INTERNATIONAL SYMPOSIUM ON

NUCLEAR DATA EVALUATION METHODOLOGY

INTERNATIONAL SYMPOSIUM ON

NUCLEAR DATA EVALUATION METHODOLOGY

Brookhaven National Laboratory, Upton, NY, USA

12 – 16 October 1992

Edited by

Charles L. Dunford

Brookhaven National Laboratory

World Scientific

Singapore • New Jersey • London • Hong Kong

Published by

World Scientific Publishing Co. Pte. Ltd.
P O Box 128, Farrer Road, Singapore 9128
USA office: Suite 1B, 1060 Main Street, River Edge, NJ 07661
UK office: 73 Lynton Mead, Totteridge, London N20 8DH

INTERNATIONAL SYMPOSIUM ON NUCLEAR DATA EVALUATION METHODOLOGY

ISBN 981-02-1285-2

Printed in Singapore.

International Symposium on Nuclear Data Evaluation Methodology

Brookhaven National Laboratory, 12 – 16 October 1992

Sponsors
Brookhaven National Laboratory
United States Department of Energy

Cooperation of
OECD Nuclear Energy Agency
International Atomic Energy Agency

International Organizing Committee

C.L. Dunford, chairman	Brookhaven National Laboratory, USA
S. Ganesan	IAEA Nuclear Data Section, Vienna, Austria
A.V. Ignatyuk	FEI, Obninsk, Russia
N.B. Janeva	INR, Sofia, Bulgaria
Y. Kanda	Kyushu University, Japan
Y. Kikuchi	JAERI, Tokai-mura, Japan
D. Larson	Oak Ridge National Laboratory, USA
Liu Tinjin	IAE, Beijing, PR China
R. Michel	University of Hannover, Germany
C. Nordborg	NEA Data Bank, Paris, France
J.J. Schmidt	Vienna, Austria
D.L. Smith	Argonne National Laboratory, USA
H. Vonach	IRK, Vienna, Austria
R.M. White	Lawrence Livermore National Laboratory, USA
P.G. Young	Los Alamos National Laboratory, USA

PREFACE

These proceedings contain the invited and contributed papers presented at the Symposium on Nuclear Data Evaluation Methodology which was held at the Brookhaven National Laboratory on October 12–16, 1992. The symposium was sponsored by Brookhaven National Laboratory and the United States Department of Energy with the cooperation of the Nuclear Energy Agency and the International Atomic Energy Agency.

With the completion and release of five evaluated data files, BROND in the Former Soviet Union, CENDL in the People's Republic of China, ENDF/B in the United States, JEF and EFF in Western Europe, and JENDL in Japan, it was appropriate to review the state-of-the-art in the nuclear data evaluation field. Of the registered participants, 41 came from the United States and 59 came from 18 other nations and 3 international organizations. The last meeting on this subject was held at Brookhaven National Laboratory on September 22–25, 1980. At the meeting there were 67 registered participants mostly from the United States. The participation in this recent meeting illustrates the internationalization of the nuclear data evaluation activity.

Several new topics have been included in the presentations at this symposium. These include medium energy data evaluation methods, charged-particle and photo-nuclear data evaluation methods, the generation of covariance information, and advances in computer-aided evaluation. I would like to thank the International Organizing Committee for preparing the technical content of the meeting and selecting the papers to be presented and also the staff of the National Nuclear Data Center of Brookhaven National Laboratory for the local arrangements for making the Symposium a great success.

The Editor

CONTENTS

INFLUENCE OF INTEGRAL DATA

MEDIUM ENERGY DATA

KEYNOTE ADDRESS

EVALUATION OF NUCLEAR DATA — PART, PRESENT AND FUTURE

J.J. Schmidt
Vienna, Austria

ABSTRACT

Evaluation of nuclear data is a scientific discipline of basic importance to the development of nuclear science and technology. It requires multidisciplinary nuclear scientists familiar with experimental methods, theoretical concepts and models, mathematical statistical methods, computers and data application fields. Its results are computer files of evaluated data which, appropriately processed, form the input data to all nuclear science and technology computations.

In the past the development of evaluated nuclear data files and evaluation techniques was predominantly driven by the nuclear data requirements for the development of nuclear fission reactor technology. The recently released updated versions of various evaluated nuclear data files demonstrate significant progress in evaluation techniques, but still reveal considerable gaps, differences and discrepancies. There is an imminent need for a careful intercomparison of and evaluation and removal of discrcpancies from the data files.

Nuclear data requirements in the nineties are shifting from conventional to advanced fission and to fusion reactor technology, nuclear waste incineration and fission reactor decommissioning, as well as to intermediate energy applications such as biomedical applications, spallation sources, accelerator shielding, space and cosmic research. For environmental protection and the estimate of comprehensive internationally accepted files of neutron and charged particle nuclear activation cross sections and associated decay data files are needed. All these requirements necessitate the development of new and improved nuclear data evaluation techniques and an extension of the existing evaluated nuclear data files up to GeV level. This enormous task will include large-scale nuclear model computations of nuclear data whose reliability must be improved by systematic intercomparisons of the physical features and results of existing and newly developed nuclear model codes and the development of standardized files of nuclear model input parameters. Finally, for reliable estimates of the impact of nuclear data uncertainties on nuclear technology parameters a significant extension of the currently available data uncertainty files will be needed.

1. Introduction

The honorable invitation to introduce this Symposium with some thoughts about the past, the presence and the future of nuclear data evaluation burdens me with the almost impossible task, to digest and summarize a wealth of developments and information accumulated over the past 10 to 20 years. The Conference on Nuclear Data Evaluation Methods and Procedures initiated by and held at the Brookhaven National Laboratory in 1980[1], i.e. 12 years ago, represents a milestone in these developments: it gathered, for the first time, nuclear data evaluation scientists from all over the world and provided important stimuli for the developments and international exchanges which have taken place since then; I shall therefore restrict my retrospective considerations to what happened in the period after this Conference.

On behalf of all of you I should like to congratulate the National Nuclear Data Center for its renewed timely initiative to convene another International Symposium on Nuclear Data Evaluation Methodology with the declared aim to achieve a critical and comprehensive review of the developments since the 1980 Conference and the future needs and problems to be solved in nuclear data evaluation, with an emphasis on the methodical aspects. I consider this Symposium very timely for the following reasons:

(i) several comprehensive and many specialized evaluated nuclear data files (ENDFs) have very recently been internationally released and call for intensive intercomparison and testing;

(ii) bilateral co-operations introduced in the developments of these ENDFs start to expand gradually to multilateral and international co-operations;

(iii) new developments emerge in nuclear science and technolgy, entailing new challenges and tasks in nuclear data evaluation in the 90ies;

(iv) finally the Symposium represents a necessary complement of the discussions at the recent International Conference on Nuclear Data for Science and Technology held at Jülich, Germany, in May 1991[2].

In fact, this Symposium can become a landmark for enhanced international work-sharing in nuclear data evaluation in the 90ies.

The period since the 1980 Brookhaven Conference is marked by

(i) a strong increase in importance of nuclear data evaluation;

(ii) a strongly increased role of nuclear theory and models in nuclear data evaluation;

(iii) an increasing tendency towards spot precision measurements to check and parameterize nuclear model calculations; and

(iv) by the development of the US ENDF format into the international format for the exchange and intercomparison of evaluated nuclear data.

2. Necessity of nuclear data evaluation

Nuclear data evaluation is based on available experimental and theoretical evidence. To prove the necessity of nuclear data evaluation let me recall you first some well known elementary facts characteristic of all experimental values (refer also John Story's conjectures in the 60ies[3]):

(i) each experimental value has an uncertainty consisting of statistical and systematic components;

(ii) two or more experimental values usually differ from each other and have different statistical and systematic uncertainties;

(iii) if, for a given quantity, many different measurements are already available, only "an order of magnitude" more accurate new measurement can improve the accuracy of the average value for this quantity;

(iv) an infinite number of measurements, though of course impossible, would lead to the "true" average value only, if there would be no systematic errors.

Similar conjectures hold "mutatis mutandis" also for the results of nuclear model calculations.

These deficiencies are basic in nature beyond human influence and control and are certainly unacceptable for the progress of science. They call for "arbiters" who must reconcile conflicting scientific results into compromise average values as "best" possible approach to the "true" values.

These elementary deliberations lead us to the definition of evaluation as a scientific discipline of its own which developes and uses well-defined and by the scientific community accepted values and procedures for deriving "best" values and their uncertainties from usually conflicting experimental and theoretical evidence. Since science aims at the "true" understanding of natural processes, evaluation is the necessary scientific complement to science to reach this aim. This defines the basic necessity of evaluation as unabdicable for the progress of science itself.

A second deliberation starts from the fact that the quantitative results of scientific research are the basic input parameters to any technological planning and design made possible by scientific discoveries.

Applied scientists and engineers need unique "best" values and uncertainties for realistic estimates of technological and economic design margins. The evaluator prepares and provides this information. In fact, without evaluated "best" values no technological planning and progress would be possible.

In conclusion, evaluation of scientific results is needed for the progress of science itself as well as for technological applications of science.

Finally, nuclear data evaluation is an activity which cannot be done "with the left hand"; it requires scientists with interdisciplinary interest and knowledge in the following domains:

(i) experimental nuclear physics and measurement techniques;

(ii) theoretical nuclear physics, nuclear models and computer codes;

(iii) laws and techniques of mathematical statistics;

(iv) programming techniques for creating, processing and testing large computerised data files;

(v) applied sciences and technologies.

Many nuclear scientists have acquired and applied these skills in an exemplary manner. In view of the future needs, it will be essential that these skills do not get lost and are being transferred to young scientists of the next generation(s).

3. Requirements, status and further developments of evaluated nuclear data for nuclear technology

3.1. Nuclear data requirements for nuclear technology

From the beginning until now the development of thermal and fast fission plus, more recently, fusion reactor technology was the driving force behind the development of the nuclear data field including nuclear data evaluation.

The nuclear data requirements were collected in national request lists feeding into the World Request List for Nuclear Data WRENDA published by the IAEA[4,5], and described in many additional detailed reviews, particularly by the NEANDC and INDC, and by national nuclear data committees. In the following I restrict my considerations essentially to nuclear reaction data.

In summary it turned out that already for "conventional" thermal and initial advanced liquid model breeder reactor development a large amount of neutron cross section data was needed for a large variety of elements and isotopes (~ 1000) over the whole period table of isotopes and over 12 decades of neutron energy from 10^{-5} to 10^{7} eV. For fusion increased or additional emphasis was placed on activation, transmutation and double-differential cross sections.

3.2. Recently released ENDFs

The voluminous nuclear data requirements stimulated the development of increasingly comprehensive and complex global and specialized evaluated nuclear data files (ENDFs) in various parts of the world. The latest versions of these files released over the past two years and their origin are listed in the following table.

(1) Global files

- ENDF/B-6[6,7] USA
- JEF-1[8] NEA countries
- JENDL-3[9,10] Japan
- BROND-2[11,12] (former) USSR
- CENDL-2[13,14] China
- EFF-1[8] CEC countries

An index to BROND-2, CENDL-2, ENDF/B-6, JEF-1 and JENDL-3 has been published by Lemmel[15]. The release of the following global files is expected shortly:

- JEF-2[8] NEA countries
- EFF-2[8,17] CEC countries
- FENDL-1[18] IAEA

(2) <u>Specialized data files</u>
(as separate files or parts of global files)

- ENDF/B-6 standards file[19] adopted as international standards file;
- several fission product cross section, yield and decay data files[8,20-24];
- three neutron dosimetry cross section files[6,9,25];
- Several neutron activation cross section data files:

- REAC*3[26]	USA
- EAF-2[17,27]	CEC countries
- ADL-90[28]	(former) USSR
- SINCROSACT[29]	Japan
- ACTV-FUS/INT[30]	IAEA

- sequential charged particle activation cross section file KFKXN[31,32] developed at Karlsruhe, Germany;
- three charged particle induced nuclear fusion reaction cross section files developed at Graz/Austria[33], Livermore/USA[34] and Arzamas/(former) USSR[35,36].

(3) In addition, <u>nuclear data handbooks</u> have been published by the IAEA on

- neutron, charged particle and photonuclear activation cross sections[37];
- transactinium isotope nuclear decay data[38]; and on
- nuclear data for nuclear geology[39].

3.3. Achievements and shortcomings in recent ENDFs

To what extent do these data files fulfill the nuclear data requirements of fission and fusion reactor technology? To answer this question I shall quote some of the achievements and shortcomings of the more important files and to draw some preliminary conclusions as to necessary actions for improvements.

(1) <u>Global data files</u>

The <u>achievements</u> can be summarized as follows:

- the released files show a substantial increase in size and reaction detail over the respective previous versions;
- extensive efforts have been devoted to the establishment and testing of interim file versions;

- **an extensive evaluation effort succeeded in substantial improvements of the ENDF/B-6 standard cross sections;**

- **the simultaneous evaluation concept in the evaluation of main actinide fission and capture cross sections led to enhanced consistency of the evaluated data;**

- **the number of resolved resonances more appropriately analysed with the Reich-Moore formalism has enormously increased for the main structural material isotopes and the main actinides allowing fast reactor Doppler coefficient calculations to required accuracies and the derivation of improved average resonance parameters useful for the optimization of optical potentials;**

- **improved nuclear models and computer codes were developed and extensively applied in the analysis and prediction of neutron cross sections in the MeV range;**

- **some uncertainty files using covariances were developed for several important nuclides;**

- **bilateral (e.g. between the USA and China, France and the UK respectively), regional (NEA; CEC; and (former) Comecon countries) and international co-operations (so far only for FENDL) were developed;**

- **several long-standing problems and discrepancies could be solved in important neutron capture cross sections for structural materials and ^{238}U and in thermal actinide cross sections (references[40-47,60]).**

In the present stage of the released global files the following <u>shortcomings</u> can be noted; my observations are based on recent file comparisons and on papers presented at the Jülich Nuclear Data Conference[2] and other recent meetings:

- **heterogeneities in and between ENDFs:**
  - **natural element cross section evaluations are not always consistent with the sum of the isotopic cross section evaluations[9,11,14];**
  - **the use of different ENDF formats hampers the comparison of File 6 double-differential cross section (DDX) data[48];**
  - **there are differences between the files in evaluation completion dates and completeness in material coverage[15];**

- the files show still significant gaps in double-differential, gamma-production and uncertainty/covariance data; also covariance data are not given for double-differential data and resonance parameters[48];

- wrong physical assumptions are still maintained in the files[48]:

  - neutron capture gamma-ray spectra are assumed to be the same at all neutron energies up to 20 MeV;

  - gamma-ray angular distributions are assumed to be isotropic at all energies.

Major discrepancies between global files are noted for the following material groups:

(i) <u>Structural materials</u>

The evaluations for Cr, Fe and Ni isotopes in ENDF/B-6, JEF-2 and JENDL-3 are all recently updated, but show still significant discrepancies[49,60] for individual isotopes in σ_T, $\sigma_{n'}$, σ_{2n}, σ_γ (see <u>figure 1</u> for $^{58}Ni(n,\alpha)$[49] in reference[171]) and gamma-production cross sections.

(ii) <u>Main actinides</u>

There are serious discrepancies between all global ENDFs[50,60] in $\sigma_{n'}$, σ_γ and σ_f for several isotopes.

(iii) <u>Minor actinides</u>

There are numerous discrepancies between all global ENDFs in σ_f, σ_γ, $\sigma_{n'}$ and σ_{2n}; expamples are shown in <u>figures 2 and 3</u> in reference[171] for σ_γ and σ_f for ^{242}Cm and ^{243}Cm respectively[51].

(iv) <u>Th-233 - U-233 fuel cycle nuclides</u>

Strong disagreements can be noted between the more recent JENDL-3 and the older ENDF/B-6 (= JEF-2) data in σ_γ, σ_f, σ_{2n} and $\sigma_{n'}$[52,60].

Limitations in financial support, manpower and time caused a restriction of re- or new evaluations to the more important materials and prevented updating of previous file versions for less important materials[15]. Finally, differences in the quality of evaluations can be noted, partly due to these limitations, partly due to different evaluation procedures and the lack of universally agreed evaluation rules and standards, and partly due to differences in evaluation skill and experience.

(2) Specialized data files

(i) Fission product nuclear data (FPND) files

Needs and achievements in FPND have been reviewed at a recent NEANSC Specialists' Meeting[53] whose conclusions are summarized in the table below.

Parameter	Accuracy (%) required	Accuracy (%) achieved
ν_d	5	10
FP contribution to fast reactor K_{eff}	6	10-15
decay heat prediction		
- short cooling times	5	10
- long cooling times	10	10

Shortcomings in σ_γ, $\sigma_{n'}$ and yield data for many FP lead to the above listed discrepancies. Also problems in decay heat determinations are not yet fully solved.

(ii) Neutron dosimetry files

For the three neutron dosimetry files IRDF-90[25], ENDF/B-6[6] and JENDL-3[9] (re)-evaluations, extensions compared to previous versions and addition of covariance data to most reactions have been achieved.

There are still some shortcomings in (n,p), (n,α) and (n,2n) reactions[54,55] which indicate that the three files do not yet fully meet the ± 5% accuracy requirements of neutron dosimetry particularly for fusion reactors.

(iii) Activation cross section files

Six large starter files (references[26-32]) covering more than 10 000 different reactions[64] are currently undergoing extensive testing, intercomparison and benchmarking. In view of the large gaps in experimental data, the crude models[56] applied in many cross section calculations and the large discrepancies particularly in σ_α (see figures 4-7[57] in reference[171]) much further experimental and theoretical research is needed.

3.4. Integral testing of recent ENDFs

Integral testing of the recently released ENDFs is still going on at various places; I can only quote some selected typical preliminary results.

From thermal U and Pu benchmark calculations with ENDF/B-6 and JEF-2 based multigroup data[58,59] it can be concluded that K_{eff} of thermal reactors can be satisfactorily predicted usually to better than 1%, whereas epithermal to thermal fission and capture rate ratios are only reproduced within several %.

Fast critical assembly test calculations of K_{eff} and spectral indices and computations of minor actinide concentrations in samples irradiated in fast reactor spectra[50,61,62] indicate deviations in K_{eff} by $\lesssim \pm$ 1%, a result which is not much better than that obtained with previous adjusted group constants, deviations in spectral indices for the main actinides by ± 5 to ± 10% and for Th cycle nuclides by ± 10 to ± 30%, and serious discrepancies in concentrations of minor actinides.

Concerning fusion, preliminary results from a variety of ongoing tests[63-66] indicate serious C/E discrepancies in activation rates, - 5% to + 20% deviations in tritium breeding rates and some still unexplained differences in Be sphere neutron multiplication measurements.

3.5. Conclusions

On the basis of the considerations in sections 3.3. and 3.4. the following conclusions can be drawn on the status of the presently available evaluated nuclear data files as compared with the requirements of nuclear fission and fusion technology.

(i) Thermal reactors

The available ENDFs appear now to be satisfactory for thermal reactor calculations from first principles; only a few problems need still to be solved.

(ii) Fast breeder reactors

The available ENDFs are considerably improved over the respective previous versions and some long-standing problems in important nuclear data could be solved. However, serious discrepancies still exist in important cross sections of main and minor actinides, structural material and fission product isotopes. It appears that fast breeder reactors can still not be calculated fully without some cross section adjustment.

(iii) Fusion systems

The available ENDFs are improved, but serious inconsistencies and gaps exist in all types of activation cross sections, particularly for (n,α) reactions, double-differential and gamma-ray production cross sections. The ENDFs are still incomplete and need many improvements. This will require longer term efforts, some aspects of which will be discussed in chapter 6.

Finally, some covariance data are now available, but only for selected important materials.

3.6. Evaluated data processing

An important caveat has to be added at the end of this chapter. Data processing codes such as NJOY[67], AMPX[68] and GRUCON[69] must fulfill a very important basic condition which in practice has turned out to be far from trivial: the accuracy reached in ENDF data must not get lost or worsened during data processing. Systematic comparison exercises with a variety of data processing codes conducted by Cullen, Ganesan, Muir[70-72] and others have identified and eventually rectified numerous inconsistencies and problems in the codes participating in the exercise. Not all problems have yet been solved, and it will be an important future task to continue and extend the comparisons of data processing codes.

4. Near-term improvement of ENDFs

Considerable efforts are currently going on or are planned on intercomparison, testing and improvements of ENDFs. Two international initiatives and activities deserve particular mentioning, the NEANSC Working Group (WG) on International

Evaluation Co-operation (IEC) and Subgroups[73] (see also references[41,43,49,74-79]) and co-ordinated research programmes (CRP) pursued by the IAEA Nuclear Data Section (NDS). The main emphasis of the IEC WG of NEANSC and its Subgroups is on (re)evaluations, intercomparisons and convergence of the most important and still deficient data in ENDF/B-6, JEF-2 and JENDL-3. In the future also experts from Non-OECD countries will be invited to participate in this work.

The IAEA CRPs are devoted to experimental and theoretical research on selected important missing or discrepant data[80,81] and are usually conducted as a co-ordinated three to five year co-operation of several nuclear data research groups in different advanced and developing countries.

Additional necessary improvements of existing ENDFs are put together in the following (incomplete) "shopping list":

- further development and unification of ENDF formats in different ENDFs[48];
- new measurements and evaluations to solve the problems with the ^{10}B (n,α) standard cross section[19];
- improvement of (n,p), (n,α) and (n,2n) cross sections for structural material isotopes[49] and for fusion related activation[54,64,65] and dosimetry reactions[55];
- addition and introduction of the energy dependence of neutron capture γ-ray spectra and of the anisotropy in γ-ray angular distributions for the more important reactor materials[48];
- solution of the remaining inconsistencies in the (n,2n) cross sections for ^{7}Li, Be and Pb[66];
- measurements and nuclear model calculations to close the gaps in neutron and charged particle emission spectra in the 7-14 MeV energy range[48,90];
- addition of missing neutron emission spectra and of gamma-ray production and charged particle emission ((n,Xp); (n,Xα)) cross-sections and spectra[17,48,82];
- removal of discrepancies in some global ENDFs between isotopic and natural element cross sections[9,11,14];
- for inelastic scattering calculations, theoretical estimates of spins and parities should be added to the Evaluated Nuclear Structure Data File (ENSDF)[48];

- a renewed edition of the Brookhaven handbook by Mughabghab et al.[128]; it should include thermal and 14 MeV cross sections and infinite dilute resonance integrals from all global ENDFs and make proper corrections for missing and p-wave resonances;

- development of accurate covariance data[83,84] for the most important materials and of rough covariance data for the less important materials; covariance data for secondary distributions and resonance parameters should also be included.

5. New nuclear data requirements in the 90ies

The presently ongoing and planned near-term efforts described in Section 3.7 will not suffice to meet all fusion requirements and the new nuclear data requirements of new application areas in the 90ies, as outlined below.

5.1. Advanced reactors and reactor decommissioning

Various advanced reactor concepts are currently developed or planned[60] with the following common concerns of inherent safety; limited radiation damage; safe disposal of nuclear waste; and yet economic design.

Below follows a selected list of new reactor concepts and (in brackets) nuclear data concerns:

- Mixed oxide (MOX) fuelled thermal/epithermal tight lattice light water reactors with variable fuel-to-moderator ratios (accurate low-lying fissile and fertile isotope resonance cross sections);

- Nitrite fuelled thermal and fast reactors (^{14}N and ^{15}N neutron cross sections);

- Lead cooled fast reactors (Pb neutron cross sections);

- Reactors with "inert matrices" (e.g. Ce instead of ^{238}U) designed to burn plutonium without its build-up through neutron capture in ^{238}U (no Ce cross sections in any ENDF);

- "Incineration" reactors for burning minor actinides and long-lived fission products (σ_f, σ_γ, $\sigma_{n'}$, σ_{2n} and nuclear decay data of the following more important minor actinides: ^{237}Np; $^{238,240,241,242}Pu$, $^{241,242,242m,243}Am$ and $^{242,243,244}Cm$ and long-lived fission-products such as ^{99}Tc, ^{129}I, ^{135}Cs and others.

For realistic and economic design and safety calculations a more accurate data base for the above mentioned actinide and fission product isotopes will be needed. For those "incineration" reactors which would be supposed to use larger amounts of minor actinides as fuel the accuracies of minor actinide neutron cross sections must be as high as for the main actinides.

- Very similar data needs exist for actinide and FP inventory calculations for reactor decommissioning[85] and for the estimation of environmental radiation levels and effects following reactor accidents such as the Chernobyl accident[86].

- ^{232}Th – ^{233}U fuel cycle[87-89]. A much improved data base is needed for the following main isotopes[52,87]: ^{232}Th, ^{231,232,233}Pa, and ^{232,233,234}U.

5.2. Fusion demonstration reactors including the ITER project

The nuclear data requirements for fusion reactor technology can only be fulfilled on a long-term basis. Three application aspects require long-term efforts:

(i) fusion material research[90-92] requires neutron and charged particle nuclear reaction cross sections and primary knock-on atom spectra for many materials up to 50 MeV.

(ii) development of "low-activation" fusion materials needs neutron and sequential charged particle nuclear activation cross sections and associated nuclear decay data for many initial, transmuted and impurity material isotopes[93-96].

(iii) fusion reactor shielding design caculations require neutron emission, gamma- and charged particle production cross-sections and double-differential spectra for magnet shield (H, He, B, C, O, W, Pb), biological shield (B, C, O, Si, Ca, Ba) and associated structure materials[96].

5.3. Intermediate energy applications

In the 90ies a variety of new application fields with new nuclear data requirements for particle energies up to the GeV range will become of growing interest such as:

- development of intermediate energy proton accelerators;

- application of such accelerators to nuclear waste incineration[97,98] and nuclear fuel breeding[107];

- high intensity neutron sources such as p,d-Li and p,d-Be for material and nuclear cross section testing[91], material analysis and neutron radiography[99];

- space and cosmic research including e.g. the planned US Mars mission[100];

- neutron and proton cancer therapy with cyclotrons[101];

- medical radioisotope production with cyclotrons including optimization of cyclotron operation[102-104].

These new intermediate particle energy applications require a huge amount of new neutron and charged particle (including light and heavy ions, μ, π and K mesons) nuclear reaction data from about 20 MeV up to several GeV not covered by the present ENDFs except for Fe, ^{208}Pb and ^{209}Bi[105,106].

While work has been started, particularly in Japan[98,108], there are so far, to my knowledge, no systematic evaluated charged particle nuclear data files available.

The first more comprehensive review of the requirements and status of nuclear data for the aforementioned intermediate energy applications has been performed by the IAEA Advisory Group Meeting on Intermediate Energy Nuclear Data for Applications in October 1990[107]. The most comprehensive IEND needs for intermediate energy proton accelerator design and application to nuclear waste incineration have been formulated for the Japanese Omega Project[98].

IEND needs for space applications and cosmic research have been formulated by Michel[109], Dudkin[110], Howe and Auchampaugh[100] and Dudkin et. al.[111]. They are very similar to those for intermediate energy accelerators. They comprize incident n,p,α, heavy ions, electrons and gamma-rays with energies from the MeV to the GeV range, interactions with target elements ranging from light to rare earths elements, production cross sections, energy spectra, energy-angle double differential distributions of secondary particles and some further integral quantities such as integral multiplicities for the production of secondaries, and a variety of stable and radioactive endproducts and recoil nuclides.

Optimization of neutron cancer therapy[101] requires:

- an extension of the presently available ENDFs from 20 to 70 MeV for H, Be, C, N, O, Ca, Fe, Ni, Cu, Zr, W, Pb etc.;

- data for the predominantly used $^{9}Be(p,n)^{9}B$ neutron source plus the adjunct ^{12}C (p,n) and $^{16}O(\alpha,n)$ reactions producing background neutrons for incident proton energies up to 70 MeV;
- an extension of the $(n,X\alpha)$ measurements on carbon from 30 to 70 MeV for Kerma determinations.

As a result of a co-ordinated research programme, IAEA-NDS will publish a handbook on nuclear data for neutron therapy in 1993[112,113].

For medical radioisotope production the most important data needs are as follows[102-104]:

- excitation functions for the formation of the more important radioisotopes (~ 50);
- cross sections for monitor reactions used for optimizing the cyclotron operation for the production of radioisotopes with optimum purity.

The experimental IEND data base is increasing, but so far rather sparse. Papers reported at the Jülich Conference[2] and at the IAEA AGM on IEND[106] give an overview of recent experimental data (specific references[114-127]).

6. Required evaluation activities in the 90ies - Some conclusions and recommendations

6.1. General conclusions and recommendations

The new areas of nuclear sciences and technologies sketched in chapter 5 require substantial new, mostly long-term, experimental, theoretical and evaluation efforts to satisfy the new nuclear data requirements and "revisiting" of available nuclear data in the light of higher accuracy requirements. My conclusions and recommendations presented below are mostly based on recent meetings and discussions; in line with the topic of this Symposium I shall concentrate on evaluation needs and development of theoretical tools required for evaluations.

(i) A more systematic compilation than in Section 5.1. of nuclear data requirements of all new reactor concepts should be performed, e.g. at the Advisory Group Meeting on Nuclear Data Requirements for Advanced Reactors planned by IAEA-NDS in 1993 and the present ENDFs be updated accordingly.

With regard to the ^{232}Th- ^{233}U fuel cycle, I propose that those countries interested in this fuel cycle such as Brazil, India, Japan, Turkey and others form a collaborative effort for the establishment of an improved nuclear data base.

(ii) For realistic and economic design and safety computations for "incineration" reactors and reactor decommissioning improved nuclear cross section and decay data bases for minor actinides and fission products should be established, in close co-operation between the NEANSC IECWG subgroups and the IAEA CRPs.

(iii) To satisfy various nuclear safety, material damage and environmental protection needs, the present activation data starter files should be extended stepwise to a universal file of neutron and charged particle induced activation cross sections (about 30 000 reactions) plus nuclear decay data (see also (iv) below). IAEA/NDS is planning to co-ordinate a collaborative effort.

(iv) Near- and long-term improvements of the present ENDFs require extensive testing, physics intercomparison and improvement of existing nuclear model codes for energies below 20 MeV; these efforts should be closely co-ordinated between NEA and IAEA (see also below under 6.2.).

(v) For intermediate energy applications the present global ENDFs for neutron cross sections must be extended and evaluated charged particle nuclear data libraries be established for a variety of nuclides for primary energies from 20 MeV to several GeV. This requires a very large effort which can only be successful if it is done in international co-operation and in the following initial steps as proposed by Pearlstein[129]:

1. Extension of the four nuclear data centre co-operation[130] in the compilation of experimental neutron and charged particle nuclear data into the international EXFOR data base[131] up to energies in the GeV range to test IEND nuclear models;

2. Establishment of a quick first version of IEND ENDFs for (p,Xn), (p,Xγ), product yields and monitor reactions to help in exploratory accelerator design and sensitivity studies to determine the crucial data.

(vi) In view of the large magnitude of the nuclear data needs, the limited experimental capacities, and the availability of powerful computers and codes, the principal and most economic ways to satisfy IEND requirements will be through nuclear model calculations of nuclear data supported by and adjusted to energy-dependent shape measurements and/or precision measurements at selected energy points. The establishment of IEND ENDFs thus requires the long-term development of new or improved intermediate energy nuclear model codes, their testing, intercomparison and large scale application for IEND calculations, as indicated e.g. by recent intercomparison studies by Pearlstein[158] and code development and testing work in Japan[98].

(vii) The large scale application of nuclear model codes envisaged for the future requires a unified reference base of nuclear parameter input data[48], such as nuclear masses, reaction Q-values, discrete nuclear level schemes, nuclear level densities, optical model parameters, gamma-ray strength functions and fission barrier parameters.

IAEA/NDS has started a long-term project for the establishment of such a reference data base with a Consultants' Meeting in November 1991[132] and plans a further meeting on project strategy and details in 1993.

6.2. Some specific conclusions and recommendations concerning evaluation methodology

The future nuclear data evaluation work requires development of new or improved evaluations and standardization of evaluation methodology. Some aspects are discussed below.

6.2.1. Nuclear model codes

Low-energy nuclear reaction theory and models related to nuclear data evaluation have been extensively reviewed at several recent meetings[133-135] and workshops[136,137] and formed a substantial fraction of the papers reported at the Jülich Nuclear Data Conference[2]. NEA, in co-operation with IAEA, has conducted several nuclear model code intercomparison exercises[138,139]. V.A. Konshin and O.V. Konshin[140,141] have performed the first systematic physics tests and intercomparisons of optical potentials and nuclear model codes.

The recent development of nuclear reaction theory and models towards more quantitative predictions and large scale applications in recent ENDFs have led to some overenthusiasm as regards the accuracy of present nuclear cross section predictions. In numerous recent publications it has been demonstrated that, due to the (usually) large number of free model parameters, perfect fits to known experimental data can be achieved. However, in the prediction of unknown data, significant deviations up to a factor 2 between the results of different nuclear model codes have been observed (see figure 8 in reference[171] presenting the results of the NEA/IAEA "blind" $^{60g}Co(n,p)^{60}Fe$ calculation exercise[142] and analysis by Konshin[141]).

Some recently suggested improvements to low-energy nuclear model codes are briefly summarized below:

(i) systematic derivation of optical potential parameters from dispersion relations (Hodgson[143-145]; see also[147]); for this task a collection of all known single-particle bound levels would be useful and required. The concept was successfully applied by Kitazawa et al.[146] to the calculation of ^{28}Si neutron cross sections.

(ii) Systematic application of the generalized temperature-dependent Lorentzian to the gamma-ray strength function for E1 transitions proposed by Uhl and Kopecky[149,150] in statistical model calculations and of the quasi-deuteron mechanism of hard photon emission proposed by Oblozinsky[151,152] in pre-equilibrium calculations of photon production data and gamma-ray spectra (see also[148]).

(iii) Systematic introduction of the collective enhancement correction of single particle level densities developed by Ignatyuk et al.[153].

(iv) Adoption of the energy dependence of the nuclear level density parameter a as proposed by Konshin[141].

(v) The fission nuclear data systematics developed by Märten et al.[154] and Madland[155] should be compared where applicable and comprehensively utilized in the analysis and/or prediction of fission nuclear data of the main and minor actinides.

(vi) Incorporation of angular momentum and parity conservation in nuclear model codes where still required.

(vii) Further development of quantum-mechanical statistical theories as pursued for example by Kalka[156,157].

(viii) Continuation and intensification of nuclear model code intercomparisons, with an emphasis on physics intercomparisons as exemplified by V.A. Konshin and O.V. Konshin[140,141], in close co-operation between IAEA and NEA.

6.2.2. Evaluation standards

To prove evaluation as a scientific discipline, to establish a consistent basis for the intercomparison of different evaluations and for educational purposes the establishment of universally accepted evaluation standards is highly desirable. Evaluation standards are meant in a twofold sense: first, high scientific standard of evaluations reflecting the contemporary status of experimental and theoretical evidence and knowledge of application needs, secondly, standard rules and procedures to be adhered to in evaluations. Many colleagues inside and outside this audience have significantly contributed to the development of both high quality evaluations and standard rules; among many outstanding examples, I may refer here to the works by D.L. Smith[159], Perey[160], Peelle[161], Poenitz[162], Fröhner[163], Vonach[83,164], Ignatyuk[153], Muir[84,85,165], Kanda[166], Hale[167], Arthur et al.[168], Pearlstein[169], and others.

It would be highly desirable and useful, particularly also for the attraction, orientation and education of newcomers in the field, to excerpt and summarize the rules, procedures and recipees from the above and other dispersed references and to publish a high level "cook-book" of evaluation methodologies and rules.

A good general basis for the improvement of evaluations which are based on experimental and nuclear model information is provided by Vonach's four step proposal[83,164] (see also Peelle[161], Kanda[166], and, for IEND evaluations, Pearlstein[169]). As evaluations are incomplete without quantifying their uncertainties, covariances should principally be included in ENDFs; the recent work by D.L. Smith[159], Muir[84,85,165] and Kanda[166] provides a good methodical basis for the derivation of covariances.

7. Final remarks

I hope to have succeeded in illustrating the substantial recent progress as well as some of the open problems in nuclear data evaluation and in demonstrating the important role which nuclear data evaluation is to play in the 90ies.

I hope also that I could present you with convincing arguments that evaluation is not only a basically necessary, but also a highly responsible profession. In this context I should like to remind you of some of the closing remarks by Alan Smith[170] at the end of the Jülich Conference in which he underlined the essential part which nuclear data physicists take in restoring public confidence in nuclear energy and its applications.

At the end let me welcome your response, your critique and comments. They will tell me what I should have said if I had known before. Thank you for your attention.

References

1. Conference on Nuclear Data Evaluation Methods and Procedures, Brookhaven National Laboratory, September 1980; Proceedings edited by B.A. Magurno and S. Pearlstein and published as report BNL-NCS-51363, Vols. I and II, 1981.

2. International Conference on Nuclear Data for Science and Technology, Jülich, Germany, May 1991; Proceedings edited by S.M. Qaim and published by Springer Verlag Berlin, Heidelberg, 1992.

3. - 170. Contained in reference 171.

171. J.J. Schmidt, report INDC(NDS)-270/L, 1993.

EVALUATION INTERCOMPARISON

TOOLS AND TECHNIQUES FOR EVALUATION INTERCOMPARISON

D. M. HETRICK
Oak Ridge National Laboratory, P. O. Box 2008, MS 6180
Oak Ridge, TN 37831, USA

ABSTRACT

A number of computer programs are described that will read the appropriate data from evaluations in ENDF format (V or VI), perform the required calculations, and plot results versus available measured data. These tools are useful in showing weaknesses in evaluations and for uncovering discrepancies. The programs are capable of graphing results for reaction cross sections, differential scattering cross sections, angle-integrated emission spectra, double-differential cross sections of neutron emission, and gamma-ray spectra. Examples are given that illustrate the numerous options presently available.

1. Introduction

Graphical comparison of cross sections computed from evaluated libraries (ENDF/B-VI, JEF-2/EFF-2, BROND, JENDL-3) versus measured data can be quite helpful in showing weaknesses in the evaluations, as well as uncovering discrepancies. A number of computer programs have been developed at the Oak Ridge Electron Linear Accelerator (ORELA) facility that will read the appropriate data in ENDF format (V or VI) from the evaluations, perform the required calculations, and plot results versus available data. These codes have been used for comparison analyses of the structural material evaluations done at ORELA for version VI and are capable of plotting results for (1) reaction cross sections, (2) differential scattering cross sections, (3) angle-integrated emission spectra for neutrons, protons, alphas, etc., (4) double-differential cross sections of neutron emission, and (5) total secondary gamma-ray spectra. The programs are written in FORTRAN, are interactive in that they prompt the user for information, and can be run either on a VAX mainframe computer with the results going to a plotter, or on an IBM compatible PC that is connected to a Hewlett-Packard (or compatible) laser printer. Plot files produced on the PC can be imported by a word processor. This paper describes these computer codes and the options that are available. Example plots are shown which compare evaluated cross sections with experimental data.

2. Methods

Over the last few years, ORNL evaluators at ORELA have been responsible for several ENDF/B-VI evaluations (C, and isotopes of Si, Cr, Mn, Fe, Ni, Cu, and Pb). In the course of doing this work, several computer programs were developed to plot cross sections either from existing ENDF/B-V evaluations or those calculated from nuclear model codes versus available measured data. These plotting programs were used extensively in helping to adjust input parameters for the nuclear model codes so that calculated results compared as favorably as possible to experimental data.

Subsequently, Subgroup 1 of the NEANDC/NEACRP Working Group on Evaluation Cooperation was assigned the task of comparing cross sections from the international evaluations (ENDF/B-VI, JEF2/EFF2, and JENDL-3). C. Y. Fu, D. M. Hetrick, and D. C. Larson from ORELA were involved in this effort and, thus, many of the existing programs at ORELA were modified to read the appropriate data in ENDF format (version V or VI)[1], perform the required calculations, and plot the results versus data (retrieved from CSISRS[2] and/or the literature). The programs have proven useful in showing weaknesses in the evaluations as well as uncovering discrepancies.

The following subsections give an overview for each of the computer programs from ORELA. For each of the plots shown in this paper, the programs were run on an IBM compatible PC that was connected to a Hewlett-Packard LaserJet IIIp printer. The plot files produced on the PC were imported by WordPerfect 5.1 for direct inclusion into the word-processing text.

2.1. *Comparison Plots for File 3 Cross Sections*

To plot results from File 3 (MF=3) of the evaluations, a program called XSECT is first used to retrieve the evaluated data from the evaluations and outputs the results into files that are named by the user. XSECT prompts the user for the file name of the evaluation and the MT number from File 3 that is desired. The results are plotted by the code PLOTXS, which prompts the user for the evaluated file names just produced by XSECT, measured data files that include data retrieved from CSISRS or from the literature, and also prompts the user for titles such as the cross section name and references for the data sets. Dashed lines are available in PLOTXS so several different evaluations as well as several data sets can be plotted in one figure (see Fig. 1 for an example).

2.2. *Plotting Differential Scattering Cross Sections*

A program called DSIG reads ENDF formatted files to retrieve the elastic or inelastic level cross section (file MF=3, MT=2 for elastic or 51-90 for inelastic levels) and the Legendre polynomial coefficients (file MF=4 or 6) in order to calculate the differential scattering cross section. The evaluated result can be compared to measured data. DSIG is not presently capable of using a transformation matrix to transform the coefficients from the CM system to the LAB system (or vice versa). An example showing the differential elastic scattering cross section for ^{63}Cu plotted by DSIG is shown in Fig. 2.

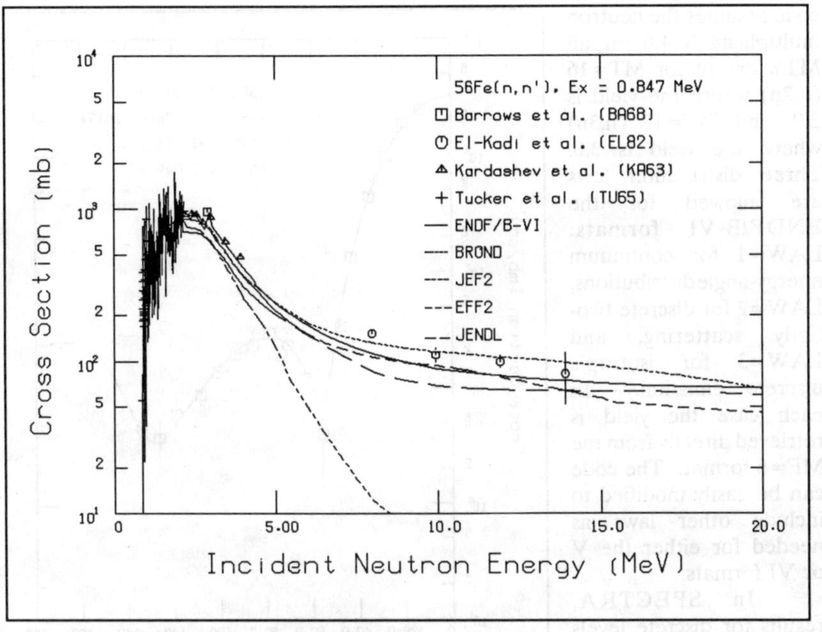

Figure 1. Example plot from the PLOTXS program.

2.3. Angle-Integrated Emission Spectra and Double-Differential Cross Sections

A program called SPECTRA can plot angle-integrated emission spectra for neutron, proton, alpha, or recoil particles, and can also plot double-differential cross sections of neutron emission. SPECTRA is capable of plotting results from the nuclear model code TNG[3], or ENDF formatted (V or VI) files, versus measured data. For both ENDF/B-V and -VI formats, the cross section is read from file MF=3, while the energy distributions are retrieved from file MF=5 for ENDF/B-V and from file MF=6 for ENDF/B-VI. At present, the code is not capable of reading an evaluation that would include both MF=5 and MF=6 distributions, but it can put two (or more) different evaluations on the same plot. For example, an ENDF/B-V spectra can be compared to an ENDF/B-VI spectra on the same plot.

SPECTRA presently allows secondary energy distribution laws LF=1 (arbitrary tabulated function), 7 (simple fission spectrum-Maxwellian), and 9 (evaporation spectrum) for ENDF/B-V formatted evaluations. For these MF=5 distributions, the

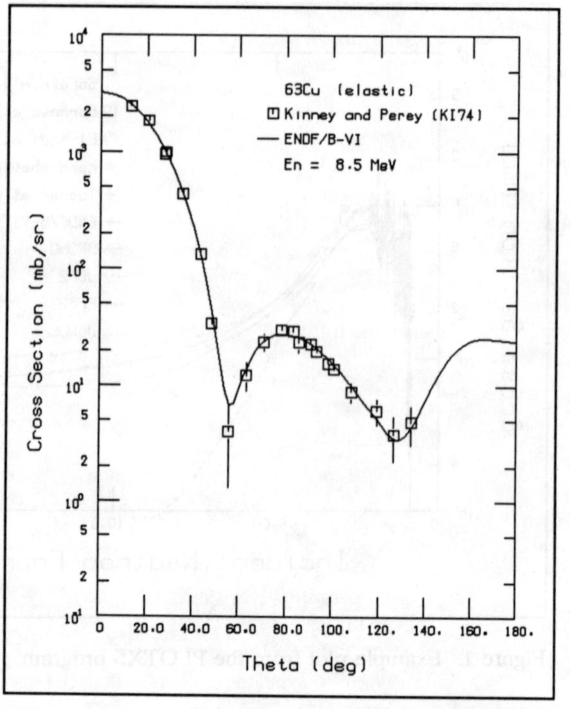

Fig. 2. Example plot from the DSIG program.

code assumes the neutron multiplicity is 1.0 for all MT's except for MT=16 (n,2n) where the yield is 2.0, and MT=17 (n,3n) where the yield is 3.0. Three distribution laws are allowed for the ENDF/B-VI formats: LAW=1 for continuum energy-angle distributions, LAW=2 for discrete two-body scattering, and LAW=3 for isotropic discrete emission. In each case the yield is retrieved directly from the MF=6 format. The code can be easily modified to include other laws as needed for either the V or VI formats.

In SPECTRA, results for discrete levels from either ENDF/B-V or -VI are converted from the CM system to the LAB system. Formulas from Evans[4] and Marion[5] are used in the conversion. At present, SPECTRA assumes that the distributions for the reactions (e.g., (n,2n), (n,np), etc.) are given in the LAB system (the evaluations for the structural materials done at ORELA all have the MF=6 distributions given in LAB). If the distribution for a reaction is in CM rather than LAB, a message is printed by the code stating that the results for the reaction will be plotted in the CM system, but for the levels will be in LAB. It is anticipated that the code will be modified at some future time to allow results to be plotted in either LAB or CM.

For the discrete levels, if an angle-integrated result is desired, the outgoing energy for the level is computed at 90 degrees and the cross section is put in a bin with a bin width BW specified by the user. A resolution function is not used. The center of this bin is the calculated energy at 90 degrees. Likewise, if a double-differential result is desired, the code computes the center of the bin for each discrete level using the angle specified by the user. The same algorithm is used for the elastic contribution.

Results can be plotted for any incident neutron energy; the code will do the required interpolations when necessary. However, for the energy distributions, SPECTRA presently assumes that INT=1 (histogram) between the secondary energies and INT=2 (linear-linear) between the incident energies for ENDF/B-V. INT=12 (corresponding-point interpolation) is allowed for ENDF/B-VI files. Partial as well as the total spectra can be plotted, and either a dashed or solid line can be used. Thus, results for the partial spectra like (n,2n) or (n,np) can be plotted with dashed lines and the total spectra (sum of the partials) can be plotted with a solid line. Or the user can plot the total spectra from ENDF/B-V with a dashed line vs. ENDF/B-VI results using a solid line (or vice versa). Results will always be plotted as histograms with the user supplied bin width BW.

SPECTRA is interactive in that it prompts the user for information. The user would need to know what reactions and levels are included in an evaluated file as given in MF=1 before running the program. Table 1 shows an example run done on an IBM compatible personal computer to plot the angle-integrated cross section of neutron emission for ^{58}Ni at 14.6 incident neutron energy. The user responds to the prompts with the information that is underlined in Table 1. Comments about the run are in bold type on the right side of the table. The plot corresponding to the example run in Table 1 is shown in Fig. 3. Total clock time to do this example on an IBM compatible 486 (50 mhz) was less than three minutes, and on an IBM PS/2 (model 80 386) was approximately four minutes.

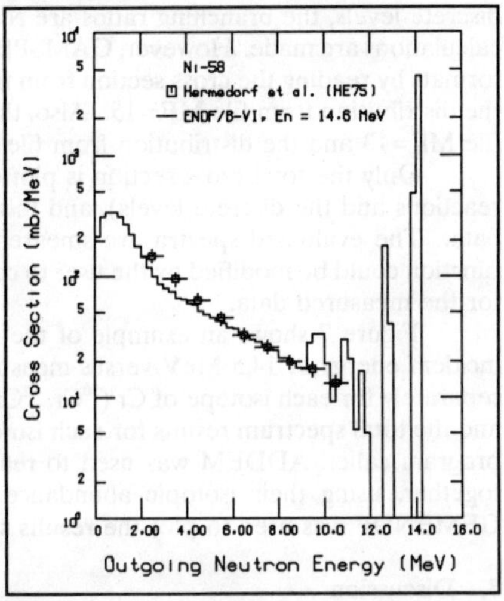

Figure 3. Example plot from the SPECTRA program.

A plot showing double-differential cross sections of neutron emission for ^{56}Fe from ENDF/B-VI and EFF-2 is presented in Fig. 4, while a proton production spectra plot is given in Fig. 5. Partial spectra (using dashed lines) from the contributing reactions are shown in the proton emission plot (Fig. 5). A plot comparing the neutron spectra for Ni from ENDF/B-V and -VI is given in Fig. 6.

If the SPECTRA program detects a case that it can not handle during a run, an error message is printed and the program will stop. For example, if the user wants a proton production spectra plot (a 2 would be typed for the second prompt in Table 1), but specifies an angle not equal to 0.0 (for the fourth prompt), the program will print a message stating that the program can not handle this scenario since, at present, only angle-integrated calculations are allowed for charged-particle spectra by SPECTRA. Also, if while reading the evaluation SPECTRA detects a ENDF format it can not handle, a message is printed and execution stops.

2.4. *Total Secondary Gamma-Ray Spectra*

The differential cross section versus outgoing gamma-ray energy can be plotted with a program called GAMSPEC. Several options are available to read ENDF formatted files. Usually, for ENDF/B-VI evaluations the cross section is read from file MF=3, and the yield and energy distribution are read from file MF=6. For the discrete levels, the branching ratios are read from file MF=12 and the appropriate calculations are made. However, GAMSPEC can handle evaluations with ENDF/B-V formats by reading the cross section from file MF=3, the yield from file MF=12, and the distribution from file MF=15. Also, the absolute cross section can be read from file MF=13 and the distribution from file MF=15.

Only the total cross section is plotted (i.e., the sum of the gammas from the reactions and the discrete levels), and this result can be compared to experimental data. The evaluated spectra are smeared by a gaussian function. If desired, this function could be modified by the user to correspond to the resolution of the detector for the measured data.

Figure 7 shows an example of the evaluated gamma-ray spectrum for Cr at incident energy of 14.5 MeV versus measured data. Note that GAMSPEC was run separately for each isotope of Cr (^{50}Cr, ^{52}Cr, ^{53}Cr, and ^{54}Cr), obtaining a plot for each and the total spectrum results for each isotope were stored in data files. An auxiliary program called ADDEM was used to read and add the results from each isotope together, using their isotopic abundance. Finally, another auxiliary code called GEMENDF was used to plot the results shown in Fig. 7.

3. Discussion

The computer programs described in this paper are useful for plotting a number of different results from ENDF formatted evaluations versus experimental data. They have been used extensively in evaluation work on the materials done at ORELA. Thus, there may be some ENDF laws and formats that are used by other evaluators that these programs will not recognize. However, modifying the programs to include other laws and formats should be straightforward.

The programs are written in standard FORTRAN-77 and have been run on a VAX mainframe and IBM compatible PC's, and therefore should be portable. The codes are available upon request from the Radiation Shielding Information Center (RSIC), Building 6025, MS-6362, Oak Ridge National Laboratory, Oak Ridge, Tennessee 37831-6362. All the programs use a local plotting package ORPLOT, written for the ORELA computing system by F. G. Perey[6], but modified by D. M.

Hetrick to remove machine dependencies. ORPLOT is a collection of callable FORTRAN subroutines and each of the codes described herein have MAIN programs and subroutines that call the ORPLOT subroutines.

A nice feature of the programs is that results can be plotted on Hewlett-Packard (or compatible) laser printers, and plot files produced by the programs on a PC can be imported by a word processor for direct inclusion into word-processing text. All the plots produced for this paper used the GRAPHCNV program of WordPerfect 5.1 to convert the plot files from the plotting programs to files that could be imported to the word-processing format.

Future work will include adding more options to the programs as the need arises and to make them more general. For example, the SPECTRA program assumes that the distributions for the reactions are given in the LAB system and should be modified to allow results to be plotted in either LAB or CM. The programs should be very useful in doing evaluation intercomparisons that will be helpful in uncovering discrepancies in evaluations.

4. Acknowledgements

The author wishes to thank D. C. Larson and C. Y. Fu for many helpful discussions held throughout the course of producing the programs described in this paper. Research sponsored by the Office of Energy Research, Division of Nuclear Physics, U. S. Department of Energy, under contract DE-AC05-84OR21400 with Martin Marietta Energy Systems, Inc.

5. References

1. P. F. Rose and C. L. Dunford, (editors), ENDF-102, Data Formats and Procedures for the Evaluated Nuclear Data File, ENDF-6, (Brookhaven National Laboratory report BNL-NCS-44945, 1991).
2. CSISRS Library, National Nuclear Data Center, (Brookhaven National Laboratory, Upton, New York 11973).
3. C. Y. Fu, *Nucl. Sci. Eng.* **100**, (1988) 61.
4. R. D. Evans, The Atomic Nucleus (McGraw-Hill, New York, 1955) pp. 420-421.
5. J. B. Marion, 1960 Nuclear Data Tables (Part 3), (United States Atomic Energy Commission, Nuclear Data Project), pp. 161-163.
6. F. G. Perey, (Oak Ridge National Laboratory, 1974, 1992).

Table 1. Example Interactive Run of the SPECTRA Program

SPECTRA H,W,SIZE,LX,LY ? 6., 4.5, .08, 0, 1	**H=height, W=width of plot; SIZE= symbol size; LX,LY indicate linear (0) or log (1) scale.**
IF N, TYPE 1; IF P, TYPE 2; IF A, TYPE 3; IF RECOIL, TYPE 4 1	**N is for neutron, A is for alpha, P is for proton.**
INCIDENT NEUTRON ENERGY EI (MEV) = 14.6	
ANGLE WANTED = (IF WANT INTEGRATED OVER ALL ANGLES, ENTER 0.0) 0.0	
TYPE BIN WIDTH WANTED (MEV) 0.25	
IF TNG TYPE 1, ENDF/B-V TYPE 5, ENDF/B-VI TYPE 6 6	
INPUT FILE = NI58V6.END	**INPUT FILE is evaluation, OUTPUT FILE will contain total spectra results when run is completed.**
OUTPUT FILE = NI58N.AIN	
FILE #S ARE 3 & 6, MT# (TYPE < 0 IF WANT SUM) = 2	**(elastic)**
FILE #S ARE 3 & 6, MT# (TYPE < 0 IF WANT SUM) = 16	**(n,2n)**
FILE #S ARE 3 & 6, MT# (TYPE < 0 IF WANT SUM) = 22	**(n,na)**
FILE #S ARE 3 & 6, MT# (TYPE < 0 IF WANT SUM) = 28	**(n,np)**
FILE #S ARE 3 & 6, MT# (TYPE < 0 IF WANT SUM) = 51	**(1st discrete level)**
TYPE FINAL LEVEL NUMBER 58	**(final discrete level)**
FILE #S ARE 3 & 6, MT# (TYPE < 0 IF WANT SUM) = 91	**(n,n') continuum**
FILE #S ARE 3 & 6, MT# (TYPE < 0 IF WANT SUM) = -1	
YMIN, YMAX = (CALC) 0.5011E+01 0.4390E+04	**YMIN, YMAX are minimum and maximum cross section results so that user can determine limits on plot.**
TYPE YMIN, YMAX, XMAX 1.,10000.,16.	
TYPE NAME FOR PLOT FILE NI58.PLT	
IF WANT DASHED LINE FOR TOTAL SPECTRUM, TYPE Y _	
IF WANT PARTIALS PLOTTED, TYPE Y _	
TYPE FILE WITH EXP PTS NI.DAT	
IF MORE DATA, TYPE Y _	
?*WIDTH, ?*HEIGHT FOR TEXT = 0.2, 0.92	**Position text on plot.**
TYPE MAT NAME: N^I^-58	**^ for lower case letter.**
TYPE REFERENCE, EXP. SET # 1 H^ERMSDORF ET AL.^ (HE75)	
TYPE INCIDENT ENERGY (MEV), EXP. SET # 1 14.6	
TYPE EVALUATION LABEL (<12 CHARACTERS) ENDF/B-VI	
IF WANT MORE ON PLOT, TYPE Y _	
MORE PLOTS DESIRED, TYPE Y _	

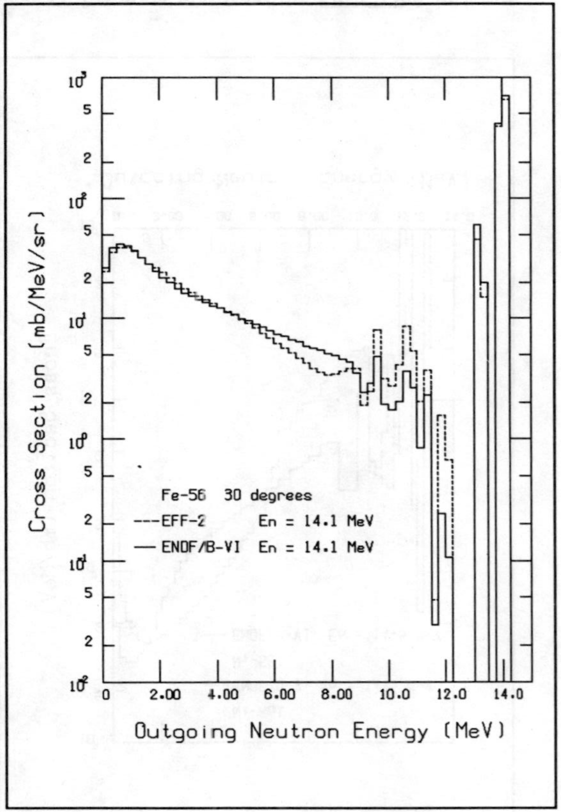

Figure 4. Example plot from the SPECTRA program.

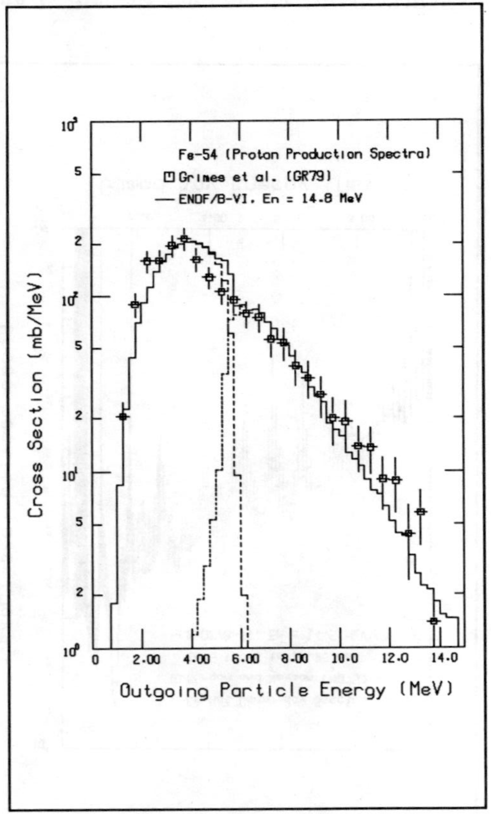

Figure 5. Example plot from the SPECTRA program.

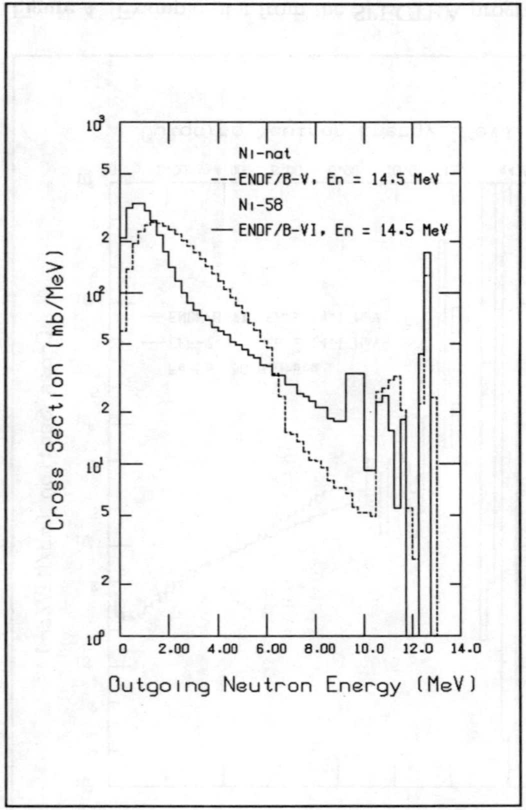

Figure 6. Example plot from the SPECTRA program.

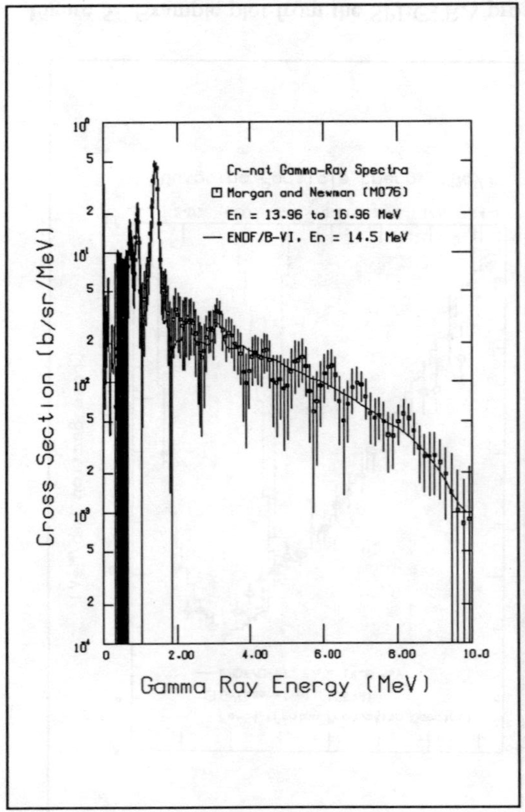

Figure 7. Example plot from the GEMENDF program.

THE INTERCOMPARISON TOOLS FOR NEUTRON NUCLEAR DATA OF STRUCTURAL MATERIALS

LIU TINGJIN   WANG CUILAN
Chinese Nuclear Data Center, CIAE, P. O. Box 275 (41)
Beijing, China

ABSTRACT

A system for intercomparison of neutron nuclear data in ENDF / B format has been developed. With the system the evaluated data in ENDF / B format and experimental data in EXFOR format can be retrieved, processed and plotted. Using the system, the neutron nuclear data of natural Fe, Cr, Ni from ENDF / B–6, JENDL–3, CENDL–2 and BROND–2 libraries have been intercompared.

1. Introduction

As the main evaluated neutron nuclear data libraries, such as ENDF / B–6, JENDL–3, CENDL–2 and BROND–2 etc., were all released in last few years, it becomes important to intercompare them, so that the best one can be recommended to users, or, as concluded at 18th INDC meeting[1], based on them develop an international standard reference evaluated neutron data library. To do the intercomparison among these libraries as well as with EXFOR experimental data, a program system was developed, with which the evaluated data in ENDF / B format and experimental data in EXFOR format can be retrieved, processed and plotted. As a test of the system and an example, using the system the intercomparison of Fe, Cr, Ni neutron nuclear data from evaluated libraries CENDL–2, ENDF / B–6, JENDL–3 and BROND–2 has been done.

2. Outline of the System

The flow chart of the system is shown in Fig. 1.

The experimental data in EXFOR format are first retrieved, processed and recorded in a interface file, which is read by codes READB or READEX and used to compare with evaluated data. The evaluated data to be compared are retrieved from the libraries, and then recorded in a interface file, which is read by codes READ3, READ4 or READ5 and used for intercomparing.

For cross section data (file 3), it is needed to interpolate in 5 interpolation modes (INT5). The data are needed not only for cross section intercomparison, but also for getting absolute differential cross section (file 4) and energy spectrum (file 5). For file 4, it is needed ot calculate the angular distribution from the Legendre coefficients (CCDS) and correspondingly to interpolate for the coefficients (INTD). Also the angular distribution needs to interpolate for incident energy E (INT2). For file 5, it is needed to interpolate for incident neutron energy E(INTY) and secondary neutron energy E′ (INTX), and to make Gaussian extension (GSS) for the single energy spectrum corresponding to inelastic scattering to discrete level. In some cases, the spectrum of natural element need to be calculated (MA4) from its isotopes′ double differential cross sections (file 6).

The data through above processing are plotted (LIBP).

3. Cross Section

At first, the complete data of a nuclide, which want to be compared, are retrieved from the corresponding evaluated libraries by using ENDF / B utility program ENDRET[2], then only cross section data (MF = 3) are retrieved by using program RETRIVE, and all cross section data from different libraries are merged into one file in standard ENDF / B format. This file is a interface file and the data for a special reaction (MT) plotting are retrieved each time. This process is to save retrieval time for each plotting as the operation speed of our available Micro–VAX–II computer is not so fast.

The experimental data in EXFOR format are first retrieved according to the required nuclide, reaction type, and then are processed in format (for example, energy point and / or region selection, unit transformation, error process etc.), analyzed and compared in physics (FORM, SIG). At last the required evaluated experimental data are recorded in a file in conveniently read format.

The retrieved evaluated data from the evaluated libraries are plotted with the corresponding measured data as well as their errors together (MF3). The coordinate scale, neutron energy and cross section region can be selected to make the figure clear. An example for cross section intercomparison is given in Fig. 2.

As there are 5 interpolation modes for cross section in the ENDF / B format, the figure could be distorted in the case that the energy interval between two adjacent energy points is large and the plot coordinate scale is not in the same as the interpolation mode. An example is given in Fig. 3. In the figure four sets of data are the same, except that there are more energy points for curves 2, 3, 4, which are given according to given linear–log, log–linear and log–log interpolation modes, while there are only two energy points for solid line, which is given in the data file. If the data are plotted in log–log (linear–log, log–linear) coordinate scale, curves 1 and 4 (2, 3) are completely the same straight line.

For angular distribution (file 4) and energy spectrum (file 5), the data retrieval, process and plotting are the same as or similar with cross section described above. So only special problems are discussed for them below.

4. Angular Distribution

Some angular distributions (for example elastic scattering) are given in Legendre coefficient format, so the relative angular distribution or absolute differential cross section must be calculated :

$$f(\mu,E) = \sum_{l=0}^{NL} \frac{2l+1}{2} f_l(E) P_l(\mu) \tag{3.1}$$

or

$$\sigma(\mu,E) = \frac{\sigma(E)}{2\pi} \sum_{l=0}^{NL} \frac{2l+1}{2} f_l(E) P_l(\mu) \tag{3.2}$$

The Legendre coefficients for given incident energy are retrieved, and if necessary, they are interpolated in linear–linear mode. In the interpolation, the coefficients for lower incident energy are supplemented to the same number as higher one by zero.

If there are experimental data to compare, the absolute differential cross sections are calculated, and in this case the cross section of the same MT number and at the same incident energy point is retrieved from file 3 (it is the same for energy spectrum).

If there are no experimental data to compare, the angular distributions are plotted at the coordinate system given by the file. If there are, they can plotted in any desired system, as the measured data can be given in any system. Also the data can be given and the desired figures can be plotted versus angle θ or their cosine μ. These information is given by four flags at the beginning of the experimental data file.

The coordinate system transformation is completed according to the following kinematics relation

$$\cos\theta_L = \frac{\gamma + \cos\theta_c}{(1 + 2\gamma\cos\theta_c + \gamma^2)^{1/2}} \tag{3.3}$$

$$\sigma_L(\theta_L) = \frac{(1 + 2\gamma\cos\theta_c + \gamma^2)^{3/2}}{|1 + \gamma\cos\theta_c|} \sigma_c(\theta_c) \tag{3.4}$$

$$\varepsilon_L = \varepsilon_c (1 + 2\gamma cos\theta_c + \gamma^2) \tag{3.5}$$

where

$$\gamma^2 = \frac{m_a m_b}{M_A M_B} \cdot \frac{K_{ci}}{K_{cf}}$$

K_{ci}, K_{cf} are the initial and final kinetic energy of the system in mass–center system respectively.

An example for differential cross section intercomparison is given in Fig .4.

5. Energy Spectrum

For energy spectrum (file 5), the interpolation is more complicated, there are three kinds of interpolation for fractional part of partial energy distribution, for secondary energy E′ and for incident neutron energy E. Usually the linear–linear mode is used for all of these interpolations and is reasonable in physics. In the interpolation for incident energy, the secondary maximum energy of the interpolated spectrum is also linearly interpolated from the maximum energies E'_{1max}, E'_{2max} of the corresponding spectra used for interpolation.

In some cases, for example there are some structures in the spectrum, the spectrum interpolated in linear–linear mode as given in the file (INT = 2) for incident energy E is unreasonable. An example is given in Fig. 5. The spectrum at E_n = 17 MeV (curve 4) is interpolated from the spectra at E_n = 16, 18 MeV (curves 1 and 2). It is clear that the shape of the curve 4 is unreasonable, it should be as curve 3, which is theoretically calculated with the same parameters as curve 1 and 2. To solve this problem, as pointed out in ENDF / B–6 format[3], "unit-base" interpolation was used. The secondary energies E'_i of the spectra used for the interpolation are first divided by their maximum energy, then the interpolation for incident energy E_n is performed in linear–linear mode as usually, and finally the "normalized" secondary energy points are multiplied by the maximum energy, which is interpolated in the same way as above.

So far, in the secondary neutron spectrum measurement it is difficult to distinguish the neutrons from different reactions, so the measured secondary neutron spectra are for all reactions, called neutron emission spectrum. To compare the evaluated (generally, calculated with model theory) secondary neutron spectrum with experimental one, the evaluated spectra for all reactions (MT numbers) of a nuclide are summed, including the contributions from the neutrons of inelastic scattering to discrete levels. The energy of these inelastic neutrons to the same level is fixt in physics, but in measurement, it is extended to a distribution due to the energy resolution of the instruments. So to compare, the

Gaussian extension is made for these single energy neutrons :

$$P_i(E_j') = \frac{\sigma_i(E)}{\sqrt{2\pi}\,\sigma} exp[-\frac{(E_j' - E_i)^2}{2\sigma^2}] \tag{5.1}$$

where i means ith discrete level, E'_j is secondary neutron energy points, σ_i (E) is the cross section of inelastic scattering to ith level at incident energy E, which is retrieved from file 3. Taking into account that the most secondary neutron spectra are measured by means of time–of–flight, the distribution width σ is taken as function of incident neutron energy $\sigma = 0.007\ E^{3/2}$ (MeV).

As the double differential cross section (MF = 6) is given in ENDF / B–6 library for each isotopes, to compare with 4π space neutron spectra of natural element in other libraries, the double differential cross sections are transform to 4π space spectra for each isotopes, and then these spectra are summed according to the abundance of the isotopes to get the spectra of natural element. In this processing, the interpolations for secondary neutron energy E′ and incident neutron energy E are needed, and the care is taken into account to the problems in the interpolation for E discussed above.

An example is shown in Fig. 6 for the intercomparison between the total neutron emission spectra with the experimental data.

Conclusion Remarks

The system has been used for intercomparison of structural materials Fe, Cr, Ni complete neutron data from latest evaluated neutron nuclear data libraries CENDL–2, ENDF / B–6, JENDL–3 and BROND–2. The plots of the neutron cross sections (MF = 3), angular distributions (MF = 4) and secondary neutron spectra (MF = 5 or 6) for each reactions (MT number) have been plotted in much detail, so that the curves can be seen clearly and the intercomparison can be made reliably.

For the system, this is a complete test. In fact, through these plotting, the system has been improved and supplemented, so that it can be used conveniently. As a whole intercomparison work, this is just a beginning, there are lot of other nuclides need to be intercompared. Even for these three elements, this is only first step, some differences have been found and need to be further analyzed in physics. An example is given in Fig. 7.

Although the system so far only used for files 3, 4, 5, as the format of files 13 (γ–production cross section), 14 (γ–production angular distribution) and 15 (γ–production spectrum) is the same as or very similar with the format of files 3, 4, 5 respectively, it can also be easily used for this files.

References

1. A. J. Deruytter, *INDC / P(91)–1* (1991). S. S. Kapoor, *Interoffice Memorandum to Prof. M. Zifferero* (1990).
2. C. L. Dunford, *ENDF utility program* (1990).
3. P. F. Rose et al., *ENDF–102* (1988).

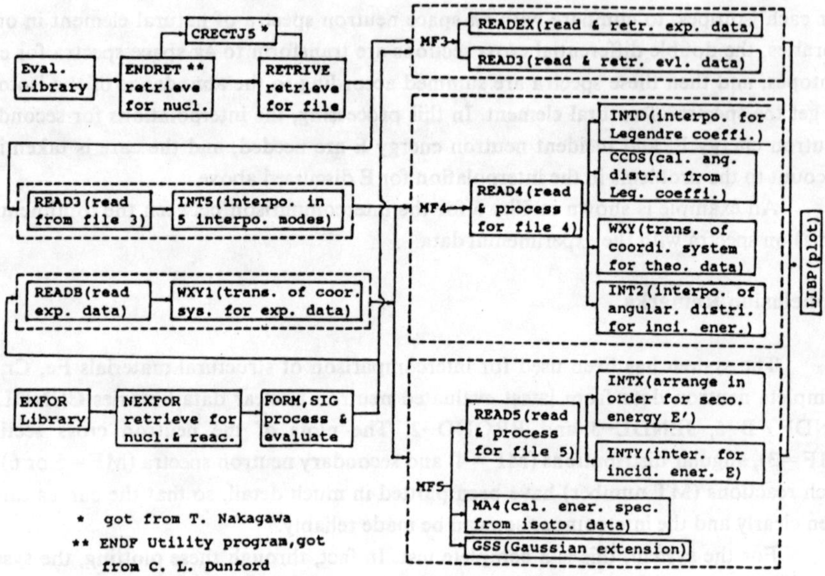

Fig. 1 Flow chart of the system for intercomparison

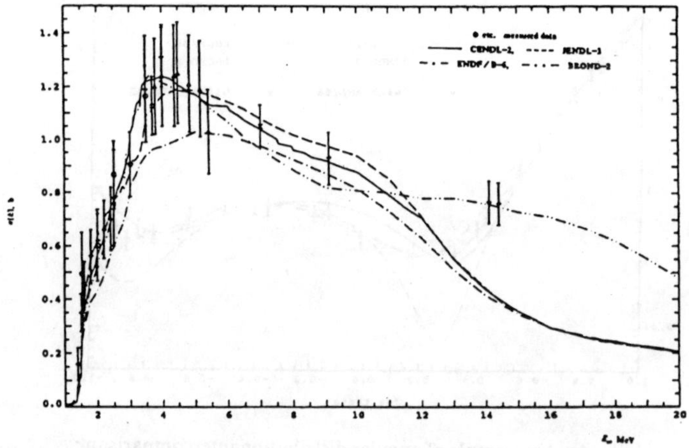

Fig. 2 An example of cross section intercomparison : Ni inelastic scattering cross section

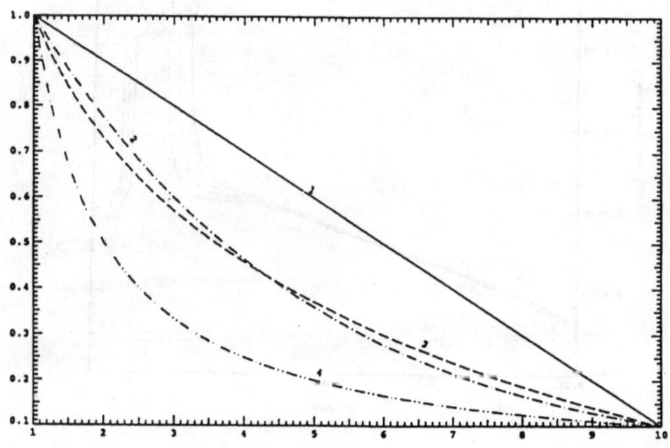

Fig. 3 Distortion form the cross section interpolation

2, linear–log; 3, log–linear; 4, log–log

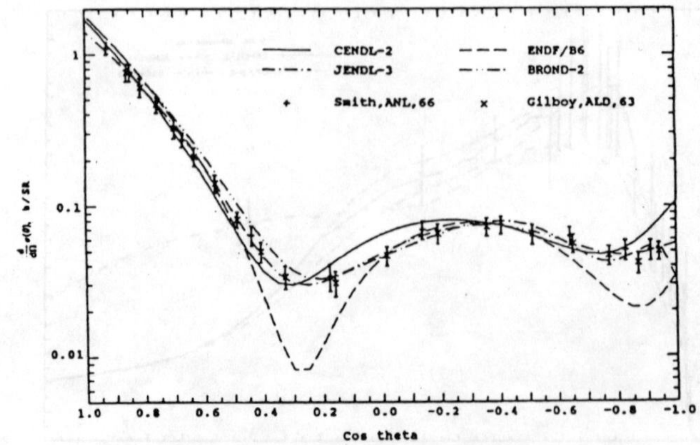

Fig. 4 An example of angular distribution intercomparison : Cr elastic differential cross section at 4.0 MeV

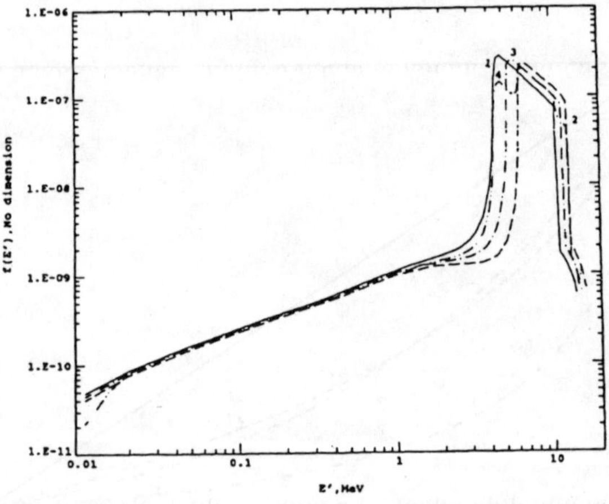

Fig. 5 An example for distortion of secondary neutron spectrum interpolation : Fe (n,n′) secondary neutron spectrum

1. E_n = 16 MeV; 2. E_n = 18 MeV;

3. E_n = 17 MeV, theoretically calculated as 1, 2;

4. E_n = 17 MeV, interpolated from 1, 2

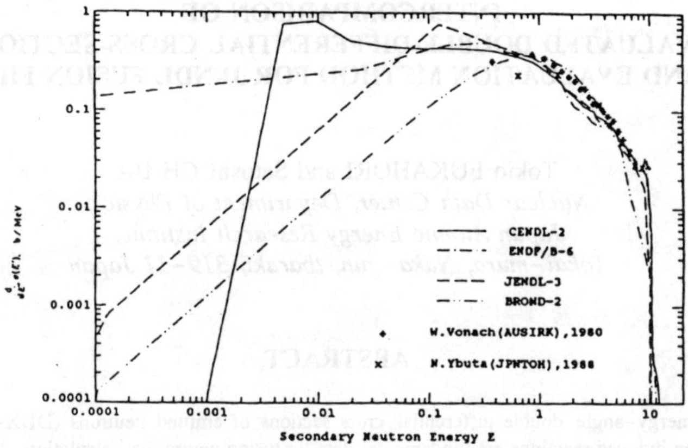

Fig. 6 An example of energy spectrum intercomparison : Ni secondary neutron emission spectrum at 14.1 MeV

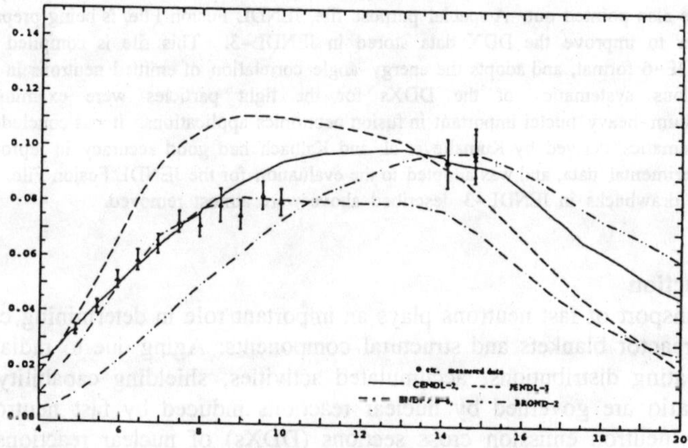

Fig. 7 An example of differences among the data from different libraries : Ni (n,α) cross section

INTERCOMPARISON OF EVALUATED DOUBLE-DIFFERENTIAL CROSS-SECTIONS AND EVALUATION METHOD FOR JENDL FUSION FILE

Tokio FUKAHORI and Satoshi CHIBA
Nuclear Data Center, Department of Physics,
Japan Atomic Energy Research Institute,
Tokai-mura, Naka-gun, Ibaraki, 319-11 Japan

ABSTRACT

The energy-angle double differential cross sections of emitted neutrons (DDXs) from neutron induced reactions are fundamental data for fusion neutronics calculation. In order to check reliability of evaluated data, the DDX data calculated from three major evaluated nuclear data libraries, JENDL-3, ENDF/B-VI, and JEF-2, were compared with experimental data measured by two Japanese groups. The comparison was made at incident energies of 1.2, 2.03, 4.2, 5.4, 6.0, 14.1 and 18.0 MeV, and at angles of 30, 60, 90, 120 and 150 deg. It was found that the DDX data calculated from the libraries could reproduce overall trend of experimentally observed values. However, some discrepancies were also pointed out. A special purpose file, JENDL Fusion File, is being prepared in order to improve the DDX data stored in JENDL-3. This file is compiled in the ENDF-6 format, and adopts the energy-angle correlation of emitted neutrons in MF=6. Various systematics of the DDXs for the light particles were examined for medium-heavy nuclei important in fusion neutronics applications. It was concluded that systematics derived by Kumabe et al. and Kalbach had good accuracy in reproducing experimental data, and was adopted to the evaluation for the JENDL Fusion File. In this file, drawbacks in JENDL-3 described above were almost removed.

1. Introduction

Transport of fast neutrons plays an important role in determining characteristics of fusion reactor blankets and structural components: Aging due to radiation damage, nuclear heating distributions, accumulated activities, shielding capability and tritium breeding ratio are governed by nuclear reactions induced by fast neutrons. Double differential neutron emission cross sections (DDXs) of nuclear reactions induced by neutrons are therefore fundamental and important data required in fusion neutronics calculations. Recent evaluated nuclear data libraries, e.g. JENDL-3[1], ENDF/B-VI[2] and JEF-2[3], have been produced to meet the above requirements. On the other hand, the DDXs have been extensively measured at Osaka university and Tohoku university[4-12].

Considering the importance of the DDX data, it is useful to compare the DDXs calculated from those libraries with the experimental data and to check reliabilities of the evaluated data. A DDX data processing code PLDDX[13] was developed to calculate the DDX from the evaluated nuclear data libraries. In Ref. 13, the comparison was made for the 32 elements at incident neutron energies of 1.2, 2.03, 4.2, 5.4, 6.0, 14.1 and 18.0

MeV, and at angles of 30, 60, 90, 120 and 150 deg. The evaluated data libraries considered were JENDL-3.1 (released in Dec. 1990), ENDF/B-VI.1 (released in Sep. 1991) and JEF-2.2 (released in Nov. 1991). It was found that the DDX data calculated from the three evaluated nuclear data libraries could reproduce overall trend of experimentally observed values. However, some discrepancies were also pointed out.

In Chapter 2, some of discrepancies are summarized. In order to improve the DDXs data in JENDL-3, a special purpose file, JENDL Fusion File, is being developed for the fusion engineering purpose. The evaluation method for the JENDL Fusion File is briefly introduced in Chapter 3. Concluding remarks are given in Chapter 4.

2. Discrepancies between Evaluated and Experimental Data

The comparison of the experimental data and the DDX calculated from the evaluated data libraries was made at the neutron incident energies of 1.2, 2.03, 4.2, 5.4, 6.0, 14.1 and 18.0 MeV, and at the angles of around 30, 60, 90, 120 and 150 deg. for the 32 elements summarized in Table 1. In this chapter, the discrepancies found by the comparisons are discussed. The discussions are restricted to limited number of elements, and only the typical graphs are shown in this paper. All the graphs are given in Ref. 13.

2.1. Lithium

The comparison made for ^{6,7}Li at the incident neutron energy range from 4.2 to 6.0 MeV indicates that the three evaluated data libraries give similar results. The DDXs calculated from the libraries can reproduce the experimental data well. However, discrepancies were found at 14.1 MeV. Figures 1 and 2 show the DDXs of ^{6}Li at the emitted angles of 30 and 120 deg. and at the incident neutron energy of 14.1 MeV. The JENDL-3 data agree with the experimental data[11] at all the angels excellently. The ENDF/B-VI data are smaller than the experimental data below 2 MeV at all the angles, and larger around 5 MeV at the backward angles. JEF-2 gives smaller values in the 6-10 MeV region at the forward angles, and larger at the backward angles. The ^{7}Li data at 14.1 MeV are compared with the experimental data for ^{7}Li[7] and ^{nat}Li[4]. It is found that the data of JENDL-3 reproduce the experimental data, but other two libraries give smaller values in the secondary neutron energy region just below the second inelastic scattering peak at the forward angles.

2.2. Beryllium

Figures 3 and 4 show the comparisons of ^{9}Be with the experimental data[6,9] at the incident neutron energy of 14.1 MeV. The JENDL-3 data agree with the experimental data satisfactorily. The ENDF/B-VI data almost have similar trend to the JENDL-3 data, except around the first inelastic scattering peak at the backward angles. The JEF-2 data slightly overestimate the DDXs below 9 MeV at the forward angles. At the backward angles, a 'shoulder' shape appears above the elastic scattering peak calculated from JENDL-3. In JENDL-3, the (n,2n) neutron spectrum corresponding to the continuum inelastic neutrons is given in MF=5 which cannot take account of the energy-angle correlation properly.

2.3. Oxygen

The DDXs of ^{16}O are compared with the experimental data[6,9] at the incident energies of 14.1 and 18.0 MeV. The JEF-2 data are equivalent to the ENDF/B-VI data.

The three libraries give similar results at 14.1 MeV and reproduce the experimental data. At 18.0 MeV, the data of the three libraries have the same trend in the secondary energy region above 10 MeV. In the energy region below 10 MeV, however, they are discrepant from each other. Below 4 MeV, the JENDL-3 data reproduce the experimental data at the forward angles, but they are significantly lower than the experimental data just below the first inelastic peak. For the elastic peak at the backward angles, JENDL-3 gives higher values than other two libraries. The other two libraries reproduce them above 4 MeV, but give much lower values than the experimental data below 4 MeV.

2.4. Aluminum

The comparisons of ^{27}Al data are made with the experimental data at the incident energies of 14.1[5,10] and 18.0 MeV[10]. The evaluated data are in almost good agreement with the experimental data. However, JENDL-3 gives smaller values in the discrete inelastic region at the forward angles at 14.1 MeV and at all the angles at 18.0 MeV. The JEF-2 data overestimates the DDXs in the low energy region at the forward angles.

2.5. Iron

The DDXs calculated from the evaluated data of ^{nat}Fe are compared with the experimental data[6,12] at the incident energy of 14.1 MeV in Figs. 5 and 6, and with the data[10] at 18.0 MeV. For ENDF/B-VI and JEF-2, the ^{56}Fe data are shown in the figures, since they include no data for the natural element. The JENDL-3 data at both the incident energies tend to underestimate the DDX slightly at the forward angles and to overestimate them at the backward angles in the middle energy region of secondary neutrons. JEF-2 gives the lower values than the experimental data in the continuum region at the forward angles. The reason of this underestimation is attributable to insufficient consideration of the direct inelastic scattering and preequilibrium processes.

2.6. Copper

The DDXs of ^{nat}Cu are compared with the experimental data at the incident neutron energies of 14.1[5,10] and 18.0MeV[10]. The comparison is not made for ENDF/B-VI, since no data for the natural element are given in ENDF/B-VI. JENDL-3 almost reproduces the experimental data at both the incident energies, while the JEF-2 data are too small in the discrete inelastic region.

2.7. Niobium

Figures 7 and 8 show the DDXs calculated from the evaluated data of ^{93}Nb with the experimental data[6] at the incident energy of 14.1 MeV. JENDL-3 almost reproduces the experimental data, while ENDF/B-VI overestimates the DDX in all the energy region at the backward angles, and JEF-2 underestimates them at the forward angles.

2.8. Lead

The comparison is made for ^{nat}Pb with the experimental data[5] at the incident neutron energy of 14.1 MeV. ENDF/B-VI is not included in the comparison, because of no data for the natural element. JENDL-3 overestimates the experimental data in the discrete inelastic scattering region at the backward angles, and underestimates them in the preequilibrium neutron energy region at the forward angles. The JEF-2 data almost reproduce the experimental data, except just below the elastic scattering peak.

2.9. Uranium

The DDXs of ^{238}U are compared with the experimental data[8] of ^{nat}U at the incident neutron energy range from 1.2 to 14.1 MeV. The JENDL–3 data almost reproduce the experimental data at 1.2 and 2.03 MeV. At the incident neutron energy of 14.1 MeV, JENDL–3 gives smaller values in the energy region from 5 to 12 MeV, since the energy spectrum of the continuum inelastic scattering in JENDL–3 is represented with only the evaporation spectrum. In the discrete inelastic scattering energy region, JENDL–3 gives much higher values than the experimental data at the incident energies above 4.2 MeV. The fission tail lying in the high energy region above the elastic scattering peak is in good agreement with the experimental data.

2.10. Summary of Discrepancies

Table 2 summarizes the discrepancies between the experimental data and the data calculated from the three evaluated nuclear data libraries. The discrepancies are caused by the following four reasons: 1) The direct process of the discrete inelastic scattering is not considered. 2) The preequilibrium process is not considered properly. 3) The energy–angle correlation (MF=6) is not considered. 4) The other discrepancies are caused by each evaluation method, such as the evaporation spectrum is not reproduced.

3. Evaluation Method for JENDL Fusion File

For applications of nuclear data in the fusion neutronics field, the accurate energy–angle correlated neutron spectra are strongly required. Therefore the drawbacks of the evaluated data described above must be removed. A special purpose file, JENDL Fusion File, is being prepared in order to solve the above mentioned problems of the DDX data stored in JENDL–3. This file is compiled in the ENDF–6 format, and adopts the energy–angle correlation of emitted particles in MF=6. Data are basically taken from JENDL–3, by making modification of the data concerning the DDX. The DDX data of charged particles are stored as well as those of neutrons. Their DDXs are evaluated by adopting suitable systematics.

Various systematics to calculate the double–differential light particle emission cross sections from nuclear reactions were examined for medium–heavy nuclei important in fusion neutronics applications[14]. They need a ratio of the multistep direct component to the total particle emission spectrum, which was calculated with a code system[15] of ELIESE–3, EGNASH2 and an auxiliary code. Results calculated by the systematics were compared with the experimental data for 14 MeV neutrons. It was concluded that the systematics derived by Kumabe et al.[16] and Kalbach[17] had good accuracy in reproducing the experimental data. In the case where JENDL–3 dose not include the contribution of the direct inelastic process, its contribution was calculated with DWUCKY which is a modified code of DWUCK–4. The JENDL Fusion File reproduces the measured DDX data better than JENDL–3, as shown in Figs. 5–8 with the thick solid line. At present, the evaluation work for the JENDL Fusion File is continued. Finally, the data for about 100 nuclides will be stored. More details of the evaluation method are given in Ref. 18.

4. Concluding Remarks

The comparisons of DDXs calculated from the three evaluated nuclear data libraries with the experimental data were performed for 32 elements important in fusion technology. The discrepancies between the experimental and evaluated data are mainly caused by the following reasons: 1) The direct process of the discrete inelastic scattering is not considered. 2) The preequilibrium process is not included properly. 3) The energy–angle correlation is not included. The comparison of DDX is useful to check the reliability of the evaluated data. According to the results of the comparison, the JENDL Fusion File is being made in order to improve the DDX data of JENDL–3 for the fusion neutronics. The evaluation method of JENDL Fusion File was described briefly.

Acknowledgements

The authors would like to thank Dr. A. Takahashi of Osaka university, Dr. M. Baba of Tohoku university and their coworkers for supplying them with their experimental data.

References

1. K. Shibata, T. Nakagawa, T. Asami, T. Fukahori, T. Narita, S. Chiba, M. Mizumoto, A. Hasegawa, Y. Kikuchi, Y. Nakajima and S. Igarasi, *JAERI*–**1319** (1990).
2. BNL/NNDC, ENDF/B–VI (1990).
3. NEA/Data Bank, JEF–2 (1991).
4. A. Takahashi, *JAERI–M* **86–029** (1986) 99.
5. A. Takahashi, E. Ichimura, H. Sugimoto and T. Kato, *JAERI–M* **86–080** (1986) 393.
6. A. Takahashi, private communication.
7. S. Chiba, M. Baba, H. Nakashima, M. Ono, N. Yabuta, S. Yukinori and N. Hirakawa, *J. Nucl. Sci. Technol.* **22** (1985) 771.
8. M. Baba, H. Wakabayashi, N. Itoh, K. Maeda and N. Hirakawa, *JAERI–M* **89–143** (1986) in Japanese.
9. M. Baba, M. Ishikawa, T. Kikuchi, H. Wakabayashi and N. Hirakawa, *Proc. of Int. Conf. on Nucl. Data for Sci. and Technol., Mito, May 30– Jun. 3* (1988) 209.
10. M. Baba, M. Ishikawa, N. Yabuta, T. Kikuchi, H. Wakabayashi and N. Hirakawa, *Proc. of Int. Conf. on Nucl. Data for Sci. and Technol., Mito, May 30– Jun. 3* (1988) 291.
11. M. Baba, S. Matsuyama, M. Fujisawa, T. Iwasaki, S. Iwasaki and R. Sakamoto, *JAERI–M* **90–025** (1990) 383.
12. M. Baba, private communication (1991).
13. T. Fukahori, S. Chiba and T. Asami, *JAERI–M* **92–053** (1992).
14. B. Yu, S. Chiba and T. Fukahori, *J. Nucl. Sci. Technol.* **29** (1992) 677.
15. N. Yamamuro, *JAERI–M* **90–006** (1990).
16. I. Kumabe, Y. Watanabe, Y. Nohtomi and M. Hanada, *Nucl. Sci. Eng.* **104** (1990) 280.
17. C. Kalbach, *Phys. Rev.* **C37** (1988) 2350.
18. S. Chiba, B. Yu and T. Fukahori, *JAERI–M* **92–027** (1992) 35.

Table 1 The summary of elements and isotopes for which the comparison was performed. The element name without atomic number means the natural element.

Incident Energies	Elements or Isotopes
1.2 MeV	Th,U
2.03 MeV	Th,U
4.2 MeV	^{6}Li,Th,U
5.4 MeV	^{6,7}Li
6.0 MeV	^{7}Li,Th,U
14.1 MeV	^{6,7}Li,^{9}Be,^{10,11}B,C,N,O,^{19}F,^{23}Na,Mg,^{27}Al,Si,S,Ca,Ti,V,Cr,^{55}Mn, Fe,^{59}Co,Ni,Cu,Zr,^{93}Nb,Mo,Sn,Sb,^{181}Ta,W,Pb,^{209}Bi,Th,U
18.0 MeV	C,O,^{27}Al,Fe,Ni,Cu,Zr

Table 2 Classification of the elements by the origin of discrepancies between the experimental data and the evaluated data

Origins of Discrepancies	JENDL-3	ENDF/B-VI	JEF-2
The Region of Direct Process for Discrete Inelastic Scattering	N,F,Ti,V,Co, Zr,Nb,Mo,Sn, Sb,Ta,W,Pb,U	Be,N,Na,Mg, Al*,Si,S,Ti,V, Mn,Zr,Nb,Mo, Ta,W,Bi,U	N,Mg,Al*,Si, Ti,V,Mn,Fe, Cu,Zr,Nb,Mo, Ta,Pb,Bi,Th,U
The Region of Preequilibrium Process	Al*,Ti,Cr,Co, Cu*,Zr,W	Li,Ti,Fe,Co, Zr,Nb,Ta,Bi	Ti,V,Mn,Fe, Co,Cu*,Zr,Nb, Mo,Ta
The Energy-Angle Correlation	Be,Mg,Ca		
The Others	B,O*,Si,S,Ca	Li,B,C,O*,F, Si,S,Ca,Zr, Nb,Mo	Li,Be,B,C,O*, F,Si,Ca,V,Fe, Zr

* The discrepancies are observed only at the incident neutron energy of 18 MeV.

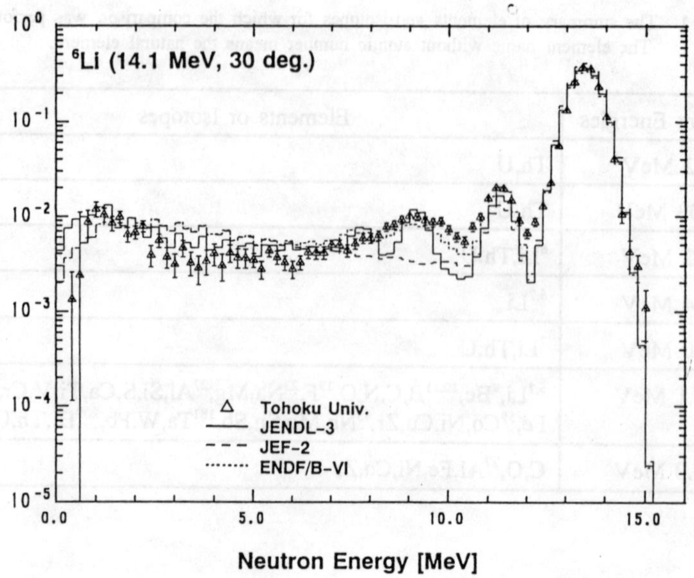

Fig. 1 The ^{6}Li DDX at 14.1 MeV, Emitted Angle = 30 deg. in Laboratory System

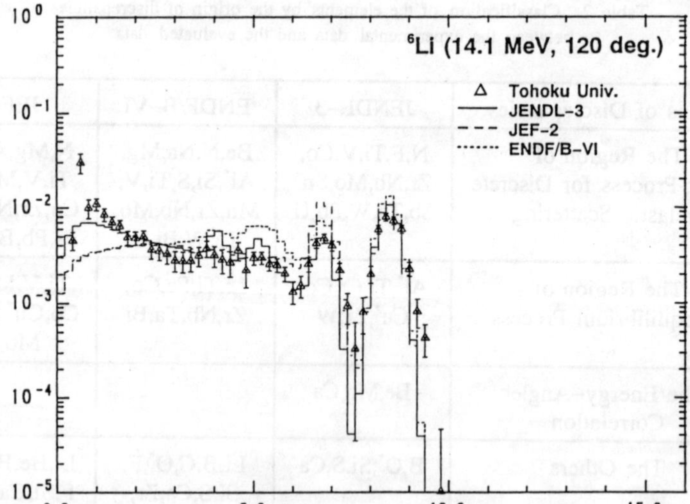

Fig. 2 The ^{6}Li DDX at 14.1 MeV, Emitted Angle = 120 deg. in Laboratory System

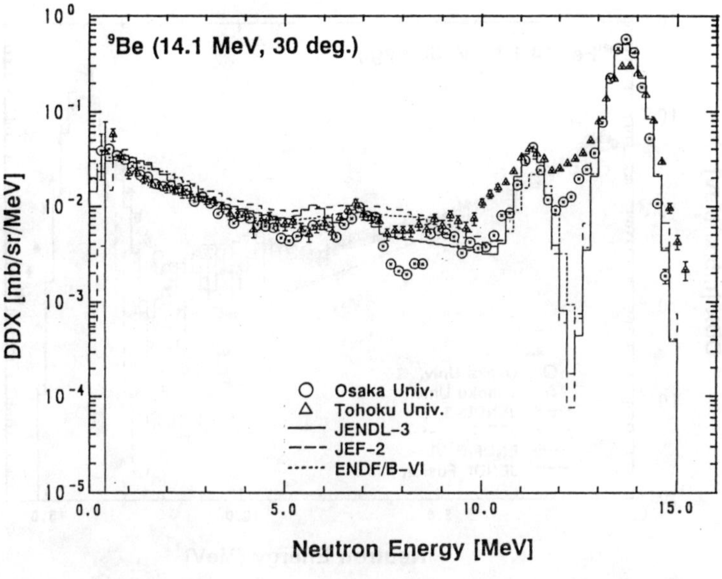

Fig. 3 The ^{9}Be DDX at 14.1 MeV, Emitted Angle = 30 deg. in Laboratory System

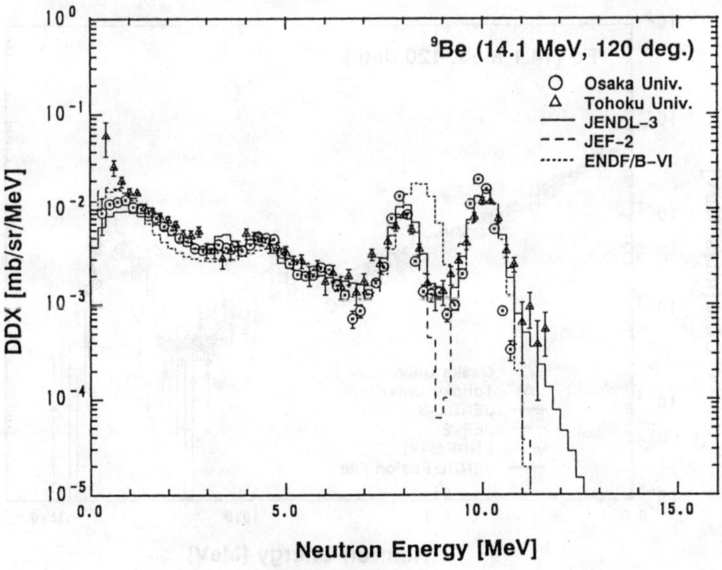

Fig. 4 The ^{9}Be DDX at 14.1 MeV, Emitted Angle = 120 deg. in Laboratory System

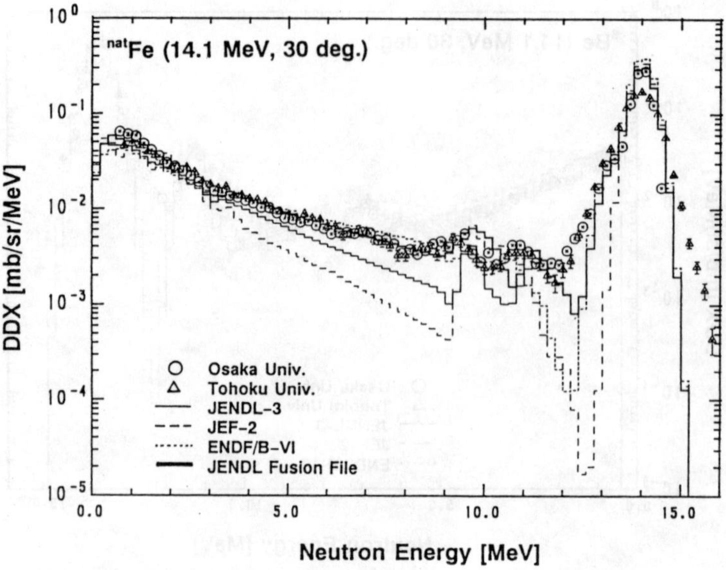

Fig. 5 The ^{nat}Fe DDX at 14.1 MeV, Emitted Angle = 30 deg. in Laboratory System

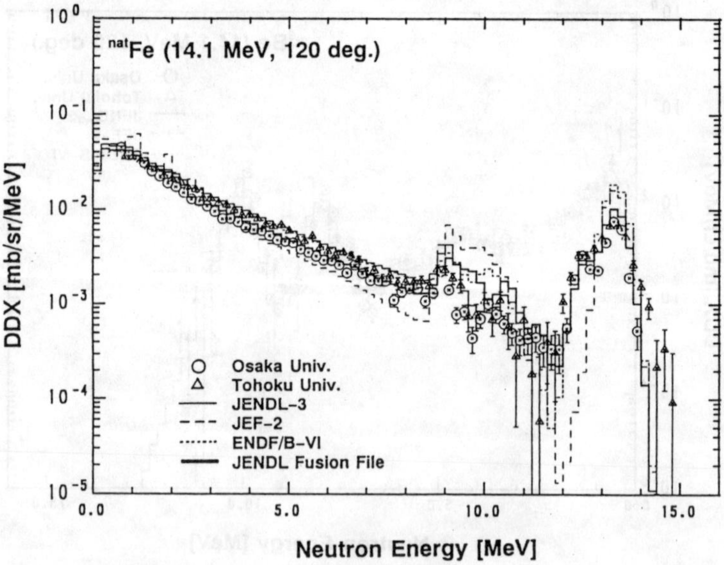

Fig. 6 The ^{nat}Fe DDX at 14.1 MeV, Emitted Angle = 120 deg. in Laboratory System

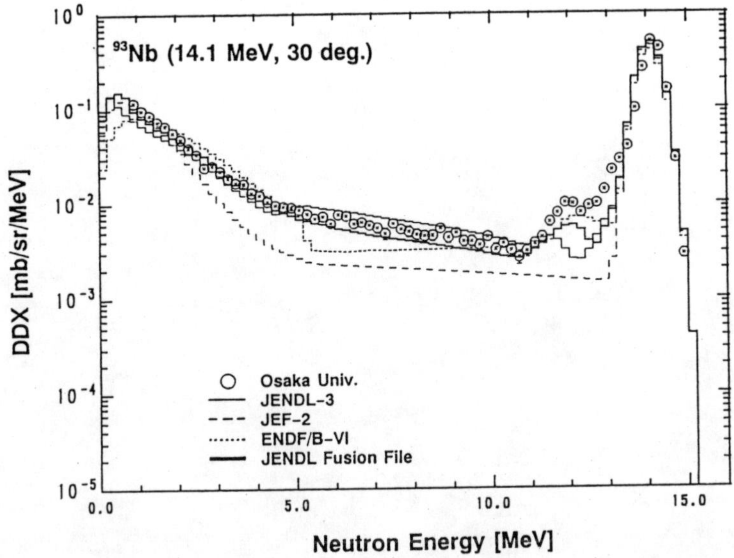

Fig. 7 The ^{93}Nb DDX at 14.1 MeV, Emitted Angle = 30 deg. in Laboratory System

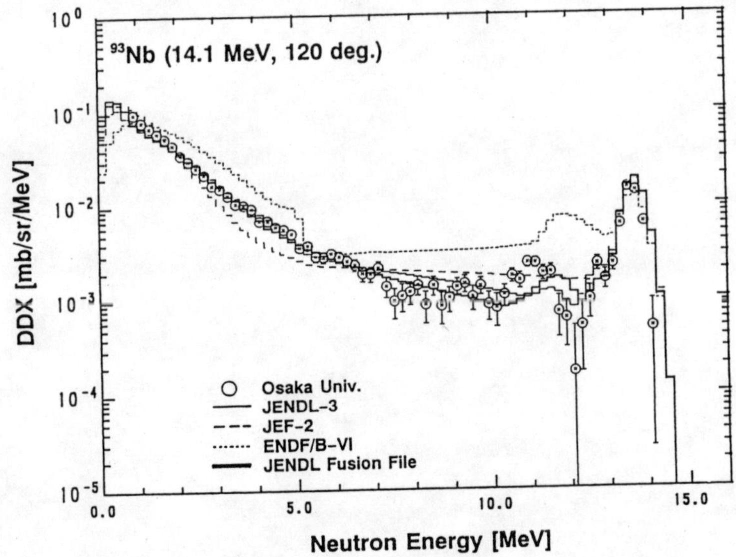

Fig. 8 The ^{93}Nb DDX at 14.1 MeV, Emitted Angle = 120 deg. in Laboratory System

ANALYSIS OF EXPERIMENTAL DATA

THE RESOLUTION OF DISCREPANCIES AMONG NUCLEAR DATA

ROBERT PEELLE
Consultant, Nuclear Data Measurement and Evaluation Section
Oak Ridge National Laboratory, P. O. Box 2008
Oak Ridge, Tennessee, USA 37831-6356

ABSTRACT

Significant differences among input data occur in the evaluation of nuclear data because it is difficult to achieve experimental results with the accuracy required for some applications. Types of "discrepancies" are classified. The means are reviewed by which an evaluator may treat discrepancies in the process of evaluation. When all means fail that are based on how the discrepant data were obtained, the perplexed evaluator must sometimes combine discrepant data based just on the stated values and uncertainties; techniques for treating such challenges are compared. Some well-known data discrepancies are examined as examples.

1. Introduction

Particularly for neutron reactions, experimenters often must strain to achieve the level of accuracy required by nuclear data users. In most such cases theory provides some guidance, but model calculations yield insufficient absolute accuracy for those cross sections technology requires be known within a few percent. Moreover, in some cases there has been difficulty in reaching consensus on whether certain cross sections should be accurately represented in file formats and, if so, how best to accomplish the goal. An example is that only in the recent data files for the MeV region has the ENDF/B system included formats to represent continuum yields that are functions of both secondary angle and energy.[1] Finally, differential data sets have been subjected to application-embedded integral tests of data and methods which themselves are difficult to perform and interpret. In this context all types of discrepancies are to be expected.

Typically, microscopic evaluated neutron data are used as input for complex transport calculations designed to estimate system parameters of practical interest. First order sensitivity theory is used to estimate the effects of cross section changes on these calculated parameters.[2] The same sensitivity coefficients allow propagation of data uncertainties and covariances to give the uncertainties in, and correlations among, the calculated parameters. Differences between calculated and measured integral parameters for test systems can be analyzed using sensitivity coefficients only to the extent that evaluators have included covariances in the data files for the reactions of importance. Covariance data are also required for integral benchmark data.[3]

Data discrepancies are resolved by thoughtful persons who find problems in experimental measurements, rather than by "canned" procedures applied to published results. Good decisions are often reached by distillation of ephemeral clues. Therefore, this paper attempts to categorize, clarify, and comment on methods used to make progress in the presence of discrepancies, rather than to provide preferred recipes for action. It is hoped to remind evaluators of the present possibilities and encourage them to seek improved strategies.

2. Types of Nuclear Data Discrepancies

Dialog about "discrepancies" would be enhanced if workers more carefully distinguished among the situations that commonly arise.

"Truly discrepant" experimental values differ so much relative to the stated uncertainties that the nominal probability is $<10^{-3}$ that such disagreement would occur by chance. (Beware non-normal probability density functions!) This is a case that pains evaluators as old records are reviewed, values and/or uncertainties are altered with the grudging permission of authors, and somewhat arbitrary techniques are sometimes employed in attempts to resolve the apparent problem.

"Important differences" among values, judged relative to application needs, occur when data are not truly discrepant but differ enough to concern users. Frequently, more accurate experiments need to be sought, for there is little point in debating which inadequate values to adopt. Careful analysis of uncertainties by experimenters allows this situation to be recognized.

"Discrepancies with integral experiments" are important differences between the results of integral experiments and the corresponding values computed from a differential cross section data base. When all data include covariance information and analysis biases and uncertainties are accounted for, true discrepancies sometimes remain. Since such comparisons usually require complicated computational models, the study of such discrepancies must go beyond the experimental values themselves.

"Discrepancies with theory" occur when experimental values for some quantities are systematically inconsistent with model calculations. Such discrepancies often occur in angle or energy distributions for a range of incident energy and a series of nuclides. Perhaps the theory being used can't account for an observed trend, or perhaps there are uncompensated systematic errors in the experiments in these regions. Such systematic discrepancies should not be treated as random uncertainties in assessing the uncertainty of a set of model parameters, but rather as an indication of the need to refine either the nuclear model or the experiments.

3. Evaluating Discrepant Data

Here we consider true discrepancies among comparable data elements, and wish to itemize the approaches used by evaluators to deal with the discrepancy to allow presentation of a combined value and its uncertainty. Assume that the experimenters have given equivalently defined numerical uncertainties as well as any pertinent covariance data.

A typical evaluator first updates values to correct apparent mistakes in analysis and to include modern values of any auxiliary information previously used in reducing the raw data. Data sets should be used according to the quantities measured, even if these are not the quantities reported by the authors.[4] For example, values reported as cross sections in journals may have been measured as unnormalized ratios to some standard cross section. By the time these initial steps have been completed, the existence of an apparent discrepancy may be recognized. If the data are important, the evaluator should ask the experimenters to comment on the discrepancy and any weak points in the documentation. The evaluator should also consider initiating a request for a measurement with improved technique or most careful execution of a previously used technique. This can be done by direct contact with experimenters and/or by submitting the item to the appropriate nuclear data request list.[5] In the interim, some means to combine the existing data will usually be required.

An evaluator may next look for <u>legitimate</u> reasons to increase uncertainties on specific data elements based on an author's failure to include numerical values for all acknowledged uncertainties, failure to recognize some inevitable sources of uncertainty, or documentation that does not justify unusually small reported uncertainties. (The "needed increase" is not a justification here.) A clue in this search may be large differences between uncertainties reported for different experiments performed using similar techniques; however, sometimes the conditions were indeed much more favorable for some of the experiments. The difficulties of this evaluation process are legendary.[6]

If the data remain discrepant, the evaluator may consider the following steps:

(a) Examine the raw data for the experiments that give the discrepant results. If the features of interest are unclear in the raw data, there may be justification to reject a data set or to enlarge the uncertainties quoted for an experiment. Examples occur in the evaluation of resonance parameters.[7]

(b) Increase the uncertainties on older data as once discussed by Poenitz,[8] preferably in some systematic way that doesn't depend on recognition of a discrepancy. Equipment and techniques were generally superior in 1985 relative to those in 1965, and experimenters had the advantage of more experience. Note the weakness, however, in general use of weighting according to newness; nuclear data facilities and scientists may well be less able to produce precise measurements in 1995 than in 1985.

(c) Assign a minimum uncertainty for any measurement of a given type, based in part on the known capabilities of the experimental technique.[9] In the limit, each experiment would be given equal weight, but it is probably self-defeating to set the minimum uncertainty too large. For multipoint data sets this idea must be elaborated to consider correlation patterns.

(d) Consider trends apparent in other data from the same author and experimental equipment, perhaps for nuclides for which a greater wealth of relevant data is available, that could justify rejection of a data set or assignment of an expanded uncertainty. Sometimes data nearly impossible to explain by theory can be rejected.

If discrepancies persist after these techniques have been used or rejected, one may perform a weighted average and accept the result. A few apparent discrepancies are indeed expected. Often, the uncertainty of the quoted weighted average is increased by scaling the variances and covariances of all input values by a common factor to make the chi-square just equal to the number of degrees of freedom.[10] This means that each input is given a relative weight in accord with its author's estimate, and the scatter among the experimental results determines the uncertainty of the average. The combined result itself remains unchanged, even though it is likely that the (unknown) proper correction of the input data would change the data correlation pattern and the average value.

If reliance on the above approaches seems inadequate, the evaluator may consider those in the next section.

4. Numerical Techniques For Addressing Discrepancies

The techniques considered below are more controversial because they do not refer to the experiments that produced the discrepant values. They can be attractive and perhaps necessary if discrepancies remain after detailed evaluation of the underlying experiments or if there is no opportunity for such expert evaluation. Use provides the data evaluator a transparently impersonal "objective" approach and thereby some protection against authors of errant data. However, primary reliance on such techniques will reduce the likelihood that the evaluator can make an intellectual contribution by pointing to corrections needed in the presented data. More important, experimenters would have less incentive to document their work carefully and thereby might be less likely to find their own previously unsuspected errors. The paragraphs below refer to some recent papers on this popular topic.

Froehner[11] exhibited a Bayesean formalism in terms of unrecognized errors on each data element. The approach has the merit that it is derived in a clear-cut way from simple assumptions. Knowledge concerning the relative "reliability" of the experiments (with respect to unrecognized errors) can be included. The evaluated results from this formulation move smoothly from the limit for data consistency to the limit for extreme data inconsistency. The former yields the usual weighted average with the original uncertainties. The latter leads (for equal reliability) to an unweighted average with uncertainty tied to the value of the parameter τ for the width of the assumed density function of the unrecognized errors. The text of ref. 11 does not prescribe a specific choice of this parameter τ. Stepping outside the spirit of Froehner's work, one could adjust τ to achieve consistency. For the case of equal reliability, the updated uncertainties on the input data would be formed by combining the original uncertainties with τ in quadrature. The uncertainty in the combined result is taken from the extended uncertainties just as in a weighted average. Note that "outlier" values are not automatically taken to be suspect. Froehner has considerably extended this approach to deal with the scale of the τ values.[12]

Statistical criteria are used by Yaborov[13] to identify which of a series of data elements are "mismatched" in that they are believed to have uncertainties larger than postulated. In the proposed scheme, the uncertainties of these mismatched data are then scaled up. Note that in this approach and the others below the uncertainty will never be scaled up on a value close to the average, even though in fact its uncertainty may have been badly underestimated.

The paper of James, Mills, and Weaver[14] considers each experiment result in terms of its "normalized residual," the difference between it and the weighted mean of the remaining measurements, divided by the standard deviation of that difference. They suggest reducing the weight of the data element that has the largest normalized residual if it is greater than a selected limiting value. The weight is reduced by the amount just needed for the modified residual to have the chosen limiting value. The defined normalized residual has nice properties for this type of analysis. The present author is concerned that this method could lead to inappropriate deweighting of outliers unless the user has previously identified and corrected all unreasonably small uncertainties.

Various techniques are discussed and compared by Rajput and MacMahon,[15] and a process termed the "Rajeval technique" is described. First, results are rejected as outliers that have residuals greater than a selected critical value. (Here, the residuals are calculated like the normalized residuals of the previous paragraph except that the unweighted mean is used.) Then the weighted residuals for the remaining values are examined to identify those deemed inconsistent according to a criterion that depends on the number of measurements. The uncertainties on these values are increased until the criterion for consistency is met.

While all these approaches are thoughtful efforts that have performed encouragingly in specific applications, some seem to require rather arbitrary selection of critical values. The present author suggests that a potential user try any prospective methods against a few test cases with anomalies of the types expected in the data to be examined, just to be sure odd cases will be handled appropriately. For example, selection of a technique might depend on whether one expects biases in some of the values rather than badly underestimated standard deviations. Overall, it seems safer to utilize a technique that has a clear if idealized theoretical foundation.

5. Examples of Resolution of Nuclear Data Discrepancies

A few examples of nuclear data discrepancies of the last 20 years are reviewed below. Talks later in this session cover other examples. A discrepancy file is maintained by international nuclear data organizations.[16]

5.1 *Fission Neutron Multiplicity of* ^{252}*Cf.*

The average number of fast neutrons emitted per fission $\overline{\nu}$ is the single parameter to which reactor physics calculations are most sensitive.[17] Fortunately it can be measured well. The value for spontaneous fission of ^{252}Cf has long been used as a reference standard. By 1970, in addition to ongoing experiments, there were two careful experimental measurements based on counting of prompt neutrons absorbed in liquid scintillator, and others equally careful based on time-uncorrelated measurements of emitted neutrons absorbed in manganese baths or analogous instruments. Depending on the corrections applied, the results were very weakly inconsistent with the quoted uncertainties. However, the scintillator tank results appearing to be larger by 1%, a few times the 1/4% uncertainty quoted on the weighted mean of about 3.73 neutrons/fission.[18]

Teams of expert experimenters combed experiment documentation for obscure errors, and authors received repeated requests for additional experimental details. Numerous re-evaluations of the data observed in particular experiments were performed and several were adopted. H. Kouts in the keynote address to the 1975 nuclear data conference labeled the situation a scandal.[19] It became apparent that improved data were required to support a needed evaluated uncertainty in the neighborhood of 0.2%, while the measurements existing then had quoted uncertainties of greater than 0.4%. Considering the importance of the data, experts were not willing to rely too much on reduction of uncertainty through averaging. In this light, it appeared that additional measurements were needed using techniques that would take advantage of all the insight gained over the years. Spencer and Gwin performed such a painstaking experiment and achieved a claimed uncertainty of about 0.2%.[20] For now the controversy has abated even though their value is 1% higher than the average of the others.

Looking back at the situation, the data set after corrections in 1972 showed important differences among the data, but only borderline discrepancy. When an improved experiment could be completed and documented to avoid many of the concerns about earlier experiments, it appeared that the need had been met. Moreover, the whole set of thermal neutron constants had become internally consistent with a ^{252}Cf $\overline{\nu}$ value of 3.768 ± .005.[21]

5.2 *Fission Cross Section of* ^{235}*U.*

Because of the direct importance of fast neutron fission in ^{235}U and its use as a standard cross section, work was strongly underway in the 1960s toward achieving 1% uncertainty in this cross section from 0.1 MeV to at least 15 MeV. Great efforts were made to improve techniques. An apparently excellent measurement of this era was performed by White[22] at several energies up to 14 MeV. His documentation was so detailed and thoughtful that other workers assumed that his results would endure even though his values were about 5% below previously favored ones. Viewed from a more recent perspective, he worked against major disadvantages including lack of a pulsed source to allow flight-time discrimination against backgrounds from fission at lower energy. Except at 5.4 MeV, his original values, quoted to 2-3%, are all larger than those in ENDF/B-VI by 0.7 to 3 times his uncertainty estimates.

The process of upgrading data continues, though only extremely precise and somewhat diverse experiments are generally encouraged. Since the ENDF/B-VI evaluation, new broad-range data were obtained including energies above 15 MeV[23] where the data base had been relatively thin, and notable differences were recorded. Another evidence that this and other standards data are not yet mature is that complete covariance data are not yet released for ENDF/B-VI, and the values of the reported evaluated uncertainties are questioned. The uncertainties given by the data combination procedure used for the evaluation do suggest that the 1% accuracy goal has been reached from 0.2 to 15 MeV for 70 average cross section values covering that interval.

This example illustrates that while skillfully obtained and documented data can be affected by systematic error, persistent effort by a variety of approaches can be rewarded by gradually convergent results. The evaluator is left with a puzzle how to fairly treat the results from the decade prior to the work of White.

5.3 Energy Balance in Evaluations.

Upon the release of ENDF/B-V, it was discovered that when evaluations were used to compute the kinetic energy released locally in a material (KERMA) there were energy balance discrepancies.[24] That is, after taking into account the Q values of reactions, the incident and outgoing energies were not equal. In the MeV region, a major source of this discrepancy was that measured reaction gamma-ray emission spectra were often placed directly in the file, while the neutron energy distributions for reactions feeding those emissions were evaluated based on other data. There were bound to be distressing differences between the two sources of information even if those differences were within known uncertainties.

All (excluding fission) or most of the neutron cross sections and secondary distributions in these regions are obtained from nuclear model calculations that are benchmarked against the available data. When these models include de-excitation gamma-ray cascade treatments they produce photon yields vs. incident energy that are included in the benchmarking against experimental observations.[25] However, much of the experimental photon data were obtained using elemental samples, while the evaluations are best performed isotope by isotope.

In many of the current evaluations,[26] energy imbalance in the evaluation has been avoided by using model codes to give the ENDF/B gamma-ray data required for applications. (In others, various adjustments have been applied to force energy balance.) Energy distributions from neither secondary neutron nor photon data are included directly in the file, but both types of data are examined during the process of setting up the model calculations to achieve a good compromise parameter set. The presence of any discrepancy between them may be deduced from the separate comparisons with model calculations. In that sense, the evaluations have been made internally consistent, but serious inconsistencies between neutron and photon data may remain.

6. Concluding Remarks

Numerical techniques such as weighted least-squares data combination have most validity with consistent or nearly consistent data; in face of discrepancy the use of technical judgement is necessary. For important data, overcoming either real discrepancies or important differences is likely to be slow and thereby expensive work.

Evaluators need to be bold though diplomatic, and should not fear to exercise and document scientific judgement related to updating values and uncertainties in the work of others. (Note that there is substantial disagreement on this point!)

Use of theory as an evaluation tool is virtually required as incident energies rise above the threshold for emission of second and third particles, and numerous reaction channels open that may not have been observed in any experiment. Evaluation theorists need to remain aware of the level of experimental verification available, and experimenters need to seek and perform critical measurements to help theorists control systematic modeling error.

7. Acknowledgement

The author acknowledges the many workers who generated nearly all the ideas compiled above. Helpful reviews were received from D. C. Larson and L. W. Weston. Only in a few cases could sources be identified. Research sponsored by the Office of Energy Research, Division of Nuclear Physics, U. S. Department of Energy, under contract DE-AC05-84OR21400 with Martin Marietta Energy Systems, Inc.

8. References

1. P. F. Rose, ed., *ENDF/B-VI Summary Documentation*, (Brookhaven National Laboratory Report BNL-NCS-17541, 1991) p. 3.
2. e.g., E. Greenspan, in *Sensitivity and Uncertainty Analysis of Reactor Performance Parameters*, vol. 14 of Advances in Nuclear Science and Technology, eds. J. Lewins and M. Becker (Plenum Press, New York, 1982) p. 193ff.
3. ibid. P. Collins, p. 159ff.
4. W. P. Poenitz, in *Proceedings of the Conference on Nuclear Data Evaluation Methods and Procedures*, eds. Magurno and Pearlstein, (Brookhaven National Laboratory Report BNL-NCS-51363, 1981) p. 254.

5. e.g., *Compilation of Requests for Nuclear Data*, eds. P. F. Rose and A. Daly, (Brookhaven National Laboratory Report BNL-NCS-52028, 1987). *WRENDA 83/84, World Request List for Nuclear Data*, ed. Wang Dahai, (International Atomic Energy Agency report INDC(SEC)-95/URSF, 1988).
6. R. F. Taschek, M. S. Moore, and J. J. Schmidt, discussion in *Nuclear Data for Reactors*, proceedings of the International Conference at Helsinki, STI/PUB/259 (International Atomic Energy, Vienna, 1970) p. 464.
7. D. K. Olsen, *Report to the ^{238}U Discrepancy Task Force on SIOB Fits to the ORNL, CBNM, and JAERI Transmission Data*, (Oak Ridge National Laboratory Report ORNL-TM-9023, 1984).
8. W.P. Poenitz, in *Neutron Standards and Flux Normalization*, ed. A. B. Smith (U. S. Atomic Energy Commission AEC Symposium Series 23, 1971) p. 334.
9. N. E. Holden, in *Nuclear Data for Science and Technology*, ed. S. Igarasi (Saikon Publishing Company, Tokyo 102, Japan, 1988) pp. 872,3.
10. R. Peelle, in *Sensitivity and Uncertainty Analysis of Reactor Performance Parameters*, op.cit., p. 42.
11. F. H. Froehner, *Principles and Techniques of Data Evaluation*, **KfK 4099**, (Kernforschungszentrum Karlsruhe, 1986), p. 26.
12. F. H. Froehner, in *Maximum Entropy and Bayesian Mehods*, ed. J. Shilling (Kluwer Academic Publisher, 1989) p. 467.
13. Yu. F. Yaborov, *Method for Evaluating Non-Consistent Data Using Two Statistical Criteria*, **INDC(CCR)-343**, (IAEA Nuclear Data Section, 1991). (Translation of original paper published in Russian in *Nuclear Constants* **2**, 1989 and as **INDC(CCR)-303**.)
14. M.F. James, R.W. Mills, and D.R. Weaver, *Nucl. Instr. and Methods*, **A313** (1992) 277.
15. M. U. Rajput and T. D. MacMahon, *Nucl. Instr. and Methods*, **A312** (1992) 289.
16. *The INDC/NEANDC Joint Discrepancy File 1990*, eds. B. H. Patrick and N. P. Kocherov (International Atomic Energy Agency Report INDC(NDS)-235, 1990).
17. e.g., M. Becker and D. R. Harris, *Sensitivity of Nuclear Fuel-Cycle Cost to Uncertainties in Nuclear Data*, (Electric Power Research Institute Report EPRI NP-1632, 1980) p. 4-5.
18. H. Condé, in *Neutron Standard Reference Data*, (International Atomic Energy Agency, Vienna, 1974) Table II on p. 280.
19. H. J. C. Kouts, in *Nuclear Cross Sections and Technology*, eds. R. A. Schrack and C. D. Bowman (U. S. Department of Commerce NBS Special Publication 425, 1975) p. 6.
20. R. R. Spencer, R. Gwin, and R. Ingle, *Nucl. Sci. and Engineering*, **80** (1982) 603.
21. E. J. Axton, *Evaluation of the Thermal Constants ...* (Central Bureau of Nuclear Measurements internal memo GE/PH/01/86) private communication, 1986.
22. P. H. White, *J. nucl. Energy*, (Parts A/B *Reactor Sci. Technol.*) **19** (1965) 325.
23. A. D. Carlson et al., in *Nuclear Data for Science and Technology*, ed. S. M. Qaim, (Springer-Verlag Berlin Heidelberg New York,1992) p. 518.

24. R. E. MacFarlane, *Transactions of the American Nuclear Soc.* 33 (1979) 681.
25. J. K. Dickens et al., in *Nuclear Data for Science and Technology*, ed. S. M. Qaim, (Springer-Verlag Berlin Heidelberg New York,1992) p. 307.
26. D. M. Hetrick et al., C. Y. Fu et al., and D. C. Larson et al., in *ENDF/B-VI Summary Documentation*, ed. P.F. Rose (Brookhaven National Laboratory Report BNL-NCS-17541, 1991) pp. 131, 149, 181, 198, and 336.

A SUGGESTION FOR SPECIFICATION OF THE NEUTRON ENERGY SCALE IN MEASUREMENTS, ANALYSES AND EVALUATIONS OF DIFFERENTIAL REACTION CROSS-SECTION DATA

DONALD L. SMITH
Engineering Physics Division, Argonne National Laboratory
9700 South Cass Avenue, Argonne, Illinois 60439, U.S.A.

ABSTRACT

Neutron energy distributions (spectra) in differential neutron cross section measurements are represented conventionally by giving the energies (not well defined) and full widths at half maximum of the spectra. In most modern experiments, it is possible for investigators to provide a much more thorough representation of neutron spectra. However, reporting such information would entail more numerical detail than appears practical to document. Therefore, it is suggested that such spectra be represented in publications and files of experimental data by providing the following four moments of each distribution: mean value (energy), standard deviation (resolution), skewness (asymmetry), and kurtosis (sharpness). Implementation of such a standard for data reporting would offer evaluators a much more complete and rational basis for comparing reported values and performing evaluations than previously possible. Some examples are provided to illustrate the concepts.

1. Introduction

Differential neutron measurements are performed in order to acquire knowledge of the energy and/or angular dependence of important physical parameters (e.g., cross sections). However, energy differential measurements are never truly monoenergetic, and no measurement exercise is ever complete due to limitations of time and technique. Thus, all differential measurements - even those with very high resolution - are somewhat inadequate and require profound assumptions about the physical processes to reach conclusions about those fundamental quantities involved. Also, basic information is lost or distorted by the effects of finite resolution. Finally, interpretation of data from experiments can be thwarted by arbitrary and/or inadequate specification of neutron spectra involved in the measurements. The objectives of this paper are to discuss these issues, make some suggestions for improvement in the current ways of handling the problem, and provoke constructive thought.

2. Energy Scale Effects

A differential measurement aims to provide a value q for a physical quantity Q, corresponding to parameters $\underline{p}$, i.e., $q = Q(\underline{p})$. Examples are $s = \sigma(E)$ or $y = d\sigma(E,\theta)/d\Omega$. Errors, biases or general confusion over the definition of the energy scale for a measurement can impact severely on the interpretation of such differential information. To understand this, consider $\sigma(E)$ to be the "true" differential cross section. Let E_m be a measured, specified or assumed energy, while E_0 is its true (but unknown) equivalent value. Then, there is a systematic error, $\Delta\sigma = \sigma(E_0)-\sigma(E_m)$, in our knowledge of σ near E_0 due to energy scale uncertainty. It can be significant for rapidly changing σ.

3. Loss of Information

If $\sigma(E)$ is the true cross section, then its measurement will lead, inevitably, to the result $\sigma'(E) = \int R(E,\epsilon)\sigma(\epsilon)d\epsilon$, where $\int R(E,\epsilon)d\epsilon = 1$ for all E (in the discussions of this paper, integration extends from 0 to ∞). The process of converting σ to σ' is irreversible, in a practical sense, even if measurements of σ' are made for many closely spaced values of E, and $R(E,\epsilon)$, the resolution function, is well known. We must accept as inevitable that information about $\sigma(E)$ is irretrievably lost by the measurement process. However, interpretation of σ' as representative of σ becomes more meaningful if we follow consistent procedures for specifying $R(E,\epsilon)$. In particular, the measured results $\sigma'(E)$ will then be free of biases and distortions so that physical principles can be used with some reliability to infer the nature of σ from the measured values of σ'. Figs. 1 and 2 show how the resolution function distorts basic information. For example, it is seen that finite resolution leads to cross sections that are too large near threshold. This can have a profound effect on C/E comparisons with integral data. Nuclear model results are often preferable to real data near threshold unless the measurements are of very high resolution.

4. Characterization of Differential Neutron Spectra

Modern experimental methodology, aided by computer analysis, allows for characterizing the neutron spectra from differential measurements with considerable accuracy and detail[1-8]. It is impractical to record all this detail in computer data files for posterity, so a consistent system for identifying and recording key spectrum parameters is needed. Contemporary practice generally offers ad hoc and inadequate characterization, e.g., differential energy ⇒ peak neutron yield, and resolution ⇒ full width at half maximum. <u>It is suggested that differential neutron spectra be treated as probability distributions and that well defined low-order moments from statistics be used to characterize these distributions for practical purposes</u>. As an example, consider Fig. 3 which is derived from printout of a data analysis code used to determine neutron activation cross sections as measured against a fission standard. The neutron energy spectrum shown corresponds to the activation sample in this experiment.

5. Moments of Localized Neutron Distributions

Let us assume that the shape of the neutron energy distribution is defined by p(E) and that it is normalized, i.e., $\int p(\epsilon)d\epsilon = 1$. For convenience, the following change of notation is made: $p(\epsilon) \Leftrightarrow R(E,\epsilon)$ for a particular E. Neutron energy for a differential measurement can be equated to the distribution mean value, $E = \int \epsilon p(\epsilon)d\epsilon$. The neutron energy resolution can be associated with the distribution standard deviation, $\sigma = [\int (\epsilon - E)^2 p(\epsilon)d\epsilon]^{1/2}$. Note that in this context, σ ought not be confused with a cross section. As an example, consider the distribution $p(\epsilon) = \exp[-(\epsilon - E)^2/(2\sigma^2)]/(2\pi\sigma^2)^{1/2}$, i.e., the normal or Gaussian distribution with mean value E and standard deviation σ. Then, FWHM $\approx 2.35\,\sigma$. Deviations of the spectrum relative to a Gaussian can be examined by considering two additional distribution moments: skewness, $\alpha_3 = \mu_3/\sigma^3$, where $\mu_3 = \int (\epsilon - E)^3 p(\epsilon)d\epsilon$, and kurtosis, $\alpha_4 = \mu_4/\sigma^4$, where $\mu_4 = \int (\epsilon - E)^4 p(\epsilon)d\epsilon$. The skewness is zero for a distribution which is symmetric about the mean value, while it is negative for one with a tail sloping to the left and positive for one sloping to the right. For a Gaussian distribution, the kurtosis equals 3. If $\alpha_4 < 3$, the distribution has drops off more rapidly to zero on the wings than a Gaussian, while $\alpha_4 > 3$ signifies a distribution which drops off more slowly than a Gaussian[9]. These four quantities embody the principle features of a localized distribution function. Therefore, it is suggested that a reasonable characterization of the spectrum associated with a differential measurement can be recorded for posterity by preserving E, σ, α_3 and α_4. This approach does not suggest a major new commitment of effort or data storage space since all that is required is a consistent determination of E and σ, plus two additional parameters beyond what is conventionally expected from an experimenter.

6. A Suggested Standard for Data Reporting and Filing

Consideration needs to be given to a means for implementing this suggestion in the reporting of experimental data submitted for publication in the literature or storage in archival files. The issue should be debated but, for simplicity, let us assume energy differential cross section data, such as a reaction cross section, are to be presented. An experimenter could attend to detailed characterization of all his measurement spectra and then calculate E, σ, α_3 and α_4, as defined earlier, for each data point. The experimenter would then report for each experimental data point: σ_{meas}, E, σ, α_3 and α_4, where σ_{meas} is the measured differential cross section. These five values could be introduced into computer files for each data point, along with any descriptive information of a conventional nature, as available.

7. To What Extent Can We be Successful in Reconstructing a Differential Spectrum Given only Four Parameters?

It is clear that generally we could not do an outstanding job because the conversion process, $p(\epsilon) \Rightarrow E$, σ, α_3 and α_4, certainly entails loss of information. In fact, it is not suggested here that provision of four parameters is a substitute for knowledge of the complete distribution. However, a user of reported data would be in a far better position to evaluate the significance of the information if he knew exactly how the energy scale were determined and how broad the spectrum used was in terms of its standard deviation, which is well defined. Furthermore, he would benefit by knowing whether or not the spectrum used was skewed and if it dropped to zero in the wings more quickly or more slowly than a Gaussian. Nevertheless, it is interesting to examine just what can be done with four parameters, and the following discussion is offered in this spirit. The approach presented here is to represent $p(\epsilon)$ by a 4th-order polynomial expansion. There are a couple of possibilities which come to mind. The first is to use a simple Taylor series, namely, $p(\epsilon) \approx \Sigma_{k=0,4}\, p^k(E)(\epsilon-E)^k$, where $p^k(E)$ is the kth-order derivative of p at $\epsilon = E$. This approach is conceptually rigorous, but not particularly useful because the parameters, $p^k(E)$, are not easily related to the moments we have chosen to consider. The second approach is to consider an empirical expansion which does directly involve E, σ, α_3 and α_4, namely, $p(\epsilon) \approx p(E)\{1 + \Sigma_{k=1,4}\, c_k[(\epsilon-E)^k/(k!\sigma^k)]\}$. The coefficients, c_k, are defined as follows (with indicated justification): $c_1 = 0$ (simplicity and it works for a Gaussian), $c_2 = -1$ (it works for a Gaussian), $c_3 = \alpha_3$ (a reasonable guess), and $c_4 \approx 0.6\alpha_4$ (it works for a Gaussian). The results for a Gaussian are shown in Fig. 4, where various values of $A = c_4/\alpha_4$ are considered. With this empirical expansion, fairly good results were obtained for both a Gaussian and a spectrum obtained from analysis of an actual neutron activation experiment[10]. In this measurement, for several spectra corresponding to the D(d,n)He-3 source reaction, the empirical polynomial expansion provides a pretty good shape representation of $p(\epsilon)$ over the approximate energy range $E-1.5\sigma < \epsilon < E+1.5\sigma$ (e.g., see Fig. 5). However, the spectra calculated from the empirical expansion are found to be shifted to lower energies relative to the true $p(\epsilon)$, by an amount ΔE that is dependent very roughly in linear fashion on α_3 (see Fig. 6). Even though the agreement is not very good for energy deviations ϵ of more than 1.5σ to 2σ (see Figs. 4 and 5), and one must account for the energy shift when the spectrum is skewed, it is encouraging that the approach does provide the user with a much better understanding of the spectrum shape than deduced from knowing only the mean energy and standard deviation alone.

8. Summary

The approach suggested in this paper capitalizes on the ability which experimentalists now possess to characterize differential neutron energy spectra quite reliably and in considerable detail. Representation of these spectra by four low-order distribution moments offers a statistically rigorous way to define energy and resolution, and also provides useful information on higher-order spectrum features, i.e., skewing and rapidity of the drop off to zero. Given these four parameters, it is possible to reconstruct spectral shapes in qualitative fashion in the vicinity of the mean energy, provided that allowance is made for a small shift in energy which roughly depends linearly on α_3.

9. References

1. D.L. Smith, *Remarks Concerning the Accurate Measurement of Differential Cross Sections for Threshold Reactions Used in Fast Neutron Dosimetry for Fission Reactors* (Report ANL/NDM-23, Argonne National Laboratory, 1976).

2. J.V. Meadows, *Determination of the Energy Scale for Neutron Cross Section Measurements Employing a Monoenergetic Accelerator* (Report ANL/NDM-25, Argonne National Laboratory, 1977).
3. J.V. Meadows and D.L. Smith, *Neutron Source Investigations in Support of the Cross Section Program at the Argonne Fast-Neutron Generator* (Report ANL/NDM-53, Argonne National Laboratory, 1980).
4. D.C. Larson, N.M. Larson, J.A. Harvey, F.G. Perey, D.E. Pierce and R.H. Seals, *Laser Measurements of Distances from the ORELA Neutron Target to Experimental Stations Along Flight Paths 1 and 6* (Report ORNL/TM-9097, Oak Ridge National Laboratory, 1985).
5. N.V. Kornilov, *Possibilities of Experimental Determination and Theoretical Prediction of Different Properties of Accelerator Based Neutron Sources* (Proceedings of an Advisory Group Meeting on Properties of Neutron Sources, Organized by the International Atomic Energy Agency and held in Leningrad, U.S.S.R., 9-13 June 1986, IAEA Vienna, Austria.)
6. A.B. Kagalenko and N.V. Kornilov, *The Differential Neutron Production Cross Sections in the D(d,np) Reaction* (ibid. Ref. 5).
7. P. Raics, S. Nagy and S. Szegedi, *Nuclear Data for Science and Technology*, ed. S.M. Qaim (Springer-Verlag, Berlin, 1992) p. 660.
8. B. Strohmaier, M. Wagner and H. Vonach, ibid. Ref. 7, p. 663.
9. D.L. Smith, *Probability, Statistics and Data Uncertainties in Nuclear Science and Technology* (American Nuclear Society, LaGrange Park, Illinois, 1991).
10. L.P. Geraldo, D.L. Smith and J.V. Meadows, *Annals of Nuclear Energy* Vol. 16, No. 6 (1989) 293.

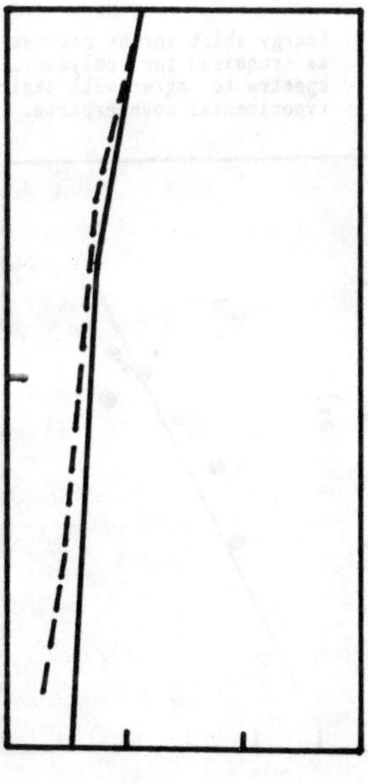

Fig. 1: Effect of finite resolution on threshold data Original (___), Broadened (_ _ _). Vertical scale is logarithmic.

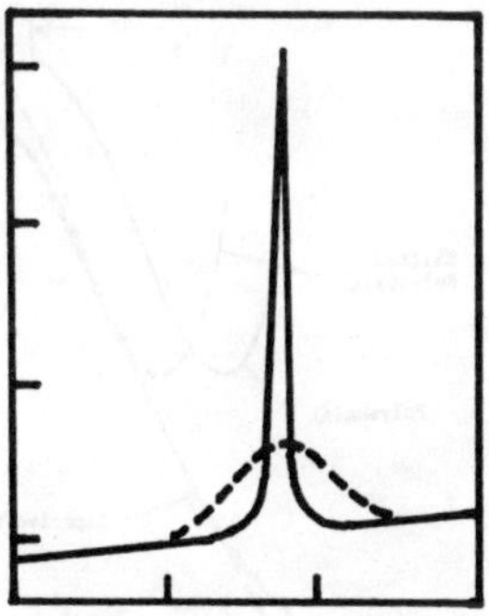

Fig. 2: Effect of finite resolution on resonance data Original (___), Broadened (_ _ _).

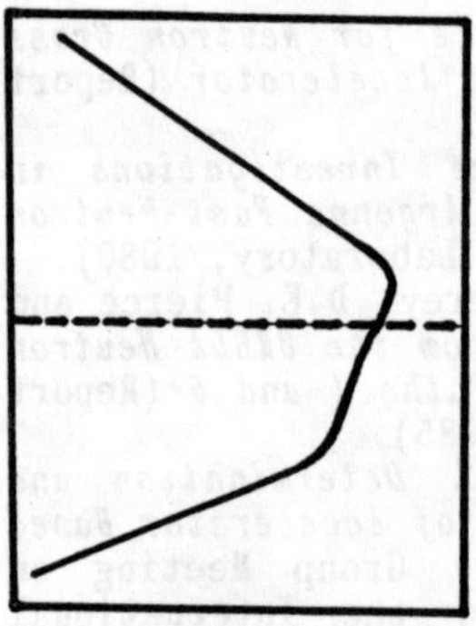

Fig. 3: Calculated neutron energy spectrum for a fast neutron activation experiment (dashed horizontal line marks mean value of the energy).

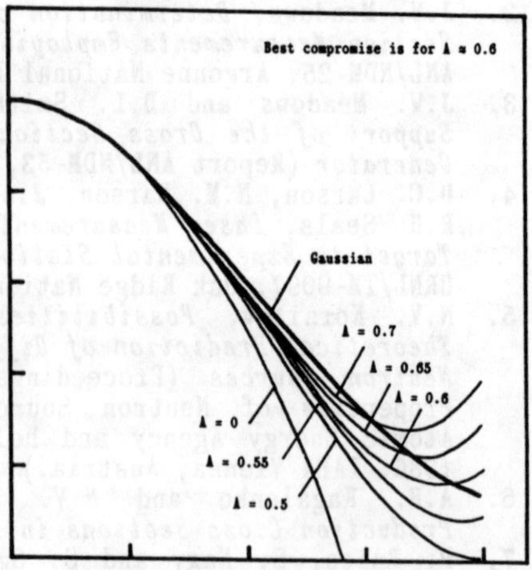

Fig. 4: Comparison of a Gaussian spectrum shape with an empirical polynomial expansion for $c_4 = A a_4$ and various values of A (see text).

Fig. 5: Various representations of an experimental "monoenergetic" neutron energy spectrum.

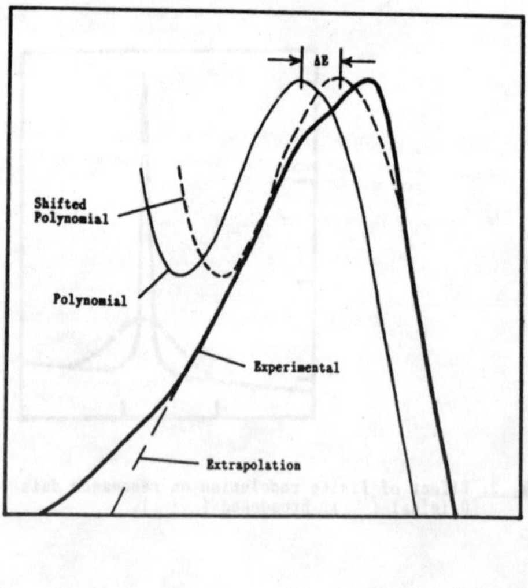

Fig. 6: Energy shift versus skewness as required for polynomial spectra to agree with their experimental counterparts.

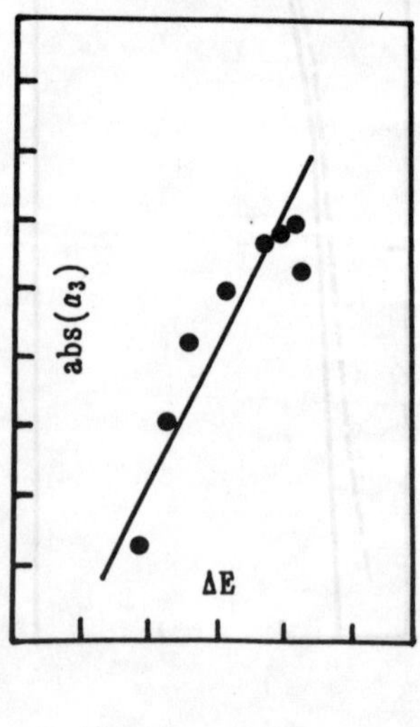

THE DATA BASE OF THE STANDARDS AND RELATED CROSS-SECTIONS AFTER ENDF/B-VI

W. P. Poenitz
Argonne National Laboratory - West, P. O. Box 2528, Idaho Falls, ID 83403
A. D. Carlson
Nat'l Institute of Standards and Technology, Rad. Phys. Bldg., Washington, DC 20234

ABSTRACT

A brief description is given of the procedure used in the global evaluation of the standards and other important cross sections for ENDF/B-VI. The results of the evaluation are compared with new or revised experimental data.

I. Introduction

Measurements of neutron interaction cross sections have been carried out most often relative to a few other neutron interaction cross sections, called standards, which were assumed to be well known. In Table 1a the cross sections are listed which have been declared "standards" by one committee or another since they are frequently used reference cross sections. Some other reference cross sections and

Table 1. Standards and Other Cross Sections Involved in the ENDF/B-VI Global Evaluation

	Cross Section or Data	Energy Range of Standard	Energy Range of Global Eval.	Involvement in Global Evaluation
a	H(n,p)	E ≤ 20 MeV	---	Independent Eval. for ENDF/B-VI[1].
	^{3}He(n,p)	E ≤ 50 KeV	---	Independent Eval. for ENDF/B-VI[2].
	C(n,n)	E ≤ 1.8 MeV	---	Independent Eval. for ENDF/B-VI[3]. Scattering Standard.
	^{6}Li(n,t)	E ≤ 1 MeV	E ≤ 2.8 MeV	Flux Measurement Standard. ^{6}Li (n,n) also Included in Evaluation.
	^{10}B(n,αγ) - - - - - - - ^{10}B(n,α)	E ≤ 250 keV	E ≤ 1.0 MeV	Flux Measurement Standard. ^{10}B(n,α_0), ^{10}B(n,α_1), ^{10}B(n,n) were Involved in the Evaluation.
	^{197}Au(n,γ)	0.2 ≤ E ≤ 2.5 MeV	E ≤ 2.8 MeV	Capture Cross Section Standard. ^{238}U(n,γ) was also Included in the Global Evaluation.
	^{235}U(n,f)	0.15 MeV ≤ 20 MeV	E ≤ 20 MeV	General Standard Cross Section. ^{239}Pu(n,f) also Included in Global Evaluation.
b	^{238}U(n,f)	---	1 MeV ≤ E ≤ 20 MeV	Used as Reference for E > 2 MeV
	^{252}Cf Spectrum, $\bar{\nu}$	All energies of the spontaneous fission spectrum	---	Independent Evaluation. $\bar{\nu}$ was Together with the Thermal Parameters Included in the Global Evaluation.

data which are occasionally used are listed in Table 1b. The neutron spectrum of the spontaneously fissioning ^{252}Cf is used as a neutron spectrum standard. It is used for the calibration of neutron detectors which, in turn, can be used for cross section measurements. The number of neutrons emitted per fission ($\bar{\nu}$) of ^{252}Cf has been used as the standard for the measurements of $\bar{\nu}$(E) for fissile and fertile nuclei.

Absolute values of the standard cross sections have been measured involving in most cases absolute neutron fluence determinations. An exception is the hydrogen scattering cross section which can be determined as a total cross section in transmission measurements and thus does not require absolute fluence measurements. As a consequence, this cross section is quite well known (uncertainties of less than 0.5%). However, the angular distribution for neutron scattering from hydrogen, which is used in some detectors for neutron fluence measurements, is less well known. The detection efficiency for the recoil protons and other aspects of the cross section measurement introduce additional uncertainties with the result that cross sections measured relative to the hydrogen scattering cross section display discrepancies similar to those which involved absolute neutron fluence determinations.

Evaluations of the existing data prior to ENDF/B-VI often were artistic in nature instead of being based on scientific procedures. The evaluations generally involved drawing a curve through the existing data. Curves were drawn which emphasized certain data sets which were perceived to be "better" than others, or because they were the result of the most recent measurements. These evaluation procedures were in part chosen in response to the poor data base. The evaluations up to version VI were hierarchical, i.e., H(n,p) was chosen first, then ^{6}Li(n,t) was evaluated, then ^{10}B(n,α), followed by ^{235}U(n,f), and ^{197}Au (n,γ). There is no reason why a ratio measurement, e.g., involving the ^{6}Li(n,t) and ^{235}U(n,f) cross sections, should be used to derive ^{235}U(n,f) cross sections. A proper evaluation should permit ratio measurements to have an impact on each of the standard cross sections in the ratio. At the First Conference on Nuclear Data Evaluation Methods and Procedures, it was demonstrated that a simultaneous evaluation of interrelated standard cross sections by generalized least-squares was feasible[5]. After development of a formulation which permitted the inclusion of the R-matrix evaluations of the light elements,[6] the Standards Subcommittee of the Cross Section Evaluation Working Group (CSEWG) decided to perform the evaluation of the standards and other important cross sections for ENDF/B-VI using these objective evaluation techniques.

The data base which was available for the ENDF/B-VI evaluation has been repeatedly discussed (for more recent examples see Refs. 7-11). Therefore, in the following we will mainly consider data which have become available since the ENDF/B-VI evaluation. Comparison with the evaluations will indicate possible problems.

II. The ENDF/B-VI Evaluation

The global evaluation of the standards and other important cross sections for ENDF/B-VI has been described previously.[12-14] Therefore, only a brief account will be given here. A schematic of the global approach is shown in Fig. 1.

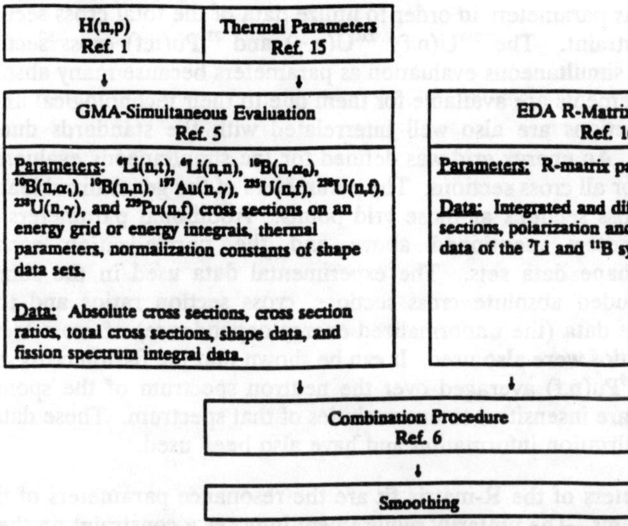

Fig. 1 Schematic of the Global Evaluation for ENDF/B-VI

The evaluation process consisted of three major parts:

1. A simultaneous evaluation of 773 cross sections and thermal parameters using the generalized least-squares fitting code GMA (for Gauss, Markov, Aitken).
2. R-matrix fits of the ^{7}Li and ^{11}B systems using the R-matrix code EDA (for Energy Dependent Analysis).
3. A procedure for combining the results from the parallel steps 1 and 2.

The evaluation involved the ^{6}Li(n,t), ^{10}B(n,α_0), ^{10}B(n,α_1), ^{197}Au(n,γ), and ^{235}U(n,f) standards as parameters. Other reference data like the ^{3}He(n,p), C(n,n) cross sections and the ^{252}Cf-neutron spectrum were evaluated independently because they have very few correlations with the data involved in the global approach. The H(n,p) cross sections had been originally included as parameters of the simultaneous evaluation[5], however, adjustments of the H(n,p) cross section were felt to represent the transfer of measurement errors and not cross section improvements. Therefore, a prior R-matrix analysis of the H(n,p) cross section by Dodder and Hale[1] was used to convert all measurements relative to H(n,p) into absolute or shape cross section data. Many data sets involve a normalization to thermal cross sections. Instead of including the thermal cross section data base in the simultaneous evaluation, a recent evaluation of the thermal parameters (cross sections, g-factors, $\bar{\nu}$), often called thermal constants, of ^{233}U, ^{235}U, ^{239}Pu and ^{241}Pu by Axton[15] with the associated covariance matrix was used as an input data set. The set of thermal parameters also contains $\bar{\nu}$ of ^{252}Cf which thereby became part of the global evaluation.

In addition to the standard cross sections, the scattering cross sections of ^{6}Li and ^{10}B were used as parameters in order to utilize data of the total cross sections as an additional constraint. The ^{238}U(n,f), ^{238}U(n,γ) and ^{239}Pu(n,f) cross sections were included in the simultaneous evaluation as parameters because many absolute cross section measurements are available for them due to their technological importance. These cross sections are also well interrelated with the standards due to ratio measurements. An energy grid was defined for the simultaneous evaluation which was the same for all cross sections. The parameters of the generalized least-squares fit were the cross sections at those grid points. Additional parameters were the thermal parameters mentioned above and the normalization constants of experimental shape data sets. The experimental data used in the simultaneous evaluation included absolute cross sections, cross section ratios and total cross sections. Shape data (the unnormalized energy dependence) of cross sections and cross section ratios were also used. It can be shown that the fission cross sections of ^{235}U(n,f) and ^{239}Pu(n,f) averaged over the neutron spectrum of the spontaneously fissioning ^{252}Cf are insensitive to uncertainties of that spectrum. These data provide valuable normalization information and have also been used.

The parameters of the R-matrix fit are the resonance parameters of the ^{6}Li+n and ^{10}B+n systems. The unitarity requirement imposes a constraint on the reaction channels. Theory describes not only the total and the integrated cross sections, but differential cross sections, polarization and inverse reaction data as well, thus experimental data for those have been used in the fits. The ^{6}Li and ^{10}B data had to be partitioned between the simultaneous evaluation and the R-matrix evaluation. Because the R-matrix fits were made separately for the ^{6}Li+n and the ^{10}B+n systems, all ratio data and data strongly correlated with other measurements remained in the GMA data base. Some data sets, which were originally in the GMA data base, were transferred to the EDA data base in order to speed up the convergence of the R-matrix fits. The partitioning of the ^{6}Li and ^{10}B data base left the simultaneous evaluation with poorer data, therefore resulting in larger uncertainties for the ^{6}Li and ^{10}B cross sections from that analysis than that obtained from the R-matrix results. As a consequence, the simultaneous evaluation results had lesser weight in the combination procedure for the ^{6}Li and ^{10}B cross sections. E.g., though the simultaneous evaluation resulted in ~12% higher ^{6}Li(n,t) cross sections around 2 MeV compared with the R-matrix solution, the combination procedure adjusted the R-matrix result only by ~3% in this range. On the other hand, for cross sections like ^{235}U(n,f), the R-matrix results had very little impact on the combination result for energies above 1-2 MeV.

Data scarceness or inconsistencies at some energies introduce arbitrary fluctuations of the pointwise cross sections where they are known to be smooth based on the statistical-optical model. Therefore, as a final step of the evaluation process some smoothing was done for the $^{197}Au(n,\gamma)$, $^{235}U(n,f)$, $^{238}U(n,f)$, $^{238}U(n,\gamma)$ and $^{239}Pu(n,f)$ cross sections by comparing the pointwise data with theoretical calculations and compensating high and low values in local regions.

III. ENDF/B-VI and New Experimental Data

III.1 The H(n,p), ^{3}He(n,p), and C(n,n) Cross Sections

The H(n,p) cross sections of ENDF/B-III through ENDF/B-V were based on an analysis by Hopkins and Breit.[17] A charge independent R-matrix analysis of n-p and p-p experimental data below 30 MeV by Dodder and Hale[1] was adopted for ENDF/B-VI. The newest measurement by Priesmeyer *et al.*[18] is ~0.24% lower than ENDF/B-VI. The largest difference between ENDF/B-VI and V up to 20 MeV is only ~0.5%. However, the center-of-mass system (CMS) scattering cross sections at 180° differ by almost 2% near 11 MeV. This is significant because the 180° CMS neutron angle corresponds to a 0° laboratory angle for the recoil protons, which is frequently used in proton recoil detectors for neutron fluence measurements. A recent measurement by Ryves and Kolkowski[19] resulted in a 180° cross section at 14.5 MeV which is between the ENDF/B-V and VI evaluations. A $\sigma(180°)/\sigma(90°)$ ratio of 1.053 $\pm$ 0.015 which also was measured in this experiment is consistent with ENDF/B-V but not with VI for which the value is 1.093 $\pm$ 0.010.

The ENDF/B-VI evaluation of ^{3}He(n,p) led to a considerable improvement in data quality for this cross section since the last ENDF evaluation of this cross section was performed in 1968 for ENDF/B-III. That evaluation was carried over to ENDF/B-IV and V. The new evaluation was an R-matrix analysis done by Hale[2] using all possible two body reactions in the ^{4}He system. Measurements have been made by Borzakov[20] from 0.26 to 142 keV neutron energy relative to the ^{6}Li(n,t) cross section. These measurements are in excellent agreement with the ENDF/B-VI evaluation in the standards energy region.

The evaluation for ENDF/B-V of C(n,n) was based on an R-matrix analysis of ^{12}C by Fu[3]. The evaluation for ENDF/B-VI was modified in order to include two resonances of ^{13}C (1.1% abundance) so that the evaluation would be appropriate for natural carbon. New measurements by Schmiedmayer and Moxon[21] in the energy region from 50 eV to 100 keV are in excellent agreement with the ENDF evaluation. Also a filtered neutron beam measurement at 2 keV by Kirilyuk[22] is in very good agreement.

III.2 The ^{6}Li(n,t) and ^{10}B(n,α) Cross Sections

The cross section data base for ^{6}Li and ^{10}B has been reviewed recently in Refs. 9 and 10. New data for the ^{6}Li(n,t) cross section have not become available. It had been the conclusion in Ref. 9 that the ^{6}Li(n,t) cross section is reasonably well known but that substantial discrepancies exist for the ^{10}B cross sections. An NEANSC endorsed international inter-laboratory collaboration group was formed in order to provide a mechanism for improving the ^{10}B(n,α) cross sections.

New measurements of the shape of the ^{10}B(n,α_1) cross section have been made by Schrack *et al.*[23] The data, normalized to ENDF/B-VI between 200 and 360 keV, are shown in Fig. 2 compared with the ENDF/B-VI global evaluation result. With the exception of the resonance at 400 keV, the agreement is very good up to 1 MeV where the global evaluation for the ^{10}B cross sections ended. This agreement is important because the ENDF/B-VI values for this cross section increased relative to by up to 20% between 400 keV and 1 MeV. Large differences occur between the new measurement and ENDF/B-VI above 1.5 MeV (+15% around 2 MeV and −30% around 3 MeV).

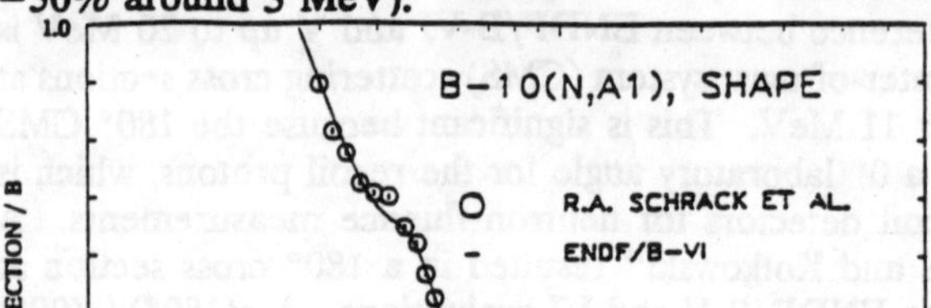

Fig. 2 Comparison of the Result for ^{10}B(n,$\alpha\gamma$) from the Global Evaluation with New Measurements by Schrack *et al.*[23]

Fig. 3 Comparison of the Result for ^{10}B(n,α_0) vs ^{10}B(n,α_1) from the Global Evaluation with New Measurements by Weston & Todd[24]

A new measurement of the cross section ratio $\sigma(n,\alpha_0)/\sigma(n,\alpha_1)$ by Weston and Todd[24] is compared with ENDF/B-VI in Fig. 3. The discrepancy is obvious. However, including the new data in the evaluation would probably have little effect on the outcome below 100 keV (because of the reasonable agreement) and above 400 keV (because of the large uncertainties of the new data). The ENDF/B-VI evaluation of ^{10}B(n,α_0) was strongly influenced by the inverse reaction cross section measurement of Olson and Kavanagh.[25]

The ratio of the ENDF/B-VI cross sections to those of V for ^{10}B(n,α) (sum of the ^{10}B(n,α_0) and ^{10}B(n,α_1)) is shown in Fig. 4. Also shown are the adjustments made to group cross sections of ^{10}B(n,α)

which are required in order to bring a large number of calculated fast reactor integral quantities into better agreement with experiment.[26] It can be seen from Fig. 4 that the change from ENDF/B-V to VI will be beneficial for the calculation of these quantities and reduce the C/E (for calculated vs. experiment) discrepancies.

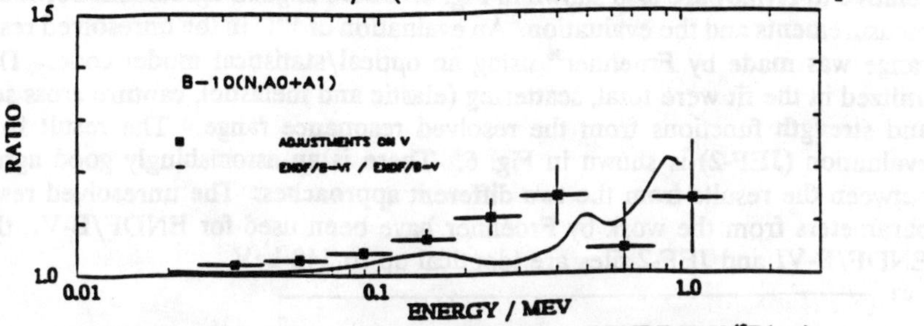

Fig. 4 Comparison of Cross Section Adjustments of ENDF/B-V $^{10}B(n,\alpha)$ Required by Integral Data and the Ratio of ENDF/B-VI to V

III.3 *The $^{197}Au(n,\gamma)$ and $^{238}U(n,\gamma)$ Cross Sections*

New measurements of the $^{197}Au(n,\gamma)$ cross section using calibrated photoneutron sources[27] are shown in Fig. 5. These data are somewhat lower at 23 keV and substantially higher (~20%) at 967 keV compared with ENDF/B-VI. Around 1 MeV the ENDF/B-VI evaluation agrees well with absolute measurements and $^{197}Au(n,\gamma)/^{235}U(n,f)$ data. A previous photoneuton source measurement[28], however, also resulted in a similarily high value at 967 keV. Revised and new data by Davletshin *et al.*[29] were measured relative to the H(n,p) cross section and are also shown in Fig. 5., together with new data by Kazakov *et al.*,[30] Voinier[4], Demekhin *et al.*[31] and some additional data by Davletshin *et al.*[29] referenced to ^{235}U (n,f) of ENDF/B-VI. The agreement with ENDF/B-VI overall is good, but there might be a problem around the inelastic levels at 270 keV.

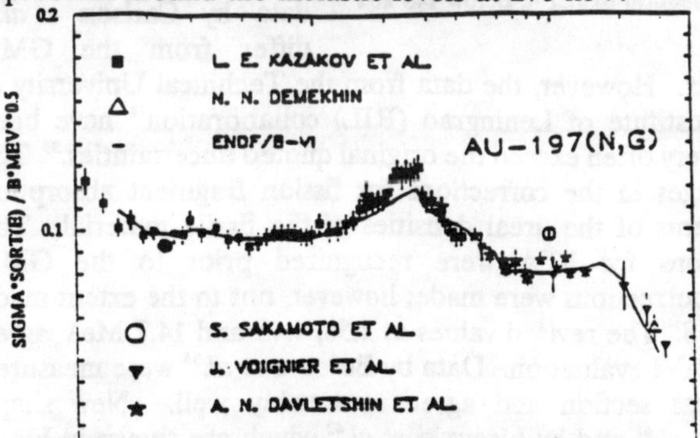

Fig. 5 Comparison of the Global Evaluation Result (ENDF/B-VI) for $^{197}Au(n,\gamma)$ with New Experimental Data

$^{238}U(n,\gamma)$ values were obtained with the same calibrated photoneutron sources[32] used for the $Au(n,\gamma)$ measurements mentioned above. They are shown in Fig. 6 together with the result from the global evaluation. The agreement is quite good at 23 keV but the 967 keV value is again higher than ENDF/B-VI. Other recent measurements by Kobayashi *et al.*[33] were made with Fe and Sc filtered beams, and normalized at 24 keV to the ENDF/B-VI value. New measurements by Adamchuk *et al.*[34] using a multiplicity detector as well as data by Buleeva *et al.*[35] relative to H(n,p) are also shown in Fig. 6. There is good agreement between these measurements and the evaluation. An evaluation of ^{238}U in the unresolved resonance range was made by Froehner[36] using an optical/statistical model code. The data utilized in the fit were total, scattering (elastic and inelastic), capture cross sections, and strength functions from the resolved resonance range. The result from this evaluation (JEF-2) is shown in Fig. 6. There is an astonishingly good agreement between the results from the two different approaches. The unresolved resonance parameters from the work by Froehner have been used for ENDF/B-VI, thus the ENDF/B-VI and JEF-2 files are identical up to 148 keV.

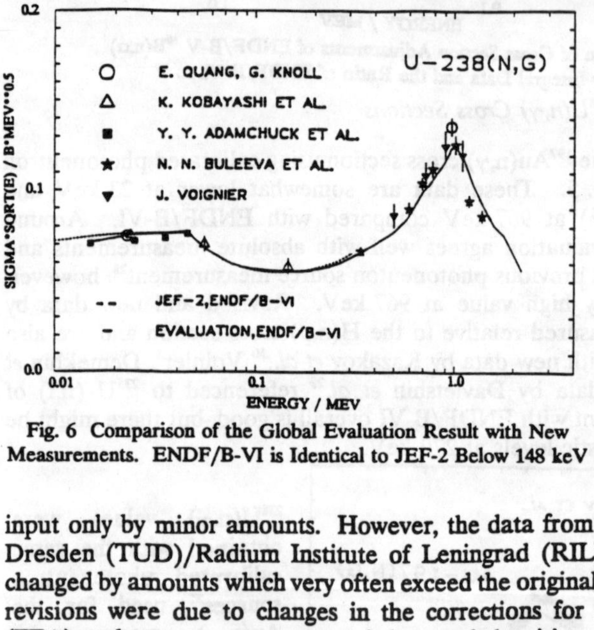

Fig. 6 Comparison of the Global Evaluation Result with New Measurements. ENDF/B-VI is Identical to JEF-2 Below 148 keV

III.4 The ^{235}U(n,f) and ^{238}U(n,f) Cross Sections

A substantial number of new measurements have been made of the ^{235}U(n,f) cross section. Figure 7 shows two data sets which recently have been finalized or revised,[37,38] but had been available and used for the simultaneous evaluation with GMA. The final data by Carlson *et al.*[37] differ from the GMA input only by minor amounts. However, the data from the Technical University of Dresden (TUD)/Radium Institute of Leningrad (RIL) collaboration[38] have been changed by amounts which very often exceed the original quoted uncertainties.[39] The revisions were due to changes in the corrections for fission fragment absorption (FFA) and new measurements of the areal densities of the fissile material. The initial insufficient corrections for FFA were recognized prior to the GMA evaluation[40] and additional corrections were made; however, not to the extent made for the final values in Ref. 38. The revised values at 2.56, 4.45 and 14.7 MeV agree very well with the ENDF/B-VI evaluation. Data by Buleeva *et al.*[35] were measured relative to the H(n,p) cross section and agree reasonably well. New shape measurements by Carlson *et al.*[41] and by Lisowski *et al.*[42] which are shown in Fig. 8 normalized around 14 MeV to ENDF/B-VI, support higher cross sections than

ENDF/B-VI above 15 MeV but agree very well with ENDF/B-VI at all other energies, specifically around 8 MeV.

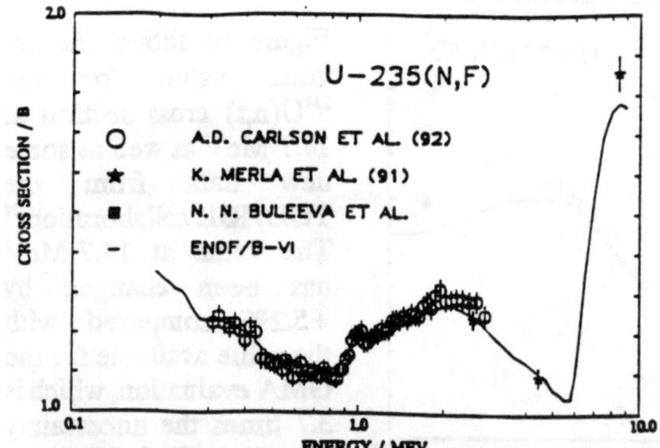

Fig. 7 Comparison of the Result of the Global Evaluation (ENDF/B-VI) for ^{235}U(n,f) with New or Revised Experimental Data

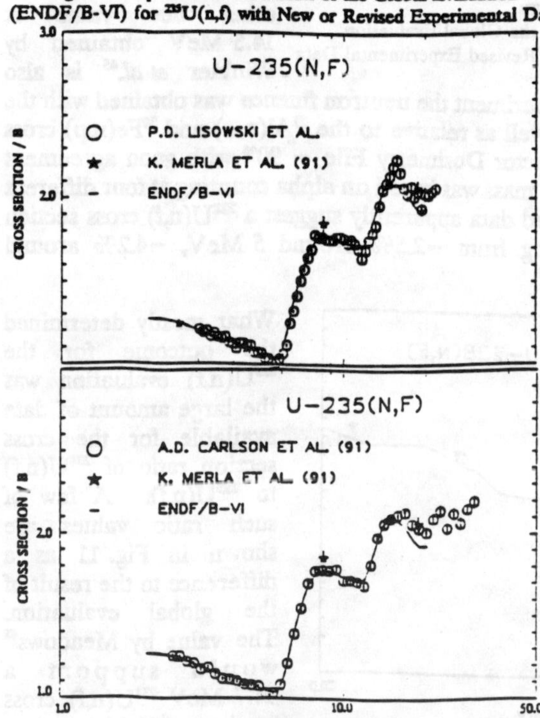

Fig. 8 Comparison of the Result of the Global Evaluation (ENDF/B-VI) for ^{235}U(n,f) with New or Revised Experimental Data

Two more ^{235}U(n,f) data sets are shown in Fig. 9. The data by Johnson *et al.*[43] are correlated with the data by Carlson *et al.*[37] because the same fission chamber was used. The agreement with ENDF/B-VI is very good. The measurements by Iwasaki *et al.*[44] were made with a recoil proton telescope for the determination of the neutron fluence. The cross sections have total uncertainties of about 2.5%. The originally quoted values were in very good agreement with the ENDF/B-VI ^{235}U(n,f) evaluation. However when the data are converted using the ENDF/B-VI hydrogen cross sections instead of V, an increase of 1.7% occurs. An experimental check was made of the telescope by determining fluence with the telescope using ENDF/B-V hydrogen cross sections and comparing this with that obtained using the time-correlated associated-particle technique. This agreement was within 1% with uncertainties of 2%. These measurements suggest that the

ENDF/B-VI hydrogen cross sections may be too high at 14 MeV for a CMS scattering angle of 180 degrees; however the uncertainties make this conclusion quite tentative.

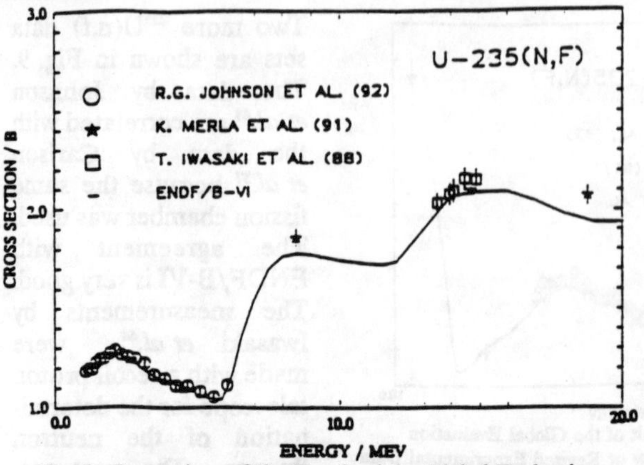

Fig. 9 Comparison of the Result of the Global Evaluation (ENDF/B-VI) for ^{235}U(n,f) with New or Revised Experimental Data

Figure 10 shows the revised value for the ^{238}U(n,f) cross section at 14.7 MeV as well as some new data from the TUD/RIL collaboration.[38] The value at 14.7 MeV has been changed by +5.2% compared with the value available for the GMA evaluation, which is 3.7 times the uncertainty quoted in Ref. 39. Another new value at 14.5 MeV obtained by Winkler *et al.*[45] is also shown in Fig. 10. In the latter experiment the neutron fluence was obtained with the associated particle technique as well as relative to the ^{27}Al(n,α) and ^{56}Fe(n,p) cross sections of the International Reactor Dosimetry File − 90[46] with good agreement among the three values. The ^{238}U mass was based on alpha counting at four different laboratories. The new and revised data apparently suggest a ^{238}U(n,f) cross section higher than ENDF/B-VI ranging from ~2.5% around 5 MeV, ~4.2% around 14 MeV to ~5.6% at 18-19 MeV.

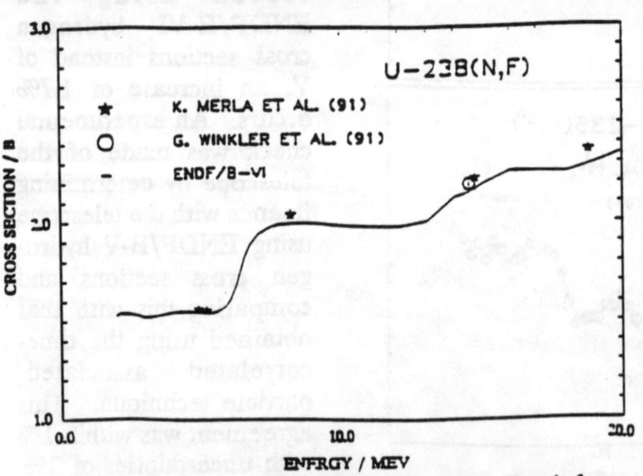

Fig. 10 Comparison of the Global Evaluation Result for ^{238}U(n,f) (ENDF/B-VI) with New and Revised Experimental Data

What mostly determined the outcome for the ^{238}U(n,f) evaluation was the large amount of data available for the cross section ratio of ^{238}U(n,f) to ^{235}U(n,f). A few of such ratio values are shown in Fig. 11 as a difference to the result of the global evaluation. The value by Meadows[48] would support a 14.7 MeV ^{238}U(n,f) cross section ~4% higher than ENDF/B-VI and had been of some concern during the evaluation.[49] However, although it has a low uncertainty, it was "swamped" by ~15 other data sets which contradict it and agree well with ENDF/B-VI. Of specific interest are the data by Behrens and Carlson[50] and by Difilippo *et al.*[51]. These data extend over the whole energy range from threshold to 20 MeV, therefore they transfer information available at lower energies from many other experiments to the range shown in Fig. 11. New measurements of the ^{238}U(n,f)/^{235}U(n,f) cross section ratio have been made by Lisowski *et al.*[52], Formichev *et al.*[53], and Baba *et al.*[54] These data are both higher and lower than the data by Difilippo *et al.*[51].

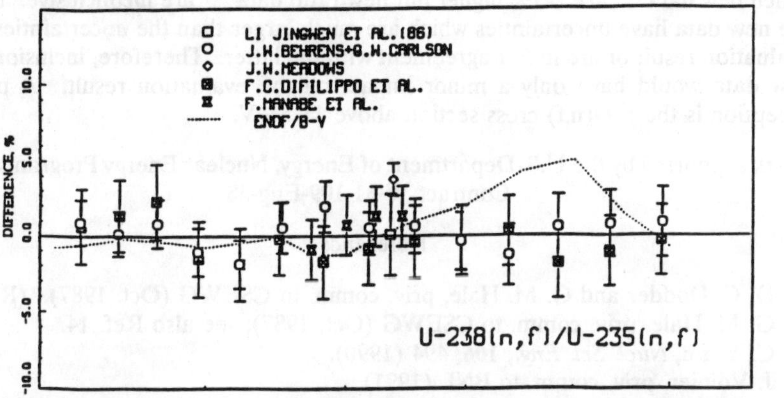

Fig. 11 Some of the Experimental Data of ^{238}U(n,f)/^{235}U(n,f) Used in the Global Evaluation. Shown is the Difference to the Evaluation Result (ENDF/B-VI)

IV. Summary and Conclusions

The evaluation of the standards and other important cross sections for ENDF/B-VI was carried out in three distinct steps: A generalized least-squares fit of various correlated cross section data, an R-matrix fit of the light element data, and a combination procedure involving the results from the two prior steps. The results from this global evaluation (ENDF/B-VI) stand up very well in the comparison with new experimental data which have become available after the completion of ENDF/B-VI. Specifically, ^{235}U(n,f) appears to be well confirmed by numerous measurements at all energies below 15 MeV. The same applies for the ^{238}U(n,γ) cross section over the energy range of the evaluation. It is interesting that the new measurements confirm the low values of the evaluation, as most of the older data are higher than ENDF/B-VI.

A few possible problems are indicated by the new or revised data. The differential H(n,p) cross section around 14 MeV which was not part of the global evaluation is in conflict with new direct and indirect (via ^{235}U(n,f)) measurements. A new measurement of the ^{10}B(n,α_0)/^{10}B(n,α_1) ratio is in conflict with the ENDF/B-

VI ratio above 100 keV, whereas a new shape measurement of ^{10}B(n,$\alpha\gamma$) agrees very well with ENDF/B-VI, and the sum of ^{10}B(n,α_0) + ^{10}B(n,α_1) of the evaluation reduces the C/E discrepancy observed for integral fast reactor quantities. The ^{10}B + n data base appears still to be inconsistent. New and revised data suggest higher cross sections for ^{197}Au(n,γ) below the inelastic cusp at ~ 270 keV by ~ 8% for a limited energy range of ~ 70 keV. At all other energies the data confirm the evaluation. Higher cross sections are suggested by new (preliminary) and revised cross section data for ^{235}U(n,f) above 15 MeV where the evaluation was based on discrepant data. The greatest problem is present in the ^{238}U(n,f) cross section for which new data[38,45] are 2-5% higher but new ratio data[52-54] are inconclusive. Most of the new data have uncertainties which are much larger than the uncertainties of the evaluation result or are in fair agreement with the latter. Therefore, inclusion of the new data would have only a minor impact on the evaluation result. A possible exception is the ^{235}U(n,f) cross section above 15 MeV.

Work supported by the U.S. Department of Energy, Nuclear Energy Programs, under Contract W-31-109-Eng-38

References

1 D. C. Dodder and G. M. Hale, priv. comm. to CSEWG (Oct. 1987). (Ref. 14)
2 G. M. Hale, priv. comm. to CSEWG (Oct. 1987); see also Ref. 14.
3 C. Y. Fu, *Nucl. Sci. Eng.*, **106**, 494 (1990).
4 J. Voinier, priv. comm. to BNL (1991).
5 W. P. Poenitz, *Proc. Conf. on Nuclear Data Evaluation Methods and Procedures*, BNL Report BNL-NCS-51363, **Vol I**, p. 249 (1981).
6 R. W. Peelle, priv. comm. to W. P. Poenitz (1981), see Appendix C of Ref. 14.
7 D. B. Gayther and B. H. Patrick, *Proc. IAEA Consultant's Meeting on the ^{235}U Fast-Neu. Fission Cross Section*, Smolenice, March, 1983, INDC(NDS)-146, p. 97.
8 V. G. Pronyaev *et al.*, ibid. Ref. 7, p. 107.
9 W. P. Poenitz, *Proc. Advisory Group Meeting on Nuclear Standard Reference Data*, Geel, 12-16 Nov. 1984, IAEA TECDOC-335, p. 112.
10 A. D. Carlson, *Progr. Nucl. Energy*, **13**, 79 (1984).
11 W. P. Poenitz, *Proc. Topical Meeting on Reactor Physics and Shielding*, ANS, Chicago, Sept. 1984, **Vol II**, p. 529.
12 A. D. Carlson *et al.*, *Proc. Conf. Nuclear Data for Basic and Applied Science*, Santa Fe, 13-17 May 1985, **Vol 2**, p. 1429; also as Radiat. Eff. 96, 87 (1986).
13 R. Peelle and H. Condé, *Proc. Conf. Nuclear Data for Science and Technology*, Mito, Japan, May 30 – June 3, 1988, p. 1005.
14 A. D. Carlson *et al.*, *The ENDF/B-VI Neutron Cross Section Measurement Standards*, ENDF-351, to be published (1992).
15 E. J. Axton, *CBNM*, GE/PH101/86; also priv. comm. (1986).
16 G. M. Hale, *Proc. Symp. on Neutron Standards and Applications*, NBS Spec. Publ. **493**, 30 (1977).
17 J. C. Hopkins and G. Breit, *Nucl. Data*, **A9**, 137 (1971).

18 H. G. Priesmeyer *et al.*, ibid, Ref. 12, p. 1463.
19 T. B. Ryves and P. Kolkowski, *Ann. Nucl. Energy*, **17**, 657 (1990).
20 S. B. Borzakov *et al.*, *Sov. J. Nucl. Phys.*, **35**, 307 (1982).
21 J. Schmiedmayer and M. C. Moxon, ibid. Ref. 13, p. 165.
22 A. L. Kirilyuk *et al.*, *Proc. Int. Conf. on Neu. Phys.*, Kiev, 1987, **Vol. 2**, p. 298.
23 R. A. Schrack *et al.*, *Nucl. Sci. Eng.*, to be published (1992).
24 L. W. Weston and J. H. Todd, *Nucl. Sci. Eng.*, **109**, 113 (1991).
25 M. D. Olson and R. W. Kavanagh, *Phys. Rev.*, **C30**, 1375 (1984).
26 W. P. Poenitz and P. J. Collins, *Proc. NEACRP Spec. Meeting on Applications of Critical Experiments and Operating Data to Core Design via Formal Methods of Cross Section Data Adjustments*, Jackson Hole, WY, Sept. 1988, Report NEACRP-L-307, p. 97.
27 S. Sakamoto *et al.*, *Nucl. Sci. Eng.*, **109**, 215 (1991).
28 J. C. Robertson *et al.*, *J. Nucl. Energy*, **23**, 205 (1969).
29 A. N. Davletshin *et al.*, *Sov. J. Atomic Energy*, **65**, 913 (1988).
30 L. E. Kazakov *et al.*, *Yad. Konstanty*, **44**, 85 (1990).
31 V. L. Demekhin *et al.*, *Proc. 36th All Union Conf. on Nucl.*, Kharkov, p. 94 (1986).
32 E. Quang and G. Knoll, *Nucl. Sci. Eng.*, **110**, 282 (1991).
33 K. Kobayashi *et al.*, *Proc. Conf. on Nuclear Data for Science and Technology*, Jülich, (1992) p. 65.
34 Yu. V. Adamchuk *et al.*, *Sov. J. Atomic Energy*, **65**, 920 (1989).
35 N. N. Buleeva *et al.*, *Sov. J. Atomic Energy*, **65**, 930 (1989).
36 F. H. Froehner, *Nucl. Sci. Eng.*, **103**, 119 (1989).
37 A. D. Carlson *et al.*, *Nucl. Sci. Eng.*, to be published (1992).
38 K. Merla *et al.*, ibid. Ref. 33, p. 510.
39 V. N. Dushin *et al.*, ibid. Ref. 8, p. 53.
40 W. P. Poenitz and J. W. Meadows, ibid. Ref. 8, p. 27.
41 A. D. Carlson *et al.*, *Proc. Spec. Meeting on Neu. Cross Section Standards for the Energy Region above 20 MeV*, Uppsala, May 1991, Re. NEANDC-305, "U", p.165.
42 P. W. Lisowski *et al.*, ibid. Ref. 41, p. 177, see also p. 165.
43 R. G. Johnson *et al.*, *Nucl. Sci. Eng.*, to be published (1992).
44 T. Iwasaki *et al.*, ibid. Ref. 13, p. 87.
45 G. Winkler *et al.*, ibid. Ref. 33, p. 514.
46 N. Kocherov and H. Vonach, *Int. Reactor Dosimetry File (IRDF-90), Proc. 7th ASTM Symp. on Reactor Dosimetry*, 27-31 Aug. 1990, Strasbourg, France.
47 Wu Jingxia *et al.*, *Chinese J. Nucl. Phys.*, **5**, 150 (1983).
48 J. W. Meadows, *Argonne National Lab. Report ANL/NDM-97* (1986).
49 W. P. Poenitz, priv. comm. to A. Carlson, NIST; G. Hale, LANL, and R. Peelle, ORNL, July 30, 1987.
50 J. W. Behrens and G. W. Carlson, *Nucl. Sci. Eng.*, **63**, 250 (1977).
51 F. C. Difilippo *et al.*, *Nucl. Sci. Eng.*, **68**, 43 (1978).
52 P. W. Lisowski *et al.*, priv. comm. (1992).
53 V. A. Formichev *et al.*, ibid. Ref. 33, p. 534.
54 M. Baba *et al.*, *J. Nucl. Sci. & Techn.*, **26**, 11 (1989).

U-238 CAPTURE AND INELASTIC SCATTERING

YUKINORI KANDA
Department of Energy Conversion Engineering,
Kyushu University,
Kasuga, Fukuoka, 816 / Japan

ABSTRACT

The capture and inelastic scattering cross sections of ^{238}U have been left as the two important data of those which are agreed to be re-evaluated in a nuclear data field. From a viewpoint of evaluation methodology, they are the extremes of conditions about experiments and theories. Concerning the capture cross sections, recently the reliable evaluation base on a few experiments has been done with a resonance analysis. The procedure of the evaluation is briefly reviewed. As contrasted with it, the evaluation of the inelastic cross section should be performed on the fact that there are scarce and scattered experimental data and some theories can be applicable even though the parameters used in the theoretical evaluation are not decisive for ^{238}U get. Referring the previous reports analyzing the inelastic scattering experiments for ^{238}U, the theory and parameter set applicable to the evaluation are searched. Both distorted wave Born approximation and coupled-channel model are suitable because the ^{238}U is strongly deformed. In practice, computing coast is important. The optical model parameters of Bruyeres group is used in all the previous report.

1. Introduction

Fuel in nuclear reactors is composed of an overwhelmingly large amount of ^{238}U. Therefore, its neutron cross section is one of the most basic data bases to determine the reactor-physics features which depend on the neutron spectrum in the reactor. The neutrons generated in a nuclear fission experience slowing-down through scattering by nuclei and annihilation through absorption in a reactor and then they form a proper neutron spectrum. The capture and inelastic scattering cross sections of ^{238}U play a decisive role in these process. Nevertheless, nuclear data evaluators have not produced the data base sufficient to answer in quality the requirement from reactor designers. The evaluators and experimenters who are interested in the problem have made efforts to solve it by developing new facilities and detectors to make accurate measurements and new techniques and methods to produce reliable cross sections from scattered experiments. Their studies continue for a few decades. A difficulty of the recent status for these works, however, is that the experimental efforts in the world are abruptly decreasing because of a lack of funds in spite of no experimental data enough to solve it. As the result, it must be solved by the evaluation which utilizes the experiments available at present, that is, to do this we are demanded not only to make full use of the available evaluation methods but also to develop ones on the basis of new ideas.

Considering the status mentioned above, the evaluation of the capture and inelastic scattering cross sections for ^{238}U were adopted as the subject at a working group in the international cooperative evaluation of NEANSC (Nuclear Energy Agency Nuclear Science Committee) in OECD. A report on the capture cross section was already presented at Juelich conference[1].

In this report, we discuss the methods applicable to the evaluation of the ^{238}U capture and inelastic scattering cross sections. In the following section, general issues in nuclear data field are discussed. In the section 3, the evaluation methods of ^{238}U capture cross section developed by the JEF-2 group are reviewed. The evaluation method for $^{238}U(n,n')$ is discussed in the section 4 taking account of referring available literatures. In the last section, a short discussion is presented.

2. General Consideration

In nuclear data evaluation, there are generally several requirements which are necessary to satisfy at all cases : It is the highest priority in these demands that the evaluated cross sections must be absolute values and it is the most difficult besides. The majority of the cross sections is measured relative to some references which are selected to match the facilities and techniques used in their experiments. As the result, it is natural that many kinds of references are adopted even for the specific cross section so that a troublesome renormalization of the individual data is needed before evaluating the cross section value itself. There is, however, no guarantee that the result is correct and no decisive means that one directly confirms its correctness. In another way, the evaluated data must reproduce integral experiments : This is the ultimate end at the present application of the nuclear data. If they are not correct it can not be realized. Therefore , the study in the nuclear data evaluation is primarily to develop the method to produce the correct absolute-value or to determine the most probable value if the evaluator can not believe it correct.

These problems in the nuclear-data field are not so important in the study of nuclear physics in which one can very frequently discuss nuclear structure and reaction mechanism without the absolute cross sections. Therefore, the data acquired in their experiments must be processed with different ways in both fields even though they are obtained with similar facilities and techniques. With regard to theoretical analysis of experiments, also, evaluators require absolutely agreement between theoretical and experimental results but physicists may not. Therefore, the theoretical calculation by the former requires absolutely decided parameters to reproduce the measurements with high reliability as possible.

From a view point of evaluation methodology, the two cross sections of ^{238}U are the extremes of conditions on experiments and theories. The capture cross section is evaluated on the basis of the measurement and not with the theoretical calculation, and the inelastic scattering one is calculated with the theory by referring the experiments to the contrary. There are a few reasons. First, the former can be measured over the whole interesting energy range with high density energy points using a white neutron source : The latter has no enough available data since it must be primarily measured at a single energy by a monoenergetic source though there are few white-neutron-source experiments in which γ-rays are detected in stead of neutrons. Secondary, the capture cross section is expressed merely as a function of the incident neutron energy : The inelastic scattering cross section consists of the partial one corresponding to the energy levels in ^{238}U and should be also given angular distributions for every partial cross section. Thirdly, applicable and appropriate theories for the inelastic scattering have been developed but not for the capture. In other words, there are the theories which are usable to calculate excitation functions of the inelastic scattering and angular distributions of the secondary neutrons for the individual energy levels

3. Evaluation Methods of $^{238}U(n,\gamma)$

The capture cross sections of ^{238}U compiled in the evaluated nuclear data files, JENDL-3, ENDF/B-VI and JEF-2 are consistent and in addition the evaluators of the three files have agreed that the respective evaluation methods adopted in their works are valid. As the report summarized briefly about it was already published[1], in this section a viewpoint of the evaluation methodology is emphasized.

Sowerby et al.[2,3] and Moxon et al.[4,5] and Froehoner[7] in the JEF evaluation group played a leading part in solving the long-pending problem that the evaluated values for ^{238}U (n,γ) must be larger than the correct ones near 100 keV neutron energy.

Sowery and Moxon's approach to the problem is said to be orthodox in a certain sense because they adopt a reliable experiment and renormalize it taking account of background in experimental data. They use, however, a new method which has been never applied to such a work and has become possible with a high-speed computer.

They said that an accurate measurement must be chosen in order to obtain the good evaluated data. In their study, Olsen et al.'s[8,9] transmission data is used to analyze resonance peaks simultaneously with the capture data of de Saussure et al.[10], Moxon[11] and Macklin et al.[12].

Both the data are measured with the time-of-flight facility of ORELA in the neutron energy range of 1eV-100keV. On the occasion of the selection of the transmission data, reliability of three experiments is examined and quantitatively compared with each other. The Macklin et al.'s capture data is chosen as the most accurate, newest and widest energy range experiment though these are a guess by the author taking account of their believe on the choice of the data base as the statement cited in the beginning of this paragraph. The both time-of-flight spectra have many resonance peaks whose shapes depend on their proper physical characteristics and are different at the same resonance between both data. The REFIT computer code[13] is used to analyze the resonance shape of the data, called the shape analysis method. So far, such the data are almost analyzed with the area analysis method which does not take account of the resonance shape but the magnitude of only the area under the resonance peak. In the latter, information including in the peak data can be reduced comparing with the former because the peak area is an integrated values as the result regardless of the height and width of the peak. Contrary to this, the former fully utilize information of the peak data. Therefore, the shape analysis method is superior to the area analysis method. In spite of it, that has not been used because that consume much of computing time. In addition, the REFIT code can simultaneously analyze both transmission and capture data : This function of the code increases the ability to resolve overlapping resonance peaks and identify a small peak by mutual supplementing of both data containing different information through distinct phenomena, namely, the transmission and capture.

On the other hand, the right analysis of the resonance peaks makes clear an amount of background including in the acquired data since the experimental resonance-shape can not be reproduced with the calculated curve if the background is not correctly subtracted. Sowerby and Moxon found that Macklin et al.'s data should be corrected with a factor of 0.9 resulting in 10% decreasing of the capture cross section and that this is similarly applicable to de Saussure et al's data[10]. Their study not only solved the discrepancy in the capture cross section of ^{238}U between the evaluated value and the expected one in the nuclear reactor analysis but also proved the ability of the shape analysis method, which was improved by simultaneous application to both transmission and capture data.

Another result concerning to the evaluation methodology is a demonstration that the normalization of a time-of-flight data to obtain an absolute capture capture cross section can be performed at ordinary resonances, and this means, that although so far the time-of-flight data for the capture cross section was normalized at a saturated resonance, which is usually limited only the 6.67eV resonance for ^{238}U, non-saturated resonances can be also used as the normalization resonances. If there are numerous resonances whose shapes are reliably determined, they can be multiple normalization points and then their absolute values can be estimated with small uncertainty.

Froehner[7] evaluated simultaneously total, inelastic scattering and capture cross sections in the energy range from 10 to 300KeV by utilizing theoretical formulae for individual cross sections. Since the formulae are incoherent. They can be easily fitted to experimental data by adjusting the parameters in them. This method is excellent from a viewpoint of the application of physically-deduced-curves fitting though the method is available to limited energy ranges and cases.

4. Evaluation Methods of ^{238}U (n, n')

The measurements and analysis of ^{238}U(n, n') has been studied as one of the even-even actinide nuclei : They are not only important in nuclear reactor development but also interesting to nuclear physics because they are so strongly deformed that the interaction of a neutron with them can not be described by the simple theories based on the spherical optical model which is preferably used in the nuclear data evaluation. In this section, the results are compared in the following paragraphs to search the guideline on a more reliable evaluation of ^{238}U(n, n'), leaving out the progress about the measurement and data analyses, because Sheldon et al.[14], which was published in 1986 as one of the reports from University of Lowell on fast-neutron inelastic scattering cross sections for the principal even-even actinide nuclei, includes the brief summary of early studies of ^{238}U(n, n') level excitation function and angular distributions.

The total cross section of ^{238}U(n, n') is quite similar in the JENDL-3 and ENDF/B-VI but their partial cross sections are largely discrepant. The reason is that the former can be evaluated conventionally by sutracting the cross sections competing with the inelastic scattering from the total cross section and by referring to existing non-elastic scattering cross section data but the methods and parameters applied to the latter are too ambiguous to result in agreeable evaluation.

The evaluation of the inelastic scattering cross section can not be discussed without considering the other types of interaction such as the elastic scattering, fission and multi-neutron emission reactions while charged-particle emission reactions are precluded owing to their negligibly small cross sections comparing with the ones mentioned above. A main competing interaction in the energy range from the threshold energy of the inelastic scattering to 5MeV is the fission, from 5MeV to 10MeV the fission and (n, 2n), from 10MeV to 15MeV (n, 3n) joins to them and from 15MeV to 20MeV (n, 4n) should be taken into account in addition to the before. Above 8MeV the inelastic scattering cross section is the smallest. Accordingly, a theory used to evaluate the inelastic scattering cross section is desirable to include the competing processes. One of available theories fitting to this requirement is the Hauser-Feshbach model (H-F), whose outstanding feature is the possibility to estimate absolute cross sections of all the available competing and multi-step processes by using the optical model parameters and level density parameters which can be obtained from many literatures. Although the H-F theory is extremely convenient and is widely utilized in the nuclear data evaluation, it is not almighty since it stands on the compound nucleus model (CN). It is known at present that direct interaction (DI) and pre-equilibrium process must be taken into account in order to reproduce experiments with theoretical calculation. The formulae available to the present interesting issues are the ones based on the distorted wave Born approximation and coupled-channel optical model in addition to the H-F.

If the nucleus is deformed, it is physically natural on the basis of the past works to use the distorted wave or coupled-channel formalism for calculation of the inelastic scattering process, in particular, of the angular distribution of the secondary neutron. It depends on the strength of the coupling between the initial and final states. If the coupling between the elastic scattering and reaction channels is strong it must be calculated with the coupled-channel formula. If, however, the coupling is weak, it may be sufficient to use the distorted wave approximations. The ^{238}U nucleus is strongly deformed, so there is strong coupling between the elastic channel and some of the inelastic scattering channels, requiring to perform the coupled-channel calculation.

The analysis methods studied recently for $^{238}U(n,n')$ can be grouped into two types though they all are taking account of the contribution from both CN and DI : In the one, both the contributions are separately calculated and the results incoherently add (hereafter called as incoherent method). In the alternative, both are coherent treated (hereafter designated as coherent method). The former was studied by the Lowell group (Beghian et al.[15], Chan et al.[16], Chan and Sheldon[17], Shao et al.[18], Sheldon et al.[14]), the Bruyeres group (Haouat et al.[19]) and the Oxford group (Hodgson and Kobos[20]), and the latter by the Lowell group (Chan et al.[16], Sheldon et al.[14]).

In the incoherent method, the CN contributions are calculated in all the groups using the Hauser-Feshbach formalism which includes the Moldauer's width fluctuation correction but may not in a group : The DI calculations are performed by the distorted wave approximation or the couple-channel formalism using the JUPITOR code[21] or its modified version. The approach of the coherent method, on the other hand, takes proper account of the possible interaction between direct and compound components by using a unified approach, employing the Engelbrechit-Weidenmueller transformation, and the statistical S-matrix ensemble.

A few examples of the results for the level excitation functions are quoted in Figs. 1 and 2. The whole results do not comprehensively agree with available experiments. Fortunately, the optical model parameters of the Bruyeres group[19] or the slightly modified ones were used in every work referred above. Therefore, all the works can be considered to be independent of the choice of the optical model set. The absolute cross section values are fixed through the proper adjustable parameters which are selected fitting the respective formula and decided by the calculator. Alternatively, the shape of the excitation function reflects the characteristics of the formalism utilized to calculate it. Therefore, which method is useful to the evaluation of the $^{238}U(n,n')$ excitation function should be judged from comparing in the shapes between the calculations and experiments. Any results do not represent the superior agreement to the others. The coherent calculation increasing inclusion of excited states rapidly raises the computing time so that its application to the nuclear data evaluation becomes practically impossible at present, though its physical potentiality is extremely high.

In Figs. 3 and 4, the comparison of the incoherent calculations for the angular distribution of $^{238}U(n,n')$ with available experiments is found partly quoting from the reports. There are no works for the angular distribution by the coherent method. The agreement in shape is acceptable to some extent. The disagreement in the absolute value is the quite same problem in the excitation function.

5. Discussion and Conclusion

The problem in the $^{238}U(n,\gamma)$ cross section are solved. The methods applied in the works have been briefly reviewed.

The incoherent method should be applied to the evaluation of ^{238}U (n,n') but the problem for the normalization is leaving as a considerable difficulty. The optical model parameter set of the Bruyeres group is quite appropriate as the set to be used in the evaluation of $^{238}U(n,n')$ because it has been used and tested in the studies of the various groups as seen above resulting in good performances and also its availability in the preliminary analysis of the new scattering data measured at Argonne National Laboratory has been confirmed[22]. The good result is expected on the basis of these consideration. The present work is a starting point.

The attention to be paid in comparison of a calculation with available experiments is that there is a systematic discrepancy in the cross section for $^{238}U(n,n')$ between the measurement with the direct (n,n') experiment detecting neutrons and with the indirect $(n,\gamma n')$ detecting γ-rays : Roughly speaking the latter is often larger ~20-30% than the former.

The double differential experiments are valuable to check the evaluated $^{238}U(n,n')$ data.

The author would like to thank Dr.T.Kawano, K.Kamitsubo, N.Fujikawa and T.Iwamoto for helpful discussions.

Reference

1. Y.Kanda, Y.Kikuchi, Y.Nakajima, M.G.Sowerby, M.C.Moxon, F.H.Froehner, W.P.Poenitz, and L.W.Weston, "A Report of Evaluated $^{238}U(n,\gamma)$ Cross Section" in Proc. International Conference on Nuclear Data for Science and Technology, 13-17 May 1991, Juelich, Germany (Springer-Verlag, Berlin, 1992) p.851
2. M.Sowerby, in Proc. International Conference on Nuclear Data for Basic and Applied Science, May 1985, Santa Fe, USA (Gordon and Breach, New York, 1986) p.1511
3. M.Sowerby and F.Corvi, in Proc. International Conference on Nuclear Data for Science and Technology, May / June 1988, Mito, Japan (Saikon, Tokyo, 1988) p.37
4. M.C.Moxon, M.G. Sowerby, Y.Nakajima, and C.Nordborg, in Proc. 1988 International Reactor Physics Conference, Sept. Jackson Hole, USA (1988) p. I-281
5. M.C.Moxon, M.G.Sowerby, and J.B.Brisland, in Proc. 1990 International Reactor Physics Conference, PHYSDR'90, April, Marseille, France(1990) p. III-41
6. M.C.Moxon, in Neutron Data of Structural Materials for Fast Reactors (Pergamon Press, Oxford (1979) p.644
7. F.H.Froehner, Nucl. Sci. Eng., **103** (1989)119
8. D.K.Olsen, G.De Saussure, R.B.Perez, E.G.Silver, F.C.Difilippo, R.W.Ingle and H.Weaver, Nucl. Sci. Eng., **62** (1977) 479
9. D.K.Olsen, G.De Saussure, R.B.Perez, F.C.Difilippo, R.W.Ingle and H.Weaver, Nucl. Sci. Eng., **69** (1979) 202
10. G.De Saussure, E.G.Silver, R.B.Perez, R.Ingle, and H.Weaver, Nucl. Sci. Eng., **5** (1973) 385
11. M.C.Moxon, "The Neutron Capture Cross-section of ^{238}U in the Energy Region 0.3 to 100KeV", UKAE Report AERE-R 6074 (1968)
12. R.L.Macklin, R.B.Perez, G.De Saussure and R.W.Ingle, Proc. Int. Conf. Nucl. Data for Sci. Tech., Mito, Japan, Saikon Pub. Co., Tokyo, p.71 (1989)
13. M.C.Moxon, Neutron Data of Structural Materials for Fast Reactors, Pergamon Press, Oxford, p.644 (1979)
14. E.Sheldon, L.E.Beghian, D.W.S.Chan, A.Chang, G.P.Couchell, J.J.Egan, G.Goswami, G.H.R.Kegel, S.Q.Li, A Mittler, D.J.Pullen, W.A.Schier, J.Q.Shao, and A.Wang, J.Phys.G **12**, 443 (1986)
15. L.E.Beghian, G.H.R.Kegel, T.V.Marcella, B.K.Barnes,

G.P.Couchell, J.J. Egan, A.Mittler, D.J.Pullen, and W.A.Schier, Nucl. Sci. Eng., **69** (1979)191
16. D.W.S.Chan, J.J. Egan, A.Mittler, and E.Sheldon, Phys. Rev., **C26** (1982)841
17. D.W.S.Chan and E.Sheldon, Phys. Rev., **C26** (1982)861
18. J.Q.Shao, G.P.Couchell, J.J.Egan, G.H.R.Kegel, S.Q.Li, A Mittler, D.J.Pullen, and W.A.Schier, Nucl. Sci. Eng., **92** (1986)350
19. G.Haouat, J.Lachkar, Ch. Lagrange, J.Jary, J.Sigaud, and Y.Patin, Nucl. Sci. Eng., **81** (1982)491
20. P.E.Hodgson and A.M.Kobos, Nucl. Sci. Eng., **89**(1985)111
21. T.Tamura, Rev. Mod. Phys., **37** (1963)679
22. S.Chiba and A.B.Smith (Private Communication,1992)

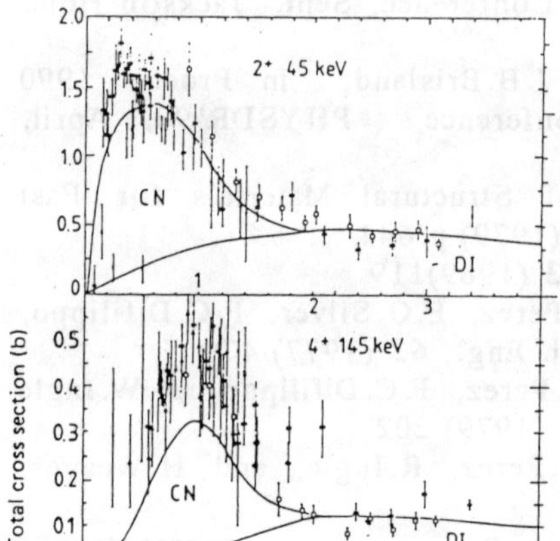

Fig 1. Example of level excitation function of inelastic neutron scattering to 2^+ and 4^+ states in ^{238}U.[20]

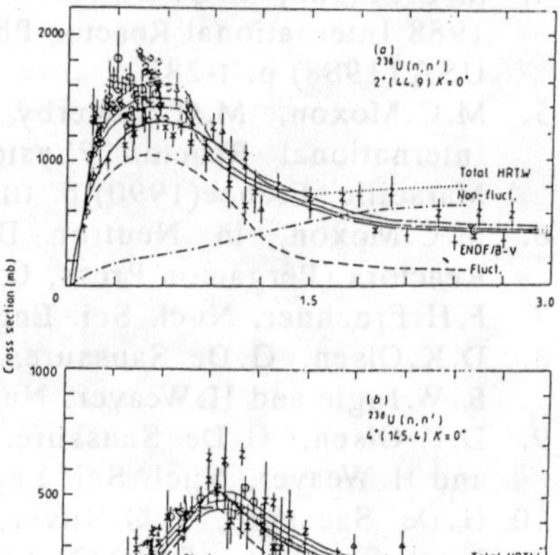

Fig 2. Other example of level excitation function of inelastic neutron scattering to 2^+ and 4^+ states in ^{238}U.[14]

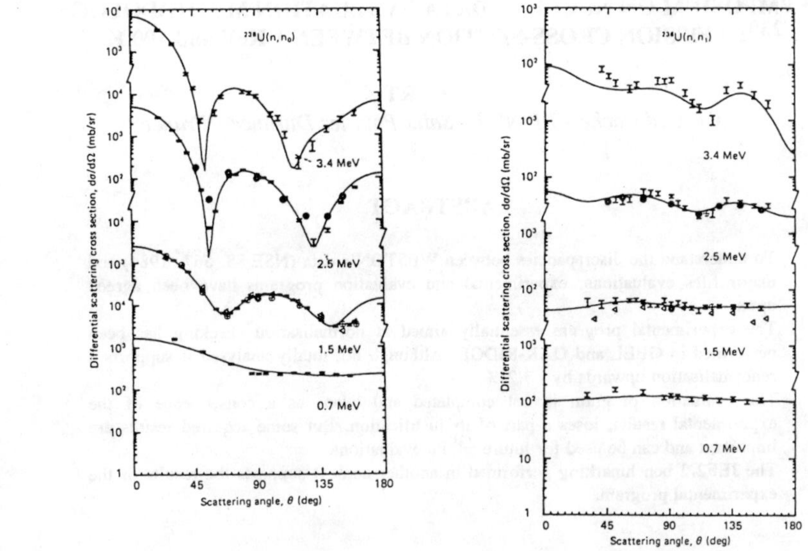

Fig 3. Angular distribution for neutron elastic scattering from ^{238}U and inelastic scattering to first 2^+ (45keV) excited state of ^{238}U at 0.7, 1.5, 2.5, and 3.4 MeV.[19]

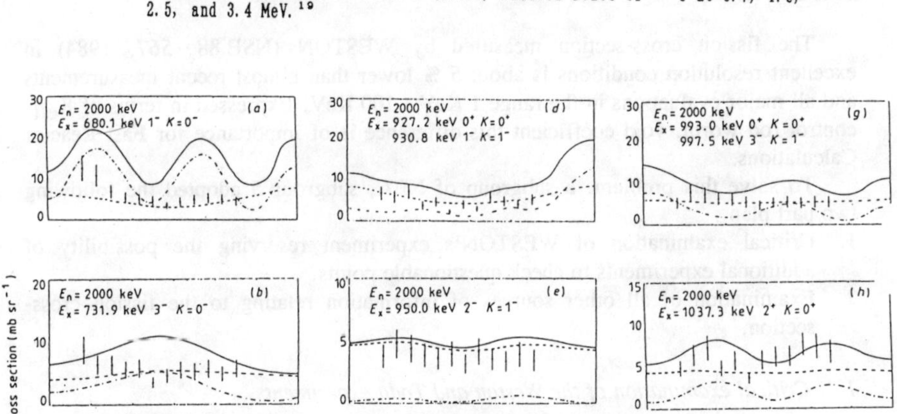

Fig 4. $^{238}U(n, n')$ angular distribution at 2.0MeV for vibrational levels up to E*=1060KeV.[14]

SYMPOSIUM ON NUCLEAR DATA EVALUATION METHODOLOGY ^{239}Pu FISSION CROSS-SECTION BETWEEN 1 KeV and 100 KeV

E. FORT

CE Cadarache - BP N° 1 - Saint Paul lez Durance - France

ABSTRACT

To understand the discrepancies between WESTON's data (NSE 88, 567, 1984) and major files evaluations, experimental and evaluation programs have been agreed upon.

The experimental program essentially aimed at normalisation checking has been performed in GEEL and OAK-RIDGE. Although not totally analysed it supports a renormalisation upwards by ~ 3.2 %.

The evaluation program is not completed and even, as a consequence of the experimental results, loses a part of its justification. But some acquired results are important and can be used for future ^{239}Pu evaluations.

The JEF2.2 benchmarking performed in another context supports the results of the experimental program.

Introduction

The fission cross-section measured by WESTON (NSE 88, 567, 1984) in excellent resolution conditions is about 5 % lower than almost recent measurements and all major evaluations in the range 1 KeV - 100 KeV. Expressed in terms of K_{eff}, control rod worth, void coefficient this difference is of importance for Fast Reactor Calculations.

To solve this problem, a subgroup of NEA, subgroup 5 adopted the following two part plan :

1. Critical examination of WESTON's experiment reserving the possibility of additional experiments to check questionable points.
2. Examination of all other sources of information relating to the fission cross-section.

1.1. Critical examination of the Weston and Todd experiment

This is a T.O.F. experiment performed with ORELA, using a multiparallel plate fission chamber ; the shape of the neutron flux was measured relative to a $^{10}BF_3$ chamber up to a neutron energy a 1 KeV and to a ^{6}Li glass scintillator a higher energies. The flux internormalisation was made in the energy interval 100 eV - 1000 eV with a statistical uncertainty less than 0.1 %. Finally the σ_f curve was normalised to the thermal value. For this experiment conceived for a measurement at high energy, the following aspects were considered to detract from such way of normalisation : sample thickness self absorption, dead time. On the other hand, the fission integral for the interval 100 eV - 1000 eV (I_f = 8996 b.eV) is know only with a rather significant uncertainty (1.9 % for normalisation, 1 % systematic, 0.15 %

statistical). The comparison with the I_f values obtained from GWIN's measurements (I_f = 9268 b.eV) [1] and (I_f = 9286 b.eV) [2] reveals a 4 % difference not in contradiction with the above quoted errors. But since GWIN's measurements are affected by a large uncertainty (11 %) in the ^{10}B content of the neutron flux counter, they have not been considered as absolutely reliable references. Therefore it has been decided to plan experimental programs both at OAK-RIDGE and GEEL to check this normalisation point.

In GEEL (1992) two different geometries have been used, by C. WAGEMANS [3] :

① Measurement with a double ionisation chamber.
The characteristics were as follows :
^{239}Pu sample : 186 μg/cm^2,
^{10}B sample : 10 μg/cm^2.
Flight path length : 8,5 m.

The normalisation has been made in two ways :

i) via the fission integral between 20 meV and 60 meV,

$\int_{20\ \text{meV}}^{60\ \text{meV}} \sigma_f\, dE = 25.36$ b.eV, from which an average value of 634 b is obtained, to be compared with 631.4 b that is an average of several measurements selected by H. DERRIEN in his most recent resonance parameter evaluation [4] ;

ii) via a linear least square fit of the $\sigma\sqrt{E}$ data between 20 meV and 30 meV from which value of 784.25 b was derived for the thermal value.
A preliminary result gives for the interval 100 eV - 1000 eV chosen for the normalisation check a value : $\int_{100\ \text{eV}}^{1000\ \text{eV}} \sigma_f\, dE = 9280$ b.eV, that is 3.15 % higher than WESTON (1984).

② Measurement in "low geometry" using surface barrier detectors.
The ^{239}Pu sample was 106 μg/cm^2 and ^{10}B sample was 10 μg/cm^2.
The neutron flight path length was : 8.3 m.
The data are being analysed.

In OAK-RIDGE a new fission cross-section measurement has been performed by WESTON and TODD with ORELA on a 19 m flight path. This is a still unpublished work and the information on it is obtained through the evaluation report of H. DERRIEN. It is indicated that a careful normalisation in the thermal range leads to a revision of the 1984 data upwards by about + 3.1 %. This evaluation is a simultaneous fit [4] of various experimental data sets according to a Bayesian method using the SAMMY Code, with allowance for background and normalisation coefficient adjustment. The following experimental data base has been used :

- absorption and fission data from GWIN et al [5], [6],
- fission data from GWIN et al [7], [8], BLONS [9], WESTON and TODD renormalised [10], [11],
- transmission data from SPENCER [12], HARVEY [13].

These experimental data sets are characterised by high energy resolution and/or low background. With respect to the ensemble from which the resolved resonance parameter set in JEF2.2 and ENDF-B6 has been derived, additional sets are included, in particular the high resolution transmission data by HARVEY et al [13] and the high resolution fission cross-section measurement by WESTON and TODD (1988) [16]. If one uses this evaluation one obtains for the range 100 eV - 1000 eV a fission integral equal to 9304 b.ev from which one deduces a renormalization by 3.4 % (see table 1).

Table 1 - Averaged fission cross-section

Energy (eV)	Calcul	Weston 1991	Weston 1984 Renorm.	Standard B6
0.010-10	80.12	79.98		
9-20	94.74	94.91		
20-40	17.52	17.76	17.97	
40-60	50.64	50.90	50.87	
60-100	54.42	53.38	54.33	
100-200	18.63	18.59	18.56	18.66
200-300	17.85		17.89	17.88
300-400	8.31		8.34	8.43
400-500	9.59		9.58	9.57
200-500	11.92	11.93	11.93	11.96
500-600	15.39		15.57	15.86
600-700	4.37		4.30	4.46
700-800	5.51		5.53	5.63
800-900	4.84		4.89	4.98
900-1000	8.33		8.38	8.30
500-1000	7.69	7.73	7.73	7.79

Although they are preliminary, all the results of the experimental program are quite consistent and support a ~ 3.2 % renormalisation.

1.2. Examination of all other sources of information relating to the fission cross-section :

A specific study made in JAPAN (NAKAGAWA, KAWAÏ, ...) has shown that other types of data such as alpha data or competitive reaction (capture, non elastic, inelastic, ...) data are too inaccurate to be used as references.

At the beginning it was decided to perform two model calculations to describe the fission data, one based on WESTON's 84 ❶, another one based on ENDF-B6 ❷. Both would be based on the same neutron channel description and validated against clean integral experiments.

Neutron channel description :

High resolution transmission data have been obtained at ORELA by J. HARVEY et al [12] using 3 samples cooled at liquid nitrogen temperature whose thicknesses were chosen so as to be a good compromise between opposite conditions to get accurate data and moderate self screening effects. From these transmission data a total cross-section has been derived that is in agreement with POENITZ et al [13] experimental data but significantly lower (3 % - 4 %) than JEF2 or ENDF-B6. The difference results, in our judgment, from the self screening correction and also from the better quality of the raw transmission data.

The data base formed by POENITZ's and DERRIEN's data have been considered as total cross-section reference data in the range 1 KeV - 500 KeV.

The scattering radius and $S^{l=0}$ and $S^{l=1}$ neutron strength functions values in the resonance range, were extracted from the simultaneous fit by H. DERRIEN to derive the JEF2 resonance parameter sets.

These were all ingredients needed to derive an optical model parametrization that has been obtained by C. LAGRANGE in 1990 (OMP 90). Compared to the previous OMP used in JEF2 (OMP 86) [14] real and imaginary parts of the potential were renormalised, the ranges of the effective interactions and the spin-orbit potential remained unchanged. With this new optical model parametrization one obtains an overall improved fit to the experimental data : the neutron scattering angular distributions (see fig. 1 below) but essentially the total cross-section above 0.5 MeV in an energy range rich in data (see fig. 2).

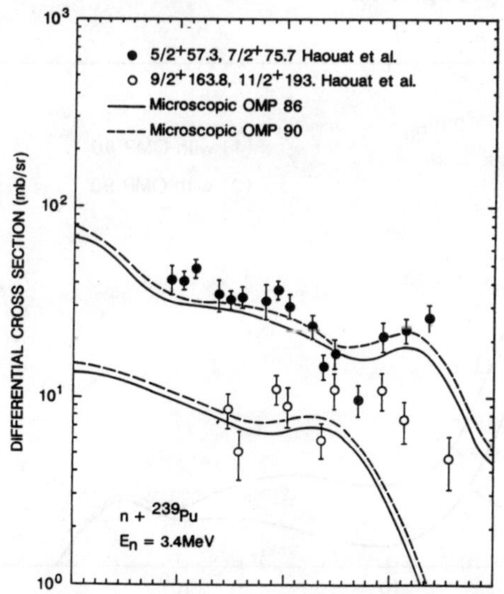

Fig. 1 - Comparison of experimental inelastic scattering cross-sections at 3.4 MeV with data calculated with OMP86 and OMP90

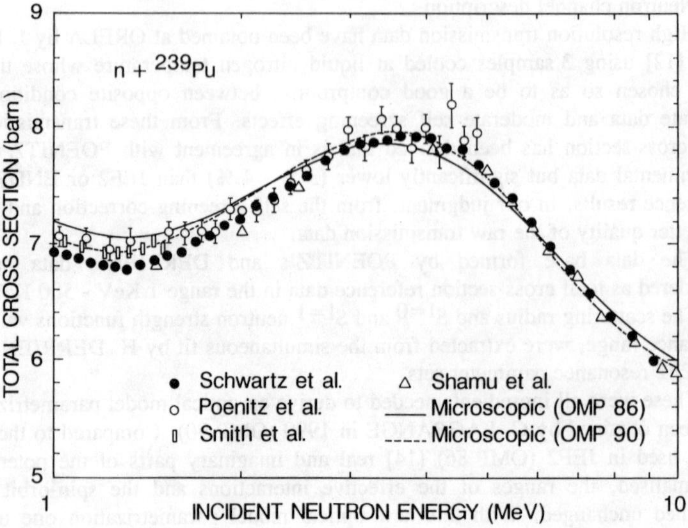

Fig. 2 - Experimental total cross-sections compared with data calculated using OMP86 and OMP90

In addition there is a significant lowering (10 % - 15 %) of the compound cross-section below 1.5 MeV (see figure 3).

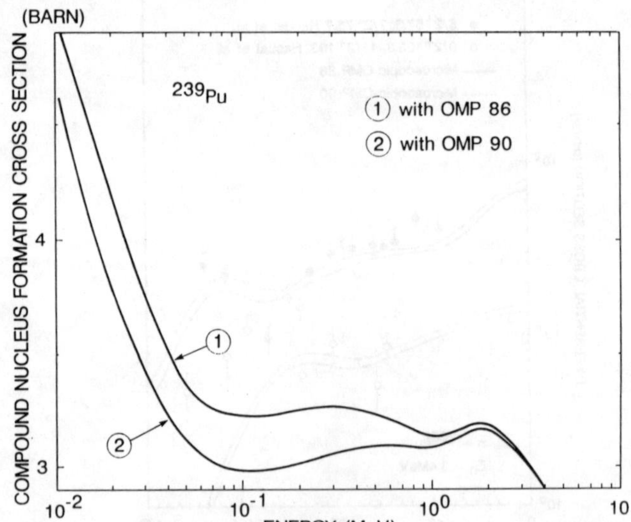

Fig. 3 - Comparison of the compound nucleus formation cross-section obtained by using OMP86 (curve 1) and OMP90 (curve 2)

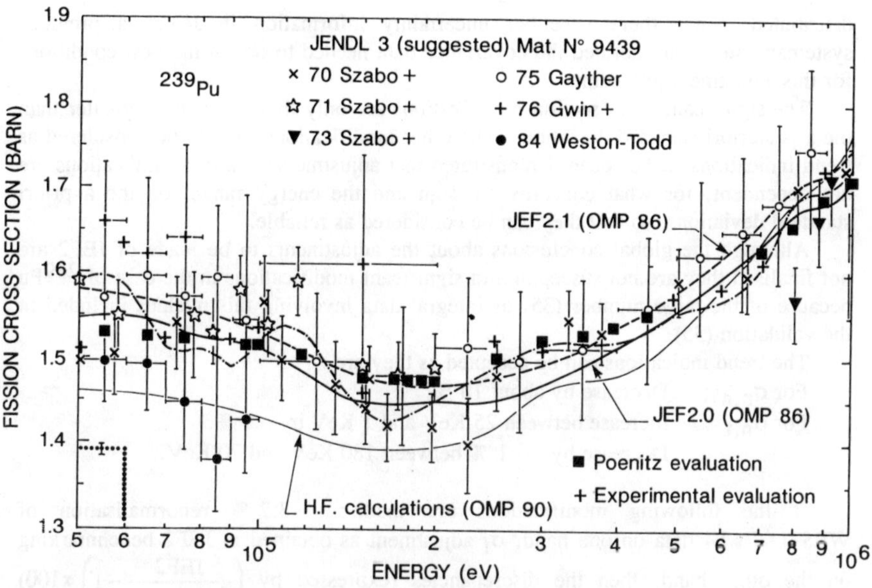

Fig. 4 - Experimental fission cross-section data compaed with the model calculations (OMP90) based on WESTON's data

This is an important feature from which direct conclusions can be drawn. The model evaluations ❶ is almost completed and shows (see figure 4) that a fissions cross-section based on WESTON 84 is low with respect to all the other evaluations up to 1 MeV, i.e., in the range of full sensitivity of Fast Reactors. One observes also with respect to the JEF2.2 evaluation a lowering of the inelastic cross-section which partially compensates, for what concerns the K_{eff} value of fast systems, the effect of a lowering of σ_f.

The model evaluation ❷ has not been performed. The model evaluation ❶ has not been yet tested against chosen clean integral data, but the JEF2.2 benchmarking that is being performed allows some conclusions to be transposed. The conditions of this benchmarking are fully described in another publication at this meeting [15]. They can be summarised as follows :

The JEF2.2 library has been processed into a 1968 energy groups library containing infinite dilute cross-section and probability tables which have been produced by using the NJOY-THEMIS and CALENDF codes respectively. A great effort has been devoted to Quality Assurance to be sure that the data processing preserved the quality and the integrity of the information. The integral data have been recalculated with this fine group library. The cross-section modifications to minimise the discrepancies between the experimental and precalculated integral data have been obtained by a statistical adjustment, assuming that all the data were normally distributed. Since there is either uncertainty information in JEF2, a previous systematic study had defined the covariance data needed to obtain the best conditions for this adjustment procedure.

The significant cross-section modifications are only those which are greater than the a posteriori standard deviations. The other modifications have to be considered as trend indications. It has been demonstrated that adjustments and trend indications are not dependent, for what concerns the sign and the energy range, on the a priori standard deviations, so that they can be considered as reliable.

Although the global conclusions about the adjustments to be made of JEF2 are not finalised they are not susceptible to significant modifications in the case of ^{239}Pu because of the large number (35) of integral data involving this nucleus included in the validation (35).

The trend indications can be accepted as they are :

For $\sigma_{n,n'}$: Decrease by about 10 %.

For $\sigma_{n,f}$: Increase between 25 KeV and 1 KeV by ~ 1 %.
Decrease by ~ - 1 % between 180 KeV and 70 KeV.

If the following modifications are made : + 3.2 % renormalisation of WESTON's 84 data on one hand, σ_f adjustment as obtained in JEF2 benchmarking on the other hand, then the discrepancies (expressed by $\left(\frac{\text{JEF2}}{\text{WESTON}}-1\right)$ x100) between WESTON revised and JEF2 data are significantly reduced when not totally cancelled below 10 KeV. Above this energy limit there is still a discrepancy of about ~ 3 %, as it, can be seen in the table 2.

Table 2 - Discrepancies between JEF2 and WESTON's data

Energy group structure (eV)	JEF2-2 (barn)	WESTON84 (barn)	Before modification (%)	After modification (%)
4.5400E+02 - 1.2340E+03	7.091	6.99	1.44	-2.76
1.2340E+03 - 3.3550E+03	3.653	3.50	4.28	1.08
3.3550E+03 - 9.1190E+03	2.246	2.18	2.75	1.3
9.11904+03 - 2.4790E+04	1.722	1.67	3.1	0.2
2.47904+04 - 6.7380E+04	1.578	1.48	6.2	2.3
6.7380E+04 - 1.1110E+05	1.544	1.42	8.7	4.2

A similar conclusion can be stated for the standard ENDF/B6. As a matter of fact if one refers to figure 5 there is still a discrepancy of ~ 2 % above 10 KeV between POENITZ's fit and the renormalised WESTON's data. Without drawing any final judgement it is possible to claim that a careful benchmarking of the evaluation would be helpful in deciding about the existence of an energy dependent effect in WESTON's experiment.

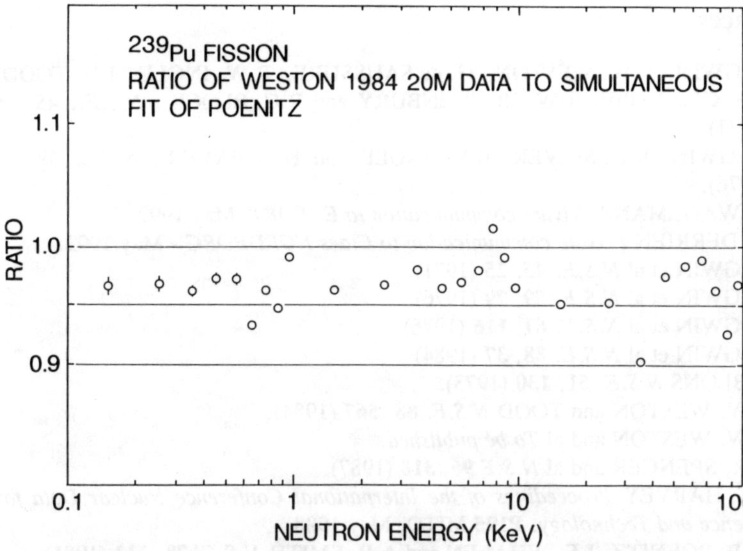

Fig. 5 - Ratio of WESTON's data (1984) to simultaneous fit of POENITZ

Conclusion

After 2 years of work the subgroup 5 has not fully completed the task it had planned. The experimental program is almost finished and although not all the results are finalised the results that are already available lead to a firm conclusion : the fission cross-section data obtained by WESTON and TODD in 1984 have to be renormalised upwards by approximately 3 %. What is left for a near future is to provide a refined figure for this renormalisation.

The question now is to evaluate as correctly as possible what still should be done to improve the situation at low cost. In particular should the validation work of the model evaluation ❶ be continued ? Or should the finalised conclusion be based on the experimental information only ?

But, in any case what has been done outside the experimental field, has been very useful since it is of prime importance for any future reevaluation of ^{239}Pu. This is in particular true for the conclusions concerning the examination of the competitive data and for the definition of the best presently available OMP for ^{239}Pu.

References

1. R. GWIN, L.W. WESTON, G. de SAUSSURE, R.W. INGLE, J.H. TOOD, F.E. GILLESPIE, R.W. HOCKENBURY and R.C. BLOCK : *N.S.E.*, **45, 25** (1971).
2. R. GWIN, E.G. SILVER, R.W. INGLE and H. WEAVER : *N.S.E.* **59, 79** (1976).
3. C. WAGEMANS *Private communication to E. FORT*. May 1992.
4. H. DERRIEN *Private communication to Claes NORDBORG* - May 1992.
5. R. GWIN et al *N.S.E.* **45, 25** (1971).
6. R. GWIN et al *N.S.E.* **59, 79** (1976).
7. R. GWIN et al *N.S.E.* **61, 116** (1976)
8. R. GWIN et al *N.S.E.* **88, 37** (1984)
9. J. BLONS *N.S.E.* **51, 130** (1973).
10. L.W. WESTON and TOOD *N.S.E.* **88, 567** (1984).
11. L.W. WESTON and al *To be published.*
12. R.R. SPENCER and al *N.S.E* **96, 318** (1987).
13. J.A. HARVEY *Proceedings of the International Conference Nuclear Data for Science and Technology*. **P195** MITO, May 1988.
14. W.P. POENITZ, J.F. WHALEN and A.B. SMITH *N.S.E.* **78, 333** (1981).
15. Ch. LAGRANGE and D.G. MADLAND *Physical Review C.* **Volume 33 number 5, p. 1616.**
16. L.W. WESTON and T.H. TODD *Unpublished work. Private communication to H. DERRIEN.*
17. E. FORT and M. SALVATORES *"Fast Reactor Benchmarks and integral Data testing and feedback into JEF2". A paper presented to this meeting.*

THERMAL NEUTRON ACTINIDE DATA

H. Tellier

Service d'Etudes de Réacteurs et de Mathématiques Appliquées
Centre d'Etudes Nucléaires de Saclay - 91191 Gif-sur-Yvette FRANCE

INTRODUCTION

During the 70's, the physicists involved in the cross section measurements for the low energy neutrons were almost exclusively interested in the resonance energy range. The thermal range was consedered as sufficiently known. In the begining of the 80's, reactor physists had again to deal with the delicate problem of the power reactor temperature coefficient, essentially for the light water reactors. The measured value of the reactivity temperature coefficient does not agree with the computed one. The later is too negative. For obvious safety reasons, it is an important problem which must be solved. Several causes were suggested to explain this discrepancy. Among all these causes, the spectral shift in the thermal energy range seems to be very important. Sensibility calculations shown that this spectral shift is very sensitive to the shape of the neutron cross sections of the actinides for energies below one electron-volt. Consequently, reactor physisists require new and accurate measurements in the thermal and subthermal energy ranges [1,2]. A part of these new measurement results were recently released and reviewed [3]. The purpose of this study is to complete the preceding review with the new informations which are now available. In reactor physics the major actinides are the fertile nuclei, uranium 238, thorium 232 and plutonium 240 and the fissile nuclei, uranium 233, uranium 235 and plutonium 239. For the fertile nuclei the main datum is the capture cross section, for the fissile nuclei the data of interest are nu-bar, the fission and capture cross sections or a combination of these data such as η or α. In the following sections, we will review the neutron data of the major actinides for the energy below 1 eV.

URANIUM 238

At low energy, the cross section shapes of uranium 238 are given by the 4.4 eV "p" wave and the 6.67 eV "s" wave resonances. These two resonances induce for capture cross section a 1/v behaviour in the thermal range. It is the shape which is universally adopted by all the evaluated files. But a part of the temperature coefficient descrepancy could be explain by a non 1/v dependence of the uranium 238 capture cross section for the low energy neutron. The cross section must decrease with the energy faster that the 1/v shape. This effect can be obtained with the assumption of a weak bound level near the zero energy. L. Erradi proposed a small resonance at - 0.005 ev [4]. This hypothesis was supported by the measurement of the fission cross section at 0.025 ev which was performed in Grenoble [5]. The expérimental value cannot be exclusively explained by the contribution of the nearby resonances. An extra resonance is needed. If this resonance is close to the zero energy as in the Erradi's assumption, it must have an impact on the cross section shape in the thermal range. The non 1/v shape which is obtained with the - 0.005 eV resonance is not incompatible with the Harwell measurement of the uranium 238 capture cross section [6] but only because the experimental incertainties of this measurement are rather large. New and more accurate measurements of the uranium 238 capture cross section were required. These measurements were performed at the Geel laboratory [7]. As it can be seen on figure 1 which represents the experimental variation of $\sigma\gamma \sqrt{E}$ as a function of the neutron energy, and a comparison with the JEF evaluation, these results confirm without ambiguity a 1/v behaviour for the capture cross section. The assumption of the resonance in the immediate vicinity of the zero energy and the explanation of a part of the temperature coefficient discrepancy by such a resonance must be dropped. Reasonably, we must consider that the uranium 238 capture cross section problem is solved in so far as the thermal energy range is concerned.

THORIUM 232

Nowadays, the use of thorium fuel cycle in thermal neutron reactors is no more a high priority and the physicist interest for thorium nuclear data is less important in the case of low energy neutron than in the fast range. Consequently the situation is fondamuntally different from the one of uranium 238. That is why no request recently appeared for the thermal neutron energy range. In this energy domain it exists only two recent differential measurements which give access to the shape of the thorium 282 capture cross section. They are the measurement performed in Brookhaven for the energies between 35 and 1000 meV [8] and the experiment of RPI which covered the neutron energies above 10 meV [9]. These two sets of experimental data are compared on figure 2 which displays the variation of $\sigma_\gamma \sqrt{E}$ versus the neutron energy. If we take into account the experimental uncertainties, the agreement between the two measurements is good enough above 50 meV. They are also in good agreement with the recommended value of the ENDF/B6 evaluation. In the very low energy domain, between 10 and 25 meV, it exist only one measurement which significantly deviates from the evaluated recommendation. The experimental cross section decreases less than a 1/v shape. We know that the temperature coefficient of a multiplying lattice is very sensitive to the shape of the fertile nucleus capture cross section below 25 meV. In the case of a thorium cycle revival, the observed discrepancy between the measurement and the evaluation must be clarified as it was recently done for uranium 238. New and accurate measurement of the thorium 282 captur cross section would be needful in the thermal and subthermal energy range.

PLUTONIUM 240

For this isotope, the cross section behaviour in the thermal energy range is mainly governed by the 1.056 eV resonance. Consequently it is necessary to have a very accurate knowledge of this resonance parameters. Only two measurements of the 1.056 eV resonance parameters were recently carried out, the experiment of Brookhaven [10] and the one of Oak Ridge [11]. In the Brookhaven experiment, total and capture cross sections with room temperature and cooled samples were used. In the Oak Ridge measurement, transmission measurements with seven thicknesses of sample were performed. Thus it was expected that the results could be very satisfactory. Unfortunately, as it can be seen in table 1, the two sets of results are significantly discrepant.

Table 1 : Parameters of the ^{240}Pu 1.056 eV resonance

Γ_n (meV)	Γ_γ (meV)	Laboratory
2.32 ± 0.06	32.4 ± 0.6	Brookhaven (81) [10]
2.45 ± 0.02	30.3 ± 0.3	Oak Ridge (87) [11]

A detailed analysis of the two experiments leads to be more confident in the second set of results. But the interpretation of spent fuel isotopic composition suggests tendencies closer to the high value of radiative capture width. The différence between the two series of resonance parameters induces a change of 1.2 percent on the contribution of the 1.056 eV resonance to the capture cross section at 0.025 eV and a change of 4.5 percent on the resonance integral value. These modifications become very important every time that the plutonium 240 is significantly involved. It is mainly the case for irradiated fuel analysis or for plutonium recycling in light water reactor. The discrepancy between the two differential measurements and the tendency deduced from the integral experiments must be clarified.

URANIUM 235

In the case of a fissile nucleus, the problem is more complex than for a fertile nucleus. In addition to the capture cross section, other fundamental nuclear data are involved, the number ν of neutron which are emitted in a fission and the fission probability. These three quantities, or a combination of them such as η or α must be investigated.As it has been shown that the temperature coefficient is very sensitive to the shape of the various nuclear parameters versus the neutron energy, several differential measurements were performed during the last years. They are relative to the fission cross section, nu-bar and eta in the very low energy range.

a) Nu-bar

One measurement of nu-bar was recently performed in the energy range which we are interested in. It is the Oak Ridge experiment which gives the ratio of the uranium 235 prompt nu-bar over the Californium ν_{sp} of spontaneous fission for the neutron energies between 5 meV and 1 eV [12]. According to these results, which are displayed on figure 3, nothing appears in the vicinity of the 0.29 eV resonance and we can reasonably keep the assumption of a constant value of nu-bar below 1 eV. This flat shape, which is adopted in all the evaluated files, has important consequences for the capture cross section behaviour at low energy.

b) Fission cross section

Since 1984, very accurate measurements was performed in the low energy neutron range [13, 14, 15]. As it can be seen on figure 4, above 20 meV and in particular in the 0.290 eV resonance all these results are in agreement between them and with the most recent evaluated files ENDF/B5 and JEF2. In the subthermal energy range, below 5 meV, only the Geel experiment gives informations. According to this measurement the uranium 235 fission cross section reaches an 1/v shape for energies higher than it was assumed in ENDF/B5. On the contrary, JEF2 which was released after the Geel experiment takes into account its results below 10 meV and adopts a fission cross section shape closer to a 1/v behaviour.

c) **Eta measurements**

To explain the temperature coefficient discrepancy of the uranium fuel reactor, it is $\eta = \nu \frac{\sigma_f}{\sigma_\alpha}$ which is the most sensitive parameter. In all the previous evaluated files, including ENDF/B5 and JEF1 eta was assumed to have a constant value below 0.1 eV. As the reactor physicists proposed to increase eta between 5 and 100 meV, measurements of this neutron parameter were necessary to validate the modification of shape. Four measurements of eta were recently performed in the range of interest.

The first one is the Geel experiment [16]. This experiment was performed with a linac and a liquid methane moderator to inhance the importance of the low energy neutrons. It covered the neutron energy between 2 and 450 meV. As it is shown in figure 5 the results suggest an increase of eta between 2 and 80 meV. It is in the good direction for the reactor physics.

A second measurement was carried out with the Harwell linac [17]. Unfortunately the number of low energy neutron was not very high. Consequently the accuracy was not good enough. Nevertheless this experiment did not show a significant shape of eta versus the neutron energy as it is displayed on figure 6. It is contradictory with the Geel results.

A third experiment was performed in Grenoble [18]. Instead of a linear accelerator, like in Geel, the neutron source was constituted by a cold neutron beam of the high flux reactor and more neutron of low energy were obtained. The accuracy was then expected to be better because the background would be lower. In the energy range between 2 and 150 meV, this experiment perfectly confirms the Geel results and the shape of eta versus energy as it is shown on figure 5.

Finally a fourth experiment was carried out with the Oak Ridge linac [19]. The preliminary results of the last experiment are compared with the Harwell results on figure 6. These two series of results seem more or less in agreement and do not show a significant shape of eta.

These four experimental results can be splitted in two sets, a first set which indicates a shape of eta (the Geel and Grenoble data) and the second set which does not (the Harwell an Oak Ridge data). As the low neutron flux was the highest in the Grenoble experiments, it is possible to given a more important weight to these results and to propose a slope for the eta shape below 100 meV. This attitude was adopted for the preliminary version of JEF2. But for the physical point of view, the desagreement between the two sets of results is not acceptable. On behalf on the NEA Nuclear Science Comittee, a working group carefully studied the various corrections (count loss, background substraction, absorption...) which were applied to the raw data of the four measurements. The final recommendations of the working group are not established yet but the preliminary results are encouraging. It seems that it may be quite possible to define a curve of eta with an energy dependent shape. Eta would increase of about 1,3 % between 3 and 80 meV and this energy dependence would be compatible with the four experimental data [20]. This shape would be close to the reactor physicist suggestion.

d) **Alpha measurement**

A complementary and important information upon the uranium 235 cross section in the thermal energy range is given by the recent Geel measurement of α [21]. It is a very interesting result because it constitutes an independent way to obtain information about $\eta = \frac{\nu}{1+\alpha}$. As it can be seen on figure 7, the experimental values of α are not reproduced by the ENDF/B5 evaluation. As all the previous files, ENDF/B5 recommend a flat shape of α below 100 meV. As ν is energy independent this shape corresponds to a constant value of η. On the contrary, it was adopted in JEF2 a slope for η and this file, which was released before the experimental values of α, is in good agreement with the measurement results. It is an important fact, because now, we have a coherent set of experimental data for ν, σ_f, η and α, which confirm the slope of η, as it was suggested by the integral experiment.

URANIUM 233

For the same reasons as for thorium, the uranium 233 nuclear data in the low energy range was not systematically studied these last years. Nevertheless it exist some scarce results, mainly for nu-bar and the fission cross section. These measurement were generally performed in the same campaign as uranium 235 when the physicists have uranium 233 samples at their disposal. For η and α, nothing new is available.

a) **Nu-bar**

The only result about nu-bar below 1 eV is the one of Oak Ridge [12] which gives the ratio of uranium 283 prompt nu-bar to the spontaneous fission nu-bar of californium 252. No significant structure was observed in this energy range, as it is shown as figure 8 and we can reasonally admit the flat shape which is adopted in ENDF/B6. But don't forget that the ENDF/B6 absolute value of nu-bar is not in agreement with the tendency which is deduced from the buckling measurements [22].

b) **Fission cross section**

At the opportunity of the campaign of measurements on fissile nuclei, an accurate determination of the shape of the uranium 233 cross section shape was carried out with the Geel linac [23]. In order to inhance the low energy neutron flux and obtain a good accuracy in the thermal and subthermal energy range, a liquid nitrogen cooled moderator was used. In these experimental conditions we can be very confident in the results which are displayed on figure 9. They are also in fair agreement with the ENDF/B6 recommendation. Reasonably we can admit that the shape of the uranium 233 fission cross section is well known below 1 eV.

PLUTONIUM 239

All the old evaluations of the plutonium 239 neutron data, including ENDF/B5, are considered as not satisfactory by the reactor physicists. As a matter of fact, in all these files it was adopted a flat behaviour of nu - bar in the low energy range, the spin of the resonances was not considered and it was used a Breit and Wigner formalism to compute the cross sections. With the high burn up fuels and the recycling, the plutonium became more and more important in the thermal neutron reactors. An updating of the plutonium 239 neutron data was strongly required. This updating was performed by Derrien et al. [24] who took into account new experimental results, concerning nu-bar [12], the fission cross section [13] and the total cross section [11], and used a Reich and Moore formalism which is more convenient for the fissile nuclei.

a) **Nu-bar**

On the contrary to the uranium 233 and uranium 235 cases, the recent measurement of plutonium 239 nu-bar [12] shows an important decrease in the vicinity of the low energy resonance at 0.3 eV. This strong structure is well reproduced by the Fort's theoritical calculation which takes into account the spin effect and the (n, γf) effect of the J = 1 resonances of plutonium 239 [25]. Figure 10 represents a comparison between the experimental values of nu-bar, normalized to the spontaneous fission nu-bar of californium 252, and the evaluation of Fort et al. As the agreement is very good this shape of nu-bar was included in JEF2.

b) **Fission cross section**

All the new evaluated files use the resonance parameter set which was deduced from the Derrien's analysis. The behaviour of the plutonium 239 cross sections in the thermal range is well reproduced by the contributions of the low energy resonances and of the bound level. After the release of the initial version of the recent evaluated files a new measurement of the plutonium 239 fission cross section became available [23]. The figure 11 shows the comparison of these new experimental values with the recommended values of JEF2 below 1 eV. The agreement is quite satisfactory and the new fission cross section measurement constitute a confirmation of the recommended values. Today, no request upon the plutonium 239 fission cross section seems needful as far as the low energy is concerned.

CONCLUSION

Since a few years, the status of the thermal neutron data for the major actinides was greatly improved. The recent measurements of the microscopic data led to a better knowledge of the cross section shapes in the low energy domain. Several problems which were of great importance for reactor physics were solved. Let us mention the 1/v dependance of the uranium 238 capture cross section and the behaviour of the uranium 235 fission cross section below 20 meV. The structure of the plutonium 239 nu-bar, which was not taken into account in the past, was well established and theoritically explained. We can reasonably expect that the uranium 235 eta discrepancy will be solved in the next future. Nevertheless it remains some problems which have to be further investigated. The most important, today, is the discrepancy between the two sets of parameters of the 1.056 eV plutonium

240 resonance. This desagreement has an important impact in high burn up fuels or plutonium recycle studies. The difference between the evaluated values and the measured values of the thorium 232 capture cross section below 20 meV and the problem of the absolute value of the uranium 233 nu - bar have a lower priority. But in the case of new interest for the thorium cycle, requests upon these two actinides will be certainly needful.

REFERENCES

[1] - J.L. Rowlands : "Neutron physics and Nuclear Data - Harwell", 7 (1978).

[2] - J. Bouchard, C. Golinelli et H. Tellier : "Neutron data for Science and Technology - Anvers", 21 (1982).

[3] - H. Tellier : "Neutron Data for Science and Technology - Jülich" (1991).

[4] - L. Erradi : These doctorat es sciences. Paris Orsay (1982).

[5] - P. D'Hont et al. : Annals of Nuclear Energy, 11, 485 (1984).

[6] - M.C. Moxon and J.E. Jolly : "Uranium and Plutonium resonance parameters - Vienna" 124, INDC-129 (1981).

[7] - F. Corvi and G. Froni : "Nuclear Data for Science and Technology- Mito", 127 (1988).

[8] - R. E. Chrien et al. : Nuclear Science and Engineering, 72, 202 (1979).

[9] - R.C. Little et al. : Nuclear Science and Engineering, 79, 175 (1981).

[10] - H.J. Liou and R.E. Chrien : "Uranium and Plutonium resonance parameters - Vienna" 438, INDC-129 (1981).

[11] - R.R. Spencer et al. : Nuclear Science and Engineering, 96, 318 (1987).

[12] - R. Gwin et al. : Nuclear Science and Engineering, 87, 381 (1984).

[13] - R. Gwin et al. : Nuclear Science and Engineering, 88, 37 (1984).

[14] - C. Wagemans and A.J. Deruytter : "Nuclear Data for Basic and Applied Science - Santa Fe" I, 499 (1985).

[15] - R.A. Schrack : "Nuclear Data for Science and Technology - Mito", 101 (1988).

[16] - J.A. Wartena et al. : Report IAEA Tecdoc 491, 123 (1987).

[17] - M.C. Moxon et al. : "Nuclear Data for Science and Technology - Mito", 75 (1988).

[18] - J.A. Wartena et al. : Report NEANDC (E) 312, III (1989).

[19] - M.C. Moxon et al. : Private communication to the NEA Data Bank (1990).

[20] - H. Weigmann : Private Communication (1991).

[21] - H. Weigmann et al. : "Nuclear Data for Science and Technology - Jülich" (1991).

[22] - H. Tellier and J. Vanuxeem : (This meeting).

[23] - C. Wagemans et al. : "Nuclear Data for Science and Technology - Mito", 91 (1988).

[24] - H. Derrien et al. : Nuclear Science and Engineering, 106, 434 (1990).

[25] - E. Fort et al. : Nuclear Science and Engineering, 99, 375 (1988).

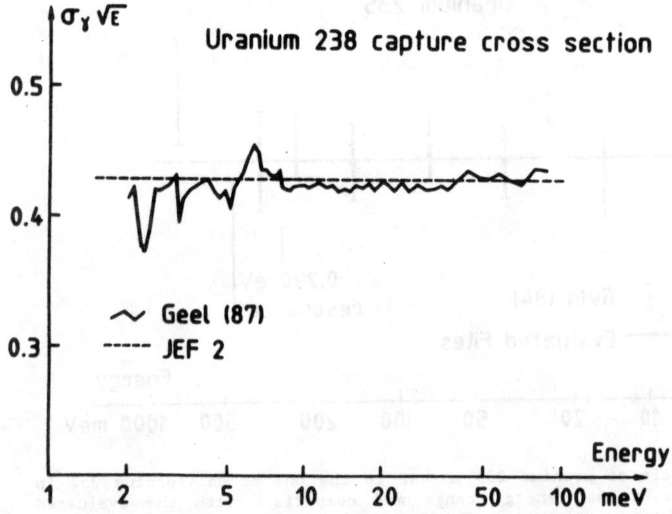

Figure 1 - Comparison between the uranium 238 capture cross section measurement of Geel, for a metal sample, and the evaluated values of JEF2.

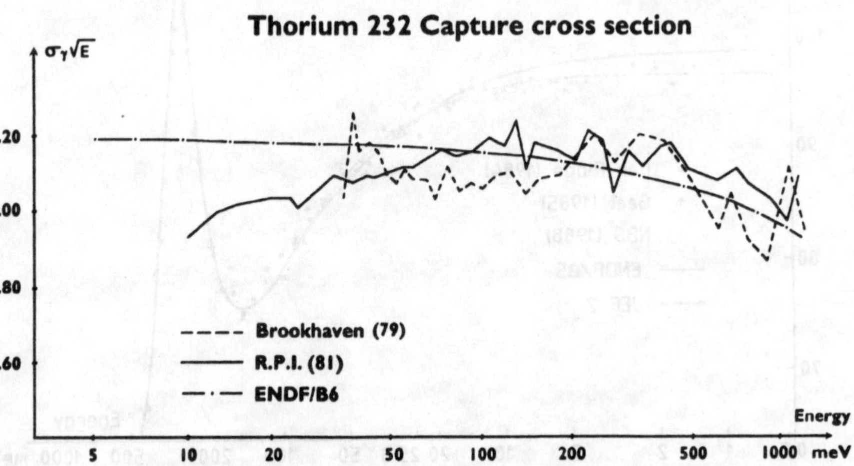

Figure 2 - Comparison of the recent measurements of the thorium 232 capture cross section and the evaluated values of ENDF/B6.

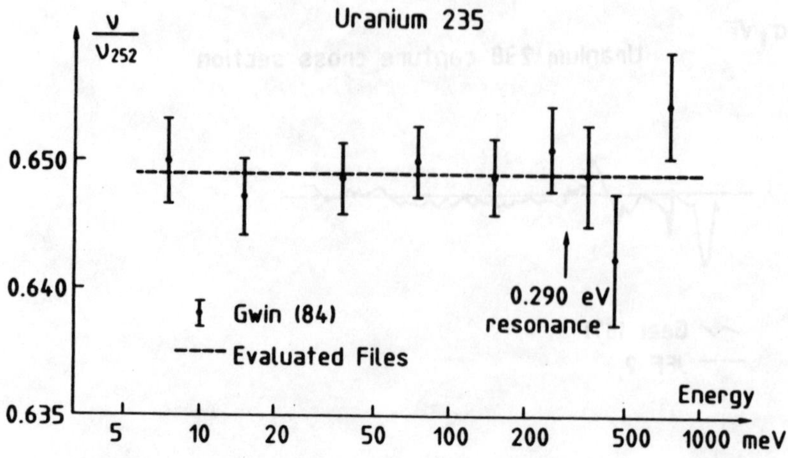

Figure 3 - Ratio of uranium 235 nu-bar to the one of californium 252 in the thermal energy range and comparison with the evaluated values.

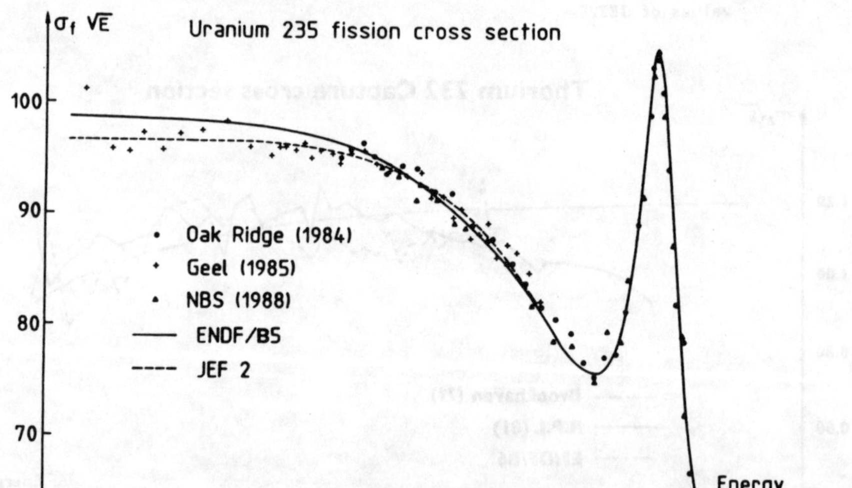

Figure 4 - Recent measurements of the uranium 235 fission cross section and comparison with ENDF/B5 and JEF2.

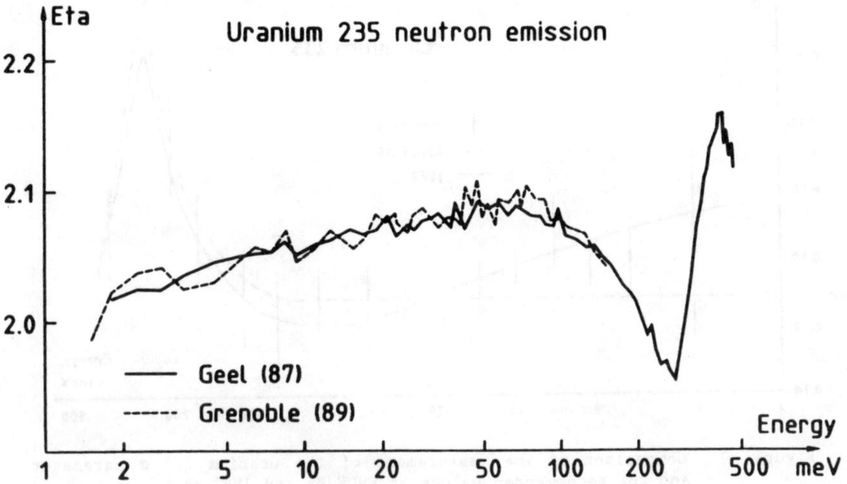

Figure 5 - Experimental results of uranium 235 eta measurements of Geel and Grenoble in the low energy neutron range.

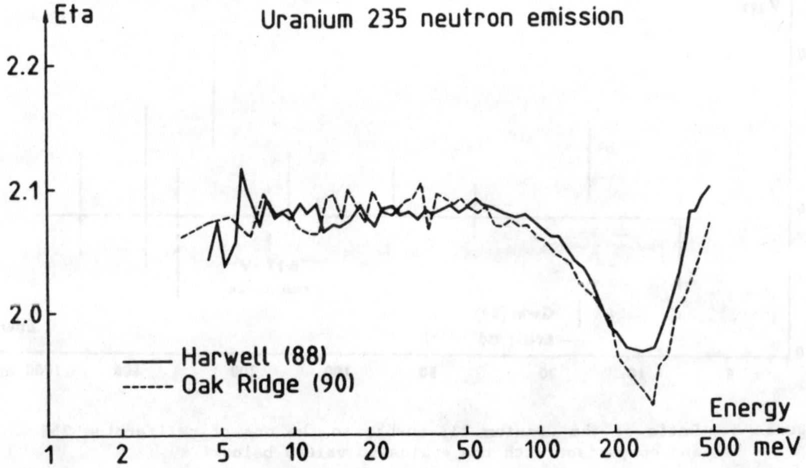

Figure 6 - Experimental results of the uranium 235 eta measurements of Harwell and Oak Ridge below 500 meV.

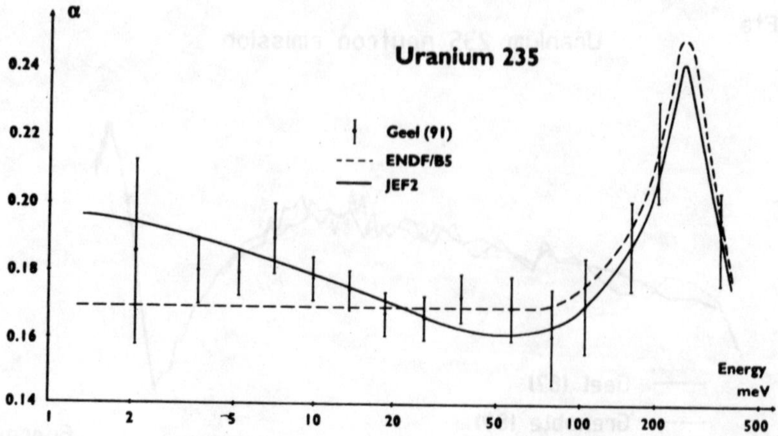

Figure 7 - Comparison of the measurement of the uranium 235 α parameter and the recommended values of ENDF/B5 and JEF2.

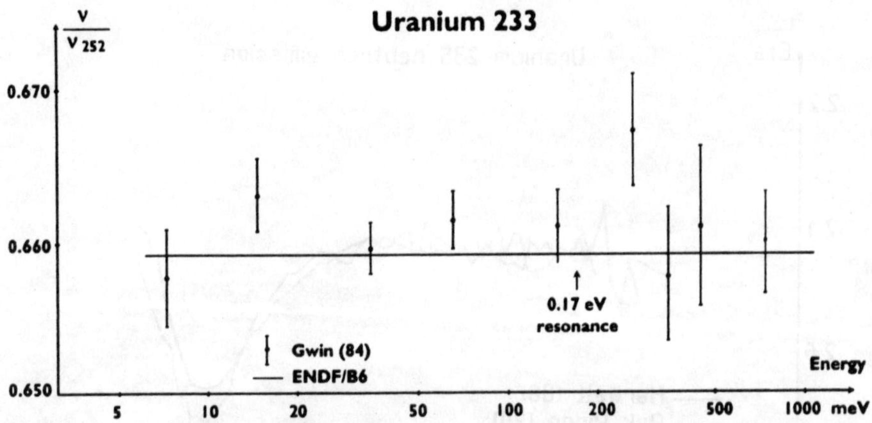

Figure 8 - Ratio of the uranium 233 nu-bar to the one of californium 252 and comparison with the evaluated values below 1 eV.

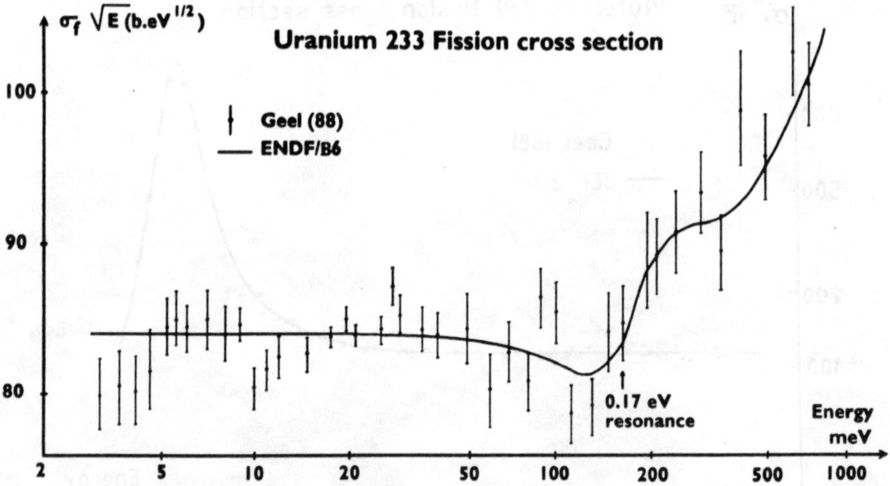

Figure 9 - Experimental fission cross section of uranium 233 and comparison with ENDF/B6.

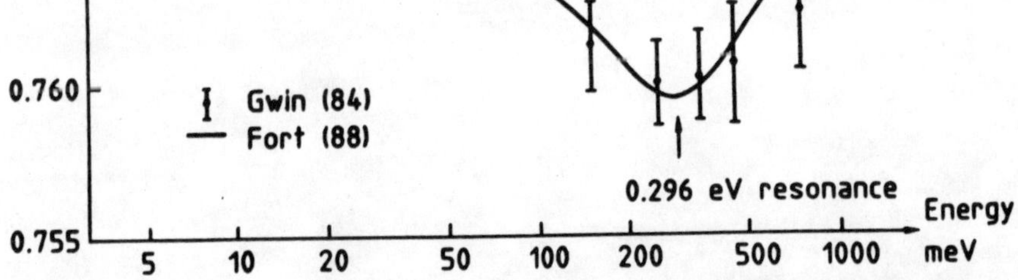

Figure 10 - Ratio of the plutonium 239 nu-bar to the one of californium 252 and comparison with the theoritical calculation of Fort.

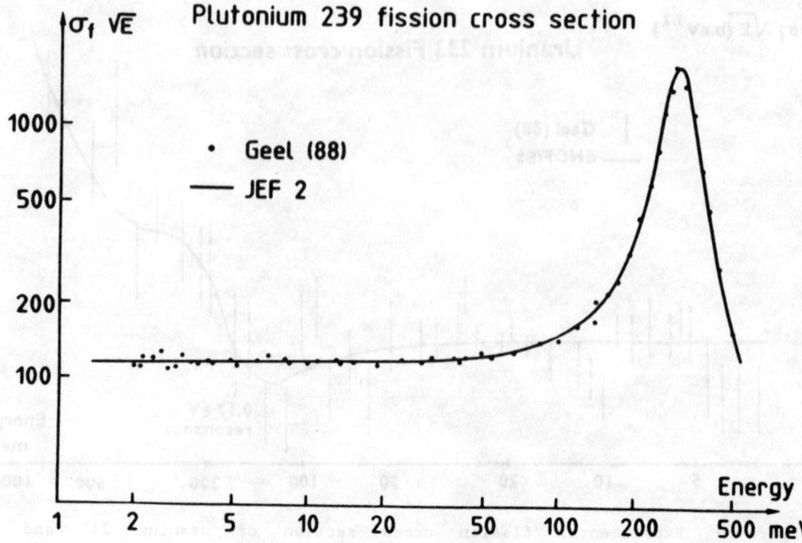

Figure 11 - Comparison of the last fission cross section of Geel and the recommended values of JEF2 for the plutonium 239 below 500 meV.

EVALUATING FISSION NEUTRON-MULTIPLICITY DATA

M. S. Zucker
Department of Nuclear Energy

and

N. E. Holden
Reactor Division
Brookhaven National Laboratory, Upton, New York 11973, U.S.A.

1. Introduction

The present work had its origins in the practical need to obtain P_ν (the probability that ν neutrons are emitted in a fission) for the purpose of calculating the theoretical response of instrumentation that used a correlation technique to assay spontaneously fissioning nuclides. The assay results are proportional to the factorial moments calculated with the P_ν distribution.[1] Obtaining experimentally derived sets of P_ν from many sources reported over several decades led immediately to the problem of evaluating the data: Aside from the trivial problem of sometimes not being properly normalized, the first moments $\langle\nu\rangle = \Sigma\nu P_\nu$ were typically not in accord with the best recent evaluations, or the calibration of detector efficiency was based on obsolete values for $\langle\nu\rangle$ for supposedly well-characterized "standard" nuclides such as ^{252}Cf. The problem of evaluating P_ν is unusual in that, compared to the usual situation where the quantities being evaluated are single numbers, the P_ν are sets of numbers, that moreover, are constrained so that $\Sigma P_\nu \equiv 1$ and $\Sigma\nu P_\nu = \langle\nu\rangle$, where $\langle\nu\rangle$ is usually determined more accurately from a separate experiment than it can be calculated from the experimentally derived P_ν distribution.

2. Basic Methodology

The method developed to reconcile different sets of P_ν proposed for the same nuclide, and to reconcile a given set with a value of $\langle\nu\rangle$ different from the first moment calculated from it, follows directly from the basic equations used to evaluate the experimental data leading to these quantities: The quantity $\langle\nu\rangle$ is usually independently determined by measuring a gross count

$$G = \epsilon\langle\nu\rangle q \tag{1}$$

where ϵ is the detection efficiency for a single neutron, and q is the source strength (in fissions). The observed neutron multiplicity Q_n, i.e. that n neutrons from a given fission are observed, is related to the P_ν by

$$Q_n = \Sigma P_\nu C(\nu, n)\epsilon^n(1-\epsilon)^{\nu-n}, \quad C(\nu, n) \equiv \nu!/n!(\nu-n)! \tag{2}$$

and the inverse relation from which the P_ν are determined,

$$P_\nu = \Sigma Q_n C(n, \nu)\epsilon^{-n}(\epsilon - 1)^{(n-\nu)} \tag{3}$$

It follows from these equations that a set P_ν may be considered as being derived from experimentally obtained Q_n using a detector of efficiency ϵ. Moreover, if the best value of the first moment is $\langle\nu\rangle'$ instead of the $\langle\nu\rangle$ implied by the data, then [from (1)] the $P_\nu{}'$ corresponding to $\langle\nu\rangle'$ would be given by

$$P_\nu{}' = \Sigma Q_n C(n, \nu)(\epsilon')^{-n}(\epsilon' - 1)^{(n-\nu)} \tag{4}$$

where $\langle\nu\rangle'\epsilon' = \langle\nu\rangle\epsilon$. (The equations (2), (3), and (4) preserve the normalization of the P_ν and also the values of quantities that can be shown to be independent of ϵ, such as $\langle\nu(\nu-1)\rangle/\langle\nu\rangle^2$ ("Diven's parameter"), and $\langle\nu(\nu-1)(\nu-2)\rangle/\langle\nu\rangle^3$, etc.) In effect the original incorrect value for the first moment is being blamed on an inaccurate value for the efficiency of the detector used in the experiment. These transforms then produce a normalized set of $P_\nu{}'$ yielding the accepted value for $\langle\nu\rangle' = \Sigma\nu P_\nu{}'$.

3. Illustration Using Spontaneous Fission

Neutron multiplicity data for spontaneous fission of ^{240}Pu, gathered from 1955 to 1984, will be used to illustrate the process.[2] Since some authors rounded off the P_ν presumably in accord with the estimated error of their measurements, the first step in the process was to renormalize the probabilities if necessary. Then using the procedure outlined above, each probability distribution was transformed to yield the best available (evaluated) value for $\Sigma\nu P_\nu = \langle\nu\rangle = 2.154 \pm .005$ (prompt neutrons per fission). These results are shown in Table I. The standard deviation was then computed using the deviations from the mean of the group of seven for each ν. The first two data sets showed significantly more deviations beyond $\pm\sigma$ than the other sets and on that basis were not used to form the "consensus" set, in which the individual terms are a simple unweighted average of the remaining five results.

TABLE I

P_ν for ^{240}Pu

	Hammel** 55	Hicks** 56	Diven 56	Baron 66	Wang 74	Zhang 84	Boldeman 84	Consensus	Std. Dev.
P_0	.0674770	.0518555-	.0598767	.0672230	.0602461	.0628325	.0657477	.0631852±	.0033
P_1	.2123022-	.2412679+	.2346677	.2296671	.2323636	.2307696	.2323544	.2319644	.0019
P_2	.3709042+	.3512038	.3290595	.3287954	.3417481	.3380100	.3290022	.3333230	.0061
P_3	.2208222-	.2298913-	.2625685+	.2535573	.2502577	.2466918	.2510283	.2528207	.0060
P_4	.1050492	.1090102	.0963776	.0996201	.0945106	.1015568	.1011654	.0986461	.0031
P_5	.0234452+	.0160176	.0167024	.0192328	.0159182	.0199289	.0183174	.0180199	.0017
P_6	.0000000	.0007536	.0007476	.0019044	.0049556	.0002106	.0023847	.0020406	.0018
$\langle\nu\rangle$	2.1540000*	2.1540000*	2.1540000*	2.1540000*	2.1540000*	2.1540000*	2.1540000*	2.1540000*	.005
$\langle\nu(\nu-1)\rangle$	3.7962	3.7328	3.7465	3.8162	3.7862	3.7797	3.8160	3.7889	.0290
$\langle\nu(\nu-1)(\nu-2)\rangle$	5.2528	5.0471	4.9803	5.2947	5.3196	5.1385	5.3193	5.2105	.1492
$\langle\nu(\nu-1)\rangle/\langle\nu\rangle^2$	.81820	.80454	.80749	.82250	.81604	.81465	.8224	.81663	.0063
$\langle\nu^2\rangle-\langle\nu\rangle^2$	1.3105	1.2471	1.2608	1.3304	1.3005	1.2940	1.3303	1.3032	.0290
$\langle\nu^2\rangle$	5.9502	5.8868	5.9005	5.9702	5.9402	5.9337	5.9700	5.9429	.0290

* data sets made to conform to this value

** data sets not used to form consensus because of several deviances beyond $\pm\sigma$

\+ means P_ν deviates by $> +\sigma$

\- means P_ν deviates by $< -\sigma$

4. Methodology for Neutron Induced Fission

Induced fission data presents additional problems, but mostly of a practical nature brought about by the relatively poor statistical precision of the existing data. There is only one available data set for $n+^{235}U$, $n+^{238}U$, and $n+^{239}Pu$, for E_n between 1.36 and 25 MeV, unpublished presumably because of the poor statistical precision.[3] Certain of the P_ν derived at some energies were negative (because of the formal way in which statistical errors in the experimental results propagated), which is mathematically and physically impossible for a probability. Moreover, considered as a function of energy, the cited P_ν often exhibit fluctuations obviously related to poor counting statistics rather than to any real physical process. Fortunately relatively precise values for $\langle\nu\rangle$ at thermal neutron energy (except for ^{238}U), and as a function of incident energy well into the fast neutron energy region, are available. Moreover the behavior of $\langle\nu\rangle$ with energy is particularly simple, allowing interpolation (and extrapolation to zero energy in the case of ^{238}U) with confidence. In principle the evaluated available data for $\langle\nu\rangle$ should be used, but in the present instance it was thought better to use the values obtained by the same experimenters furnishing the multiplicity data.[4] The value of $\langle\nu\rangle$ for the three nuclides is shown in Figure 1.

FIGURE 1

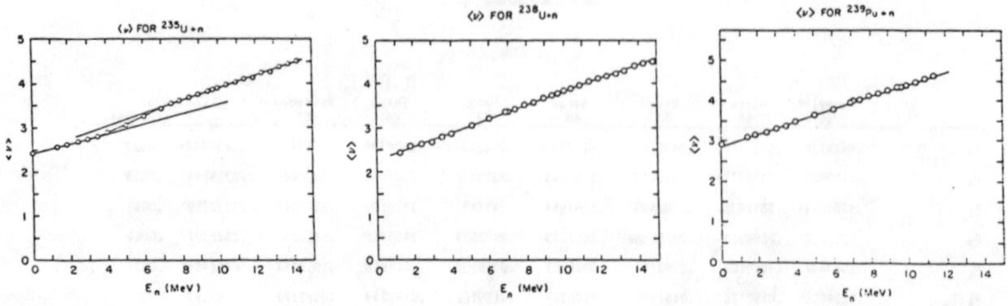

Any of the P_ν , for a given ν as a function of neutron energy E_n, is a bell shaped curve, or a portion of one. For the smaller ν values only the decreasing tail of the bell appears, starting at $E_n = 0$. For intermediate values of ν the whole bell shaped curve appears except the ascending part cut off at $E_n = 0$; the function rises to a maximum and then declines, as though it might approach 0 asymptotically at high energies. The P_ν , being probabilities, must sum to unity at any energy. Therefore the only way for $\langle\nu\rangle$ to increase monotonically with energy, as can be seen to be the case from Figure 1, is for there to be an increase in the P_ν for higher values of with a corresponding decrease in the relative importance of those P_ν for smaller values of ν. Thus the maxima of the bell-shaped functions appear at successively higher energies the higher the ν value. The data show no evidence for any more complicated behavior than this for any of the P_ν as a function of neutron energy.

5. Illustration for Induced Fission

The above qualitative picture of the behavior of each multiplicity probability as a function of energy furnished a guide for processing the data. As illustrated for ^{239}Pu in Figure 2, the lowest possible orders of polynomials were used for least-squares fitting each multiplicity. It might be preferable in principle to use Hermite functions for the fitting since they have the possibility of going asymptotically to zero at either extreme and certainly reproduce the desired bell-shape characteristic desired.

However over the limited range of energy 0 to 10 MeV, simple polynomial fits were adequate and certainly easier to implement. Though the reported multiplicities extended up to 25 MeV, it was decided that only the data up to about 14 MeV merited being treated, with the data from the last 4 MeV being used mainly as a guide to indicate how best to fit functions over the lowest 10 MeV. In the case of P_8 it was also necessary to average groups of points with respect to energy to reduce the scatter to a tolerable amount. After smoothing by fitting polynomials the P_ν will in general no longer sum to unity, failing by about a percent in the present case, and so the numbers were then renormalized. Equations (2), (3), and (4) were then applied as in the spontaneous fission example, with the condition that the value of $\langle\nu\rangle$ at any energy must equal the interpolated evaluated data for that parameter (Figure 1) at that energy.

FIGURE 2

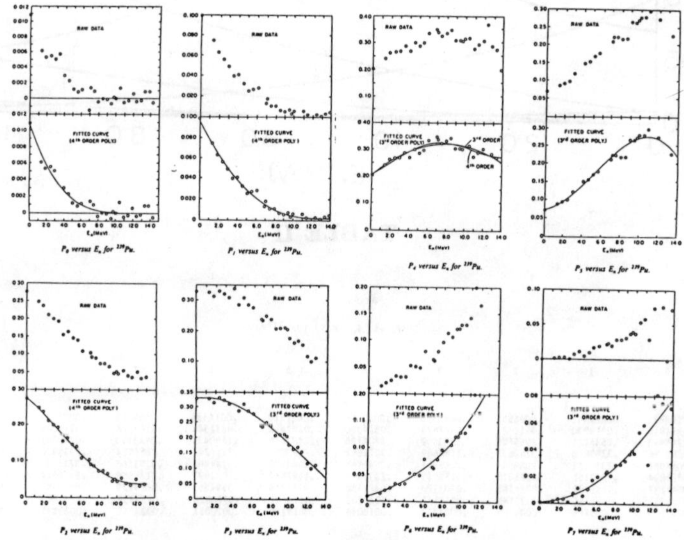

This procedure then produces a normalized set of $P_\nu{}'$ as a function of E_n which yields the proper value of $\langle\nu\rangle'$ at that energy. In principle a new least-squares fit with respect to energy can be made for the $P_\nu{}'$, the results renormalized, and again reconciled with the best available values for $\langle\nu\rangle$. However this process is

rapidly convergent, so that only a marginal improvement would result. A plot of the P_ν resulting from the above process, for neutrons bombarding ^{239}Pu, for neutron energies from 0 to 10 MeV, is shown in Figure 3; Table II lists the same information.

FIGURE 3

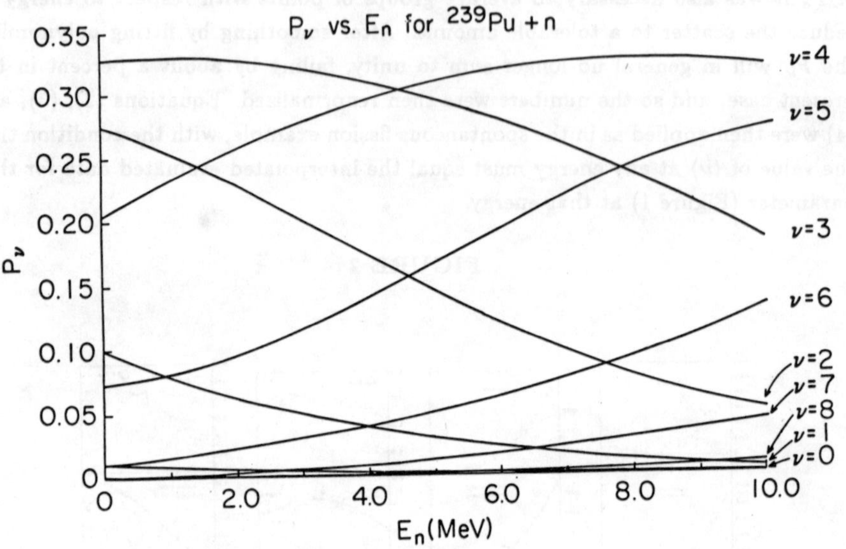

TABLE II

P_ν vs. E_n (MeV) for ^{239}Pu

E_n / ν	0	1	2	3	4	5	6	7	8	9	10
0	.0108826	.0084842	.0062555	.0045860	.0032908	.0022750	.0014893	.0009061	.0004647	.0002800	.0002064
1	.0994916	.0790030	.0611921	.0477879	.0374390	.0291416	.0222369	.0163528	.0113283	.0071460	.0038856
2	.2748898	.2536175	.2265608	.1983002	.1704196	.1437645	.1190439	.0968110	.0775201	.0615577	.0492548
3	.3269196	.3289870	.3260637	.3184667	.3071862	.2928006	.2756297	.2558524	.2335926	.2089810	.1822078
4	.2046061	.2328111	.2588354	.2792811	.2948565	.3063902	.3144908	.3194566	.3213289	.3200121	.3154159
5	.0726834	.0800161	.0956070	.1158950	.1392594	.1641647	.1892897	.2134888	.2356614	.2545846	.2687282
6	.0097282	.0155581	.0224705	.0301128	.0386738	.0484343	.0597353	.0729739	.0886183	.1072344	.1295143
7	.0006301	.0011760	.0025946	.0048471	.0078701	.0116151	.0160828	.0213339	.0274895	.0347255	.0432654
8	.0001685	.0003469	.0005205	.0007233	.0010046	.0014140	.0020017	.0028246	.0039531	.0054786	.0075217
$\langle\nu\rangle$	2.8760000	3.0088800	3.1628300	3.3167800	3.4707300	3.6246800	3.7786300	3.9325800	4.0865300	4.2404900	4.3944400
$\langle\nu(\nu-1)\rangle$	6.7479888	7.4107740	8.2397008	9.1241373	10.0544377	11.0243551	12.0311784	13.0754119	14.1602527	15.2911453	16.4750997
$\langle\nu(\nu-1)(\nu-2)\rangle$	12.5894114	14.5928444	17.3210445	20.4416846	23.9063613	27.6864348	31.7771206	36.1974684	40.9883672	46.2105526	51.9416469
$\nu(\nu-1)\rangle/\langle\nu\rangle^2$	.8158242	.8185663	.8236823	.8293894	.8346726	.8391002	.8426349	.8454740	.8479331	.8503700	.8531413
$\langle\nu^2\rangle-\langle\nu\rangle^2$	1.3526129	1.3662954	1.3990374	1.4398876	1.4792012	1.5107299	1.5317640	1.5428066	1.5470550	1.5498799	1.5584365
$\langle\nu^2\rangle$	9.6239888	10.4196539	11.4025308	12.4409173	13.5251677	14.6490351	15.8098084	17.0079919	18.2467828	19.5316353	20.8695397

6. Critique of the Results and Future Directions

Table I indicates that for spontaneous fission, once sets of P_ν are normalized and transformed to yield the same value of $\langle \nu \rangle$, they become comparable and allow a subset of them to be chosen for inclusion in a consensus set and uncertainties in the consensus set to be evaluated objectively.

For induced fission there are other indications of the plausibility of the procedure outlined above: P_ν data for $E_n = 0$ for $(^{235}U + n)$ and $(^{239}Pu + n)$ exist that were derived in experiments independent of those for $E_n \geq 1.36$ MeV. The latter data extrapolated to $E_n = 0$ gives a P_ν which agrees well with the consensus of measured P_ν at $E_n = 0$. This is illustrated in Table III using $(^{235}U + n)$ for example. Column (a) has the extrapolated data resulting from the fitting process; it predicts a $\langle \nu \rangle = 2.41052$. Column (b) lists this data transformed so that it predicts a value $\langle \nu \rangle = 2.41400$, the best value for this nuclide at $E_n = 0$. Column (c) lists the consensus P_ν arrived at from experiments with thermal neutrons, which give $\langle \nu \rangle = 2.414 \pm 0.007$. The three alternatives are consistent with one another, given the listed uncertainties in the thermal data. Similar agreement was obtained for the evaluation of the $(^{239}Pu + n)$ system multiplicities.

TABLE III

Comparison of $P_\nu(E_n = 0)$ for $(^{239}Pu + n)$, derived in different ways

a) curve fitted, $\langle\nu\rangle = 2.87737$.
b) curve fitted, normalized to $\langle\nu\rangle = 2.87600$.
c) consensus data, normalized to $\langle\nu\rangle = 2.87600$.

	(a)	(b)	(c)	
P_ν	P_ν	P_ν	P_ν	std. dev.
0	.0108819	.0108593	.0108601	.00003
1	.0993848	.0995994	.0993044	.0028
2	.2747747	.2749800	.2748737	.0003
3	.3269322	.3268549	.3270500	.0041
4	.2046922	.2044754	.2047660	.0087
5	.0727458	.0726004	.0727720	.0133
6	.0097395	.0097138	.0097430	.0027
7	.0006463	.0006452	.0006310	.0009
8	.0002726	.0002716	--	--
$\langle\nu\rangle$	2.8773708	2.8760000	2.8760000	.009
$\langle\nu(\nu-1)\rangle$	6.7569620	6.7505254	6.7435	.0184
$\langle\nu(\nu-1)(\nu-2)\rangle$	12.6350212	12.6169716	12.5447	.0539
$\langle\nu(\nu-1)\rangle/\langle\nu\rangle^2$	.8161309	.8161309	.81528	.00223
$\langle\nu^2\rangle-\langle\nu\rangle^2$	1.3550705	1.3551498	1.3481	.0184
$\langle\nu^2\rangle$	9.6343328	9.6265254	9.6195	.0184

Two distributions may be considered equivalent if their corresponding moments are equal. In the above work only the lowest two moments, the normalization (the "zeroth" moment) and the first moment, were forced to be equal. Speculating, it would seem better if a way could be found to have higher moments involved also, say the second factorial moment, $\langle\nu(\nu-1)\rangle$, or the ratio $\langle\nu(\nu-1)\rangle/\langle\nu\rangle^2$. The latter is especially attractive since it is insensitive to the efficiency of the detector used and is a measure of the width of the P_ν distribution. However, at this writing the formalism for this has yet to be developed.

Recently there has been an upsurge of interest in spallation neutron sources. It is thought that the similarities between the evaporation part of such sources and fission sources (aside from the fact that $\langle\nu\rangle$ is an order of magnitude larger) will allow similar treatment of the multiplicity data for such sources.

7. References

1. (a) K. Bohnel, *Nuclear Science and Engineering 90*, 75 (1985).

   (b) W. Hage and D. M. Cifarelli, *Nuclear Science and Engineering 85*, 159 (1985).

   (c) R. Dierckx and W. Hage, *Nuclear Science and Engineering 85*, 325 (1983)

   (d) N. Ensslin, *Proc. 7th Annual Symp. on Safeguards and Nuclear Material Management*, Liege, Belgium, 23 May 1985.

   (e) W. Hage and D. M. Cifarelli, *Nucl. Inst. and Methods A236*, 165 (1985).

2. We have to date evaluated the P_ν data for thermal, spontaneous, or fast neutron induced fission for the following nuclides based on what data could be gleaned from the literature or private communications (all the data is available at least in the form of BNL reports): ^{257}Fm, ^{252}No, $^{242,\,244,\,246,\,248}Cm$, $^{246,\,250,\,252,\,254}Cf$, $^{236,\,238,\,239,\,240,\,241,\,242}Pu$, $^{233,\,235,\,238}U$.

   (a) N. E. Holden and M. S. Zucker, Proc. ANS/INMM Topical Meeting on Safeguards and Technology, Trans. Am. Nucl. Soc. 45, Suppl. 1 23 (1983), BNL 33965.

   (b) M. S. Zucker and N. E. Holden, *Proc. 6th ESARDA Symp. on Safeguards and Nuclear Material Management*, Venice, Italy, May 1984, p.341, BNL 34804.

(c) N. E. Holden and M. S. Zucker, *IAEA Advisory Group, Meeting on Nuclear Standard Reference Data*, Geel, Belgium, Nov. 1984, IAEA-TECDOC-335, p.248, BNL 35513.

(d) M.S. Zucker and N. E. Holden, *Trans. Amer. Nuc. Soc. 49*, 195, June 1985 BNL 35829

(e) N. E. Holden and M.S. Zucker, Int. Conf. on Nucl. Data for Basic and Applied Science, Santa Fe, N.M., May 1985, *Radiation Effects*, 96 (1986) 1631; BNL 36379.

(f) M. S. Zucker and N. E. Holden, *Trans. Am. Nuclear Society 52*, 636, June 1986, BNL 37575.

(g) N. E. Holden and M. S. Zucker, *ANS Topical Meeting on Advances in Reactor Physics and Safety*, Saratoga Springs, N.Y. Sept. 1986, BNL 37362.

(h) N. E. Holden and M. S. Zucker, *7th Annual Symposium on Safeguards and Nuclear Material Management*, Liege, Belgium, May 1985, p.229, BNL 36467.

(i) M. S. Zucker and N. E. Holden, *Proceedings, International Symposium on Nuclear Material Safeguards*, Vienna, Nov. 1986, (1987) Vol. 2: p.329-347, BNL 38665.

3. We are indebted to J. Frehaut and coworkers at the Commissariat a l'Energie Atomique, Centre d'Etudes de Bruyeres le Chatel, for kindly furnishing their unpublished data.

4. M. Soleilhac, J. Frehaut, and J. Gauriau, Journal of Nuclear Energy, 23, 257 (1969).

INFLUENCE OF INTEGRAL DATA

FAST REACTOR BENCHMARKS AND INTEGRAL DATA TESTING AND FEEDBACK INTO JEF2

E. FORT, M. SALVATORES

CE Cadarache - BP N° 1 - Saint Paul lez Durance - France

ABSTRACT

Studies relating to the validation of the JEF2.2 library in the KeV and MeV energy ranges are described. The processing of the file into group constants and probability tables has been made in a 1968 group scheme with the care required for Quality Assurance. The selection for inclusion in the Integral Data Base, limited for the moment to Reactor core data, has been made on the basis of experimental environment cleanness and suitability for accurate neutronic calculation. The uncertainty information is almost absent in JEF2. Covariance data have been elaborated on the basis of personal judgment and systematic studies have defined the condition for a statistical adjustment procedure to preserve the consistency of a maximum of useful integral data information with microscopic information. Although the results are not final since not all available integral data have been considered the conditions of the adjustment are such that the present conclusions are unlikely to be modified : ^{235}U, ^{238}U, ^{58}Ni, ^{16}O evaluations need revisions.

Introduction

In the past, when the information in the evaluation was of relatively poor quality the integral data were considered as absolute references to correct the nuclear data which were considered to be affected by significant systematic errors. The covariance matrices, when existing, played little role in the adjustments.

The somewhat recent improvements of the quality of the evaluations and the development of theories and tools for data adjustment have made it possible to consider now the microscopic and the integral data as two complementary sources of information about the same basic physical phenomenon. The validation of evaluated data becomes more and more the research into the extent of the consistency in the complementarity. In this operation the uncertainty information plays a key role, provided that some conditions are fulfilled.

General Conditions for Adjustments

Nuclear data adjustment is a long process involving several steps :

1. Processing the evaluated data into parameter or group constants in a way that preserves the quality and the integrity of the information. This requires that fully validated tools are used for the processing and that methods sophisticated enough exist to use the processed data.

2. A choice of integral data according to criteria of cleanness of the experimental environment. The integral data should be derived from the raw data by applying corrections of small amplitude so that the integral values are not

sensitive to uncertainties in the corrections (Doppler, heterogeneity, self-shielding, ...) nor to the methods used to calculate them. It has to be noted that in recent years these methods have received considerable improvements so that a number of additional experiments can be considered now as valuable benchmarks.

On the other side, in order to obtain the energy dependence of the nuclear data corrections the integral data should have differing energy dependent sensitivities.

This is the reason why the benchmark integral data set should contain as many experiments with different spectral indices as possible, and for a given integral experiment as many parameters (K_{eff}, B_m^2', reaction rate ratios) as possible in order to separate the effects of the different cross-section types.

3. Neutronics calculations which must be as exact as possible so that the calculated integral parameters are not dependent on the calculational method nor on the calculational scheme. If these conditions are satisfied the differences between experimental data and calculated data are the exact measures of the imperfect quality of the input data.

The sensitivity coefficient calculations are based on perturbation theory or variational methods. Both types of methods are equivalent provided that the amplitude of the nuclear data modification in the adjustment is small enough for the linearity principle to apply.

But here again the condition of exact calculations imposes an energy scheme that is adapted to the fine structure of the flux $\phi(E)$ as a function of energy. In fact, sensitivity calculations using perturbation theory imply the calculation of expressions like $<\phi^*.dx.\Phi>$ which are integrals over space, angle, energy of the product of the vector $dx.\Phi$ by the adjoint flux ϕ^*, dx being the variation of the Boltzman operator which acts on the vector flux. Keeping in mind that the components Φ_g (g stands for a given group number in the energy scheme) are average values, incorrect values of ϕ_g induce incorrect values for ϕ^*_g and the errors are "squared" in the integral.

This condition is easily satisfied with fast and very fast systems, but one has to be careful in analysing thermal and epithermal systems.

4. The main objective in nuclear data adjustment is to check the consistency of Integral and microscopic data so as to create data set able to calculate any integral parameter with the best reliability and with an accuracy at least equal that of the measurements.

Everything has already been said about adjustments, their general theoretical bases, the conditions and limitations in application.

We will incidentally recall some special features of relevance to the present topics and used in the work described here below as an illustration.

A set of measured integral quantities, combined in a vector E with covariance matrix V is considered. The nuclear data implied in these integral parameters are combined also in a vector T with a covariance matrix M. The integral parameters calculated from T are denoted by the vector C.

If the sensitivity matrix S expresses the dependence of $\overline{R}$ on relative variations of T one has :

$$\frac{dC}{C} = S\frac{dT}{T} \tag{1}$$

The use of such logarithmic variates (with possibility of variation between -∞ and +∞) is the most appropriate to be used in the operation of entropy maximisation.

Actually, in the adjustment the objective is to minimise the quantity :

$$\frac{E-C}{C} = S\frac{dT}{T}, \tag{2}$$

that is obtained by maximising a likelihood function corresponding to the system C+T.

A vector T' is obtained. This vector is recommended as the best estimate of the unknown true vector since it minimises the expected consequences of choosing a vector deviating from the true vector [1].

T' needs to be close to T in order <u>to satisfy the linearity assumption</u>. Statistical adjustments are meaningful only when the a priori evaluated data are of sufficient quality, i.e. when non excessive values are obtained for any component of :

$$\frac{E-C}{C} \tag{3}$$

(let say less than 4% - 5%).

The final vector C' for the integral data is :

$$C' = C + S\left(\frac{T'-T}{T'}\right) \tag{4}$$

with a new covariance matrix V', while the covariance matrix for T' is denoted by M'.

Simplifications are obtained if the assumption is made that the probability distributions for T and C are Gaussian. The Gaussian distributions are kept in the adjustment for T' and C' and it is supposed that the true unknown distributions are also Gaussian.

Significant deviations from this assumption (existence of systematic errors) are detected by large :

$\left(\frac{E-C'}{C'}\right)$ values, or/and by adjustments to cross-sections much greater than the assigned standard deviations. The existence of such deviations lead to a deviation of the χ^2 parameter from its expected normal value that should equal N, the number of integral data (whether they are correlated or not).

If $\chi^2 > N$ the consistency between integral and microscopic data is not good. It is well known that the solution is to correct the discrepant data when possible, or to increase by judgment the related uncertainties or to reject the data when justified.

Finally, to keep the goodness of the fit all the "a posteriori" standard deviations have to be enhanced by the factor :

$$\left(\frac{\chi^2}{N}\right)^{1/2} \tag{5}$$

that will be referred to, in the following as the enhancement factor.

It follows that, for a given integral observable i the standard deviation after adjustment is :

$$\left(V_{ii}\,\frac{\chi^2}{N}\right)^{1/2} = \left[S_i\; M'\; S_i^T\,.\,\frac{\chi^2}{N}\right]^{1/2} \quad \text{when } \chi^2 > N. \tag{6}$$

Practical Conditions of JEF2 Validation

- *Data processing :*

The JEF2 file has been processed, for the purpose of benchmarking, in different group schemes : 33g for the analysis of the LOS ALAMOS criticals, 172g for thermal systems and minor constituents, 1968g for fast systems. All have common boundaries and are therefore consistent.

The processing work has been performed using the NJOY version 89.62* with agreed updates to specific modules (treatment of the unresolved region, fine flux calculation in GROUPR ...). As agreed by the JEF project the same parameter values were used by all the Laboratories participating in the data processing (CADARACHE, BOLOGNE, WINFRITH) : Control of the number of points for cross-section reconstruction ($ERMAX=10^{-3}$), the numerical accuracy ($NDIGITS=8$), accuracy on the resonance integral RI, ...

The 33g library, called FASTLIB [3], has been processed on an IBM machine, while the other Libraries have been processed on a CRAY. There is consistency between both procedures provided that the double length word version of NJOY is used on the IBM. The 33g library contains only infinite dilution cross-section since it has been demonstrated for the hard spectrum Los Alamos criticals [3] that the selfshielding effect is negligible. It has been developed in the P5 order although a P3 development was sufficient.

The 1968g multitemperature library JECCOLIB2 [4] has been established with systematic controls for quality assurance. It contains infinite dilute cross-sections and probability tables. These probability tables were produced by using CALENDF [5] and are used in selfshielding calculation for heterogeneous media. They have been established by using the so-called "Moments" method whose basis is the following relationship which defines Mn as the nth order moment of the total cross-section and Mx_n as the moments of the partial cross-section $[\sigma_{xi}]$:

$$Mn = \frac{1}{\Delta E}\int \sigma_t^n(E)dE = \sum_{i=1}^{I} p_i\, \sigma_{t_i}^n \tag{7}$$

$$Mx_n = \frac{1}{\Delta E}\int \sigma_x(E)\, \sigma_t^n(E)\, dE = \sum p_i\, \sigma_{xi}\, \sigma_{t_i}^n \tag{8}$$

The set of values $\{p_i, \sigma_{ti}, [\sigma_{xi}]\}$ is the probability table or the subgroup parameters for the energy group of reference. For a given i^{th} subgroup p_i represents the probability, σ_{ti} the ith subgroup value for the total cross-section and $[\sigma_{xi}]$ stands for every partial cross-section. The cross-sections are recalculated in CALENDF with the same basic formalism as NJOY (with a few exception, e.g. ^{58}Ni) but in different conditions for the energy grid and the numerical approach (ladders of resonances being used in the unresolved region ...).

BONDARENKO (NJOY) and subgroup parameter (CALENDF) formalisms can be related so that one has for any energy group, for a given grid of dilution values σ_d :

$$\sigma_{x,eff}^{(\sigma_d)} = \frac{\int \frac{\sigma_x(E)dE}{\sigma_t(E)+\sigma_d}}{\int \frac{dE}{\sigma_t(E)+\sigma_d}} = \frac{\sum_i \frac{p_i\, \sigma_{x_i}}{\sigma_{t_i}+\sigma_d}}{\sum_i \frac{p_i}{\sigma_{t_i}+\sigma_d}} \tag{9}$$

($\sigma_{x,eff}^{(\sigma_d)}$ is the effective value of the cross-section σ_x corresponding to the "dilution" σ_d value).

This intercomparison provided a good opportunity to check the quality of NJOY production. Isolated differences of the order of 2% per group were tolerated because of differences in the numerical treatment. Systematic differences were examined and explanations found.

Very many errors have been detected in the evaluations, in NJOY and in CALENDF. But such rigorous and systematic controls are a guarantee that the information in the evaluated data has been group averaged without any distortion.

In addition, checks have been performed at each step of the transformation of the group constants into formatted data for use by the cell code.

- *Choice of integral experiments :*

It is reasonable to think that all the experiments considered for the JEF2 benchmarking a priori fulfil the conditions of cleanness of the experimental environment.

Most of them were extensively used in the past for this purpose of validation, mainly with the objective to produce data sets for specific applications (Fast BREEDER CORE Formulaire, as an example) and were (and still are) proprietary. With the few exceptions of the experiments in the very fast range (LOS ALAMOS spheres) the chosen experiments have been performed on the MASURCA (FRANCE) or RB2 (ITALY) critical facilities, or the power Reactor PHENIX (FRANCE). These experimental programs were part of the EUROPEAN FAST REACTOR project and were designed to check the data involved in specific applications : major actinides for clean cores or start up cores (^{235}U, ^{238}U, ^{239}Pu : Program RZ on MASURCA, importance of higher Pu isotopes in reprocessed fuels (^{240}Pu, ^{241}Pu, ^{242}Pu : Program PLUTO on MASURCA), contribution of the structural material (Fe, Cr, Ni) to the neutron balance measured relative to ^{235}U fission or ^{10}B capture program on RB2. Data from other programs (ZEBRA, SNEAK) will be included in future adjustments.

The integral data of reference are : bücklings, critical masses, K^+ for $K_\infty=1$ experiments, reaction rate ratios. If the parameter :

$$r = \frac{<\nu \Sigma_f>}{<\xi \Sigma_s>} \tag{10}$$

(where the numerator stands for the total number of neutrons produced by fission and the denominator for the integral of the slowing down density) is used as the spectral index, one notes a very large spread of the values ($0.2<r<$), a situation that satisfies the condition of energy dependent sensitivities in the integral experiments.

- *Neutronics calculations :*

The integral data in the very fast range (GODIVA, ...) need a very accurate spatial and angular treatment. This has been performed in the P^3 S^{32} approximations with S^N corrections made to Keff calculations, corrections obtained from the following relationship :

$$\left(K_{4n} = \frac{4.K_{2n} - K_n}{3} \right) \tag{11}$$

Mesh and energy grid effects were investigated in order to specify the optimal calculational conditions.

The integral data of the fast reactor range were assumed to have been obtained in the conditions of the asymptotic flux. Therefore, the calculations have been limited to cell calculations, using the recent European cell code ECCO [6].

The essential approximations [7] are related to the anisotropy treatment :

a. By using a transport cross-section, in particular for collision probability calculations and the treatment of leakage, defined as :

$$\sigma_{tr} = \sigma_t - \bar{\mu}\, \sigma_s$$

In this relationship σ_t refers to the total cross-section, $\bar{\mu}$ to the average cosine, σ_s to the scattering cross-section. The correct anisotropy is replaced by a forward scattering anisotropy without energy change.

b. The so called P_1 inconsistent approximation is used for flux calculation. It is not clear whether the B1 approximation which is better in principle, would be more appropriate.

c. The leakage is assumed to be isotropic

In addition one has to mention that no correlation is represented between the probability tables, even in the fine group Library. This is an approximation in the resonance structure, of very small importance for fast and very fast systems.

All the approximations here above mentioned have impacts which have not yet been estimated, but assumed to be small.

The uncertainties on the integral parameters should ideally be increased to allow for these approximations.

The sensitivity coefficients have been calculated using the ERANOS code system.

The Standard Perturbation theory has been applied for the K_{eff} sensitivity calculations. For a given energy group g the sensitivity coefficient to a nuclear constant variation inducing variations δF and δA on the neutron production and removal operators respectively is expressed as :

$$\left(S^{K_{eff}}\right) = \frac{<\varnothing^*, \delta F \varnothing> - <\varnothing^*, \delta A \varnothing>}{<\varnothing^*, F' \varnothing'>} \tag{12}$$

(∅ is the scalar flux, ∅* the adjoint flux and ∅' is the perturbated flux). In this expression the denominator represents the fission normalization integral.

For a spectral Index I the sensitivity coefficient to the cross-section Σ_x has two components :

- the "direct effect" component whose expression is obvious,
- the component corresponding to the modification of the neutron energy spectrum. It is easy to demonstrate, in the frame of the Generalized Perturbation Theory, that the expression for the "spectral effect" is :

$$\left(S_x^I\right) = \frac{dI / I}{d\Sigma_x / \Sigma_x} = - <\varnothing^*, \Sigma_x \varnothing> \tag{13}$$

- *Adjustments :*

The adjustments have been performed using the code AMERE [8].

The a posteriori χ^2 parameter value has been used as a test of goodness of the fit, and consequently as a test of <u>reliability</u> of integral parameter prediction, by the adjusted data set.

The "a posteriori" average uncertainty on integral parameters served as a criterion of <u>accuracy</u> of prediction.

In practice a sequential procedure has been adopted to implement the system of integral data so as to detect the introduction of any inconsistent integral datum by a sharp increase of the χ^2 value.

The different experimental programs have been considered first separately (i.e., the "LOS ALAMOS spheres" data only, or the "MASURCA" data only, ...) in order to test their internal consistency. Some experiments have been rejected on a statistical criterion, but the elimination has been found justified by arguments of experimental technics. Finally a system of N=50 integral data has been selected.

The JEF2 evaluations are not provided with covariance matrices, with the exception of ^{239}Pu [9] and ^{238}U [10]. Covariances have been generated on the basis of personal judgement for what concerns both the standard deviations and the range of correlations (medium range). (A covariance matrix can be written as $B=\Sigma P \Sigma$ in a way that displays explicitly the standard deviations (diagonal matrix Σ) and their correlations (matrix P)).

As a compromise between constraints relevant to physics and to have common boundaries with the 3 above mentioned schemes a 15 group scheme has been established to perform the sensitivity calculations and the cross-section adjustments (see the table 1). The covariance matrices have been collapsed into this 15 group scheme.

Table 1 - 15 groups scheme used for adjustment calculations

Group boundaries upper limit (eV)	Group	Scheme 1968 g	Scheme 172 g	Scheme 33 g	Lethargy
19.64033 + 6	1	1	1	1	-0.675
6.06531 + 6	2	142	9	3	0.500
2.231302 + 6	3	262	15	5	1.500
1.353353 + 6	4	322	18	6	2.000
4.978707 + 5	5	442	26	8	3.000
1.831564 + 5	6	664	32	10	4.000
6.737947 + 4	7	686	36	12	5.000
2.478752 + 4	8	808	42	14	6.000
9.118820 + 3	9	928	46	16	7.000
2.034684 + 3	10	1108	53	19	8.500
4.539993 + 2	11	1268	61	22	10.000
2.260329 + 1	12	1648	80	28	13.000
4. + 0	13	1837	93		14.731
5.3158 - 1	14	1920	137	33	16.75
1. - 1	15	1952	156		18.42068
1.1 - 4		1968	172		

This system was well adapted for systematic studies so as to define the optimal conditions for the JEF2 validation.

The gains (expressed in percent) on the accuracy of K_{eff} and the spectral indices I_s resulting from adjustments are defined as : $\left(1-\frac{<\Delta K_{eff}>^*}{<\Delta K_{eff}>}\right)$ x 100 and $\left(1-\frac{<\Delta I_s>^*}{<\Delta I_s>}\right)$ x 100 respectively. The brackets denote the average value and the star * distinguishes the "a posteriori" value from the "a priori" value. The variations in the gains have been investigated as functions of the enhancement factor for different standard deviations obtained from this nominal a priori values by a multiplication by factors 0.3, 0.6, 1.3, 1.5, and for different values of the range of the "a priori" correlations labelled by R = 1, 2, ..., 15.

Figure 1 shows that the gains are generally improved by an increase of the standard deviations in the case of no energy correlation (R=0). This situation is practically unchanged by introducing correlations of any range (R = 1, 2, ..., 12).

Figure 1- Gains (%) in accuracy on K_{eff} and I_s as functions of enhancement factor variations resulting from changes in a priori standard deviations

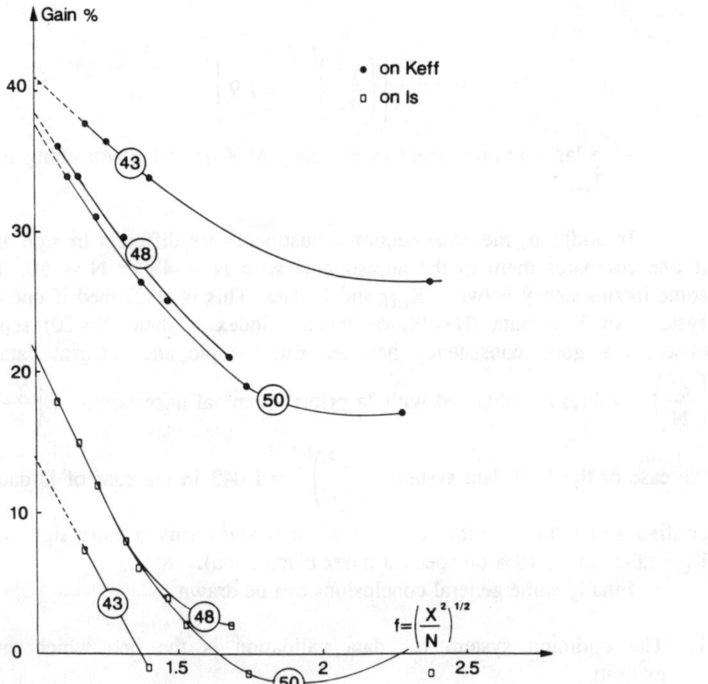

Two experiments have been rejected from the N = 50 system because of adjustment $\left(\left|\frac{E-C'}{C'}\right| > 3\sigma\right)$.

With this new system of N=48 experiments the situation is globally improved : better consistency with integral experiments is obtained with the nominal a priori standard deviations ($\left(\frac{\chi^2}{N}\right)^{1/2} = 1.314$, against $\left(\frac{\chi^2}{N}\right)^{1/2} = 1.37$ in the case of N=50). Both gains in Keff and Is are increased but the importance of "a priori" correlations remains small (effect of ~ 5% on $<\Delta Keff>*$ for example).

If one reduces the system to the N = 43 integral data such that $\left|\frac{E-C'}{C'}\right| < 1.5\,\sigma$, one notes (see fig. 1) when using the nominal a priori standard deviations :

- a less good consistency :

$$\left(\left(\frac{\chi^2}{N}\right)^{1/2} = 1.9\right) \qquad (14)$$

- a large improvement in accuracy of K_{eff} and a worsening in accuracy of I_S.

In addition, the cross-section adjustments are different in sign and amplitude if one compares them to the adjustments with N = 48 or N = 50. This suggests some inconsistency between K_{eff} and Is data. This is confirmed if one considers the systems of K_{eff} data (N=28) or spectral index Is data (N=20) separately. One observes a good consistency between microscopic and integral data since good $\left(\frac{\chi^2}{N}\right)^{1/2}$ values are obtained with "a priori" nominal uncertainties [$\left(\frac{\chi^2}{N}\right)^{1/2}$ = 1.11 in the case of the Keff data system ; $\left(\frac{\chi^2}{N}\right)^{1/2}$ = 1.043 in the case of Is data]. For these configurations the importance of the off diagonal terms is more significant (14% on K_{eff} calculation, 10% on spectral index calculation).

Finally some general conclusions can be drawn :

1. The optimum system for data validation is the one which simultaneously exhibits :
   - Good consistency between integral and microscopic data (enhancement factor $\left(\frac{\chi^2}{N}\right)^{1/2}$ as close to 1 as possible).
   - Most accurate prediction for each integral parameter.
2. Concerning the microscopic data, the very sophisticated covariance matrices with very small standard deviations are not of practical use, at present, because of the existence of some degree of inconsistency between integral and microscopic data, which limits the importance of the "a priori" off diagonal terms.
   In inconsistent situations the weight of the integral data is dominant in the "a posteriori" microscopic correlations. This is visible since the accuracies on the integral parameters are improved when the "a posteriori" standard deviations are increased.

This suggests :

- The correlations existing between integral data of the same type (I_s essentially), in a given experimental program should be taken into account in the adjustment. In the present study their effects have not been investigated.
- Uncertainty calculations for practical configurations should be performed only with adjusted data sets.

- *Results :*

According to the conclusions of the systematic studies the system of N = 48 experiments is the more adapted, for the moment, to validate JEF2. For what concerns the microscopic data the range of the correlations has been fixed to 3 groups, unless it has been differently assigned in the evaluated covariances (^{238}U, ^{239}Pu).

We have considered as valuable adjustments only those which are greater than the a posteriori standard deviations. As demonstrated in the systematic study, the sign and the energy range of these adjustments are independent of the input standard deviations, so that they can be considered as significant.

For this first approach in the general process of JEF2 validation, 5 nuclear constants have been considered :

- the average number of neutrons emitted per fission : ν,
- the fission cross-section : σ_f,
- the neutron absorption cross-section : $\sigma_a = \sigma_c + \sigma_{n,p} + \sigma_{n,\alpha} + \ldots$,
- the elastic and the inelastic scattering cross-sections : $\sigma_{n,n}$ and $\sigma_{n,n'}$.

For these 2 last nuclear constants the sensitivity coefficients correspond to the global energy transfer so that both the cross-sections and the secondary angular and energy distributions are concerned by the adjustments.

The significant adjustments involve very few nuclei :

^{235}U : σ_f ; 67 KeV < E < 497 KeV
decrease by about -1.5 %

σ_c ; 450 eV < E < 2 KeV
increase by ~ 12 %

^{238}U : $\sigma_{n,n}$; 180 KeV < E < 1.35 MeV
decrease by ~ - 7 %

^{58}Ni : σ_a ; 25 KeV < E < 2.23 MeV
decrease by ~ - 28 %

The capture and probably the (n,p) cross-section are affected by adjustments. Further studies are needed to separate the effects on both cross-sections.

^{16}O : $\sigma_{n,n}$; 25 KeV < E < 500 KeV
decrease by ~ 10 %

This is a surprising conclusion of the present benchmarking considering the quality of the evaluation. Essentially the concerned energy range is the one below the first resonance at ~ 450 KeV (see figure 2). It is worthwhile noting that the transport cross-section is higher than the total cross-section below the resonance energy and lower above because of the sign of $\overline{\mu}$ that changes at the resonance energy.

Figure 2 - Adjustment of the elastic cross-section of ^{16}O as requested by the benchmarking

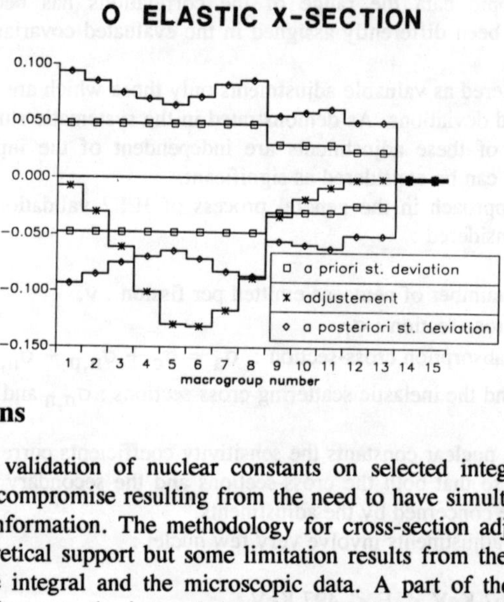

Conclusions

The validation of nuclear constants on selected integral data is a difficult exercise of compromise resulting from the need to have simultaneously complete and consistent information. The methodology for cross-section adjustment is based on a strong theoretical support but some limitation results from the relative inconsistency between the integral and the microscopic data. A part of the responsibility in this situation belongs to the integral data (too optimistic uncertainties, in particular for some types of spectral index ?) and to the calculational methods.

In this situation, the constraints on the quality of uncertainty information on the nuclear data are less stringent and it seems, for the time being, that for valuable data adjustments relatively simple covariance are sufficient : standard deviations on large energy bands which can match the current multigroup energy structures of applications libraries and correlation coefficients both in energy and between nuclei.

In addition to limited revisions to cross-sections of some nuclei such as ^{235}U, ^{238}U, ^{58}Ni, the present JEF2 benchmarking has revealed a severe conflict between the integral and the microscopic data concerning the elastic cross-section and/or the related angular distributions of ^{16}O.

References

1. F. FROHNER *"Sermon on the basic simplicity of least squares"*.
2. S. CHIBA and D. SMITH *ANL/NDM* - 121 September 1991.
3. E. FORT, C. MOUNIE, S. OKAJIMA *JEF Report to be published*.
4. G. RIMPAULT, P. RIBON, J.L. ROWLANDS, E. FORT, P. LONG, S. RALPHS, R. SOULE, C. EATON, C. DEAN *The ECCO/JEF2 library. N.T. SPRC/LEPh 92/231 (1992)*.
5. P. RIBON *Seminar on Njoy and Themis. OECD NEA DATA BANK. Saclay. June 1989*.
6. M.J. GRIMSTONE, J.D. TULLET, G. RIMPAULT *PHYSOR 90, International Conference on the Physics of Reactors* Vol. 2, IX, 24.
7. J. ROWLANDS *Private communication*.
8. French version of the AMARA code by A. GANDINI, M. PETILLI. *CNEN Report RT/FI* (73) 39.
9. *"Covariance Matrices for the ^{239}Pu evaluation proposed for the JEF2 library"*. E. FORT, Y. KANDA, Y. NAKAJIMA *Preliminary JEF Report*.
10. M. SOWERBY *Private communication*

Reactor Benchmarks and Integral Data Testing and Feedback into ENDF/B-VI

R. D. McKnight
Reactor Analysis Division
Argonne National Laboratory
Argonne, Illinois 60439

and

M. L. Williams
Nuclear Science Center
Louisiana State University
Baton Rouge, Louisiana 70803

ABSTRACT

The role of integral data testing and its feedback into the ENDF/B evaluated nuclear data files are reviewed. The use of the CSEWG reactor benchmarks in the data testing process is discussed and selected results based on ENDF/B Version VI data are presented. Finally, recommendations are given to improve the implementation in future integral data testing of ENDF/B.

1. The Role of Integral Data in Nuclear Data Evaluation

To the reactor designer, integral data testing defines how well target accuracies are met. To the data evaluator, integral data testing measures the performance of the evaluated file. Within the evaluation process, integral data can be used in two distinct ways, depending on whether the data files are to be "adjusted" or "validated". The former means is clearly the more elegant and the integral results play an active role in the data specification. The latter is more of an afterthought in which the integral data play a passive role -- simply providing a means of testing the evaluated differential data. The use of integral measurements in the production of adjusted libraries has been the subject of numerous scholarly papers.[1-4] Within the structure of CSEWG (the Cross Section Evaluation Working Group), the use of integral measurements in the production of the ENDF/B data files is that of quality assurance of the finished product. Traditionally with ENDF/B, integral data testing has not been an afterthought portion of the evaluation process, but rather the culmination of the process. Nevertheless, for either an adjusted or evaluated data library, the integral data test represents the "bottom line"; it produces bias factors and uncertainties associated with using the data libraries with specified methods and models.

There are four components of integral data testing: evaluated differential data files (and their uncertainties); data processing methods (and their uncertainties); integral measurements (and their uncertainties); and analysis of integral experiments (and its uncertainties). All four components are required to produce a C/E (calculation/experiment) ratio or bias. Likewise, the uncertainties of all four components are required to determine if the C/E value is meaningful. Obviously, a C/E value close to unity with a large uncertainty might be either good or fortuitous and might bear very little information regarding the quality of the evaluated data files.

1.1. *Evaluated Differential Data Files*

The first of these components is the subject of this Symposium and, though obviously essential to the process, need not be discussed herein. It should perhaps be emphasized that the covariance data are also an essential ingredient in understanding the C/E values. Although these data did not receive formal testing with ENDF/B-V, it is highly desirable to process and test these data in Version VI.

1.2. *Data Processing Methods*

With regard to the second component (cross section processing), suffice it to say "You don't really appreciate the importance of something until you lose it!" Such has been the case with ENDF/B-VI. Foiled by insufficient support and the new ENDF/B-6 formats, when the time came to "data test" ENDF/B-VI data, it was found that very little data testing could be done. Only the NJOY code system[5] was able to produce some limited multigroup libraries. Thus, the ENDF/B-VI data files were released with a little integral data testing, a little sensitivity analysis, and more than a little trepidation. Historically, ENDF/B data testing has also included sufficient methods comparisons to understand the "differences" in data processing. Lacking that, current data testing results still feel preliminary, because it is difficult to quantify the processing uncertainties. This statement is not intended to slight the NJOY code system, which has produced most of the ENDF/B-VI data testing results to date and is doing very well with ENDF/B-VI; but rather, it is intended to highlight the need for additional independent processing methods such as continuous-energy Monte Carlo.

1.3. *Integral Measurements*

There are several requirements for integral experiments for data testing. The integral experiments should span a wide range of data sensitivities. This can be achieved with a diverse range of material compositions and spectra. The integral experiments should contain minimal complexity (i.e., clean geometry and simple heterogeneity) in order to minimize modeling assumptions. The use of integral measurements obtained from various critical facilities and/or with different experimental methods will minimize the effect of systematic uncertainties. Finally, the integral measurements must include uncertainty data and covariances in order to be useful.

1.4. *Analysis of Integral Measurements*

The uncertainty in the calculation of an integral quantity must be small enough to allow conclusions about the evaluated nuclear data. The calculation methods must include an appropriate technique for determining the resonance absorption when it is significant and accurate neutron transport approximations which adequately account for scattering anisotropy, leakage and heterogeneities. This implies application of an accurate analytical method to a precise model. In some cases, a homogeneous sphere or an infinite lattice may be a very precise model of the integral experiment. In such cases, application of a rigorous analytical method is both accurate and convenient. On the other hand, a homogeneous spherical model is an analytical leap of faith for a three-dimensional plate-type critical assembly. Although it might be practical to calculate an integral parameter with negligibly small methods' uncertainties using the simplified model, the true methods' uncertainties in the analysis of the integral parameter must include the methods' uncertainties in generating the simplified model. That is, all of the modeling approximations and uncertainties in those methods which were applied to the as-built critical experiment in the derivation of the simplified model become imbedded in the use of that model, and must be combined with the additional uncertainties (presumably small) resulting from the analysis of the simplified model. These uncertainties "inherent" in the use of the simplified model may be large, small or difficult to assess. The present availability of more rigorous methods (e.g., three-dimensional transport theory and continuous-energy Monte Carlo with combinatorial geometry) allow more rigorous modeling in the generation of a benchmark model today, thereby greatly simplifying the that task and eliminating those uncertainties inherent in the use of a simplified model.

2. CSEWG Benchmark Models

The CSEWG benchmark models used for integral data testing are specified in ENDF-202.[6] This compilation, which contains benchmark models for fast reactor, thermal reactor, shielding, fission product and dosimetry data testing, was produced almost 20 years ago and includes integral measurement data gathered almost 40 years ago. Although known to be deficient in some areas, it continues to be one of the most useful single sources of integral data for nuclear data testing, methods' comparisons, and code validation.

2.1. *Fast Reactor Benchmarks*

The fast reactor benchmarks include fourteen plutonium fueled assemblies (JEZEBEL, VERA-11A, ZPR-3/48, ZEBRA-3, ZPPR-2, ZPR-6/7, ZPR-3/56B, SEFOR, SNEAK-7A, SNEAK-7B, ZPR-9/31, JEZEBEL-PU, FLATTOP-PU and THOR) and eleven uranium fueled assemblies (GODIVA, VERA-1B, ZPR-3/6F, ZPR-3/11, ZPR-3/12, ZEBRA-2, ZPR-6/6A, JEZEBEL-23, BIG TEN, FLATTOP-25 and FLATTOP-23). This collection includes a diverse range of compositions and geometries and affords the testing of nuclear data over a range of neutron spectra. A critical review of this set of benchmarks was presented by Collins et al[7] in 1982. Because of the limited resources applied to data testing in the decade since that review, their conclusions are still valid today; some bear repeating. Although many of the integral measurements in ENDF-202 remain of value today (particularly those in the homogeneous cores built at Los Alamos), much of the data (particulary in the LMFBR criticals) were obtained by methods now considered inadequate and have lost their usefulness. Examples of such measurements include fission rates measured by inserting fission chambers into cavities in the assembly and small sample reactivity worth measurements obtained by oscillating a small sample through a cavity. Uncertainty data provided with the benchmark values are often inadequate, and correction factors from the as-built assemblies to the idealized models are not always complete.

2.2. *Thermal Reactor Benchmarks*

ENDF-202 contains a series of 36 thermal reactor data testing benchmark assemblies for the principal fissile and fertile materials. This series includes homogeneous aqueous assemblies and simple rod lattices. These data are of similar vintage and quality as the fast reactor benchmark data. However, in most cases the benchmark testing employs Monte Carlo or transport theory methods, thereby minimizing modeling approximations. In low enriched uranium benchmarks the resonance calculation is especially important and is usually responsible for most of the methods' uncertainty. The report of the benchmark data testing of ENDF/B-V assessed this integral data base to be good for ^{235}U-H_2O systems, but to be poor for the D_2O and plutonium experiments.[8] The benchmark specifications are incomplete in characterizing experimental uncertainties and do not include measurements of the moderator temperature coefficient of reactivity and Doppler reactivity. A fuel depletion benchmark would also be useful.

2.3. *Shielding Benchmarks*

The shielding data testing of ENDF/B relies principally on a series of 12 shielding data testing benchmarks (the SDT series). The first five of this series are ORNL "broomstick" experiments which provide tests of total cross sections of iron, oxygen, nitrogen, sodium and stainless steel, respectively. Several of this series are mockup experiments which are relevant to LMR design and provide tests of transport through combinations of sodium, iron and steel. Some of the LLL pulsed-sphere measurements, which test data for neutrons at 14 MeV and are of value for fusion applications, are included in the shielding data testing series. Additional shielding data testing benchmarks would be desirable, particularly measurements in concrete and air.

2.4. *Other Benchmark Testing*

ENDF/B data testing also includes integral tests of fission product, actinide and dosimetry data.[8] T. R. England will be presenting results of integral tests of fission product and decay data at this Symposium.[9]

Analysis of the data testing benchmarks in ENDF-202 is certainly the center, though not the circumference, of the integral testing of ENDF/B. Increasingly, feedback to evaluators is based on analysis of other (than CSEWG benchmark) integral measurements. For example, measurements of the ^{10}B(n,He) reaction rate were made in ZPPR using ^{10}B helium accumulation fluence monitors (HAFMs).[10] Calculations using ENDF/B-IV and -V data yielded C/E values of $\sim$0.93. Furthermore, C/E discrepancies for ^{10}B(n,He) in five very different benchmark spectra ranged from 0.86 - 0.91 with ENDF/B-V.[11] (It appears this discrepancy will be reduced but not eliminated with ENDF/B-VI data.)

3. Integral Data Testing with ENDF/B-VI

It is good that integral results with ENDF/B-VI are not the subject of this Symposium, for data testing results have been very slow in arrival and very limited in scope. However, because the participants in this Symposium do wish to know the performance of the Version VI evaluations, a brief synopsis of results is included.

3.1. *Fast Reactors*

Preliminary results for a series of fast reactor benchmark assemblies have been generated by R. E. MacFarlane (LANL)[12] and R. Q. Wright (ORNL).[13] Both sets utilized Version VI constants prepared by NJOY. These results are in general agreement with each other (except for the Big-10 assembly) and are in general slightly improved with the Version VI data (relative to Version V.2). Eigenvalues for all the reflected assemblies are larger than for the bare assemblies. The eigenvalue for FLATTOP-23 is very high and unchanged from Version V (^{233}U was not changed for Version VI). For the ZPR (LMR-type) assemblies, the eigenvalue bias between ^{235}U and ^{239}Pu fuel is not eliminated, but reversed between Versions V.2 and VI. Changes in the C/E values for central reaction rate ratios obtained with Version VI (relative to Version V.2) in these fast assemblies are slightly improved and are more self-consistent. The overprediction of capture in ^{238}U relative to fission in ^{235}U (or ^{239}Pu) is reduced by $\sim$4% (or $\sim$2%, respectively). The threshold fission ratios tend to be underpredicted ($\sim$2-3%), particularly in the harder spectra.

3.2. *Thermal Reactors*

Initial benchmark results with ENDF/B Version VI indicate results generally close to the Version V results, with some results somewhat worse and some improved.

Specifically, the Version VI eigenvalues for low leakage $^{235}U/H_2O$ criticals are underestimated by ~0.5% (whereas Version V values are close to unity). This appears to be due to the ^{235}U thermal data which is less reactive in Version VI. The thermal eta for ^{235}U droops in the low thermal range, which should improve the calculation of the moderator temperature coefficient. The epithermal alpha value for ^{235}U remains low in ENDF/B-VI. It has decreased for the Version V value, which was low compared to integral measurements. Furthermore, Version VI data testing results have a trend of increasing eigenvalues for $^{235}U/H_2O$ criticals with increasing epithermal fission fraction, suggesting the ^{235}U resonance evaluation may be too reactive. However, the new ENDF/B-VI oxygen evaluation increases fast leakage relative to Version V. This change greatly improves the tendency of calculated eigenvalues to increase with increasing leakage. The ρ^{28} parameter is changed only slightly with Version VI (relative to Version V) and is still overestimated (by ~2%).

3.3. *Shielding*

Several of the ORNL Broomstick measurements, including iron, oxygen, nitrogen, and sodium, have now been analyzed with Version VI data. For each of these, the results are very similar to the Version V results and generally indicate adequate total cross section data for these materials. The Version VI oxygen data produce some slightly more discrepant values at higher energies (especially above 6 MeV), whereas the Version VI nitrogen data display significant discrepancies (as did Version V) in the valleys of the spectrum at 1.5 and 3.5 MeV.

The Version VI data for iron show dramatic improvements in LWR pressure vessel dosimetry analysis. Significant differences in the the neutron flux in the reactor cavity are caused by the new iron data.[14] This change in the calculated flux transmission (shown in Figure 1) with Version VI iron data also produces significant and consistent improvement in dosimetry measurements made in the reactor cavity.

Preliminary results for the LMR Radial Shield Benchmark from the US-Japan JASPER Program Radial Shield Experiment also indicate dramatic improvements in C/E values (for flux) using Version VI data for iron and ^{11}B.

4. Summary and Recommendations

The results of ENDF/B-VI data testing are still somewhat preliminary and limited in scope. This is due both to the general lack of support (funding) for the processing and testing of the Version VI files and to the considerable difficulty which the changes in Version VI data and formats have created for the processing codes. Further development on the processing codes is continuing and additional data testing results are planned. In the mean time, these initial integral results are already being considered in plans for future revisions of the Version VI evaluations.

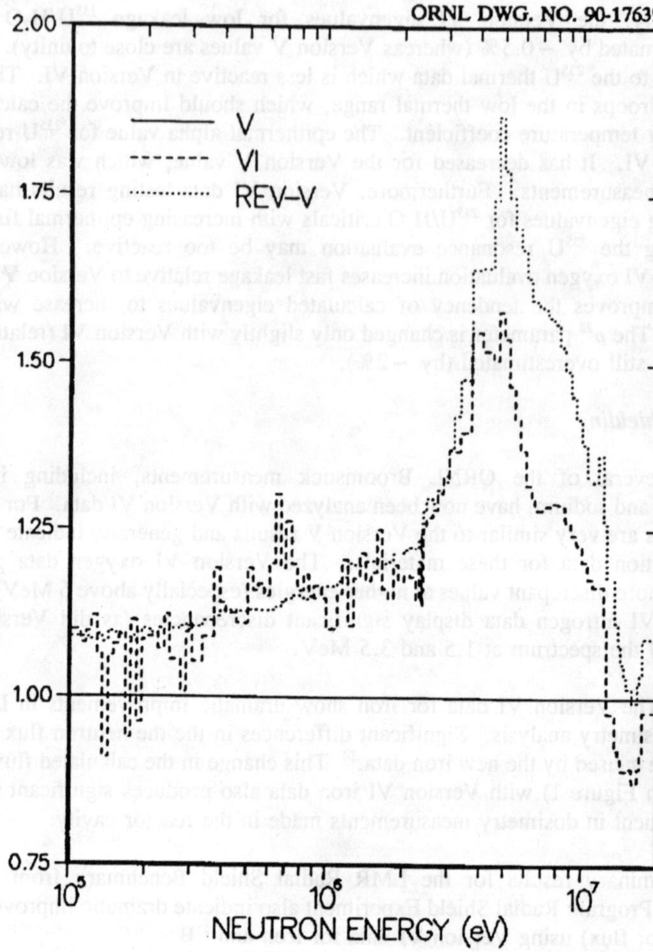

Figure 1. Ratio of Group Fluxes Calculated in the Reactor Cavity of Arkansas Nuclear One Unit 1 (taken from Ref. 14).

In addition to resolving the observed discrepancies in the integral comparisons discussed above, there are several nuclear data areas of current high interest to reactor designers:

(i) Nuclear data and branching ratios for the minor actinides are very important for the transuranic (TRU) burning evaluations being performed worldwide. Delayed neutron properties for the minor actinides are also important in these studies.

(ii) The current graphite thermal scattering data are known to be poor in ENDF. The thermal scattering kernels for graphite and the covariance data on their temperature dependence is an important requirement for MHTGR design studies.

(iii) Scattering data for fission products are more important in reactor designs which recycle fission products along with reprocessed fuel.

(iv) For unknown reasons, many criticality safety studies show a larger discrepancy between measured and calculated eigenvalues of plutonium systems (than uranium systems).

(v) New fuel designs and burnable poison nuclides are continually being studied in power reactors. This will require improved data for the erbium burnup chains, hafnium, gadolium, and possibly other new absorber nuclides in the future.

(vi) "Burnup credit" for spent fuel reactivity studies require more accurate data for some fission products after extended storage time.

Current requests from reactor designers are often not for improved data for the "Big 3" isotopes, but for data to fill a "niche" for a particular application. This omission reflects not only the progress made over the 6 major releases of the ENDF files for these most important data, but also the current "climate" of the nuclear industry. There is presently no demonstration reactor in final safety review or commercialization project trying to set fuel enrichments. At such a time there will be a need to know the bias and uncertainty in predicting integral parameters and a need to reduce those uncertainties. There will be a call for the best available differential data (and its covariances), for the best available integral data (at its covariances) and for the best design methodologies (with verification and validation). The feedback of data testing results to the evaluator is an important step toward that goal.

5. References

1. G. Cecchini et al., *Analysis of Integral Data for Few Group Parameter Evaluation of Fast Reactors*, Proc. International Conference on the Peaceful Uses of Atomic Energy, Geneva, August 1964, **2**, p 388 (1964).

2. J. L. Rowlands and J. D. Macdougall, *The Use of Integral Measurements to Adjust Cross Sections and Predict Reactor Properties*, Proc. International Conference on the Physics of Fast Reactor Operations and Design, London, June 1969, The British Nuclear Energy Society, p 180 (1969).

3. A. Gandini and M. Salvatores, *Nuclear Data and Integral Measurement Correlations for Fast Reactors*, CNEN Report RT/FI (74) 3 (1974).

4. W. P. Poenitz and P. J. Collins, *Utilization of Experimental Integral Data for the Adjustment and Uncertainty Evaluation of Reactor Design Quantities*, Proc. NEACRP Specialists' Meeting on Application of Critical Experiments and Operating Data to Core Design via Formal Methods of Cross Section Data Adjustment, Jackson Hole, Sept. 1988, NEACRP-L-307, p 97 (1988).

5. R. E. MacFarlane and R. M. Boicourt, *NJOY: A Neutron and Photon Photon Processing System*, Trans. Am. Nucl. Soc., **22**, 720 (1975).

6. *Cross Section Evaluation Working Group Benchmark Specifications*, BNL-19302 (ENDF-202), Brookhaven National Laboratory (1974).

7. P. J. Collins, R. D. McKnight and S. G. Carpenter, *Critical Review: Fast Reactor Benchmarks*, Trans. Am. Nucl. Soc., **41**, 573 (1982).

8. *ENDF-311 Benchmark Data Testing of ENDF/B-V*, BNL-NCS-31531 (ENDF-311), Brookhaven National Laboratory (1982).

9. T. R. England, *Decay Data Evaluation for ENDF/B-VI*, these proceedings (1992).

10. B. M. Oliver et al., Measurements of the 10B(n,He) Reaction Rate in ZPPR, Proc. International Conference on Nuclear Data for Basic and Applied Science, Sante Fe, May 1985, **1**, 625 (1985).

11. J. A. Grundl, *Examination of 10B(n,He) and 6Li(n,He) Cross Section Measurements in Reactor Physics Benchmarks*, Proc. International

Conference on Nuclear Data for Basic and Applied Science, Sante Fe, May 1985, 1, 471 (1985).

12. R. E. MacFarlane, private communication (1992).

13. R. Q. Wright and J. E. White, *Fast Reactor Data Testing of ENDF/B-VI at ORNL*, Trans. Am. Nucl. Soc., 64, 566 (1991).

14. M. L. Williams et al., *Transport Calculations of Neutron Transmission Through Steel Using ENDF/B-V*, Revised ENDF/B-V, and ENDF/B-VI Iron Evaluations, ORNL/TM-11686 (NUREG/CR-5648), Oak Ridge National Laboratory (1991).

QUALIFICATION AND IMPROVEMENT OF IRON ENDF/B-VI AND JEF-2 EVALUATIONS BY INTERPRETATION OF THE ASPIS BENCHMARK

S.H. ZHENG - I. KODELI - C. RAEPSAET
C.M. DIOP - J. C. NIMAL - A. MONNIER

COMMISSARIAT A L'ENERGIE ATOMIQUE
CENTRE D'ETUDES NUCLEAIRES DE SACLAY
DMT/SERMA/LEPP
91191 GIF SUR YVETTE CEDEX
FRANCE

INTRODUCTION

The aim of the present study is to contribute to the validation of the new evaluated nuclear data files like ENDF/B-VI (distr. 90) or JEF-2.2.

The new cross-section evaluations for iron isotopes are of particular interest for the nuclear community, since it is well known that the ENDF/B-IV data underestimate the neutron flux on deep penetration problems. The performances of the new nuclear data libraries are compared with those of ENDF/B-IV. The ASPIS benchmark, where the neutron transports through more than one meter iron plate, was chosen for this study.

The cross-section libraries were produced by the THEMIS/NJOY (ref 1) processing system and the transport calculations were carried out using the 3D Monte Carlo code TRIPOLI (ref 2) . The influence of different multigroup cross-section representations was investigated.

Finally, sensitivity, uncertainty and data adjustment analyses were carried out to obtain some additional informations about the quality of the cross-section data in ENDF/B-VI files. The analyses were performed using the code package set up of different modules, either developed at CEA or obtained from the NEA Data Bank. The adjustment indicated that some modifications have to be introduced to the neutron cross-sections of iron and the whole calculations were repeated with the adjusted set of cross sections.

The comparison of the results of the uncertainty and the adjustment analyses applied to ENDF/B-IV and ENDF/B-VI iron data permits to establish the progress made and gives some indications about the state-of-the-art of the cross section data.

DESCRIPTION OF THE EXPERIMENT

The ASPIS experiment on neutron deep penetration in iron has been carried out at Winfrith to contribute to the OECD (NEA) Collaborative Benchmark Exercise on assessing the nuclear data requirements for radiation shielding.

The details of the ASPIS experiment are described in ref 3 and 4. It consists of a low-power natural uranium converter plate, driven by the source reactor NESTOR, a graphite moderator and an extensive iron shield. This iron shield is composed by 24 mild steel plates, each 1830 mm wide, 1910 mm high and 50.8 mm thick. A 6.35 mm simplified air gap between adjacent plates allows the location of the detectors.

A cylindrical model (from ref 3) with rectangular members shown as equal area cylinders is used for the calculation (fig 1); in the iron shield, the steel has been stretched to fill the 6.35 mm air gap and the steel density has accordingly been reduced by a factor 0.889. The detectors, $Rh_{103}(n,n')Rh_{103}^{m}$, $In_{115}(n,n')In_{115}^{m}$ and $S_{32}(n,p)P_{32}$ are used to measure the reaction rates on the axes of iron shield.

CACULATIONAL METHOD AND CROSS SECTIONS

The calculations were realized using the 3 dimensional Monte Carlo code TRIPOLI (ref 2), which solves the Boltzmann integral transport equation, which governs the neutron propagation in the matter.

In the TRIPOLI code, the general geometry is exactly described using equations of the first and second order. The biasing technics are used to accelerate the simulation, specially in the neutron deep penetration cases.

The neutron cross sections are given in a multigroup representation by the THEMIS/NJOY system (ref 1) as follows :

$$\sigma_{x_g} = \frac{\int_{E_g}^{E_{g+1}} \sigma_x(E)\, \mathcal{F}(E)\, dE}{\int_{E_g}^{E_{g+1}} \mathcal{F}(E)\, dE}$$

where x is the reaction type, g the group number and $\mathcal{F}$, the weight function.

The cross section libraries ENDF/B4, ENDF/B6 (distr. 90) and JEF 2.2 have been used and processed with the following multigroup representations :

- 315 group structure ($\Delta u \approx 1/20$ in the region of interest) weighted by a fission spectrum for E>1MeV, and by $1/(E^*\Sigma_t(E))$ for the energy below 1 MeV, where $\Sigma_t(E)$ is the total macroscopic cross section of the composition.
- 3857 group structure ($\Delta u \approx 1/480$ for E>9 keV) with the same weight function.
- 315 group structure represented by a probability table method (ref 5 and 6).
- 315 group adjusted cross sections.

The cross sections of the 3 detectors used during the calculation are taken from the IRDF85 library. Their thresholds are 100 keV for Rh_{103}, 300 keV for In_{115} and 1 MeV for S_{32}.

SENSITIVITY AND ADJUSTMENT ANALYSES

The principal sources of the uncertainty in the neutron fluence calculations are :

- mathematical methods (numerics, group collapsing, transport calculation),
- geometry and material composition uncertainties and simplifications used in the calculational model,
- cross sections and other nuclear data uncertainties,
- neutron / gamma source strength uncertainties.

Assuming the error introduced by the calculational method itself is negligible, the uncertainty in the calculated flux can be reduced by adjustment. A least square method is normally adopted. Data adjustment requires the measured and the calculated integral results, the sensitivity profiles of the integral results to the nuclear data, as well as a covariance matrix of both the basic nuclear data and the measured integral data. In so far as the data are well compatible, the expected uncertainties in the adjusted parameters are substantially smaller than those in the unadjusted library.

The sensitivity and adjustment analyses were performed using a code package (ref 7) set up of different modules, either obtained from the NEA-Data Bank or developed at CEA. The calculational scheme used is presented in another paper at this conference (ref 8, Table 2). The sensitivity profiles are calculated using the SUSD code (ref 9) in conjunction with the one- and two-dimensional S_N transport codes ANISN and TWODANT (ref 10, 11, 12). ZOTT code (ref 12) is used for the adjustment. The covariance matrices were taken from the ZZ-VITAMIN-J/COVA library (ref 13). The linear perturbation method is adopted.

To limit the covariance matrices and the number of parameters used in the adjustment to a reasonable size, rather broad 30 group ENDF/B-IV based cross-sections were adopted in the ANISN and TWODANT calculations. It is obvious that for an accurate description of the absolute flux in a deep penetration problem like ASPIS this is not sufficient, because the errors due to the cross-section processing (group collapsing, self-shielding treatment) become increasingly important with the distance. On the other hand the sensitivity profiles as relative quantities were expected to be less affected by these errors.

The 30 group transport calculations served therefore only to determine the relative sensitivity profiles of the Rh, In and S reaction rates at the different penetrations in iron, to the iron cross-sections and the response functions. The TRIPOLI Monte-Carlo code calculations, performed in 315 and 3857 energy groups, provided the reference calculated reaction rates for the adjustment. The adjustment was nevertheless restricted to the penetration of less than 65 cm, since the data processing error has increasing impact on the sensitivity calculations, as well as on the 315 group neutron flux calculations.

The principal objective was to adjust the iron cross-sections on the basis of the calculated and measured Rh, In and S reaction rates. The TRIPOLI code was then rerun using the adjusted sets of cross-sections and the results were compared to the reaction rates from the original calculation.

The adjustment was first applied to the C/E values obtained using the standard 315 energy group calculation. The comparison to the 3857 group and probability table calculations (table 1) shows that these values (in particular In reaction rates) are subject to a non-negligible data processing error at large distances. The adjustment gives nevertheless reasonable results for the In and S reaction rates which are significantly improved. The Rh reaction rates on the other hand look marginally better at larger distances, but only on the cost of an overestimation at smaller distances. It is though evident that the adjustment factors partly account for the cross-section processing error. In this way these adjustments should be considered as valid only for the specific applications, which do not differ significantly from ASPIS experiment and where a similar energy group structure and cross-section processing are used. The adjustment is mathematically not rigorous, since the C/E values are not completely consistent with the input covariance data, which contain only the data evaluation uncertainties.

The adjustment performed on the basis of the reaction rates, calculated using the probability tables, clearly demonstrates this fact. As the cross-section processing errors have been reduced considerably, the adjustment factors correct primarily the data evaluation errors. The nonelastic cross-section adjustments above 3 MeV remain practically the same as in the 315 group adjustment, but are substantially different below. It is immediately clear that the difference is due to the fact that the 315 group adjustment factors in this region reflect the errors due to the use of the coarse group structure and the absence of self-shielding.

Figure 2 and 3 present the adjustments for the iron ENDF/B-IV and ENDF/B-VI cross-sections, obtained adopting the same procedure to the ENDF/B-VI probability table results. The factors are preliminary, since they are based on the sensitivity profiles from the ENDF/B-IV calculations. The covariance matrices of the Fe-56 isotope were used.

RESULTS

The comparison between the experimental and calculational results are given in table 1 and figure 4. The experimental systematic errors are 3% for Rh, 2% for In and 4% for S. The statistical errors in the calculational reaction rates are 4% for Rh, 5% for In and 3% for S.

ENDF/B4 results : from the results of the calculation with non adjusted ENDF/B4, we can conclude that a 315 group structure is not sufficient to process deep penetration problems for neutrons. The results are improved with probability tables and 3857 group structure. The 3857 group calculation gives a negligible error due to the energy group collapsing. The results and the adjusted coefficient show that the nonelastic cross section was overestimated in this library over 1 MeV.

ENDF/B6 results : for the 3857 group structure and the probability tables, the In and S results are in good agreement with the experiment, but the Rh reaction rates are really different. A 3857 group structure could somehow be unsufficiently precise to represent the Rhodium cross section in that energy range.

JEF-2 results : the results obtained are different from measurements. If we compare the JEF-2 and ENDF/B6 cross sections for Fe^{56}, we can see that the first excited level of the inelastic scattering is systematically 0,1 barn less in JEF-2 than in ENDF/B6. This has a direct influence on the S32 response, for E>2 MeV. The results for Rh and In are more difficult to explain. However, the source is described as a fission spectrum (E≈2MeV). As the first inelastic level is underestimated in JEF-2, the flux in the energy range $E' = E_{source} - E_{first\ level}$, which is In and Rh energy range, is underestimated too.

ACKNOWLEDGMENTS

Thanks are due to P. Ribon for his appreciable contribution and his help, specially concerning the probability tables.

CONCLUSIONS

The uncertainty and adjustment analyses proved to be of great value for the validation of the neutron transport calculations. The adjustment indicated the needs for some modifications to be introduced to the neutron cross sections of iron. The results obtained with this adjusted sets of data show considerably improved agreement between the calculation and the experiment.

Before adopting the adjusted results, one must be well aware of their limitations. In particular, one must pay attention to the following:
- The fundamental restriction is the quality of the covariance (uncertainty) information. It is evident that the quality of the adjusted data cannot be better than the quality of the input covariance data and that the deficiencies in the covariance information can lead to completely erroneous conclusions.
- The processing of the data (e.g. cross section group averaging, self shielding treatment, etc.) introduces additional errors, which are sometimes difficult to take into account properly. These errors are in addition peculiar to a specific problem, which makes the results of the adjustment to some extend problem dependent. In order to obtain the good set of data for general use it is therefore of prime interest to reduce this data processing errors to a minimum.
- Linear perturbation methods are usually used for the sensitivity and adjustment analyses. This implies the linear dependence between the target quantity and the basic data. This approximation is valid if the uncertainties and the data correction factors are relatively small.

References

1 - C. Raepsaet, C. Diop et al : *The THEMIS/NJOY processing system*, CEA to be published

2 - J. C. Nimal, A. Baur, D. Dejonghe, J. Gonnord, A. Monnier, T. Vergnaud : *TRIPOLI : Energy dependant three dimensional Monte Carlo program* ORNL - OLS - 80 110 (1980)

3 - J. Butler at al. : *Results and calculational Model of the Winfrith Iron Benchmark*, NEACRP- A-629

4 - A. K. McCracken : *The establishment of a shielding experimental benchmark at the NEA Data Bank*, NEACRP-A-1044, June 1990

5 - P. Ribon et al. : *Les tables de probabilités, application au traitement des sections efficaces pour la neutronique*, CEA to be published

6 - S.H. Zheng et al. :*The Monte Carlo transport calculations using the probability tables method for cross sections*, CEA to be published

7 - I. A. Kodeli, et al., *Uncertainty and Data Adjustment Analyses for PWR Shielding*, ANS Topical, M&C, Pittsburg, 1991.

8 - I. A. Kodeli, E. Sartori, *Experience with Collapsing and Expanding Multi-group Cross-section Covariance Matrices*, presented at this conference.

9 - K. Furuta et al., *SUSD: A Computer Code for Cross Section Sensitivity and Uncertainty Analyses Including Secondary Neutron Energy and Angular Distributions*, UTNL-R-0185, 1986.

10 - W. W. Engle Jr., *A Users Manual for ANISN*, Report ORNL/K-1693, ORNL, 1967.

11 - R. E. Alcouffe, et al., *Users Guide for TWODANT: A Code Package for Two-Dimensional, Diffusion-Accelerated, Neutral-Particle Transport* , LA-10049-M, LANL, 1989.

12 - D. W. Muir, *Evaluation of Correlated Data using Partitioned Least Squares: A Minimum-Variance Derivation*, Nucl. Sci. Eng: 101, p88 (1989).

13 - I A. Kodeli, E. Sartori, *Covariance Data Library ZZ-VITAMIN-J/COVA*, NEA Data Bank package 1264/03, 1990.

315 group structure						3857 group structure			Probability tables 315 group structure						d
Rh 103		In 115		S 32		Rh 103	In 115	S 32	Rh 103		In 115		S 32		(cm)
n adj	adj	n adj	adj	n adj	adj	n adj	n adj	n adj	n adj	adj	n adj	adj	n adj	adj	
ENDF/B4															
0.89	0.92	0.97	1.03	0.92	0.99	0.89	1.00	0.95	0.85	0.87	0.97	1.00	0.92	0.95	5.72
0.96	0.98	0.94	0.99	0.89	1.01	0.94	0.99	0.93	0.92	0.92	0.98	0.98	0.91	0.97	11.43
0.99	1.04	0.90	0.98	0.91	1.01	0.96	0.92	0.96	0.96	0.93	0.96	0.92	0.92	0.98	17.15
0.99	1.01	0.89	1.00	0.84	1.01	0.96	0.99	0.88	0.93	0.94	0.95	1.00	0.88	0.96	22.86
1.03	1.07	0.95	1.05	0.82	1.00	1.03	0.99	0.85	0.98	1.01	1.00	1.00	0.87	0.95	28.58
1.04	1.10	0.88	1.07	0.79	1.06	1.06	1.04	0.84	1.05	1.02	1.09	1.06	0.85	0.95	34.29
				0.79	1.05			0.85					0.86	1.00	40.01
1.06	1.18	0.81	1.05	0.72	1.05	1.09	0.98	0.81	1.12	1.07	1.05	1.03	0.82	0.97	45.72
1.02	1.09	0.76	0.91	0.72	1.02	1.07	0.95	0.84	1.09	1.04	1.01	0.98	0.84	0.96	51.44
1.00	1.09	0.75	0.95	0.62	0.93	1.06	0.97	0.72	1.11	1.04	1.10	1.04	0.77	0.90	57.15
0.92	1.05	0.72	0.93	0.63	0.99	1.04	1.02	0.78	1.06	0.99	1.12	1.02	0.82	0.97	62.87
0.91	1.06			0.52	0.86	1.07		0.73	1.10	1.07			0.73	0.86	68.58
0.89	0.99					1.05			1.12	1.05					74.30
0.74	0.86					1.02			1.03	1.01					85.73
0.69	0.81					0.96			1.03	0.99					91.44
0.51	0.65					0.84			0.88	0.88					102.87
0.43	0.56					0.82			0.85	0.85					114.30
ENDF/B6															
						0.92	1.05	1.02	0.86	0.85	0.99	0.97	1.00	0.98	5.72
						0.96	1.02	1.05	0.95	0.91	1.05	0.97	1.03	1.02	11.43
						0.97	0.95	1.09	0.95	0.96	0.98	0.95	1.09	1.06	17.15
						0.99	1.02	1.06	0.94	0.93	1.03	0.99	1.05	1.02	22.86
						1.06	1.08	1.07	0.97	0.97	1.03	1.06	1.10	1.01	28.58
						1.09	1.09	1.05	1.09	1.04	1.12	1.10	1.08	1.03	34.29
								1.08					1.16	1.08	40.01
						1.16	1.03	1.05	1.11	1.06	1.08	1.03	1.10	1.01	45.72
						1.16	1.02	1.05	1.05	1.06	1.02	1.01	1.14	1.03	51.44
						1.19	1.06	0.97	1.10	1.06	1.15	1.04	1.02	0.95	57.15
						1.17	1.01	1.10	1.08	1.05	1.07	1.03	1.15	1.06	62.87
						1.28		0.93	1.15	1.12			1.00	0.96	68.58
						1.26			1.10	1.12					74.30
						1.30			1.13	1.10					85.73
						1.36			1.14	1.13					91.44
						1.22			1.00	0.99					102.87
						1.26			1.05	1.06					114.30
JEF 2															
						0.90	1.05	1.01	0.86		0.98		1.02		5.72
						0.90	0.99	1.06	0.88		0.94		1.10		11.43
						0.90	0.93	1.13	0.91		0.91		1.19		17.15
						0.91	0.95	1.11	0.93		0.89		1.19		22.86
						0.93	0.94	1.13	0.89		0.89		1.18		28.58
						0.94	0.93	1.15	0.88		0.85		1.23		34.29
								1.26					1.32		40.01
						0.96	0.80	1.22	0.90		0.72		1.31		45.72
						0.93	0.75	1.26	0.88		0.63		1.35		51.44
						0.92	0.72	1.20	0.89		0.66		1.28		57.15
						0.89	0.71	1.34	0.85		0.62		1.49		62.87
						0.92		1.19	0.89				1.29		68.58
						0.89			0.88						74.30
						0.86			0.87						85.73
						0.85			0.88						91.44
						0.75			0.77						102.87
						0.74			0.77						114.30

Table 1:

C/E values for the 3 detectors versus the thickness of the iron slab

C: TRIPOLI results
E: Experimental results
d: thickness
n adj: calculation with non adjusted cross sections
adj: calculation with adjusted cross sections

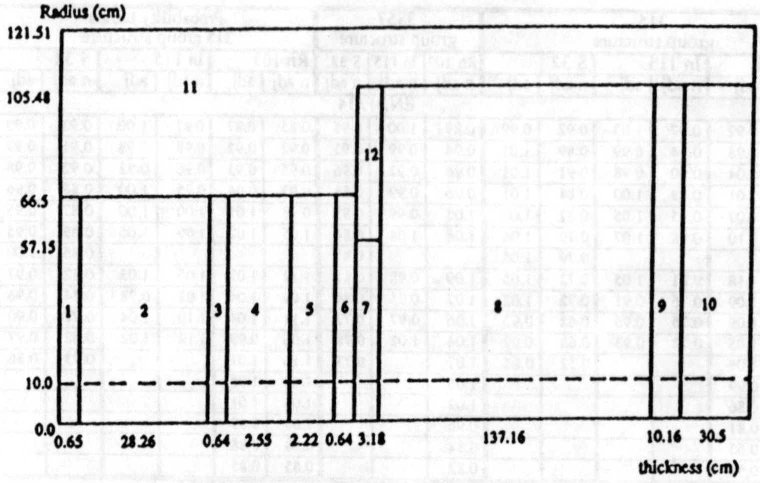

Figure 1 : Calculational geometry of the Benchmark ASPIS
1, 3, 7 : Aluminium; 2 : Carbone; 4, 6 : Air; 5 : Converter plate
8 : Iron shield; 9,12 : Mild Steel; 10, 11 : Concrete

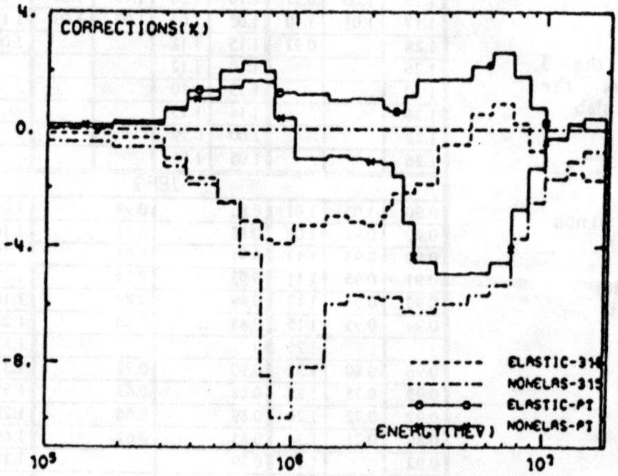

Figure 2 : Adjustment of the iron cross section
(ASPIS experiment)

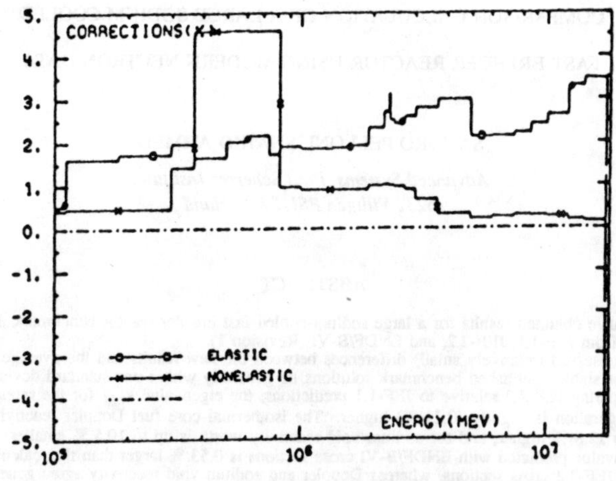

Figure 3 : **Adjustment of Fe56 ENDF/B6 cross sections (ASPIS experiment - probability tables)**

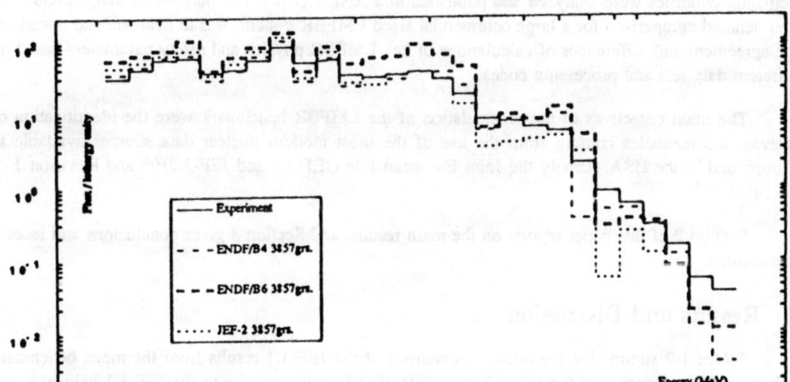

Figure 4 : **TRIPOLI and experimental neutron spectra at 114,3 cm**

COMPARISON CALCULATION OF A LARGE SODIUM-COOLED FAST BREEDER REACTOR USING MODERN NEUTRON DATA

SANDRO PELLONI, SHAHID AHMED

Advanced Systems, Paul Scherrer Institute,
5232 Villigen PSI, Switzerland

ABSTRACT

We have obtained results for a large sodium-cooled fast breeder reactor benchmark using data from JEF-1.1, JEF-2.2, and ENDF/B-VI (Revision 1).
The observed (relatively small) differences between the new results and the average values of sixteen published benchmark solutions lie generally within one standard deviation. Comparing JEF-2.2 relative to JEF-1.1 predictions, the eigenvalue k_{eff} for the reference configuration is slightly (0.21 %) higher. The isothermal core fuel Doppler reactivity is up to 12.8 % higher, and the sodium void reactivity worth is up to 10.5 % smaller. The eigenvalue predicted with ENDF/B-VI cross sections is 0.53 % larger than that calculated with JEF-2.2 cross sections, whereas Doppler and sodium void reactivity agree generally within 5 %. The main contributions to these differences are from the fission production cross section of ^{239}Pu and from the inelastic scattering cross section of ^{238}U and ^{23}Na below 1 MeV.

1 Introduction

Results of an international calculation of a large (1250 MWe) sodium-cooled fast breeder reactor benchmark model (LMFBR benchmark) were presented and discussed at a specialist's meeting at Argonne National Laboratory (ANL) in February 1978. Sixteen solutions from ten different participating countries were analyzed and published in a ANL report[1]. The purpose of that exercise, the first detailed comparison for a large commercial sized LMFBR system, was to evaluate and document the agreement and differences of calculations of key LMFBR physics and safety parameters based on different data sets and processing codes.

The main objectives of the recalculation of the LMFBR benchmark were the identification of relevant discrepancies coming from the use of the most modern nuclear data sources available in Europe and in the USA, namely the Joint European File (JEF-1.1 and JEF-2.2)[2,3] and Revision 1 of ENDF/B-VI[4].

Section 2 of this paper reports on the main results, and Section 3 gives conclusions and recommendations.

2 Results and Discussion

Tables 1-2 summarize the relative deviations of the JEF-1.1 results from the mean benchmark values, and the variation of the JEF-2.2 and ENDF/B-VI results relative to the JEF-1.1 values.

The tables show that the new results generally lie within one standard deviation of the average benchmark values.

Specific differences are:

- The eigenvalue k_{eff} for Configuration 1 (reference configuration) predicted with JEF-2.2 is slightly (0.21 %) larger than that calculated with JEF-1.1. Comparing ENDF/B-VI relative to JEF-2.2 predictions, k_{eff} is 0.53 % higher.
- Breeding and conversion ratios predicted with JEF-2.2 and JEF-1.1 agree well within 1 %. Comparing ENDF/B-VI relative to JEF-2.2 predictions, the maximum, 4.30 % increase (i.e. 0.417 versus 0.399) of the breeding ratio is in the blanket region.
- The sodium void reactivity worths for the inner core and whole reactor calculated with JEF-2.2 are significantly (i.e. about 10 %) smaller than the corresponding JEF-1.1 values. The ENDF/B-VI and JEF-2.2 predictions differ by a maximum of 5 %.
- The reactivity worths of the central control rod with sodium calculated with JEF-2.2 increases by 5.5 % when compared with JEF-1.1. The ENDF/B-VI and JEF-2.2 predictions agree well within 2 %.
- The JEF-2.2 isothermal core fuel Doppler-reactivity coefficient exceeds the JEF-1.1 value by 3.87 % for Configuration 1 (i.e. -0.00751 versus -0.00723), and by as much as 12.77 % for the voided Configuration 3 (i.e. -0.00452 versus -0.00401). The ENDF/B-VI and JEF-2.2 predictions agree within 4 %.
- The effective delayed neutron fraction and the inhour of reactivity for Configuration 1 calculated with JEF-2.2 are about 7 % larger than the respective JEF-1.1 values. The ENDF/B-VI values for these parameters are significantly (4.61 % and 11.2 %) lower than the JEF-2.2 values. These differences are primarily due to the different delayed neutron yields of ^{238}U and ^{239}Pu (see Table 2).
- There is a significant difference in the central sodium void reactivity worth. The rather big sodium void worth in Configuration 1, calculated with JEF-1.1 data (about 1.3 standard deviations larger than the mean benchmark value), decreases by about 0.65 standard deviations when using JEF-2.2 cross sections, and by about 0.25 standard deviations when using ENDF/B-VI cross sections.

Table 3, which is indicative for the above differences, shows the relative variation of the JEF-2.2 eigenvalue k_∞ for the inner core of Configuration 1 when substituting data for individual isotopes with data from ENDF/B-VI and JEF-1.1. The main contributions to this variation are ^{239}Pu from ENDF/B-VI (-0.66 %), ^{238}U from ENDF/B-VI (+0.71 %), and ^{23}Na from JEF-1.1 (-0.553 %).

The decrease of k_∞ due to ^{239}Pu from ENDF/B-VI comes primarily from the fission production cross section. However, there are two opposite effects, a strong decrease coming from the fission cross section itself, and an increase coming from nubar and from the capture cross section[5]. The increase of k_∞ due to ^{238}U from ENDF/B-VI, and the decrease due to ^{23}Na from JEF-1.1 are from the inelastic scattering data.

The fast fission and fission production cross sections of ^{239}Pu between 10 keV and 1 MeV are up to 6 % larger in JEF-1.1 than in ENDF/B-VI. The maximum differences occur at about 50 keV (1.6 versus 1.5, respectively 4.7 versus 4.4 barns). The JEF-2.2 cross section is even larger than the JEF-1.1 cross sections at energies below 50 keV.

The inelastic scattering ENDF/B-VI cross section for ^{238}U is up to 15 % larger than the JEF-1.1 cross section at energies between 200 and 400 keV (i.e. 1.24 versus 1.07 barns at about 240 keV). The JEF-2.2 cross section is slightly smaller than the JEF-1.1 cross section in this energy range.

There is a considerable discrepancy between JEF-1.1 and ENDF/B-VI inelastic scattering cross section for ^{23}Na at energies below 400 keV. The JEF-2.2 cross section is rather similar to the ENDF/B-VI cross section.

	Configuration 1		Conversion Ratio		Breeding Ratio		
	k_∞ Inner Core	k_{eff}	Inner Core	Outer Core	Core	Blanket	Reactor Total
BENCHMARK Mean (of 16 Solutions) Std. Deviation (%)	1.12428 1.8	1.00509 1.3	1.09262 4.2	0.82144 4.4	0.98984 4.2	0.40227 2.4	1.39211 3.5
JEF-1.1 Deviation from mean in (%)	1.12791 0.32	1.01103 0.59	1.06431 -2.59	0.79834 -2.81	0.96323 -2.68	0.39536 -1.72	1.35858 -2.41
JEF-2.2 % Diff. Compared to JEF-1.1	1.13320 0.46	1.01316 0.21	1.06152 -0.26	0.79563 -0.34	0.96111 -0.22	0.39933 1.00	1.36044 0.13
ENDF/B-VI % Diff. Compared to JEF-1.1	1.14352 1.38	1.01852 0.74	1.08005 1.48	0.80904 1.34	0.98438 2.19	0.41652 5.35	1.40091 3.11
	Central Control Rod Worth			Na-Void Worth		Fuel Doppler Reactivities	
	$(k_{rod}-k_{ref})/k_{ref}$			$(k_{void}-k_{ref})/k_{ref}$		$(k_{2200}-k_{1100})/k_{1100}$	
	Na-In Relative to Fuel	Na-In Relative to Na	Na-Void Relative to Fuel	Na-Void Inner Core	Na-Void Outer Axial Blanket	Na-In Conf.1	Na-Void Conf.3
BENCHMARK Mean (of 16 Solutions) Std. Deviation (%)	-0.00355 13.2	-0.00298 14.4	-0.00460 10.2	0.02116 12.1	0.02098 16.8	-0.00729 10.8	-0.00433 14.3
JEF-1.1 Deviation from mean in (%)	-0.00336 -5.35	-0.00282 -5.37	-0.00445 -3.26	0.02471 16.77	0.02592 23.54	-0.00723 -0.82	-0.00401 -7.39
JEF-2.2 % Diff. Compared to JEF-1.1	-0.00354 5.35	-0.00298 5.67	-0.00456 2.47	0.02230 -9.75	0.02320 -10.49	-0.00751 3.87	-0.00452 12.77
ENDF/B-VI % Diff. Compared to JEF-1.1	-0.00362 7.73	-0.00304 7.80	-0.00461 3.59	0.02336 -5.46	0.02463 -4.97	-0.00734 1.52	-0.00435 8.48

Table 1: Results for the LMFBR Benchmark Obtained Using JEF-1.1, JEF-2.2, and ENDF/B-VI (Rev. 1) Cross Sections

	JEF-1.1	JEF-2.2	ENDF/B-VI	Std. Dev. (%)	JEF-1.1 Dev. Mean Value (%)	JEF-2.2 % Incr. Compared to JEF-1.1	ENDF/B-VI % Incr. Compared to JEF-1.1
β_{eff}	0.003823	0.004113	0.003924	2.62	-1.16	7.59	2.65
Inh. Reac.	1.1398	1.2193	1.0823	1.62	-1.02	6.97	-5.04
^{238}U D. N. Y.	0.0460	0.0481	0.0440	/	/	/	/
^{239}Pu D. N. Y.	0.0061	0.0065	0.00645	/	/	/	/
^{23}Na R. W.	-7.7156	-7.2150	-7.5303	9.75	11.69	-6.49	-2.40

Table 2: Effective Delayed Neutron Fraction β_{eff}, Inhour of Reactivity (x10^5), and Central Sodium Reactivivity Worth Expressed in $\delta k_{eff}/k_{eff}/(10^{31}$ Atoms)for Configuration 1, and Their Deviations from the Mean Benchmark Values, and Delayed Neutron Yields for ^{238}U and ^{239}Pu

Evaluation	Isotope	Eigenvalue k_∞	% Diff. Compared to (Pure) JEF-2.2
ENDF/B-VI	^{16}O	1.13320	0.000
	^{23}Na	1.13506	0.164
	Natural Iron	1.13638	0.281
	^{235}U	1.13315	-0.004
	^{238}U	1.14126	0.711
	^{239}Pu	1.12575	-0.657
	^{240}Pu	1.13575	0.225
	^{241}Pu	1.13421	0.089
	^{242}Pu	1.13332	0.011
JEF-1.1	^{23}Na (JEF-1.1)	1.12694	-0.553
	^{238}U (JEF-1.1)	1.12915	-0.357
	^{239}Pu (JEF-1.1)	1.13418	0.086

Table 3: Eigenvalues k_∞ for the Inner Core (Configuration 1), Calculated Using JEF-2.2 Data, and Their Percent Increase Due to the Use of ENDF/B-VI and JEF-1.1 Data for Individual Nuclides

3 Conclusions and Recommendations

The present analyses dealt with neutronics calculations of a large sodium-cooled fast breeder reactor performed using various nuclear data.

From this study the following conclusions can be reached, and some recommendations can be made.

- Good agreement, generally within one standard deviation, was achieved between the new results and the average values over the sixteen benchmark solutions obtained in the past. However, comparing JEF-2.2 with JEF-1.1 predictions, the isothermal core fuel Doppler reactivity is significantly, up to 12.8 % higher, and the region sodium void reactivity worth is up to 10.5 % smaller.

  The eigenvalues predicted with ENDF/B-VI are about 0.5 % larger than those calculated with JEF-2.2 cross sections.

  There are additionally large discrepancies up to 10 % in the delayed neutron fractions and sodium reactivity worths predictions.
- The k_∞ deviations are mostly coming from the different fission production cross sections of ^{239}Pu, from the different inelastic scattering cross section of ^{238}U and ^{23}Na between 10 keV and 1 MeV, and from the different delayed neutron yields of ^{238}U and ^{239}Pu. Particularly, it is found that ^{238}U and ^{239}Pu from JEF-2.2 and JEF-1.1 behave neutronically in a similar way, whereas ^{23}Na from JEF-2.2 behaves rather like ^{23}Na from ENDF/B-VI.
- Therefore, we recommend to further analyze this data in order to improve the accuracy of fast breeder calculations.

4 References

1. L. G. Sage, R. D. McKnight, D. C. Wade, K. E. Freese, P. J. Collins, ANL-80-78/NEA-CRP-L-243 (1980)

2. J. Rowlands, N. Tubbs, The Joint Evaluated File: a New Data Resource for Reactor Calculations, Int. Conf. on Nuclear Data for Basic and Applied Science, Vol. 2, p. 1493, (Santa Fe, New Mexico, 1985)

3. E. Fort, M. Salvatores, Fast Reactor Benchmarks and Integral Data Testing and Feedback into JEF-2, Paper Presented at this Conference

4. C. L. Dunford, Evaluated Nuclear Data File, ENDF/B-VI, Int. Conf. on Nuclear Data for Science and Technology, p. 788-792, (Jülich, Germany May 1991)

5. S. Pelloni, Comparison Calculation of a Large Sodium-Cooled Fast Breeder Reactor Using the Cell Code MICROX-2 in Connection with ENDF/B-VI and JEF-1.1 Neutron Data, PSI-118 (1992)

BENCHMARK EXPERIMENTS FOR FUSION NEUTRONICS AND DATA TESTING OF JENDL-3

MAEKAWA, Hiroshi
Fusion Neutronics Laboratory, Department of Reactor Engineering
Japan Atomic Energy Research Institute
Tokai-mura, Naka-gun, Ibaraki-ken, 319-11 Japan

ABSTRACT

The most suitable experiments for the verification of data and methods relevant to fusion neutronics are clean benchmark experiments on a simple geometry with simple material compositions using a D-T neutron generator. As there is no good parameter such as criticality, spectrum indices have been measured as the absolute values along with spectra of neutron and gamma-ray. Most of the benchmark experiments used for the data testing of JENDL-3 were performed using FNS/JAERI and OKTAVIAN/Osaka University. The accuracy of nuclear data in JENDL-3 seems to be good enough for applying to the fusion reactor design. Minor changes are, however, desirable for many nuclei in JENDL-3 to improve their accuracy.

1. Introduction

In May 1983, a tentative nuclear data file for fusion neutronics study was requested to the Japanese Nuclear Data Committee. Because JENDL-2 was prepared mainly for fission reactors and its accuracy was not enough for fusion neutronics application. Especially, there are problems in the nuclear data above 5 MeV. In December 1983, JENDL-3PR1, having 8 nuclei of ^{6}Li, ^{7}Li, ^{9}Be, ^{12}C, ^{16}O, Cr, Fe and Ni, was completed. Three nuclei of ^{6}Li, ^{7}Li, and ^{12}C were re-evaluated as JENDL-3PR2 in March 1985. Both nuclear data files were used for the analyses of experiments relevant to fusion neutronics at FNS/JAERI and OKTAVIAN/Osaka University. In July 1985, the first Specialists' Meeting on Nuclear Data for Fusion Neutronics was held at Tokai Research Establishment, Japan Atomic Energy Research Institute.[1)] The meeting was very fruitful and full of suggestions.

The Working Group of Fusion Neutronics Integral Test, belong to both the Japanese Nuclear Data Committee and the Japanese Committee on Reactor Physics, was starting in April 1987 for the data testing of JENDL-3. A tentative version was released in April 1987 as JENDL-3T Rev. 0, which had 32 nuclei. The results of integral test were reported at the 1987 Seminar on Nuclear Data.[2)] At the Seminar, many groups pointed out the problem of Be and Pb nuclear data. A technical meeting was held to discuss the problems in mainly Be and Pb data in December 1987.

In March 1988, the revised version was released as JENDL-3T Rev. 1. Additional integral tests were presented at the 1988 Seminar on Nuclear Data.[3] JENDL-3 was released domestically in October and world-widely in December 1989.[4] Integral test results for JENDL-3 were also reported at the 1989 and 1990 Seminars on Nuclear Data.[5,6] The second Specialists' Meeting on Nuclear Data for Fusion Reactors was held at Tokai Research Establishment, Japan Atomic Energy Research Institute.[7] The main object of the Meeting was to review the evaluated data of JENDL-3 and the results of integral tests.

Specific feature of D-T fusion neutronics is discussed in **Section 2**. Benchmark experiments for the data testing of JENDL-3 are shown in **Section 3**. The procedure of data testing and results are presented in **Section 4**.

2. Specific Feature of D-T Fusion Neutronics

Specific features of D-T fusion neutronics are, (1) source neutron energy is 14 MeV and exceeds the binding energy of nucleons, (2) neutrons are generated in only core plasma region, (3) tritium is fuel but exists negligibly small amount in natural, and so on. They make new problems in the study of neutrons and related behavior in the fusion reactor blanket and shield. Anisotropic scattering and induced activity are important for such high energy neutrons. Most part of nuclear heating is generated in the super-conductive magnet region. Since the source neutons distribute unevenly at the plasma region, anisotropic scattering should be considered in the transport calculation and nuclear heating-rate distribution changes steeply from the first wall to the super-conductive magnet. Finally, tritium self-sustain is essential in the case of D-T fusion power reactors. The tritium breeding rate should be exceed unity with an enough margin. They also request additional accuracy in the evaluated nuclear data files, i. e., nuclear data of threshold reactions [(n,2n), (n,p), (n,α), etc.], anisotropic scattering, tritium production [^{6}Li(n,t)^{4}He, ^{7}Li(n,n't)^{4}He] and neutron multiplication [^{9}Be(n,2n), Pb(n,2n), Zr(n,2n), etc.], and nuclear data accuracy over wide-energy range.

Problems in the benchmark experiments for the data testing are summarized as follows: (1) There is no good parameter such as the criticality in the case of critical experiments for fission reactors. The experimental error is expected to be 3 ~ 5 % in a good case, while the criticality can be measured with the error of less than 0.1 %. Therefore it is desirable to measure neutronic parameter as many as practical, i.e., to measure spectrum indices such as reaction rates along with spectra of neutron and gamma-ray. (2) Both neutrons and gamma-rays exist simultaneously in the experimental field. The n-γ discrimination is always necessary for the detectors such as NE213 liquid scintillator and proton-recoil proportional counter. Gamma-insensitive detectors such as fission counter and activation foils are effective. (3) Nuclear heating and induced activities are important as well as tritium production rate. An intense D-T neutron generator is required to measure these parameters. Developments of new techniques are also needed in the available resource and condition for the benchmark experiments. (4) The neutron flux attenuates largely by 4 ~ 6 decades, especially in the case of shielding experiments. This requests the large dynamic range to both the neutron generator and detectors.

Problems in the calculation for the analysis of benchmark experiments and the nuclear design of fusion reactors are divided into mainly following three items: (1) Anisotropic scattering is dominant and important, especially for high energy neutrons. That is, the transport calculation should be necessary and the diffusion calculation is inadequate any more. (2) The wide energy range should be considered from 15 MeV to thermal. At least 100 groups for neutron is desirable in the group-wise calculation. (3) Fusion reactors such as tokamak-type have a complex geometry and configuration. Three-dimensional calculations are needed.

3. Benchmark Experiments for Data Testing

In 1981, two powerful D-T neutron generators, FNS/JAERI and OKTAVIAN/Osaka University were constructed in Japan to apply the fusion neutronics study. Following benchmark experiments have been performed using mainly FNS and OKTAVIAN. The first type is integral experiment on spherical or cylindrical assemblies made of Li, Li_2O, C, Be, Fe, Cu, etc. Measured items are tritium production-rate distribution, reaction rate distribution, in-system neutron spectrum and so on. The second type is measurement of neutron and gamma-ray leakage spectra from spherical or slab assemblies. The third type is very useful for data testing of secondary neutron energy and angular distributions, i.e., double-differential cross section (DDX). Measured DDX have been used directly in the data testing of DDX in JENDL-3. Recently two new type experiments have been conducted at FNS. One is measurement of induced activity in a fusion environment to assess the JENDL Activation File. The other is measurement of nuclear heating rate by a micro-calorimetric method for data testing of kerma factor.

4. Data Testing of JENDL-3

Integral test plan of JENDL-3 nuclear data relevant to the fusion neutronics is summarized in **Table 1**. Two types of cross section libraries based on JENDL-3, FUSION-J3[8)] and FSXLIB-J3[9)] were prepared for the data testing. The FUSION-J3 library has 125- and 40-groups for neutron and gamma-ray, respectively, and is for ANISN and DOT. The FSXLIB-J3 library is used for the continuous energy Monte Carlo code MCNP.

Table 2 summarizes the status of JENDL-3 nuclear data based mainly on the discussion at the second Specialists' Meeting mentioned above. The discussion was proceeded nuclide by nuclide according to the data tests presented at the meeting. The accuracy requested is not the same for all nuclei. Some nuclei such as ^{7}Li, ^{9}Be and Fe are major elements for the fusion reactor and the high accuracy is requested for the data of these nuclei. The status of data for these nuclei in the table is judged critically. As space is limited here, discussions are made below for some interest elements.

Figure 1 shows the measured and calculated spectra at 24.9-degree for 50.8-thick Be slab.[7] It is clear that the JENDL-3 is improved much in the energy range 2 ~ 10 MeV comparing to the JENDL-3T. From the C/E (ratio of calculation to experiment) trends with energy, angle and thickness, and from the results of integral experiments on Be[10] and Be-Li_2O assemblies[11], additional examinations are needed especially in (n,2n) and DDX data.

In the data of ^{10}B and ^{11}B, no integral test was presented. These data seem to be not so bad from the comparison between measured and evaluated DDX data.

Figure 2 shows the gamma-ray flux distributions in a graphite assembly of 63 cm-diameter and 61 cm-long.[12] A good agreement is observed between measured and calculated fluxes within 10 %. Both neutron and gamma-ray data of C in JENDL-3 seem to be good.

A problem of nitrogen data in JENDL-3 was pointed out in the previous test[5]. The data have been modified as JENDL-3M (tentative name)[6], the DDX data calculated from JENDL-3 are quite different from measured data. This is due to the lack of the energy-angle correlation of continuum inelastic scattering. The calculated angular spectra based on this JENDL-3M agree well with measured ones except angular distribution of elastic reaction peak.[13] Recently some strange structure is discovered in total cross section. This problem will be resolved in JENDL-3.2.

Neutron and gamma-ray emission spectra from a Si sphere are shown in **Fig. 3**.[5] From the analysis of neutron leakage spectrum, it was pointed out that there was some discrepancy in the energy range of 0.2 ~ 2 MeV and around 10 MeV. In the case of gamma-ray, the JENDL-3 calculation agrees well with the measurement in whole energy region and reproduce the peaks of 0.97 MeV from ^{28}Si(n,p), 1.8 and 2.8 MeV from ^{28}Si(n,n'). The Si data of both neutron and gamma-ray in JENDL-3 seem to be good enough for application to the fusion neutronics.

Figures 4 and **5** show neutron spectra at 0.41 m[14] and gamma-ray heating rate distributions[7] in a iron assembly of 1 m-diameter and 0.95 m-long. The MCNP calculation gives a good agreement with the measurement. The DOT calculation overestimates the measurement by more than a factor of 2 for the flux below 500 keV in the front part of the iron assembly, and underestimates by a factor of 2 for the flux above 500 keV in the rear part. This discrepancy is attributed to the lack of self-shielding correction. A reasonable agreement is observed for the gamma-ray heating.

Figure 6 shows a sample of angular spectra from lead slabs.[15] The spectra calculated with JENDL-3 agree well with the measured ones except the region between 5 and 11 MeV. In the C/E trends of integrated flux over energy, the underestimation is observed for the energy region of 5 ~ 10 MeV. This is caused by a lack of some inelastic levels. A good agreement should be expected when the number of inelastic level increases more than present limit in JENDL-3 or bunched levels are used.

5. Concluding Remarks

Various types of benchmark experiments for data testing have been accumulated in Japan using mainly FNS/JAERI and OKTAVIAN/Osaka University. These data were applied to the data testing of preliminary version of JENDL-3 (JENDL-3PR1, -3PR2, -3T) and JENDL-3. Most of them were reported and discussed at the two "Specialists' Meeting on Nuclear Data for Fusion Neutronics" published in JAERI-M 86-029 and JAERI-M90-123, respectively.

The accuracy of nuclear data in JENDL-3 seems to be good enough for applying to the fusion reactor design. Minor changes are, however, desirable for many nuclei in JENDL-3 to improve their accuracy. The range of these minor changes, especially in the cases of total and elastic cross sections, are now within the uncertainty of data evaluation and the error of differential experiments. Namely, the accuracy for a large part of nuclear data in JENDL-3 becomes close to the limit of errors of both differential and integral experiments. This fact suggests that we should be careful to examine and discuss the status of evaluated nuclear data considering the accuracy and/or uncertainty of data evaluations and experiments.

Japanese fusion neutronics community expects that the problems pointed out through the data tests will be solved in the next JENDL-3.2 version and in the JENDL Fusion File including MF-6. The Fusion File was requested at the Specialists' Meeting in order to judge the usefulness of the file-6.

Acknowledgement

The author would like to thank the member of the Working Group of Fusion Neutronics Integral Test for their kindly cooperation for the data testing of JENDL-3. He also wish to thank the members of Subcommittee on Fusion Reactor of the Japanese Research Committee on Reactor Physics and the Japanese Nuclear Data Committee for their supports and mutual communication.

References

1. S. Igarashi and T. Asami (Eds.), *Proc. Specialists' Meeting on Nuclear Data for Fusion Neutronics*, JAERI-M 86-029 (March 1986).
2. T. Nakagawa and A. Asami (Eds.), *Proc. 1987 Seminar on Nuclear Data*, JAERI-M 88-065 (March 1988).
3. T. Nakagawa and A. Zukeran (Eds.), *Proc. 1988 Seminar on Nuclear Data*, JAERI-M 89-026 (March 1989).
4. K. Shibata, et al., "Japanese Evaluated Nuclear Data Library, Version-3 - -JENDL-3 --", JAERI 1319 (June 1990).
5. Y. Nakajima and M. Igashira (Eds.), *Proc. 1989 Seminar on Nuclear Data*, JAERI-M 90-025 (Feb. 1990).
6. M. Igashira and T. Nakagawa (Eds.), *Proc. 1990 Seminar on Nuclear Data*, JAERI-M 91-032 (March 1991).
7. Y. Nakajima and H. Maekawa, *Proc. Second Specialists' Meeting on Nuclear Data for Fusion Reactors*, JAERI-M 91-062 (March 1991).
8. K. Maki, et al., "Nuclear Group Constant Set FUSION-J3 for Fusion Reactor Nuclear Calculations Based on JENDL-3," JAERI-M 91-072 (May 1991) (In Japanese).
9. K. Kosako, Y. Oyama and H. Maekawa, "FSXLIB-J3: MCNP Continuous Energy Cross Section Library Based on JENDL-3," JAERI-M 91-187 (1991); *Proc. Topical Mtg. on New Horizons in Radiation Protection and Shielding*, Apr. 26-May 1, 1992, Pasco, Washington, U.S.A., pp357-363.
10. H. Maekawa, et al., *Fusion Technol.*, 19 [3], pp1949-1954 (1991).
11. H. Maekawa, et al., ibid., 15 [2] pp1283-1286 (1989).
12. F. Maekawa, et al., To be published.
13. Y. Oyama, H. Maekawa and K. Kosako, "Measurements and Analyses of Angular Neutron Flux Spectra on Liquid Nitrogen, Liquid Oxygen and Iron Slabs," *Proc. Int'l Conf. on Nuclear Data for Science and Technology, Jülich, Germany, 13-17 May 1991* pp337-340.
14. C. Konno, et al., *Fusion Eng. Design*, 18 pp297-303 (1991).
15. H. Maekawa and Y. Oyama, ibid., 18 pp287-291 (1991).

Table 1 Integral test plan of JENDL-3 for fusion neutronics.

Organization	Benchmark Problem	Personnel	Standard analysis	Additional analysis
JAERI	• Integral experiments on Li_2O, C, Li_2O-C, Be-Li_2O, Be, Fe, assemblies	H. Maekawa K. Kosako C. Konno	DOT 3.5	
	• TOF experiments on Li_2O, C, Li, Be, LN_2, LO_2, Fe & Pb slabs	Y. Oyama H. Maekawa	DOT 3.5	MCNP MORSE-DD
	• γ-ray, heating Be, C, Li_2O, Fe, W	Y. Oyama S. Yamaguchi	DOT 3.5	
	• Pulsed Sphere Program (LLNL), ^{6}Li, ^{7}Li, C, O, Fe, Pb, P.E., H_2O, Be	T. Mori M. Nakagawa	MORSE	ANISN-DD DOT-DD MORSE-DD
University Union	• Integral experiments on Li, Pb-Li, Pb-Li-C & Be-Li-C spheres assemblies	T. Iguchi A. Takahashi K.Sugiyama	ANISN	MCNP
Osaka Univ.	• TOF experiments on Li, Pb, SS316 concrete assemblies	A. Takahashi J. Yamamoto	ANISN	NITRAN
Kyoto Univ.	• Backscattering experiments on concrete & SS304 assemblies	K. Shin	ANISN	MCNP
KURRI Osaka Univ.	• TOF and γ-ray integral experiments on LiF, Teflon, Li, Al, SI, Ti, Cr, Mn, Co, Cu, Zr, Nb, MO &W sphere assemblies	C. Inchihara J. Yamamoto	ANISN	ANISN-DD NITRAN MCNP
TIT	• Spectra in LiF, C, H_2O & LiF-C assemblies	H. Sekimoto	MORSE	
Tohoku Univ.	• TOF experiments on P.E.& Pb slabs	S. Iwasaki		MCNP
JAERI &SRI	• Experiment on SS304 assembly at ORNL	K. Sakurai K. Ueki		MORSE MCNP
Toshiba	• Pulsed Sphere Program (LLNL)	M. Kawai M. Uematsu	ANISN MORSE	

Table 2 Status of Nuclear Data in JENDL-3

Nuclide	Priority	Status		Comment
		neutron	gamma	
Li-6	High	△	○	There is a problem in DDX.
Li-7	High	○	○	There is no large problem.
Be-9	High	△		If the accuracy of TBR is requested to be less than 5%, the status in unsatisfactory. Additional examination is required especially for (n, 2n) and DDX.
C-12	High	□	○	Re-evaluation is desirable for total and elastic cross section. Information of spectrum in (n,n'3α) is not enough and is important for calculation of DPA. Data for nuclear heating is good.
N-14	Medium	×(J-3) □(J-3M)		JENDL-3 is not good but newly evaluated JENDL-3M is almost good.
O-16	High	○		JENDL-3 seems to be enough.
F-19	Low	×	×	Data is not good but priority is low.
Al	High	□	○	
Si	High	○	○	Evaluation is fine.
Ti	Medium	×	△	
Cr	High	□	△	There are some problems.
Mn	High	□		
Fe	High	□		
Ni	High	□		
Co	Medium	×		
Cu	High	○	○	There is a problem on ^{63}Cu (n, 2n) for dosimetory.
Zr	Medium	×		
Nb	High	△		
Mo	High	×		Evaluators understand the status.
W	High	×		
Pb	High	△		Addition of more inelastic levels is required. Re-evaluation is desirable for total, elastic and inelastic cross section.

* Above status is judged considering the importance and priority of each nuclide.

○ : good, □ : almost good, △ : fairly good, × : not good.

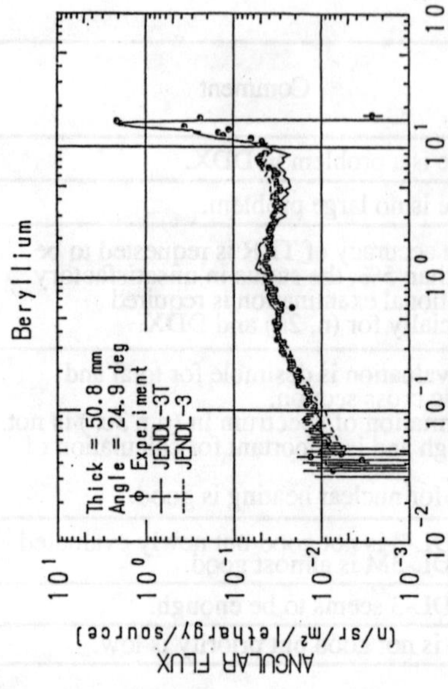

Fig. 1 Angular neutron flux spectrum on Be slab.

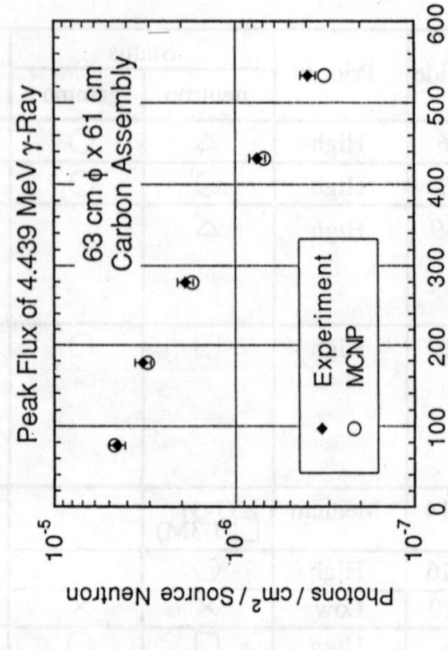

Fig. 2 Gamma-ray flux distribution in graphite assembly.

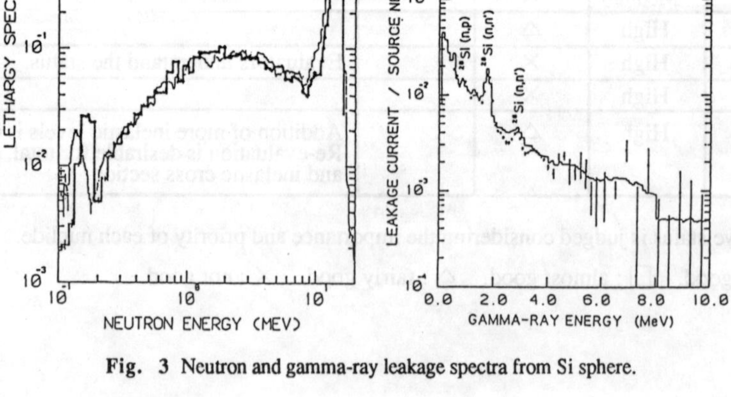

Fig. 3 Neutron and gamma-ray leakage spectra from Si sphere.

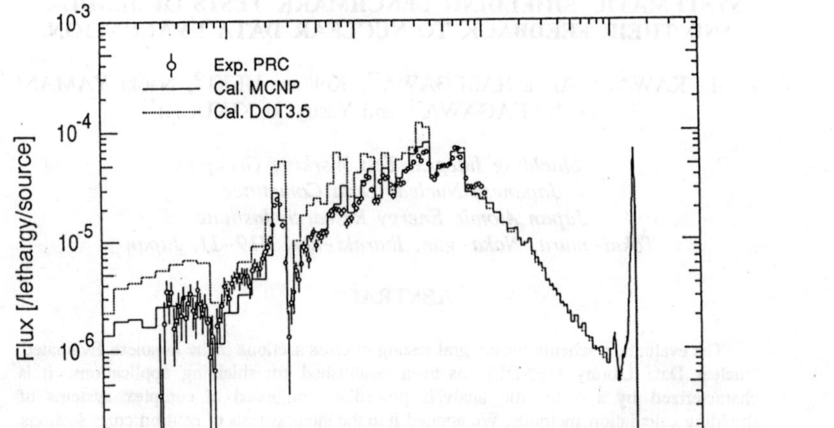

Fig. 4 Measured and calculated neutron spectra at 0.41 m from front surface of iron assembly.

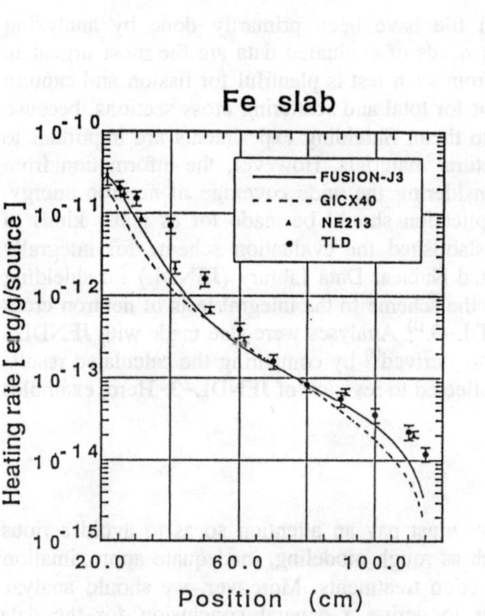

Fig. 5 Gamma-ray heating rate distribution in Fe assembly.

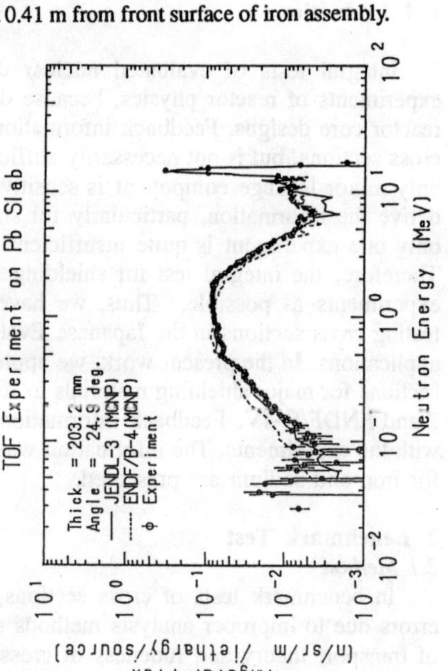

Fig. 6 Angular neutron flux spectrum on Pb slab.

SYSTEMATIC SHIELDING BENCHMARK TESTS OF JENDL–3 AND THEIR FEEDBACK TO NUCLEAR DATA EVALUATION

Masayoshi KAWAI[*1], Akira HASEGAWA[*2], Kohtaro UEKI[*3], Naoki YAMANO[*4], Tsuneo NAKAGAWA[*2] and Yasuyuki KIKUCHI[*2]

Shielding Integral Test Working Group
Japanese Nuclear Data Committee
Japan Atomic Energy Research Institute
Tokai-mura, Naka-gun, Ibaraki-ken, 319-11, Japan

ABSTRACT

The evaluation scheme for integral testing of cross sections in the Japanese Evaluated Nuclear Data Library (JENDL) has been established for shielding applications. It is characterized by a systematic analysis procedure composed of complex systems of shielding calculation methods. We applied it to the integral tests of neutron cross sections for major shielding materials in JENDL–3 to obtain feedback information. General confirmation of the reliability of JENDL–3 was obtained, but several kinds of problems for the total and inelastic scattering cross sections were pointed out. In this report, the example for sodium and iron is described including the data revision for JENDL–3.2.

1. Introduction

Integral tests of evaluated nuclear data file have been primarily done by analyzing experiments of reactor physics, because data needs of evaluated data are the most urgent in reactor core designs. Feedback information from such test is plentiful for fission and capture cross sections, but is not necessarily sufficient for total and scattering cross sections, because only minor leakage component is sensitive to them. Shielding experiments are important to derive the information, particularly for structural materials. However, the information from only one experiment is quite insufficient considering the wide coverage of neutron energy. Therefore, the integral test for shielding application should be made for as many kinds of experiments as possible. Thus, we have established the evaluation scheme for integrally testing cross sections in the Japanese Evaluated Nuclear Data Library (JENDL) for shielding applications. In the present work, we applied the scheme to the integral tests of neutron cross sections for major shielding materials in JENDL–3.[(1)] Analyses were also made with JENDL–2 and ENDF/B–IV. Feedback information was derived[(2)] by comparing the calculated results with the experiments. The information was reflected to revision of JENDL–3. Here, examples for iron and sodium are presented.

2. Benchmark Test

2.1 Method

In benchmark tests of cross sections, we must pay an attention so as to avoid serious errors due to improper analysis methods such as rough modeling, inadequate approximation of transport theory and rudeness of cross section treatments. Moreover, we should analyze the experiments of different types in order to derive a general conclusion for the data

---

1) Toshiba Corp., Kawasaki–ku, Kawasaki–shi, Japan, 2) JAERI, Tokai–mura, Ibaraki–ken, Japan, 3) Ship Research Institute, Mitaka–shi, Tokyo, Japan, and 4) Sumitomo Atomic Energy Industries, Ltd., Chiyoda–ku, Tokyo, Japan.

reliability. Accordingly, we have prepared the comprehensive system for shielding benchmark test composed of three kinds of shielding analysis systems: (a) the most popular system of Sn transport codes and Monte Carlo code using "ANISN" type cross sections expressed with the spherical harmonics expansion series (such as DOT3.5), (b) RADHEAT–V4 system[3] using the energy–angle double differential group cross sections (such as DIAC), and (c) a pointwise Monte Carlo code, MCNP. In the systems (a) and (b), resonance self–shielding effects are treated with the Bondarenko procedure to calculate effective macroscopic cross sections. The most appropriate system was selected by considering the experimental geometry and the source distribution of each benchmark.

Analyzed shielding benchmark experiments are generally characterized as having simple geometries, well described experimental methods and results. In the present work, the ORNL Broomstick experiment[4] and four kinds of transmission experiments have been analyzed.

2.2 Analyses of Broomstick Experiments

Broomstick experiments were performed using long stick samples to investigate the effect of minima in the total neutron cross sections in MeV energy region for the typical shielding materials. Since this experiment is of good geometry, no transport calculations are needed for the analysis. The calculation consists from two parts: (1) to determine a transmitted uncollided spectrum and (2) to fold the uncollided spectrum, obtained from the first step calculation, with the resolution function of the spectrometer system used in the experiments.

The results for sodium in 60.6 cm thickness (actually stick length) are shown in Fig. 1. From the figure, we can see the spectrum by all evaluations are overestimated systematically: C/E=1.243±0.189, 1.242±0.188 and 1.217±0.215 for JENDL–3, JENDL–2 and ENDF/B–IV, respectively. The results of JENDL–2 is almost the same as JENDL–3. On the other hand, there are clear differences between JENDL and ENDF/B–IV in the energy range from 6 to 10 MeV. In this range, ENDF/B–IV has better agreement than JENDL.

For iron, there are two experiments with different sample thickness, 20.3 cm and 30.5 cm. The typical result for thin sample is given in Fig. 2. The C/E values of uncollided fluxes calculated with JENDL–3 are 0.927±0.126 and 0.936±0.242 for thin and thick samples, while those for ENDF/B–IV are 0.936±0.086 and 0.986±0.176. From these results, we can draw following observations: For the thick sample case, the agreement by ENDF/B–IV is better than JENDL. The values of C/E of JENDL–3 is too small (about 0.7) in the energy range between 1 to 3 MeV compared with the results of ENDF/B–IV. They are too large (about 1.2) in the energy range between 8 MeV and 10 MeV, although the discrepancy observed in the higher energy part with JENDL–2 is diminished in JENDL–3 results. Such tendency of the C/E value increasing with energy is much milder for the ENDF/B case.

2.3 Analyses of Transmission Experiments

The following benchmark experiments were analyzed: (1) ORNL neutron transmission experiments[5,6] for sodium and iron, (2) ASPIS deep penetration experiments[7] for iron, (3) KfK leakage spectrum measurements[8] from iron spheres, (4) LLNL pulsed sphere experiments[9] for iron using D–T neutron source. In the analyses for the first two experiments, DOT3.5 and MCNP were used. For the latter two experiments, DIAC and MCNP were used, respectively. As for DOT3.5 calculations, anisotropy of cross section was taken up to P5. From the analyses, the following results were obtained.

(1) Sodium

In the analyses of the ORNL deep penetration experiments for sodium, the results were influenced by the contents of hydrogen impurity in the sodium sample, as shown in Fig. 3

for example of Bonner ball response for 15–ft thick sample. Here, we assume that the actual hydrogen content is near to the reported maximum value. For 10–ft and 15–ft long samples, JENDL–3 reproduced fast neutron spectrum above 70 keV with C/E = 1.0 to 1.3, and the responses for 3 to 10 inches diameter Bonner ball detectors and cadmium covered BF_3 counter with the average C/E = 0.79 to 0.98. Comparison of the results by JENDL–3 with those by JENDL–2 and ENDF/B–IV was tentatively made in the case without hydrogen. JENDL–3 gives larger C/E values (0.91 to 1.10 and 1.07 to 1.33 for the 10–ft and 15–ft samples, respectively) than in the case with the maximum content. Even in this case, ENDF/B–IV and JENDL–2 underestimated the Bonner ball responses for the 10–ft sample by 10–20% and 10–30%, respectively.

(2) Iron

Figure 4 compares the calculated spectra with the experimental data at 57.1 cm and 85.7 cm penetration positions in the iron stack of the ASPIS experiments. For small penetration, general agreement is observed in all libraries, but for deep penetration, the large discrepancy exists in the energy range between 600 keV and 2 MeV. The C/E values of the reaction rates for the threshold reactions of ^{103}Rh and ^{115}In also show a trend of decreasing from 1.2 to 0.4, 1.1 to 0.6 for JENDL–3, 1.4 to 0.6 and 1.1 to 0.9 for ENDF/B–IV, respectively. ENDF/B–IV gives better results than JENDL in these high energy region, while the agreement between calculation and experiment in the spectrum just below the 24 keV window is superior for JENDL–3. For the ORNL iron benchmark, the underestimation also appears above 700 keV.

For relatively thin transmission problem of the KfK benchmark using different size of iron sphere (10 cm to 40 cm in diameter), the neutron leakage spectra calculated with JENDL–3 and ENDF/B–IV show good agreement in detailed spectrum shape due to resonances in the energy region between 100 keV and 1 MeV. The C/E values for the various energy ranges distribute from 0.95 to 1.35 with JENDL–3 and from 0.95 to 1.24 with ENDF/B–IV.

In the calculation with MCNP for the LLNL pulsed sphere benchmark, a good agreement was obtained with JENDL–3 to the elastic peak including uncollided source neutrons and the multi–scattering region, except for the energy region between 7.5 and 12.5 MeV.

3. Feedback Information and JENDL–3 Revision

3.1 Sodium

For the ORNL transmission experiments, JENDL–3 gives the C/E values of closest–to–unity and is concluded better than ENDF/B–IV and JENDL–2. However, there are some problems in sodium cross sections of JENDL–3. The overestimation of MeV neutron transmission is commonly observed in the calculated neutron spectra for the Broomstick and ORNL benchmarks. These fast neutron fluxes are the most sensitive to the total cross section. The analyzed results suggest that total cross sections of JENDL–3 and ENDF/B–IV evaluated on the basis of the old measurements are too large; this suggestion was confirmed by the recent data of Larson et al.,[10)] as shown in Fig. 5. Therefore, revision of total cross section was made by tracing the data of Larson et al. with eye–guide technique. Revised data will be published in JENDL–3.2.

In Fig. 1 for the result of Broomstick benchmark, it is clearly observed that JENDL–3.2 diminishes the overestimation of transmitted neutron spectrum.

3.2 Iron

Comparison of the JENDL–3 results on the Broomstick benchmark with those of JENDL–2 shows that the modification of total cross section above about 5 MeV improved to diminish the overestimation of fluxes by JENDL-2. However, the underestimation (C/E=about 0.7) of fluxes in the energy range between 1 and 3 MeV suggests that the total cross section of JENDL-3 is too large or cross-section minima are too shallow. JENDL-3 data were evaluated by tracing the experimental raw data with eye-guide technique as shown in Fig. 6. However, the measured data are usually smeared because of the finite energy resolution of detector system and electronics. Accordingly, the total cross section was reanalyzed by unfolding the energy resolution function of measured data. The unfolded result, denoted by JENDL-3.2, is compared with the previously evaluated data and the experimental data. It has shaper resonance structure than JENDL-3. Figure 2 shows that the result of JENDL-3.2 was remarkably improved from JENDL-3 for the Broomstick benchmark.

For deep penetration problems, the ASPIS and the ORNL benchmarks give good information. From these results, overall applicability of JENDL-3 for natural iron has been confirmed for the deep penetration in the shielding calculation. The agreement between calculation and experiment in the spectrum just below the 24 keV window is superior for JENDL-3 evaluation. It means the evaluation of the 24 keV window is very good for JENDL-3. The higher energy above 3 MeV, JENDL-3 gives better results than ENDF/B-IV.

However, underestimation between 600 keV and 1.2 MeV in deep penetration for the ASPIS benchmark, and 600 keV and 3 MeV for the ORNL benchmark suggests some problems still remain in the assignments between elastic and inelastic scattering cross sections in the energy range between 600 keV and about 1 MeV where group constants for the total cross section are not so different between JENDL-3 and ENDF/B-IV.

For D-T neutron source problem of the LLNL pulsed sphere benchmark, good result was obtained, except for the energy region between 7.5 and 12.5 MeV. In this region, a slight underestimation was shown in the calculation results with JENDL-3. It means that the preequilibrium and direct components of inelastic scattering cross section may be underestimated in JENDL-3.

4. Concluding Remarks

Shielding benchmarks are very effective to integrally test the total and scattering cross sections for structural materials. Systematic benchmark tests with various integral data on the same material can give substantial information about the validation of nuclear data, so that the evaluation scheme established in this study is essential for the nuclear data evaluation.

5. References

1) K. Shibata, et al., *JAERI-1319* (1990).
2) M. Kawai, et al., "*Shielding Benchmark Test of JENDL-3,*" to be published as JAERI-report.
3) N. Yamano, et al., *JAERI-1316* (1989).
4) E.A. Straker, *ORNL-TM-2242* (1968).
5) R.E. Maerker, et al., *Nucl. Technol.,* **22**, 275 (1974).
6) R.E. Maerker and F.J. Muckenthaler, *Nucl. Sci. Eng.,* **52**, 227 (1973).
7) M.D. Carter and A. Packwood, *Proc. of the Specialists' Meet. Sensitivity Studies and Shielding Benchmarks*, Paris, October 1975, p.119 (1975), OECD.
8) H. Werle, et al., *NEACRP-U-73* (1976).
9) L.F. Hansen, et al., *Nucl. Sci. Eng.,* **51**, 278 (1973).
10) D.C. Larson, et al., *ORNL-TM-5614* (1976).

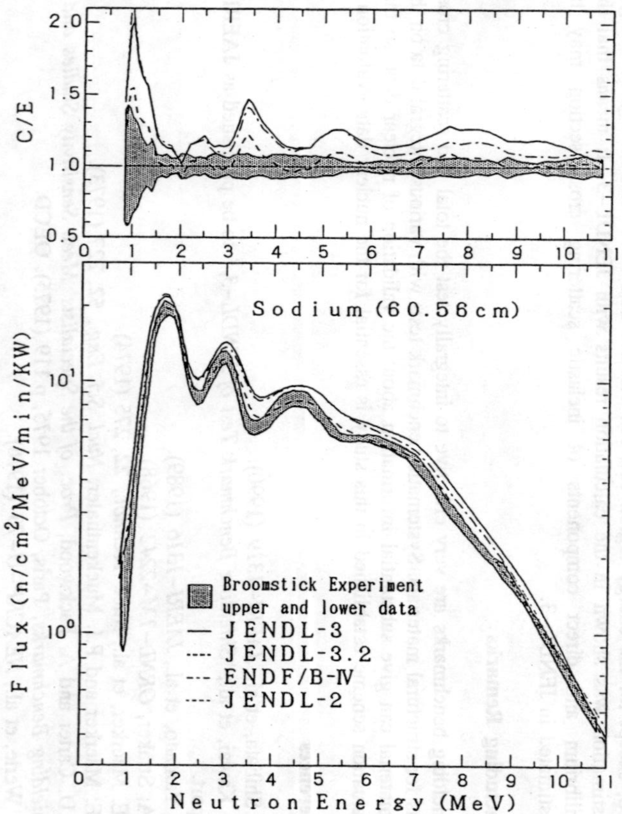

Fig. 1 Comparison of Measured and Calculated Uncollided Flux for Sodium Benchmark.

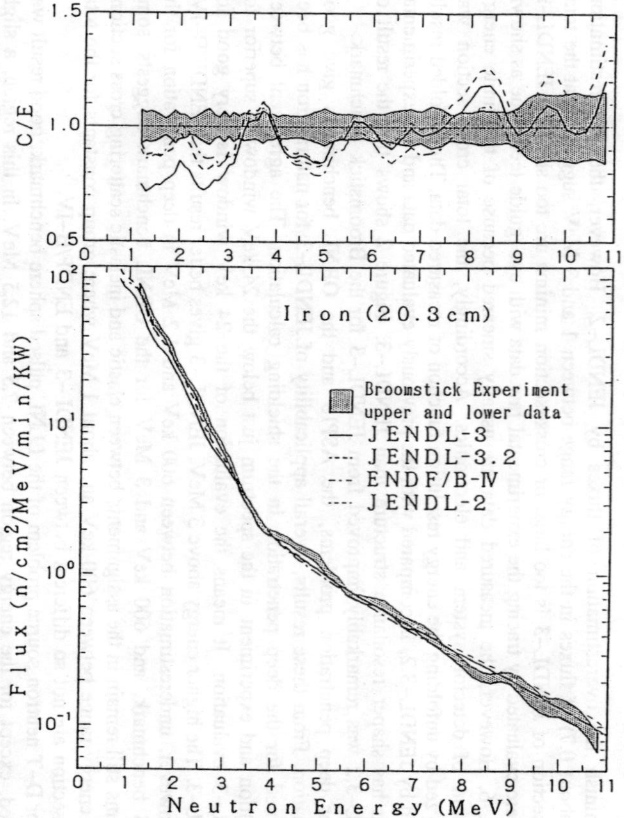

Fig. 2 Comparison of Measured and Calculated Uncollided Flux for Iron Benchmark.

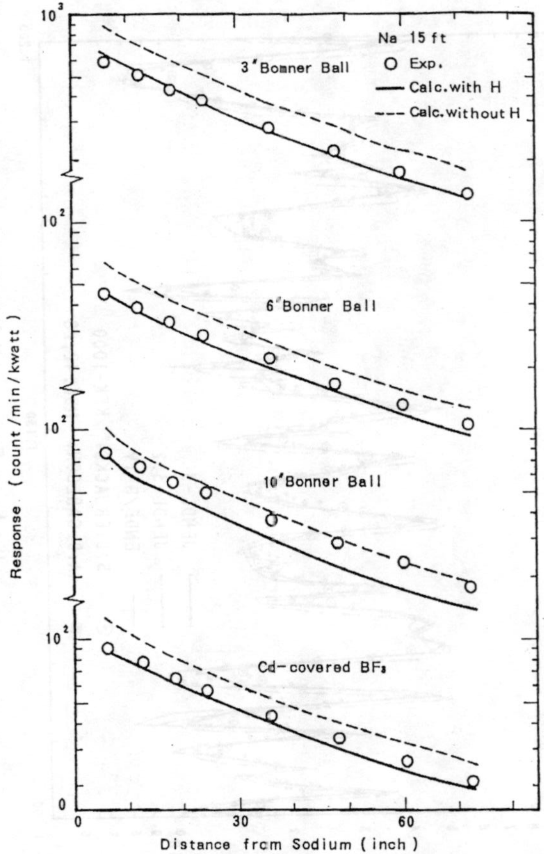

Fig. 3 Comparison of Bonner Ball Responses for Sodium Benchmark

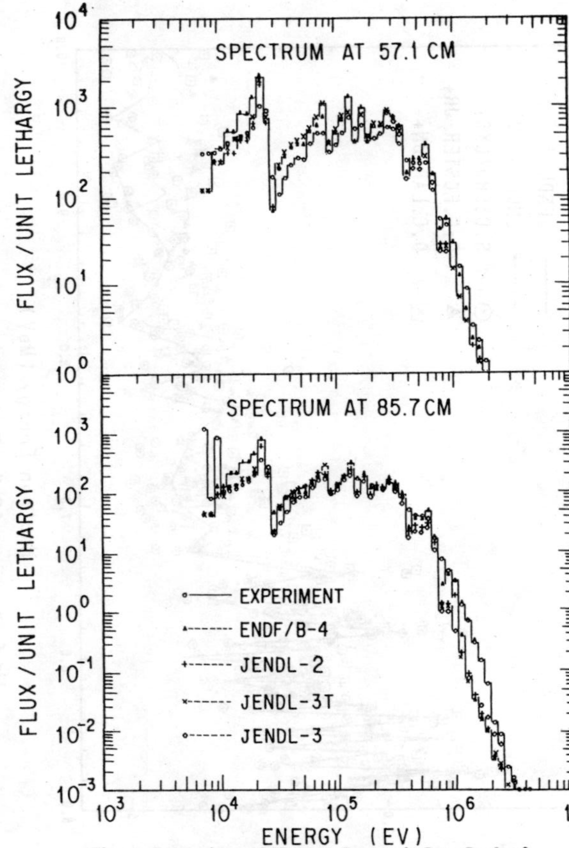

Fig. 4 Comparison of Neutron Spectra in Iron Stack of ASPIS Experiment.

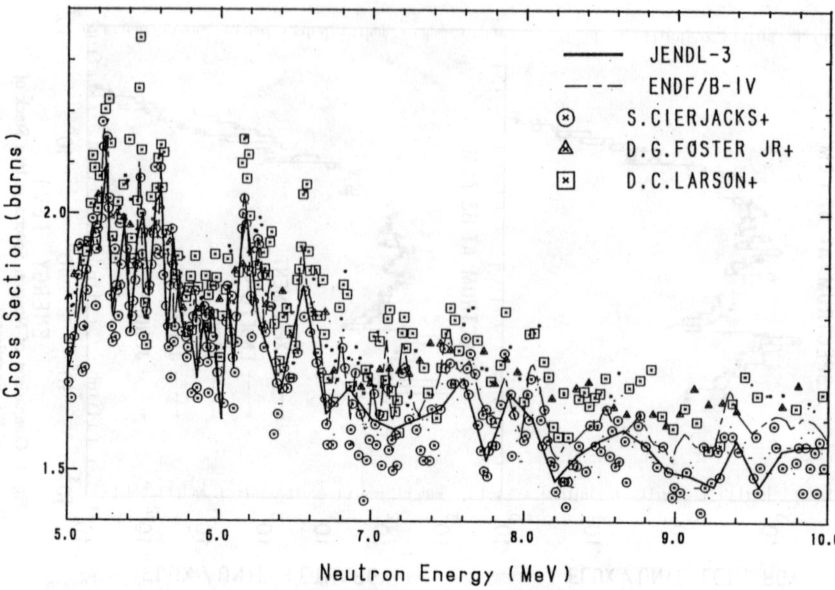

Fig. 5 Comparison of Sodium Total Cross Sections.

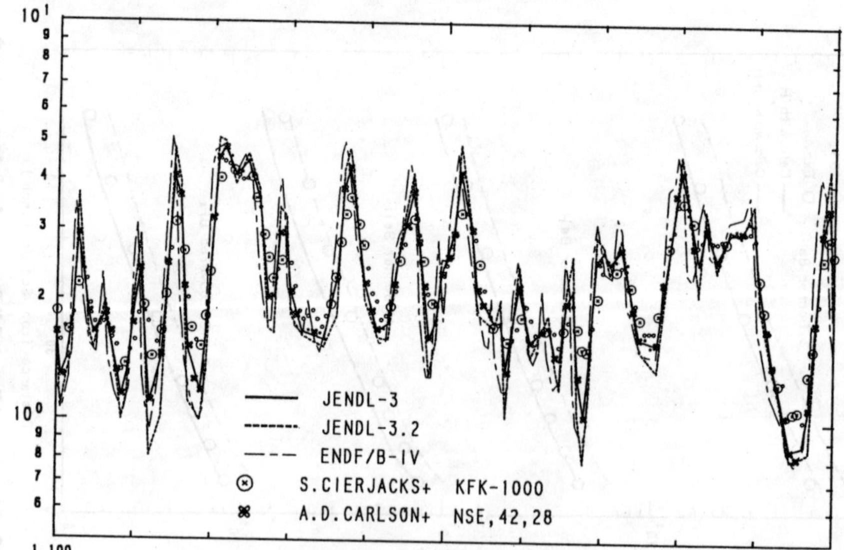

Fig. 6 Comparison of Iron Total Cross Sections.

INTEGRAL DATA TESTING FOR THERMAL REACTORS AND FEEDBACK INTO JEF2

H. Tellier and J. Vanuxeem

Service d'Etudes de Réacteurs et de Mathématiques Appliquées
Centre d'Etudes Nucléaires de Saclay, 91191 GIF-sur-YVETTE, FRANCE

INTRODUCTION

Nowadays, the calculation of a nuclear reactor core is generally performed by solving the Boltzmann equation. According to the computer code we solve the integral form or the integro-differential form of the transport equation. But in any case we need numerous numerical data : the geometrical and chemical data, and the neutron and nuclear data. The former data represent the dimensions of the cell and the core, the chemical and isotopic composition of the fuel, the structures and the moderator. They are generally known with a good accuracy. The latter data represent the neutron cross sections and the nuclear properties of the various nuclides. They are not always known with an accuracy as good as the reactor physicist wishes. The neutron data are generally deduced from direct nuclear measurements. These measurements give the variation of the nuclear properties of the nuclides versus the incoming neutron energy : they are the differential experiments. Generally, the shape of the cross sections can be more or less well achieved. But very often, it is difficult to measure the absolute magnitude of the nuclear data with the accuracy required by the physicist needs. Some of the best estimated values of the evaluated files can have an uncertainty which is too large. To improve the knowledge of the nuclear parameters we must use another type of experiment : the integral experiments. In this kind of experiments we use a mock-up of a reactor or the reactor itself and we measure some synthetic parameters which are representative of the neutron properties of the cell or of the reactor for the actual neutron spectrum. For example, we can measure criticality factor, buckling, reaction rates or irradiated fuel composition.

If we choose integral experiments with very simple geometries and asymptotic spectrum such as uniform lattices or homogeneous media we can perform the calculation of these experiments without numerical approximations. In these conditions, it is possible to use very sophisticated options of the reactor computation codes.

So the difference between the computed values of the synthetic parameters and the experimental ones can be attributed to the inaccuracy of the neutron data used in the calculation. These neutron data depend on the evaluated files but also on the processing. It is very important to handle very carefully the processing codes which generate the multigroup cross sections used by the reactor physics. The second important thing is to solve the Boltzmann equation without bias. For this it is necessary to check the geometrical description of the cell and the energy mesh of the multigroup cross sections for each integral experiment. Only if all these conditions are achieved, we can deduce modifications of the neutron data from the difference between the computed and the measured values of a set of integral experiments. For this purpose, we can use the tendency research method [1,2]. In this method, the difference between the computed and the measured values of an integral parameter, effective multiplication coefficient for example, is expressed as a linear function of the cross section modifications. The cross section modifications are obtained by a least square computation. Therefore, it is necessary to have different sensitivity coefficients. For the thermal neutron reactor this condition is achieved by using multiplying media with different kind of moderator : heavy water, light water and graphite, and for each moderator we use several fuel over moderator ratios and various fuel enrichments. In addition, the number of integral experiments must be much higher than the number of sensitive neutron data.

In the following sections, we will describe the main synthetic parameters which govern the physics of the thermal neutron reactors, the set of integral experiments which were used in this study and the results obtained in the case of the JEF2 benchmarking. The fundamentals of the tendency research method are briefly recalled in the appendix.

SENSITIVE NEUTRON DATA

To validate the neutron data, with integral experiments, we must use multiplying media with a very simple geometry. Consequently, we can perform the computation of a unique cell or of a fuel assembly in its asymptotic neutron spectrum. In these conditions we can use the more sophisticated options of the reactor physics code and this without geometrical approximation. For the thermal reactor benchmarking of JEF2, we used the recent Apollo2 code [3]. In Apollo2, which solves the integral Boltzmann equation by the collision probability method in the multigroup approximation, the resonance range is treated with an improved self-shielding formalism [4,5]. Now, it is possible to use several self-shielding areas inside the fuel pin and the equivalence procedure between the heterogeneous lattice and an homogeneous medium can be made group by group with various slowing down models. For this work, we used six different self-shielding areas in the rods and the equivalence was performed with the statistical model for the high energy resonances and the wide resonance approximation for the low energy resonances. These new possibilities have a significant impact on the benchmarking results in the resonance energy range because the effective integrals are computed with a better accuracy than in the past, where we were able to use only one self-shielding domain [6].

The total energy range is described by 172 groups. That is to say, that in the neutron library, each cross section of each isotope is represented by 172 values and that the various transfer matrices have a very high order. With integral experiments, it is not possible to obtain informations for all the cross sections and all the groups and for each matrix element. But the differences between the computed values of integral results and the measured ones can be explained with a smaller number of parameters which globally represent the neutron behaviour of a multiplying medium. They are the synthetic parameters and it is for these parameters that we will be able to deduce informations. The nature and the number of the synthetic parameters depend on the reactor types which are studied. As far as the heavy nuclei are concerned and in the case of thermal neutron reactors, we can divide the energy range in three domains :

- The energy range above 50 keV which is characterized by smooth variations of the cross section. It is mainly the fission cross sections which are important in this domain.

- The resolved and unresolved resonance energy range between 1 eV and 50 keV, in which both the resonance parameters and the self-shielding formalism are involved. The effective integrals constitute the convenient parameters in this domain.

- The thermal and subthermal energy range below 1 eV where the cross section have also a smooth behaviour. For these energies, the cross section shapes are generally well known through the differential measurements. Consequently, we have adopted the cross section shapes of the evaluated file, JEF2 in our case, and have chosen the magnitudes of the cross section for a 0.025 eV neutron as sensitive parameters.

- For the plutonium 240, the low energy range is governed by the 1.056 eV resonance and the most important parameters are the radiative capture width of the resonance and the thermal absorption cross section.

Another quantity is important for the fissile nuclei, the average number of neutrons which are emitted in a fission, nu-bar. As in the thermal neutron reactor it is mainly the low energy range which is preponderant, we have adopted the nu-bar shape of the evaluated file and chosen as synthetic parameter the nu-bar value for a 0.025 eV neutron.

For the moderator (light water, heavy water or graphite), the low energy behaviour can be characterized by the thermal absorption cross section and the slowing down and diffusion properties by the migration area. The thermal absorption of the heavy water is too few sensitive to be kept as a synthetic parameter.

Finally 29 synthetic parameters were taken into account for the validation of JEF2. They are given in table I.

INTEGRAL EXPERIMENTS

Three essential conditions must be satisfied to obtain accurate tendencies on the basic neutron data : very simple experiments with an asymptotic neutron spectrum, different sensitivity coefficients and the possibility to disconnect the effects of uranium from the ones of plutonium and thorium. The asymptotic neutron spectrum can be observed in critical facilities which used cells with only one kind of fuel. In this case, the buckling can be define without ambiguity and the reactor calculation can be made with a one cell computation. If this buckling is measured with a very good accuracy, this type of clean experiment is very interesting for the tendency research. Unfortunately they are scarce. The homogeneous experiments for which we known exactly the critical composition and the geometrical dimensions are also interesting because some of them contain only plutonium and for that they are not perturbed by uranium. At last various sensitivity coefficients can be obtained with lattices moderated by heavy water, light water or graphite and several moderating ratios.

The integral data set which we used for the JEF2 benchmarking can be split into three parts : the multiplying media which only contain uranium (uranium 235 alone or a mixture of uranium 235 and uranium 238), the ones which contain plutonium and finally the lattices representative of the thorium cycle. In each case we had at our disposal experimental value of the buckling which corresponds to the fondamuntal mode. The various multiplying media which cover a wide range of neutron spectra, from the well thermalized lattices (95% of this neutron are slowed down below 2.8 eV) to the hard spectra of the tight pitch lattices (only 25% of the neutron are slowed down below 2.8 eV) are briefly described below.

a) **Uranium**

Independantly of the moderating ratio, the uranium lattices depends on the nature of the moderator and the enrichment in uranium 235. They are :

- 7 heavy water lattices with natural uranium metallic fuel. These very thermalized media are mainly sensitive to the low energy range of the cross sections.

- 4 natural uranium and graphite lattices. They are also sensitive to the low energy range but in association with the preceding experiments they allow to separate the effect of the heavy water from the one of the graphite.

- 1 light water moderated lattice with metallic fuel containing 1,3% of uranium 235.

- 9 uranium dioxide lattices containing from 1,4 % to 2,7 % of uranium 235 and moderated by light water. In these experiments are included the well known TRX benchmarks.

- 6 uranium dioxide lattices for which the epithermal range was enhanced by using various tight pitches.

- 9 recent french critical experiments which were performed in the frame of the light water reactor studies.

- 2 homogeneous multiplying media of Oak Ridge which used 93% enriched uranium. These experiments are very interesting because they are very few sensitive to uranium 238.

- 1 critical facility with bucklings mesured between 20°C and 210°C : the swedish experiment KRITZ.

b) **Plutonium**

The plutonium fuels are characterized by two important parameters : the amount of plutonium or the plutonium over uranium ratio and the isotopic composition of the plutonium. The fuel can contain a more or less important quality of plutonium 240 and the lattices can be more or less sensitive to this isotope. Consequently we have used multiplying media with various plutonium over uranium ratios and various isotopic compositions.

The integral experiments used for this study are :

- 6 heavy water moderated lattices with metallic fuel which only contain 0,4% of plutonium with 6% of isotope 240.

- 14 light water moderated lattices which contain from 1,5 to 4,0% of plutonium. The proportion of plutonium 240 is equal to 8 or 22%.

- 5 homogeneous media with 3 and 5% of plutonium 240. These multiplying media are important for the plutonium qualification because they do not contain uranium.

- 3 tight pitch lattices with 11% of plutonium This experiments were performed in France, in the frame of the plutonium recycling studies.

c) **Thorium cycle**

Since several years the thorium cycle has not a high priority. The experiments which are related to this cycle are generally old, scarce and not always accurate enough. The uranium 233 and thorium validation is based on four sets of critical or exponential experiments.

- 2 light water moderated lattices the fuel of which is a mixture of uranium 235 and thorium. They are critical experiments.

- 7 exponential experiments in which the fuel is composed of uranium 233 and thorium. The moderator is light water.

- 5 exponential experiments which used the same uranium 233 and thorium fuel but heavy water as moderator.

- 2 homogeneous media composed by a solution of uranium 233 nitrate in light water.

Finally we had at our disposal 83 buckling measurements. To these integral experiments we added results coming from spent analysis and spectrum indices which gave informations about the radiative capture cross sections or the radiative capture over fission ratios of various fertile and fissile nuclei.

RESULTS AND ANALYSIS

All the eighty-three critically factor computations were used in a single least square calculation. For a first run, the twenty-nine synthetic parameters were considered as free parameters. A detailed analysis of this preliminary result showed that the observed tendencies were small and inaccurate in the case of several parameters. The statistical uncertainties deduced from the least square calculation were much greater than the magnitude of the suggested changes. Consequently, these small modifications can be considered as having no physical meaning and there is no inconvenience to fix these parameters and to keep their initial JEF2 values. Then a second least square calculation was performed with a reduced number of free parameters. In comparison with the first computation we observe a very small increase of the chi-square. This small increase justifies the reduction of the free parameter number.

As far as the thermal neutron reactors are concerned the tendency research suggests meaningful modifications for only a small number of sensitive parameters : the migration area of the light water and of the heavy water, the effective integral of the fertile nuclei uranium 238 and thorium 232 and the thermal data of uranium 233. For all the other parameters, the required modifications are very small and the best estimated values are very close to the initial values of the file. If we take into account the suggested modifications, the computation of the whole set of integral experiments is satisfactory as it can be seen on figures 1 through 3. On these figures we have displayed the difference between the computed value k_{eff} and experimental value which is equal to unity. The error bars represent the experimental uncertainty. Each integral experiment is identified by the slowing down density q. It is the number of neutrons which become thermal for one emitted fission neutron. This quantity is representative of the neutron spectra. The high values of q correspond to the soft spectra of the well thermalized lattices (mainly with heavy water or graphite as moderator) and the low values to the hard spectra of the tight pitch lattices (light water with a low moderator to fuel ratio). The figure 1 is relative to the multiplying media which contain only uranium. The average discrepancy of K_{eff}-1 is equal to -14.10^{-5} and the dispersion is equal to 482.10^{-5}. We do not observe any shift as a function of the spectrum hardness. On figure 2, we have the same representation for the multiplying media which contain, only plutonium or a mixture of plutonium and uranium. In this second case, the average k_{eff}-1 is $(182 \pm 535).10^{-5}$. For the integral experiment relevant to the uranium 233-thorium cycle the comparison is displayed on figure 3. The average k_{eff}-1 is equal to $(-275 \pm 640)10^{-5}$. For this case the dispersion is higher than for the uranium and plutonium cycles. But the integral experiments are old and not always accurate enough because they are mainly exponential experiments. For the whole set of critically factors the average deviation is $(-5 \pm 550)10^{-5}$. The statistical distribution of the reduced residual deviations $|k_{eff}-1|/\Delta K_{eff}$ is given in figure 4.

The dotted line is the Gauss distribution. We can observe a slight difference with a pure Gauss distribution. There is an excess of large deviations. This can be explained by the fact that some experimental error bars are likely too small. Nevertheless the results of the tendency research seem very satisfactory. They are the following ones.

For the major fissile nuclei uranium 235 and plutonium 239 the requested modifications are very small and within the error bars. Consequently, it is possible to keep the initial values of JEF2 for the thermal neutron reactors. The tendency research values are given in table II and we can see that the agreement with the initial values is very good. In the case of uranium 233, the situation is less satisfactory because the analysis of the integral experiments suggests important modifications of nu-bar and of the radiative capture cross section. For these two parameters the obtained values are incoherent with the initial ones.

For the fertile nuclei, uranium 238 and thorium 232, the problem is a little bit more complicated than for the fissile nuclei. The sensitive parameters are the effective integrals and the computed value of these quantities is strongly linked to the formalism and approximations which are used in the reactor core calculations. Even with the very sophisticated options of Apollo2 that we have used, the analysis of the integral experiments suggests a slight decrease of the fertile nuclei effective integral as it can be seen in table III. Let us notice that the uncertainties which are associated with these modifications remain important. They do not allow to conclude with assurance that the basic data of the fertile nuclei must be revised. Further investigations are needed before to endorse such a requirement.

For the other actinides, the sensitivity of the integral experiments are not important enough to obtain significant modifications of the neutron data.

CONCLUSION

The use of integral experiments and of the tendency research method, or a similar one, is a very efficient tool to improve the basic neutron data. It is a complementary approach to the microscopic cross section measurements and it gives important informations to the evaluators. In the case of JEF2 it was taken into account the most recent differential measurements of the major actinides and the benchmarking results of the preliminary version of the file was also used. We have now at our disposal a file which gives entire satisfaction for uranium 235 and plutonium 239 as far as thermal neutron reactors are concerned. No modification of the low energy part of these two evaluations is needed. For uranium 233 we observe a significant discrepancy between the thermal data deduced from the differential measurements and the values obtained with the integral experiments. In case of thorium cycle revival, a revision of the uranium 233 evaluation will be probably necessary, jointly with new and more accurate integral experiments.

REFERENCES

[1] **P. REUSS** : "La méthode de recherches de tendances". Report CEA-N 2222 (1988)

[2] **P. REUSS** : "The Search of tendencies", Thermal Reactor Benchmark calculation, EPRI-NP 2855,10 (1983)

[3] **R. SANCHEZ, J. MONDOT, Z. STANKOVSKY, A. COSSIC** and **I. ZMIJAREVIC** : "Apollo2 : A User Oriented Portable Modular Code for multigroup Transport Assembly Calculations", Advances in Reactor Physics Mathematics and Computations. Paris, III, 1563, (1987).

[4] **M. COSTE, R. SANCHEZ, Z. STANKOVSKY, C. VAN DER GUCHT** and **I. ZMIJAREVIC** : "Apollo2 Assembly Spectrum Code : New features", Advances in Mathematics, Computations and Reactor Physics-Pittsburg, II, 10.1.4 (1991).

[5] **M. COSTE, H. TELLIER, C. BRIENNE-RAEPSAET and C. VAN DER GUCHT** : "Heavy Nucleus Resonance Absorption in Heterogeneous lattices", Advances in Reactor Physics, Charleston, II, 363 (1992).

[6] **H. TELLIER, C. VAN DER GUCHT AND J. VANUXEEM** : "Integral validation of the JEF2 major actinides for thermal neutron reactors", Advances in Mathematics, Computations and Reactor Physics - Pittsburg, IV, 19.1.2 (1991).

APPENDIX

FUNDAMENTALS OF THE TENDENCY RESEARCH METHOD

For each integral experiment (criticality factor, reaction rate ...), we know the experimental result Y_i and the measurement uncertainty E_i. In any case, we can compute the same quantity which is a function of the neutron parameters x_k. The result of this calculation is F_i (..., x_k, ...). If we change the value of the neutron parameter x_k, which becomes $x_k + \Delta xk'$, the result of the computation is now $F_i(\ldots, xk + \Delta x_{k'}, \ldots)$.

The principle of the tendency research method is to choose the modification Δx_k of the neutron parameters in such a way that the quantity

$$Q = \sum_i \frac{1}{E_i^2} [Y_i - F_i(\ldots, x_k + \Delta x_k, \ldots)]^2$$

for all the set of integral experiments becomes minimum. Nowadays the magnitude of the main neutron cross sections are more or less well known. So, the modifications Δx_k are expected to be small and we can make a first order expansion of the computed value.

$$F_i(\ldots, x_k + \Delta x_k, \ldots) = F_i(\ldots, x_k, \ldots) + \sum_k \Delta x_k \frac{\partial F_i}{\partial x_k}$$

We can also replace the partial derivatives by the sensibility coefficients

$$S_{ik} = \frac{\Delta f_i}{\Delta x_k}$$

These sensibility coefficients (variation of the integral quantity F_i for a one per cent change of the parameter x_k) can be computed by the perturbation theory or a variational method.

With these assumptions we must now minimize the quantity

$$Q = \sum_i \frac{1}{E_i^2} [Y_i - F_i(\ldots, x_k, \ldots) - \sum_i S_{ik} \Delta x_k]^2$$

or if ΔY_i represents the difference between the experimental result and the computed value for the integral experiment i

$$Q = \sum_i \frac{1}{E_i^2} [\Delta Y_i - \sum_k S_{ik} \Delta x_k]^2$$

The minimization is done with the least square method. That is why, if we want to determine the modification Δx_k with a good accuracy, it is absolutely necessary to use a set of integral experiments for which the sensitivity coefficients are as different as possible. We can obtain different sensibility coefficients by using integral experiments corresponding to various types of reactors.

From the mathematical point of view, the least square calculation leads to the Δx_k values which minimize the quantity Q. But we must take two remarks into account. First, the Δx_k are assumed to be small (don't forget that we have made a first order expansion of F_i). Secondly the cross sections are measured by differential experiments with an experimental uncertainty ϵ_k. The Δx_k must be lower or of the same order than ϵ_k. This is why, instead of minimizing the Q value, we prefer minimize the following quantity :

$$Q' = \sum_i \frac{1}{E_i^2} [\Delta Y_i - \sum_k S_{ik} \Delta x_k]^2 + \lambda \sum_k \left(\frac{\Delta x_k}{\epsilon_k}\right)^2$$

In this expression λ is a free parameter which represent the importance given to the microscopic data in the tendency research.

TABLE I

SENSITIVE PARAMETERS FOR THERMAL NEUTRON REACTORS

Nucleus	Thermal Range	Resonance Range	High energy range
U 233	ν, σ_f, σ_a	$I^a{}_{eff}$, $I^f{}_{eff}$	-
U 235	ν, σ_f, σ_a	$I^a{}_{eff}$, $I^f{}_{eff}$	-
U 238	-	$I^a{}_{eff}$	σ_f
Pu 239	ν, σ_f, σ_a	$I^a{}_{eff}$, $I^f{}_{eff}$	-
Pu 240	σ_a	Γ_γ	-
PU 241	ν σ_f, σ_a	-	
Th 232	-	$I^a{}_{eff}$	σ_f
H_2O	σ_a	M^2	-
D_2O	-	M^2	-
C	σ_a	M^2	-

TABLE II

FISSILE ACTINIDE THERMAL DATA

		Initial JEF2 value	Tendency Research result
U 235	ν	2.437	2.439 ± 0.004
	σ_f	582.5	582.2 ± 1.0
	σ_c	98.8	99.1 ± 1.0
Pu 239	ν	2.877	2.872 ± 0.007
	σ_f	747.2	745.4 ± 2.0
	σ_c	270.2	273.8 ± 3.9
U 233	ν	2.498	2.480 ± 0.004
	σ_f	525.1	525.4 ± 1.1
	σ_c	45.9	50.8 ± 2.0

TABLE III

FERTILE ACTINIDE EFFECTIVE INTEGRAL MODIFICATIONS

	ΔI_{eff} (barn)
U 238	- 0.20 ± 0.11
Th 232	- 0.16 ± 0.08

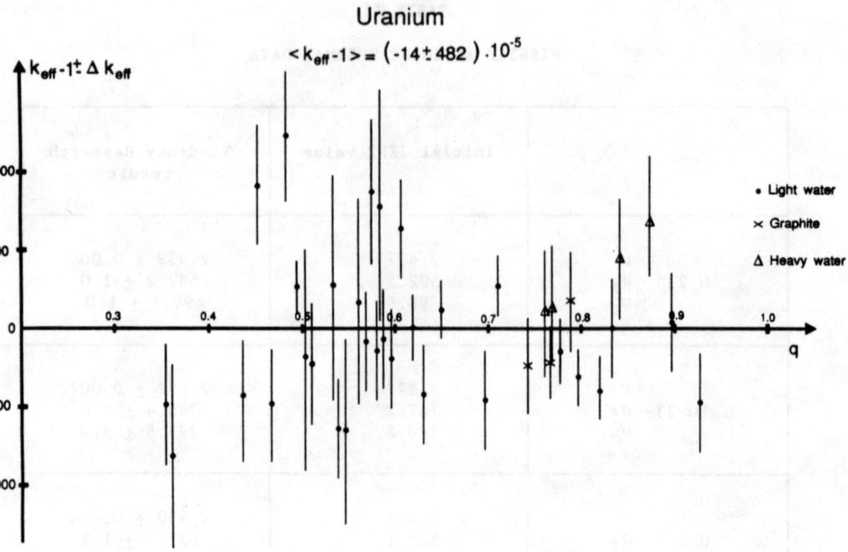

Figure 1

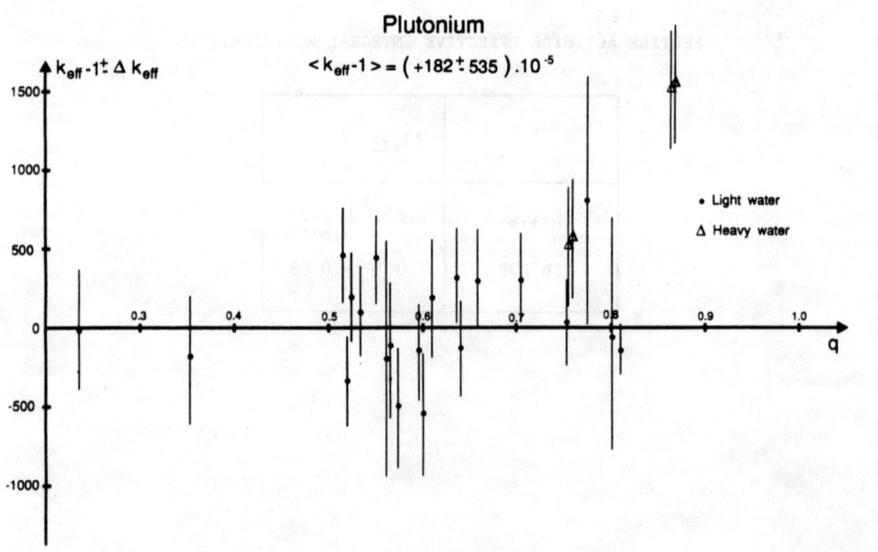

Figure 2

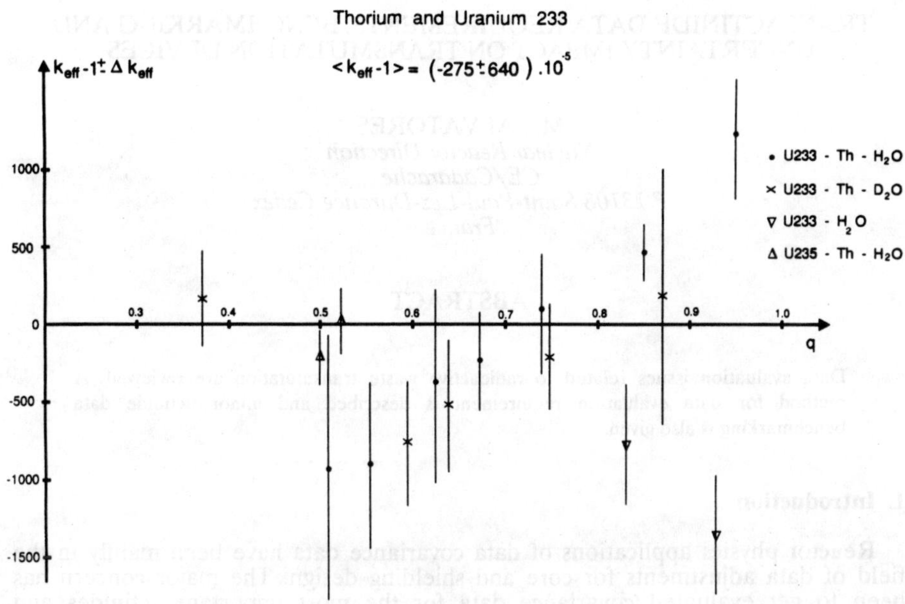

Figure 3 -

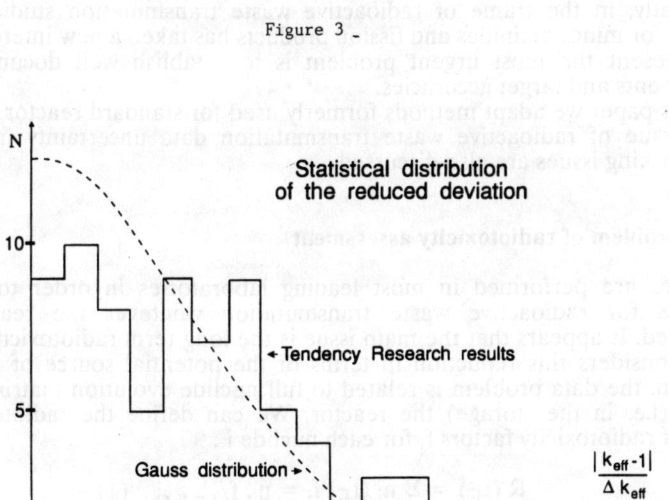

Figure 4

TRANSACTINIDE DATA REQUIREMENTS, BENCHMARKING AND UNCERTAINTY IMPACT ON TRANSMUTATION DEVICES

M. SALVATORES
Nuclear Reactor Direction
CE/Cadarache
F 13108 Saint-Paul-Lez-Durance Cedex
France

ABSTRACT

Data evaluation issues related to radioactive waste transmutation are reviewed. A method for data evaluation requirement is described and minor actinide data benchmarking is also given.

1. Introduction

Reactor physics applications of data covariance data have been mainly in the field of data adjustments for core and shielding design. The major concern has been to get evaluated covariance data for the most important actinides and structural materials.

Recently, in the frame of radioactive waste transmutation studies, the data problem for minor actinides and fission products has taken a new interest.

At present the most urgent problem is to establish well documented data requirements and target accuracies.

In this paper we adapt methods formerly used for standard reactor applications to the issue of radioactive waste transmutation data uncertainty impact. Data benchmarking issues are also discussed.

2. The problem of radiotoxicity assessment

Studies are performed in most leading laboratories in order to implement strategies for radioactive waste transmutation whatever the reactor system considered. It appears that the main issue is the long term radiotoxicity reduction. If one considers this reduction in terms of the potential source of radiotoxicity reduction, the data problem is related to full nuclide evolution matrix, inside and outside (i.e. in the storage) the reactor. We can define the radiotoxicity R by means of radiotoxicity factors f_i for each nuclide i :

$$R\,(t_F) = \sum_i n_i\,(t_F)\,f_i = \underline{n}\,.\,\underline{f}/_{t=t\,F} \quad (1)$$

where $n_i(t)$ is the nuclide i density at time t_F. The time interval to be considered starts from t_o, initial time of irradiation in a reactor. In the time interval (t_o,t_F) one has to consider the irradiation time steps, the cooling times before possible

reprocessing and fabrication (e.g. in the homogeneous actinide recycling mode), and storage time.
The $n_i(t)$ can be obtained by the standard nuclide evolution equations :

$$\frac{d\,\underline{n}\,(t)}{dt} = A\,\underline{n}\,(t) \qquad (2)$$

where A is the time evolution matrix and $\underline{n}(t)$ is the nuclide vector of components $n_i(t)$.

The equation for the component n_i is the classical one :

$$\frac{dn_i}{dt} = -(\sigma_{ai}\phi + \lambda_i)n_i + \sum_j (\sigma_{j->i}\phi + \lambda_{j->i})\,n_j$$

where σ represent (one-group) cross-sections and λ, decay constants with the appropriate branching ratios.
The problem to define data requirements and target accuracies for R can be formally stated as follows (Réf. 1 and 2). A "cost" function F is defined :

$$F \cdot \Sigma \left(\frac{a_i}{y_i}\right)^2 \quad i=1.....I \text{ (I total number of nuclear data of relevance)} \quad (3)$$

where a_i are "cost" parameters related to data uncertainties y_i. The function F is minimized (i.e. the y_i values which minimize F are determined), with conditions on several integral parameters x_j (like radiotoxicity defined in equation (1)) :

$$\text{St dev } x_j = \sqrt{\sum_{i,k} S_i^{+j} y_{ik} S_k^j} \leq d_j \qquad (3')$$

where S_i^j are the sensitivity coefficients of the integral parameter j to the data of index i, y_{ik} is the data covariance matrix, and d_j are the designer target accuracies on parameter x_j.

3. The radiotoxicity sensitivity calculation

The sensitivity of radiotoxicity to a basic data p_k is defined as :

$$S_k^R = \frac{dR}{R} \Big/ \frac{dp_k}{p_k} \qquad (4)$$

To p_k's are associated uncertainties y_k and a covariance matrix y_{ik}.

The sensitivity coefficients Eq. (4) can be obtained (at first order) with a specific application of the generalized perturbation theory in the nuclide field (reference (3)) :

$$S_k^R = \frac{p_k}{R} \int_{t_o}^{t_F} \underline{n}^* \frac{\delta A}{\delta p_k} \underline{n}\, dt \qquad (5)$$

where $\underline{n}^*$ (usually defined as "nuclide importance vector") is obtained as solution of :

$$-\frac{d\underline{n}^*}{dt} = A^* \underline{n}^* \qquad (6)$$

with boundary condition at $t = t_F$

$$\underline{n}^* (t = t_F)) = \underline{f} \qquad (7)$$

where $\underline{f}$ is the vector or components f_i.

Applications of the above formalism have been made for reactor burn-up problems (see for example ref. 4 and 5).

The linearity of the operator A (and then the possibility to add time dependent contributions to R) allow to consider very simply functionals R of the type :

$$R_I = \int_{t_x}^{t_F} \underline{f} \cdot \underline{n}\, dt \qquad (8)$$

(this can be the case of the radiotoxicity integrated over a given interval of time). For such a case, it is only necessary to replace equation (6) with the more general formulation :

$$-\frac{dn^*}{dt} = A^* \underline{n}^* + \underline{f} \qquad (6')$$

Time dependent perturbations can also easily be dealt with. In that case, if $\delta A/\delta p_k$ is constant over a short time interval (t_{l-1}, t_l), equation (5) can be rewritten :

$$S_k^R = \frac{p_k}{R} \sum_l \int_{t_{l-1}}^{t_l} \underline{n}^* \frac{\delta A}{\delta p_k} \underline{n}\, dt \qquad (5')$$

where the sum over l extends from t_o to t_F.
Finally, it is worth to note that, as pointed out in reference 4, "nuclide contribution densities" can be defined as :

$$c_i(t) = n_i^*(t)\, n_i(t) \qquad (9)$$

The knowledge of $c_i(t)$ allows to determine at each time t which isotopes contribute most to the response of interest. Beside the use for the purpose of data requirements assessment, this quantity can be exploited to find optimum strategies for radiotoxicity reduction.
The interest of the c_i quantities defined by equation (9) is due to the fact at each time t the total importance is conserved :

$$\underline{n}^* \cdot \underline{n} = \text{constant} = \underline{n}^* \cdot \underline{n} \,\Big|_{\,t=t_k} = R \quad (10)$$

All the results given in the present paragraph, are related to the first order approximation. Higher order developments already obtained for classical burn-up problems (see reference 6), can be readily extended to the present case.

4. Data requirements for transmutation and related fuel cycles

The very general procedure outlined in paragraph 2 for data requirement assessment can be specialized to specific radioactive transmutation strategies and the associated fuel cycles.

In fact, if one considers the radiotoxicity issue as the most relevant, the consequences on the fuel cycle of each transmutation strategy using for example fission reactors or accelerator-based systems, have to be accounted for.

For the data requirement point of view, this implies that more "integral" parameters x_j have to be considered, as formally indicated in equation 3'.

A typical example is the radioactivity or the neutron emission of actinide-bearing fuel, which has to be reprocessed, with the related transport and fabrication problems. Data requirements should then take into account supplementary target accuracies.

For example, in the case of neutron emission the integral parameter to be introduced in the minimisation procedure of equations 3 and 3', is defined as :

$$E = \underline{n}.\underline{Y} \,\Big|_{\,t=t_E}$$

where $\underline{Y}$ is the neutron emission yield vector and t_E is a specifically defined time after some cooling period outside the neutron flux.
All what has been said for the R functional in paragraph 3, can be used to find the appropriate sensitivity coefficients S_k^E, to be used in equation 3'.

In general, the interest of the approach outlined here, is to consider globally the problem of data requirements, taking into account explicitely correlations among data, both in energy and among data of different types, when they exist.
It is clear that simplified evaluations of uncertainties and correlations will play an important role in the present phase of transmutation-related studies, to give credibility to data requirements, and to quantify, at least as a first approximation, the impact of data uncertainties on the performance of the different transmutation strategies, in terms of the most relevant, long-term, effects.

5. Data uncertainties impact on actinide burner reactor performances

Data uncertainties can play an important role also in the more classical field of the reliability of the announced neutronics caracteristics of the different types of actinide burner reactors which are presently proposed.
This analysis is of the same type of the analysis performed in the past for standard reactor types. It concerns the critical mass, the reactivity swing during irradiation, control rod worth and, in particular, reactivity coefficients.
Since there is at present a lack of data uncertainty assessment in the field of the so-called minor actinides, a systematic effort to intercompare at least the major evaluated files for minor actinides has started, as part of the International Evaluation Cooperation, sponsored by the Nuclear Science Committee of the Nuclear Energy Agency of the OECD. Np-237 and Am-241 data have been considered first.
To gain some insight on the impact of the data discrepancies observed among the different files, we have made a first attempt to apply them to the concept of a fast reactor with homogeneous recycling of Np and Am (reference 7). We have been looking to the critical mass (i.e. the K_{eff}) and the sodium void reactivity effect of such a system. The aim was to see if, even in the case of the introduction of a limited amount of minor actinides, the additional uncertainties due to these isotopes, would significantly affect the prediction of integral parameter, in such a way that one should consider data improvement efforts or, possibly, "ad hoc" integral experiments. Two typical sensitivity profiles are shown in figures 1 and 2.
A $\pm$ 10 % uncertainty on the σ_c and σ_f of Am-241 and Np-237 and a $\pm$ 30 % uncertainty on their total inelastic cross-sections have the following impact on the K_{eff} and the Na-void reactivity coefficient of a large LMFBR core (of the EFR type) when ~ 3 % of each of the two isotopes are introduced homogeneously in the core, replacing U-238 :

Percentage variation of	Np-237 σ_c	Np-237 σ_f	Np-237 σ_{in}	Am-241 σ_c	Am-241 σ_f	Am-241 σ_{in}
K_{eff}	$\pm$ 0.3 %	$\pm$ 0.15 %	$\pm$ 0.05 %	$\pm$ 0.4 %	$\pm$ 0.16 %	$\pm$ 0.3 %
Na-void	$\pm$ 2.7 %	$\pm$ 0.5 %	$\pm$ 0.3 %	$\pm$ 3.5 %	$\pm$ 0.5 %	$\pm$ 0.4 %

The impact of uncertainties of the order of those indicated above are not negligeable, but not large enough to prevent preliminary calculations and parametric analysis.

Finally, as a contribution to data benchmarking, a re-evaluation has been made of the pure sample irradiation experiments PROFIL in PHENIX [8], using the recently released JEF-2 data. In the following table the C/E results for some minor actinide reaction of interest, are indicated :

	C/E with	
	JEF-1 data	JEF-2 data
σ_c (Pu-238)	0.95	0.97
σ_c (Np-237)	0.90	0.94
$\sigma_{n,2n}$ (Np-237)	1.19	1.15
σ_c (Am-241)	1.03	1.05
σ_c (Am-243)	0.94	0.97

6. Conclusions

The problem of data evaluation and associated uncertainties can play a significant role in long-lived radioactive waste transmutation studies and strategies.

Well founded data requirements are still needed and an approach has been discussed, relevant to the potential radiotoxicity source in the deep storages.

Data uncertainties play a role also in the actinide-burner reactors, but the present data situation, at least for Np-237 and Am-241, is sufficient for most preliminary studies.

7. References

1. L.N. Usachev et al., Proc. Int. Conf. Neutron Physics, Harwell, Sept 1978, CONF-780921, p. 181, International Atomic Energy Agency (1978).
2. G. Palmiotti and M. Salvatores, Nucl. Sci. Eng., 87, 333 (1984).
3. A. Gandini, "Time dependent Generalized Perturbation Method for Burn-up Analysis" CNER RT/FT (75) 4 - Comitato Nazionale per l'Energia Nucleare (1975).
4. M. Williams, C.R. Weisbin, "Sensitivity and Uncertainty Analysis for Functionals of the Time Dependent Nuclide Density Field" ORNL-5393 (1978).
5. A. Gandini, M. Salvatores, L. Tondinelli, Nucl. Sci. Eng. 62, 339 (1977).
6. G. Palmiotti, M. Salvatores, Nucl. Sci. Eng. 84, 150 (1983).
7. H. Küsters, M. Salvatores et al. "Recent Investigations in France and Germany on the Transmutation of Actinides and Fission Products in Reactors", ANP'92 International Conference, Tokyo, October 1992.
8. A. D'ANGELO et al., Nucl. Sci. Eng., 105, 244 (1990).

8. Acknowledgement

Paragraph 5 is based on not yet published results, provided by R. SOULE. His contribution is gratefully acknowledged.

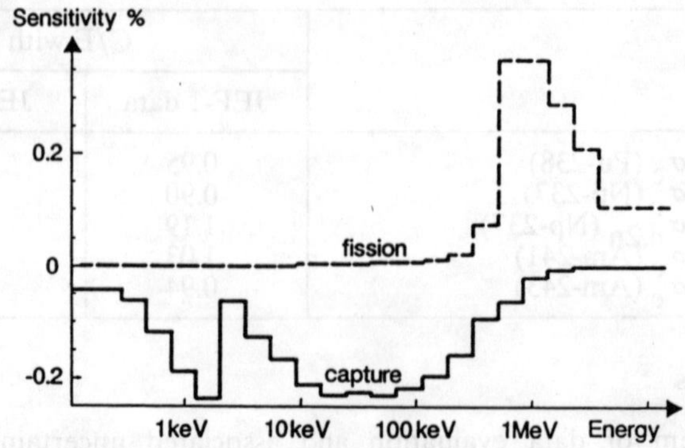

Effect of introducing 3% ^{237}Np in a large LMFBR core :

- **2700pcm due to capture**

+ 1450pcm due to fission

- **150pcm due to inelastic scattering**

Figure 1 : Keff SENSITIVITY TO ^{237}Np 100% SIGMA VARIATION

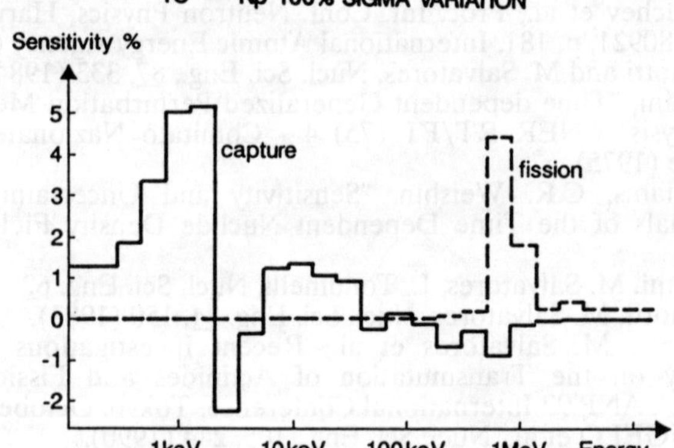

Total core void effect (large LMFBR core)	**standard core (no Np) : 1650pcm** **with 3% ^{237}Np : 2000pcm**
Sensitivity integral (core with Np)	$\sigma_c(^{237}Np) \rightarrow$ **+ 17% (+22%-5%)** $\sigma_f(^{237}Np) \rightarrow$ **+ 8%**

Figure 2 : SODIUM VOID SENSITIVITY TO ^{237}Np 100% SIGMA VARIATION

DATA FITTING METHODS

MODERN PRINCIPLES OF DATA FITTING AND EVALUATION

F. H. FRÖHNER
Kernforschungszentrum Karlsruhe
Institut für Neutronenphysik und Reaktortechnik
Postfach 3640, W-7500 Karlsruhe 1, Germany

ABSTRACT

The probabilistic foundations of data fitting and evaluation are reviewed, with due account of the progress achieved over the past five decades: Restrictions imposed by the frequency definition of probability were removed, and objective rules for the assignment of a-priori probabilities came from group and information theory. As a result, Bayesian estimation of the parameters of physical theories and models is now the most powerful method for the evaluation of experimental, hence error-afflicted and incomplete, data. Bayesian hypothesis testing yields a formal equivalent of Ockham's razor. Bayesian curve fitting deals readily with point data where both ordinate and abscissa are uncertain, and suggests improvements to current techniques even for negligible abscissa uncertainties.

1. Introduction

Much of current data evaluation and fitting relies on more or less plausible rules about first-order error propagation, least squares and hypothesis testing. Many physicists have learnt them during laboratory courses and on the job, finding most books on probability theory geared more to the needs of statisticians interested in trends and correlations in large "random" samples of economic, biological, medical etc. data, than to their own need to find the parameters of physical theories from often sparse and always uncertain experimental data. Conceptual difficulties arose (a) from an overly restrictive definition of probabilities as relative frequencies, (b) from perpetual warnings throughout the literature against Bayes' theorem with its "subjective" a-priori probabilities. During the last decades, however, progress concerning prior probabilities has been such that today the Bayesian method of inductive inference, or learning from error-affected and incomplete observations, is more objective, more coherent, more powerful, yet simpler, than conventional sampling theory with its unsatisfactory ad-hoc elements.

2. Probability as Quantitative Measure of Rational Expectation

Probability theory unfolds from the fundamental sum and product rules,

$$P(A|B) + P(\bar{A}|B) = 1\,, \tag{1}$$

$$P(AB|C) = P(A|BC)\,P(B|C) = P(B|AC)\,P(A|C)\,, \tag{2}$$

where A, B and C represent propositions such as "the coin shows head" or "the cross section is larger than 210 b", AB means "both A and B are true", $\bar{A}$ means "A is false", and $P(A|B)$ denotes the probability of A given B which indicates that all probability assignments are conditional, based on either empirical or theoretical information or on assumptions. Following J. Bernoulli and Laplace we interpret probabilities as degrees of rational expectation on a numerical scale ranging from 0 (impossibility) to 1 (certainty). The two forms of the product rule follow from the symmetry $AB = BA$.

"Frequentists" criticise the interpretation of the P as degrees of plausibility (not the equations among them). They insist that by probability one must mean only "relative frequency in a random experiment" such as coin tossing, in the limit of very many repetitions. They argue that physical constants are not random variates assuming various possible values with certain frequencies, therefore they should not have probability distributions associated with them. For scientists in general, and data evaluators in particular, this outlook is too narrow. It would not allow them to say that, according to measured data, a physical constant has a certain probability to lie between given limits. Now the task to infer the values of half-lives or cross sections from error-affected and incomplete data is not a random trial that can be repeated at will, but an exercise in inductive inference (reasoning in the face of uncertainty). *Laplace's probability concept is therefore the appropriate one for the evaluation of scientific data.* Probabilities are not frequencies, but they permit prediction of frequencies in repeated trials.

Actually, R.T. Cox (1946) proved that degrees of plausibility can always be represented by real numbers between 0 and 1, within any logically *consistent* formal system of inductive reasoning. The most general consistency conditions can be written as two functional equations whose solutions are equivalent to the sum and product rules. Cox was criticised for assuming his probability functions to be twice differentiable, but L. Rényi (1954) gave a proof without that assumption. In quantum mechanics the idea that probabilities quantify incomplete knowledge has always been implicit, and E. Schrödinger (1947) arrived independently at the same conclusions as Cox: *Any consistent scheme of logical inference must be equivalent to probability theory as derived from the basic sum and product rules, with Laplace's interpretation of probability.* Most proponents of Fuzzy Sets and Artificial Intelligence, on the other hand, seem to be unaware of Cox's achievement.

3. Bayes' Theorem as the Rule for Updating Knowledge with New Data

The simplest form of Bayes' theorem,

$$P(A|BC) = \frac{P(B|AC)\,P(A|C)}{P(B|C)}, \tag{3}$$

is a direct consequence of the symmetry of the product rule (2) with respect to A and B. Laplace generalised this to the case of several distinct, mutually exclusive alternatives A_i:

$$P(A_i|BC) = \frac{P(B|A_iC)\,P(A_i|C)}{\sum_i P(B|A_iC)\,P(A_i|C)}, \qquad i = 1, 2, \ldots n. \tag{4}$$

For continuous alternatives we replace the finite probabilities $P(A|C)$, $P(B|AC)$ etc. by infinitesimal probabilities $p(A|C)dA$, $p(B|AC)dB$ etc. with probability densities $p(A|C)$, $p(B|AC)$ etc., and sums over alternatives by integrals. The result,

$$p(A|BC)\,dA = \frac{p(B|AC)\,p(A|C)\,dA}{\int p(B|AC)\,p(A|C)\,dA}, \qquad A_L \le A \le A_H, \tag{5}$$

can be considered as the cornerstone of modern data evaluation. The usual situation is that we have data B depending on the value of an unknown physical quantity A and on other circumstances C. If we know the mathematical form of the so-called likelihood function $p(B|AC)$, in which the unknown quantity A occurs as parameter, and also the a-priori probability ("prior" for short) $p(A|C)dA$, then we can form their product which is proportional to the a-posteriori probability ("posterior") $p(A|BC)dA$ of the unknown parameter lying in an infinitesimal interval dA at A. The normalising denominator is readily calculated by integration (numerically if necessary) but often we shall not need this constant at all.

As an example consider determination of the decay constant λ of some short-lived radioisotope from decays registered at times $t_1, t_2, \ldots t_n$. Obviously we must identify λ with A, and the data $t_1, \ldots t_n$ with B, while C consists of all other relevant information such as purity of the sample, sufficiently long observation time for all observable decays to be recorded, and applicability of the decay law

$$p(t_i|\lambda)\, dt_i = \exp(-\lambda t_i)\, \lambda dt_i\,, \qquad 0 < t_i < \infty\,. \tag{6}$$

According to the product rule the joint probability of observing the mutually independent data $t_1, \ldots t_n$, given λ, is

$$p(t_1, \ldots t_n|\lambda) dt_1 \ldots dt_n = \exp\Big(-\lambda \sum_{i=1}^{n} t_i\Big)\lambda^n\, dt_1 \ldots dt_n\,. \tag{7}$$

Multiplying the likelihood function $p(t_1, \ldots t_n|\lambda)$ by a prior $p(\lambda)d\lambda$ we get

$$p(\lambda|t_1, \ldots t_n)\, d\lambda \;\propto\; \exp\Big(-\lambda \sum_{i=1}^{n} t_i\Big)\lambda^n p(\lambda) d\lambda\,, \qquad 0 < \lambda < \infty\,, \tag{8}$$

which is an example of the Bayesian model for learning from experience: *The posterior is proportional to the product of likelihood function and prior.* The likelihood function conveys the impact of the data, while the prior summarises what we knew already beforehand. But what is the correct prior for the decay constant?

4. Least Informative Priors from Invariance Arguments

The basic rules tell us how various probabilities are related but not how we can assign numerical values. For the latter purpose Laplace used the *principle of insufficient reason: If there is no reason to prefer one alternative to another one ought to assign the same probability to both.* This works well for discrete alternatives, for coins, dice, particle spins etc., but not for continuous ones. Ought we to assume all decay constants as equally probable a priori, $p(\lambda)d\lambda \propto d\lambda$, or all mean lives $\tau = 1/\lambda$, in which case the prior is different, $p(\tau)d(\tau) \propto d\tau \propto d\lambda/\lambda^2$?

For more than a century this seeming arbitrariness about priors for continuous variates has led most statisticians to abandon and even attack the Bayesian approach and to seek alternatives that circumvent priors. Others, comparing this with attempts to do arithmetic without zero, used "subjective" priors or invoked invariance arguments to find unique priors (H. Jeffreys 1939). The breakthrough was achieved by E.T. Jaynes (1968) who demonstrated that for the simplest, practically most important cases the symmetry of the problem uniquely determines the prior that reflects total ignorance about numerical values. If a *location parameter* is to be estimated, for instance the centre (mean) μ of a Gaussian, the form of the prior must be invariant under an arbitrary shift of location, $p(\mu)d\mu = p(\mu+c)d(\mu+c)$. Otherwise some locations would be preferred to others, contrary to the assumption of total ignorance. The functional equation is satisfied by the uniform distribution

$$p(\mu)\,d\mu \;\propto\; d\mu\,, \qquad -\infty < \mu < \infty\,. \tag{9}$$

The prior for a *scale parameter*, for instance the standard deviation σ of a Gaussian, is perhaps less obvious. If there is no preferred scale, one expects invariance under rescaling, $p(\sigma)\,d\sigma = p(c\sigma)\,d(c\sigma)$. The solution of this functional equation is

$$p(\sigma)\,d\sigma \;\propto\; \frac{d\sigma}{\sigma}\,, \qquad 0 < \sigma < \infty\,, \tag{10}$$

as advocated already by H. Jeffreys (1939). In the language of group theory the prior corresponding to invariance under a group of transformations is the right invariant Haar measure on this group (see Berger 1985, ch. 6.6)

The non-normalisability of such "least informative" priors has been criticised, and they are called "improper". Now one may consider the priors (9) and (10) as limits of extremely broad, normalised distributions on the linear ($d\mu$) and on the logarithmic ($d\sigma/\sigma = d\ln\sigma$) scale, just as Dirac's delta function is the limiting case of extremely narrow, normalised distributions. Mathematical difficulties do not arise for those who keep in mind that both *the least informative priors and the "most informative" delta function are just convenient shorthand notations for distributions that are extremely broad or extremely narrow compared to the distributions with which they are convolved* (Fröhner 1986).

The decay constant in our example is a scale parameter, multiplying all times in the "sampling distribution" (6), hence Jeffreys' prior is appropriate. It satisfies $p(\lambda)\,d\lambda \propto d\lambda/\lambda \propto d\tau/\tau$ which removes all ambiguity: Whether we estimate the scale parameter λ or the scale parameter τ, we always get the same posterior,

$$p(\lambda|n\bar{t})\,d\lambda = \Gamma(n)^{-1}\,e^{-x}\,x^{n-1}\,dx\,, \qquad 0 < x \equiv \lambda n\bar{t} < \infty\,, \tag{11}$$

a gamma distribution depending on the average recording time of the sample, $\bar{t} \equiv \sum_i t_i/n$.

5. Recommended Values from Estimation under Quadratic Loss

Most users of evaluated data are not interested in the details of a posterior distribution. They want, for example, a recommended decay constant and its uncertainty, and nothing else. So we calculate the expectation value of the posterior (11),

$$\langle\lambda\rangle = \int_0^\infty \lambda\, p(\lambda|n\bar{t})\,d\lambda = \frac{1}{\bar{t}}\,, \tag{12}$$

and its root-mean-square error (square root of the variance, also called standard deviation or 1σ uncertainty),

$$\Delta\lambda = \left[\int_0^\infty (\lambda - \langle\lambda\rangle)^2\, p(\lambda|n\bar{t})\,d\lambda\right]^{1/2} = \frac{1}{\bar{t}\sqrt{n}}\,, \tag{13}$$

and state our recommendation summarily as $\langle\lambda\rangle \pm \Delta\lambda$. What we just did is called *estimation under quadratic loss* in decision theory. The idea is that there is always a penalty for bad estimates $\hat{\lambda}$, describable by a *loss function* that vanishes at the true value and is positive elsewhere. Near the true value any reasonably smooth loss function can be taken as proportional to the squared error, $(\hat{\lambda} - \lambda)^2$, since its Taylor expansion about zero error begins with the quadratic term. In this approximation the expected loss is minimised,

$$\int_0^\infty (\hat{\lambda} - \lambda)^2 \, p(\lambda|n\bar{t}) \, d\lambda = \min, \tag{14}$$

if we choose $\hat{\lambda} = \langle\lambda\rangle$, which makes the expected square error minimal and equal to the variance, and justifies our recommendation. We note that the relevant "statistic", $\bar{t}$, appeared automatically, without guesswork. Thus *an evaluator who does not know the users' loss functions ought to recommend posterior means and standard deviations* rather than modes, medians or other "estimators" and "confidence intervals" suggested in the older literature.

6. Assignment of Probabilities by Entropy Maximisation

E.T. Jaynes (1957) recognised that entropy maximisation, employed by J.W. Gibbs in his axiomatic approach to thermodynamics, is a powerful logical tool for inductive inference in general, permitting assignment of probabilities (prior or any other) in a well defined way whenever "macroscopic" numerical information is available. This discovery became possible after C.E. Shannon (1948) had proved that the missing information implied by a probability distribution is given by the information entropy. For of a discrete probability distribution p_i, with mutually exclusive alternatives i, the information entropy is $S = -\sum_i p_i \ln p_i$ (up to a constant). For continuous distributions with probability density $p(x)$ one must take (Jaynes 1968)

$$S = -\int dx \, p(x) \ln \frac{p(x)}{m(x)}, \tag{15}$$

where $m(x)$ is a possibly uniform (least informative) prior ensuring invariance under change of variable. Let us now assume that we do not know $p(x)$ but that we have global information about it in the form of expectation values for several known functions $f_k(x)$,

$$\langle f_k \rangle = \int dx \, p(x) \, f_k(x), \qquad k = 1, 2, \ldots K. \tag{16}$$

What probability density $p(x)$ satisfies these constraints without implying any other, spurious information or hidden assumptions? The answer is provided by the *principle of maximal entropy: If we want compatibility with given information, but minimal information content otherwise, we must choose $p(x)$ such that the information entropy is maximised subject to the constraints.* This variational problem is readily solved by the technique of Lagrange multipliers, with the result

$$p(x) = \frac{m(x)}{Z} \exp\Big(-\sum_{k=1}^{K} \lambda_k f_k(x)\Big), \tag{17}$$

$$Z = \int m(x) \exp\Big(-\sum_{k=1}^{K} \lambda_k f_k(x)\Big). \tag{18}$$

The Lagrange multipliers λ_k must be found from the constraints (16) or from the equivalent equations

$$\langle f_k \rangle = -\frac{\partial}{\partial \lambda_k} \ln Z \,, \qquad k = 1, 2, \ldots K \,. \tag{19}$$

If, for instance, the variate is inherently positive, such as a life time t, and we know only the mean life τ, the maximum entropy principle (with uniform prior) tells us that the exponential decay law, $p(t|\tau)dt = \exp(-t/\tau)\,dt/\tau$, is optimal for predictions. For a thermodynamic system with given mean energy we find Gibbs' canonical ensemble (i. e. distribution) with partition function Z, etc. A fascinating review of the maximum entropy method, including a wide variety of applications from imperfect dice to irreversible statistical mechanics, was given by Jaynes (1978).

The evaluator is mainly confronted with data reported in the form $\langle x \rangle \pm \Delta x$. As our notation indicates, these figures are to be interpreted as the measurer's best estimate under quadratic loss. For known first and second moments, $\langle x \rangle$ and $\langle x^2 \rangle = (\Delta x)^2 + \langle x \rangle^2$, and uniform prior, the maximum entropy algorithm yields $p(x) \propto \exp(-\lambda_1 x - \lambda_2 x^2)$ as the most conservative, most objective probability density for further inference (if the range of the variate is $-\infty < x < \infty$). Obviously this is the Gaussian distribution

$$p(x|\langle x \rangle, \Delta x)\, dx = \frac{1}{\sqrt{2\pi(\Delta x)^2}} \exp\left[-\frac{1}{2}\left(\frac{x - \langle x \rangle}{\Delta x}\right)^2\right] dx \,, \qquad -\infty < x < \infty \,. \tag{20}$$

Similarly, if measurements of several quantities x_μ are reported in the form of data $\langle x_\mu \rangle$ with covariance matrix elements $C_{\mu\nu} \equiv \langle (x_\mu - \langle x_\mu \rangle)(x_\nu - \langle x_\nu \rangle) \rangle = C_{\nu\mu}$, the maximum entropy algorithm with uniform priors yields $p(\mathbf{x}) \propto \exp(-\sum_\mu \lambda_\mu x_\mu - \sum_{\mu\nu} \lambda_{\mu\nu} x_\mu x_\nu)$ with $\lambda_{\mu\nu} = \lambda_{\nu\mu}$. We recognise this as the multivariate Gaussian

$$p(\mathbf{x}|\langle \mathbf{x} \rangle, \mathbf{C})\, d(\mathbf{x}) = \frac{1}{\sqrt{\det\,(2\pi\mathbf{C})}} \exp\left[-\frac{1}{2}(\mathbf{x}-\langle \mathbf{x} \rangle)^\dagger \mathbf{C}^{-1}(\mathbf{x}-\langle \mathbf{x} \rangle)\right] d(\mathbf{x}) \,, \qquad -\infty < x_\mu < \infty \,, \tag{21}$$

where $\mathbf{x}$ is the vector with coordinates x_μ, $d(\mathbf{x}) = \prod_\mu dx_\mu$ is the associated volume element (to be distinguished from the infinitesimal vector $d\mathbf{x}$), $\mathbf{C} = \langle \delta\mathbf{x}\delta\mathbf{x}^\dagger \rangle$ with $\delta\mathbf{x} \equiv \mathbf{x} - \langle \mathbf{x} \rangle$ is the given covariance matrix, and the dagger denotes transposition. If only variances are reported, without information about correlations, the maximum entropy algorithm yields automatically the expected diagonal matrix $\mathbf{C}$ with $C_{\mu\mu} = (\Delta x_\mu)^2$.

Gaussian distributions are often considered appropriate only if measurements are affected by many independent "random" errors so that the central limit theorem can be applied, or they are invoked for mere convenience, with stern warnings about the consequences in case the true distribution is not Gaussian. The maximum entropy principle cannot eliminate those consequences, but it reassures *an evaluator, who is given only best values and their (co)variances, that the corresponding Gaussian is his best choice for all further inference*, whatever the unknown true error distribution may happen to be.

7. Bayesian Parameter Estimation and Least Squares

Let us now assume that observables y_i (e. g. neutron transmission data) depend on parameters x_μ (e. g. those of an optical-model potential) according to some theory or model $\mathbf{y}(\mathbf{x})$. A-priori estimates (or perhaps previous measurements) of the parameters, given in the form $\boldsymbol{\xi}$, $\mathbf{A}$, imply a Gaussian maximum-entropy prior. Similarly, new data on the observables, in the form $\boldsymbol{\eta}$, $\mathbf{B}$, imply a Gaussian maximum-entropy likelihood. Multiplication of prior and likelihood function yields the posterior

$$p(\mathbf{x}|\boldsymbol{\xi},\mathbf{A},\boldsymbol{\eta},\mathbf{B})\,d(\mathbf{x}) \;\propto\; \exp\Big[-\frac{1}{2}(\boldsymbol{\xi}-\mathbf{x})^{\dagger}\mathbf{A}^{-1}(\boldsymbol{\xi}-\mathbf{x})-\frac{1}{2}(\boldsymbol{\eta}-\mathbf{y}(\mathbf{x}))^{\dagger}\mathbf{B}^{-1}(\boldsymbol{\eta}-\mathbf{y}(\mathbf{x}))\Big]\,d(\mathbf{x}). \tag{22}$$

Introducing the hypervectors $\mathbf{z}(\mathbf{x}) \equiv \{\mathbf{x},\mathbf{y}(\mathbf{x})\}$ and $\boldsymbol{\zeta} \equiv \{\boldsymbol{\xi},\boldsymbol{\eta}\}$, and the associated covariance matrix $\mathbf{C} \equiv \langle\delta\boldsymbol{\zeta}\,\delta\boldsymbol{\zeta}^{\dagger}\rangle$, we can write more concisely

$$p(\mathbf{x}|\boldsymbol{\zeta},\mathbf{C})\,d(\mathbf{x}) \;\propto\; \exp\Big[-\frac{1}{2}(\boldsymbol{\zeta}-\mathbf{z}(\mathbf{x}))^{\dagger}\mathbf{C}^{-1}(\boldsymbol{\zeta}-\mathbf{z}(\mathbf{x}))\Big]\,d(\mathbf{x}). \tag{23}$$

The mean vector $\langle\mathbf{x}\rangle$ and the covariance matrix $\langle\delta\mathbf{x}\delta\mathbf{x}^{\dagger}\rangle$ are easily found if this posterior is a pure Gaussian in $\mathbf{x}$, i. e. if the model $\mathbf{y}(\mathbf{x})$ is linear. If not, one must either integrate numerically (by Monte Carlo if necessary) or employ the *method of steepest descent* (saddle point integration) which means that the posterior is replaced by a Gaussian with the same maximum and the same curvatures at maximum, so that the approximation is best in the domain that contributes most to the integrals: Expansion of the exponent in (23) around the maximum, $\mathbf{x} = \hat{\mathbf{x}}$, and truncation after the quadratic term,

$$Q(\mathbf{x}) \equiv (\boldsymbol{\zeta}-\mathbf{z}(\mathbf{x}))^{\dagger}\mathbf{C}^{-1}(\boldsymbol{\zeta}-\mathbf{z}(\mathbf{x})) \simeq Q(\hat{\mathbf{x}}) + (\mathbf{x}-\hat{\mathbf{x}})^{\dagger}\mathbf{A}'^{-1}(\mathbf{x}-\hat{\mathbf{x}}), \tag{24}$$

shows that $\langle\mathbf{x}\rangle \simeq \hat{\mathbf{x}}$, $\langle\delta\mathbf{x}\delta\mathbf{x}^{\dagger}\rangle \simeq \mathbf{A}'$, the approximations becoming exact for linear models. At the maximum one has $Q(\mathbf{x}) = \min$, i. e.

$$\frac{1}{2}[\nabla Q]_{\mathbf{x}=\hat{\mathbf{x}}} = 0 = \mathbf{S}(\hat{\mathbf{x}})^{\dagger}\mathbf{C}^{-1}(\boldsymbol{\zeta}-\mathbf{z}(\hat{\mathbf{x}})), \tag{25}$$

where $\nabla_{\mu} \equiv \partial/\partial x_{\mu}$, and $\mathbf{S}$ is the rectangular matrix of sensitivity coefficients $S_{i\mu} = \partial z_i/\partial x_{\mu}$. One can solve this system of "normal" equations for instance by Newton-Raphson iteration starting from $\hat{\mathbf{x}} \simeq \boldsymbol{\xi}$ (see Fröhner 1986). The covariance matrix $\mathbf{A}'$ is then obtained from

$$\frac{1}{2}[\nabla\nabla^{\dagger}Q]_{\mathbf{x}=\hat{\mathbf{x}}} = \mathbf{A}'^{-1} = \mathbf{S}(\hat{\mathbf{x}})^{\dagger}\mathbf{C}^{-1}\mathbf{S}(\hat{\mathbf{x}}). \tag{26}$$

One can also work with relative errors, $(\zeta_i - z_i)/\zeta_i$, relative covariance matrix elements $\langle\delta\zeta_i\,\delta\zeta_j\rangle/(\zeta_i\zeta_j)$, and relative sensitivity coefficients, $S_{i\mu}/\zeta_i$. The replacement

$$\boldsymbol{\zeta}-\mathbf{z} \to \mathbf{D}(\boldsymbol{\zeta}-\mathbf{z}), \qquad \mathbf{C} \to \mathbf{DCD}, \qquad \mathbf{S} \to \mathbf{DS} \qquad \text{with} \qquad D_{ij} \equiv \delta_{ij}/\zeta_i, \tag{27}$$

leaves the equations invariant since the diagonal matrices $\mathbf{D}$ are canceled everywhere by their inverses. The advantage is that relative (percentage) errors and sensitivities are more easily grasped, remembered and compared than absolute ones, especially for physically distinct parameter and data types.

Without correlations ($\mathbf{C}$ diagonal) the quadratic form $Q(\mathbf{x})$ is just a sum of squares, so we recognise the condition $Q(\mathbf{x}) = \min$ as the least squares principle, generalised to nonuniform priors and nonlinear models. (The most general form admits correlations even between prior estimates and new data.) Thus *generalised least squares is essentially Bayesian parameter estimation under quadratic loss in saddle point approximation, for the special case that only best values and covariance matrices are given (which implies Gaussian maximum-entropy distributions).* A separate least-squares principle is not needed.

The Bayesian least-squares formalism is employed, for example, in the neutron resonance analysis code SAMMY (Larson and Perey 1980) and in the program FITACS that fits resonance-averaged cross sections (Fröhner et al. 1982). Gaussian priors often improve convergence dramatically, and increase the number of parameters that can be adjusted simultaneously. The data vector $\boldsymbol{\eta}$ may contain data of various types, from different experiments, and the Bayesian approach permits stepwise fitting of large bodies of data, the last posterior always becoming the prior of the next step.

8. Construction of Covariance Matrices from Statistical and Systematic Errors

Experimental uncertainties and correlations are usually not reported as covariance matrices, but evaluators can construct them if measurers document their various error sources and assign uncertainties to each one. *Statistical errors* are caused by finite count rates. The "direct" probability of n counts registered during time t, for a mean rate λ, is given by the Poisson distribution. The "inverse" probability of λ, for given t and n, is obtained with Bayes' theorem and Jeffreys' prior $p(\lambda)d\lambda \propto d\lambda/\lambda$ as the gamma distribution

$$p(\lambda|n,t)\,d\lambda = \Gamma(n)^{-1}\,e^{-x}x^{n-1}\,dx\,, \qquad 0 < x \equiv \lambda t < \infty\,. \tag{28}$$

This yields (compare eqs. 11-13) the familiar rule for the estimation of statistical uncertainties

$$\frac{\Delta\lambda}{\langle\lambda\rangle} = \frac{1}{\sqrt{n}}\,. \tag{29}$$

Common ("systematic") errors induce correlations, in contrast to statistical errors: If, for instance, the reported data were obtained from raw count rates a_i by application of common corrections $b, c, \ldots$, the total errors are $\delta\eta_i \equiv \eta_i - y_i \simeq (\partial\eta_i/\partial a_i)\delta a_i + (\partial\eta_i/\partial b)\delta b + (\partial\eta_i/\partial c)\delta c \ldots$. The statistical errors, δa_i, δa_j, are always uncorrelated, and the common errors, δb, $\delta c, \ldots$, are normally uncorrelated, too, in which case the covariances between all error components vanish, $\langle\delta a_i \delta a_j\rangle = \langle\delta a_i \delta b\rangle = \langle\delta b \delta c\rangle = \ldots = 0$. The overall covariance matrix elements depend then only on mean-square errors (variances) for the components and on sensitivity coefficients (derivatives),

$$B_{ij} \equiv \langle\delta\eta_i\delta\eta_j\rangle \simeq \left(\frac{\partial\eta_i}{\partial a_i}\right)^2(\Delta a_i)^2\delta_{ij} + \frac{\partial\eta_i}{\partial b}(\Delta b)^2\frac{\partial\eta_j}{\partial b} + \frac{\partial\eta_i}{\partial c}(\Delta c)^2\frac{\partial\eta_j}{\partial c} + \ldots\,. \tag{30}$$

This shows that *for the construction of covariance matrices the evaluator needs (i) (root-)mean-square errors for the various error components, (ii) information about the data reduction so that he can calculate the sensitivity coefficients.* If he is told that the data reduction involved subtraction of a common background b and division by a calibration factor c, i. e. $\eta_i = (a_i - b)/c$, he can calculate $B_{ij} \simeq [\delta_{ij}(\Delta a_i)^2 + (\Delta b)^2 + \eta_i\eta_j(\Delta c)^2]/c^2$. Combination of statistical and systematic errors is sometimes considered as incorrect and dangerous, but in first-order error approximation it is seen to be perfectly straightforward.

9. Goodness of Fit: Ockham's Razor

If a theory does not fit the data well one is tempted to refine it. When is this justified? We consider a simple model, $\mathbf{y}_a(\mathbf{x}_a)$, and a competing, more complicated one, $\mathbf{y}_b(\mathbf{x}_b)$, that has e. g. a higher-dimensional parameter vector $\mathbf{x}_b$ (more parameters). Bayes' theorem tells us that $p(\mathbf{y}_a|\boldsymbol{\eta}, \mathbf{B}) \propto p(\boldsymbol{\eta}, \mathbf{B}|\mathbf{y}_a)p(\mathbf{y}_a)$, where $p(\mathbf{y}_a)$ is the prior probability of model a, and $p(\boldsymbol{\eta}, \mathbf{B}|\mathbf{y}_a) = \int d(\mathbf{x}_a)p(\mathbf{x}_a|\boldsymbol{\xi}_a, \mathbf{A}_a)p(\boldsymbol{\eta}, \mathbf{B}|\mathbf{x}_a)$ is the integral over all possible parameter vectors. With posteriors of the type (22) and the conventional notation $\chi^2 \equiv Q(\hat{\mathbf{x}})$ (see Eq. 24), one gets for the odds of a against b

$$\frac{p(\mathbf{y}_a|\boldsymbol{\eta}, \mathbf{B})}{p(\mathbf{y}_b|\boldsymbol{\eta}, \mathbf{B})} = \frac{p(\mathbf{y}_a)}{p(\mathbf{y}_b)}\,\frac{\exp(-\chi_a^2/2)}{\exp(-\chi_b^2/2)}\,\frac{\det\,[1 + \mathbf{A}_b\mathbf{S}_b(\hat{\mathbf{x}}_b)^\dagger \mathbf{B}_b^{-1}\mathbf{S}_b(\hat{\mathbf{x}}_b)]^{1/2}}{\det\,[1 + \mathbf{A}_a\mathbf{S}_a(\hat{\mathbf{x}}_a)^\dagger \mathbf{B}_a^{-1}\mathbf{S}_a(\hat{\mathbf{x}}_a)]^{1/2}}\,. \qquad (31)$$

The three ratios on the right represent (i) the prior odds, (ii) the goodness of fit (the model with the smaller chi-square is favoured), (iii) the adequacy of the models for the given prior information (many parameters with appreciable sensitivities, i. e. large $\mathbf{S}$, and very diffuse priors, i. e. large $\mathbf{A}$, are unfavourable). *There is a trade-off between goodness of fit and model simplicity.* The last ratio is called the Ockham factor, an allusion to William of Ockham's (14-th century) *principle of parsimony: Keep explanations simple, do not introduce unnecessary entities.*

A widely used fix for bad least-squares fits is multiplication of the input covariance matrix $\mathbf{C}$ by a factor c, so as to make chi-square per degree of freedom equal to its expected value, $\langle\chi^2\rangle/\nu = 1$, where ν is equal to the number of data η_i if a Gaussian prior is used (instead of that number minus the number of adjusted parameters in non-Bayesian least squares fits without prior). If we compare the original model a, with $c = 1$, and a model b, with c adjustable but constrained by $\langle c\rangle = 1$, which implies the maximum entropy prior $p(c)dc = e^{-c}dc$, the odds are found to be

$$\frac{p(\mathbf{y}_a|\boldsymbol{\eta}, \mathbf{B})}{p(\mathbf{y}_b|\boldsymbol{\eta}, \mathbf{B})} = \frac{p(\mathbf{y}_a)}{p(\mathbf{y}_b)}\,\Gamma\Big(\frac{\nu}{2}+1\Big)^{-1}\,\exp\Big(-\frac{\chi^2}{2}\Big)\,\Big(1+\frac{\chi^2}{2}\Big)^{\nu/2+1}\,. \qquad (32)$$

The prior odds ratio is typical for Bayesian hypothesis testing, while the remaining factors imply, except for very small ν, the rule familiar from the conventional least-squares test: model a is supported by the data as long as $1 - \sqrt{2/\nu} \lesssim \chi^2/\nu \lesssim 1 + \sqrt{2/\nu}$ (see Fig. 1).

If the vector $\boldsymbol{\zeta}$ contains data from several sources, indiscriminate application of the same factor c to all uncertainties, whether realistic, overoptimistic, or overly conservative, seems unjust. More justice is ensured in case of discrepancies if individual diffuse priors for unrecognised errors are introduced. The resulting Bayesian procedure downweights data that deviate widely from sample averages, like the much advertised James-Stein estimators but without their bizarre discontinuities (Fröhner 1989). Fig. 2 shows a real-life example.

10. Curve Fitting with Uncertainties in Two Dimensions

Conventional curve fitting is restricted to point data with uncertain ordinates, abscissa uncertainties are usually neglected. Let the curve equation in Cartesian coordinates be $y = f(x, \boldsymbol{\theta})$, where $\boldsymbol{\theta}$ is the vector of curve parameters, with prior $p(\boldsymbol{\theta})\,d(\boldsymbol{\theta})$. For data without abscissa uncertainties, $\xi_i = x_i$, $\eta_i = \langle y(x_i)\rangle$, $\Delta\eta_i = \sigma_i$, the posterior for the curve parameters is

$$p(\boldsymbol{\theta}|\{\xi_i, \eta_i, \sigma_i\})\,d(\boldsymbol{\theta}) \propto d(\boldsymbol{\theta})\,p(\boldsymbol{\theta}) \prod_i \exp\Big[-\frac{1}{2}\Big(\frac{\eta_i - f(\xi_i, \boldsymbol{\theta})}{\sigma_i}\Big)^2\Big]\,, \qquad (33)$$

from which recommended values can be obtained as described already. If both ξ_i and η_i are uncertain one has, instead of an error bar, an error ellipse defined by a covariance matrix $\mathbf{C}_i$ with elements $C_{11} = \langle(\delta\xi_i)^2\rangle$, $C_{12} = \langle\delta\xi_i\delta\eta_i\rangle$, $C_{22} = \langle(\delta\eta_i)^2\rangle$. We can write down the probability for true points (x_i, y_i) given the data and the curve parameters,

$$p(\{x_i, y_i\}|\{\xi_i, \eta_i, \mathbf{C}_i\}, \theta) d^n x d^n y \propto \prod_i [p(\xi, \eta|x, y, \mathbf{C}, \theta) p(x, y|\theta) dx dy]_i \tag{34}$$

where $p(\xi, \eta|x, y, \mathbf{C}, \theta)$ is the maximum entropy likelihood function, a Gaussian. The prior

$$p(x, y|\theta)\, dx\, dy = p(y|x, \theta)\, dy\, p(x|\theta)\, dx \propto \delta(y - f)\, dy \sqrt{1 + f_x^2}\, dx \tag{35}$$

with $f \equiv f(x, \theta)$ and $f_x \equiv \partial f(x, \theta)/\partial x$, expresses the facts that a true point lies on the curve, and that the probabilities for its location are proportional to the length of the line elements $ds = \sqrt{1 + f_x^2}\, dx$. A second application of Bayes' theorem yields the joint distribution $p(\theta, \{x_i, y_i\}|\{\xi_i, \eta_i, \mathbf{C}_i\}) d(\theta)\, d^n x\, d^n y \propto p(\{x_i, y_i\}|\{\xi_i, \eta_i, \mathbf{C}_i\}, \theta)\, d^n x\, d^n y\, p(\theta) d(\theta)$ for parameters and true points. Integrating out the true coordinates x_i, y_i one gets, instead of Eq. 33,

$$p(\theta|\{\xi_i, \eta_i, \mathbf{C}_i\}) d(\theta) \propto d(\theta)\, p(\theta)$$
$$\times \prod_i \Big\{ \int dx \sqrt{1 + f_x^2} \exp\Big[-\frac{1}{2} (\xi - x \quad \eta - f) \begin{pmatrix} C_{11} & C_{12} \\ C_{12} & C_{22} \end{pmatrix}^{-1} \begin{pmatrix} \xi - x \\ \eta - f \end{pmatrix} \Big] \Big\}_i . \tag{36}$$

The integrations over the x_i can be performed in saddle point approximation, and also the calculation of means and (co)variances. The square root, giving more weight to steeper curve sections, remains present even in the limit of very small abscissa uncertainties, in contrast to Eq. 33 where it is absent. *This suggests improvements in existing fitting programs* that must deal with sharp peaks such as resonances, or with data exhibiting steep rises e. g. at reaction thresholds.

The simplest example for two-dimensional errors is a straight-line fit ("linear regression"). The least informative prior must be invariant under parallel translations of the straight line, and also under rotations about the origin, i. e. $p(\theta) d\theta \propto dr_0\, d\phi_0$ for straight lines given in polar coordinates by $r = r_0/\sin(\phi - \phi_0)$ (Fig. 3). One finds, in saddle point approximation,

$$p(r_0, \phi_0|D)\, dr_0\, d\phi_0 \propto dr_0\, d\phi_0 \prod_i \frac{1}{\sigma_i} \exp\Big[-\frac{1}{2} \Big(\frac{r_0 - r_i \sin(\phi_i - \phi_0)}{\sigma_i} \Big)^2 \Big] \tag{37}$$

where the polar coordinates r_i, ϕ_i correspond to the Cartesian coordinates ξ_i, η_i, and where $\sigma_i^2 \equiv [C_{11} \sin^2 \phi_0 - 2C_{12} \sin \phi_0 \cos \phi_0 + C_{22} \cos^2 \phi_0]_i$. For the special case that all error ellipses are circles, $C_{11} = C_{22}$, $C_{12} = 0$, the best estimate under quadratic loss is

$$\langle r_0 \rangle = \sqrt{\bar{\xi}^2 + \bar{\eta}^2}, \qquad \langle \phi_0 \rangle = \frac{1}{2} \arctan \frac{2\overline{(\xi - \bar{\xi})(\eta - \bar{\eta})}}{\overline{(\xi - \bar{\xi})^2} - \overline{(\eta - \bar{\eta})^2}}, \tag{38}$$

overbars denoting weighted averages over the data, with weights $1/\sigma_i^2$.

11. Concluding remarks

The present brief review shows that the modern Bayesian approach provides straightforward solutions to the basic probabilistic problems of data evaluation and model fitting. For more details see Fröhner (1986, 1987). The connection with the fundamental sum and product rules of probability theory is always quite direct. Ad hoc assumptions are not needed once prior and loss function are specified, nor is there any room for guesswork about statistics that can serve as estimators. Properties like bias, efficiency, sufficiency, admissibility etc. are more or less incidental aspects of the unique Bayesian estimates: means and (co)variances, under quadratic loss. The least-squares method emerges naturally if only data with uncertainties and, perhaps, correlations are given. Bayesian hypothesis testing, not tied to arbitrary significance levels, generates odds for hypotheses directly, and produces a formal equivalent of Ockham's razor. Bayesian curve fitting deals readily with two-dimensional uncertainties, in contrast to older methods. Bayesian techniques have also been applied to the thorny problem of discrepant data, but much work remains to be done.

REFERENCES

J.O. Berger, *Statistical Decision Theory and Bayesian Analysis*, Springer, New York etc. (1985)

R.T. Cox, "Probability, Frequency and Reasonable Expectation", *Am. J. Physics* **14** (1946) 1

F.H. Fröhner, B. Goel and U. Fischer, *Proc. Meeting on Fast Neutron Capture Cross Sections*, Argonne, ANL-83-4 (1982) p. 116

F.H. Fröhner, "Principles and Techniques of Data Evaluation", Karlsruhe report KfK 4099 (1986)

F.H. Fröhner, "Once again, how many reactor accidents", *Nature* **326** (1987) 834

F.H. Fröhner, "Bayesian Evaluation of Discrepant Experimental Data", in *Maximum Entropy and Bayesian Methods, Cambridge, England, 1988*, Kluwer, Dordrecht (1989)

E.T. Jaynes, "Information Theory and Statistical Mechanics I, II", *Phys. Rev.* **106** (1957) 620; **108** (1957) 171

E.T. Jaynes, "Prior Probabilities", *IEEE Trans. Syst. Cybern.* **SSC-4** (1968) 227; reprinted in Jaynes (1983) p. 114

E.T. Jaynes, "Where Do We Stand on Maximum Entropy", in *The Maximum Entropy Formalism*, R.D. Levine and M. Tribus, eds., M.I.T. Press, Cambridge, Massachusetts (1978), reprinted in Jaynes (1983) p. 210

E.T. Jaynes, *Papers on Probability, Statistics and Statistical Physics*, R.D. Rosenkrantz, ed., Reidel, Dordrecht (1983)

H. Jeffreys, *Theory of Probability* , Oxford (1939)

N.M. Larson, F.G. Perey, Oak Ridge Report ORNL/TM-7485 (1980);

L. Rényi, *Valószinűségszámítás* (= Probability Calculus), Budapest (1954); presented in slightly generalised form by J. Aczel, *Lectures on Functional Equations and Their Applications*, Academic Press, New York (1964), Sect. 7.1.4

E. Schrödinger, "The Foundation of the Theory of Probability", *Proc. R. Irish Acad.* **51 A** (1947) 51, **51 A** (1947) 141; reprinted in *Gesammelte Abhandlungen*, Wien (1984) vol. 1, pp. 463, 479

C.E. Shannon, *Bell Syst. Tech. J.* **27** (1948) 379 and 623; C.E. Shannon and W. Weaver, *The Mathematical Theory of Communication*, U. of Illinois Press, Urbana (1949)

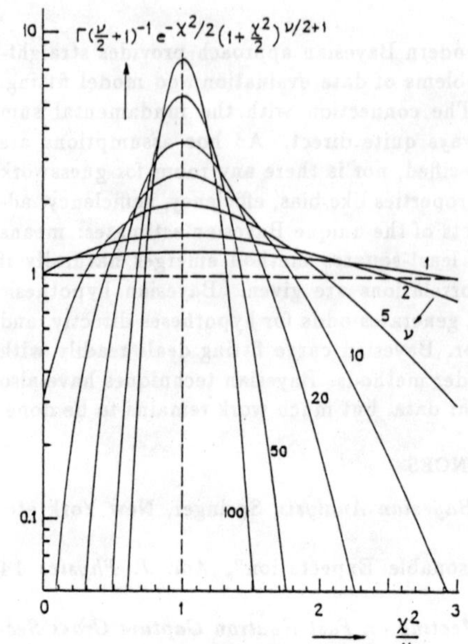

Fig. 1. Hypothesis test: Likelihood ratios for two least-squares models, (a) with fixed uncertainties (no adjustment, $c = 1$) versus (b) with uncertainties taken as adjustable estimates (expected adjustment factor $\langle c \rangle = 1$), see Eq. 32.

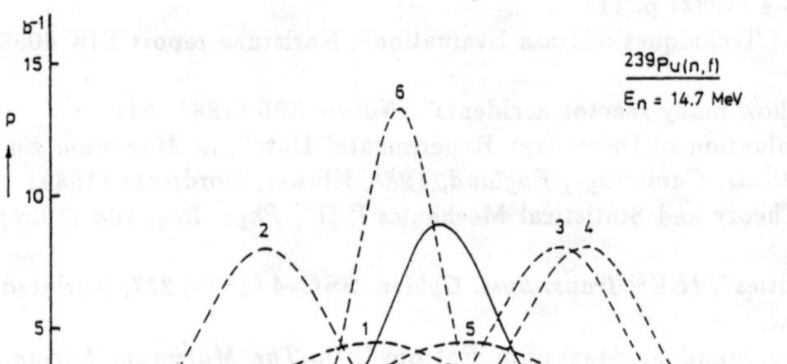

Fig. 2. Discrepancies: Gaussians representing experimental, partially discrepant data from six sources (dashed lines), and posterior density (solid line) from Bayesian estimation including vague prior information about unrecognised errors (from Fröhner 1989).

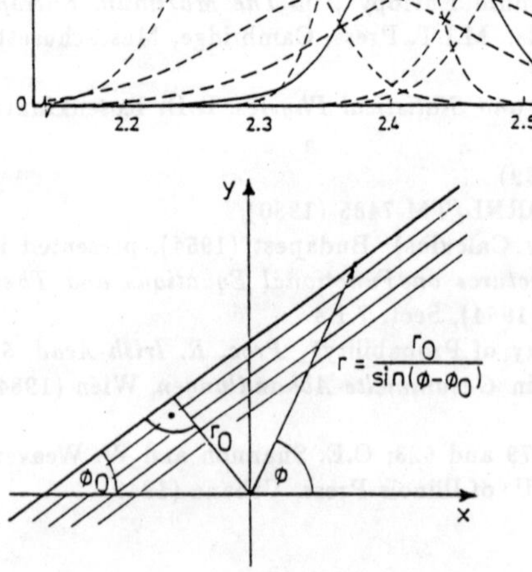

Fig. 3. Definition of coordinates r, ϕ and straight-line parameters r_0, ϕ_0 in Eqs. 37, 38.

PERSPECTIVES ON PEELLE'S PERTINENT PUZZLE AND ITS SIGNIFICANCE IN DATA FITTING AND EVALUATION

Satoshi CHIBA

Nuclear Data Center, Japan Atomic Energy Research Institute
Tokai-mura, Naka-gun, Ibaraki-ken 319-11, Japan

Donald L. SMITH

Engineering Physics Division, Argonne National Laboratory
9700 South Cass Avenue, Argonne, Illinois 60439, U.S.A.

Abstract

The least-squares method is widely used as a method of statistical inference. In applying this method to analysis of actual data, however, the least-squares method sometimes yields strange results which is strongly against our intuition. For example, the average of the two data (1.5, 1.0) becomes 0.88 in a certain combination of the elements of the covariance matrix, an answer outside the range of the input data! This problem, as it was introduced by R.W. Peelle in 1987, is called the "Peelle's Pertinent Puzzle (PPP)". The purposes of this paper are to explain what the PPP is, what its origins are and what the correct answer is. However, whether or not such an answer is correct depends strongly on the history how the data were derived. An approximate but practical method to achieve the correct answer in actual data evaluation will be also described.

1. Introduction

The least-squares method (LSM) is commonly used in nuclear data evaluation as a method of parameter estimation and curve fitting because its solution, the least-squares solution (LSS), has such desirable properties as being consistent, unbiased and minimum variance. The formalism of the LSM seems to be well established. However, the LSM sometimes gives strange results. For example, an average of a set of data becomes outside the range of the data under a certain combination of elements of the covariance matrix. In 1987, R. W. Peelle of ORNL found a typical example of such an anomaly and named the problem as "Peelle's Pertinent Puzzle (PPP)"[(1)]. The problem, which is to seek an average of two data, seemed very easy first. In 1990, the authors proposed a method to avoid such a fallacy result, and distributed drafts of a report which was published later as Ref. 2 to possibly interested readers. After this, a world-wide big discussion emerged[(3)]; some claimed that the answer outside the range of the data was correct, while other methods to avoid such a result were also proposed.

The problem indicated by PPP turned out to express very important aspects of the LSS which affects the whole process of parameter estimation through the LSM. For example, the authors have made an investigation on, being inspired by PPP, properties of the LSS under transformation of data spaces[(4)]. The important consequences are; if a set of data is transformed into a set of another having the same dimension, the covariance matrix and the design matrices must be transformed according to special criterion in order to get the same LSS obtained before the transformation is done. Furthermore, if the data space is truncated, the LSSs before and after the truncation are not necessarily equal. Sometimes impacts of these points are serious, and PPP offers a good example to understand these properties. They will be universally important in every field in which the LSM is applied.

In this paper, two definitions of PPP are explained first. Then, answers to both definitions are given. These arguments are mostly based on the thoughts described in Ref. 4. Through this discussion, origins of the PPP will be made clear to be a neglect of the invariance criterion and improper truncation of data spaces. In the particular problem known as PPP, the answer outside the range of the data was proved to be incorrect. However, such an answer can be regarded to be correct in other circumstances. Whether or not this kind of answer is correct depends strongly on the history how the data were processed. Then, an approximate method to avoid such a bad answer is presented which is applicable in actual nuclear data evaluation.

2. Peelle's Pertinent Puzzle – as originally introduced

The following statement summarizes the problem known as PPP originally introduced by Peelle[1].

"Suppose we are required to obtain the weighted average of two experimental results for the same physical quantity. The first result is 1.5, and the second result 1.0. The full covariance matrix of these data is believed to be the sum of three components. The first component is fully correlated with standard error 20% of each respective value. The second and third components are independent of the first and of each other, and correspond to 10% random uncertainties in each experimental results.

The weighted average obtained from the least–squares method is 0.88 ± 0.22, a value outside the range of the input values! Under what conditions is this the reasonable result that we sought to achieve by use of an advanced data reduction technique?

One's first reaction is to blame the non–intuitive result on the discrepant nature of the input data. This is not the reason, because the whole input data covariance matrix can be scaled up without changing its 'shape' until the data are consistent. The input data are indeed strange, but a similar if muted effect would occur if a less–odd example of this type were offered."

Let us trace this result below. Let the two data be the two components of a vector **y**, i.e., $\mathbf{y}=(y_1,\ y_2)^t = (1.5,\ 1.0)^t$. The covariance matrix associated with **y** is calculated from the above information as

$$V_y = \begin{bmatrix} 1.5^2\cdot(0.1^2 + 0.2^2) & (1.5\cdot0.2)\cdot(1.0\cdot0.2) \\ (1.5\cdot0.2)\cdot(1.0\cdot0.2) & 1.0^2\cdot(0.1^2 + 0.2^2) \end{bmatrix} = \begin{bmatrix} 0.1125 & 0.06 \\ 0.06 & 0.05 \end{bmatrix} \tag{1}$$

If X be the average of the two data to be sought, the observation equation relating the data and the parameter is written as;

$$\mathbf{y} = \begin{bmatrix} y_1 \\ y_2 \end{bmatrix} = \begin{bmatrix} 1.5 \\ 1.0 \end{bmatrix} \sim \begin{bmatrix} X \\ X \end{bmatrix} = \begin{bmatrix} 1 \\ 1 \end{bmatrix} X. \tag{2}$$

Therefore, the design matrix "a" (which is referred to as "sensitivity matrix" as well) is determined to be $\mathbf{a} = (1,\ 1)^t$. By putting these quantities into the formula of the least–squares method[5], the result is obtained as;

$$p = X = (a^t V_y^{-1} a)^{-1} a^t V_y^{-1} y = 0.88 \tag{3}$$

$$V_X = (a^t V_y^{-1} a)^{-1} = 0.22^2 \tag{4}$$

where **p** denote the parameter vector, i.e., **p** = (X). We could thus successfully reproduce the result obtained by Peelle. In this definition, therefore, the PPP is defined as **getting a LSS outside the range of the input data.**

A lot of different opinions emerged about this result from the participants of a world–wide discussion performed through letters and E–mails[3], e.g.,

1. Because we believe in LSM, we should accept this result to be correct.
2. Even though we feel this is incorrect, we have to accept this result because we have nothing else.
3. Because the data are statistically inconsistent (chi–square turned out to be more than 5), it is no wonder to get such an incorrect answer. We should not try to estimate a parameter based on such a fallacy data base.
4. This answer can be considered to be correct because it does agree with the lower data within the quoted uncertainties.
5. More reasonable result must be sought.

Before giving our explanations on this answer, we have to modify the PPP as was first proposed by Zhao[6,7] in order to understand the significance contained in this puzzle in more detail.

3. Second Definition of PPP

Let a vector **y** and its covariance matrix V_y are related with the physical quantity X and C by the following observation equation;

$$y = \begin{bmatrix} y_1 \\ y_2 \\ y_3 \end{bmatrix} = \begin{bmatrix} 1.5 \\ 1.0 \\ 1.0 \end{bmatrix} \sim \begin{bmatrix} X \cdot C \\ X \cdot C \\ C \end{bmatrix} \equiv g(p) = \begin{bmatrix} g_1(p) \\ g_2(p) \\ g_3(p) \end{bmatrix}, \; V_y = \begin{bmatrix} 0.15^2 & 0 & 0 \\ 0 & 0.1^2 & 0 \\ 0 & 0 & 0.2^2 \end{bmatrix} \tag{5}$$

where $p = (X,C)^t$ denotes the parameter vector to be sought, and **g** the model relating the data and the parameters. The elements of design matrix (**a**) of this non–linear problem is calculated to be

$$a_{ij} = \frac{\partial g_i}{\partial p_j}, \; \therefore \; a = \begin{bmatrix} C & X \\ C & X \\ 0 & 1 \end{bmatrix} \tag{6}$$

The LSS under the non–linear problem is written down as

$$p = p_0 + (a^t V_y^{-1} a)^{-1} a^t V_y^{-1} [y - g(p_0)] \tag{7}$$

where p_0 denotes an initial parameter vector. Because of the non–linearity of the problem, iterations are required. Expression for V_p is the same as Eq. (4). From these equations, the LSS is obtained after a few iterations as

$$p = \begin{bmatrix} X \\ C \end{bmatrix} = \begin{bmatrix} 1.15 \\ 1.0 \end{bmatrix}, \; V_p = \begin{bmatrix} 0.06 & -0.046 \\ -0.046 & 0.04 \end{bmatrix} \tag{8}$$

Now let us transform the data and the covariance matrix according to the following rule;

$$y' = f(y) = \begin{bmatrix} y_1/y_3 \\ y_2/y_3 \\ y_3 \end{bmatrix} = \begin{bmatrix} 1.5 \\ 1.0 \\ 1.0 \end{bmatrix}, \; \left[V_{y'_{ij}}\right] = \left[\sum_{k,m} \frac{\partial y'_i}{\partial y_k} \cdot \langle \Delta y_k \cdot \Delta y_m \rangle \cdot \frac{\partial y'_j}{\partial y_m}\right] = \begin{bmatrix} 0.1125 & 0.06 & -0.06 \\ 0.06 & 0.05 & -0.04 \\ -0.06 & -0.04 & 0.04 \end{bmatrix} \tag{9}$$

Here, the covariance matrix was transformed based on the law of error propagation. According to this transformation, the "model" should be also transformed as

$$\boldsymbol{g}(\boldsymbol{p}) = \begin{bmatrix} X\cdot C \\ X\cdot C \\ C \end{bmatrix} \rightarrow \boldsymbol{g}'(\boldsymbol{p}) = \boldsymbol{f}(\boldsymbol{g}(\boldsymbol{p})) = \begin{bmatrix} \frac{X\cdot C}{C} \\ \frac{X\cdot C}{C} \\ C \end{bmatrix} = \begin{bmatrix} X \\ X \\ C \end{bmatrix} = \begin{bmatrix} 1 & 0 \\ 1 & 0 \\ 0 & 1 \end{bmatrix} \cdot \begin{bmatrix} X \\ C \end{bmatrix} = \boldsymbol{a}\cdot\boldsymbol{p} \tag{10}$$

Because of the data transformation, the model was thus linearized. By putting these quantities into Eqs (3) and (4), the results now become;

$$\boldsymbol{p} = \begin{bmatrix} X \\ C \end{bmatrix} = \begin{bmatrix} 0.88 \\ 1.24 \end{bmatrix}, \boldsymbol{V}_{\boldsymbol{p}} = \begin{bmatrix} 0.22^2 & -0.035 \\ -0.035 & 0.031 \end{bmatrix} \tag{11}$$

The answer X = 0.88 ± 0.22 is the one obtained in Eq. (3), i.e., the "strange" result obtained in the ordinal form of PPP. In this modified problem, X was obtained to be 1.15 in Eq. (8) first. On the contrary, by applying a transformation, we have a different answer given in Eq. (11). Therefore the LSSs before and after the transformation seemed disagree seriously, showing another paradox of the least–squares method. Zhao and Perey argued that we should call this difference of LSSs before and after a transformation as PPP. This is the second definition of PPP. Indeed, if we cut the third element of y' given in Eq. (9), the data vector and covariance matrix become exactly the same as those given in the original PPP, i.e., Eqs. (2) and (1), respectively.

In the next section, answers to these two definitions of PPP are explained based on the results described in Ref. 4.

4. Explanations

4–1 Invariance Criterion under Isomorphic Transformations

Let us first consider the problem of different LSSs caused by a data transformation explained in section 3. Because the transformation considered in Section 3 conserves the number of data, it corresponds to the "isomorphic transformation" defined in Ref. 4. According to Ref. 4, when the data y is transformed into another by using a set of n–functions f as y'=f(y) in an isomorphic way, the covariance matrix and the design matrix must be transformed in the covariant manner in order to get the same LSS before and after the transformation, i.e., $\mathbf{V}_{\mathbf{y'}} = \mathbf{c}\cdot\mathbf{V}_{\mathbf{y}}\cdot\mathbf{c}^t$ and $\mathbf{a}' = \mathbf{c}\cdot\mathbf{a}$, where c must be either one of the following three matrices;

$$c_{i,j}^{(-)} = \left[\frac{\partial f_i}{\partial y_j}\right],\ c_{i,j}^{(+)} = \left[\frac{\partial f_i}{\partial g_j}\right],\ \boldsymbol{c}^{(1/2)} = \frac{1}{2}\cdot(\boldsymbol{c}^{(-)} + \boldsymbol{c}^{(+)}) \tag{12}$$

These three matrices correspond to the backward, forward and central differentiations in evaluating the quantity (y'–g') in the form c·(y–g). Corresponding to the three choices of matrix c, there are three physically meaningful criterion to make the LSSs before and after the transformation invariant. Let us apply these criterion to the problem described in Section 3.

a) Criteria–1

In the transformation performed in Eq. (9), the function f is defined as

$$y' = f(y) = \begin{bmatrix} f_1(y_1,y_2,y_3) \\ f_2(y_1,y_2,y_3) \\ f_3(y_1,y_2,y_3) \end{bmatrix} = \begin{bmatrix} y_1/y_3 \\ y_2/y_3 \\ y_3 \end{bmatrix} \tag{13}$$

From this definition, the matrix $c^{(-)}$ is calculated as;

$$c^{(-)} = \left[\frac{\partial f_i}{\partial y_j}\right] = \begin{bmatrix} \frac{1}{y_3} & 0 & -\frac{y_1}{y_3^2} \\ 0 & \frac{1}{y_3} & -\frac{y_2}{y_3^2} \\ 0 & 0 & 1 \end{bmatrix} = \begin{bmatrix} 1 & 0 & -1.5 \\ 0 & 1 & -1 \\ 0 & 0 & 1 \end{bmatrix} \tag{14}$$

The covariance matrix and the design matrix in the transformed data space are calculated to be;

$$V_{y'} = c^{(-)} V_y c^{(-)t} = \begin{bmatrix} 0.1125 & 0.06 & -0.06 \\ 0.06 & 0.05 & -0.04 \\ -0.06 & -0.04 & 0.04 \end{bmatrix}, \; a' = c^{(-)} \cdot a = \begin{bmatrix} C & X-1.5 \\ C & X-1 \\ 0 & 1 \end{bmatrix} \tag{15}$$

By putting these quantities and y' given in Eq. (9) into Eq. (7), we get the same answer after a few iterations as given in Eq. (8), i.e., the same answer obtained before the transformation. The covariance matrix of transformed data obtained here is identical with that obtained by the usual law of error propagation, i.e., Eq. (9).

b) Criteria–2

According to this criteria, the covariance matrix and the design matrix must be transformed by using a matrix $c^{(+)}$;

$$c^{(+)} = \left[\frac{\partial f_i}{\partial g_j}\right] = \begin{bmatrix} \frac{1}{C} & 0 & -\frac{X}{C} \\ 0 & \frac{1}{C} & -\frac{X}{C} \\ 0 & 0 & 1 \end{bmatrix} \tag{16}$$

The covariant form of the transformed quantities is

$$V_{y'} = c^{(+)} V_y c^{(+)t}, \; a' = c^{(+)} \cdot a = \begin{bmatrix} 1 & 0 \\ 1 & 0 \\ 0 & 1 \end{bmatrix} \tag{17}$$

By putting these quantities into Eq. (7) and iterating a few times, the answer is again the same as given in Eq. (8).

c) Criteria–3

According to this criteria, the covariance matrix and the design matrix must be transformed in a covariant manner by using a matrix $c^{(1/2)} = 1/2 \cdot (c^{(+)} + c^{(-)})$. The answer is again the same as obtained in Eq. (8).

From the above calculations, it was shown that if the covariance matrix and the design matrix are transformed in a consistent manner according to either one of the three criterion, the LSS after the transformation is equal to the one obtained before the transformation. Although the covariance matrix of Eq. (15) was criticized to be incorrect[7], the correct answer can be still sought if the design matrix is also transformed in a correct manner. It will be worthwhile to note that the transformed quantities obtained according to the criteria–2 are equivalent with what was proposed by Zhao and Perey[7] which is to obtain these quantities with "averaged data" by using the inverse function $\mathbf{f}^{-1}$. This point will be explained later in more detail.

Then, let us transform the covariance and design matrices by neglecting these invariance criterion. For example, if

$$V_{y'} = c^{(-)} V_y c^{(-)t} = \begin{bmatrix} 0.1125 & 0.06 & -0.06 \\ 0.06 & 0.05 & -0.04 \\ -0.06 & -0.04 & 0.04 \end{bmatrix}, \; a' = c^{(+)} \cdot a = \begin{bmatrix} 1 & 0 \\ 1 & 0 \\ 0 & 1 \end{bmatrix} \tag{18}$$

the LSS becomes (X,C) = (0.88,1.24). It must be recognized that the covariance and design matrix used in this solution are exactly the same as Eqs. (9) and (10) which were used in obtaining Eq. (11). Therefore, the anomaly appeared in Section 3 was completely understood to be an improper combination of transformed covariance matrix and design matrix. This anomaly can be resolved by satisfying the invariance criterion proposed in Ref. 4. This is the answer to the second definition of PPP.

4–2 Impact of Truncation of a Data Space

In Ref. 4, a term "truncation" was defined as a transformation which does not conserve the dimensions of the data vector, its covariance matrix and the design matrix. By defining the operators α_i and β_i which delete the i–th raw and column of a matrix, respectively, the truncation is schematically written as

$$y' = \alpha_i y, \; V_{y'} = \alpha_i \beta_i V_y \tag{19}$$

It was shown in Ref. 4 that the least–squares solutions before and after this type of transformation were generally not equivalent and sometimes caused serious disagreement which is equivalent with the one appeared in PPP. Let us apply this result to PPP. The data vector **y** and covariance matrix V_y defined in Eqs. (1) and (2) are obtained by truncating the data vector **y**' of Eq. (9) and covariance matrix of Eq. (15) as follows;

$$\begin{gathered} y = \begin{bmatrix} 1.5 \\ 1.0 \end{bmatrix} = \alpha_3 \begin{bmatrix} 1.5 \\ 1.0 \\ 1.0 \end{bmatrix} = \alpha_3 [y' \textit{ of Eq. } (9)], \\ V_y = \begin{bmatrix} 0.1125 & 0.06 \\ 0.06 & 0.05 \end{bmatrix} = \alpha_3 \beta_3 \begin{bmatrix} 0.1125 & 0.06 & -0.06 \\ 0.06 & 0.05 & -0.04 \\ -0.06 & -0.04 & 0.04 \end{bmatrix} = \alpha_3 \beta_3 [V_y' \textit{ of Eq. } (15)] \end{gathered} \tag{20}$$

The design matrix **a** of Eq. (2) is obtained from that of Eq. (15) after replacing C by its estimate 1.0 (this is required because we cannot estimate two parameters out of two data) as;

$$a = \begin{bmatrix} 1 \\ 1 \end{bmatrix} = \alpha_3 \beta_2 [a' \textit{ of Eq. } (15)] \tag{21}$$

Therefore, it was proved that the quantities appeared in the original form of PPP are obtained by truncating the third dimension of a three–dimensional problem. According to the conclusions obtained in Ref. 4, a truncation of a data space totally destroys the "causality" of the LSS before and after the truncation. The problem of dimension is thus very important and sometimes serious.

In PPP as originally introduced, the off–diagonal term of the 2×2 covariance matrix is not zero. This fact indicates an existence of the third data which introduced the correlation between the two data. Therefore, the minimum dimension of this problem must be identified to be 3, and the correct solution must be sought in the 3–dimensional parent space. Once this is recognized, the correct answer is obtained as Eq. (8). This is the answer to the original definition of PPP.

4–3 A Method Proposed by Zhao and Perey

In Ref. 7, Zhao and Perey argued that the law of error propagation must be applied with the derivatives calculated with the "average" of the arguments, i.e.,

$$\frac{\partial y'_i}{\partial y_j} \rightarrow \frac{\partial y'_i(y_1,y_2,\ldots,y_n)}{\partial y_j}\Big|_{y=\langle y\rangle} \tag{22}$$

In Zhao and Perey's method, the estimate <y> is to be obtained iteratively by using the inverse transformation as $\langle y\rangle = f^{-1}(g'(p))$ which is inevitably equal to $f^{-1}(f(g(p))) = g(p)$. By replacing the average <y> by its estimate g, the above equation can be modified as

$$\frac{\partial y'_i(y_1,y_2,\ldots,y_n)}{\partial y_j}\Big|_{y=\langle y\rangle} = \frac{\partial f_i}{\partial y_j}\Big|_{y=g} = \frac{\partial f_i(g_1,g_2,\ldots,g_n)}{\partial g_j} \equiv c^{(+)}_{i,j}$$

$$\therefore\ V_{y'} = \left(\frac{\partial y'}{\partial y}\right)\cdot V_y \cdot \left(\frac{\partial y'}{\partial y}\right)^t = c^{(+)}\cdot V_y \cdot c^{(+)t} \tag{23}$$

The transformed covariance matrix obtained by their method is thus equivalent with that of Eq. (17). Although Zhao and Perey's derivation is based on a special form of transformation, it was proved here that their proposal was completely equivalent with the invariance criteria–2 derived in Ref. 4, which in turn is applicable to any type of isomorphic transformations.

5. Important Notice

So far, the authors have clarified the problem known as PPP and their interpretations of it. In this problem, the answer 0.882 which is outside the range of the underlying data could be proved to be incorrect for both definitions of PPP. However, this remark is not necessarily true.

Let a data vector **y** and its covariance matrix V_y be given below;

$$y = \begin{bmatrix}1.5\\1.0\end{bmatrix},\ V_y = \begin{bmatrix}0.15^2 & 0\\ 0 & 0.1^2\end{bmatrix} \tag{24}$$

The average of the two elements of the vector **y** is obtained by setting the design matrix **a** as $\mathbf{a} = (1,1)^t$. Putting these quantities into Eqs. (3) and (4), the average is obtained to be X = 1.15 ± 0.25. This is inside the range of the data. Then, by applying a linear–isomorphic transformation **F** defined as

$$F = \begin{bmatrix}0.1 & 0.9\\ 0 & 1\end{bmatrix}, \tag{25}$$

these quantities are transformed into

$$\boldsymbol{y}' = \boldsymbol{F}{\cdot}\boldsymbol{y} = \begin{bmatrix} 1.05 \\ 1 \end{bmatrix}, \boldsymbol{V}_{y'} = \boldsymbol{F}{\cdot}\boldsymbol{V}_y{\cdot}\boldsymbol{F}^t = \begin{bmatrix} 0.008 & 0.009 \\ 0.009 & 0.01 \end{bmatrix}, \quad (26)$$

$$\boldsymbol{a}' = \boldsymbol{F}{\cdot}\boldsymbol{a} = \begin{bmatrix} 1 \\ 1 \end{bmatrix}$$

The two elements of the transformed data vector y' has the same physical meaning as those of y because the design matrix is unchanged. This is a result of the special form of the transformation F adopted here. The LSS obtained from the transformed data is again X = 1.15 ± 0.25, which is **outside the range of the transformed data**, but **inside the range of the original data**. Therefore, it is no wonder to obtain a LSS outside the range of the underlying data if they are obtained from a set of other data by this kind of isomorphic transformation.

The above argument can be traced inversely to evaluate whether or not a LSS outside the range of the data is correct. If we can find a linear–isomorphic transformation F which diagonalizes the covariance matrix (i.e., $V_{y'} = F{\cdot}V_y{\cdot}F$ is a diagonal matrix) without changing the design matrix (i.e., a' = F·a = a), the data before and after the transformation have the same physical meaning. Because the transformed data (y' = F·y) is independent of each other, the average of these data is always inside them. Then, the problem is : "Are those transformed data reasonable?"

In the case of PPP in the original 2–dimensional problem, a linear transformation

$$\boldsymbol{F} = \begin{bmatrix} 0.622 & 0.378 \\ -1.544 & 2.544 \end{bmatrix} \quad (27)$$

diagonalizes the matrix V_y of Eq. (1) without changing the design matrix, i.e.,

$$\boldsymbol{y}' = \boldsymbol{F}{\cdot}\boldsymbol{y} = \begin{bmatrix} 1.311 \\ 0.229 \end{bmatrix}, \boldsymbol{V}_{y'} = \boldsymbol{F}{\cdot}\boldsymbol{V}_y{\cdot}\boldsymbol{F}^t = \begin{bmatrix} 0.281^2 & 0 \\ 0 & 0.346^2 \end{bmatrix}, \quad (28)$$

$$\boldsymbol{a}' = \boldsymbol{F}{\cdot}\boldsymbol{a} = \begin{bmatrix} 1 \\ 1 \end{bmatrix} = \boldsymbol{a}$$

In this interpretation, the original PPP is equivalent with a problem of seeking an average of the two data 1.311 ± 0.281 and 0.229 ± 0.346. Therefore, if the two dimensional data space given in Eq. (28) is the original data space of PPP, we must accept the answer of 0.882 to be correct because it lies inside the range [0.229, 1.311]. However, the covariance matrix of Eq. (28) has only two error components while in the original definition of PPP it is clear, from the statement of Peelle, that the <u>number of error components must be three</u>. This fact implies an existence of the third data, and then the complete data space of this problem must have a dimension of three. Therefore, the two dimensioal data space appeared in the original definition of PPP cannot be justified as a data space derived, by a linear–isomorphic transformation F^{-1}, from the uncorrelated two–dimensional space given in Eq. (28).

6. An Approximation

In the previous sections, it was shown that the problem of data transformation has significant impacts on the least–squares solutions. Even though a LSS outside the range of the underlying data may be sometimes considered to be correct, these "odd" results are more often obtained by a neglect of the invariance criterion and truncation of the data spaces. Whether or not such an answer is correct depends strongly on the history how the underlying data are derived.

In actual nuclear data evaluation, the final results are usually reported after truncating a lot of information. For example, if data y was obtained by subtracting the background from the foreground, i.e., if y = F – B, only the result y will be reported. In this example, the original data space has a dimension of 2 (F and B), while the transformed data has only a dimension of 1 (y). The final quantities, e.g., cross sections, found in the literatures and EXFOR are usually obtained by truncating a considerable amount of the data spaces in this way. Therefore, the correct methods described in Ref. 4 cannot be usually applied, and then there will be no guarantee to get the correct results from the LSM by using these "reported" data and their covariance matrices alone. Because of this reason, the LSS applied in actual nuclear data evaluations had often yielded strange results[(2)] which is outside the range of the data. A similar if muted effect would occur if a less-odd example of this type were offered. However, because the necessary information is lost, we cannot extend the dimension of data spaces which was possible in PPP fortunately. Then, when we get a result which is apparently incorrect, how can we handle it? Rigorously speaking, it is impossible to get the correct answer. However, we can have a better answer by assuming the relative error instead of absolute one to be correct. Let the relative error of a data y_i be r_i. Then, the error associated with y_i is written as $\Delta y_i = r_i y_i$. The (i,j) element of the covariance matrix is thus written to be $V_{ij} = \langle \Delta y_i \cdot \Delta y_j \rangle = \langle r_i \cdot r_j \rangle y_i \cdot y_j$, where $\langle r_i \cdot r_j \rangle$ denotes the (i,j) element of the relative covariance matrix. The authors proposed to replace this by $V_{ij} = \langle r_i \cdot r_j \rangle g_i \cdot g_j$ where g_i denote the least-squares estimate of y_i. By taking the least-squares estimate, fluctuation of the data is smoothed out partly. Therefore, this is a way to realize the invariance criteria-2 of Ref. 4 approximately. The details of our method, e.g., formulation, convergence criteria and numerical examples, are given in Ref. 2. For example, if this method is applied to PPP in the original form, the answer is 1.25 ± 0.28, which is much closer to the correct value 1.15 ± 0.25 than the original answer 0.88 ± 0.22. In the cases where the relative errors are more reliable than the absolute errors, this method is indeed rigorously correct.

We have applied this method to the evaluation of $^{115}In(n,n')^{115m}In$ reaction cross section. The curve A in Fig. 1 shows the result of usual LSM, which is lower than the underlying data in almost all energy interval, while curve B shows the result obtained by the method proposed in Ref. 2. The latter curve goes middle of the data base, and the integral measurement does support curve B more than curve A. The method proposed in Ref. 2 may be only an approximation. However, it is applicable to actual data evaluation, and as long as the authors know, gives better result than the usual LSM. It seems to be the only method to approximately realize the invariance criteria after considerable amount of information has been lost and unrecoverable.

7. Conclusions

Origins of a problem known as PPP were made clear to be an improper truncation of a data space and neglect of the invariance criterion which is required to make the LSSs before and after a data transformation equal. These remarks are also applicable to actual nuclear data evaluation, because experimentalists will report their results after transforming and truncating raw data. If the truncated information is not lost, the correct answer of the least-squares method can be obtained by recovering the lost dimensions and satisfying either one of the invariance criterion. On the contrary, if the truncated information is lost, the least-squares method may cause such strange results as shown in PPP. If such a result is obtained, an approximate method must be adopted. This paper described the PPP, its interpretations, the correct method, and an approximation. These consequences must be understood well by nuclear data evaluators and data analysists. From all of these discussions, it can be concluded that **the least-squares method is completely correct even if the number of data is as small as two and they are seriously discrepant, if the covariance matrix and design matrix are consistently treated and if the dimension of the problem is correct.**

Acknowledgements

The authors are grateful Dr. Z. Zhao of CNDC on recent communications on PPP. They also appreciate the comments from members of JAERI Nuclear Data Center.

References

1. R. W. Peelle, *"Peelle's Pertinent Puzzle"*, Informal memorandum dated October 13, 1987, Oak Ridge National Laboratory, Oak Ridge, Tennessee, U.S.A. (1987).
2. S. Chiba and D. L. Smith, *"A Suggested Procedure for Resolving and Anomaly in Least-Squares Data Analysis Known as Peelle's Pertinent Puzzle and the General Implications for Nuclear Data Evaluation"*, ANL/NDM-121 (1991).
3. Private communications among R. W. Peelle (ORNL), F. G. Perey (ORNL), Z. Zhao (CIAE), H. K. Vonach (IRK), F. H. Fröhner (KfK), D. L. Smith (ANL) and S. Chiba (JAERI), (1990).
4. S. Chiba, *"Impacts of Isomorphic Transformations and Truncations of Data Spaces on the Least-Squares Solutions"*, Submitted to NEANSC Specialists' Meeting on Evaluation and Processing of Covariance Data, Oak Ridge, Tennessee, U.S.A., October 7 – 9, 1992.
5. D. L. Smith, *"Probability, Statistics, and Data Uncertainties in Nuclear Science and Technology"*, An OECD Nuclear Energy Agency Nuclear Data Committee Series, Neutron Physics and Nuclear Data in Science and Technology, Vol. 4, American Nuclear Society (1991).
6. Z. Zhao, private communication (1990).
7. Z. Zhao and F. G. Perey, *"The Covariance Matrix of Derived Quantities and Their Combination"*, ORNL–TM 12106 (1992)

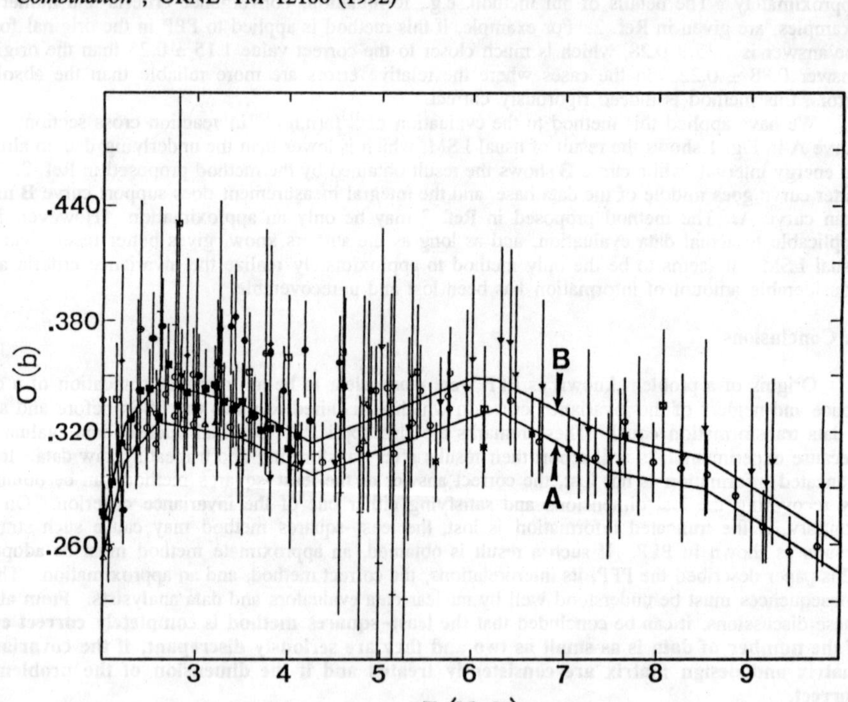

Fig. 1 The $^{115}In(n,n')^{115m}In$ reaction cross section. The curve A denotes a LSS obtained by the usual LSM. The curve B is a result obtained by a method proposed in Ref. 2. The points correspond to experimental data.

THE LEAST-SQUARES COMBINATION OF CORRELATED DATA IS ALIVE AND WELL

R.L.PEREL, J.J.WAGSCHAL and Y.YEIVIN
Racah Institute of Physics, The Hebrew University of Jerusalem, Israel

We discuss the least-squares procedure for averaging any number of correlated data. In fact, we consider the procedure in its general formulation which serves to determine m parameters by n correlated experimental data on different functions of the parameters. If m>n, the same formulation applies to the adjustment of a given library of m correlated parameters by a set of n relevant correlated data which may even be correlated to the given parameters. There are, however, certain restrictions on the application of the least-squares procedure. Combining averages over two partially overlapping sets of data, for instance, yields a result which is not equal to the average over the union of the two sets. We offer a general comment on the applicability of the least-squares method.

1. Background, Motivation and Goals

In 1987 R.W.Peelle distributed an informal memorandum in which he discussed the least-squares combination of two correlated data. The essential part of the memorandum is reproduced in Ref.1. In Peelle's particular example $x_1=1.0 \pm \sqrt{.05}$, $x_2=1.5 \pm \sqrt{.1125}$, and $\text{cov}(x_1,x_2)=.06$. The average of these data is $x=.882$. The fact that x falls outside the range (x_1,x_2) led Peelle to question the applicability of least-squares to combinations of correlated data. But the conventional average of two positively correlated data, for substantial correlation coefficients, is more likely to fall outside the data range than within it. And in Peelle's example the correlation coefficient is an impressive $\rho=.8$. Yet the question as to whether or not the standard averaging algorithm applies, arises as one considers the situation which entails the two correlated data. The scenario suggested by Peelle is that of an experiment in which two independent activation measurements are performed on the same foil. The *relative* uncertainties in the activation measurements are both 10%, and that in the foil mass is 20%. We may assume that the measured mass is m=1.0. Its variance is thus 0.04. The activations a_1 and a_2 are then 1.0 and 1.5, and their respective variances 0.01 and 0.0225. The activation cross section is x=a/m. In the foregoing example x is estimated by $\langle x_1,x_2\rangle$. It could just as well be estimated by $\langle a_1,a_2\rangle/m$, which is 1.154. The *questions*, therefore, are (a) why the two results are different; (b) if either is correct, which one is it; and (c) if neither is, what is the logically consistent estimate of x.

Many have participated in informal discussions and an extensive correspondence on what became known as Peelle's Puzzle. We are aware of two *publications* on the subject. The earlier is a report by Chiba and

Smith,[1] who argue that one should, in effect, employ the relative uncertainty matrix associated with x_1 and x_2. This leads to $\langle x_1,x_2\rangle=1.25$. However, the argument for preferring the relative uncertainties is not very convincing. The other report, by Zhao and Perey,[2] attributes the discrepancy between the two results to the nonlinearity of the derived quantity (x) as a function of the directly measured quantities (a,m). In this argument the authors are speciously right. And by means of a clever iterative procedure they demonstrate that the result $\langle x_1,x_2\rangle$ converges to $\langle a_1,a_2\rangle/m$. Yet, in our humble opinion, there is no need to fall back on the non-linearity of x. We will later return to this matter. We should also note the treatise on Data Uncertainties by Smith,[3] who mentions the matter as a demonstration of "certain hazards associated with attempting to circumvent the rigorous Bayesian approach to parameter estimation."

In view of all the above, our two-fold purpose is, first, to present the general formulation of the *conventional* least-squares procedure which serves to determine a number of parameters by a given set of experimental values of quantities which are functions of the parameters, to which quantities we will refer as *responses*. By conventional we mean that, whatever functions the responses are, we approximate them by their first-order Taylor expansions about the *given* values of the parameters. Our purpose, further, is also to discuss situations in which the procedure might lead to erroneous results, and should therefore be carefully applied. We recall that in considering correlated data, one can ascertain the applicability of the standard procedure only after the circumstances which give rise to the correlations are understood. We will discuss examples to which the procedure can be safely applied, and others which should be cautiously reformulated before the respective problems can be properly handled.

2. Some Elementary Observations

It is rather amusing that, once the correlation of the two data in Peelle's example is accounted for, the apparent Puzzle is hopefully laid to rest. However, before embarking on our main objective, we would still like to elaborate two of the above comments. We will first show that the *formal* average of two positively correlated data can easily fall outside the data range. In the absence of information on how the correlation came about, such an average may be the only possible estimate. And secondly, we will discuss an elementary example in which the experimental data are measurements of the parameter whose value we wish to determine. And yet we show that it is possible to produce two correlated estimates of the unknown parameter of which the *formally-correct* combination is clearly wrong.

2.1 *The averages of two positively correlated data*

The simple prescription to combine two experimental data, $x_1 \pm \sigma_1$ and $x_2 \pm \sigma_2$, in order to obtain their proper average, is to weigh each datum by the inverse of its variance. The average then falls within the range (x_1,x_2). However, when the data are correlated, the proper relative weights become the inverses of the respective-variance-minus-the-covariance, i.e.

$$x = \frac{(\sigma_2{}^2-\sigma_1\sigma_2\rho)x_1 + (\sigma_1{}^2-\sigma_1\sigma_2\rho)x_2}{\sigma_1{}^2+\sigma_2{}^2-2\sigma_1\sigma_2\rho} . \tag{1}$$

When the correlation coefficient ρ is positive, one weight might become negative, and the average x of the two input data might fall outside the range (x_1, x_2): it would lie above the range if $\sigma_2/\sigma_1<\rho$, and below it if $\sigma_1/\sigma_2<\rho$.

2.2 *The Combination of Partial Averages*

Consider three disjoint sets of independent experimental data on a quantity x, respectively consisting of n_1, n_2 and n_3 data. All the individual measurements are associated with identical variances, which we denote by v. The averages of x over each of the three sets are

$$m_i = \frac{1}{n_i}\sum x_j \quad . \quad . \quad . \quad . \quad i=1,2,3 \tag{2}$$

where the summation index j runs over the data in the respective set. The best estimate of the average of x, however, is based on all the data at our disposal, and is therefore

$$m_{123} = \frac{n_1 m_1 + n_2 m_2 + n_3 m_3}{n_1 + n_2 + n_3} , \tag{3}$$

and its variance is $v_{123}=v/(n_1+n_2+n_3)$.

And now let us consider the average of x over the union of sets 1 and 3, and its variance,

$$m_{13} = \frac{n_1 m_1 + n_3 m_3}{n_1 + n_3} , \quad v_{13} = \frac{v}{n_1 + n_3} , \tag{4}$$

and, similarly, the average m_{23} of x over the union of set 2 and set 3, and its variance v_{23}. The covariance of the two partial averages can be easily evaluated, and evidently is

$$c = \mathrm{cov}(m_{13}, m_{23}) = \frac{n_3 v}{(n_1+n_3)(n_2+n_3)} . \tag{5}$$

Hence, by Eq.(1), the properly weighted average of m_{13} and m_{23} is

$$M = \langle m_{13}, m_{23}\rangle = \frac{(v_{23}-c)m_{13} + (v_{13}-c)m_{23}}{v_{13} + v_{23} - 2c} . \tag{6}$$

Substituting the explicit expressions for the two variances and the covariance, the expression for M surprisingly reduces to

$$M = \frac{n_1 m_{13} + n_2 m_{23}}{n_1 + n_2} . \tag{7}$$

Thus M (provided $n_1, n_2, n_3 \neq 0$) is definitely different from m_{123}. The last equation, however, is most instructive in that it demonstrates that in particular circumstances M nevertheless coincides with m_{123}. First, if $n_3=0$, then trivially $M=m_{12}$. But secondly, if $n_2=0$, i.e. if m_{23} is actually m_3 so that the second "combined" set is a partial set of the first, then $\langle m_{13}, m_3\rangle = m_{13}$. Thus, when one is faced with the need to average two correlated data, one can never be sure whether the formal result makes any sense, that is unless one knows the story behind the correlation.

3. The General Formalism of Correlated Data Analysis

We now turn to our main objective: to present the general least-squares formalism for the analysis of experimental data. A central proposition of which we make use is that the conventional adjustment prescriptions are also most efficient in handling classical data-fitting problems. In order to determine N parameters by I given data on relevant responses, one should first use N of the given responses to evaluate a "first guess" for the parameters, which one could then adjust by the remaining I-N responses. Such analysis is applied to Peelle's example. The formalism, finally, is employed to demonstrate the validity of successive adjustments by independent data.

3.1 *The Adjustment of a Parameter Library by a Set of Measured Responses*

Consider a set of N parameters, such as the entries in a table of microscopic cross sections, to which we will refer as a parameter library, and which we regard as components of an N-dimensional vector p. Note that non-physical correction factors, introduced to account for calculational[4] or other uncertainties, might be also included in a parameter library. We denote the uncertainty matrix associated with the parameter vector p, the matrix $\langle \delta p \delta p^{\dagger} \rangle$, by C_p. Consider also a set of I measured responses. Responses, by definition, are functions of the parameters. We regard the responses as components of an I-dimensional vector r, and denote its associated uncertainty matrix by C_r. The given parameters and the measured responses might be correlated, and we denote their possible N×I cross-covariance matrix by C_{pr}.

Along with the measured responses we further consider their respective calculated values, which we denote by $\bar{r}_i = \bar{r}_i(p)$. The uncertainty matrix associated with the vector $\bar{r}$ is derived by the well-known Sandwich Rule (the Law of Uncertainty Propagation). It is $SC_pS^{\dagger}$, where S is the I×N sensitivity matrix $\{\partial \bar{r}_i / \partial p_n\}$.

The calculated responses, as a rule, deviate from their respective measured values, which after all is the primary motivation of Adjustment. The deviations $d=\bar{r}-r$ and their associated uncertainty matrix,

$$C_d = \langle \delta(\bar{r}-r)\delta(\bar{r}-r)^{\dagger} \rangle = \ldots = SC_pS^{\dagger} - SC_{pr} - C_{rp}S^{\dagger} + C_r \ , \quad (8)$$

play an important role in the adjustment formalism. The straightforward derivation of the adjustment prescription is briefly reviewed in Ref.5. In any event, the adjusted parameters p' are given by

$$p' = p + (C_{pr} - C_pS^{\dagger})C_d^{-1}d \ , \quad (9)$$

and their associated uncertainty matrix is

$$C_{p'} = C_p - (C_{pr} - C_pS^{\dagger})C_d^{-1}(C_{rp} - SC_p) \ . \quad (10)$$

Attention should be paid to the fact that the given responses r are also being adjusted. The adjusted responses r' are given by

$$r' = r + (C_r - C_{rp}S^{\dagger})C_d^{-1}d \ , \quad (11)$$

and their associated uncertainty matrix is

$$C_{r'} = C_r - (C_r - C_{rp}S^{\dagger})C_d^{-1}(C_r - SC_{pr}) \ . \quad (12)$$

Finally, the adjusted parameters and the adjusted responses are obviously correlated, and their cross-covariance matrix is

$$C_{p'r'} = C_{pr} - (C_{pr} - C_pS^{\dagger})C_d^{-1}(C_r - SC_{pr}) \ . \quad (13)$$

3.2 *The Proper Solution of Peelle's Original Example*

In Peelle's example the measured "responses" are a_1, a_2 and m_1. The parameter we wish to determine is $x=a/m$. The calculated responses, however, depend on both x *and* m. Furthermore, m is responsible for the correlation, so it must be considered as an explicit parameter. As we have already explained, we first use two of the given responses, say a_1

and m_1, to evaluate a first guess for the two parameters, so that

$$p_1 = x = a_1/m_1 = 1.0 \ , \quad p_2 = m = m_1 = 1.0 \ . \tag{14}$$

The variance of p_1 is $var(a_1)/m_1^2 + a_1^2 var(m_1)/m_1^4 = .05$, and that of p_2 is 0.04. The covariance of the parameters is $-a_1 var(m_1)/m_1^2 = -.04$. Thus

$$C_p = \begin{bmatrix} 0.05 & -0.04 \\ -0.04 & 0.04 \end{bmatrix} . \tag{15}$$

This given "library" should then be adjusted by the response $r = a_2 = 1.5$, of which the variance C_r is 0.0225, and which is *independent* of the parameters. The calculated response $a = p_1 p_2$, and so $\bar{r} = 1.0$, and $d = \bar{r} - r = -.5$. To evaluate C_d, which with no parameter-response correlations is just $SC_pS^\dagger + C_r$, we first evaluate the sensitivity profile of $\bar{r} = p_1 p_2$:

$$S = \left(\frac{\partial \bar{r}}{\partial p_1}, \frac{\partial \bar{r}}{\partial p_2}\right) = (p_2, p_1) = (1.0, 1.0) \ . \tag{16}$$

Therefore $SC_pS^\dagger$ is just the sum of all the elements of C_p which is given in Eq.(15), that is $SC_pS^\dagger = .01$, and thus $C_d = .0325$.

All in all, and noting that $C_d^{-1} d = -1/0.065$, we conclude that

$$p' = p - C_pS^\dagger C_d^{-1} d = \begin{bmatrix} 1.0 \\ 1.0 \end{bmatrix} + 15.38 \begin{bmatrix} 0.01 \\ 0 \end{bmatrix} = \begin{bmatrix} 1.1538 \\ 1.0 \end{bmatrix} . \tag{17}$$

It is also important to note that

$$r' = r + C_r C_d^{-1} d = 1.5 - \frac{0.0225}{0.065} = 1.1538 \ . \tag{18}$$

This means that $\bar{r}(p') = r'$, which *a posteriori* justifies the application of the standard procedure, the non-linearity of $\bar{r}(p)$ notwithstanding. Needless to say these *are* the correct values of x, m and a.

3.3 *The Applicability of the Formalism to Successive Adjustments*

Eqs.(8),(9) and (10), with no parameter-response correlations, reduce to

$$C_d = SC_pS^\dagger + C_r \ , \quad p' = p - C_pS^\dagger C_d^{-1} d \ , \quad C_{p'} = C_p - C_pS^\dagger C_d^{-1} SC_p \tag{19}$$

If we now consider the adjustment of a parameter library by a set of responses r, independent of the library, and which consists of two uncorrelated subsets r_1 and r_2, then the adjusted library p' is given by

$$p' = p - C_p(S^\dagger_1 \ S_2^\dagger) \begin{bmatrix} C_{d1} & S_1C_pS_2^\dagger \\ S_2C_pS_1^\dagger & C_{d2} \end{bmatrix}^{-1} \begin{bmatrix} d_1 \\ d_2 \end{bmatrix} , \tag{20}$$

where $C_{d1} = C_{r1} + S_1C_pS_1^\dagger$ and $C_{d2} = C_{r2} + S_2C_pS_2^\dagger$.

We will now prove the equivalence of this operation to that of first adjusting p by the subset r_1, which results in

$$p_1 = p - C_pS_1^\dagger C_{d1}^{-1} d_1 \ , \quad C_{p1} = C_p - C_pS_1^\dagger C_{d1}^{-1} S_1 C_p \ , \tag{21}$$

and then adjusting this library by the subset r_2, leading to

$$p_2 = p_1 - C_{p1}S_2^\dagger (C_{r2} + S_2C_{p1}S_2^\dagger)^{-1} [\bar{r}_2(p_1) - r_2] \ . \tag{22}$$

To demonstrate that $p' = p_2$, we rely on the matrix-inversion identity

$$\begin{bmatrix} C_{d1} & S_1C_pS_2^\dagger \\ S_2C_pS_1^\dagger & C_{d2} \end{bmatrix}^{-1} = R \begin{bmatrix} C_{d1} + S_1C_pS_2^\dagger T S_2C_pS_1^\dagger & -S_1C_pS_2^\dagger \\ -S_2C_pS_1^\dagger & T^{-1} \end{bmatrix} R \tag{23}$$

where

$$R = \begin{bmatrix} C_{d1}^{-1} & 0 \\ 0 & T \end{bmatrix} , \quad T = (C_{d2} - S_2C_pS_1^\dagger C_{d1}^{-1} S_1 C_p S_2^\dagger)^{-1} . \tag{24}$$

We also rely on the auspicious observation that this matrix T reduces to

$$T = (C_{r2} + S_2C_{p1}S_2^\dagger)^{-1} \ , \tag{25}$$

which is just the inverse of the C_d needed in the second-step adjustment. Substituting Eqs.(23),(24) and (25) in Eq.(20), and performing some straightforward though tedious algebra, the explicit expression for p' eventually reduces to p_2.

4. Summary and Conclusions

Our discourse started with the elementary exercise suggested by R.W.Peelle as an illustration of problems involved in averaging correlated data. As a matter of fact, correlated data should *always* be handled with great care. Unhappily not much can be said, and nothing should therefore be done, before the sources of the correlations are known *and* understood. While Peelle's particular example is by now adequately solved, the general problem he has raised, concerning the simplistic and carefree application of the least-squares method, persists. We have mentioned the formal combination of averages over partially-overlapping sets of data, of which the outcome is just wrong, as an example of such an application. A deeper analysis of the problem would show how to use the available data in order to produce the best estimate of the average. This already hints at the general proposition, on the applicability of least-squares, we eventually wish to formulate.

We have listed the general prescriptions of Adjustment. More specifically, we wrote down the explicit expressions for the final parameter library resulting from the adjustment of an initial library by experimental data on a set of responses, and for the (reduced) uncertainties associated with the adjusted library. We explained how this *formulaire* also applies to data-fitting problems, and we applied it to Peelle's example. We then discussed successive adjustments of a library by additional data on responses, and proved that it works as long as the added data are absolutely independent of the data which have already participated in the adjustments. Further, our experience with response-uncertainty matrices which are not block diagonal, showed that successive adjustments failed miserably.

There obviously remains a lot to be done. For instance, we have tried to imagine a situation which leads to a true "average" outside the range of the given data, and now believe it does not exist. Can this be rigorously proven?

We do not, at this time, offer a formal proposition on the applicability of least-squares, yet we would like to share our experience in cross-section adjustment with The Evaluators: A given evaluation of a set of parameters can be safely improved (i.e. adjusted) by additional data only if these data are truly independent of any information used in the evaluation.

References

1. S.Chiba and D.L.Smith, U.S. Report ANL/NDM-121 (1991).
2. Z.Zhao and F.G.Perey, U.S. Report ORNL/TM-12106 (1992).
3. D.L.Smith, *Probability, Statistics, and Data Uncertainties in Nuclear Science and Technology* (American Nuclear Society, 1991). pp.205-9
4. Y.Yeivin et al., in *Proc. Conf. 1980 Advances in Reactor Physics and Shielding*, ANS (1980). pp.705-718
5. D.Shalitin et al., in *Reactor Dosimetry: Methods, Applications and Standardisation*, H.Farrar and E.P.Lippincott (Eds.), ASTM-STP-1001 (1989). pp.399-404

FITTING AND EVALUATION TECHNIQUES FOR DELAYED NEUTRON DATA AND FISSION YIELDS

D R WEAVER
School of Physics and Space Research, The University of Birmingham, Edgbaston, Birmingham, B15 2TT, England

R W MILLS
School of Physics and Space Research, The University of Birmingham, Edgbaston, Birmingham, B15 2TT, England

and

M F JAMES
Roevin Management Services Ltd, Park Road, Timperley, Altrincham, Cheshire, WA14 5QH

ABSTRACT

Delayed neutron data in evaluated files originate from two sources: (i) experimental results of measurements on delayed neutron parameters themselves, and (ii) from summation of results for individual fission products. One example from each route is discussed: (i) Derivation of parameters and, in particular, covariance matrices from delayed neutron spectrum measurements. (ii) Evaluation of fission yields, including the use of the normalized residual technique in averaging experimental data and in treating outliers. In addition some thoughts are presented on coping with covariance between large files of input data, as is the case in fission yields.

1. Introduction

The parameters of delayed neutron emission useful for technological application such as the total number of delayed neutrons per fission, $\overline{\nu}_d$, the Keepin six-group time dependence descriptors, a_i and λ_i, and the six-group spectra are usually derived from one of two sources:

(i) direct experiments (e.g. Keepin[1]),

(ii) from summation calculations based on data for individual delayed neutron precursors (e.g. Brady and England[2], and Mills et al[3]).

A third technique uses empirical models based on systematics (e.g. Tuttle[4]); this is employed when insufficient experimental data are available for the use of the other two methods - often the case with the more exotic fissioning system.

This paper will draw on examples of evaluation techniques drawn from experience of the first two of these sources.

2. Determination of covariance matrices from delayed neutron spectral measurements

2.1. The *Spectrometry System*

Measurements have been made in Birmingham of the spectra of delayed neutrons from fast neutron fission in ^{235}U; see Walker et al[5]. The technique for determining the full response function of the ^{3}He spectrometer and the covariance matrix for the resultant spectra was described in detail in a paper by Owen et al[6]. The relationship between the true neutron spectrum **Φ(I)** and the measured pulse-height spectrum, **C(J)** may be written as follows:

$$\mathbf{C(I)} = \sum \mathbf{R(I,J)\,\Phi(J)} \qquad (1)$$

where **R(I,J)** is the response function of the detector, and **I** = 1 to **I'** and **J** = 1 to **J'** represent the energy spectra channels of the true spectrum and the measured pulse height respectively. On the basis of knowledge of the response function from separate experiments and a measurement of the pulse height spectrum derived from a delayed neutron measurement, one can hope to invert Eq.(1) and derive an estimate of the true spectrum. Clearly the problem cannot be solved for a greater number of channels in the true spectrum than in the measured spectrum (**I' > J'**). While solutions for cases with **I' < J'** can be obtained by least squares analysis, it is often the case that the maximum amount of information on the true spectrum shape is desired; thus the limiting case of **I' = J'** is employed in which case the matrix **R(I,J)** is square and the unfolding procedure consists of determining the inverse of **R(I,J)** either analytically or, more usually, iteratively (see Owen et al[6] and Rudstam et al[7]).

The uncertainty in the derived spectrum, **Φ(I)**, depends upon the measurement uncertainties of the pulse-height spectrum **C(J)**. It also depends, however, on the uncertainties in the response function **R(I,J)** which was determined in the separate experiments referred to earlier. Many reports of spectral measurements have concentrated upon the effects of the uncertainties in **C(J)**, but relatively few allow fully for the effects of uncertainties in the response function though Perey[8] and McCracken et al[9] had noted the problem. Indeed McCracken had introduced a mechanism into the code RADAK which allowed for the effects of the response function by a Monte Carlo sampling technique but, while this provided an approach to the problem, it was felt by the Birmingham group that a more complete method was required. This concern led to the method of analysis described in Owen et al[6].

The response of a ^{3}He neutron spectrometer can be described by the combination of components. First there is a near Gaussian shaped peak at an energy corresponding to the incident neutron energy, $\mathbf{E_n}$, plus the Q-value of the ^{3}He(n,p)T reaction, 764keV. In practice two Gaussians are usually used in order to represent the slightly asymmetric shape of the observed spectrum but the combination is still clearly identifiable as an obvious single peak. It is the fact that the reaction gives this well-defined peak that makes this spectrometer so useful, however there are a number of competing reactions that have to be taken into account when describing the full response function.

First the cross section for the $^3He(n,p)T$ reaction is greatest at low energies; thus even with significant thermal neutron shielding, there is always a component in the spectrum at about 764keV corresponding to thermal neutron events. This problem is usually solved by limiting the range of $\mathbf{E_n}$ for which the spectrum is unfolded to that part >20 keV.

Secondly, there is a finite chance that one of the charged particles from the $^3He(n,p)T$ reaction may collide with the wall of the detector before giving up all its energy to the filling gas. This leads to a wall-effect continuum at energies less than the full energy peak.

Thirdly, elastic collisions of the incident neutrons with the components of the filling gas lead to the usual rectangular recoil counts spectra. The major contributors to this effect in the ^3He spectrometers used in Birmingham are the ^3He itself and protons from the small component of methane in the counter.

Several different parameterisations have been used over the years for the two Birmingham spectrometers, however for consistency with Owen et al[6] the following discussion will use the parameterisation described in that paper; changing the number of parameters used leads to no fundamental difference to the technique for determining the covariance matrix of the final neutron spectrum. In Owen et al [6] the response was described by

$$\mathbf{Y(E)} = \mathbf{P}(1)\left(\exp\left[-0.5\frac{(\mathbf{E}-\mathbf{E_n})^2}{\mathbf{P}(2)}\right] + \mathbf{P}(3)\exp\left[-0.5\frac{(\mathbf{E}-\mathbf{E_n}-\mathbf{P}(4))^2}{\mathbf{P}(5)}\right] + \mathbf{P}(6)\,\omega(\mathbf{E})\right) \quad (2)$$

where $\mathbf{Y(E)}$ is the magnitude of the pulse height spectrum at energy $\mathbf{E}$; $\mathbf{E_n}$ is the energy of the incident neutrons. In Owen et al[6] $\omega(\mathbf{E})$ is the wall effect spectrum alone, however later work added components to allow for the recoil spectra.

P(1) is an overall normalisation factor related to the efficiency of the detector for neutron incident at energy $\mathbf{E_n}$ and was described by two separate functions η_L and η_H for the low and high energy ends; these were obtained by fitting polynomials to the efficiency curve measured by use of a series of mono-energetic test spectra. **P(2)** through **P(6)** are the parameters describing the shape of the response function and these were determined by non-linear least squares fitting of responses to mono-energetic test spectra. The variations of **P(2)** through **P(6)** as a function of the test energy were plotted and fitted to polynomials of up to fourth order (later work used even higher order fits). Take a couple of examples: the parameter **P(5)** might be described with a third order polynomial with respect to the energy E of the monoenergetic test spectrum, thus:

$$\mathbf{P(5)\ at\ energy\ E} = q_{50} + q_{51}\mathbf{E} + q_{52}\mathbf{E}^2 + q_{53}\mathbf{E}^3 \quad (3)$$

Similarly **P(4)** might be described by a second order polynomial

$$\mathbf{P(4)\ \text{at energy}\ E = q_{40} + q_{41}\,E + q_{42}\,E^2} \tag{4}$$

The procedure has been described in more detail in a laboratory report, Weaver [8].

Thus it was possible to predict the spectrometer's response for any incident energy by interpolating with these polynomials to derive the five parameters **P(2) - P(6)**, plus the efficiency term **P(1)** and from these to determine the distribution using Eq.(2).

2.2. *Covariance Matrix determination*

The relationship in Eq.(1) may be represented in matrix terms by

$$\mathbf{C = R\,\Phi} \tag{5}$$

from which the result, Φ, is derived by the spectrum unfolding.

Consider for a moment the normal propagation of errors in the relationship:

$$\mathbf{y = f(x_1, x_2)} \tag{6}$$

The variance of y is given, again in matrix terms, by

$$\sigma_y^2 = \mathbf{D^T\,V_x\,D} \tag{7}$$

where $\mathbf{V_x}$ is the covariance matrix for the **x** parameters:

$$\mathbf{V_x} = \begin{pmatrix} \sigma_{x_1}^2 & \mathrm{cov}(x_1, x_2) \\ \mathrm{cov}(x_1, x_2) & \sigma_{x_2}^2 \end{pmatrix}, \quad \mathbf{D} = \begin{pmatrix} \partial f / \partial x_1 \\ \partial f / \partial x_2 \end{pmatrix}$$ and $\mathbf{D^T}$ is the transpose of **D**.

If, instead, **y** is a vector with **k** dimensions which is derived from $\mathbf{x_i}$ where **i** = 1 to **n**:

$$\mathbf{y = y(x_1 \ldots x_i \ldots x_n)} \tag{8}$$

the covariance matrix for **y** can still be written as:

$$\mathbf{V_y = D^T\,V_x\,D} \tag{9}$$

where **D** is now a matrix of dimension **k** by **n**

$$D = \begin{pmatrix} \frac{\partial y_1}{\partial x_1} & \frac{\partial y_2}{\partial x_1} & \cdots & \cdots & \frac{\partial y_k}{\partial x_1} \\ \cdots & \cdots & \cdots & \cdots & \cdots \\ \frac{\partial y_1}{\partial x_n} & \cdots & \cdots & \cdots & \frac{\partial y_k}{\partial x_n} \end{pmatrix} \tag{10}$$

Now the unfolding problem can be seen as deriving the neutron spectrum as a function of the input parameters:

$$\Phi = \Phi(C, Q_1, \ldots, Q_m) \tag{11}$$

where the Q_i represent the polynomial coefficients that describe the efficiency and response function shape. Thus the terms of the count spectrum vector, **C**, and the polynomial coefficients Q_i are the equivalents of the x_i in Eq.(8). Hence, as long as the equivalents of V_x and **D** in Eq.(9) can be determined, the covariance matrix for Φ, Var Φ, can be obtained.

As a result of the way that the components were obtained V_x is in fact equal to:

$$\begin{vmatrix} \text{Var } C & & & & & & & \\ & \text{Var } \eta_H & & & & & 0 & \\ & & \text{Var } \eta_L & & & & & \\ & & & \text{Var } P(2) & & & & \\ & & & & \text{Var } P(3) & & & \\ & & & & & \text{Var } P(4) & & \\ & 0 & & & & & \text{Var } P(5) & \\ & & & & & & & \text{Var } P(6) \end{vmatrix}$$

Var P(2) through **Var P(6)** represent the covariance matrices derived from the polynomial fits made as in Eq.(3) and Eq.(4); thus **Var P(5)** would be a 4 by 4 matrix from Eq.(3), and **Var P(4)** would be a 3 by 3 matrix from Eq.(4). Because these fits were done separately they are effectively independent of each other and the off diagonal terms are zero, as shown.

Ideally a grand overall fit would have been done to the response function shape in Eq.(2) over the range of test energies E_n, and this would have identified the off diagonal values for the covariance between the individual parts (but this was not practicable with the computing facilities available at the time).

The **D** matrix was estimated by performing the unfolding operation with all the parameters set to their predefined values. Then a further unfolding was performed with one parameter, $\mathbf{Q_i}$, modified by $\mathbf{f}\,\sigma_{\mathbf{Q_i}}$ where **f** was some fraction between +1 and -1 and $\sigma_{\mathbf{Q_i}}$ is the standard deviation of $\mathbf{Q_i}$. The differentials in **D** were then estimated by approximating with:

$$\partial\boldsymbol{\Phi}_1 / \partial\mathbf{x}_i \approx \Delta\boldsymbol{\Phi}_1 / \mathbf{f}\,\sigma_{\mathbf{Q}i} \quad \text{etc.}$$

A similar process was then followed whereby the counts' spectrum was perturbed one channel at a time in order to determine the effect on **Var** $\boldsymbol{\Phi}$ of the uncertainties in the individual channels of the counts' spectrum.

2.3. Comments

The technique could be used successfully as long as a few points were observed. First, it was necessary to ensure that the appropriate value of **f** was used for each parameter. For some, quite a large fraction could be used, indeed had to be used, so that the size of $\Delta\boldsymbol{\Phi}_i$ was significant. For others, very small values of **f** had to be used because of the sensitivity of the resulting spectrum to that particular parameter. Secondly it was necessary to test that the perturbation was equally well predicted by positive or negative values of **f**; this ensured that the approximation was reasonably valid being in a region where the perturbation in was linear with the change in the parameter value.

The overall result was a covariance matrix that could describe the uncertainties in the individual channels of the unfolded neutron spectrum, and the correlations between them. It transpired that the major effects of the uncertainties in the response function were at the low energy end of the spectrum, where they amounted to about 50% of the effect of the uncertainties in the counts. In the medium to high energy portions of the spectrum the response function uncertainties contributed only 20% or less to the overall uncertainty.

An interesting observation from the unfolding of spectra was that some channels could have predictions of **negative fluxes**. These are clearly unphysical but are simply the manifestation of the randomness of the distribution of best estimate values about a zero true mean. It is felt that unfolding programs that constrain spectra to be positive in all channels will introduce a bias into the results. This arises since the response to a monoenergetic incident neutron is not a delta function and thus there has to be a contribution in lower energy parts of the counts' spectrum if a high energy component exists. If the effect of constraining an unfolded spectrum to be always positive is, in fact, to add neutrons to a particular part of the spectrum to eliminate a "false" negative prediction, then this must result in the corresponding wall effect also being added to the lower energy part. This immediately introduces a bias in these lower energy channels. It is our view that constraining the results of unfolding to be always positive is unwise.

3. The Normalised Residual and Its Use in Averaging Fission Product Yield Data

The second method of determining delayed neutron parameters is to sum the effects of the physical precursors. As there are some 271 known precursors, it is necessary to be able to evaluate both their cumulative yields and their **Pn** values if summation is to be performed; for example the total delayed neutron yield, $\overline{\nu}_d$, can be obtained through:

$$\overline{\nu}_d = \sum_i Pn_i \, Y_{ci} \qquad (12)$$

where Pn_i is the **Pn** value of the **i**-th precursor and Y_{ci} is its cumulative yield.

A problem that exists in evaluating data is that experimental results do not always conform to the requirements for the application of normal statistical averaging rules. If data is well behaved it follows that:

(i) Each measurement is unbiased,

(ii) The measurements are independent and uncorrelated,

(iii) The standard deviation of each measurement is accurately known so that it can be used in the weighting of the result in the averaging process.

Results for a recent evaluation of fission yields [James et al[11] – James et al[14]] have shown that on the whole measurements of fission yields are generally unbiased, independent and with reasonably well known uncertainties. There are, however, a significantly large fraction of outliers; see James et al[15] and Figure 1.

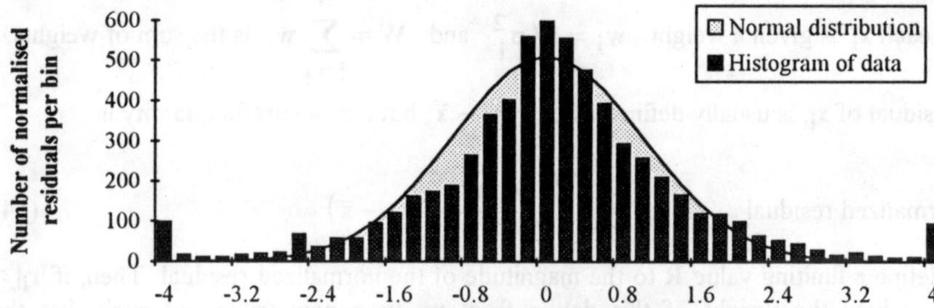

Figure 1. Normalised residuals for the dataset of all chain yields compared with the normal distribution N.B. Residuals of magnitude greater than 4 are shown at either -4 or +4 as appropriate

Ideally, a review of the original experimental data should lead to the revision of the discrepant results; however, this may often have been attempted to no avail, and anyway may be impractical when there are very large numbers of data values, as in fission yield evaluation.

Traditional techniques for removing outliers rely on (i) removal of data which contribute more than a specified amount to the χ^2, or (ii) calculation of the internal and external standard deviations of the mean and the larger taken as the recommended value - if the external is the larger the process is equivalent to multiplying the standard deviation of each measured value by the square root of χ^2/degrees of freedom (the Birge ratio). The first method has the disadvantage that data are actually eliminated from the set and there have been cases where outlying data have ultimately been accepted as correct - see references in [James et al [12]]. In addition there is a certain arbitrariness in the cut-off value beyond which a datum is regarded as discrepant. The second method has the disadvantage that all data are treated equally and have their standard deviations renormalized irrespective of whether they are outliers or not.

Thus the method of the "normalized residual" was introduced to the evaluation process for the fission yield data sets. Suppose **N** measurements have been made of a quantity **X**. Let the **i**-th measurement be $\mathbf{x_i}$ with a standard deviation σ_i, and suppose that an estimation of **X** is made by calculating a weighted mean $\bar{\mathbf{x}}$:

$$\bar{\mathbf{x}} = \sum_{i=1}^{N} \mathbf{w_i \, x_i} \, / \, \mathbf{W} \tag{13}$$

where each $\mathbf{x_i}$, is given a weight $\mathbf{w_i} = 1/\sigma_i^2$ and $\mathbf{W} = \sum_{i=1}^{N} \mathbf{w_i}$ is the sum of weights.

The residual of $\mathbf{x_i}$, is usually defined as $\rho_i = \mathbf{x_i} - \bar{\mathbf{x}}$, but a more useful quantity is

the normalized residual $\quad \mathbf{r_i} = \sqrt{\dfrac{\mathbf{w_i W}}{(\mathbf{W} - \mathbf{w_i})}}\,(\mathbf{x_i} - \bar{\mathbf{x}}) \qquad (14)$

Now define a limiting value **R** to the magnitude of the normalized residual. Then, if $|\mathbf{r_i}| >$ R, we reduce the weight of this datum from $\mathbf{w_i}$ to a new value $\mathbf{w_i}'$ such that the normalized residual has a magnitude **R**. The adjusted weight is:

$$\mathbf{w_i}' = \mathbf{w_i}\left[1 - \mathbf{W}\left(\mathbf{r_i}^2 - \mathbf{R}^2\right) / \left(\mathbf{W}\,\mathbf{r_i}^2 - \mathbf{w_i}\,\mathbf{R}^2\right)\right] \tag{15}$$

The mean becomes $\quad \bar{\mathbf{x}}' = \bar{\mathbf{x}} - \left(\mathbf{w_i} - \mathbf{w_i}'\right)\left(\mathbf{x_i} - \bar{\mathbf{x}}\right) / \mathbf{W_{new}} \qquad (16)$

where $\mathbf{W_{new}} = \mathbf{W} - \mathbf{w_i} + \mathbf{w_i}'$ is the new sum of weights. The value of χ^2 is reduced by $\left(\mathbf{r_i}^2 - \mathbf{R}^2\right)$.

The values of the normalized residuals for the other data have to be recalculated. Of course, this procedure may still be followed by a second adjustment using the Birge factor.

It should be immediately obvious that the technique cannot be used when there are just two measurements as their normalised residuals will be equal and opposite. It is also important to note that the method cannot be applied if the distribution of data falls into two or more subsets as this indicates there are systematic effects that cannot be reconciled by the application of statistical averaging. It must be emphasised that, if at all possible, reference should be made to the original sources of the data in order to resolve discrepancies **before** the application of methods such as the normalized residual.

4. Some Thoughts About the Evaluation of Large Sets of Data Which Have Correlations Between Individual Elements.

Standard statistical techniques provide the formulation for the least squares analysis of data which are correlated. However, in some cases - and fission yields clearly fall into this category, the size of the data files and the sparseness of the correlation information in the original literature makes the task of formulating the full covariance matrix of the input data beyond the realms of possibility. There are however, significant degrees of correlation in the data. For example, fission yield measurements divide into three categories:

(i) absolute

(ii) ratio - where one yield has been measured relative to another, supposedly, well-known yield

(iii) ratio of ratio - where the ratio of two yields for fission induced one nuclide is compared with the same ratio for fission in a well known nuclide (usually thermal fission in ^{235}U).

So far it has been impractical to attempt to combine all these data into one grand least squares analysis; there are 11750 absolute, 1300 ratio and 1500 ratio of ratio results in our dataset. As a result, a step by step approach has been applied whereby the results of averaging only the absolute values have then been used to convert the ratio results to an absolute basis prior to a second round of averaging, and then finally the ratio of ratio results have been included once the "well known" fission system has been finalised. Fortunately the standard results from the absolute measurements and the "well known" fission system are by definition known accurately, their values hardly change in the second stage and no further iterations are required.

We believe, however, that there is a possibility that we can incorporate into our next evaluation an attempt at performing a least squares analysis of the whole dataset which will have the effect of including both the ratio and ratio of ratio results at the first stage. We are encouraged by the analysis of Smith[16], particularly that in his example 12.3. However, there is a significant difference in the size of the fission yield problem compared with that example.

Alternative concepts being explored are (i) that the ratio measurements be used as constraints on a least squares analysis of the absolute results, or (ii) that consideration of the relative magnitudes of the correlations between some of the weakly related values may allow the simplification of the problem by ignoring small correlation terms. Whatever the outcome of that investigation, we would reiterate the plea that measurers record full information, in particular on the values and uncertainties of standards employed in their work. so that the evaluators' task does not require undue recourse to assumptions about what may or may not have been done.

5. Acknowledgements:

The work was funded at various times by SERC, CEGB, BNFL and UKAEA. The contributions of J Walker, J G Owen, and S Chilton to the ^{3}He spectrometry are gratefully acknowledged. as are the helpful discussions with many colleagues, in particular A Wahl, T R England, and M Brady.

6. References

1. G R Keepin, *Physics of Nuclear Kinetics*, (Addison-Wesley, 1965).
2. M C Brady and T R England, *Nucl. Sci. and Eng.* **103**,(1989) 129.
3. R W Mills, M F James, and D R Weaver, *Proceedings of the International Conference on Nuclear Data for Science and Technology, Jülich, "Study of the delayed neutron yield and its time dependence by the summation method"*, (1991) 86-88
4. R J Tuttle, *Proceedings of the Consultants' Meeting on Delayed Neutron Properties, Vienna*, (1979) 26-30
5. J Walker, D R Weaver, J G Owen and S Chilton, *Radiation Effects* **94** (1986) 101
6. J G Owen, D R Weaver and J Walker, *Nuclear Instruments and Methods* **188**, (1981) 579-593.
7. G Rudstam, *The Swedish Councils' Laboratory Report LF-57*, Studsvik, (1974).
8. F G Perey, *Oak Ridge National Laboratory Report ORNL/TM-6267 (ENDF-259)* (1978)
9. A K McCracken and M J Grimstone, *OECD-NEA Specialists' Meeting on Sensitivity Studies and Shielding Benchmarks,* (1975).
10. D R Weaver, *Birmingham Radiation Centre Report BRC 82-02 "A Covariance Cookbook"* (1982).
11. M F James, R W Mills and D R Weaver, *Progress in Nuclear Energy*, **26**, (1991) 1-29
12. M F James, R W Mills and D R Weaver, *UKAEA report AEA-TRS-1015* (1991)
13. M F James, R W Mills and D R Weaver, *UKAEA report AEA-TRS-1018* (1991)
14. M F James, R W Mills and D R Weaver, *UKAEA report AEA-TRS-1019* (1991)
15. M F James, R W Mills and D R Weaver, *Nuclear Instruments and Methods*, **A313**, (1992) 277-282.
16. D L Smith, *Probability, Statistics and Data Uncertainties in Nuclear Science and Technology* (OECD 1991)

DATA FITTING AND EVALUATION TECHNIQUES FOR NEUTRON SPECTRA

W. MANNHART
Physikalisch-Technische Bundesanstalt
Bundesallee 100
D-3300 Braunschweig
Fed. Rep. of Germany

ABSTRACT

The methodoloy of a neutron spectrum evaluation is reviewed. The various steps in the spectrum evaluation process are discussed and possible alternatives are shown. Details of the procedure are demonstrated in the example of an evaluation of the prompt fission neutron spectrum of californium-252. The substantial differences between model-free and model-dependent evaluations are outlined.

1. Introduction

The aim of an evaluation is the generation of a "best set" of recommended data with realistic uncertainty estimates which represent the actual status of the physical knowledge of a quantity of interest based on the information gained from experiments and, if possible, from theoretical models. In the evaluation process of energy-dependent quantities, for example of neutron energy spectra N(E), a distinction must be made between two different cases. The first case is given when besides a variety of experimental data points $N(E_i)$, each taken at a different and experiment-specific energy E_i, a realiable physical model of N(E) is available. Then the evaluation process is essentially a data-fitting process which uses the original experimental data and their uncertainties to derive a set of parameters belonging to the physical model. This means that all the available information is concentrated on a few parameters. It is obvious that the quality of the model describing the underlying physical process is an essential point in such a strong data reduction procedure. The second case is when a credible physical model is lacking. In such circumstances, which quite often occur, the evaluation process is confined to a reduction of the experimental data base alone and requires the additional introduction of a common energy grid valid for all the different experiments involved in the evaluation procedure.

In the present work, the example of a past evaluation of the neutron spectrum of spontaneous fission of californium-252[1,2] is used to outline the typical methodology of a neutron spectrum evaluation. The evaluation procedure corresponds essentially to the "case 2" method, but also some aspects of a "case 1" evaluation are discussed. The numerical examples given exceed that of the previous evaluation and contain additional recent results[3].

For none of the fission neutron spectra is such a broad experimental data base available as for the neutron spectrum of Cf-252. However, when scanning the literature one is often surprised to find how sparse the available documentation is even for experiments which claim a high level of precision for their data. The recent development of evaluation methodology (see Ref. 4, for example) gave us very powerful tools for evaluation, but the quality of each evaluation depends mainly on the exactness and the correct interpretation of the input data and not so much on the evaluation technique. Indeed, the requirements on the input data have become more stringent, because today's evaluation codes need a quantified uncertainty covariance matrix for the input. Only where there is such information can the full potential of these codes be used for an objective evaluation based on sound statistical methods.

2. Steps of the Evaluation Procedure

2.1. Experimental Data Base

Measurements of the Cf-252 neutron spectrum have been published since 1955. Many of these measurements were analyzed in terms of a Maxwellian distribution:

$$N(E) = (2/\pi^{1/2})\, T^{-3/2}\, E^{1/2}\, e^{-E/T} \tag{1}$$

In the early eighties there was a peak in new experimental efforts accompanied by intense attempts to describe the spectrum on the basis of nuclear reaction theories (see Refs. 5-7). The new experimental investigations identified a few critical corrections which very strongly influence the final result of the data analysis. A good overview of these aspects is given in Refs. 8 and 9. The selection of the experimental data finally used in the evaluation was very much governed by two points, whether the advanced data correction techniques were taken into account in the experiments, and whether the documentation of the experimental procedure, which is mandatory for realistic estimates of the data covariance matrix, was complete. The data base and details of the individual experiments are listed elsewhere[1].

2.2. Generation of the Data Covariance Matrices

The generation of these matrices, which comprise on a large scale a description of the details of the experimental procedure, only makes sense when it is based on fundamental facts. Each experiment was therefore carefully checked for the completeness and validity of the experimental method.

The experimental data were related to a common set of constants and were modified in the case of incomplete or missing corrections. Most of the experimental data were quoted as numerical values of the spectrum at discrete energies, i.e. as $N(E_i)$. All experiments used in the evaluation were based on time-of-flight (TOF) measurements. The energy distribution N(E) of the neutron spectrum was derived from a measured time distribution n(t) with

$$N(E) = \frac{|\,\partial t/\partial E\,|\; n(t)\, C(E,\Omega)}{\nu\, (N_f/\epsilon_f)\, (\Omega/4\pi)\, \epsilon(E)} \tag{2}$$

The neutron source strength is given by ν, the average number of neutrons per fission, and by N_f the number of counted fission events and ϵ_f the efficiency of the fission detector. Ω is the solid angle and $\epsilon(E)$ the efficiency of the neutron detector. $C(E,\Omega)$ is a complex direction dependent correction function, and the transformation between time and energy is given by $| \partial t/\partial E |$ with

$$E = m_0 \{ [1 - L^2/(c^2t^2)]^{-1/2} - 1 \} \tag{3}$$

m_0 being the neutron rest mass, L the length of the neutron flight path and c the value of the velocity of light. The quantity n(t) of Eq. (2) has already been corrected for random coincidences and background effects. But n(t) is also a derived quantity. Due to the final width (2τ) of the TOF channel, the measured quantity is an integral of n(t) which is equated to n(t) with

$$\frac{1}{2\tau} \int_{t_i-\tau}^{t_i+\tau} n(t)\, dt = f_{BIN}\, n(t_i) \tag{4}$$

f_{BIN} is the bin-width correction which takes into account that n(t) is a function with a rapidly changing slope. A further correction is necessary for the time resolution of the experiment. The actually measured quantity is

$$\bar{n}(t) = \int n(t+\lambda)\, g(\lambda)\, d\lambda \tag{5}$$

where g (λ) describes the time resolution function of the experiment. $g(\lambda)$ comprises the time width of the gamma peak, the finite flight time inside the neutron detector and the width of the TOF window. This means that the determination of n(t) requires the deconvolution of Eq. (5). The corrections of Eqs. (4) and (5) vary in their magnitude with the neutron energy, and therefore for each experiment, when not explicitly given, they were re-calculated to obtain the necessary details for the covariance matrix.

Another quantity which was taken into account was the uncertainty, dE, of the neutron energy scale. This uncertainty was transformed into a spectrum uncertainty, dN, with

$$(dN/N)^2 = [\, (1/2 - E/T) + (E/\epsilon)\, (\partial\epsilon/\partial E)\,]^2 \ (dE/E)^2 \tag{6}$$

The first term on the right-hand side of Eq. (6) is the relative derivative of the spectrum with respect to the neutron energy, assuming a spectral shape as given in Eq. (1), and the second term is the corresponding derivative of the efficiency of the neutron detector. The energy scale uncertainty comprises the uncertainty of the flight path, that of the TOF window and that of the zero time definition.

The uncertainty components of these corrections (and of all other quantities which influenced the result of Eq. (2)) and their correlations, formed the final covariance matrix. For each of the experiments such a matrix was calculated based on the details of the specific experimental procedure.

2.3. *Common Energy Grid*

The data used in the evaluation covered the energy range between 25 keV and 20 MeV. To have a common basis for the evaluation process, a grid of 70 fixed neutron energies was established. Each experimental data point in the energy interval ΔE between $E_g - \Delta E/2$ and $E_g + \Delta E/2$ was moved to the grid energy E_g. For this purpose a Maxwellian distribution (Eq. (1)) was fitted to each experimental data set. The individual slope of this fit was used to transform the data to the grid energies. In this procedure the energy shifts were usually small and the uncertainty contribution was almost negligible.

2.4. *Data Evaluation*

During recent years, enormous progress has been achieved in the description of the Cf-252 neutron spectrum based on nuclear reaction theory[5-7]. In the energy range between 0.5 and 8 MeV the results of the theoretical models are very similar and have been successfully used[5] to approximate single experimental data sets. However, up to now the models are somewhat deficient in that they do not adequately describe the whole energy range between a few keV and 20 MeV. The present evaluation did not therefore include a theoretical model and was confined to the generation of a best set of experimental data.

The evaluation was performed with a general least-squares method and combined the individual experiments with regard to their variances and covariances. Not all of the experiments were "absolute" ones. Some had been normalized by the authors, mainly in such cases where the neutron source strength or the neutron detector efficiency of Eq. (2) was not given in absolute units. These experiments were regarded as shape data and the determination of the normalization factor was a part of the evaluation process. For two of the experiments used in the past evaluation[1], data which had meanwhile been modified[9,10] were used and two additional experiments[11,12] were included in the evaluation.

The evaluation comprised more than 400 data points and resulted in a value of 0.94 for the chi-square per degree of freedom, indicating no real inconsistencies between the experimental data. The result of the evaluation, given as point data in Figs. 1 and 2, represents the present knowledge of the Cf-252 neutron spectrum based on experiments. To scale the large dynamic range of the spectrum of five orders of magnitude, the data are given as ratios to a Maxwellian with T = 1.42 MeV. Compared with the spread of the original data (see Ref. 2), the evaluated result is fairly smooth. One should bear in mind that the evaluation procedure did not introduce any smoothing other than that originating from the data correlations. Relative standard deviations were derived from the evaluated covariance matrix. They are 10% at 25 keV, decrease to $\leq$ 2% between 0.3 and 8 MeV, being actually $\leq$ 1.5% between 1 and 6 MeV, are $\leq$ 5% between 8 and 16 MeV, and reach 25% at 20 MeV.

2.5. *Data Smoothing*

The "point-wise" evaluation is quite useful to test theoretical models, but for practical applications and for the interpolation between the data a smooth form is

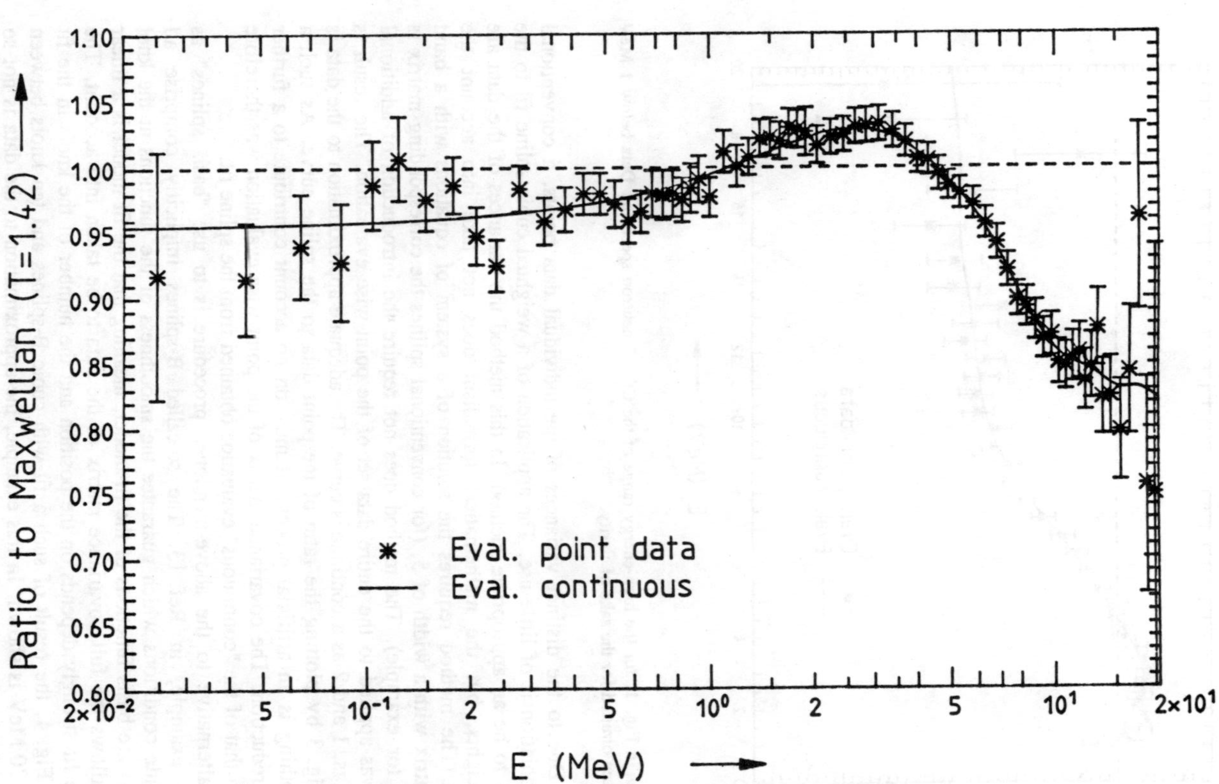

Fig. 1 Result of the "point-wise" (data with error bars) and the "continuous" (solid line) evaluation of the Cf-252 neutron spectrum. The data are plotted relative to a reference Maxwellian with T = 1.42 MeV.

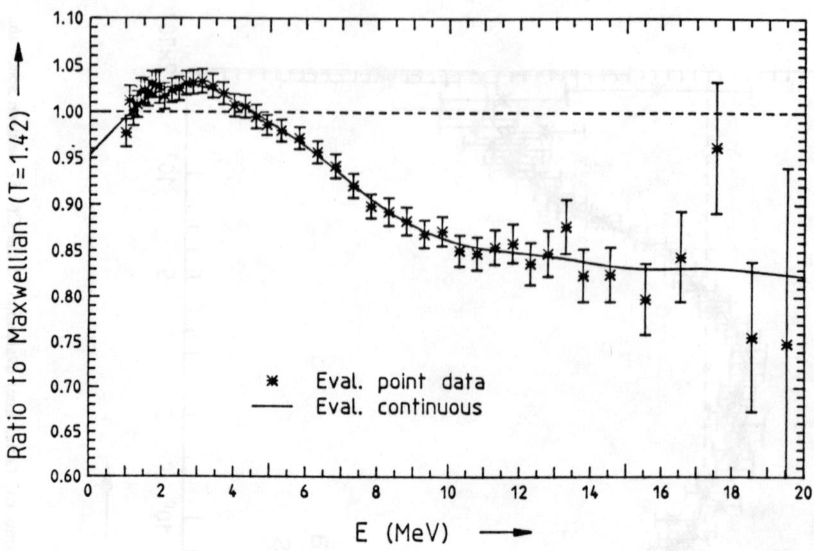

Fig. 2 Same as in Fig. 1, for the high-energy range of the Cf-252 neutron spectrum. Data below 1 MeV have been omitted for the sake of clarity.

preferable. Due to the distinct variances of the individual data points, a conventional spline interpolation is of little use. The application of a weighted cubic spline fit to the data proved to be an appropriate method. In this method the variances of the data are used as weights, but the mathematical formalism does not take into account the correlations. The method requires the solution of a system of equations with a band diagonal matrix with a width of 5, (for conventional splines the corresponding matrix is tridiagonal, for example). The method does not require the introduction of additional knots and was applied to the entire data set of the point-wise evaluation. The result is shown in Figs. 1 and 2 as a continuous curve. The adequate approximation to the data is shown in Fig. 3 by plotting the ratio of the point data to the spline curve. As such a spline smoothing is an artificial process, it must on no account contribute to a further uncertainty reduction. The covariance matrix of the point-wise evaluation was therefore equated with that of the "continuous" evaluation obtained from the spline fit.

An alternative to the above-mentioned procedure is to use "basis splines" as shown, for example, in Ref. 13. The so-called B-splines implicitly comprise all continuity side conditions which guarantee the smoothness of the functions at the knot positions and are basis functions of the fit model, meaning that the fit model is a linear one which allows the full covariance matrix of the data to be taken into account. The result of the fit strongly depends on the position and the number of the knots in the fit interval. In Fig. 4, the result of such a fit with cubic B-splines and ten knots between 20 keV and 20 MeV is shown. In this example, the approximation to the data is not so good as that of the weighted spline fit in Fig. 3. However, the method is quite promising

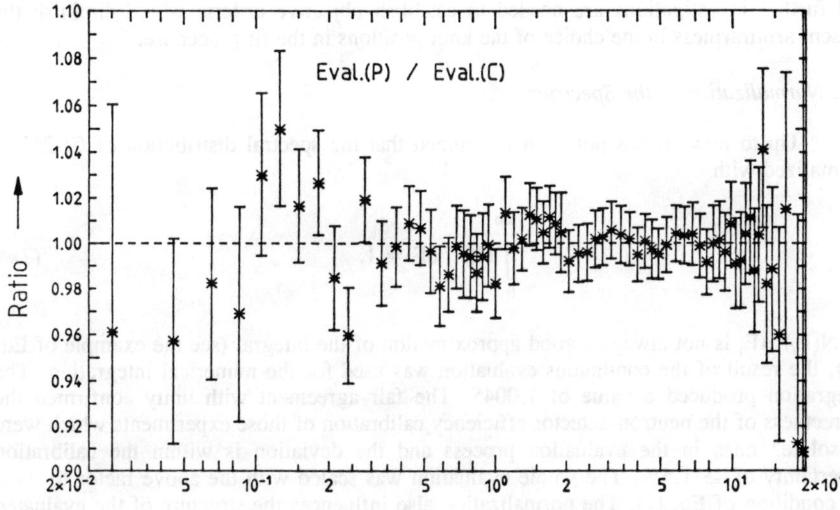

Fig. 3 Ratio of the point-wise evaluation to the continuous one (spline fit).

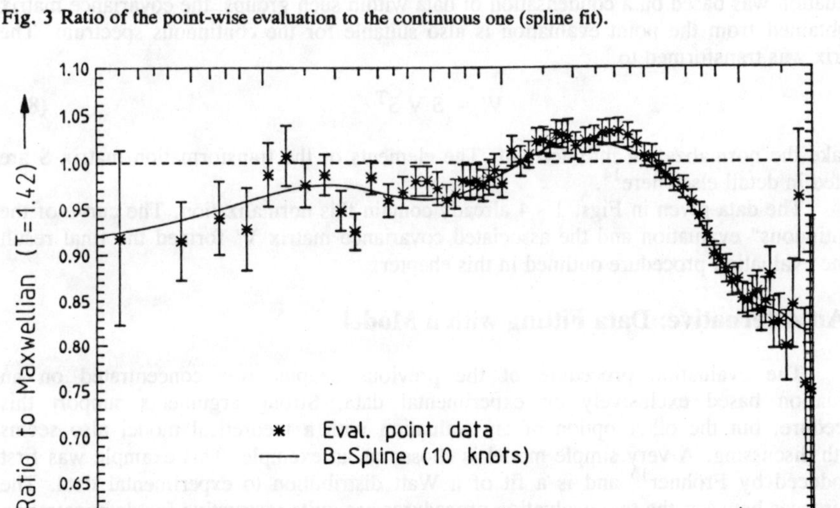

Fig. 4 B-spline fit to the evaluated point data. The fit was done with ten knots at 0.02, 0.3, 0.8, 1.5, 2.5, 4.0, 6.0, 10.0, 14.0 and 20.0 MeV.

and further investigations are needed to establish objective criteria which diminish the present arbitrariness in the choice of the knot positions in the fit procedure.

2.6. Normalization of the Spectrum

Up to now, it has not been mentioned that the spectral distribution of Cf-252 is normalized with

$$\int_0^\infty N(E)\, dE = 1. \tag{7}$$

As $N(E_i)\, \Delta E_i$ is not always a good approximation of the integral (see the example of Eq. (4)), the result of the continuous evaluation was used for the numerical integration. The integration produced a value of 1.0045. The fair agreement with unity confirmed the correctness of the neutron detector efficiency calibration of those experiments which were "absolute" ones in the evaluation process and the deviation is within the calibration uncertainty of $\geq 1.5\%$. The whole evaluation was scaled with the above factor to meet the condition of Eq. (7). The normalization also influences the structure of the evaluated covariance matrix of the spectrum. For a continuous spectrum the matrix represents energy groups and the associated uncertainties within these groups. As the point-wise evaluation was based on a condensation of data within such groups, the covariance matrix V obtained from the point evaluation is also suitable for the continuous spectrum. The matrix was transformed to

$$V' = S\, V\, S^T \tag{8}$$

to take the normalization into account. The elements of the transformation matrix S are quoted in detail elsewhere[14].

The data given in Figs. 1 - 4 already contain this normalization. The curve of the "continuous" evaluation and the associated covariance matrix V' formed the final result of the evaluation procedure outlined in this chapter.

3. An Alternative: Data Fitting with a Model

The evaluation procedure of the previous chapter was concentrated on an evaluation based exclusively on experimental data. Strong arguments support this procedure, but the other option of an evaluation with a theoretical model also seems worth discussing. A very simple model is chosen as an example. This example was first introduced by Fröhner[15] and is a fit of a Watt distribution to experimental data. The differences between the two evaluation procedures are quite instructive for demonstration purposes. However, it is not considered that the example is a real alternative to the result of chapter 2.

The advantage of using a theoretical model is that it is possible to avoide the grid transformation procedure described in chapter 2.3 and the smoothing process in chapter 2.5. The risk is a substantial loss of information contained in the evaluation

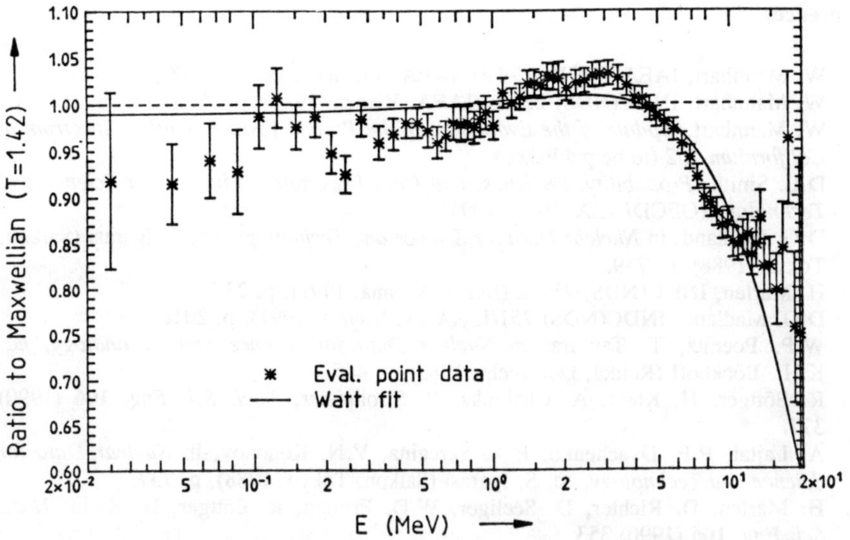

Fig. 5 Fit of a Watt distribution to the evaluated point data.

process, which always happens when the model is not an adequate description of the underlying physics. An example of this is knowledge of the uncertainty of the spectrum at certain energies, which is quite important for application purposes. With regard to the spread of the experimental data (see Fig. 4 of Ref. 15) the Watt fit gave quite a good approximation of the spectrum[15], but such a fit has only two adjustable parameters and the value of the chi-square is of a global nature. A more crucial test is made by applying the fit to the pre-evaluated point data of chapter 2. The result of such a fit is shown in Fig. 5. One recognizes the deficiencies of the Watt distribution in adequately describing the structure of the spectrum around 3 MeV and also above 7 MeV. It has already been shown that an optimum fit result can only be obtained by ignoring the correlations between the experimental data[2].

A serious disadvantage of the model of a Watt distribution is the complete loss of the uncertainty information gained from the experiments. As shown in Ref. 15, the propagated uncertainties of the Watt fit are at least one order of magnitude smaller than those obtained with the method in chapter 2. It remains questionable whether the proposed introduction of additional model uncertainties[15] is an adequate substitute. To summarize, when introducing models into the evalation process it is highly advisable to have the result of a prior model-free evaluation at hand. Only then can one obtain realistic estimates of the adequacy of the model for the evaluation process.

References

1. W. Mannhart, IAEA-TECDOC-410 (IAEA, Vienna, 1987), p. 158.
2. W. Mannhart, INDC(NDS)-220/L (IAEA, Vienna, 1989), p. 305.
3. W. Mannhart, *Update of the Evaluation of the Prompt Fission Neutron Spectrum of Californium-252* (to be published).
4. D.L. Smith, *Probability, Statistics, and Data Uncertainties in Nuclear Science and Technology* (OECD/NEA, Paris, 1991).
5. D.G. Madland, in *Nuclear Data for Science and Technology*, ed. S. Igarasi (Saikon, Tokyo, 1988), p. 759.
6. H. Märten, INDC(NDS)-251/L (IAEA, Vienna, 1991), p. 23.
7. D.G. Madland, INDC(NDS)-251/L (IAEA, Vienna, 1991), p. 201.
8. W.P. Poenitz, T. Tamura, in *Nuclear Data for Science and Technology*, ed. K.H. Böckhoff (Reidel, Dordrecht, 1983), p. 465.
9. R. Böttger, H. Klein, A. Chalupka, B. Strohmaier, *Nucl. Sci. Eng.* **106** (1990) 377.
10. A. Lajtai, P.P. Dyachenko, E.A. Seregina, V.N. Kononov, in *Nuclear Data for Science and Technology*, ed. S. Igarasi (Saikon, Tokyo, 1988), p. 737.
11. H. Märten, D. Richter, D. Seeliger, W.D. Fromm, R. Böttger, H. Klein, *Nucl. Sci. Eng.* **106** (1990) 353.
12. A. Chalupka, L. Malek, S. Tagesen, R. Böttger, *Nucl. Sci. Eng.* **106** (1990) 367.
13. Y. Kanda, Y. Uenohara, INDC(NDS)-192/L (IAEA, Vienna, 1988), p. 81.
14. W. Mannhart, INDC(NDS)-146/L (IAEA, Vienna, 1983), p. 229.
15. F.H. Fröhner, *Nucl. Sci. Eng.* **106** (1990) 345.

DATA FITTING AND EVALUATION TECHNIQUES FOR RADIOACTIVE DECAY DATA

G. WINKLER
Institut für Radiumforschung und Kernphysik
University of Vienna, Boltzmanngasse 3, A-1090 Vienna, Austria

ABSTRACT

Aspects of decay data evaluation are discussed with an emphasis on the point of view of the user of such data, in particular with respect to radionuclide metrology. Eventual requirements are outlined and strategies suggested to meet possible needs referring to already existing tools or procedures.

1. Introduction

Methods for the evaluation of nuclear decay data, of special importance for the application in different disciplines using techniques for quantitative radioactivity measurements[1], are in principle not different from those used for the evaluation of, e.g., neutron (reaction) data, particle properties[2], or in order to find the "best" values of the fundamental physical constants. In particular, from the techniques used for obtaining the latter and following the history of this interesting development (see, e.g. refs. 3-8) one can gather a lot of useful hints. Problems of data treatment which are still topical, were discussed already many decades ago, for example the burning question how to deal with discordant data (e.g. by Birge[4] and Edgeworth[9]). It has turned out that the relatively simple computations and some graphical procedures used in the past before the advent of computers were quite effective.

The accurate quantitative assay of decay-scheme data for a particular nuclide may become a very complex and elaborate task, comprising several classes of data (e.g. level properties, Q-values, types, energies and intensities of radiations emitted, half-lives),[10,11] requiring a step-by-step procedure. It often involves the knowledge or beforehand evaluation of several auxiliary data, for which experimentally determined values are often not readily available. They often have to be calculated or estimated with the help of nuclear or atomic structure theory or from systematics of the relevant properties. Typical examples of such auxiliary data are:

- internal-conversion coefficients for gamma-ray transitions and branching ratios for orbital electron capture from different subshells (for both see, e.g., ref. 12 and refs. cited therein),
- fluorescence yields (see, e.g., ref. 13 and refs. cited therein),
- beta-transition branching ratios (log ft values) or
- branching ratios for low-intensity gamma-ray transitions or transitions which are very difficult to measure (e.g., E0 transitions).

Their reliable determination and evaluation may need a good nuclear physicist's expertise. A survey on existing radionuclide-data compilations and evaluations was given recently by Bambynek.[14] A concise description of methods of analysis for decay data can be found in refs. 15,16. Only a few aspects of nuclear decay data evaluation will be highlighted, some of which are particularly relevant to the practical work in radionuclide metrology. For a particular radionuclide often special features of the decay have to be considered and there is obviously no universal recipe just to follow. The most recent meeting to discuss problems of decay-data evaluation concentrating on open questions for certain radionuclides was just about one week ahead of this symposium.[17]

2. What is needed

The determination of the decay rate of a radionuclide on the basis of an emission-rate measurement requires information about the decay scheme and a quantitative knowledge of its parameters, particularly the yield of the detected radiation per disintegration. For a sample assay by γ-ray spectrometry choosing single peaks, one often only needs the respective gamma-ray emission probability and its uncertainty.[18] In case of complex decay schemes or when employing integral counting the knowledge of the decay scheme is also necessary in order to estimate the influence of other radiations emitted (e.g. in coincidence) on event counting.

The uncertainties in the various emission probabilities may contain strongly correlated components due to common features of the experimental techniques and therefore common sources of uncertainty, e.g. via the efficiency curves, the peak-evaluation method, or via reference sources, used to determine the emission probabilities, furtheron due to normalization, or due to any other physical constraints. A fitting procedure of branch intensities as described below will generally also lead to correlations described by the covariance matrix of the evaluated output parameters. A correct uncertainty estimate of the source strength obtained should take this into account.

One example is the total counting efficiency of a high-efficiency detector (e.g. 4π-NaI(Tl) gamma-ray well-type[19] or 4π-CsI(Tl) sandwich[20] detector), particularly intended for high-precision metrology of radionuclides with complex decay schemes. If a decay scheme can be split into n independent decay paths, each with the partial abundances h_i per decay and k_i coincident transitions within a branch, the total counting efficiency ϵ is

$$\epsilon = \sum_{i=1}^{n} h_i \left[1 - \prod_{j=1}^{k_i} (1-\epsilon_i^j) \right], \tag{1}$$

where ϵ_i^j is the efficiency for the jth transition in the ith decay path. Beside the efficiencies for gamma-ray transitions including internal conversion (producing also bremsstrahlung and x-rays), those for β^--ray and β^+-ray emission (producing also bremsstrahlung and annihilation quanta), and for electron capture (producing x-rays) have to be explicitly included in Eq. (1). Accuracies approaching 0.1% can be achieved, depending on the radionuclide, just by simple integral counting above a low and reproducible discrimination level.[21] Taking into account correlations or not may produce a significant difference in the final uncertainty (compare, e.g., refs. 18,22-26). Using the general law of propagation of variances, a consistent treatment of uncertainties is possible if one starts from the relative gamma-ray emission probabilities g_{ij} as primary experimental data and uses values for the conversion coefficients and for the abundance β_o' of the direct beta branch to the ground state in connection with an uncertainty statement of each of these data.[19,27] The relative intensities of beta-decay branches β_i feeding exited levels of the daughter nucleus are strongly related to the relative intensities p_{ij} of electromagnetic transitions (total conversion coefficient α_{ij}) between the levels i and j

$$p_{ij} = g_{ij}(1+\alpha_{ij}) \qquad i = 1 \text{ to } n,\ j = 0 \text{ to } (n-1) \tag{2}$$

via the balance between deexciting and feeding

$$\beta_i = \sum_{j=0}^{i-1} p_{ij} - \sum_{j=i+1}^{n} p_{ji} \quad . \tag{3}$$

Using Eqs. (9) or (10) below and assuming the (equivalent) standard deviations $s(p_{ij})$ to be uncorrelated, the covariances of the β_i can be derived to be

$$\mathrm{cov}(\beta_i,\beta_j) = -\,\mathrm{var}(p_{ij}) = -\,s^2(p_{ij}) \quad \text{for } i > j \tag{4}$$

whereas the variances are

$$\mathrm{var}(\beta_i) = s^2(\beta_i) = \sum_{j=0}^{i-1} s^2(p_{ij}) + \sum_{j=i+1}^{n} s^2(p_{ji}) . \tag{5}$$

Normalized beta-ray branching ratios β_i' per decay can be obtained by means of the relation

$$\beta_i'/\beta_i = (f - \beta_o')/\sum_1^n \beta_i \tag{6}$$

where f is the total beta-decay fraction (f may be less than unity if several decay modes are possible).

Thus a test of the physical consistency of the chosen decay-scheme data (e.g. from compilations such as ref. 11) could be carried out for several applications within the limits of uncertainty.[21] We have found several cases where the gamma-ray branching data are discordant with beta-branching data, sometimes even giving negative feeding.[21] Uncritical mixing of beta and gamma data should be avoided.

The calculation of average-energy-per-decay data, important in nuclear engineering, is another example where correlations play a role for the estimate of the uncertainties of these data. Since average values such as $<E_\alpha>$ or $<E_\gamma>$ are the sums of individual-radiation energies or group energies E_i weighted by their intensities per decay I_i,

$$<E> = \sum_i E_i I_i \ , \tag{7}$$

correlations amongst energies and intensities have to be considered. Beside those introduced via the experimental procedures (systematic uncertainties), these correlations will depend on how the energy and intensity values have been consistently determined. Alpha-particle energies, for instance, may all be expressed as differences of level energies and the energy of the ground-state alpha-particle group. Particle-branch intensities are constrained to sum up to 100% per decay or to the total branching fraction, which automatically produces correlations. Gamma-ray intensities are not subject to this particular constraint, but there may be other balance constraints dependent on the decay scheme.

3. The basic tools

There are two basic calculation procedures showing up in almost every evaluation:

a) The generalized error-propagation law (*propagation of variances*) considering also off-diagonal elements of the involved variance-covariance matrices which permits to take into account correlations amongst uncertainties: If the (m×1) column vector of variables, **y**, is related with the vector of variables, **x**, with the dimension n, by a linear transformation

$$\mathbf{y} = \mathbf{T}\,\mathbf{x} + \mathbf{a}\ , \tag{8}$$

then the corresponding covariance matrices $\mathbf{C}_y$ and $\mathbf{C}_x$ transform according to

$$\mathbf{C}_y = \mathbf{T}\,\mathbf{C}_x\,\mathbf{T}^t \tag{9}$$

with **T** being the transformation matrix and $\mathbf{T}^t$ its transpose. A nonlinear relationship can be linearly approximated (within small uncertainty intervals) by a Taylor series expansion such that the transformation matrix becomes

$$\mathbf{T} = \left(\begin{array}{ccc} \frac{\partial y_1}{\partial x_1} & \cdots & \frac{\partial y_1}{\partial x_n} \\ \vdots & & \vdots \\ \frac{\partial y_m}{\partial x_1} & \cdots & \frac{\partial y_m}{\partial x_n} \end{array} \right)_{\mathbf{x} = \bar{\mathbf{x}}} . \tag{10}$$

For the elements of the covariance matrix $\mathbf{C}_y$ we get

$$\mathrm{cov}(y_k,y_l) = \sum_{i=1}^{n} \sum_{j=1}^{n} \frac{\partial y_k}{\partial x_i} \frac{\partial y_l}{\partial x_j}\ \mathrm{cov}(x_i,x_j). \tag{11}$$

$$\mathrm{cov}(x_j,x_j) = s^2(x_j),\quad \mathrm{cov}(y_k,y_k) = s^2(y_k). \tag{12}$$

The diagonal elements in $\mathbf{C}_x$ (i=j) or in $\mathbf{C}_y$ (l=k) represent the variances, i.e., squares of (equivalent) standard deviations. In general, $\mathbf{C}_y$ will not be diagonal even if $\mathbf{C}_x$ is a diagonal matrix.

b) The procedure to *solve an overdetermined system of equations* which relate the observed data (each subject to uncertainties) to the desired parameters: The favoured reduction method invokes the principle of least squares (for reasons, conditions of applicability, and possible alternatives see, e.g., refs. 29-32) in order to obtain a consistent set of recommendable values for the parameters and its uncertainties and correlations.[30,33,24,34,25]

Let the values of the estimated parameters be a column (m×1) vector **p** with elements $p_1,..p_j,..p_m$ and $\mathbf{p}_o$ an initial estimate of these values, i.e., $(\mathbf{p} - \mathbf{p}_o) = \Delta\mathbf{p}$, where $\Delta\mathbf{p} \ll \mathbf{p}$. Let the elements of the (n×1) vector **d** represent the measured quantities, which can be expressed as

$$d_i = f_i(p_1,..p_j,..p_m) + r_i,\quad i = 1 \text{ to } n,\quad n > m,\ m \geq 1 \tag{13}$$

where r_i is the residual, or difference between the "true" value $f_i(p_1,...p_m)$ and the observed value d_i. The functions f_i can be of the same kind or of different kinds (e.g.

sums, ratios, more complicated combinations of decay-scheme parameters, combinations of these parameters with other known variables or simply just a single parameter itself) dependent on what has been measured by the technique applied. Using the vector and matrix notation, Eq.(13) reads

$$\mathbf{d} = \mathbf{f}(\mathbf{p}) + \mathbf{r} \quad . \tag{14}$$

With initial estimates $\mathbf{p}_o$, to a first-order Taylor expansion is

$$\mathbf{f}(\mathbf{p}) = \mathbf{f}(\mathbf{p}_o) + \mathbf{G}_o(\mathbf{p} - \mathbf{p}_o) , \qquad G^{o}_{ij} = \left(\frac{\partial f_i}{\partial p_j}\right)_{\mathbf{p}_o} \tag{15}$$

and

$$\mathbf{d} - \mathbf{f}(\mathbf{p}_o) = \mathbf{G}_o(\mathbf{p} - \mathbf{p}_o) + \mathbf{r} \tag{16}$$

where the elements of the (nxm) matrix $\mathbf{G}_o$ are the partial derivatives of the functions f_i taken at the estimates $\mathbf{p}_o$.

If the covariance matrix **V** belonging to the n input data is known, then the minimum variance unbiased estimates of **p** are obtained by minimizing

$$\mathbf{S} = \mathbf{r}^t \mathbf{W} \mathbf{r} \tag{17}$$

where **W** is the (nxn) weight matrix, derived as the inverse of the (nxn) covariance matrix **V**.

Via the normal equations the solution is

$$(\mathbf{p} - \mathbf{p}_o) = (\mathbf{G}_o^{\,t} \mathbf{W} \mathbf{G}_o)^{-1} \mathbf{G}_o^{\,t} \mathbf{W} (\mathbf{d} - \mathbf{f}(\mathbf{p}_o)) \ , \tag{18}$$

and the covariance matrix of the desired output parameters is

$$\mathbf{C}_p = (\mathbf{G}_o^{\,t} \mathbf{W} \mathbf{G}_o)^{-1} , \qquad \text{where} \quad \mathbf{W} = \mathbf{V}^{-1} . \tag{19}$$

In general, the uncertainties of the parameters from the fit will be correlated ($\mathbf{C}_p$ not diagonal) even when the uncertainties in the input values are not correlated (i.e., **V** and **W** are diagonal).

The estimate of the parameter vector **p** can be improved by refitting the data using the values $(\mathbf{p} - \mathbf{p}_o) = \Delta\mathbf{p}$ from the previous fit to give more precise starting values instead of $\mathbf{p}_o$. By means of this iterative process one tries to reach convergence in the parameter values. The fit and eventually the quality of the input data can be judged by means of the goodness-of-fit parameter

$$\chi^2 = \mathbf{S}_{min}(\mathbf{p}) \tag{20}$$

which should approach (n - m), the expected number of "degrees of freedom" independent of any assumption of the underlying statistical distribution function.[28,30]

Incompleteness or uncertainty in the functional relationships f_i, and consequently in the design matrix **G** are neglected. Invoking a Bayesian approach, the formalism can be generalized permitting to combine already obtained a priori information on **p** and $\mathbf{C}_p$ with new observations **d'** and their covariance matrix **V'** as long as there are no correlations between the new data and the old data (having been used to obtain **p**).[24,25,29,35]

Taking a weighted average of the observations of a single quantity is the simplest special case of the above formulated multivariate least-squares procedure having just a single variable p (m=1) at the right side of Eqs. (13), the design equations strictly linear with no initial estimate and iterations necessary.

4. Isolated data - unconstrained (weighted) averaging

The typical example of isolated data are half-life data, with no constraints on the physical quantity. Generally, the following way of data treatment applies, when not more than one quantity is adjusted at a time, when a single-path approach can be followed in the evaluation (usually with no ambiguity which path to take). This approach is used to evaluate very simple decay schemes[36] or, e.g., to produce a set of (weighted) relative gamma-ray intensity values from literature data on a common scale as a start of the decay-scheme evaluation, in most cases considering the input data as being uncorrelated.

Usually, the weighted mean x_w of independent data x_i is calculated using as weights the inverse squares $(1/s_i^2)$ of the measurement uncertainties given as equivalent standard deviations. As was already pointed out by Birge[4] in 1932, there are two ways to assign an uncertainty to the weighted mean, namely via the variances

$$s_{int}^2 = 1 \,/\, \sum_i (1/s_i^2)\,, \qquad s_{ext}^2 = [\, \sum_i (x_i - x_w)^2/s_i^2 \,] \,/\, [\, (n-1) \sum_i (1/s_i^2) \,]\,. \quad (21)$$

The internal variance s_{int}^2 is the expected uncertainty of the mean based on the individual a priori variances s_i^2 (by uncertainty propagation), while the external variance s_{ext}^2 includes the scatter of the data and is based on the amount by which each x_i deviates from the mean, measured as a fraction of its given uncertainty s_i. The Birge ratio

$$s_{ext}/s_{int} = [\, \chi^2/(n-1) \,]^{1/2} \quad (22)$$

should be consistent with unity if the data are actually distributed as expected from the individual a priori uncertainties and is thus a measure of the consistency of the data. If s_{ext}/s_{int} is significantly greater than unity it implies that particular s_i may have been underestimated.

This touches the general question what to do with *discordant data* and how to assign uncertainties of recommended values, a broad topic itself.[29,30] The procedure will depend on whether the data set is large or small. Several recipes were given in the literature which more or less adjust the weights by inflating the original uncertainties moving the weighted mean closer to the unweighted mean. The simplest procedure would be to assign s_{ext} as the final uncertainty which is equivalent to reducing all weights uniformly leaving the weighted mean unchanged. Other procedures try to adjust the uncertainties of only those data that may be considered to be discrepant.

The method "limitation of the relative statistical weights"[37] is quite commonly used to determine a weighted mean:[38,36] After computing the sum of the individual weights, in case of three or more data points the uncertainty of any experimental result is adjusted as necessary so that no datum contributes more than

50% to the sum of weights (according to the principle that at least two measured values must contribute significantly to any proper evaluation). Furthermore, if after this adjustment still $\chi^2/(n-1) > 2$, the weighted or unweighted mean may be chosen dependent on whether or not the standard deviations of the respective means overlap, since it seems unreasonable to take a weighted mean of clearly inconsistent data.[38,36]

With the goal to find an "objective" data evaluation procedure based on a Bayesian approach, Gray et al.[32] proposed a way to find a function specifying the extent to which experimentalists missestimate the uncertainties in their measurements. The relative merits of various procedures were discussed by Rajput and McMahon[39] who applied them also to different subsets of a large discrepant data set (e.g. half-life values for ^{137}Cs) therewith testing "resistance" to the effects of including "outliers" by looking at the consistency of the obtained group results. Anyway, one should be aware of the fact that each uncertainty derived by statistical methods has its own uncertainty.

Although the techniques of obtaining best values have been gradually refined over the years, now taking special advantage of the capability of computers permitting more sophisticated and faster analyses than hitherto, no amount of statistical sophistication can make bad results any better. Therefore the first and primary task is to sort out which data should be included in the evaluation. In this sense any evaluation is subjective. It must be recommended that the persons involved in this compilation and screening phase really study the original papers in detail, check critically the measuring techniques for flaws and drawbacks, and the validity of the calculations leading to the reported values, and the realistic assignment of uncertainties, taking into account random and non-random components. And yet there may still exist unrecognized errors causing experimental data to be biased (see, e.g., ref. 40). To decide whether a work is sufficiently reliable as a whole, enough information must be present whether all significant effects have been adequately considered and necessary corrections correctly applied. Sometimes data must be excluded due to insufficient information.

5. Interrelated data - constrained fits

The decay of a radionuclide is described by many parameters some of which form an interrelated and constrained system. This is especially true of the energies of the emitted radiations (level-scheme correlations) and their intensities (balance conditions). Experiments and evaluations often utilize these constraints to infer parameters which are difficult to measure directly (e.g. β_o'); in some instances even parent-daughter-granddaughter equilibrium relations can be employed.[10] Often observations determine not a single quantity, but rather some combination of physical quantities. Differently to Section 4, all the selected data are looked at simultaneously as an entity in order to deduce a unique consistent set of corresponding parameters using the fitting procedure outlined in Section 3.

Just as an example let me mention the determination of the branching-ratio parameters in the decay scheme of ^{64}Cu, which decays by the emission of positrons, electrons, and also by electron capture. Using several distinct, but partially correlated measurements for one individual parameter, for the sum and the ratio of two of these, "best" estimates were deduced using the generalized least-squares technique starting with a full input covariance matrix of the measured data.[41]

For the analysis of particle and gamma-ray emission probabilities, the feeding-deexcitation balances can be used for a fully constrained fit.[42] Let the desired transition-probability parameters be P_{kl} for transitions between level k and l, by the aid of which all transitions P_r (r...consecutive transition number) in the decay scheme can be described by means of a linear relationship. The observed intensities p_{ij} of electromagnetic transitions between levels i and j, usually primarily obtained relative to a prominent transition and determined according to Eq. (2), are related to the P_{kl} (taken on the same relative scale) by a set of linear equations analogous to Eq. (3)

$$p_{ij} = \begin{pmatrix} \text{matrix with} \\ \text{elements} \\ \ldots +1 \ldots 0 \ldots -1 \ldots \\ \\ \end{pmatrix} \cdot \begin{pmatrix} \vdots \\ P_{ki} \\ \vdots \\ P_{kl} \\ \vdots \\ P_{il} \\ \vdots \end{pmatrix} \begin{matrix} k = (i+1) \text{ to } n \\ \\ k \neq i, l \neq i \\ \\ l = 0 \text{ to } (i-1) \; . \end{matrix} \qquad (23)$$

The matrix element is +1 for parameters P_{kl} feeding the i-th level, -1 for parameters deexciting the i-th level, and 0 if the parameter does not involve the level, assuming a total of n levels. Experimental particle-emission probabilities are usually measured on a different scale. In order to include them in Eqs. (23) and place them on a common scale, a relative normalization factor has to be introduced, which is to be treated as an additional fit parameter (see the method by Tepel[43,15]). Free normalization parameters have also to be introduced when different sets of γ-ray measurements are included without averaging them beforehand. Emission probabilities per decay for any transition (P_r') are obtained from the fitted relative parameters (P_r) by summing the latter over a set of radiations (transitions) which represent the full decay intensity of the radionuclide, possibly distributed in several decay modes, which gives the normalization denominator N in analogy to Eq. (6) (e.g. summing all ground-state γ-ray transition intensities if there is no beta feeding of the ground state). Dependent on the decay scheme and the available information, a variety of variations of the evaluation procedure are possible (compare refs. 44,45). When evaluating the uncertainties of the absolute intensities according to

$$P_r' = P_r / N = P_r / \Sigma P_r \, , \qquad (24)$$

generalized propagation of variances including the covariance terms in every step must be followed. In order to calculate, for instance, the uncertainty of a quotient like the right-hand side of Eq. (24), where some variables show up both in the numerator and the denominator, it may be more lucid to calculate first the uncertainties of the numerator and the denominator separately and then combine them including the covariance term. Uncertainty-cancellation effects are then clearly seen.

A constrained fit with a set of equations like Eqs. (23) can analogously be applied in order to fit measured γ-ray energies into a level scheme,[46] since all transition energies (after correction for recoil effects) can be represented by differences of level energies.

6. Conclusion

Processing the full covariance information is still not the common procedure in decay data evaluation, both with respect to the input data and to the evaluated output parameters. As far as the input is concerned, it is often due to a lack in available information or due to not paying proper attention to the necessity of preparing full uncertainty information. The available procedures would relatively easily permit to produce automatically full covariance information for the evaluated output data, especially since the input information (possibly except for uncertainty correlations) is present in many existent files such as ENSDF.[47] The question is whether there is really a common need to have this complete output information, justifying the additional effort particularly with respect to the documentation. Some examples where it may play a significant role have been mentioned. This point should be further discussed.

7. References

1. C.W. Reich and R.G. Helmer, *Radioactive-nuclide decay data in science and technology*, in Proc. Conf. on Nuclear Cross Sections and Technology, Washington, D.C. (1975), NBS Spec. Publ. 425, Vol. 1, p. 14
2. Particle Data Group, *Review of Particle Properties, Rev. Mod. Phys.* **56**, No. 2, Part II (1984) S1
3. R.T. Birge, *Rev. Mod. Phys.* **1** (1929) 1
4. R.T. Birge, *Phys. Rev.* **40** (1932) 207
5. R.T. Birge, *Phys. Rev.* **40** (1932) 228
6. R.T. Birge, *Rep. Prog. Phys.* **8** (1941) 90
7. E.R. Cohen, K.M. Crowe and J.W.M. Dumond, *The Fundamental Constants of Physics* (Interscience, New York, 1957)
8. *Metrology and Fundamental Constants*, ed. A.F. Milone , P. Giacomo and S. Leschiutta (North-Holland, Amsterdam, 1980)
9. F.Y. Edgeworth, *Phil. Mag.* **23** (1887) 364
10. R.L. Bunting and C.W. Reich, *Evaluation procedures for experimental decay data*, Proc. Conf. Nuclear Data Evaluation Methods and Procedures, Sep. 22-25, 1980, BNL-NCS-51363 (National Nuclear Data Center, Brookhaven National Laboratory, 1981) Vol. 1, p. 163
11. *Tables of Isotopes*, ed. C.M. Lederer and V.S. Shirley, 7-th edition (Wiley, New York, 1978)
12. W. Bambynek, *Production of inner-shell vacancies by orbital electron capture of the nucleus and by internal conversion of gamma rays*, in *Low-Level Measurements of Man-made Radionuclides in the Environment*, ed. M. Garcia-Leon and G. Madurga (World Scientific, Singapore, 1991) p. 138
13. W. Bambynek, *Reorganization of atomic shells after radioactive decay*, in *ibid.*, p. 156
14. W. Bambynek, *Radionuclide data*, in *Low-Level Measurements and their Applications to Environmental Radioactivity*, ed. M. Garcia-Leon and G. Madurga (World Scientific, Singapore, 1988) p. 37
15. E. Browne and R.B. Firestone, *Table of Radioactive Isotopes*, ed. V.S. Shirley (Wiley, New York, 1986)
16. F. Lagoutine, N. Coursol and J. Legrand, *Table de Radionucléides* (Commissariat à l'Energie Atomique, LMRI, Saclay, 1983-1987)
17. ICRM Data Evaluation Workshop, Oct. 5-7, 1992, Idaho Falls, USA, organized by A.L. Nichols and R.G. Helmer (private communication)
18. K. Debertin and R.G. Helmer, *Gamma- and X-ray Spectrometry with Semiconductor Detectors* (North-Holland, Amsterdam, 1988)

19. G. Winkler and A. Pavlik, *Int. J. Appl. Radiat. Isot.* **34** (1983) 547
20. B. Denecke, *Appl. Radiat. Isot.* **38** (1987) 823
21. A. Pavlik and G. Winkler, *Int. J. Appl. Radiat. Isot.* **34** (1983) 1167
22. C. Ballaux, *Int. J. Appl. Radiat. Isot.* **34** (1983) 493
23. K. Debertin, *Nucl. Instr. and Methods* **226** (1984) 566
24. W. Mannhart, *A Small Guide to Generating Covariances of Experimental Data*, Report PTB-FMRB-84 (Phys. Techn. Bundesanstalt, Braunschweig, 1981)
25. D.L. Smith, *Probability, Statistics, and Data Uncertainties in Nuclear Science and Technology* (American Nuclear Society, LaGrange Park, Ill., 1991)
26. W. Bambynek, *Uncertainty assignment in radionuclide metrology*, in *Low-Level Measurements and their Applications to Environmental Radioactivity*, ed. M. Garcia-Leon and G. Madurga (World Scientific, Singapore, 1988), p. 3
27. A. Pavlik, *Sitz. Ber. Österr. Akad. Wiss.* **191** (1982) 253
28. R.L. Branham, *Scientific Data Analysis - an Introduction to Overdetermined Systems* (Springer, New York, 1990)
29. F.H. Fröhner, *Principles and Techniques of Data Evaluation*, Report KfK 4099 (Kernforschungszentrum Karlsruhe, 1986)
30. E.R. Cohen, in *Metrology and Fundamental Constants*, ed. A.F. Milone et. al. (North-Holland, Amsterdam, 1980), p. 581
31. B.R. Martin, *Statistics for Physicists* (Academic Press, London, 1971)
32. P.W. Gray, T.D. MacMahon and M.U. Rajput, *Nucl. Instr. and Methods* **A286** (1990) 569
33. A. Allisy, in *Metrology and Fundamental Constants*, ed. A.F. Milone et. al. (North-Holland, Amsterdam, 1980), p. 20
34. W.L. Zijp, *Treatment of Measurement Uncertainties*, Report ECN-154 (Netherlands Energy Research Foundation, Petten, 1987)
35. M. Kitamura, *Advanced techniques of information processing*, Proc. 1991 Symp. on Nuclear Data, Nov. 28-29, 1991, JAERI, Tokai, Ibaraki-ken, Japan, Report JAERI-M 92-027 (Japan Atomic Energy Research Institute, 1992)
36. International Atomic Energy Agency, *X-ray and Gamma-ray Standards for Detector Calibration*, IAEA-TECDOC-619 (IAEA, Vienna, 1991)
37. W.L. Zijp, *On the Statistical Evaluation of Inconsistent Measurement Results Illustrated on the Example of the ^{90}Sr Half-Life*, Report ECN-179 (Netherlands Energy Research Foundation, Petten, 1985)
38. M.J. Woods and A.S. Munster, *Evaluation of Half-Life Data*, NPL Report RS(EXT)95 (National Physical Laboratory, Teddington, 1988)
39. M.U. Rajput and T.D. MacMahon, *Nucl. Instr. and Methods* **A312** (1992) 289
40. S. Tagesen and G. Winkler, *Troubles, traps and tricks in fitting exponential decay data* (this symposium)
41. P. Christmas, S.M. Judge, T.B. Ryves, D. Smith and G. Winkler, *Nucl. Instr. and Methods* **215** (1983) 397
42. R.B. Firestone, *Nucl. Instr. and Methods* **A286** (1990) 584
43. J.W. Tepel, Report INDC(NDS)-115/NE (International Atomic Energy Agency, Vienna, 1980), p. 121
44. E. Browne, *Nucl. Instr. and Methods* **A249** (1986) 461
45. E. Browne, *Nucl. Instr. and Methods* **A265** (1988) 541
46. R.G. Helmer, R.C. Greenwood and R.J. Gehrke, *Nucl. Instr. and Methods* **155** (1978) 189
47. Evaluated Nuclear Structure Data File (ENSDF), National Nuclear Data Center, Brookhaven National Laboratory, U.S.A.

TROUBLES, TRAPS AND TRICKS IN FITTING EXPONENTIAL DECAY DATA

S. TAGESEN and G. WINKLER
Institut für Radiumforschung und Kernphysik
University of Vienna, Boltzmanngasse 3, A – 1090 Vienna, Austria

ABSTRACT

A systematic study of two widely used fit procedures for the analysis of exponential decay data has provided much insight into the behaviour of those procedures. Parameters with extremely strong influence on the results of a fit have been identified and limits for safe applicability can be deduced. Discrepancies in published results for nuclear half–lifes may have their origin in underestimating these constraints.

1. Introduction

Two examples may illustrate the problems related to deducing half–lifes from measured decay data:

1.1. Walz, Debertin and Schrader from PTB Braunschweig pointed out[1], that calculated uncertainties from fitting decay data might considerably underestimate reality. This was deduced from strong fluctuations of the results of fitting subgroups in a larger data set. The authors attributed these fluctuations to unknown systematic errors and seemed to proof this by comparing to results calculated from an artificial decay curve generated by a computer code, which did not exhibit fluctuations much beyond "normal statistics".

1.2. Results on the half–life of ^{212}Po, given in Table of Isotopes[2] are rather discrepant.

$T_{1/2}$ [ns]	Ref.
296 ± 2	PR C12 318 (1975)
304 ± 4	PPSL 62A 253 (1949)
305 ± 5	NP 45 49 (1963)
304 ± 8	IIso 23 527 (1972)

This finding again rises the question, whether the given uncertainties are underestimated, or the methods for analysis bear some hitherto unrevealed traps.

2. Common Techniques for the Analysis of Decay Data

Two methods are widely applied in analyzing decay curves:

– Transformation of the data to a logarithmic scale, fit of transformed data by a linear function.

Disadvantage: Any background has to be determined separately and subtracted before fitting.
Advantage: This method has a straightforward unique solution.

– Nonlinear fitting of a curve by multidimensional minimizing of chi squared.
Disadvantage: High expense of computer time, convergency slow and not generally guarantied.
Advantage: Method should be capable of fitting background simultaneously.

Both techniques are well established and assumed to be fully understood. So, why do we bother – and why are there discrepant results?

To permit a detailed study under well–defined conditions we have generated two different test data sets:

– A Po–212 decay curve with a rather low background component and moderate statistical uncertainty (Fig.1).

– An experimentally generated "pure" exponential decay curve without any "background". These data are in fact the measured distribution of time intervals between pulses of a statistical sequence of events (radioactive source with a long life–time) (Fig.2).

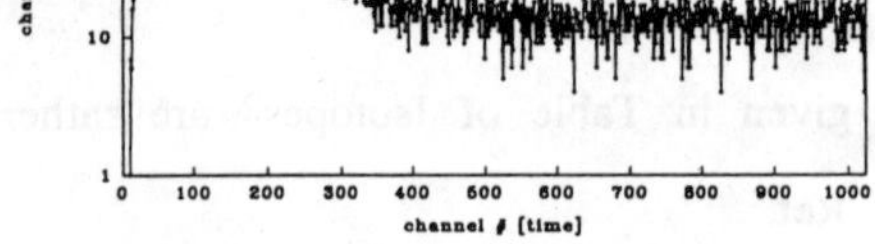

Fig. 1: Po–212 decay curve

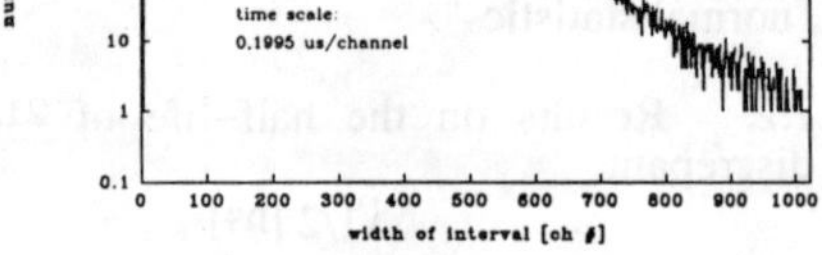

Fig. 2: "Ideal" decay curve

3. Method 1: Linear Fit to Data in Logarithmic Scale

This procedure is apparently simple and should be the method of choice due to its straightforward solution.

The function to be fitted is $y = a \exp(b\ t)$ (1)

which transforms to $\ln y = \ln a + b\ t$ (2)

with uncertainties

$$\sigma (\ln y) = (d (\ln y) / d y)\ \sigma (y) = \sigma (y) / y \quad (3)$$

which leads – in case of purely statistical uncertainty – to weights of

$$g = 1 / \sigma^2 = y \tag{4}$$

Three aspects are of main concern for the result of such a fit:

3.1. Background Determination

As can be seen in Fig.1, any measurement of a physical quantity is usually disturbed by some unavoidable background. One might think, that keeping the background small would be a safe measure to prevent influences on the desired result for the half-life. Fig.3 demonstrates the influence of background changes on the outcome of the fit to the Po-data, where the background component is in fact very small. More intriguing: A fit to the area between channels 500 and 1000 (Fig.1) gave a background of 13.49±0.16 counts per ch. Nevertheless the fit to the decay curve had best results for a background of approximately 12.4 c.p.ch.

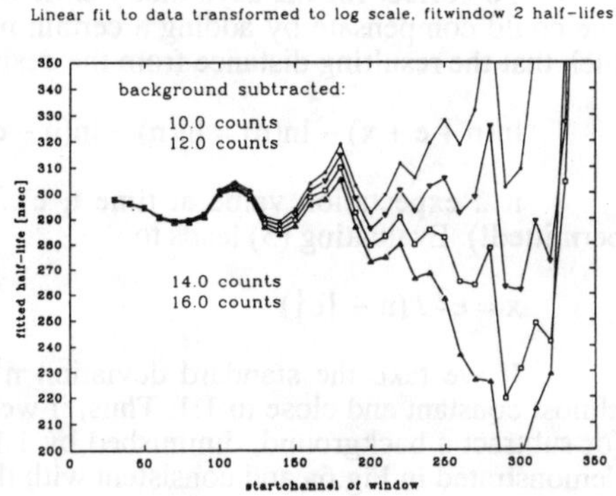

Fig. 3 sensitivity of the fit to background

Do we loose background counts while we register genuine events from our source? We will come back to this finding in the discussion of 3.3.

3.2. Assigning Correct Weights to the Data Points

It is clear that we have to assign weights to our data points, because our data set usually will contain points ranging over at least one order of magnitude and we have to account for the varying statistical uncertainty along the decay curve. However, each point of our set represents a single sample out of a (possible) distribution in the corresponding time interval. Thus, if we would assign weights calculated from the uncertainty of the measured data, we would systematically favor points above the mean and disfavor neighboring points below the mean. The result may be seen in Fig.4. The correct procedure is therefore to iterate the calculation, taking the first result as a guess of the expectation value and calculate weights for further loops from the respective expectation value at time interval t.

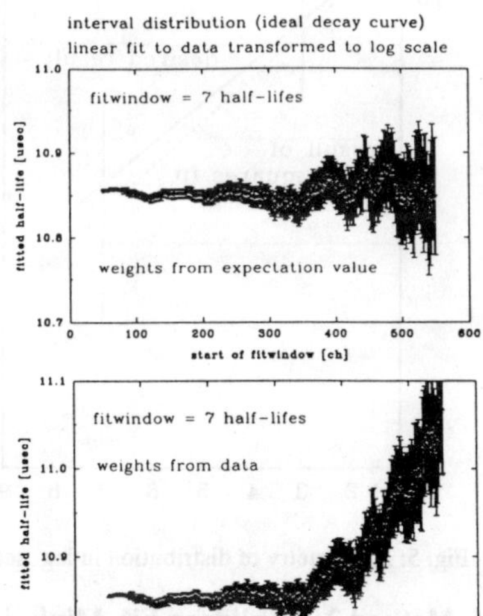

Fig. 4: Influence of the assignment of weights on the result of the fit

3.3. Distribution of the Data Points in Logarithmic Scale

A least squares fit assumes that data points are distributed evenly on both sides of the expected curve. As is obvious from Fig.5, this is not the case in a log. scale. Consequently the outcome of a fit neglecting this asymmetry is a half–life which is too short (see also Fig.6b). Since this effect acts in the opposite direction than the above mentioned incorrect weight assignment, both nearly cancel. This is the reason that method 1 produces nevertheless rather acceptable results.

To correct for the asymmetry shift we have tried a simple procedure: Obviously one could compensate by adding a certain number of counts x to each channel content, such that the resulting distance from the desired curve would be equal on both sides, i.e.

$$\ln(n + e + x) - \ln(n) = \ln(n) - \ln(n - e) \tag{5}$$

n .. expectation value at time t, e .. residuum of actual data point (both signs permitted!). Evaluating (5) leads to

$$x = e^2 / (n - |e|) \tag{6}$$

If we take the standard deviation $n^{1/2}$ for e, the resulting x for any $n > 50$ is almost constant and close to 1.1. Thus, if we add 1.1 to the content of each time channel (or subtract a background, diminished by 1.1), we should get a correct fit result. This is demonstrated in Fig.6a and consistent with the numerical outcome in 3.1.

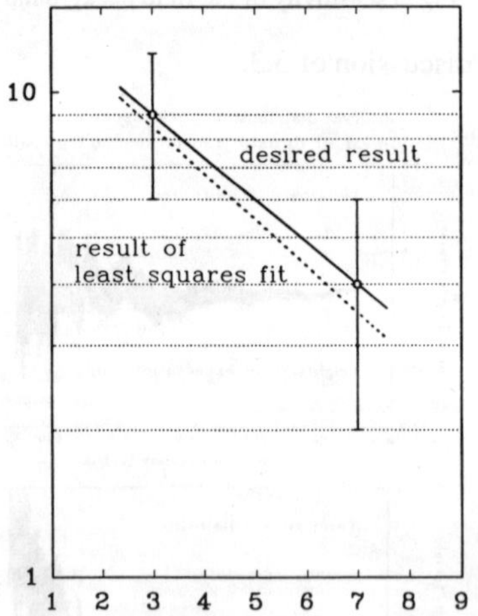

Fig. 5: Asymmetry of distribution in log. scale

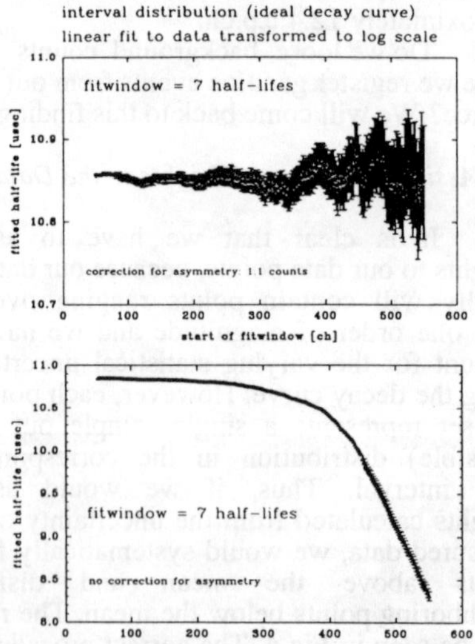

Fig. 6: Influence of asymmetry correction on fitresult

4. Method 2: Nonlinear Fit, Minimizing Chi Squared

A function of type $y = a + b \exp(c\ t)$ (7)

can only be fitted by searching the minimum on a multidimensional chi squared surface. Several solutions to this problem are known from numerous textbooks. We have chosen a two step procedure: First determine a fairly good approximation by means of a grid search algorithm, then use the result for a fine search by approximating the chi squared surface close to the minimum by a three-dimensional paraboloid. The uncertainty in each fitted parameter is then given by the variation that increases chi squared by 1.0 while maintaining the minimum for the other parameters.

4.1. Background

Background is in principle determined correctly, but with rather large uncertainty. Corresponding fluctuations result in large anticorrelated fluctuations of the other parameters, especially the fitted half-life (Fig.7). This very large sensitivity is caused by the fact that the chi squared minimum is extremely shallow with respect to background.

4.2. Weights

Since we do not apply any transformation, the weights would be 1/y, the channel content. However, due to the same reasons given in 3.2. this assignment would tend to favor low-lying datapoints and thus decrease the calculated half-life. This effect is overcome again by using the expectation values after the first approximation (Fig.8).

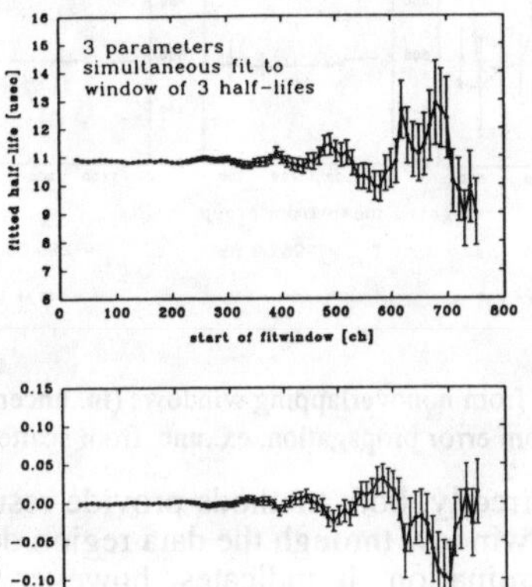

Fig. 7: Anticorrelated results backgr.vs.half-life

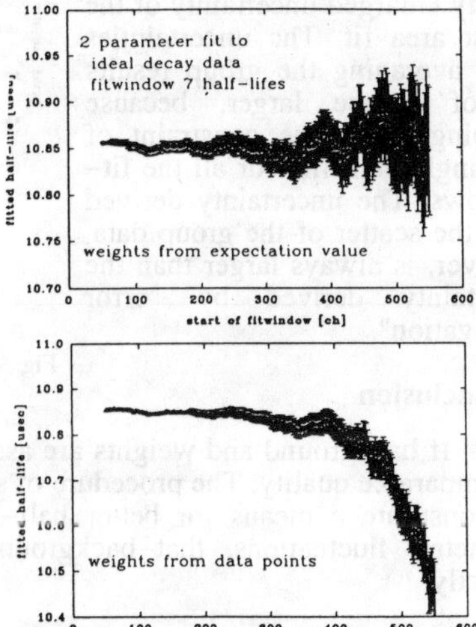

Fig. 8: Influence of assignment of weights

No problem arises with symmetry, since we are on a linear scale. To compare methods 1 and 2 it can be said, that results derived under comparable conditions are close to identical (within numerical precision in iterations).

5. Fluctuations

In order to study this phenomenon we have predominantly used the "artificial" decay curve, which provides data over more than 18 half-lifes. The results can be summarized in the following statements:

– Fluctuations are present with both methods. Due to the sensitivity to the background fit, they are slightly stronger with method 2.

– If the fit area spans less than one half–life (which was the case in the PTB computer simulation), uncertainties are large enough to make it appear purely statistical.

– When the fit area is increased fluctuations are smoothed, the uncertainties, however, decrease faster. With an area of two to three half–lifes, it appears that 50% of the datapoints are outside one standard deviation. A study, carried out by a Hungarian group[3] sheds some light on this effect. If one analyses the statistical distribution of events in a time window much larger than the half–life, the distribution function is much wider than a Gaussian and asymmetric. The consequence is, that we may calculate a variance, but the deduced standard deviation does no longer represent a 68% confidence level as with a normal distribution. Our results indicate that one would have to increase the calculated uncertainty qualitatively by a factor of square root of 2 to 3, to reach again consistency with the 68% confidence level.

– The last statement is also supported by a comparison of results, derived by calculating the weighted mean over group results from nonoverlapping fitwindows (Fig.9). The means of all calculations are consistent with a slightly enlarged uncertainty of the widest area fit. The uncertainties when averaging the group results are of course larger, because grouping drops the constraint of one single half–life for all the fit–windows. The uncertainty derived from the scatter of the group data, however, is always larger than the uncertainty derived by "error propagation".

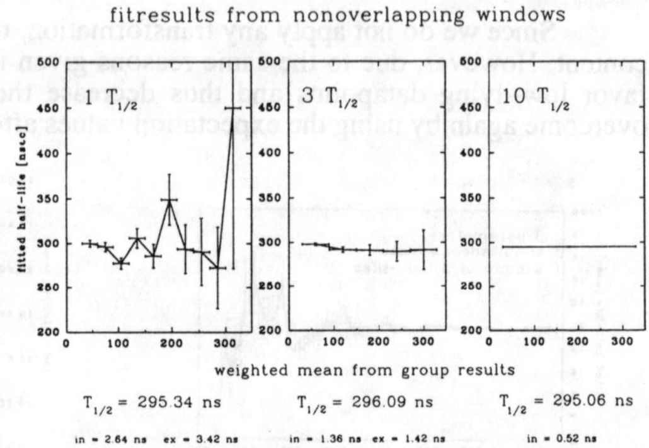

Fig. 9: Fitresults from nonoverlapping windows (in..uncertainty from error propagation, ex..unc. from scatter)

6. Conclusion

If background and weights are assigned correctly, both methods provide results of comparab!e quality. The procedure of shifting a window through the data region does not constitute a means for better half–life determination. It indicates, however, by symmetric fluctuations, that background and other corrections have been treated correctly.

7. Acknowledgement

The authors would like to express their special thanks to Jörg W. Müller from BIPM, Sèvres, for his enlightening hints with respect to the assignment of weights.

8. References

1. K. F. Walz, K. Debertin and H. Schrader, *Int. J. Appl. Radiat. Isot.* **34(8)** (1983) 1191
2. C. M. Lederer and V. S. Shirley (Eds.), *Table of Isotopes, seventh edition* (1978)
3. I. Salma and E. Zemplen–Papp, *J. Radioanal. Nucl. Chem. Letters* **153(4)** (1991) 245

On some unused capabilities of Evaluated Nuclear Structure Data File (ENSDF)

F.E.Chukreev

Nuclear Structure and Reaction Data Center,
Kurchatov's Institute, Moscow, Russia.

There is no need to argue in favor of ENSDF usefulness for fundamental and applied investigations because most serious researchers engaged in nuclear spectroscopy cannot do without Nuclear Data Sheets or full ENSDF file.

However ENSDF contains some still unused information which could be of special interest to fundamental researches but is not available in the evaluated data.

ENSDF contains many input experimental data from which the evaluator obtains the evaluated data. For the process of this work the evaluator usually uses some assumptions but in the majority of cases he does not inform the reader on the effect of these assumptions on a final result, thus producing a false impression of completeness of a level and transition pictures.

If a section where the evaluator should indicate clearly the assumptions requiring experimental verification were introduced into ENSDF or NDS, this data file would present not only a data reference book but also a collection of unsolved problems. And one may suppose that this would increase the interest of nuclear physicists for ENSDF.

We restrict ourselves to two examples to illustrate the helpfulness of such extension.

The first example refers to a barium-133 decay. This decay is important because this nucleus is on the list of those recommended for Ge(Li) detector calibration [1]. The Working Group established by the IAEA Nuclear Data Section, at preparing recommended values of gamma-ray intensity emitted at barium-133 decay used evaluated data from [2] where anomalously high penetration factors were attributed to 79.623 and 80.997 transitions. So the internal conversion coefficients (ICC) for these transitions are differing noticeably from tabulated data. The penetration factor, as it is known, cannot be determined in a direct experiment. To determine it one should know an absolute ICC value and transition multipolarity. The absolute ICC value were obtained by the authors of [2] in the assumption that a 356.017-keV gamma-transition is a pure E2 transition.

Is this assumption true? To what extent depend the other data on this assumption? To analyse the experimental data on this decay the methods involving solution of linear programming problems [3,4] were applied. These methods have the advantage that allow one the write down the results of different kind experiments in the form of a system of linear restrictions to check their consistency and unambiguity of interpretation. The details of this analysis can be found in [5].

The result of the analysis shows that three possibilities of interpretation differing in 276-and 356-keV transition multipolarities could be consisted with experimental data. These versions describe experimental data essentially well.

Table 1 shows these versions while Table 2 contains gamma-ray intensities for these three versions. When one abandons the assumption that the 356-kev transition is pure E2, then anomalous ICC for barium-133 decay would not be necessary.

Table 1

Quantum energy (keV)	Version 1	2	3
276	0.267M1+0.733E2	E2	0.267M1+0.733E2
356	0.003M1+0.997E2	0.365M1+0.635E2	0.207M1+0.793E2

Table 2

Quantum energy (keV)	Version 1	2	3
53.16	2.123	2.116	2.125
79.62	2.538	2.545	2.539
81.00	33.12	33.13	33.12
160.61	0.6204	0.6227	0.6202
223.23	0.4337	0.4400	0.4378
276.40	6.993	6.942	6.997
302.85	17.88	17.83	17.83
356.02	60.33	60.16	60.39
383.85	8.674	8.649	8.683

The difference between versions appears to be larger than the uncertainty in the quantum intensity adopted in [1].

The authors of the evaluation [2] probably know about a strong influence of the basic assumption (that the 356-keV transition is pure E2) on the interpretation of experimental data. But the reader does not know that rejection of this assumption would seriously alter the interpretation of experiments and believes that everything is well known. Though, this very reader may have suitable instrumentation to make the data on this decay more precise.

The second example refers to a plutonium-239 alpha-decay. It was shown in [3] that the experimental data by Baranov [6], Gunnik [7] and Tretyakov [8] contradict the adopted plutonium-239 decay scheme, but this fact was not reflected in a corresponding evaluation [9].

Plutonium-239 and uranium-235, the product plutonium decay, are more important for nuclear power. The precise knowledge of lower uranium-235 levels is quite necessary to make a correct evaluation of the cross-section of neutron interactions with uranium-235.

References

1. "X-ray and gamma-ray standards for detector calibration", *IAEA-TECDOC-619*, September 1991.
2. Yu.V. Sergeenkov, V.M. Sigalov, *NDS*, 49, 639 (1986).
3. F.E. Chukreev, *INDC(CCP)-309/N*, IAEA, December 1989.
4. F.E. Chukreev, *Voprosy atomnoj nauki i tekhn., Ser. "Jadernye konstanty"*, №2, 34 (1990).
5. F.E. Chukreev, *Voprosy atomoj nauki i tekhn., Ser. "Jadernye konstanty"*, №2, 1992.
6. S.A. Baranov et al. *ZETP*, 43, 1135 (1962).
7. P.Gunnik et al. *UCRL-52139* (1976).
8. E.F.Tretyakov, L.N.Kondratyev, *Izv. AN SSSR, Ser. fiz.*, XXIX, 2, 242 (965).
9. M.R. Schmorak, *NDS*, 40, 1, (1983).

NUCLEAR MASSES: DATA EVALUATION AND DATA TREATMENT

Georges AUDI
CSNSM, IN2P3-CNRS, Laboratoire René Bernas
Bâtiment 108, F-91405 Orsay Campus, France

ABSTRACT

The evaluation of nuclear masses share with other evaluations many procedures, but has also some particularities. I shall describe here how the evaluation is conducted and introduce the reader to two new tools that are important in this evaluation: the Flow-of-Information matrix which is of general use everytime a least-squares procedure is being developed, and the Interactive Graphical display of the Surface of Masses.

I cannot start this talk without recalling that the evaluation of nuclear masses has been carried on since many decades by Pr. A. H. Wapstra with the help of various collaborators. For more than 10 years now, we have been working together and with great pleasure in this mostly interesting field.

1. Introduction

Mass measurements and the evaluation of nuclear masses are in the very heart of fundamental physics. In nuclear physics, for testing nuclear models, for exploring the properties of newly produced nuclei and for observing new structures as they stem from breakings in the regularity of the surface of masses. In weak interaction physics, for the determination of the neutrino mass, for testing the vector current conservation hypothesis, and when searching for neutrinoless double beta decay (0ν-$\beta\beta$). In astrophysics, nuclear masses are essential in predicting level densities, β-decay half-lives, β-delayed neutron emissions and fission rates, in describing the r-process nucleosynthesis pathes, in locating the limits of nuclear stability, and in evaluating the detectable fluxes of solar neutrinos. In metrology, measurement of the mass of ^{28}Si is one of the keystones in the process of redefining the mass unit on a microscopic scale, as has been already achieved for the length, time and electrical units.

The *Evaluation of Masses* share probably with most other evaluations many procedures like litterature scan, data collection and critical reading. While checking for the calibration procedures, spectra and precision of the measurements, the *Evaluation of Masses* search for the primary information in the data, which do not necessarily appear always as clearly as they should. Then, new data on masses can be compared to previous existing results, either directly or through a combination of other results. If there are no previous results, comparison could be done with extrapolations, exploiting the above mentioned smooth behaviour of the *Surface of Masses*, or to estimates from mass formulae.

One special character in the treatment of data in the *Evaluation of Masses* is that all measurements are relative measurements. Each experimental data is thus represented by a link connecting two or three nuclei. The set of connections results in a complex canvas where data of different type and origin are entangled. Here lies the very challenge to extract values of masses from the experiments.

Some tools are going to help. The main one is the weighted least-squares method which is briefly recalled in section *2*. But this method, despite its high power, had a serious drawback: it didn't allow to *see*, except in very simple cases, how information could flow between the experimental data (inputs) and the adjusted masses (outputs). Yet, it is extremely useful to know the *"influence"* of each piece of data on each of the adjusted masses, and also the total influence, or *"significance"*, of one datum. I shall show in section *2.2* how a *"Flow-of-Information Matrix"* has been constructed[1], whose elements are the above *influences* and one projection yields the *significances*. As a result, the least-squares method has now become transparent and this new tool happens to provide a very important help in the Evaluation of Masses as illustrated in section *2.3*.

Another very recently developed tool is the *Interactive Graphical Display* of the mass surface allowing to see how a change in one mass or the creation of a new mass may show up in any type of projection (separation energies S_n, S_{2n}, S_p, S_{2p}, decay energies Q_α, Q_β, $Q_{2\beta}$, pairing energies Δ_{nn}, Δ_{np}, Δ_{pp}). Such a tool is essential for analysis every time a new data disagrees strongly in the comparisons above. It is also very useful in estimating still unknown masses. A short description of this new tool is given in section *3*.

2. The Flow-of-Information Matrix

2.1. Least-Squares Method: recall and notations

Each piece of data has a value $l_i \pm dl_i$ with the accuracy dl_i (one standard deviation) and makes a relation between 2, 3 or 4 masses with unknown values m_λ. An overdetermined system of M data to N masses ($M > N$) can be represented by a system of M linear equations to N parameters:

$$\sum_\lambda k_i^\lambda m_\lambda = l_i \pm dl_i \tag{1}$$

or, in matrix notation, $\mathbf{K}$ being the matrix of coefficients: $\mathbf{K}|m\rangle = |l\rangle$. A diagonal weight matrix $\mathbf{W}$ can be defined by its elements $w_{ii} = 1/(dl_i dl_i)$. The solution of the least-squares method leads to a very simple construction:

$$^t\mathbf{KWK}|m\rangle = {}^t\mathbf{KW}|l\rangle \tag{2}$$

the normal matrix $\mathbf{A} = {}^t\mathbf{KWK}$ is a square matrix of order N, positive-definite, symmetric and regular and hence invertible[2]. Thus:

$$|m\rangle = \mathbf{A}^{-1}\, {}^t\mathbf{KW}|l\rangle \quad or \quad |m\rangle = \mathbf{R}|l\rangle \tag{3}$$

The rectangular (N, M) matrix **R** is called the "response" matrix.

A computer program in 4 phases and 7300 lines will **i)** decode and check the data file; **ii)** build up a representaion of the connections between masses, allowing thus to reduce drastically the size of the system of equations to be solved (9000 equations in 3050 masses reduced to 1570 equations and 818 unknowns) without any loss of information; **iii)** do the matrix calculations above and **iv)** deduce the masses, the adjusted values for the input data, the reactions energies, the influences below, and give information on the inversion errors, the correlations coefficients, the distribution of the χ^2s, etc... Making use of parallel processing on the IBM 9021-820 in our institute, the computing time has been drasticaly decreased to 22 seconds CPU, while it used to be a couple of hours a few years ago.

2.2. Construction of the Flow-of-Information Matrix

Construction of the Flow-of-Information Matrix **F** is very simple: **K**, the matrix of coefficients, is a rectangular (M, N) matrix, the transpose of the response matrix $^t\mathbf{R}$ is also a (M, N) rectangular one. The (i, λ) element of **F** is defined as the product of the corresponding elements of $^t\mathbf{R}$ and of **K**. It has been demonstrated[1] that each element of this matrix represents the **influence** of datum i on parameter (mass) m_λ. A column of **F** will thus represent all the contributions brought by all the data to a given mass m_λ, and a line of **F** will represent all the influences given by a single piece of data. The sum of these influences will be called the **significance** of that datum. It has also been proven[1] that the *influences* and *significances* have all the expected properties, namely that the sum of all the influences on a given mass is unity, that the significance of a datum is always less than unity and that it always decreases when new data are added. Moreover, as a further proof, the *significances* as defined here are exactly the quantities obtained by squaring the ratio of the adjusted over the input uncertainty. The Flow-of-Information Matrix **F**, constructed as shown above, provides thus insight on how the information from datum i flows into each of the masses m_λ.

There is a simple way to understand the *influences* and the *significances*. Once the masses are adjusted to $|m>$, one can calculate, using Eq.(1), the adjusted data:

$$|l> = \mathbf{KR}|l> \tag{4}$$

The ith diagonal element of **KR** represents then the quantity of datum i used in the determination of l_i (same datum): this quantity is exactly what is called above the *significance* of datum i. This i^{th} diagonal element of **KR** is the sum of the products of line i of **K** and column i of **R**. The terms contributing to this sum are the individual *influences*.

2.3. The Flow-of-Information Matrix in the Evaluation of Masses

The transparency in the least-squares method brought by the Flow-of-Information Matrix can best be illustrated in the following example on ^{205}Tl and ^{205}Pb. For astrophysical purposes[4] it is highly desirable to know the energy difference between these two nuclei, accurately (a few keV) and reliably. Unfortunately a conflicting situation exists here (10 standard deviations) between the direct con-

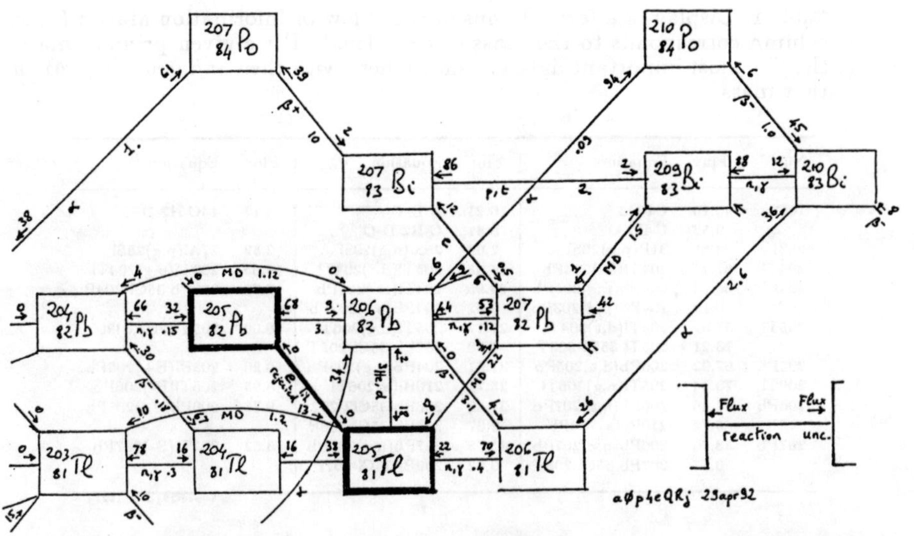

Figure 1. **Integrated connection diagram around ^{205}Tl and ^{205}Pb. Each box represents the mass of a nucleus, and each connection between boxes represents a piece of data. The arrows and numbers along a connection indicate the direction of the flux and the *influence* (in percent) of the corresponding datum.**

nection with an electron-capture measurement and several combinations of other data, connecting these two nuclei in a complex pattern. The Flow-of-Information Matrix allows to see immediately these patterns and their relative importance in the connection diagram in figure 1. Here a choice has been made as not to use the direct connection in the least-squares adjustment. Main pathes connecting ^{205}Tl to ^{205}Pb are then, on the high-A side:

^{205}Tl - ^{206}Pb - ^{205}Pb
^{205}Tl - ^{206}Tl - ^{210}Bi - ^{210}Po - ^{206}Pb - ^{205}Pb
^{205}Tl - ^{207}Pb - ^{206}Pb - ^{205}Pb

and on the light-A side:

^{205}Tl - ^{204}Tl - ^{204}Pb - ^{205}Pb

plus a small contribution from: ^{205}Tl - ^{203}Tl - ^{204}Tl - ^{204}Pb - ^{205}Pb

The Flow-of-Information Matrix can be observed along lines, displaying then for each datum which are the nuclei influenced by this datum and the values

Table 1. Display of a few columns of the Flow-of-Information Matrix (each column corresponds to the mass of a nucleus). For a given primary mass, the five most important data are shown here with their *influences* (×100) on that mass.

Nucl.	Flux	Equation	Flux	Equation	Flux	Equation
16O	45.15	C4-O3	16.21	N D-O	14.17	14C H2-O
	9.52	C D2-O	8.41	C H2 D-O		
28Si	94.85	31P(p,a)28Si	2.63	28Si(n,g)29Si	2.52	27Al(p,g)28Si
204Tl	67.47	204Tl(B-)204Pb	16.48	205Tl(d,t)204Tl	16.05	203Tl(n,g)204Tl
204Pb	66.14	204Pb(n,g)205Pb	29.95	204Tl(B-)204Pb	3.70	206Pb 35Cl-204P
	0.18	204Pb(p,d)203Pb	0.02	204Pb(p,t)202Pb		
205Tl	37.92	205Tl(d,t)204Tl	22.17	205Tl(n,g)206Tl	18.00	205Tl(3He,d)206
	13.21	205Tl 35Cl-203T	7.58	207Pb 35Cl-205T		
205Pb	67.92	206Pb(d,t)205Pb	31.81	204Pb(n,g)205Pb	0.26	205Bi(B+)205Pb
206Tl	70.34	205Tl(n,g)206Tl	25.71	210Bi(a)206Tl	3.95	206Tl(B-)206Pb
206Pb	44.46	206Pb(n,g)207Pb	34.40	206Pb 35Cl2-202	9.39	206Pb(d,t)205Pb
	5.74	210Po(a)206Pb	2.07	208Pb 35Cl-206P		
207Pb	53.30	206Pb(n,g)207Pb	41.58	207Pb(n,g)208Pb	3.52	207Tl(B-)207Pb
	0.77	207Pb 35Cl-205T	0.62	209Bi 35Cl-207P		

a0p4eQRJ 29apr92

of these *influences.* We can observe also the Flow-of-Information Matrix along columns, as illustrated by table 1, and get for each primary mass the four or five most important data with their influence on that mass. In the *Evaluation of Masses* the data-equations could be truncated to their first 15 characters without leading to ambiguities, and such table may be published, with some 800 items, in the next issue of the Table of Masses.

3. The Interactive Graphical Display of the Surface of Masses

The experimental Surface of Masses is characterized by the very smooth behaviour of its four sheets (one for each type of parity in N and Z) manoeuvering together in a 3-dimensional space (N, Z and Mass). This smoothness is interrupted in some places by "accidents" that are either associated with an experimental bias that is later shown to be so, or with a relatively violent change in one or more physical parameters (e.g. shell closures, shape transitions). Unfortunately, visualizing and handling these sheets in a tridimensional space together with the constraints concerning their relative distance is not an easy task. It is for this reason that, for the *Evaluation of Masses* and for close extrapolations, the "derivatives" of the surface of masses, like separation energies, decay energies or pairing energies were considered in various projections[3]. These derivatives, not only exhibit the same smoothness property as the surface itself, but still more important, they enhance the structures.

In order to improve the reliability of the examination of the Surface of Masses, we want to observe the continuity property in several representations of the various existing derivatives. Therefore an interactive graphic computer program was developed on a VAX workstation to display simultaneously any four derivatives (separation energies S_n, S_{2n}, S_p, S_{2p}, decay energies Q_α, Q_β, $Q_{2\beta}$, pairing energies Δ_{nn}, Δ_{np}, Δ_{pp}) plotted as a function of any of the parameters: Z, N, A, N-Z or 2Z-N. Continous lines connect (optionally) in each plot nuclei having any of the iso-properties: Z, N, A, N-Z or 2Z-N. Recentering and zooming of any of the 4 pictures allows to better focus on a zone of interest. Furthermore, the result of any mass formula can be superimposed on the same figures. Interactively, any point in any diagram can be displaced and all the subsequent changes in the chosen diagram (parentages included) as well as in the others will follow. Also new points may be created in any of the four chosen representations and will show up everywhere. Modifications and creations are recorded in a special file and can be replayed when required.

In the evaluation of data on masses, when irregularities are observed, this **Interactive Graphical Tool** is powerful in helping to trace down the mass that is responsible for any given irregularity. Besides, it is also used to extrapolate[5] the Surface of Masses and get a good estimate of still unknown masses.

4. References

[1] G. Audi, W. G. Davies and G. E. Lee-Whiting, *N.I.M.* **A249** (1986) 443.

[2] Y. V. Linnik, *Method of Least Squares* (Pergamon, New York, 1961); *Méthode des Moindres Carrés* (Dunod, Paris, 1963)

[3] K. Bos, G. Audi and A. H. Wapstra, *Nucl. Phys.* **A432** (1985) 140.

[4] K. Yokoi, K. Takahashi and M. Arnould, *Astron. Astrophys.* **145** (1985) 339.

[5] C. Borcea and G. Audi, *CSNSM-9238* report, Orsay 1992.

RESONANCE REGION

CROSS-SECTION PARAMETERIZATION IN THE RESOLVED RESONANCE REGION

N. M. Larson
Oak Ridge National Laboratory
Oak Ridge, Tennessee, USA

ABSTRACT

Experimental techniques, methods, and equipment have evolved to provide more accurate neutron cross section data with better energy resolution. Keeping pace with those developments has been a challenge for data analysts; commensurate improvements in analysis tools are required. In this paper, analysis techniques for neutron time-of-flight data in the resolved resonance region are discussed, with emphasis on contemporary needs.

1. Introduction

Cross section parameterization in the resolved resonance region is most accurate with the use of a multilevel multi-channel R-matrix formalism, although in some cases a more restrictive approximation may provide an acceptable substitute. Within the evaluations in ENDF/B-VI, a variety of formalisms can be found; single- and multi-channel Reich-Moore, single and multilevel Breit Wigner, and Adler-Adler are perhaps the most common. Applications of these formalisms for structural material evaluations are surveyed in Section 2.1 of this paper. Section 2.2 is devoted to a discussion of one computational problem that may arise in generating cross sections from ENDF/B resonance parameters.

In Section 3, specific requirements for analysis programs are discussed. Analysis of large amounts of data can tax even modern computers unless some technique (e.g., Bayes' theorem) is used to facilitate analysis of data subsets while yielding results valid for the entire data set. Simultaneous analysis of several different types of data can be accomplished in like manner; this is described in Section 3.1.

Data types which analysis programs should be able to handle include (but are not limited to) transmission, elastic and inelastic scattering, capture, fission, and angular distribution data, as discussed in Section 3.2. Improved accuracy and experimental resolution requires improvements in the mathematical descriptions of the resolution broadening functions used in the analysis, as discussed in Section 3.3. Multiple scattering and other finite-size corrections needed for calculation of many types of cross sections are discussed in Section 3.4. Finally, results obtained by the analysis process should include meaningful estimates of uncertainties and correlations on parameter values, as discussed in Section 3.5.

The multilevel R-matrix analysis code SAMMY[1] is used to illustrate one approach to these problems for medium and heavy nuclei.

2. Calculation of the cross section

2.1 Parameterizations

ENDF/B-VI[2] permits several different formulations for cross section parameterization: Single-level Breit Wigner (SLBW), multilevel Breit-Wigner (MLBW), Reich Moore (RM), Adler-Adler (AA), the General R-matrix and the Hybrid R-function. Table 1 summarizes the parameterizations used in resonance analyses reported in ENDF/B-VI for the structural materials. Note that the single- and multilevel Breit Wigner have been most commonly used; recent evaluations, however, tend toward the more accurate multilevel R-matrix formulations, in particular to the Reich Moore approximation.

Table 1. Parameterizations used in analyses for structural materials reported in ENDF-B/VI.

Nucleus	Isotopes	Laboratory	Date	Formalism
22Ti	47,48	BRC, NNL	1977	pointwise
23V		ANL-LLNL	1988	MLBW
24Cr	52,53	ORNL	1988	RM
25Mn	55	JAERI,ORNL	1988	MLBW
26Fe	56,57	ORNL	1989	RM
27Co	59	ANL-LLNL	1989	MLBW
28Ni	58,59,60,61,62	ORNL	1989	RM
29Cu	63,65	ORNL	1989	RM
39Y	89	ANL-LLNL	1986	MLBW
40Zr	90,91,92,94,96	BNL	1981	MLBW
41Nb	93	ANL-LLNL	70's?	SLBW
42Mo	95,96	LLNL	1977	SLBW
72Hf	178	SAI	1976	SLBW
74W	184	LANL	1982	MLBW
79Au	197	LANL	1984	MLBW
82Pb	206,207,208	ORNL	1989	RM
83Bi		ANL-LLNL	80's	MLBW

Advantages of the SLBW and MLBW are well known: primarily, they are simple to implement and facilitate use of psi and chi functions for broadening. However, SLBW is accurate only in situations where interference between levels is neglible; therefore its utility is limited to regions with well-isolated p- and d-wave resonances.

Figure 1 illustrates the danger inherent in the rather common practice of reconstructing pointwise cross sections with a formalism different from that used in the original data analysis, in this case RM. Note that the SLBW yields negative cross sections. While the MLBW does not produce negative cross sections, it also does not provide the correct interference between close-lying levels with the same J^{π}. It should be noted that if the data had originally been analyzed with SLBW or MLBW, the resulting reconstruction using the same formalism would generally produce cross sections in closer agreement with experiment than obtained by mixing formalisms.

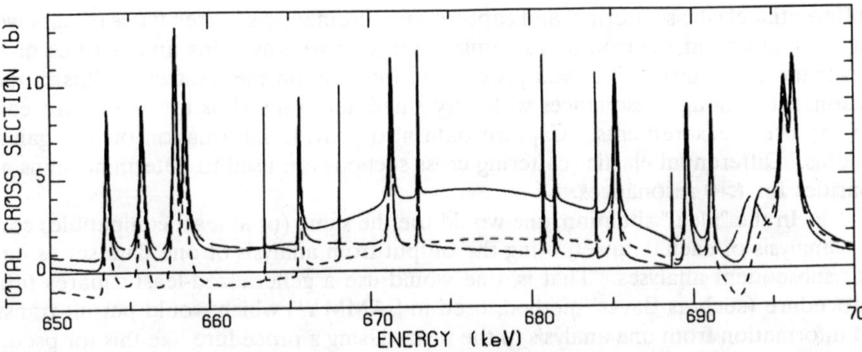

Figure 1. Calculated total cross section for ^{56}Fe; resonance parameters were obtained from a RM analysis. The solid line gives the calculated values using Reich Moore, the long dashes multilevel Breit Wigner, and the short dashes single level Breit Wigner.

2.2. Reformulation of Reich Moore approximation

A word of caution is perhaps in order here regarding calculations of cross sections via the Reich-Moore (or any other) formulation: Computers calculate exactly what they are programmed to calculate, to the precision with which they are programmed. If the formulae as given in Appendix D.1.3 of the ENDF-B/VI manual[2] are programmed directly, it is possible that unexpected results may be obtained.[3] Equation (2) on page D.8 of that manual defines the cross sections in terms of a quantity ρ, which is given by

$$\rho = I-(I-K)^{1} \tag{1}$$

where I is the identity matrix and K is defined in terms of the resonance parameters. For small energies K varies directly with the square root of the energy. Thus the expression $(I\text{-}K)$ is of the form $(1\text{-}\epsilon)$, where ϵ is small relative to 1; the inverse of $(I\text{-}K)$ then is approximately $(1+\epsilon)$, so the evaluation of ρ involves small differences of large numbers.

An equivalent, but less error-prone, formulation is to rewrite ρ as

$$\rho = (I-K)^{-1}(I-K) - (I-K)^{-1} = (I-K)^{-1}(I-K+I) = -(I-K)^{-1}K. \tag{2}$$

and write coding for this equation rather than for the original given in the ENDF/B-VI manual. Users of SAMMY may choose to use either this or the original formulation.

3. Requirements for Analysis Programs

3.1. Simultaneous Analysis

Recent studies have shown the benefit of concurrent analyses of different types of data. For example, the ^{58}Ni analysis of C. M. Perey et al.[4] considers transmission, differential elastic scattering, and capture measurements; all three types of data were needed to obtain neutron and capture widths as well as spins and parities of the resonances. Transmission data provide information on the neutron widths of most resonances, though resonances with very small neutron widths are seen most easily in capture measurements. Capture data also provide information on the gamma widths. Differential elastic scattering cross sections are used to determine spins and parities for $\ell>0$ resonances.

In the "ideal" situation, one would use the same (or at least compatible) codes for analysis of each data set, using the output from analysis of one data set as input for subsequent analyses. That is, one would use a generalized-least-squares fitting procedure (such as Bayes' method, used in SAMMY[1]) which would permit transfer of information from one analysis to the next. Using a procedure like this for pseudo-simultaneous (actually sequential) analyses yields results comparable to what one would get with a truly simultaneous fit (subject to certain limitations on linearity, computer word size, etc.).

A simple example serves to illustrate how this works: Fig. 2a shows the fission cross section for ^{239}Pu[5]. Suppose this data set could not all be analyzed at once, but instead had to be broken into the two separate pieces shown in the figure. By using the output parameter values and covariance matrix from the analysis of the first region as input to the analysis of the second region, results are obtained which are indistinguishable within the data uncertainties from those obtained by analyzing the entire data set at once, or by analyzing the second region prior to the first.

The same is true for a four-part division such as that shown in Fig. 2b, provided judicious choices are made for the division points and for the order in which the various pieces are to be analyzed. One would not, for example, choose a division

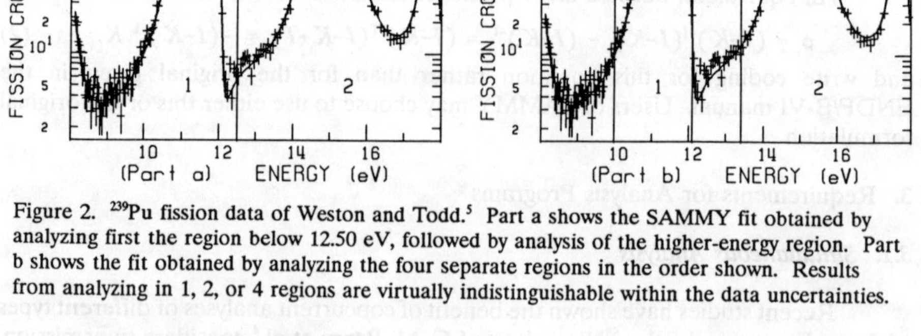

Figure 2. ^{239}Pu fission data of Weston and Todd.[5] Part a shows the SAMMY fit obtained by analyzing first the region below 12.50 eV, followed by analysis of the higher-energy region. Part b shows the fit obtained by analyzing the four separate regions in the order shown. Results from analyzing in 1, 2, or 4 regions are virtually indistinguishable within the data uncertainties.

point along the sharp rise of a resonance, or analyze first a flat region which contains information only about contributions from far-away levels and hard sphere scattering.

The obvious method of testing the validity of such a "broken" analysis is to re-analyze the entire region at once. However, if this were practical one would not have divided the data into regions to begin with! To test validity in a more realistic case, simply make other choices for division points and/or ordering of the partial analyses, and repeat the analysis.

Clearly in the simple example just considered, one would not analyze the data in four (or even two) separate pieces. However, it is often necessary to do exactly that: either there are simply too many data points, or there are too many kinds of data.

An ongoing analysis of silicon, for example, involves at least seven distinct measurements: four measurements of total cross sections for natural silicon, plus measurements of ^{29}Si oxide and ^{30}Si oxide total cross sections and inelastic scattering from ^{28}Si. Also, isotopic thermal capture and elastic values are available. An analysis program to fit all of these simultaneously would have to be rewritten for every application, and rerun whenever new data became available. Instead, we use the more attractive alternative of sequential fitting via SAMMY while communicating information obtained from earlier analyses. This yields final results which are equivalent to those which would be obtained from the truly simultaneous analysis, if it were practical to do.

3.2 Data Types and Corresponding Analysis Codes

In the ^{58}Ni study of Perey et al. alluded to above, three separate computer codes were used for the analysis. Capture data were analyzed with the code LSFIT, written by R. L. Macklin.[6] Transmission data were analyzed with SAMMY, written by this author.[1] Differential elastic cross sections were compared to calculations made by the code RFUNC, written by F. G. Perey.[7] Three separate codes were required because no one code was capable of analyzing the three different kinds of data. Unfortunately, input and output from the three codes are not mutually compatible. However, work is in progress to upgrade one of the codes (SAMMY) to correct this situation, so that future analyses will not be similarly hampered.

The capability of analyzing differential elastic scattering data is now available in SAMMY, for samples of arbitrary spin. Finite-size corrections are included in appproximate fashion (comparable to that in RFUNC): Incident-neutron attenuation is described by multiplication of the differential scattering cross sections by the factor $exp(-n_i \sigma_T(E))$, where $\sigma_T(E)$ is the Doppler-broadened total cross section at laboratory energy E, and n_i is a free parameter whose initial value is half the thickness of the sample. Outgoing-neutron attenuation is described by a multiplicative exponential of the form $exp(-n_0\ \sigma_T(E'))$; the total cross section used here is evaluated at the laboratory energy of the scattered neutron, and has been resolution-broadened to compensate for the finite size of the detectors. The thickness n_0 is a free parameter whose value can be chosen to crudely account for multiple-scattering effects.

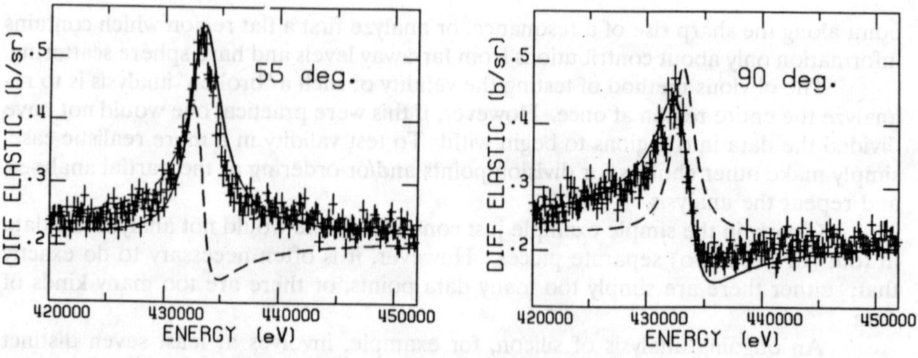

Figure 3. SAMMY fit to preliminary differential scattering data on ^{14}N at 55 and 90 degrees. The solid curve represents the d7/2 fit and the dashed curve represents the f7/2.

As illustration, a fit to preliminary ^{14}N differential elastic scattering data of J. A. Harvey et al.[8] is shown in Fig. 3. Here the sample is spin 1, the 433-keV resonance is known from high-resolution transmission measurements[9] to be J=7/2. Analysis of the differential elastic data suggests that this resonance has ℓ of 2 rather than 3, since ℓ=3 does not provide the correct angular distribution to reproduce the data.

In addition to transmission, differential elastic, and capture data, analysis codes should be applicable to fission, inelastic scattering, integral elastic scattering, and any other neutron reactions that experimentalists choose to measure in the resonance region.

3.3 *Resolution Broadening*

The excellent energy resolution of many of the recent resonance region experiments has led to refinement in the resolution-broadening techniques in analysis codes. The original resolution-broadening functions available in SAMMY were Gaussian, exponential, and a convolution of the two; these are not capable of properly describing many conditions found in ORELA and other data. Therefore, a "realistic" broadening function has been added, which attempts to correctly describe contributions to the resolution function both from the neutron production/moderation in the target and from the neutron detector. M. C. Moxon's single-channel Reich Moore code REFIT[10] has a similar realistic resolution function.

SAMMY's "Oak Ridge Resolution Function" is the convolution of relatively simple analytic functions describing the various components of the resolution function. The electron burst from the accelerator is modeled as a square function in time, as is the time-of-flight channel width (although that width can change in different time-of-flight regions). The resolution due to the water moderator part of the neutron-producing target is closely approximated by a chi-square distribution[11]

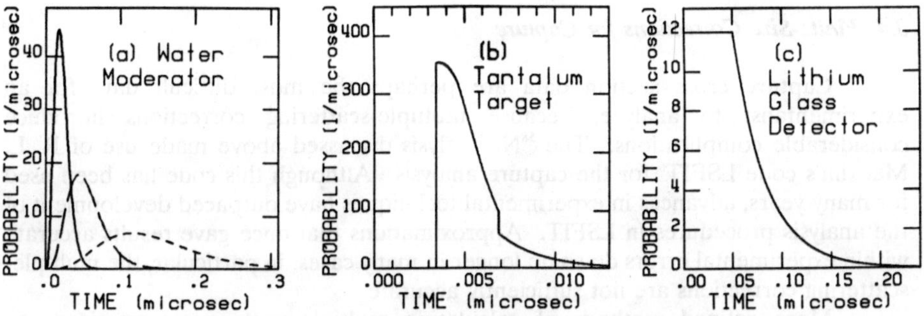

Figure 4. Components of the realistic resolution function in SAMMY. Part a shows the chi-square distribution describing the contribution to the resolution function from the water moderator; short dashes are appropriate for energies near 100 eV, long dashes near 1 keV, and the solid curve for energies near 10 keV. Part b shows the contribution from the tantalum target, and part c from the lithium glass neutron detector. The functions are normalized so that the area under each curve is unity.

(see Fig. 4a); alternatively the resolution function contribution for neutrons coming directly from the tantalum target has a more complicated (but still manageable) form shown in Fig. 4b. The resolution function appropriate for a plastic scintillator proton recoil detector is approximated by a decaying exponential, and the resolution function for a lithium glass neutron detector is shown in Fig. 4c.

The resulting "total" resolution function is a convolution of the four components described above. The result can appear highly non-Gaussian and non-exponential, depending on values of parameters in the various components. Examples are shown in Fig. 5.

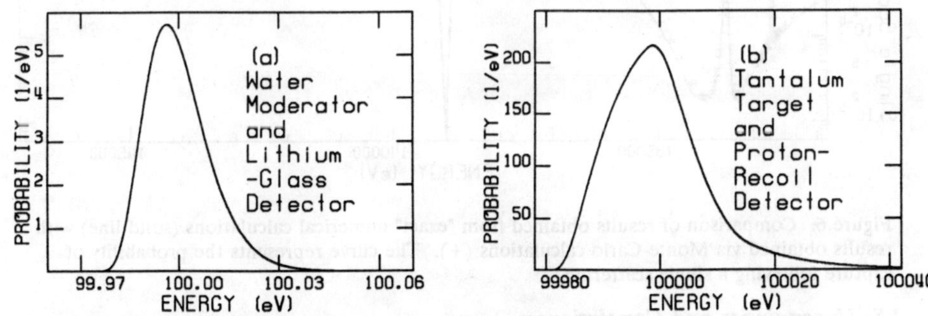

Figure 5. Realistic resolution function for water moderator with lithium-glass detector (part a) and for tantalum target with liquid scintillator detector (part b).

3.4 Finite-Size Corrections for Capture

Capture cross section data are perhaps the most difficult data for an experimentalist to analyze, because multiple-scattering corrections introduce considerable complications. The ^{58}Ni analysis discussed above made use of R. L. Macklin's code LSFIT[6] for the capture analysis. Although this code has been used for many years, advances in experimental techniques have outpaced development of the analysis procedures in LSFIT. Approximations that once gave results accurate within experimental errors do so no longer in many cases; in particular, the multiple-scattering corrections are not sufficiently accurate.

More refined methods of calculating multiple-scattering corrections are therefore now being implemented within SAMMY. The procedure, developed by F. G. Perey, is to assume a precise geometry for the sample (a flat circular disk held perpendicular to the neutron beam) and, within that geometry, to calculate the single-scattering correction exactly including the correct angular distribution, and double-scattering approximately. Comparison of results obtained from calculations via this procedure with results obtained via Monte-Carlo calculations are shown in Fig. 6. Generally, results appear more-or-less insensitive to the radius of the sample.

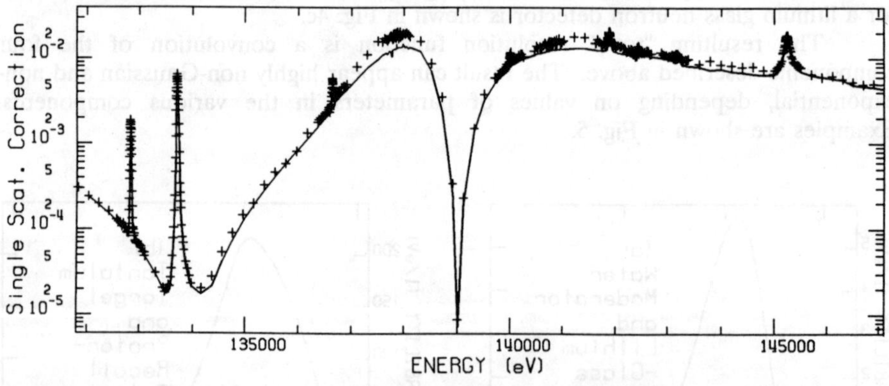

Figure 6. Comparison of results obtained from "exact" numerical calculations (solid line) with results obtained via Monte-Carlo calculations (+). The curve represents the probability of capture following a single scattering.

3.5 Uncertainties and Correlations

As nuclear design applications have progressed to using sensitivity analysis, uncertainties in nuclear parameters have become an important part of such analyses. Formats for resonance parameter uncertainties have been developed and refined.

A complete resonance parameter analysis must now include uncertainties and correlations among parameters. Obtaining correct values for these quantities requires that the experimental data which is input to the analysis also include good estimates of uncertainties *and correlations*.

This is where difficulties often arise. Most reported "experimental data" is not actually raw data, but has been subjected to an extensive data reduction procedure including such operations as background subtraction and normalization. Such data reduction operations introduce correlations among the data points, and the resultant covariance matrix is very large (N × N, where N is the number of data points). Storage and manipulation of such large arrays is beyond the capabilities of many computers, and/or of the computer programs employed to analyze the data. The "standard" solutions until recently have been to give covariances for binned cross sections or even to ignore the problem and use only the diagonal portion of the covariance matrix in the analysis codes.

An alternative approach (one which is actually correct in all cases!) is to include the data-reduction operations within the analysis itself. That is, instead of (or, perhaps, in addition to) subtracting background from and normalizing the *data*, one can normalize and add background to the *theoretical calculation*. In addition to eliminating the need for a very large data covariance matrix, this procedure has the advantage that values and uncertainties for the data reduction parameters can be refined during the course of the resonance analysis.

SAMMY is moving is this direction; a constant normalization, and several specific types of energy-dependent backgrounds, are options currently available to be added to the theoretical calculation in SAMMY. Another option available in that code (but, alas, not yet used extensively) is to include data-reduction parameters for which the user supplies the necessary partial derivatives.

4. Conclusions and Recommendations

Modern data analysis techniques in the resolved resonance region require the use of (1) accurate approximations to multi-channel, multilevel R-matrix theory (notably, the Reich-Moore approximation), (2) sophisticated and realistic resolution broadening functions, (3) careful calculations of finite-size effects, (4) statistical procedures (such as Bayes' method) which permit incorporation of prior information and pseudo-simultaneous analysis of different types of data, and (5) inclusion of data reduction parameters such as normalization and background in the data analysis process. Analysts should exercise great care in using computer codes not utilizing these techniques, to see that approximations used in those codes are adequate for the data being analyzed.

While RM is allowed in ENDF/B-VI for structural as well as fissionable materials, many important materials still have SLBW and MLBW resonance parameters. These evaluations should be reviewed to see if the analysis is adequate.

Further development of new data reduction/analysis techniques as outlined in this paper will provide meaningful uncertainties and correlations among resonance parameters in a concise, useful manner.

5. Acknowledgements

The author wishes to express her gratitude to the many SAMMY users for their frequent urgings to update the code with useful new features. Research sponsored by the Office of Energy Research, Division of Nuclear Physics, U. S. Department of Energy, under contract DE-AC05-84OR21400 with Martin Marietta Energy Systems, Inc.

6. References

1. N. M. Larson, *Updated Users' Guide for SAMMY: Multilevel R-Matrix Fits to Neutron Data Using Bayes' Equations*, ORNL/TM-9179/R2 (1989).
2. *Data Formats and Procedures for the Evaluated Nuclear Data File, ENDF*, ENDF-102, ed. P. F. Rose and C. L. Dunford (1990).
3. N. M. Greene (ORNL), private communication (1992).
4. C. M. Perey, F. G. Perey, J. A. Harvey, N. W. Hill, N. M. Larson, R. L. Macklin, and D. C. Larson, "^{58}Ni + n Transmission, Differential Elastic Scattering and Capture Measurements and Analysis Up to 813 keV," *Phys. Rev. C* (to be published).
5. L. W. Weston and J. H. Todd, *Nucl. Sci. Eng.* 88 (1984), 567.
6. R. L. Macklin, *Nucl. Sci. Eng.* 59 (1976) 12.
7. F. G. Perey, *RFUNC - A Code to Analyze Differential Elastic-Scattering Data*, ORNL/TM-11112 (1989).
8. J. A. Harvey (ORNL), private communication (1992).
9. J. A. Harvey, N. W. Hill, N. M. Larson, and D. C. Larson, *Proceedings of an International Conference on Nuclear Data for Science and Technology*, ed. S. M. Qaim (Springer-Verlag, Berlin 1992) 729.
10. M. C. Moxon, *REFIT, a Least Square Fitting Program for Resonance Analysis of Neutron Transmission and Capture Data*, AEA-InTec-0470 (1991).
11. S. N. Cramer and F. G. Perey, *Nucl. Sci. Eng.* 111 (1992) 102.

MULTILEVEL PARAMETERIZATION OF FISSILE NUCLEI CROSS-SECTIONS

ALEXANDRE LUKYANOV
Sofia University, Department of Physics,
J.Boucher NO 5, 1126 Sofia, Bulgaria

ABSTRACT

In the frame of s-matrix formalism and using the collision matrix unitarity property a method for determination of self consistent parameter set has been developed and realized in practice. The parameters of total and fission cross section, obtained by multilevel analysis of experimental data are used for creation of resonance radiative capture and neutron elastic scattering cross sections in this region.

1. General

At the calculation of fissile nuclei cross sections in the resolved resonance region are used the methods of multilevel parameterization based on general results of resonance reaction theory [1]. The difference in methods is connected with the choice of presentation of collision matrix $S^J(E)$ and parameters are found by fitting of experimental data.

The formalism of R-matrix theory gives rather general expression of collision matrix elements [1]:

$$S^J_{nc}(E)=\exp\left[-i(\varphi_n+\varphi_c)\right]\left\{\delta_{nc}+2iP_n^{1/2}\sum_{\lambda\mu}\gamma_{\lambda n}\gamma_{\mu c}(A^{-1})_{\lambda\mu}P_c^{1/2}\right\}, \qquad (1)$$

where φ_c - is potential scattering phase of the channel c, $P_c(E)$ - penetration factors, $(A^{-1})_{\lambda\mu}$ - elements of inverse matrix -

$$A_{\lambda\mu}=(E_\lambda-E)\delta_{\lambda\mu}-i\Gamma_{\lambda\mu}/2 \qquad (2a)$$

with

$$\Gamma_{\lambda\mu}=\sum_c\Gamma^{1/2}_{\lambda c}\Gamma^{1/2}_{\mu c}\ ,\quad \Gamma_{\lambda c}=2P_c\gamma^2_{\lambda c}\ .$$

Usually the sum over radiative channels $c(\gamma)$ in $\Gamma_{\lambda\mu}$ is suggested to be diagonal and then

$$A_{\lambda\mu}=(E_{\lambda}-E-i\Gamma_{\gamma\lambda}/2)\delta_{\lambda\mu}-i\Gamma'_{\lambda\mu}/2 \ , \qquad (2b)$$

where $\Gamma'_{\lambda\mu}$ is a sum only over neutron and fission channels. Accordingly the rank of collision matrix diminishes and also the number of free parameters $\gamma_{\lambda c'}$ which are real and independent on energy values.

The use of $S^J(E)$ with excluded radiative channels in $A_{\lambda\mu}$ Eq.(2b) corresponds to the most general variant of Reich-Moore scheme. However, in practice up to now is supposed that $\Gamma'_{\lambda\mu}$ does not depend on energy [2-6]. The restriction of that approximation is obvious at least for neutron widths - $\Gamma_{\lambda n}\sim E^{1/2}$. More rigorous consideration of S(E)-matrix structure leads in this case to the pole-expansion over $E^{1/2}$ and corresponding splitting of resonances [7]. However for fissile nuclei, where neutron widths are relatively small, this effect is not significant, at least in limited energy intervals.

The other methods of multilevel parameterization of resonance cross sections correspond to different variants of level matrix $A_{\lambda\mu}$ presentation in $S_{nc}(E)$ [1]. Neglecting the off-diagonal terms in Eq.(2a) leads to the result of multilevel Breit-Wigner approximation [8]. When we neglect in $\Gamma'_{\lambda\mu}$ at $\lambda\neq\mu$ only neutron channels we obtain Vogt approximation [9].

Rather attractive from practical point of view is the scheme of Adler using pole-expansion for $S^J(E)$ [1,10]. The resonance cross sections here have the form of interfering Breit-Wigner terms, which make the accounting for Doppler broadening rather simple. The transition to corresponding form of collision matrix is realized formally by level matrix A diagonalisation Eq.(2a):

$$TAT'=\tilde{\varepsilon}-E, \quad TT'=1, \quad A^{-1}=T'(\tilde{\varepsilon}-E)T \ , \qquad (3)$$

where T is the matrix of orthogonal complex transformation, $\tilde{\varepsilon}$ - diagonal matrix with elements $\tilde{\varepsilon}_k=\mu_k-i\nu_k$. As a result the collision matrix Eq.(1) comes to the form [1]:

$$S^J_{nc}(E)=\exp\left[-i(\varphi_n+\varphi_c)\right]\left\{\delta_{nc}+2iP_n^{1/2}\sum_k \tilde{\gamma}_{kn}\tilde{\gamma}_{kc}(\tilde{\varepsilon}_k-E)^{-1}P_c^{1/2}\right\}, \quad (4)$$

where $\tilde{\gamma}_{kc}=\sum_\lambda T_{k\lambda}\gamma_{\lambda c}$. The matrix element $T_{k\lambda}$, determined by the system of equations Eq.(3) -

$$E_\lambda\delta_{\lambda\mu}-i\Gamma_{\lambda\mu}/2=\sum_k T_{k\lambda}\tilde{E}_k T_{k\mu}, \quad \sum_k T_{k\lambda}T_{k\mu}=\delta_{\lambda\mu},$$

are complex values, depending in general on energy. However if this dependence in $\Gamma_{\lambda\mu}$ is not significant the parameters $\tilde{\gamma}_{kc}$ and $\tilde{E}_k$ in Eq.(4) can be considered as complex constants.

In the method for multilevel analysis of cross sections proposed below and realised practically for ^{239}Pu and ^{235}U are determined self consistent parameter sets in Adler formalism. The main result is the calculation of the resonance cross sections of radiative capture and elastic scattering by using parameters from multilevel analysis of experimental data for total and fission cross sections. This possibility follows from the unitarity property of collision matrix $S^J(E)$ for the system of levels with determined total momentum J.

2. Resonance Cross Section Parameterization

The fissile nuclei resonances, observed in the resolved region, are related to two independent level systems with J=I±1/2, where I is the spin of target nucleus. To every J value corresponds one neutron channel with L=0. The total cross section with given J is

$$\sigma^J(E)=\sigma^J_p+2\pi\lambda\!\!\!^{-2}g(J)\left[1-\mathrm{Re}\ S^J_{nn}(E)\right]$$

and using here diagonal element of collision matrix Eq.(4) we obtain $\sigma^J(E)$ as a sum of resonance terms[1,10]:

$$\sigma^J(E) = \sigma^J_p+\pi k^{-2}\sqrt{E}\ g(J)\sum_{k(J)}\frac{G_k\nu_k-H_k(\mu_k-E)}{(\mu_k-E)^2+\nu_k^2}, \quad (5)$$

where σ^J - is potential scattering cross section and the values of G_k and H_k are determined by relation -

$$(G_k - iH_k)\exp(2i\varphi_n) = 2\tilde{\Gamma}^0_{kn} , \tag{6}$$

where $\tilde{\Gamma}^0_{kn} = 2P_n \tilde{\gamma}^2_{kn}/\sqrt{E}$ are supposed to be independent on energy complex constants.

The fission cross section can be presented in similar form:

$$\sigma^J_f(E) = \pi k^{-2} g(J) \sum_{c(f)} |S^J_{nc}(E)|^2 = \pi k^{-2}\sqrt{E}\; g(J) \sum_{k(J)} \frac{G^f_k \nu_k - H^f_k(\mu_k - E)}{(\mu_k - E)^2 + \nu_k^2} , \tag{7}$$

where

$$G^f_k - iH^f_k = 2i \sum_{c(f)} \sum_{k'(J)} (\tilde{\Gamma}^0_{kn} \tilde{\Gamma}^{0*}_{k'n}\; \tilde{\Gamma}_{kc} \tilde{\Gamma}^*_{k'c})^{1/2} / (\tilde{E}^*_{k'} - \tilde{E}_k) \tag{8}$$

and summation over c(f) is related to fission channels ($\tilde{\Gamma}_{kc} = 2p_c \tilde{\gamma}^2_{kc}$).

The high resolution experimental data on total and fission cross sections in the resonance region [11-14] offer the possibility of multilevel analysis up to ~ 500 eV for ^{239}Pu and ~ 100 eV for ^{235}U. As a result now the multilevel parameter sets of both methods Reich-Moore [3-6] and Adler [10,15-17] are available and those are include in ENDF/B format [8].

In our works also has been performed simultaneous analysis of total and fission cross sections of ^{239}Pu and ^{235}U [16,17] for determination of parameters in formulae Eq.(5) and Eq.(7). The special code with non linear searching with successive expansion of analyzing interval has been used. Dopler broadening and experimental resolution were accounted for in cross sections with the standard functions Ψ and X [1].

As a result have been obtained self consistent parameter sets of Adler scheme containing 220 levels of ^{239}Pu below 600 eV and 145 levels of ^{235}U below 100 eV. These reproduce observed energy dependence of total and fission cross sections (Fig. 1).

Parameters μ_k, ν_k, $g\tilde{\Gamma}^0_{kn}$, gG^f_k, gH^f_k for separate levels are found by analysis of experimental data for cross sections, which are sums of two independent systems of resonances. However, now are known spins of the most resonances in considered energy intervals [3-6,11-14,18-20]. The identification of levels by spin value offers the possibility of creation the neutron absorption by using the unitarity property of collision matrix $S^J(E)$ Eq.(4):

$$\sigma^J_a(E) = \pi k^{-2} g(J)\left[1-|S^J_{nn}(E)|^2\right] =$$

$$= \pi k^{-2}\sqrt{E}\, g(J)\sum_{k(J)} \frac{G^a_k \nu_k - H^a_k(\mu_k - E)}{(\mu_k - E)^2 + \nu_k^2}\,, \tag{9}$$

where parameters

$$G^a_k - iH^a_k = 2\tilde{\Gamma}^0_{kn}\left[1 - i\sqrt{E}\sum_{k'(J)} \tilde{\Gamma}^{0}_{k'n}{}^{*}(\tilde{E}^*_{k'} - \tilde{E}_k)^{-1}\right] \tag{10}$$

can be calculated directly, if we know from total cross section analysis the parameters $\tilde{\Gamma}^0_{kn}$ and $\tilde{E}_k$.

The cross sections of radiative capture and elastic scattering are found from differences:

$$\sigma^J_\gamma(E) = \sigma^J_a(E) - \sigma^J_f(E)\,,$$

$$\sigma^J_n(E) = \sigma^J(E) - \sigma^J_a(E)\,. \tag{11}$$

It is to be noted that the positive values of these differences at all energies is important condition for correction of resonance parameters and identification of spin values of some levels. The energy dependence $\sigma_n(E)$ and $\sigma_\gamma(E)$ is analogous to Eq.(5) and Eq.(9), correspondingly, where parameters are determined as

$$G^n_k = G_k - G^a_k,\quad H^n_k = H_k - H^a_k,\quad G^\gamma_k = G^a_k - G^f_k,\quad H^\gamma_k = H_k - H^f_k\,. \tag{12}$$

Radiative capture and elastic scattering cross sections, which we constructed using described technique are consistent with average values, obtained experimentally in relatively small intervals (Tables 1,2). Let us note that

Table 1

Cross Sections and Self Shielding Factors of ^{239}Pu

Group N		23	22	21	20	19	18	17	16
					$\langle\sigma_\gamma\rangle$				
ENDF/B-V	[23]	7.3	1.2	26.4	66.9	34.2	37.3	17.2	12.7
JENDL	[24]	4.6	1.4	25.0	71.1	36.1	37.9	17.3	13.0
Konshin	[20]	5.2	1.6	23.4	64.9	33.3	37.2	15.1	–
Ours	[25]	4.8	1.1	23.7	66.2	32.5	36.7	16.1	12.3
					$\langle\sigma_n\rangle$				
ENDF/B-V	[23]	10.4	9.2	8.4	10.8	11.0	17.0	15.3	16.0
JENDL	[24]	9.9	9.3	8.0	10.2	10.5	17.1	14.2	14.4
Konshin	[20]	9.6	9.0	7.3	10.1	10.4	16.6	14.1	–
Ours	[25]	8.8	8.3	7.9	11.4	11.7	17.5	15.0	15.2
					$f_\gamma=\langle\sigma_\gamma/\sigma\rangle/\langle\sigma_\gamma\rangle\langle 1/\sigma\rangle$				
ENDF/B-V	[23]	.903	.930	.149	.133	.107	.177	.202	.200
JENDL	[24]	–	–	.154	.123	.129	.176	.307	.362
Konshin	[20]	.916	.973	.158	.125	.093	.167	.210	–
Ours	[25]	.965	.969	.162	.106	.089	.194	.288	.348

Table 2.

$\langle\sigma_\gamma\rangle$ and Neutron Transmission Values $\langle T(n)\rangle$ of ^{235}U

Energy interval	10-15	15-20.5	20.5-33	33-41	41-60
ENDF/B-IV	47.40	35.30	24.02	34.41	23.86
Gwin [13]	55.20	40.90	27.52	39.62	26.42
Ours	53.53	42.19	29.94	37.62	25.18

T(n)

$\Delta E \backslash n\ 10^{-2} b^{-1}$		0.257	0.386	1.02	2.14
10-21.5	1	.806	.735	.529	.322
	2	.819±.017	.739±.017	.550±.016	.356±.009
21.5-46.5	1	.834	.722	.561	.358
	2	.835±.010	.752±.017	.566±.016	.369±.011
46.5-93	1	.850	.789	.570	.356
	2	.859±.010	.788±.014	.582±.017	.379±.011

1. Calculation with our resonance parameters.
2. Experimental results [26].

the assumption about arbitrary distribution of radiative channels amplitudes is not necessary here, as it is in the Reich-Moore scheme, and there is no limitations of the fission channels number.

3. Sum Rules

The comparison of averaged over energy cross sections illustrates the precision of parameters G_k, G^f_k, G^γ_k, ν_k but it is not informative to the interference parameters H^f_k, H^γ_k, which are of the main interest in multilevel analysis. So, the integral over energy interval containing N levels, cross section of reaction c is $\int \sigma^J_c(E)dE \approx 0.652\pi g(J) \sum_{k(J)=1}^{N} E^{-1/2} G^c_k$ and generally does not depend on H^c_k.

For correction of parameters H^f_k, H^γ_k are important firstly the conditions $\sigma^J_\gamma(E)>0$ and $\sigma^J_n(E)>0$ in all interference minima and also sum rules, which follow from the unitarity of $S^J(E)$ matrix and the formal consistency between parameters of various methods. The orthogonal transformation (Eqn. 3) leads to the sums over levels with given J:

$$\sum_{k(J)} \tilde{\gamma}_{kc}\tilde{\gamma}_{kc'} = \sum_{\lambda(J)} \gamma_{\lambda c}\gamma_{\lambda c'}, \quad \sum_{k(J)} \tilde{E}_k = \sum_{\lambda(J)} (E_\lambda - i\Gamma_{\lambda\lambda}/2)$$

or

$$\sum_k \mathrm{Re}\tilde{\Gamma}^0_{kn} = \sum_\lambda \Gamma^0_{kn}, \quad \sum_k \mathrm{Im}\tilde{\Gamma}^0_{kn} = 0, \quad \sum_k \mu_k = \sum_\lambda E_\lambda, \quad 2\sum_k \nu_k = \sum_\lambda \Gamma_{\lambda\lambda} \; . \qquad (13)$$

From this particularly follows, that ν_k in general case depends on energy as $\Gamma_{\lambda\lambda} = \Gamma_{\lambda a} + \sqrt{E}\Gamma^0_{\lambda n}$. The approximation $\nu_k \approx \nu_k(\mu_k)$, which we used at cross section parameterization, is limiting the possibilities of our analysis in the same degree as the practical scheme of Reich-Moore [2-7].

The condition of collision matrix unitarity Eq.(4) leads to the relation between parameters of levels with given spin Eq.(10). Using here the assumption of Reich-Moore scheme for amplitudes of resonance widths of radiative capture channels ($\sum_{c(\gamma)} \Gamma^{1/2}_{\lambda c}\Gamma^{1/2}_{\mu c}=\Gamma_{\gamma}\delta_{\lambda\mu}$), the condition of unitarity can be presented as

$$\tilde{\Gamma}^{0}_{kn}\left[1-i\sqrt{E}\sum_{k'(J)}\tilde{\Gamma}^{0*}_{k'n}(\tilde{E}^{*}_{k'}-\tilde{E}_{k}-i\Gamma_{\gamma})^{-1}\right] =$$

$$= i\sum_{c(f)}\sum_{k'(J)}(\tilde{\Gamma}^{0}_{kn}\tilde{\Gamma}^{0*}_{k'n}\,\tilde{\Gamma}_{kc}\Gamma^{*}_{k'c})^{1/2}(\tilde{E}^{*}_{k'}-\tilde{E}_{k}-i\Gamma_{\gamma})^{-1}\,, \tag{14}$$

where in the right side is a sum over fission channels. For one channel fission this condition determines, at known total cross section parameters, the system of equations for parameters of fission cross section. In general case here obviously is necessary additional information about the number of channels and their relative contribution in the sum over c(f). Rather useful for practical analysis are rules which follow directly from the definition of parameters by Eq.(8) and Eq.(10):

$$\sum_{k(J)} H^{a}_{k} = \sum_{k(J)} H^{f}_{k} = \sum_{k(J)} H^{\gamma}_{k} = 0\ . \tag{15}$$

The sum rules Eq.(13-15) are useful for the choice of initial approximation in the code for searching and controlling of our parameters by using the available information about the results of parameterization by Reich-Moore technique. Formally, the knowledge of R-matrix parameters permits to calculate directly the Adler scheme parameters using the code POLLA [21] or Fröhner method [22]. However such conversion does not guarantee that the requirement for positive values of differences Eq.(11) is fulfilled in all interference minima .

4. Results and Conclusions

The general advantage of Adler method at multilevel fitting of resonance cross sections in the resolved region is the linear dependence on interference parameters H^{c}_{k}, which are responsible for the correctness of estimations in the cross sections minima. The scheme admits variation of these parameters, correlating with μ_k, without noticeable change of enery averaged cross sections. The use of collision matrix unitarity property in construction of resonance cross sections of radiative capture and elastic scattering permits in principle to provide the quality of these cross sections estimations on the level of knowledge of the total and fission cross sections. The comparisons with the evaluations of average cross sections for the BNAB system groups and self shielding factors, which are sensitive to the interference effects, is satisfactory (Table 1).

The results of our analysis can be adapted easily to ENDF/B format, where in Adler parameters section is convenient to use the values of $Re\tilde{\Gamma}^0_{kn}$ and $Im\tilde{\Gamma}^0_{kn}$ for total cross section and for fission cross section G^f_k and H^f_k as previous. The parameters of radiative capture and elastic scattering must be calculated by the proposed method. It is useful for this to keep in memory the helpful combinations -

$$\sum_{k'(J)} \tilde{\Gamma}^{0*}_{k'n}(\tilde{E}^*_{k'}-E_k)^{-1},$$

which determine the absorption cross section Eq.(10).

It must be noted at the and, that the great volume of information about the multilevel parameterization of resonance cross sections of fissile nuclei on the base of both methods - Reich-Moore and Adler-Adler, which are alternative presentations of the same energy dependence, is well enough to establish the direct consistency between different schemes parameters. The quality of the calculated radiative capture and elastic scattering cross sections by using the parameters of total and fission cross sections can be a good criteria of such consistency.

References

1. A.M.Lane and R.C.Thomas, *Rev. Mod. Phys.*, **30** (1958) 257. A.A.Lukyanov, *The Neutron Cross Sectoins Structure*, (Atomizdat, Moscow, 1978).
2. C.W.Reich and M.S.Moore, *Phys.Rew.*, **111** (1958) 929.
3. H.Derrien, J.Blons and A.Mishaudon, *Proc. Conf. Nuclear Data for Reactors, Helsinki*, (IAEI, Vienna, 1970) 481.
4. G.de Saussure, R.B.Perez, R.L.Maklin, *Proc.Conf.Nuclear Data for Basic and Applied Science, 1985*, (Gordon and Breach Sc. Publ., USA, 1986) v.2,p.1635.
5. J.A.Farrel, *Phys. Rew.* **165** (1968) 1371.
6. H.Derrien, G.de Saussure, N.M.Larson, L.C.Leal and R.B.Perez, *Proc.Conf.Nuclear Data for Science and Technology, Mito,1988*,ed.S.Igarasi (Saikon Publ. Company, 1988).
7. R.N.Hwang, *Proc.Conf.Nuclear Data for Science and Technology,Juelich,1991)*,ed.S.M.Qaim, (Springer-Verlag,1992),p.205.
8. D.Graber, C.L.Dunford, S.Pearlstain, *BNL-NCS-56496, ENDF-B-102*, 1975. R.E.Rose and C.L.Dunford, ed. *ENDF-102 Data Formats and Procedures for the Evaluated Nuclear Data File ENDF-6, BNL-NCS-*44945 (1990).
9. E.Vogt,*Phys.Rev.***118** (1960) 724.
10. F.T.Adler and D.B.Adler, *Proc.Conf.Neutron Cross Section Technology, Washington, 1966, Conf-*660303,Book 2, p.873
11. H.Derrien, J.Blons et al.,*Proc. Conf. Nucl. Data for Reactors, Paris* (1966) v.2, p.195, (IAEA, Vienna, 1967).see also Ref.3.,v.1, p.513.
12. C.Wagemans and A.J.Deruyter,*Ann.Nucl.Energy* **3** (1976) 437
13. R.Gwin et al., *Nucl. Sci. Eng.***59** (1976)79.
14. M.S.Moore, I.D.Moses and G.A.Keyworth, *Phys. Rew.***C18** (1978) 1328.
15. P.Lambropoulos, *Nucl. Sci. Eng.***40** (1970) 342.
16. V.V.Kolesov, A.A.Lukyanov, S.Toshkov, and N.Janeva,Ref. 4, v.2, p.1619, see also Bulg. J. of Phys.,**13** (1986) 335.
17. V.V.Kolesov and A.A.Lukyanov, Ref. 7, p.107.
18. J. Troshon, H.Derrien, B.Lucas and A.Mishaudon, Ref. 3, v. 1, p.495.
19. G.A.Keyworth et al. *Phys. Rev. Lett.* **17** (1973) 1077
20. G.V.Antsipov, V.A.Konshin, and E.S.Suchovitski, *Nuclear Constants for Pu -Isotopes* (Minsk, Nauka, 1982).
21. G. de Saussure and R.B.Perez *ORNL-2599* (1969).
22. F.H.Fröhner, *Proc. Conf. Nuclear Data Evaluation, Methods and Procedures, BNL, 1980.* (BNL-NCS-51363, 1981)
23. D.E.Cullen and O.Schwerer *Proc.Cons.Meeting on Uranium and Plutonium Isotopes, Vienna*, IAEA, (1981) p.197
24. T.Nakagawa, Y.Kikuchi et al. *Evaluation of Resonance Paramewters*, JAERI-M 9823 (1981) p.62.
25. A.A.Lukyanov, V.V.Kolesov and V.V Sinitza, *Proc.Conf. on Neutron Phys., Kiev, 1987*, (Atominform, 1988) v.2, p.279
26. A.Van'kov et al. Nucl. Sci. Eng., **96 (1987) 122.**

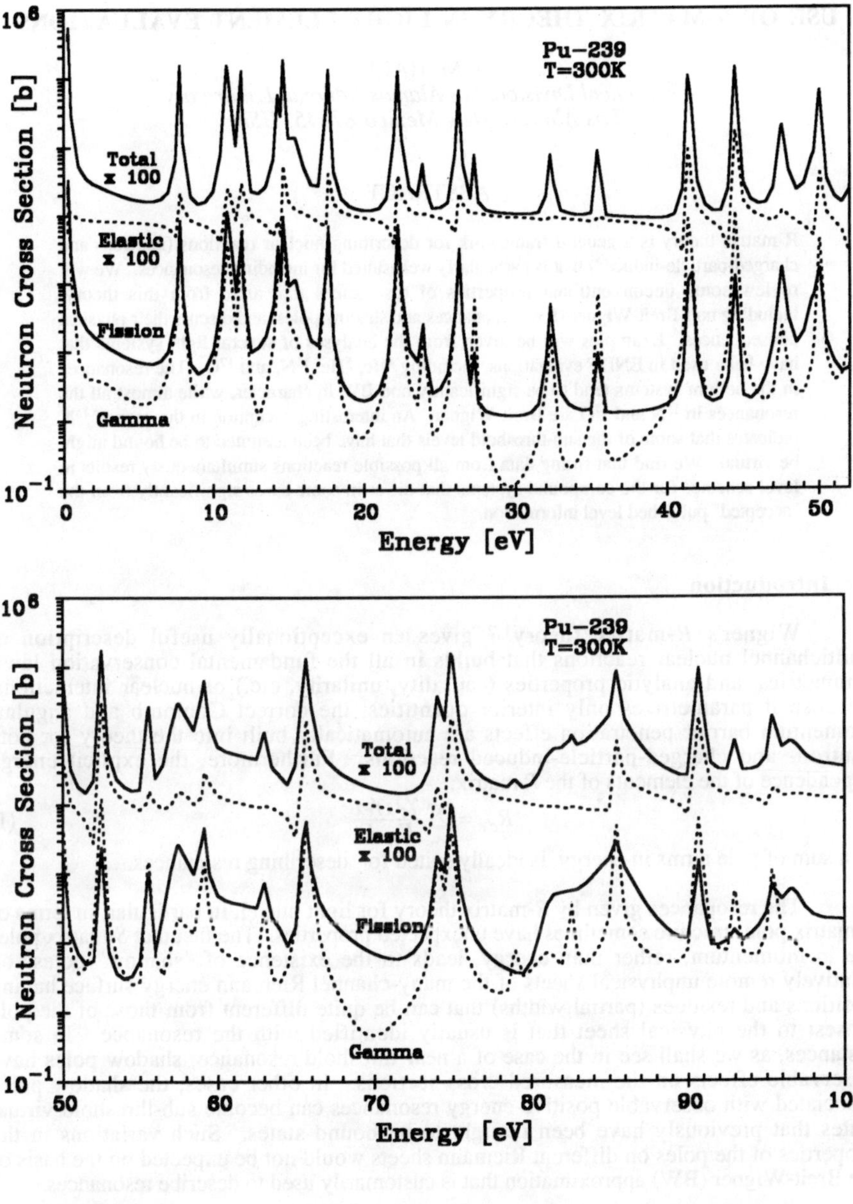

Fig. 1. Pu-239 resonance cross sections

USE OF R-MATRIX THEORY IN LIGHT ELEMENT EVALUATIONS

G. M. HALE
Theoretical Division, Los Alamos National Laboratory
Los Alamos, New Mexico 87545, USA

ABSTRACT

R-matrix theory is a general framework for describing nuclear reactions (neutron- and charged-particle-induced) that is particularly well-suited for including resonances. We will review some unconventional properties of resonances that arise from this theory, including non-Breit-Wigner (BW) resonances and shadow poles, and discuss their physical consequences. Examples will be given from the analyses of several light systems that have been used in ENDF evaluations, including ^{4}He, ^{5}He, ^{15}N, and ^{17}O. The resonances in the helium systems tend to be significantly non-BW in character, while almost all the resonances in ^{15}N and ^{17}O are Breit-Wigner. An interesting exception in the case of ^{15}N indicates that some of the sub-threshold levels that have been assumed to be bound might be virtual. We find that fitting data from all possible reactions simultaneously results in level schemes for the compound systems that differ in some cases significantly from the "accepted" published level information.

1. Introduction

Wigner's R-matrix theory[1,2] gives an exceptionally useful description of multichannel nuclear reactions that builds in all the fundamental conservation laws, symmetries, and analytic properties (causality, unitarity, etc.) of nuclear interactions. Because it parametrizes only interior quantities, the correct Coulomb and angular-momentum barrier penetration effects are automatically built into the theory for both neutron- and charged-particle-induced reactions. Furthermore, the explicit energy dependence of the elements of the R matrix,

$$R_{c'c} = \sum_{\lambda} \frac{\gamma_{c'\lambda}\gamma_{c\lambda}}{E_{\lambda}-E} \qquad (1)$$

as a sum of pole terms in energy, is ideally suited for describing resonances.

The resonances given by R-matrix theory for light nuclei, in particular, in terms of S-matrix pole structure sometimes have unexpected properties. The fact that S-matrix poles are in momentum, rather than energy, leads to the existence of "shadow" poles[3] on relatively remote unphysical sheets of the many-channel Riemann energy surface having positions and residues (partial widths) that can be quite different from those of the pole closest to the physical sheet that is usually identified with the resonance. In some instances, as we shall see in the case of a near-threshold resonance, shadow poles have observable effects on the measured cross sections. In other cases, the shadow poles associated with observable positive-energy resonances can become sub-threshold virtual states that previously have been thought to be bound states. Such variations in the properties of the poles on different Riemann sheets would not be expected on the basis of the Breit-Wigner (BW) approximation that is customarily used to describe resonances.

We will discuss in the next section aspects of resonances in the light compound systems ^{4}He, ^{5}He, ^{15}N, and ^{17}O. Calculations from the R-matrix analyses of all these systems have been used over at least part of the energy range in ENDF evaluations of neutron and charged-particle cross sections. Our approach in these analyses is to use data of all types from all possible two-body reactions. The effects of a three-body breakup channel on the two-body channels of interest can be approximated by including two-body pseudo-channels, as was done in the ^{5}He analysis. The data from all reactions are then fit simultaneously with a single set of multichannel, multilevel R-matrix parameters (the reduced-width amplitudes $\gamma_{c\lambda}$ and eigenenergies E_λ from Eq. (1), and sometimes also the channel radii a_c) using the general Los Alamos R-matrix code EDA.[4] This code features an automated variable-metric least-squares fitting algorithm that allows renormalization factors and energy shifts to be applied to the measured data included in the analysis. The calculated values can also be averaged over any of several different forms of energy resolution functions.

2. Examples

2.1 The ^{5}He System

The ^{5}He system contains one of the most famous resonances in nuclear physics: the J^π=3/2$^+$ resonance responsible for the large ^{3}H(d,n)^{4}He reaction cross section that peaks at E_d=107 keV. Values of the cross section at energies below the resonance are useful in a variety of fusion applications, and at higher energies, the differential cross section is of interest as a neutron source reaction.

The channels and data included in our analysis of the ^{5}He system for excitation energies up to 21.5 MeV are summarized in Table 1. In addition to the physical two-body channels d+t and n+α, an effective n+^{4}He*(0$^+$) channel was added to represent the effects of deuteron breakup (n+p+t). More than 2600 data points from 23 different types of measurements (cross sections, polarizations, *etc.*) were described in terms of 108 free R-matrix parameters that give a minimum in chi-square space for which χ^2 per degree of freedom is 1.48. We note that a generalized phase-shift fit would require 89 real parameters to achieve the same sort of description *at a single energy.*

Table 1. Channel configuration and data summary for the ^{5}He system analysis.

Channel	l_{max}	a_c (fm)
d+t	4	5.1
n+^{4}He	4	3.0
n+^{4}He*	1	5.0

Reaction	Energy Range	# Observable Types	# Data Points	χ^2
^{3}Hd,d)^{3}H	E_d=0-8 MeV	6	704	1164
^{3}H(d,n)^{4}He	E_d=0-8 MeV	14	1121	1379
^{3}H(d,n)^{4}He*	E_d=4.8-8 MeV	1	10	26
^{4}He(n,n)^{4}He	E_n=0-28 MeV	2	793	1150
Totals:		23	2628	3719

Examples of fits to some of the integrated cross-section data are shown in Fig. 1. At the left side of the figure are shown the fits to the reaction cross section (top) and neutron total cross section (bottom) over the whole energy range of the analysis. The right side of the figure shows essentially the same data enlarged over the region of the resonance. One sees that the calculations generally represent the data within their error bars, which, particularly in the case of the LEFCS measurements,[5] are reasonably small.

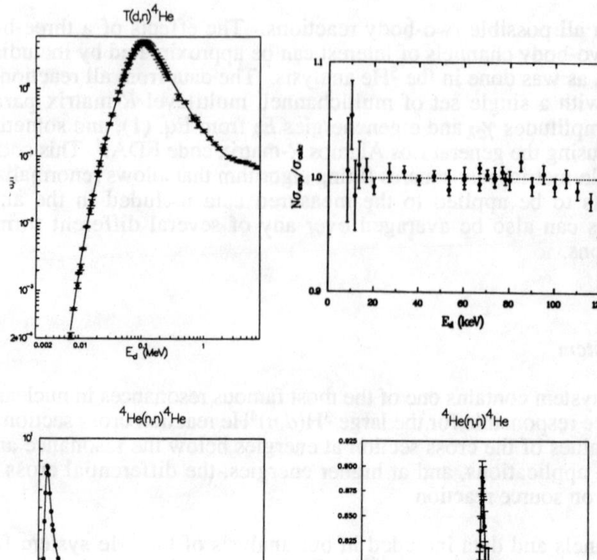

Fig. 1. ^{3}H(d,n) reaction (top) and n-α total (bottom) cross sections .

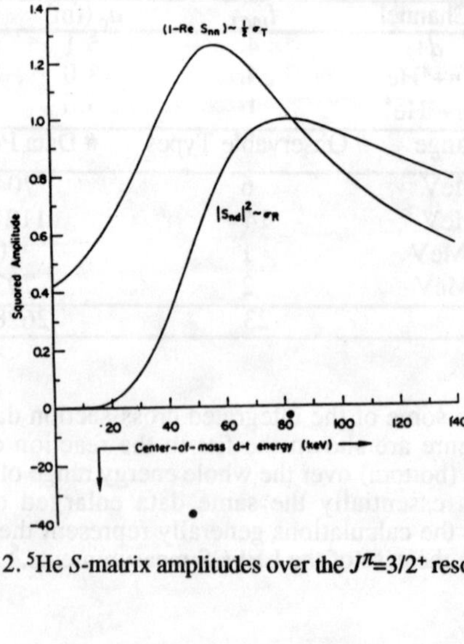

Fig. 2. ^{5}He S-matrix amplitudes over the J^{π}=3/2$^+$ resonance.

The energy dependence of the calculated S-matrix elements for $J^\pi=3/2^+$ that are related directly to the reaction (σ_R) and total (σ_T) cross sections are shown over the resonance in Fig. 2. Unlike the behavior that would be expected for an isolated Breit-Wigner (BW) resonance, the squared amplitudes corresponding to σ_R and σ_T do not peak at the same energy, and have a different energy dependence for the high-energy tail. Our analysis[6] shows that this behavior results from a two-pole structure of the resonance. One of the poles is a "conventional" BW pole that occurs on the (-,-)* Riemann sheet, and the other is a "shadow" pole on the (+,-) sheet. The total cross section is affected more by the conventional pole, shown as a circled "x" at the bottom left of Fig. 2, while the reaction cross section is influenced more by the shadow pole, shown by the same symbol near the real axis at 80 keV.

The resonance parameters for the poles are given in Table 2. The real and imaginary parts of the pole positions give the resonance energies E_r and total widths Γ, respectively, while the partial widths are defined in terms of the residues.[6] Note that for the conventional pole, the partial widths sum approximately to the total width, while for the shadow pole, the sum of the partials greatly exceeds the total. That is another manifestation of the non-BW character of the shadow pole, and indicates that its small value of Γ does not produce any narrow structure in the cross sections. The real reason for the lack of narrow structure associated with this pole is that one has to go around the d+t threshold in order to access the (+,-) sheet on which it lies, whereas the (-,-) sheet can be reached by dropping straight through the channel cuts along the real energy axis. The shadow pole associated with this famous resonance is the first one identified in a nuclear reaction, although a prior example was known for an elementary-particle resonance.

Table 2. Pole parameters for the ^{5}He resonance.

Sheet: (d,n)	(+,+)	(+,-)	(-,-)	(-,+)
E_r (keV)		81.57	46.97	
Γ (keV)		7.28	74.20	
Γ_d (keV)		2861.6	25.10	
Γ_n (keV)		68.77	39.83	

2.2 *The ^{4}He System*

Our R-matrix analysis of the ^{4}He system is part of a larger charge-independent (CI) analysis of all the two-body reactions in the four-nucleon (A=4) system. In this case, the $\gamma_{c\lambda}$ and E_λ in Eq. (1) for the R matrix are labeled with definite values of isospin quantum number T, with T=0 or 1. Furthermore, the reduced-width amplitudes for channels that belong to the same isospin multiplet are related by simple Clebsch-Gordan factors if their channel radii and boundary values are taken to be the same. In this way, the parameters of the (1,3) arrangement channels (p+^{3}He, n+t, p+t, and n+^{3}He) are interrelated, while those of the (2,2) arrangement (d+d, which is pure T=0) are unrelated to the others.

* The Riemann sheets are labeled by the signs of the imaginary parts of the momenta in the d+t and n+α channels, respectively.

The scheme of the CI analysis, as is shown in Fig. 3, was first to determine the T=1 parameters from an analysis of $p+^3$He scattering data at proton energies below 20 MeV. The fit gave a good representation of the 1004 data points included (χ^2 per degree of freedom is 1.23), using 47 R-matrix parameters determined at a channel radius of 4.9 fm for partial waves up through l=3. The validity of the isospin-1 parameters was then checked by comparing with the latest measurements of the $n+^3$H total cross-section and S-wave scattering lengths. All the $p+^3$He E_λ's were shifted by the internal Coulomb energy difference ΔE_C=-0.86 MeV in order to match the total cross section in the peak of a prominent P-wave resonance at E_n=3 MeV. Good agreement was found with the total cross section at all energies below 20 MeV, and with the coherent scattering length.[7]

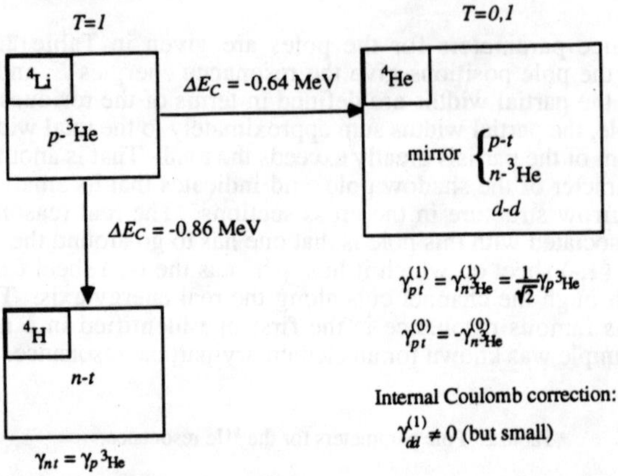

Fig. 3. Scheme of the charge-independent A=4 R-matrix analysis.

The T=1 parameters described above have been incorporated, essentially fixed, in a larger analysis of the ^{4}He reactions in which the T=0 parameters are allowed to vary. The analysis contains data from all the independent reactions among the channels $p+t$, $n+^3$He, and $d+d$ at energies corresponding to ^{4}He excitations below about 29 MeV, as is summarized in Table 3. Again, an overall Coulomb energy shift allows the T=1 E_λ's to adjust to the new compound system, and the (1,3) arrangement widths are constrained by the isospin relations given in Fig. 3.

Table 3. Channel configuration and data summary for the ^{4}He system analysis

Channel	l_{max}	a_c (fm)
$p+t$	3	4.9
n+3He	3	4.9
$d+d$	3	7.0

Reaction	Energy Range	# Observable Types	# Data Points
^{3}H(p,p)^{3}H	E_p=0-11 MeV	3	1382
^{3}H(p,n)^{3}He + inv.	E_p=0-11 MeV	5	726
^{3}He(n,n)^{3}He	E_n=0-10 MeV	2	126
^{2}H(d,p)^{3}H	E_n=0-10 MeV	6	1382
^{2}H(d,n)^{3}He	E_d=0-10 MeV	6	700
^{2}H(d,d)^{2}H	E_d=0-10 MeV	6	336
Totals:		28	4652

While the fit to the 4652 data points included using 87 free parameters is not yet at a minimum of the chi-square surface (with χ^2 per degree of freedom about 4.1), the representation of the low-energy *d+d* reactions is quite good. The differences between the two reaction branches in the unpolarized cross sections and deuteron analyzing powers are well reproduced by allowing a small amount of internal isospin mixing to enter through the non-zero width amplitudes $\gamma_{dd}^{(T=1)}$. Although these amplitudes give widths that are less than 0.1% of the single-particle value, their effects on the *P*-wave transitions, in particular, are greatly amplified in the external Coulomb field by the presence of broad, overlapping negative-parity levels of opposite isospin near the *d+d* threshold. This is shown in Fig. 4, where the "old" level scheme[8] for ^{4}He is compared with the "new" one[9] obtained from the CI *R*-matrix analysis. The new *T*=1 levels are much lower than before, and there is a new T=0, J^π=1$^-$ level just above the *d+d* threshold.

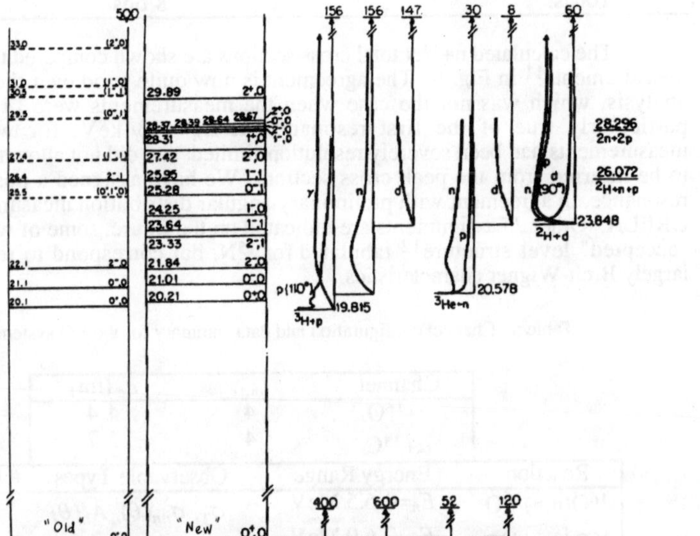

Fig. 4. Present ^{4}He level scheme (right) compared to the previous one (left).

The effects of these strongly-mixed *P*-wave levels can be seen most directly in branching-ratio measurements for muon-catalyzed *d+d* fusion at room temperature, since for identical particles, these measurements select only the *P*-wave part of the nuclear reaction. The experimental *n/p* branching ratio[10] is 1.39±0.04, but the value predicted by the analysis, 1.43±0.03, is in quite good agreement with this surprisingly large result.

2.3 *The ^{15}N and ^{17}O systems*

New *R*-matrix analyses were begun a few years ago for the ^{15}N and ^{17}O systems because of renewed interest in neutron transport through air in connection with a new study of radiation exposure during the Hiroshima and Nagasaki bombings. The analyses were generally limited to neutron energies below the first excited state of the target nucleus. Summaries of the channels and data included in the analyses are given in Tables 4 and 5.

Table 4. Channel configuration and data summary for ^{15}N system analysis.

Channel	l_{max}	a_c (fm)
n+^{14}N	2	2.5
p+^{14}C	3	4.3
α+^{11}B	2	5.1

Reaction	Energy Range	Observable Types	# Data Points
^{14}N(n,n)^{14}N	E_n=0-2.4 MeV	σ_T, $\sigma_{nn}(\theta)$	932
^{14}N(n,p)^{14}C + inv.	E_n=0-2.4 MeV	σ_{np}, $\sigma_{pn}(\theta)$, $A_p(\theta)$	685
^{14}N(n,α)^{11}B	E_n=1.3-2.4 MeV	$\sigma_{n\alpha}$	104
^{11}B(α,p)^{14}C	E_α=1.4-2.6 MeV	$\sigma_{\alpha p}$, $\sigma_{\alpha p}(\theta)$	110
Totals:		8 obs.	1831

The calculated n+^{14}N total cross sections are shown compared to the recent ORELA measurements[11] in Fig. 5. The agreement is now quite good over the whole range of the analysis, which was not the case when the measurements were first made. This was particularly true of the first resonance at E_n=433 keV, for which the previous measurements had been severely resolution-limited, and did not allow the true J-value (7/2) to be inferred from the peak cross section. We have assigned a positive parity for this resonance, in agreement with preliminary angular distribution measurements also made at ORELA. Other J^π assignments are indicated on the figure, some of which differ from the "accepted" level structure[12] tabulated for ^{15}N, but correspond to resonances that have largely Breit-Wigner characteristics.

Table 5. Channel configuration and data summary for the ^{17}O system analysis.

Channel	l_{max}	a_c (fm)
n+^{16}O	4	4.4
α+^{13}C	4	5.7

Reaction	Energy Range	Observable Types	# Data Points
^{16}O(n,n)^{16}O	E_n=0-6.5 MeV	σ_T, $\sigma_{nn}(\theta)$, $A_n(\theta)$	2421
^{16}O(n,α)^{13}C	E_n=0-6.0 MeV	$\sigma_{n\alpha}$, $\sigma_{n\alpha}(\theta)$, $A_n(\theta)$	904
^{13}C(α,α)^{13}C	E_α=0-4.6 MeV	$\sigma_{\alpha\alpha}(\theta)$	207
Totals:		7 obs.	3532

Calculated n+^{16}O cross sections are shown in Fig. 6 compared to measured data. Again, the agreement is good over the entire energy range of the analysis, and the J^π values indicated on the resonances do not always agree with the "accepted" values.[13] Most of the resonances again have BW character, with shadow poles in about the same positions on all unphysical sheets, having partial widths that sum to the total width. An interesting exception is a shadow pole associated with the lowest-energy 1/2$^-$ resonance at E_n=1.9 MeV that is a virtual state located very close to the energy of the second excited state (1/2$^-$) in ^{17}O. This sub-threshold state is a prediction of the analysis, based on fitting only the scattering data.

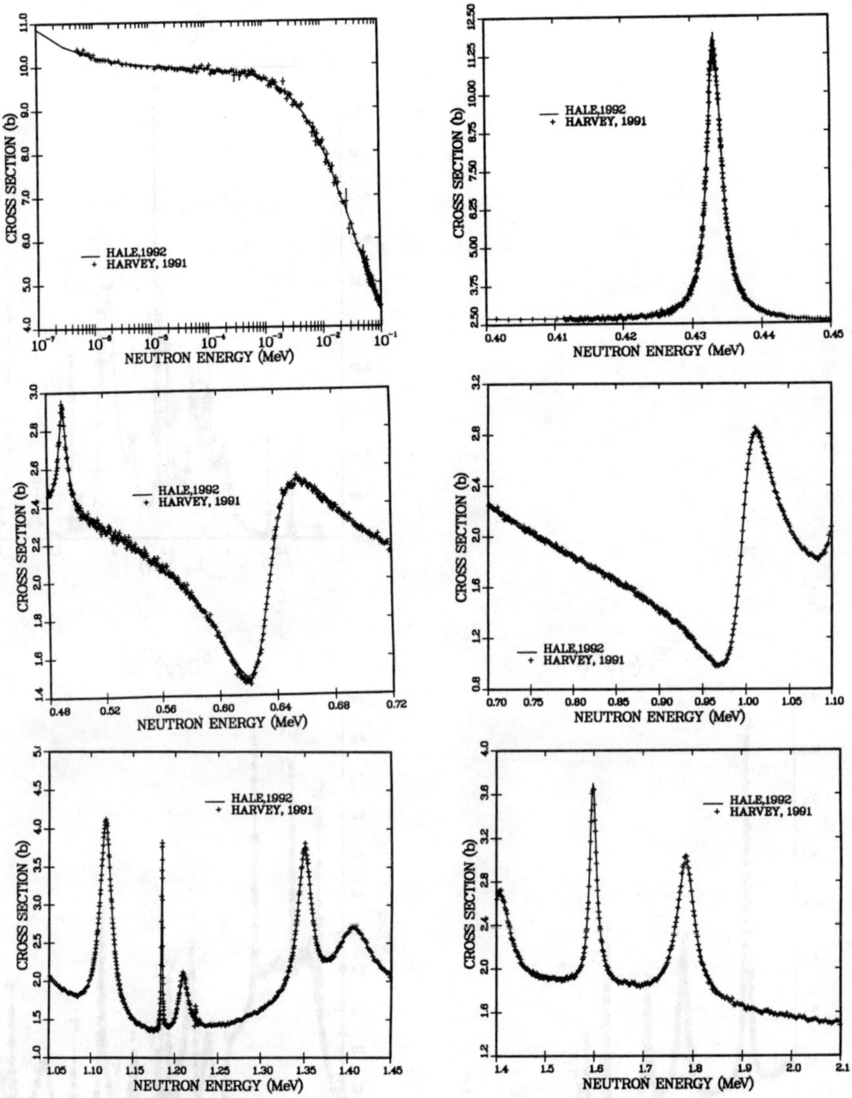

Fig. 5 Calculated n+^{14}N total cross sections compared to the data of Harvey[11] at energies below 2.1 MeV.

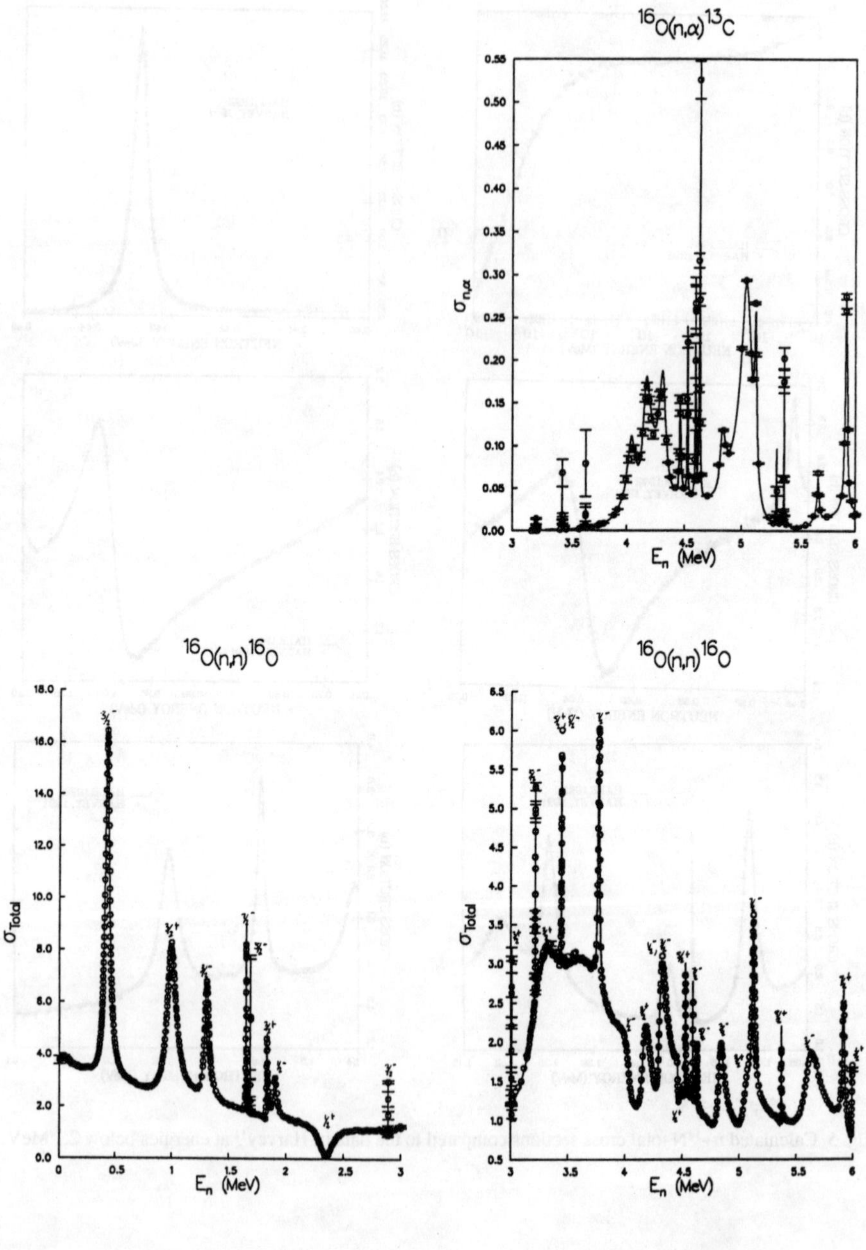

Fig. 6. Neutron total (bottom) and (n,α) reaction cross sections for neutrons on ^{16}O.

3. Summary and Conclusions

We have given several examples of using *R*-matrix theory to fit data for the light systems. Our experience has convinced us that there is no better method for representing data for multichannel systems, especially if resonances are present. It is important, however, to include data of many types from all possible reactions in order to reduce the possibility of parameter ambiguities in the chi-square fitting, and to insure the reliability of extrapolated results.

R-matrix theory predicts that all multichannel nuclear resonances have shadow poles that, in principle, have different properties on different Riemann sheets, contrary to the expectations of BW resonance forms. As a practical matter, these differences occur mainly for the broad resonances in very light nuclei, and appear to be important only near thresholds, where several unphysical sheets are nearly equally accessible from the physical sheet. The first example of an observable shadow pole in a nuclear reaction is the famous $J^\pi=3/2^+$ resonance in the ^{5}He system, where the presence of the pole on the (+,-) sheet near the real energy axis drives the reaction amplitude almost to its unitary limit, resulting in the large ^{3}H$(d,n)^4$He reaction cross section observed at low energies. Shadow poles also play a role in other ^{4}He and ^{5}He resonances, especially in producing threshold effects.

The charge-independent analysis of the A=4 reactions illustrates the utility of being able to constrain *R*-matrix parameters with approximate symmetries of the strong interactions. In this case, strong isospin mixing of *P*-wave levels near the d+d threshold that comes rather naturally from the description is able to account for the anomalous branching ratios seen in d+d muon-catalyzed fusion experiments. Other ^{4}He levels that result from the analysis do not agree with the previous scheme, but correspond well with those from a microscopic resonating-group-model calculation of the system.

The resonances resulting from analyzing the heavier ^{15}N and ^{17}O systems tend more often to satisfy the BW conditions, although they differ in detail with the "accepted" level structures for these nuclei. An interesting result for ^{17}O is the virtual state that occurs as a "shadow" to the $1/2^-$ scattering resonance some 3 MeV higher in excitation energy. This suggests that more examples of virtual states like the well-known "singlet deuteron" may exist among the sub-threshold levels of light nuclei now thought to be bound states.

4. References

1. E. P. Wigner and L. Eisenbud, *Phys. Rev.* **72** (1947) 29.
2. A. M. Lane and R. G. Thomas, *Rev. Mod. Phys.* **30** (1958) 257.
3. R. J. Eden and J. R. Taylor, *Phys. Rev.* **133** (1964) B1575.
4. D. C. Dodder, G. M. Hale, and K. Witte, "Energy Dependent Analysis Code for Nuclear Reactions," Los Alamos National Laboratory (unpublished).
5. N. Jarmie, R. E. Brown, and R. A. Hardekopf, *Phys. Rev. C* **29** (1984) 2031; R. E. Brown, N. Jarmie, and G. M. Hale, *Phys. Rev. C* **35** (1987) 1999.
6. G. M. Hale, R. E. Brown, and N. Jarmie, *Phys. Rev. Lett.* **59** (1987) 763.
7. G. M. Hale, D. C. Dodder, J. D. Seagrave, B. L. Berman, and T. W. Phillips, *Phys. Rev. C* **42** (1990) 438.
8. S. Fiarman and W.E. Meyerhof, *Nucl. Phys.* **A206** (1973) 1.
9. D. R. Tilley, H. R. Weller, and G. M. Hale, *Nucl. Phys.* **A541** (1992) 1.
10. D. V. Balin et al., *Phys. Lett.* **141B** (1984) 173; *JETP Lett.* **40** (1984) 112.
11. J. A. Harvey, N. W. Hill, N. M. Larson, and D. C. Larson, in *Nuclear Data for Science and Technology*, ed. S. M. Qaim (Springer-Verlag, Berlin, 1992) 729.
12. F. Ajzenberg-Selove, *Nucl. Phys.* **A523** (1991) 1.
13. F. Ajzenberg-Selove, *Nucl. Phys.* **A460** (1986) 1.

R-Matrix Parameters in Reactor Applications

by

R. N. Hwang
Argonne National Laboratory
9700 South Cass Avenue
Argonne, IL 60439

1. Introduction

The key role of the resonance phenomena in reactor applications manifests through the self-shielding effect. The basic issue involves the application of the microscopic cross sections in the macroscopic reactor lattices consisting of many nuclides that exhibit resonance behavior. To preserve the fidelity of such a effect requires the accurate calculations of the cross sections and the neutron flux in great detail. This is clearly not possible without viable resonance data.

The recently released ENDF/B VI resonance data[1] in the resolved range especially reflect the dramatic improvement in two important areas; namely, the significant extension of the resolved resonance ranges accompanied by the availability of the R-matrix parameters of the Reich-Moore type. The improvements all but eliminate the long standing difficulties in conjunction to the use of the Breit-Wigner representation particularly inadequate for the closely spaced resonances and at the same time, removed at least in part the ambiguity of having to extend the unresolved treatment too far into the low energy region where the self-shielding effect is apparently deterministic in nature. The benefit of such a development is not without costs to the users. Aside from the obvious increase in computing time required for the significantly greater number of resonances, the main concern is the compatibility of the Riech-Moore representation to the existing reactor processing codes which, until now, are based on the traditional cross section formalisms.

As described previously[2], one convenient way to alleviate the compatibility problem is to convert these Reich-Moore parameters into the pole and residue parameters whereby the cross sections can be cast into the similar form of the traditional cross section formalisms readily amenable for the ENDF/B format based code systems. The purpose of this paper is to summarize our recent efforts to facilitate implementation of the proposed methods into the production codes at ANL.

2. Brief Review of the Proposed Pole Representation

As reported previously[2], the rigor of the R-matrix cross sections based on the Reich Moore formalism[1] and the traditional feature of Doppler-broadening via the Voigt profile essential to many existing methods in reactor applications can be preserved by the use of either a rigorous pole representation or a simplified alternative based on pole and residue parameters numerically converted from any given set of Reich-Moore parameters[2]. By taking into account all energy dependent quantities in the R-matrix representation, one obtains, via partial fraction, the pole representation with $2(\ell+1)$ simple pole terms in $\sqrt{E}$-domain per resonance for a given ℓ- and J-state. The genuine energy-independent pole and residue parameters can be obtained by using the WHOPPER-code[2] developed for this purpose. Of the $2(\ell+1)$ poles per resonance, only two exhibit the usual sharp fluctuations that resemble the Breit-Wigner resonances as a function of $\sqrt{E}$ while the remaining 2ℓ poles are characterized by the extremely large half 'width' corresponding to the imaginary part of the pole. Upon Doppler-broadening using the rigorous kernel defined by Solbrig, one has[2]

$$\sigma_x(\sqrt{E},T) = \frac{1}{E}\sum_{\ell} \Re e\left\{\sum_{J}\sum_{\lambda=1}^{N}\sum_{j=1}^{2} R^{(x)}_{\ell,J,\lambda,j}\left[\frac{\sqrt{\pi}}{\Delta_m}W\left[\frac{\sqrt{E}-P_\lambda^{(j)*}}{\Delta_m}\right]\right] + q_\ell^{(x)}(\sqrt{E})\delta_\ell\right\} \quad (1)$$

where $R^{(x)}_{\ell,J,\lambda,j}$, $p_\lambda^{(j)}$ and Δ_m are residues (or its product with hard sphere shift factor where appropriate), poles and the Doppler width in $\sqrt{E}$-domain respectively, and the complex probability integral, W(z), is related to the usual Doppler-broadened line shape functions as

$$\sqrt{\pi}\,y\,W(z) = \psi(x,y) + i\,\chi(x,y) \quad (2)$$

The quantity $q_\ell^{(x)}(\sqrt{E})$ here represents the superposition of all poles with 'giant width' for a given ℓ not affected by the Doppler effect. One obvious concern from practical point of view is that the evaluation of Eq-1 will undoubtedly require excessive computing cost although it may not be prohibitive. Hence, a simplified version of this approach was developed to provide an alternative more attractive to reactor applications. The procedure is equivalent to the direct extraction of the Humblet-Rosenfeld[3]-type parameters and the associated 'background' term with weak $\sqrt{E}$-dependence explicitly from the converted pole and residue parameters described above. This can be accomplished by regrouping the sum of rational terms into a 'fluctuating' term and a 'smooth' term. The latter consisting of ensemble of pole terms, attributable to the outlying poles and poles with extremely large 'width' defined in Eq-1, can be approximated effectively by a rational function of low order via the non-linear least square fit via the use of the WHOPPER.JUNIOR code[2]. These fitted poles can be viewed as the 'pseudo' poles analogous to resonances assigned to regions outside of the effective resolved energy range. It has been demonstrated[2] that only three 'pseudo' pole terms for each smooth term are required for a given ℓ to provide sufficient accuracy for all major nuclides processed so far.

Upon Doppler-broadening, it leads immediately to the familiar form defined by the traditional formalisms[2]:

$$\sigma_x(E,T) = \frac{1}{E}\sum_{\ell} \Re e\left\{\sum_J \sum_{\lambda=1}^{N} 2\sqrt{E}R^{(x)}_{\ell,J,1,\lambda}\frac{\sqrt{\pi}}{\Delta_E}W\left[\frac{E-\epsilon_\lambda}{\Delta_E}\right] + \hat{S}^{(x)}_{\ell}(\sqrt{E},T) + \hat{q}^{(x)}_{\ell}(\sqrt{E})\cdot\delta_\ell\right\} \tag{3}$$

$$\text{for } \sqrt{E_\ell} \le \sqrt{E} \le \sqrt{E_u}$$

where

$$\Delta_E = 2\sqrt{E}\ \Delta_m$$

$$\hat{S}^{(x)}_{\ell}(\sqrt{E},T) = \sum_{k-1}^{NN} r_k^{(x)}\left[\frac{\sqrt{\pi}}{\Delta_m}W\left[\frac{\sqrt{E}-\alpha_k^*}{\Delta_m}\right]\right] \tag{4}$$

$$\epsilon_\lambda = [p_\lambda^{(1)*}]^2 \tag{5}$$

and $r_k^{(x)}$ and, α_k^* are the fitted residue and pole respectively.

Thus, the only difference between Eq-3 and the traditional formalisms is the presence of the 'background' terms with weak energy and temperature dependence. Its compatibility to all ENDF/B format[1] based codes is quite evident.

3. Cross Section Representations and Self-Shielding Effects

One representative quantity that characterizes the self-shielding effect is the self-shielding factor. With no loss of generality, it suffices to view it as a measure of correlation between the microscopic cross section and neutron flux in energy as well as in space at a given temperature[2]

$$f_x = \frac{<\sigma_x\phi>_{E\&\vec{r}}}{<\sigma_x><\phi>_{E\&\vec{r}}} = 1 + \frac{COV(\sigma_x,\phi)}{<\sigma_x><\phi>_{E\&\vec{r}}} \tag{6}$$

The averages involved can be cast into the form of either the usual Riemann integral or Lebesgue integral. It is quite apparent that a concise description of the cross section as a function of energy and temperature is essential to both the theoretical modeling and the practical computation of this potentially complex problem involving neutron slowing-down and transport in a reactor cell. For the resolved energy of interest here, the degree of correlation is clearly deterministic and multivariant in nature.

All traditional cross section representations[1] prior to the availability of the R-matrix parameters were essentially expressible as the linear combinations of the Breit-Wigner - like terms supplemented by the pointwise smooth data. Aside from the obvious consideration of the inherent advantage of the Doppler-broadening, the overriding consideration to preserve the traditional form in terms of the Voigt profiles[1] is its concise mathematical behavior upon which many fundamental concepts and numerical methods are based. In particular, such a description made possible the coexistence of two fundamental concepts based on the resonance integral approach and continuous energy approach respectively. The former treats the resonance integral for individual resonances separately while the later retains the continuous nature of the flux in energy over the entire resolved range. Various methods based on these concepts differ only in the rigor by which the neutron flux is treated. Thus, by preserving the traditional form, the proposed method is readily amenable to most of the reactor codes in existence with minor modifications.

As long as the resonance cross sections are represented as the linear combination defined by Eq-1 or Eq-3, the W-function alone dictates their behavior as a function of energy and temperature. The analytical characteristics of practical interest can be summarized in Table I. To be consistent with the traditional description of ψ and χ, the variables x and t will henceforth be taken to be $2(E-E_0)/\Gamma$ and $(\Delta/\Gamma)^2$ respectively instead of the generic z.

Table 1: Analytical Behavior of Voigt Profiles of Practical Interest

	ψ (x,t)	χ (x,t)
Lorentzian Limits ($\lvert z \rvert >>> 1$)	$1/(1+x^2)$	$x/(1+x^2)$
Extreme-Broadening ($t >> 1$)	$\frac{1}{2}\sqrt{\pi/t}\ \exp(-x^2/4t)$	$(1/\sqrt{t})\ D(x/2\sqrt{t})$ (Dawson integral)
Derivatives: 1st order 2nd order	 $[\chi - x \cdot \psi]/2t$ $[1-(1+2t-x^2)\ \psi - 2\chi \cdot \psi]/(2t)^2$	 $[1-\psi-x\chi]/2t$ $-[x(1-2\psi)+(1+2t-x^2)\chi]/(2t)^2$

The Lorentzian limits define the short range nature of fluctuations exhibited by these functions. The practical importance will be further addressed in the next section. The explicit knowledge of the derivatives not only provides the analytical insight of how the cross section slope varies as a function of energy but also specifies various extreme points exactly. The latter include the maximum, minimum and points of inflection for each pole term, which are generally difficult to pin-point by means of the brute force approach. For sharply varying function in question, their exact locations will have significant impact on the accuracy of the interpolation of cross sections that follows if the pointwise entry is used. It is worth noting that the points of inflection represent the locations where the temperature derivative vanishes. This stems from the fact that both ψ and χ must satisfy the heat equation of the form

$$\frac{\partial^2 f}{\partial x^2} = \frac{\partial f}{\partial t} \tag{7}$$

They are especially important from the point of view of the Doppler effect calculations.

Numerically, these extreme points for ψ, χ, and $(\psi+a\chi)$ can be determined readily as a function of t. A computer code based on the Newton-Raphson scheme has been developed to compute these extreme points as a function of t (and a where appropriate) using the zero temperature limits as initial guesses. It turns out that these points generally exhibit behavior closely resemble a hyperbola with asymptos identifiable with the expected limits inferred by two limiting cases given in Table I. Figure 1 illustrates the minimum $|X_{min}|$ of $(\psi+a\chi)$ as a function of $\sqrt{t}$ for a = 0.20. Two asymptos correspond to $|X_{min}| = |(-1-\sqrt{1+a})/a|$ for the zero temperature limit and $|X_{min}| \approx .54104\,(2\sqrt{t})$ for the extreme broadening limit reflected by the Dawson integral respectively. The relationship can be readily characterized by a family of hyperbolas with coefficients as a function of a. Such information can be incorporated into the scheme to construct the pointwise cross sections to be described.

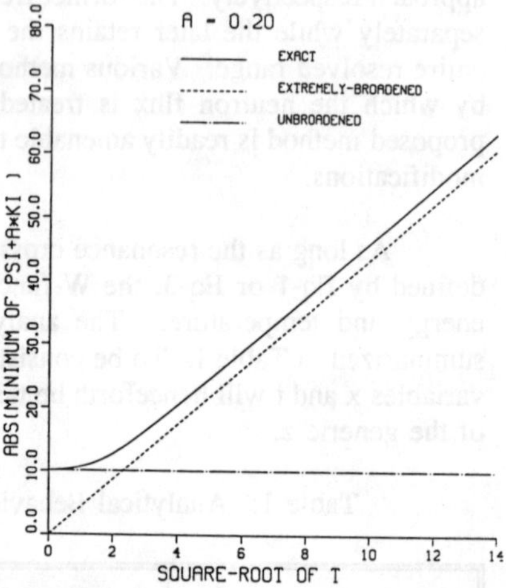

Figure 1. $|X_{min}|$ vs $\sqrt{t}$

4. Facilitation of the Proposed Method in Progress

From the practical point of view, the most essential consideration is the potential trade-offs among accuracy, computing cost and computer storage requirement that are suitable to one's objectives and computing tool available. For our purpose here, the discussions will be focused on issues pertinent to the entry of the basis resonance data to the existing deterministic and Monte Carlo codes at ANL. There are only two ways that the basic resonance data can be brought into the reactor applications; namely, the direct entry at run time and the entry with pre-computed pointwise cross sections as a function of energy and temperature. In conjunction to such applications two unique features of the proposed method of particular interest are the prior knowledge of the analytical behavior of the cross sections and the accuracy of the Doppler-broadening which is clearly difficult to match by any finite difference methods implicitly based on the Taylor's expansion without the recourse of using an exceedingly large number of mesh points.

In the following discussion, our efforts to implement the proposed method in the processing codes with two completely different data-entry philosophies will be described.

4.1. Entry at Run-Time

This type of data entry is obviously attractive for the deterministic approach in which the rigor of the composition and temperature dependence is required. It is particularly important to the code such as the MC^2-2/RABANL in which both resonance integral approach and continuous energy integral transport theory approach coexist.

However, the rigor clearly can not be achieved without the accompanying computing cost and simplifications are obviously required. The points in question is how to enhance the computational efficiencies with acceptable sacrifice in accuracy.

The question of how to manage the numerous number of resonances at run time is nothing new for the MC^2-2/RABANL code[4]. It has been demonstrated how this problem can be alleviated by taking advantage of the ultra-fine group structure of the code essential for fast reactor applications. The basic structure consists of 2000 energy group with the width approximately equal to 1/2 of the maximum energy loss per collision incurred to a neutron by the elastic scattering of the heaviest nuclide of practical interest. The basic logic of the code involves the creation of a base library of 'smooth' cross sections of all nuclides of interest beforehand but the resonance cross sections are computed for each ultra-fine group at run time. Two main cost-saving features already in place prior to the availability of the R-Matrix data are summarized briefly as follows.

(A) Pre-screening procedure

One obvious way to take advantage of the ultra-fine group structure is the use of a pre-screening process while construction the base library, whereby all resonances with the total width much greater than both the ultra-fine group width and the Doppler width corresponding to the highest temperature anticipated will be processed into the 'smooth' library. Many s-wave resonances of structural material apparently will fall into this category. Thus, it helps reduce the needs to compute them at run time.

(B) Treatment of distant resonances

As described earlier, each pole term characterized by the W-function exhibits strong fluctuations only in the energy region near the peak of the resonance. Thus, with exception of a handful of resonances within and close to a given ultra fine group in question, it suffices to evaluate the contributions of the distant levels at the ultra-fine group mesh instead of the hyper-fine meshes required to ensure the adequate treatment of the self-shielding effect.

The viability of these schemes have been successfully demonstrated especially when the continuous energy integral transport option is used[4]. The same techniques are readily amenable to the proposed method[2] based on the Hamblet-Rosenfeld[3] type of parameters. It is worth noting that the contributions of the distant tails of the latter are generally more important than those based on the traditional formalisms according to our experience with the WHOPPER code[2]. Modification and further explorations of the known analytical properties to expedite the calculations are still under way.

4.2 *Entry As Pre-Computed Pointwise Cross Section*

For calculations based on continuous energy Monte Carlo approach, the entry of resonance data at run time is obviously out of question. One must rely on the pre-computed pointwise cross sections as a function of energy and temperature. Although the storage is no longer a prohibitive constraint for the new generation computers, nevertheless, some viable criteria for the mesh spacings are obviously required to ensure the accuracy of the subsequent interpolation at run time.

For the cross section expressed as the linear combination of $\psi(x_i, t_i)$ and $a_i\chi(x_i, t_i)$, it suffices to investigate the problem in two stage: (a) Determine the mesh points adequate to cover any individual term for various values of a_i and t_i within $-\infty \leq x_i \leq \infty$; (b) optimization of the union of the individual sets within the resolved energy boundaries. In the following discussion, various steps pertinent to such a procedure will be described along with some preliminary results.

(A) Rational Transformation and Preliminary Grids

One necessary but not sufficient condition for the pointwise cross section and energy pairs generated for each term must preserve the expected average upon numerical integration. It is, therefore, reasonable to conjecture that the preliminary grids required may be taken to be the quadrature points adequate to preserve the rigor of the integration. As described in Ref-5, one of the most efficient way is via the use of the Gauss-Jacobi quadrature. It is equivalent to carry out a rational transformation of the form

$$f(x)\,dx = \frac{f(u)}{1-u^2}\,\frac{du}{\sqrt{1-u^2}}\,; \qquad -\infty \leq x \leq \infty \tag{8}$$

where

$$u^2 = \frac{(Kx)^2}{1+(Kx)^2}\,; \qquad -1 \leq u \leq 1 \tag{9}$$

and K is a constant dependent only on the Doppler width and total width under consideration. For our purpose here it was found that $K=\frac{1}{5}[\Gamma/(\Delta+\Gamma)]$ is satisfactory for the cases considered.

In terms of the Gauss-Jacobi quadrature a set of mesh points $\{u_i\}$ is specified once the total number of points is given. Extensive investigations indicate that 13 points are sufficient to reproduce the expected value adequately. A set of points $\{E_i\}$ in energy domain corresponding to $\{u_i\}$ can be obtained readily and vice versa. The transformation fully utilizes the Lorentzian behavior of ψ and χ functions, because $\lim_{u\to\pm1}\left(\psi/(1-u^2)\right) = \text{const.}$; and $\lim_{u\to\pm1}\left(x/1-u^2\right) = \text{const.}/\sqrt{1-u^2}$. Thus, the transformation amounts to optimization of number of points needed to describe each term in question from the perspective of integration.

(B) Addition of Points

Although the preliminary grids provide a good start, they are by no means adequate to ensure the accuracy of values to be interpolated at run time. In particular, those extreme points described earlier can play an important role in the interpolation process that follows. Hence, these points are included into the set determined by the quadrature.

The addition of extreme points, however, still does not guarantee the accuracy of the interpolated values especially between points where the gradient changes significantly. To ensure the accuracy of values to be interpolated, the spacing of predetermined meshes obviously must be correlated to their higher order derivatives as defined explicitly by the remainder associated with the interpolation scheme used. The precise knowledge of the derivatives for every preliminary mesh provides the needed criteria. Extensive investigations are under way and some preliminary results will be presented in the later section.

(C) Union of Individual Sets and Further Addition or Reduction of Points

The individual $\{u_i\}$ or $\{E_i\}$ sets can be combined to form an union that covers the entire resolved region. Two common scenarios that one encounters are: (1) the ensemble consists of either the well-isolated poles (or resonance where appropriate); or (2) closely-spaced poles. The former may result in an union of disjoint sets while the latter tends to produce the ensemble of strongly overlapping sets. Hence, additions or reduction of points are required accordingly.

The intersections of various sets may be defined at the average of two end points (in energy) of the neighboring sets. These intersecting points define the subset center around a given pole consisting of points that happen to fall into this interval. The table may consist of either $\{\sigma_{xi}\}$ and $\{E_i\}$ pairs or $\{\sigma_{xi}/(1-u_i^2)\}$ and $\{u_i\}$ pairs. If the subsequent interpolation scheme is to be carried out in the u-domain, all energy points within the interval must be converted into the u-mesh points based on the pole in question. For the nuclides with a dominant s-wave sequence (such as ^{238}U), the mesh points generated for the p-wave poles can be included as subsets in the interval defined by the s-wave pole in which they happen to fall.

(D) Preliminary Results

An experimental code capable of generating pointwise cross sections using the scheme described above along with plotting and interpolation routines has been developed. Although, many criteria pertinent to addition and reduction of points are still being explored, the preliminary results presented here are believed to be useful to illustrate the general features of the proposed method.

Figure 2 and Figure 3 show the U^{238} absorption and scattering cross sections vs energy respectively as obtained by the direct calculations using 'fine' energy meshes and the preliminary grids quadrature points for individual s-wave as well as p-wave poles with the minimum points added. They clearly indicate that the preliminary grid can already provide reasonable resolution for the energy region with many sharp resonances prior to the addition of extra points. The difficulties of pin-pointing the peaks and valleys by the brute force means inspite of the extremely fine points used are quite evident as illustrated by the solid lines. The problem can be alleviated readily by the proposed method when the extreme points are added.

For illustration purposes, some preliminary results were obtained via some ad-hoc conditions in lieu of the finished criteria at this time. One area of particular interest in the optimal choice between interpolation scheme and the number of mesh points required.

Figure 4 and Figure 5 show the relative errors of U^{238} absorption cross sections in percent resulting from the interpolation of the pointwise data based on the 4-point and 2-point LaGrange interpolation scheme respectively in the low energy region where the high accuracy is the most difficult to achieve. Since the results were obtained by using the same set of pointwise cross section data, the significant difference in the relative errors is directly attributable to the interpolation scheme. These figures clearly illustrate the difficult choice between the accuracy and computing cost that one must make whenever the entry of the pointwise cross sections is considered. It is believed that the accuracy of the proposed scheme can be significantly improved by removing a handful of peaks and valleys in the relative error via the use of the more viable criteria.

It is worth noting that the inherent errors attributable to the pointwise entry of cross section data are seldom noticeable in the cursory reactor bench-mark calculations based on the average quantities over considerably large energy span due to the error cancellations. Nevertheless, they do provide a point of reference as to the potential

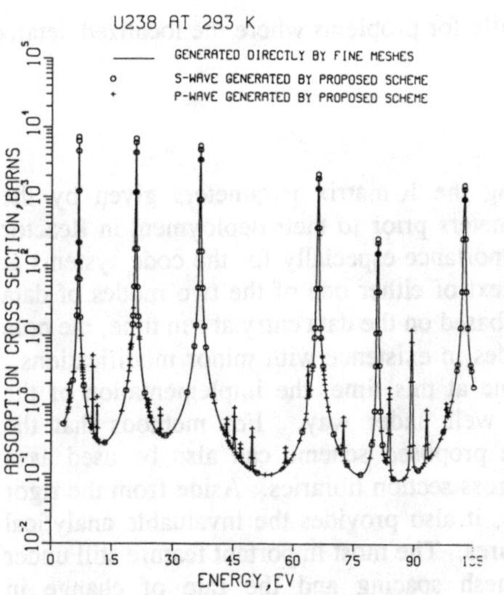

Figure 2. σ_a of ^{238}U vs E

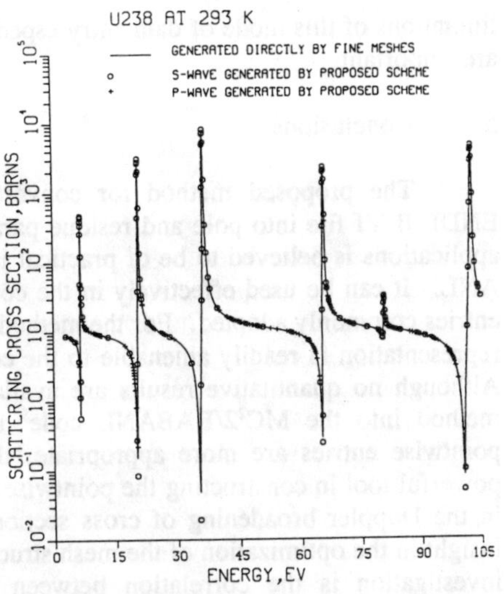

Figure 3. σ_s of ^{238}U vs E

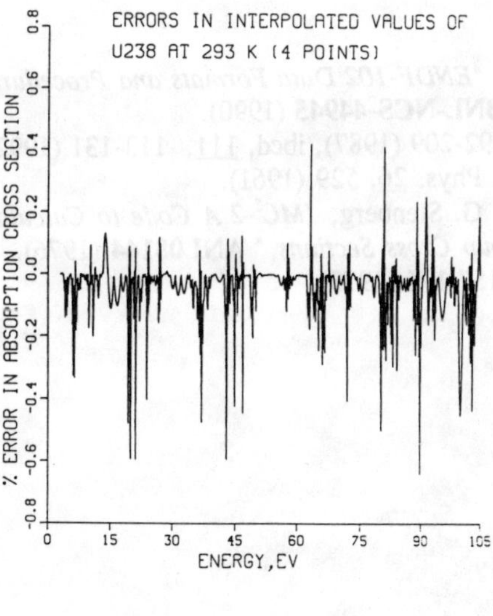

Figure 4. % Error in σ_a (4-point)

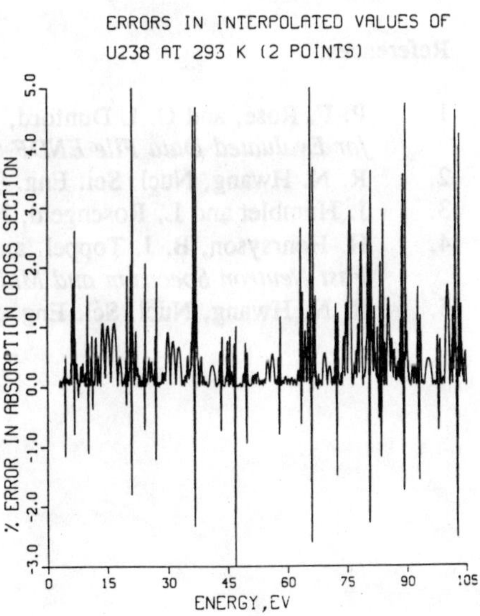

Figure 5. % Error in σ_a (2-point)

limitations of this mode of data entry especially for problems where the localized details are important.

5. Conclusions

The proposed method for converting the R-matrix parameters given by the ENDF/B VI file into pole and residue parameters prior to their deployment in Reactor applications is believed to be of practical importance especially for the code systems at ANL. It can be used effectively in the context of either one of the two modes of data entries commonly adopted. For the methods based on the data entry at run time, the pole representation is readily amenable to the codes in existence with minor modifications. Although no quantitative results are available at this time, the implementation of the method into the MC²2/RABANL code[5] is well under way. For methods that the pointwise entries are more appropriate, the proposed scheme can also be used as a powerful tool in constructing the pointwise cross section libraries. Aside from the rigor in the Doppler broadening of cross sections, it also provides the invaluable analytical insight in the optimization of the mesh structures. The most important feature still under investigation is the correlation between mesh spacing and the rate of change in derivatives between points that obviously determines the accuracy of any interpolation scheme.

References:

1. P. F. Rose, and C. L Dunford, Ed., *"ENDF-102 Data Formats and Procedures for Evaluated Data File ENDF-6,"* **BNL-NCS-44945** (1990).
2. R. N. Hwang, Nucl. Sci. Eng. 96, 192-209 (1987), ibcd, 111, 113-131 (1992).
3. J. Humblet and L. Rosengeld, Nucl. Phys. 26, 529 (1961).
4. H. Henrsyson, B. J. Toppel, and C. G. Stenberg, *"MC²-2 A Code to Calculate Fast Neutron Spectrum and Multigroup Cross Sections,"* **ANL08144** (1976).
5. R. N. Hwang, Nucl. Sci. Eng. 52, 157-175 (1973).

BAYESIAN APPROACH TO EVALUATE AVERAGE LEVEL SPACING

Zhixiang Zhao
Chinese Nuclear Data Center, P. O. Box 275 (41), Beijing, China

Zhongfu Huang
Physics Department, Guangxi University, Nanning, China

I. Introduction

In general, average level spacing is obtained from the analysis of a set of resonance parameters. The problem involved in the method to estimate average level spacing is that usually observed sample of resonance parameters is not perfect one. Many weak levels which have small neutron width are missed because of finite resolution and sensitivity of neutron experiments. This has no practical consequence on determination of strength function, but the loss of weak levels must be corrected to obtain a good estimate of average level spacing. To correct level missing and obtain better estimate of <D>, various methods have been used (see the review by Froehner[1] and by Liou[2]). Most of these method are based on a cut–tailed Porter–Thomas distribution of reduced neutron width and have not used the information in level position because of the fact that the level missing distores the whole range of level spacing distribution. The Bayesian approach simultaneously use the information involved in both neutron width and level position, therefore it could improve the estimate of <D>.

II. Method for Target with $I=0$[3]

In 1956, Wigner[4] suggested his famous level spacing distribution

$$p(t)dt = \frac{\pi}{2} exp(-\frac{\pi}{4} t^2) \tag{1}$$

for single population with a given L, J. The Wigner distribution is a good enough approximation compared to the more exact one obtained by Gaudin[5]. For S–wave neutron resonance of the target with $I=0$, there is only one resonance spin series, $J=1/2$. Let us assume that N_{obs} the number of resonances were observed in energy region from E_1 to E_2 and define

$$D_{obs} = (E_2 - E_1)/N_{obs} \tag{2}$$

Let us denote by N_0 the number of resonances estimated from measured neutron width based on a cut–tailed Porter–Thomas distribution. According to the Bayesian theorem, the posterior probability of a given level having been missed is expressed as

$$P(M|D_i) = \frac{Pr(M)P(D_i|M)}{Pr(M)P(D_i|M) + Pr(\overline{M})P(D_i|\overline{M})} \tag{3}$$

where M denotes the event "given level is missed" and $\overline{M}$ is the inverse event of event M. A priori probability Pr(M) is given by the estimate of <D> based on cut–tailed Porter–Thomas distribution

$$Pr(M) = 1 - N_{obs} / N_0 \tag{4}$$

and

$$Pr(\overline{M}) = 1 - Pr(M) \tag{5}$$

The conditional probability $P(D_i|\overline{M})$ is simply given by Wigner distribution. The $P(D_i|M)$ can be derived as the probability density of the sum of two Wigner varibles and given by[6]

$$P(D_i|M) = \frac{2\pi^2 D_i^3}{(4+\pi)^2 D_{obs}^4} exp(-\frac{\pi}{\pi+4}\frac{D_i^2}{D_{obs}^2}) \tag{6}$$

The percentage of level missing, P_M, is given by averaging $P(M|D_i)$ over all resonances

$$P_M = \frac{1}{N_{obs}} \sum_{i=1}^{N_{obs}} P(M|D_i) \tag{7}$$

Finally we have Bayes–estimate for <D>,

$$D_B = (E_2 - E_1) / N_B \tag{8}$$

where

$$N_B = N_{obs} / (1 - P_M) \tag{9}$$

III. Method for Target with I≠0[7]

We now generalize the method above to the case of S–wave resonance for target with I≠0. In this case, there are two resonance spin series, J = I+1 / 2 and J = I–1 / 2. Usually, the resonance spin are available only for few resonance which have large neutron width so that the sample to be analyzed is a mixed sample of two Wigner samples. To obtain the analytical form for mixed distribution of two Wigner populations, Monte–Carlo method was employed to create an empirical distribution. The Wigner variable $t = D_J / <D_J>$ is simply obtained from uniformed random number ζ by using

$$t = [-\frac{4}{\pi} ln(1-\zeta)]^{1/2} \tag{10}$$

and

$$D_J = t < D_J > = \frac{2(2I+1)}{2J+1} t < D > \tag{11}$$

where D_J and $<D_J>$ denote Wigner variable and its mean for J–series; $<D>$ denotes the mean of mixed distribution. In obtaining Eq. (12), the level density of J–series was assumed being proportional to (2J+1),

$$< D_J > = \frac{2(2I+1)}{2J+1} < D > \tag{12}$$

The comparisons of Monte–Carlo calculation with those calculated from Wigner distribution were carried out.The good agreements between them lead to accept the Monte–Carlo calculation as reliable. A typical example for $J=3$ series of $I=7/2$ is shown in Fig.1.

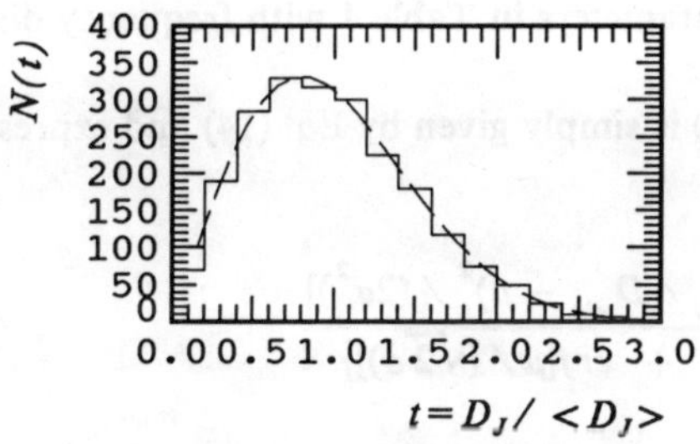

Fig.1 Distribution of Single Population with $I=7/2, J=3$
dashed:Wigner Distribution
histogram:M–C Calculation

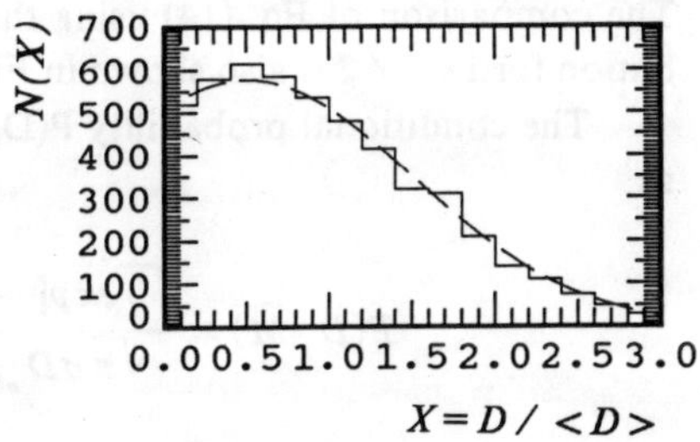

Fig.2 Mixed Distribution of Two Wigner Population with $I=7/2$
dashed:from Eq.(14)
histogram:M–C Calculation

Using Wigner variable generated in this way, two random level position series,

$$E_k^J = \sum_{i=1}^{k} D_J^i, \quad J = I \pm 1/2 \tag{13}$$

were calculated in same energy region. Then the level positions and level spacing for mixed population were obtained by mixing two level position series. A typical result of Monte–Carlo calculation for mixed distribution of I = 7 / 2 is shown in Fig. 2. It is found from Fig. 2 that the level repulsion effect inhibiting small level spacing is largely removed.

The frequency distribution obtained from Monte Carlo calculation is empirically described by a upper cut–tailed normal distribution

$$P(x)=\frac{\sqrt{2}\,exp[-(x-\mu)^2/(2\sigma^2)]}{\sigma\sqrt{\pi}\{1+erf[\mu/(\sigma\sqrt{2})]\}} \tag{14}$$

where $0 < x = D/\langle D\rangle < \infty$ and σ^2, μ represent the variance and mean, respectively. The parameters σ and μ for various I–values were obtained by fitting frequency distribution and given in Table 1.

Table 1 Parameters in Cut–tailed Normal Distribution

I	1/2	1	3/2	2	5/2	3	7/2
μ	0.60	0.58	0.57	0.55	0.53	0.52	0.50
σ	0.92	0.96	0.97	0.99	0.98	1.00	1.00

The comparison of Eq. (14) using the parameters in Table 1 with frequency distribution for I = 7 / 2 is also shown in Fig. 2.

The conditional probability $P(D_i|\overline{M})$ is simply given by Eq. (14) and expressed as

$$P(D_i|\overline{M})=\frac{\sqrt{2}\,exp[-(D_i/D_{obs}-\mu)^2/(2\sigma^2)]}{\sqrt{\pi}\,\sigma D_{obs}\{1+erf[\mu/(\sqrt{2}\,\sigma)]\}} \tag{15}$$

If a level were missed between E_i and E_{i+1}, D_i should be the sum of two random variables obeyed the distribution in form Eq. (14) and its distribution is approximately derived as

$$P(y)=\frac{f(y)}{\int_0^\infty f(y)dy} \tag{16}$$

where

$$f(y) \approx [1 - exp(-\frac{0.799\sqrt[2]{2}}{\sigma}y - \frac{0.71}{\sigma^2}y^2)]exp[-(y-\mu)^2/\sigma^2] \tag{17}$$

and

$$y = (D_1 + D_2)/(2D_{obs}) \tag{18}$$

The Monte Carlo calculation shows that Eq. (16) is a good enough approximation (see Fig. 3).

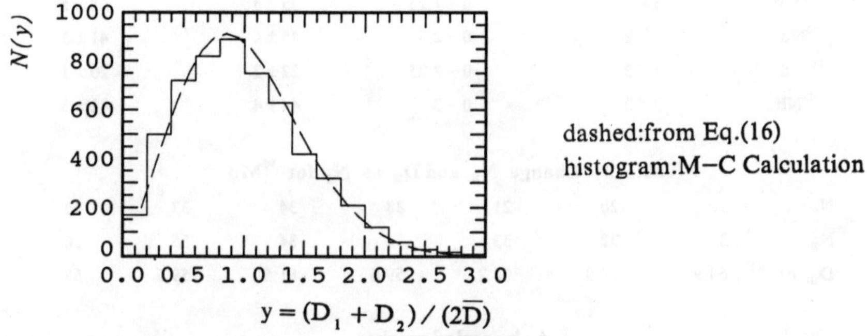

Fig.3 Distribution of Sum of Two Cut−Tailed Normal Variables with I = 7 / 2

Substituting the Eqs. (4), (5), (15) and (16) into Bayesian theorem (3), the posterior probability $P(M|D_i)$ can be obtained.

The Bayesian estimate D_B can be obtained by using Eqs. (7)−(9).

IV. Rcsults and Discussion

The method described above was used to estimate S−wave average level spacing for many target. The prior estimates for the number of resonance, N_0, were calculated by AVRPES code[8] based on neutron width data. The samples to be analyzed were taken from BNL−325[9]. Some results for target with $I \neq 0$ together with those of BNL−325 are given in Table 2. Two sets of estimates are consistent within the associated errors. The errors of estimates given in Table 2 are statistical error for sampling.

It is found that Bayesian estimates N_B or D_B always convergence even if very different prior estimate N_0 is used. A typical example for ^{95}Mo is given in Table 3. In the Table 3, seven prior values N_0 were calculated from different methods by AVRPES based on neutron width data.

Table 2 Estimates of Average Level Spacing

Nuclide	I	Energy Interval (keV)	Estimates (eV) BNL−325	This Work
^{57}Fe	1/2	1~194	6000±1000	6911±730
^{103}Rh	1/2	0~4	16±1	16.2±0.6
^{87}Rb	3/2	0.2~48.6	2640±880	2591±334
^{197}Au	3/2	0~2.611	16.5±0.9	16.3±0.7
^{85}Rb	5/2	0.1~18.5	200±45	154±8
^{95}Mo	5/2	0~2.25	55±8	58±5
^{143}Nd	7/2	0~2.5	45±4	41±3
^{145}Nd	7/2	0~2.25	22±2	20±1
^{93}Nb	9/2	0~3	44±4	42±3

Table 3 Change N_B and D_B as N_0 for ^{95}Mo

N_0	20	20	21	38	34	33	43
N_B	32	32	33	35	34	35	36
D_B, eV	64.9	64.9	64.2	59.6	61.5	59.6	58

Acknowledgments

The authors would like to give their sincere thanks to Prof. Delin Zhou of CNDC for valuable discussions.

References

1 F. H. Froehner, "Applied Neutron Resonance Theory", KFK−2669 (1978)

2 H. I. Liou, " Evaluation of Average level spacing", BNL−NCS−51363, Vol. 1, 463(1981)

3 Zhixiang Zhao, "Bayesian Approach to Estimate Average Level Spacing", *Annual Report of IAE,* (1988)

4 E. P. Wigner, ORNL−2309, 59(1959)

5 M. Gaudin, *Nucl. Phys.*,25, 447(1961)

6 Zhixiang Zhao, " Statistical Estimate for Average Resonance Parameter", *Neutron Resonance Theory and Neutron Resonance Parameters,* Atomic Energy Publishing House, p. 120(1982) (in Chinese)

7 Zhongfu Huang and Zhixiang Zhao, " Application of Bayesian Approach to Estimate Average Level Spacing", CNDP, No. 5, p. 59(1991)

8 Zhongfu Huang,*J. Chinese Atomic Sci. and Tec.,*Vol. 22, 3(1988)

9 S. F. Mughabghab et al.," Neutron Cross Section, Vol. 1, Neutron Resonance Parameters" (1981)

NEEDS FOR IMPROVEMENT OF ENDF-6 RESONANCE REGION FORMATS AND PROCEDURES

Pierre RIBON and Martine MOREAU

Centre d'Etudes Nucléaires de Saclay
91191 GIF sur YVETTE, CEDEX, FRANCE

ABSTRACT

The evaluations format does not allow the transmission of all the available information to the nuclear data processing codes. The lost informations are generally considered as having little importance : but there is an evolution to more exactness. We consider only the lacks linked with the self-shielding calculations, give a few examples of their effects and some more general considerations.

INTRODUCTION

The content of evaluations and their quality are highly dependent on experimental data and on our knowledge of nuclear physics. But there are many informations that evaluators cannot include due to restrictions of the format definition even when they are known with a reasonable degree of accuracy. Are these information of importance? Trying to answer this question implies both a knowledge of the numerical consequences of these limitations and of the required accuracy ; furthermore the answer is time dependent !.

Some other restrictions are introduced by the processing codes, sometimes in contradiction with the formal specification of ENDF - 6[1] ·

The numerical results reported here have been calculated with the latest edition of the JEF-2 evaluation by using the code CALENDF[2] and applying to this code the modifications necessary to obtain the required effect ; details on the mehods and more results are given in an internal report[3].

1 UNRESOLVED RESONANCE PARAMETERS

1-1 NUMBER OF DEGREES OF FREEDOM

1-1-1 According to the ENDF specification, the neutron width Γn is distributed according to a χ^2 law with n degrees of freedom :

$$1. \leq \nu_n \leq 2.$$

There are strong theoritical and experimental evidences that ν is very close to 1. (or, for some cases, to 2.); but it can be slightly different from 1. The current evaluations always assumed an integer value for ν_n and it seems that most of processing codes would not be able to process non-integer values. We tested the influence of this departure for ^{238}U assuming ν=1.1 ; we obtained a 0.5% variation of the self-shielding coefficient between 10. and 40. keV for a 10 barn background cross section (dilution).

1-1-2 The ENDF specification states that "at present ν_γ=0.", which physically corresponds to an infinite value of ν_γ. That implies that the radiative width is constant.

We generated $\Gamma_{\gamma\lambda}$ values distributed according to a normal law having a variance corresponding to ν_γ = 100. Figure 1 represents the results obtained for ^{238}U. We got, for a 10 barn background cross section, corrections of 1.5%.

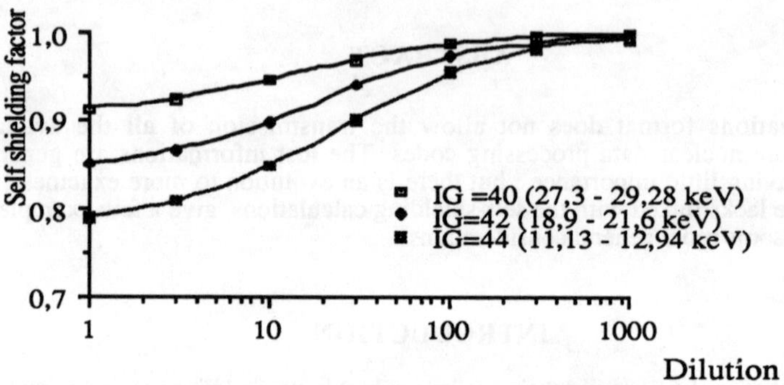

Figure 1 - *Values of the self-shielding factor as a function of the background cross section (dilution) for ^{238}U.*

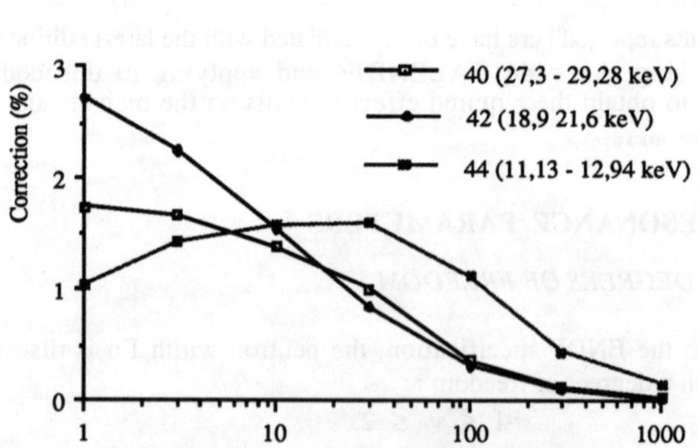

Figure 2 - *^{238}U : correction on the self-shielding factor induced by fluctuations of the radiation widths (ν_γ=100).*

1-1-3 The description of the fission widths distribution by a χ^2 law with a few degrees of freedom is an approximation because a χ^2 law is obtained only if the various average partial fission widths contributing to Γ_f are equal, which is generally false ; what we may first require is that this law describes properly the actual variance, which generally implies non integer values of ν_f. This is not in contradiction with the ENDF-6 specification, which states :

$$1.0 \leq \nu_f \leq 4.0$$

But the most used processing code[4] (at least in Europe) assumes integer values of ν_f and apparently all evaluators also define integer values for ν_f.

We tried to determine the effect of ν_f=2.5 instead of 2.0 for the ^{235}U; we concluded that this effect is <0.2 % for a 10 barn background cross section between 3. and 8. keV.

1-2 CORRELATION BETWEEN PARTIAL WIDTHS

Such a correlation has been proved between partial radiative widths and the neutron width. However it exists between the total radiation width and the neutron width, but its amplitude is probably small for nuclei such as ^{238}U; we assumes that :

$$\Gamma\gamma_\lambda = 0.98 \ \langle\Gamma\gamma\rangle \ + \ 0.02\ \Gamma_{n\lambda}\ \langle\Gamma\gamma\rangle / \langle\Gamma_n\rangle$$

The effect is small : 0.2 % between 10. and 30. keV for a 10 barn dilution.

But it may be more important for structural materials, whose $\Gamma\gamma$ is known to vary more.

1-3 FISSION PROCESS

The existence of the (n,γf) process is well established : as an example, a 1^+ resonance of ^{239}Pu has a very small fission width because the fission threshold is high for this spin state; the emission of a low energy γ ray will change the spin of the nucleus with a small decrease of the excitation energy : the probability of the fission process may then be hightly enhanced. For applications this process is a fission, but the fission width distribution is not any more a χ^2 law : it would be better described by the sum of a constant (representing the (n,γf) process) and of the actual fission width.

For ^{239}Pu, J=1$^+$, we added the fission cross sections due to :

$\Gamma_{f1} = \alpha\,\Gamma_f$, distributed according a χ^2 law with ν=1,

$\Gamma_{f2} = 0.1\,\Gamma_\gamma$, constant value $(\Gamma_\gamma = 0.0407$ eV).

The coefficient α was adjusted differently from one group to another in order to preserve the value of the cross section.

For a 10 barn dilution we obtain variations of 0.1 to 0.6 % of the self shielding factor of the ^{239}Pu fission cross section between 1. and 3. keV.

2 MULTILEVEL FORMALISM IN THE UNRESOLVED ENERGY RANGE

The ENDF-6 specification states that "at the present time, only the SLBW formalism is allowed in ENDF/B, for the reason that no significant multilevel effect can be demonstrated, when SLBW is properly handled". What means "significant"?

2-1 In 1973 G. de Saussure and R.B. Perez[5] studied the multilevel effects in the unresolved resonances region for ^{235}U. They concluded that even the fully shielded group constants are not very sensitive to the formalism at 500 eV. The difference between the single-level and multilevel value of $<\sigma_f>$ is smaller than 6% above 300K and smaller than 3% above 1000K.

2-2 We did some calculations on the ^{239}Pu : we calculated the self-shielded fission cross section for s-wave resonances with the SLBW and with the RM formalisms. For a 10 barn background cross section, between 1. and 3. keV, the formalism modifies the self-shielding factor by 0.2 to 1.6 % according to the groups (the computation was done on the APOLLO-2 energy mesh, which has 6 groups between 1.01 and 3.35 keV); it has to be noticed that the infinite dilution fission cross section also varies by 1%.

2-3 We often processed nuclides such as ^{58}Ni or ^{56}Fe, and concluded that it is not possible to treat properly such nuclei in the unresolved energy range without using the RM formalism : with SLBW formalism we obtained, for some random sets of resonances, negative values of the total cross section. Of course, it could be possible to reject the random set providing negative values, but such a choice of the random sets would introduce a bias.

As a matter of fact the importance of resonance interferences is increasing with the ratios , $<\Gamma_n/D>$ or $<\Gamma_f/D>$, which increase with energy (as $<\Gamma_n>$ and $<\Gamma_f>$ increase while $<D>$ decreases), and it is easily understandable that the multilevel formalism is required in the unresolved energy range when it is required in the resolved resonance range, at least as long as the Doppler broadening is not prevailing.

3 MISCELLANEOUS PROBLEMS

We shall just mention few lacks of the ENDF format which does not allow exact or easy description of the nuclear information, but for which we have no numerical estimation of their importance.

We remind that the self shielding, dependent on the average resonance parameters, then on the spin and orbital momentum of resonances, may induce changes in average properties as a function of the dilution cross section.

3-1 Variation of ν with energy

E. FORT[6] showed that there is a significant variation of ν within the first few eV of ^{239}Pu ; it is likely that these fluctuations are due to the nuclear states properties of $^{240}Pu^*$, then they exist at higher energy and are resonance dependent.

In the JEF2 evaluation, E. FORT introduced 2135 points below 650 eV to described the fine structure of ν ; but this representation is rather cumbersome and cannot be easily generalized.

3-2 Angular anisotropy

The angular anisotropy of scatterred neutrons depends on the orbital momentum ; then the average distribution, given for an infinite dilution cross section, will not be valid for any dilution.

3-3 Number of partial widths

The present format allows the definition of 3 or 4 partial widths ; a more accurate description of nuclei would require more partial widths with a flexible definition ; this feature would solve several of the above mentionned lacks.

3-4 Overlap of resolved and unresolved resonances

The only present solution proposed to solve this difficulty is to introduce fictitious resonances. But it would be again cumbersome, and these resolved resonances would be a false information. A better solution would be to accept the overlap with the specification of a special treatment : for example, when the unresolved energy range is processed by a resonance ladder method, it may be stipulated that the greatest random resonances have to be replaced by the resolved resonances.

3-5 . Energy interpolation.

It appears that energy interpolation of cross sections has been prefered in ENDF-6 specification to the interpolation of average parameters. This choice is less physical (the energy variation of resonance parameters is much less important that the variation of the average cross sections), and it cannot be fully applied in processing codes[7].

4 THE REQUIRED ACCURACY

Are these effects of importance ? Twenty years ago, de SAUSSURE and PEREZ concluded that it was not the case; their argument was that the effects (3 to 6 %) were small for fast reactors at high temperature.

We can submit two very different arguments to try to assess values for requested accuracy in self-shielding calculations :

a) as accuracy of microscopic data is not better than 2%, it is useless to calculate the self-shielding with an accuracy better than 1% ;

b) as in practice the cross sections are adjusted on integral experiments (at least for major nuclide), the actual accuracy on infinite dilution cross sections is of the order of 0.2 %, thus requiring a good accuracy on the calculated self-shielding factor.

Anyway we feel that the requirements on numerical accuracy have greatly increased recently, probably because the cost of computations has greatly decreased, and it is required to get free of this source of error ; but this implies that we take into account facts which were previously neglected.

CONCLUSION -

Some of the mentionned lacks can be very easily rectified at the level of the format specification.This is the case for non-integer values of ν and for the use of multilevel formalism in the unresolved resonances energy range : one has just to modify the text of ENDF-102 ; the modifications will only concern the processing codes.

Some other modifications imply the introduction of new data, as a fact of more resonance parameters (for both the resolved and unresolved energy ranges) : this imply a modification of the ENDF-B format, but is certainly feasible.

Other improvements require more studies, and will be questionnable as their interest : this is the case of the correlation between partial widths.

For the basic question is : do we require that the evaluation format allows the introduction of any nuclear information, leaving to the responsability of the evaluator (in 5 or 10 years) to decide if it is worth to include this information, and to the nuclear data processor (in 15 or 20 years) to decide if this information is worth to be processed for his specific use ?

Acknowledgements. We thank Caroline RAEPSAET for her help in preparing this paper and Henry TELLIER to have accepted to present it at this Symposium.

REFERENCES -

1. P.F. ROSE and C.L. DUNFORD, *ENDF-102*, BNL-NCS-44945 (1990)
2. P. RIBON, in the *Proceedings of the Seminar on NJOY and THEMIS*, OECD/NEA Data Bank, June 1989.
3. M. MOREAU, *internal report* SERMA 1423 (1992).
4. R.E. MaAC FARLANE et al, *The NJOY Nuclear Data Processing System*, LA9303 (1982,1985,1987).
5. G. de SAUSSURE and R.B. PEREZ, *Nucl. Sci. Eng*. 52-3 (1973) 382.
6. E. FORT et al, *Nucl. Sci. Eng*., 99(1988) 375.
7. R.E. MAC FARLANE, *Seminar on NJOY -91 and THEMIS*, Saclay, April 1992

NEUTRON CROSS-SECTIONS MODELING IN THE UNRESOLVED RESONANCE REGION

N.B.Janeva and N.T.Koyumdjieva
Inst. of Nucl. Research and Nucl. Energy,
Tzarigradsko shose No 72, 1784 Sofia, Bulgaria

A.A.Lukyanov
Sofia University, Department of Physics,
J.Boucher NO 5, 1126 Sofia, Bulgaria

Abstract: The statistical R-matrix is used for analytical modelling of resonance cross sections in the unresolved region. The characteristic function for the R-matrix elements distributions has been studied.

The modelling of the resonance structure in the unresolved region is needed for the examinations of the resonance self-shielding effects in reactor physics. For a consistent treatment of this problem both for isolated and partially overlapping resonances it is necessary to use models, which include the typical interference peculiarities of the detailed resonance cross section structure. The most convenient approach appears to be the Wigner's R-matrix theory with real and independent parameters, which obey known statistical distributions[1,2]. Different approximations for a cross section modelling have been considered in the unresolved resonance region and the method by using the reduced statistical R-matrix is a common one in this region[1].

For an independent system of resonances with fixed values of the total momentum J and parity π we use the following expression for the R-matrix elements[2]:

$$R_{nc}=S\sum_{\lambda} \zeta_{\lambda n}^{1/2} \zeta_{\lambda c}^{1/2} /(x+\Delta_{\lambda}-iy) \qquad /1/$$

where $S=\pi\overline{\Gamma}_n/2\overline{D}$, $y=\pi\Gamma_\gamma/2\overline{D}$, $x=\pi(E_0-E)/\overline{D}$, $\zeta_{\lambda n}=\Gamma_{\lambda n}/\overline{\Gamma}_n$, $\Delta_\lambda=\pi(E_\lambda-E_0)/\overline{D}$. In the region of resolved resonances the full set of the R-matrix parameters is in principle defined and the exact description of the detailed energy cross section structure is possible. In the unresolved region the only average resonance parameters ("strength functions") are defined with the corresponding laws for the statistical distributions. Those are the well known Porter-Thomas and Wigner distributions, which are defined in the random number statistics[1,2]. The random character of the resonance parameter distributions in the R-matrix Eq.(1) leads to the idea to calculate the average cross sections and their functionals by a modelling of the resonance structure and using the Monte-Carlo method[3-5]. Here we present some variants of analytical modelling of resonances in the unresolved region for the one channel R-matrix.

1. A Model of Identical Resonances.

In the one channel case the R-function ($R=R_{nn}$) is used and the total and the radiative capture cross sections are defined. Let us assume that the neutron widths are equal $\zeta_{\lambda n}=1$ and levels are equidistant $\Delta_{\lambda}=\pi\lambda$. The sum in Eq.(1) to infinity $[-\infty<\lambda<\infty]$ is[2]:

$$R=S\ \mathrm{ctg}(x-iy) \qquad /2/$$

The total and radiative capture cross sections in this case can be described as periodical functions:

$$\sigma=\sigma_p+\sigma_0\frac{\cos 2\varphi-p^{-1}\mathrm{tg}x.\sin 2\varphi}{1+p^{-2}\mathrm{tg}^2x} \qquad /3/$$

where: $p=\mathrm{th}(y+\delta)$, $S=\mathrm{th}\delta$,

$$\sigma_p=4\ \frac{\sin^2\varphi+S\ \mathrm{th}y.\cos^2\varphi}{1+S\ \mathrm{th}y},\quad \sigma_0=\frac{S(1-\mathrm{th}^2y)}{(\mathrm{th}+S)(1+S\ \mathrm{th}y)} \qquad /4/$$

and $$\sigma_\gamma=\sigma_{0\gamma}\frac{1+\mathrm{tg}^2x}{1+p^{-2}\mathrm{tg}^2x},\quad \sigma_{0\gamma}=\frac{4S\ \mathrm{th}y}{(S+\mathrm{th}y)^2} \qquad /5/$$

(the kinematical coefficients are omitted).

The averaging over a period $-\pi/2\le x\le\pi/2$ in Eq.(3,5) gives the known results of the resonance theory[1,2]:

$$\langle\sigma_\gamma\rangle=\frac{1}{\pi}\int_{-\pi/2}^{\pi/2}\sigma_{0\gamma}\ \frac{1+\mathrm{tg}^2x}{1+p^{-2}\mathrm{tg}^2x}\,dx=\sigma_{0\gamma}p=\frac{4S\ \mathrm{th}y}{(S+\mathrm{th}y)(1+S\ \mathrm{th}y)} \qquad /7/$$

$$\langle\sigma\rangle\ =\ 4\left[\sin^2\varphi+\frac{S\cos 2\varphi}{1+S}\right] \qquad /8/$$

The periodical structure of our expressions allows to present the various average functionals of cross section as mean values over the period $[-\pi/2\le x\le\pi/2]$. So in this model we have found:

$$\langle\sigma^2\rangle-(\langle\sigma\rangle)^2=8\frac{S^2}{(1+S)^2}\frac{(1-\mathrm{th}y)^2}{(1+S\ \mathrm{th}y)(S+\mathrm{th}y)} \qquad /9/$$

$$\langle\sigma_\gamma\sigma\rangle-\langle\sigma_\gamma\rangle\langle\sigma\rangle=2\langle\sigma_\gamma\rangle\cos 2\varphi\ \frac{S(1-S)}{1+S}\ \frac{(1-\mathrm{th}y)^2}{(1+S\ \mathrm{th}y)(S+\mathrm{th}y)} \qquad /10/$$

The functionals $\langle 1/\sigma\rangle$ and $\langle\sigma_\gamma/\sigma\rangle$, which are in practical importance, have been found in this scheme :

$$\langle\sigma_\gamma/\sigma\rangle=\ \langle\sigma_\gamma\rangle/(\sigma_m\sigma_M)^{1/2} \qquad /11/$$

$$\langle 1/\sigma\rangle = \frac{1}{(\sigma_m\sigma_M)^{1/2}} \frac{(p\sigma_m^{1/2}+\sigma_M^{1/2})(p\sigma_M^{1/2}+\sigma_m^{1/2})}{(p\sigma_M^{1/2}+\sigma_m^{1/2})^2+(\sigma_M-\sigma_m)\sin^2\varphi(1-p^2)} \qquad /12/$$

Here $\sigma_M=\sigma_m+\sigma_o$ is the maximal value of the total cross-section and σ_m is its minimal value[2]. We can found analytical formulae for the self indication function and average transmission:

$$\langle\sigma_\gamma e^{-n\sigma}\rangle=\langle\sigma_\gamma\rangle I_o(n\sigma_o/2)\exp[-n\sigma_o/2 - n\sigma_m] \qquad /13/$$

$$\langle e^{-n\gamma}\rangle=e^{-n(\sigma_m+\sigma_o/2)}\left[I_o(n\sigma_o/2)+2\sum_{k=1}^{\infty}\left(\frac{1-p}{1+p}\right)^k I_k(n\sigma_o/2)\cos 2k\varphi\right]$$

For $n\sigma_o\gg1$ the asymptotical limit is:

$$\langle e^{-n\sigma}\rangle \to \frac{p}{\sin^2\varphi+p\cos^2\varphi}\,\frac{1}{(\pi n\sigma_o)^{1/2}}\exp(-n\sigma_m) \qquad /14/$$

We have foundan useful functional for our further presentation:

$$\langle e^{iRt-iR^*t'}\rangle=\int_\gamma^\infty e^{-u\,\mathrm{cth}2y} I_o\left(\frac{1}{\mathrm{sh}2y}\sqrt{u^2-\beta^2}\right)du \qquad /15/$$

where $\gamma=S(t+t')$, $\beta=S(t-t')$. For $t'=0$ we have obtained[6]:

$$\langle e^{iRt}\rangle=e^{-St} \qquad /16/$$

which looks like a generalization of the well known result of the R-matrix theory $\langle R\rangle=iS$ [1]. The same result is also valid for y=0.

Simple analytical expressions of various cross section functionals can be obtained in the framework of the model of identical resonances in the unresolved resonance region. The model contains the well known results for the average cross sections and does not lead to qualitative contradictions in the transmission and self-indication functions. It is important that there are no special assumptions about values of the average resonance parameters. The resonance and potential scattering interference and also the inter-resonance interference are accounted in this model.

2. A Statistical Model of Resonance Cross Sections.

A general approach for modelling of the resonance structure in the unresolved region is that the statistical fluctuations of the R-matrix resonance parameters are taking into account. The fluctuating parameters $\xi_{\lambda n}$ and Δ_λ enter only as a definite combination in R_{nn} Eq.(1) and the

R-function enters into the cross sections as a module of $(1-iR)^{-1}$. These allows to untroduce the distribution function for real and imaguinar parts of R_{nn}.

Let Φ is any physical functional of cross sections. It can be written as the Laplace transform of a certain function φ:

$$\Phi\left[\frac{1}{1-iR},\frac{1}{1-iR^*}\right]=\int_0^\infty dt\int_0^\infty dt'\varphi(t,t')e^{-t(1-iR)}e^{-t'(1+iR^*)} \qquad /17/$$

An averaging over the resonances leads to the result:

$$\langle\Phi\rangle=\int_0^\infty dt\int_0^\infty dt'\varphi(t,t')e^{-(t+t')}\langle e^{iRt-iR^*t'}\rangle \qquad /18/$$

where: $\langle e^{iRt-iR^*t'}\rangle=$

$$=\frac{1}{\Delta E}\int dE\prod_\lambda\int d\zeta_\lambda d\delta_\lambda P(\xi_\lambda)P(\delta_\lambda)\exp i\left[S\xi_\lambda\left(\frac{t}{x-iy+\delta_\lambda}-\frac{t'}{x+iy+\delta_\lambda}\right)\right] \qquad /19/$$

Let us suppose there are equidistant levels; e.g. $\delta_\lambda=\pi\lambda$. Then the infinite product Eq.(19) is[6]:

$$\left[\frac{\sin(x-iy)\sin(x+iy)}{\sin(x-ip)\sin(x+iq)}\right]^{1/2}$$

where $p=f+\beta$, $q=f-\beta$, $f^2=\beta^2+2\gamma y+y^2$, $\gamma=S(t+t')$, $\beta=s(t-t')$. Make an average over the period $[-\pi/2\le x\le\pi/2]$ we define the characteristic function F, which corresponds to the statistical distribution of Re R and Im R:

$$F=\langle e^{iRt-iR^*t'}\rangle=\frac{1}{\pi}\int_{-\pi/2}^{\pi/2}dx\left[\frac{\sin(x-iy)\sin(x+iy)}{\sin(x-ip)\sin(x+iq)}\right]^{1/2} \qquad /20/$$

It is a real function of three parameters and we have obtained a number of presentations by using the complete elliptical integrals[7]. For y=0 and for z=0 the above expresion becomes:

$$F_0(0,\beta)=e^{-\beta}=e^{-S(t-t')} \qquad /21/$$

which coincides with our result for identical resonances Eq.(16). It means that in the one-channel case, when the radiative capture is negligible , the fluctuations of neutron widths do not appear in average functionals of cross sections Eq.(18) and here the model of identical resonances gives the "accurate" result. The distribution function for R, defined as the Fourier cosine-transformation of Eq.(21), is:

$$P(R)=\frac{1}{\pi}\int_0^\infty e^{-S\tau}\cos(R\tau)d\tau=\frac{1}{\pi}\frac{S}{S^2+R^2} \qquad /22/$$

This result has been obtained by Fröhner[8].

Let us suppose that $F_y(z,\beta)=\exp[-(z+\beta)]$. Then the distribution function is going to the form:

$$P(R_1,R_2)=\frac{1}{\pi}yR_2^{-2}\exp\left[-\frac{y}{2S}\frac{R_1^2+(R_2-S)^2}{R_2}\right] \qquad /23/$$

The universe character of the function F Eq.(19) for any average functionals Eq.(18) allows to formulate a requirement to the R-matrix modelling: equality of the characteristic function for all values of the parameters $y=\pi\Gamma_\gamma/2D$, $\beta=S(t-t')$ and $\gamma=S(t+t')$. The accuracy of the different average functionals calculations depends on the choice of the model, which must to have a characteristic function closed to our F function Eq.(20).

References:

1. A.M.Lane and R.C.Thomas, *Rev. Mod. Phys.*, **30** (1958) 257
   J.E.Lynn, *The Theory of Neutron Resonance Reactions*, (Clarendon Press, 1968)
2. A.A.Lukyanov, *Neutron Cross Sections Structure*, (Atomizdat, 1978)
3. P.A.Moldauer, *Nucl. Phys*, **A344** (1980) 185
4. N.T.Koyumdjieva and N.B.Janeva, *Bulg.J. of Nucl. Energy,* **18** (1983) 38
5. L.S.Nefedieva, V.F.Ukraintzev, N.B.Janeva, *Rep. JINR* БД-*10-83-698,* Dubna, 1983
6. N.T.Koyumdjieva, N.Savova, N.B.Janeva, A.A.Lukyanov, *Bulg. J. Phys.*, **16** (1989) 16
7. *Handbook of Mathematical Functions*, Ed. M.Abramovitz and I.A.Stegun (NBS, 1964)
8. F.H.Frohner,in *Proc.Int.Conf. on Nucl.Data for Basic and Applied Sci.*,(Gordon and Breach, New York, 1986) p.1541

NUCLEAR MODELS AND THEIR INPUT PARAMETERS

USE OF COUPLED-CHANNEL OPTICAL MODEL CALCULATIONS IN NUCLEAR DATA EVALUATIONS FOR INCIDENT NUCLEON ENERGIES UP TO 1 GeV

Jean Paul DELAROCHE

Commissariat à l'Energie Atomique,
BP 12, 91680 Bruyères-le-Châtel, France

ABSTRACT

The main features of the optical model potential (OMP) used in coupled-channel calculations are reviewed. Since there has been a recent interest in the incineration of nuclear waste materials, this paper covers a very broad energy range extending up to 1 GeV. It is shown that the OMP parameterizations currently used in neutron data evaluations below 20 MeV may be smoothly extrapolated up to 100 MeV when neutron and/or proton scattering and reaction data are available. At higher incident energies, neutron data are scarce and one should heavily rely upon proton data to build neutron OMP's. This raises problems which are discussed in the context of the Dirac equation, a proper tool for scattering and reaction OMP analyses at intermediate energies.

1 . Introduction

The optical model (OM) plays a major role in neutron nuclear data evaluation, and it is essential to know it very well. In its simplest form, the model only treats the elastic channel. This is the spherical OM (SOM), a very useful tool in evaluations of neutron data for spherical nuclei. The success met by the spherical optical model among users stems from its simplicity and fast computing times. This simplicity, however, makes it inappropriate for treating with enough accuracy many (if not all) deformed nuclei. For accurate calculations on open shell nuclei, the coupled-channel (CC) formalism is most appropriate.

The importance of performing coupled-channel calculations (CCC) has been discussed at length during the eighties, for instance when dealing with accurate evaluations in the actinide region at incident neutron energies below 20 MeV[1]. The eighties was a formative period for users of CC codes, during which many practical problems had to be solved ![2]. Many of these questions were new ones for the applied physics community. They were new, too, at a more basic level because i) the properties of neutron interactions with deformed nuclei generally were not so well known below 10 MeV, and ii) not much information on the OMP parameters could be inferred from proton scattering at low energies, the region where the interaction is dominated by Rutherford scattering. As a matter of fact, progress in the quality of nuclear data evaluations has paralleled our improved understanding of the neutron-nucleus interactions at a fundamental level.

This remark is quite general, even though it is intended to characterize some aspects of the research activities in the past decade. New challenges are now emerging for the nuclear data evaluation community. These mainly stem from needs related to the incineration of nuclear waste materials. In short, the present day evaluations covering the limited energy range E = 0 - 20 MeV may have to be extended up to about 1 GeV. This extension indeed raises fundamental problems, in particular when use is made of the coupled-channel optical model for E ≥ 100 MeV.

This paper covers the whole energy range E = 0 - 1 GeV, and deals with the OM potential properties for incident protons and neutrons. Since these potentials are related through an isovector term and a Coulomb correction, it may be convenient to build neutron OM potentials from proton potentials especially at energies E ≥ 40 MeV for which neutron scattering measurements rarely exist. In Section 2, neutron OMP parameters established for incident energies up to approximately 100 MeV are discussed. These parameterizations are obtained i) from fits to neutron scattering and reaction measurements using the Lane model[3], or ii) from a joint analysis of both proton and neutron observables.

The intermediate energy region covers the incident energies E ≥ 100 MeV. In this domaine, the phenomenological optical model analyses based on the Schrödinger equation generally break down[4], and instead the Dirac equation[5] must be used. This failure of the Schrödinger phenomenology and the success of OMP analyses based on Dirac equation are briefly described in Section 3.

2 . OMP Parameterizations up to 100 MeV

The coupled-channel formalism for scattering and reaction studies is well known and will not be described again in this paper. For a detailed presentation, see the review article by T. Tamura[6].

2.1 Preliminary considerations

(1) Relativistic kinematics

The CC equations of Ref. 6 are based on the assumption that relativistic effects are negligible. Therefore these equations may directly be used in CC calculations up to about 30 MeV. Beyond this energy, the relativistic effects gradually become more important. Minimum relativity may be considered by implementing relativistic kinematics in the CC equations. The code ECIS written by Raynal treats this kinematics[7]. Since turning on and off this option in CC calculations produces discontinuities[8] in the predictions, it may be convenient to systematically use the relativistic kinematics.

(2) Deformation lengths

The deformed optical potentials U (r) display a radial dependence r-R (α), where r is the spatial coordinate of the incident particle and R(α) the collective radius of a deformed nucleus. In the intrinsic system of coordinates, this radius is defined for deformed nuclei[6] :

$$R(\alpha) = R_0 \left[1 + \sum_{\lambda} \beta_{\lambda} Y_{\lambda}^{0} (\Omega') \right] \quad (1)$$

where β_λ is a deformation parameter that measures the permanent, static deformation. For all nuclei, the quadruple deformation β_2 plays a leading role in the interaction process. Higher order deformations β_λ (with λ = 4,6,...) are systematically smaller in magnitudes than β_2. The inclusion of quadrupole (β_4) and hexadecapole (β_6) deformations in CC calculations produces alterations of predicted observables. In general, only β_2 and β_4 deformations are important in describing (n,n') scattering to the first few levels in a rotational band.

In many CC calculations, the deformed potential U $(\vec{r})$ includes a real volume potential V $(\vec{r})$, a volume absorption W_V $(\vec{r})$, a surface absorption W_D $(\vec{r})$, and a real spin-orbit term V_{so} $(\vec{r})$. For evaluation purpose, the spin-orbit potential generally is not deformed. Since U (r) now includes three deformed components, each possibly with a different geometry, the definition of the β_λ's may become ambiguous. For the sake of simplicity, it is recommended to adopt a deformation length $\delta_\lambda = \beta_\lambda \; R_{o\lambda}$ common to all three OMP components. Moreover, δ_λ has more physical meaning than β_λ and should not depend upon incident energies. The δ_λ value is mainly determined at low energy from the real central potential while at intermediate energy it is the volume absorption that mainly fixes its value. These comments on the deformation lengths are also valid when treating vibrational nuclei in CC calculations.

(3) Coupling basis

For deformed nuclei, the number of states to be included in the coupling scheme depends upon the scattering phenomena that one may need to describe in details. For most applications, a (0^+, 2^+, 4^+) coupling basis is appropriate for even-even nuclei. In the studies of odd-even nuclei, one may either adopt the "equivalent" coupling scheme[9], which means very lengthy CC calculations, or rely upon the so-called "fictious even-even nucleus" method[10]. Finally, nobody knows for sure how to treat the nucleon interaction with odd-odd nuclei.

For vibrational nuclei (i.e., those at or near closed shells), the minimum number of states to couple is two. For even-even nuclei, one usually uses the (0^+, 2^+) coupling, where the 2^+ level is a one-quadrupole phonon excitation. If the 3^- and 5^- states are strongly collective excitations, that is when β_3 and β_5 are larger than 0.1, these levels may be included in the coupling scheme. For even-odd systems, one may use the particle-core model[11]. No prescription exists for treating inelastic scattering from odd-odd vibrational nuclei.

(4) Partial wave expansion

The CC equations are solved using partial wave expansion in terms of the angular momentum l. At low incident energy, only a few partial waves are needed (i.e.,$l_{max} \leq 20\hbar$) to reach convergence. The l_{max} value roughtly increases with incident energy as 2.4 kR_o (where k is the wave number, and $R_o = r_o \; A^{1/3}$ with $r_o \sim 1.25$ fm) and many partial waves have to be considered at high energies. It is thus suggested to check carefully the convergence of the numerical solutions at various incident energies by increasing l_{max} beyond the empirical estimate : $\lambda_{max} \approx 2.4\,kR_o$ established from SOM analyses. In CC calculations, the convergence is met using more partial waves : $l_{max} \approx 2.4\,kR_o + \lambda_{max}$, where λ_{max} is the maximum order in the expansion of the OM potential. For rotational nuclei, it is suggested to use $\lambda_{max} = 8$ when a (0^+, 2^+, 4^+) coupling is adopted[9]. For vibrational nuclei, $\lambda_{max} = 2$ if the (0^+, 2^+) coupling is considered, $\lambda_{max} = 3$ if the (0^+, 2^+, 3^-) basis also includes one-octupole phonon, etc.

One should bear in mind all these considerations before running CC calculations over a broad energy range, e.g. extending from E = 0 up to E ~ 100 MeV.

2.2 Energy dependences

In phenomenological OM analyses, one always uses local and energy dependent potentials. Because of the Coulomb field experienced by the incident protons, the nuclear part of the protron-nucleus interaction differs from that of the neutron-nucleus field. According to the Lane model[3], the nucleon-nucleus OM potential is expressed

$$U(r,E) = U_0(r,E) + (4/A)\, U_1(r,E)\, \vec{t}.\vec{T} \tag{2}$$

where $\vec{t}$ is the isospin of the incident particle and $\vec{T}$ the isospin of the target. The diagonal terms of $\vec{t}\,.\,\vec{T}$ display the difference between proton-nucleus and neutron-nucleus potentials :

$$U(r,E) = U_0(r,E) \pm \epsilon\, U_1(r,E) + \Delta U_c \begin{bmatrix} +protons \\ -neutrons \end{bmatrix} \tag{3}$$

where ϵ = (N - Z)/A is the asymmetry, U_o and U_1 the isoscolar and isovector potentials, respectively, and ΔU_c the Coulomb correction term ($\Delta U_c = 0$ for neutrons). Eq.(2) was originally introduced in the SOM framework. It may also be used in coupled-channel calculations.

The proton-nucleus OM potential contains both isovector and Coulomb correction terms, and it is essential to know ΔU_c before attempting an isoscalar-isovector decomposition. This is accomplished by comparing proton and neutron scattering from nuclei with N = Z (i.e., T = 0). Illustrations are given in Sec. 3. Once ΔU_c is known, the isoscalar-isovector decomposition is accomplished by comparing proton and neutron scattering from the same T ≠ 0 nucleus.

In most OM analyses at low energy (E ≤ 100 MeV), it is generally assumed that U_0, U_1, and ΔU_c have the same geometry. With this approximation, the deformed OM potential can be written in a condensed form :

$$\begin{aligned} -U(\vec{r},E) = {} & V_R(E)\, f(r,R_V,a_V) \\ & + 4i\, a_D W_D(E) \frac{d}{dr} f(r,R_D,a_D) \\ & + iW_V(E)\, f(r,R_W,a_W) \\ & + 2\, \lambda_\pi^2 V_{so}(E)\, \frac{1}{r}\frac{d}{dr}\, f(r,R_{so},a_{so})\, \vec{l}.\vec{s} \end{aligned} \tag{4}$$

where it is understood that R_i (i = V, D, W) are collective radii, Eq. (1), which include the deformation parameters $\beta_{\lambda i}$ (λ = 2, 4) as mentioned above. In Eq. (4), f (r, R_i, a_i) is a radial Woods - Saxon shape :

$$f\left(r,R_i,a_i\right)=\left[1+\exp\left(\frac{r-R_i}{a_i}\right)\right]^{-1} \quad (5)$$

and V_R (E), W_D (E), W_V (E), and V_{so} (E) are the potential depths for the real central potential, imaginary central surface and volume absorptive terms, and real spin-orbit interaction, respectively. The E-dependence of these OMP components are discussed below.

(1) Real central potential

Early OMP analyses of nucleon elastic scattering indicated that V_R decreases with increasing incident energy. For neutron energies less than approximatively 50 MeV, it may be assumed that V_R decreases linearly, that is : V_R (E) = V_0 (1 - α E) with $\alpha \approx$ 0.3/MeV. However, there exist indications at higher incident nucleon energies that V_R decreases less rapidily. Some analyses suggest a logarithmic E - dependence for V_R, and some others a parabolic E - dependence. More recent OMP analyses[4] that include the intermediate energy region confirm the gradual decrease of V_R with energy.

(2) Central absorptive potentials

Nuclear absorption may be expressed as the sum of a volume W_v (E) and a surface W_D (E) term with different radial dependences. The surface term has a potential depth W_D (E) that increases with incident energy, up to an energy E_o and then decreases. This E - dependence may be represented as a superposition of linear segments, with typical W_D values of about 8 MeV for E_o in the 8 - 15 MeV range. For E > E_o, the potential depth W_D (E) decreases to very small values and eventually vanishes for E $\geq$ 80 MeV. More sophisticated functional forms may also be adopted for W_D (E). The empirical volume term has a potential depth W_v that i) is small up to the energy E_O, ii) gradually increases with E typically to a value of about 10 MeV at the incident energy E ~ 80 MeV, and iii) remains roughtly constant up to about 100 MeV. Examples of E - dependences used for W_D (E) and W_v (E) are given in Ref.8.

(3) Spin-orbit interaction

Recently a large set of neutron analyzing power A_y (θ) data was measured at TUNL (Durham, North Carolina) and was used to pin down the uncertainties on the spin-orbit terms in the OMP. It has been found that V_{so} has a very weak E - dependence[12]. The spin-orbit interaction may be complex, with a sizeable imaginary component W_{so} for E above 50 MeV[12]. Since W_{so} mainly affects the A_y (θ) predictions, it may be set equal to zero when evaluating nuclear data.

(4) Coulomb correction term

This term is complex, too. Its real part $\Delta V_c(\vec{r}, E)$ is a volume shape, and its strength ΔV_c is approximately constant. Many authors adopt the value $\Delta V_c = 0.4\ Z/A^{1/3}$ (MeV), while the larger estimate $\Delta V_c = 0.48\ Z/A^{1/3}$ (MeV) is probably more realistic[13]. The imaginary component $\Delta W_c(\vec{r}, E)$ is not very well known. It is a surface shape, and its strength $\Delta W_c(E)$ is weak. The (p, p) and (n, n) elastic scattering analyses for T = 0 nuclei suggest[13, 14] that $\Delta W_c(E)$ decreases with increasing energy and vanishes for $E \geq 60$ MeV.

(5) Isovector potential

It is difficult to obtain the radial dependence of the isovector term $U_1(\vec{r}, E)$ just from single nucleon scattering data alone. However, most OM analyses are performed using a volume shape for its real part $V_1(\vec{r}, E)$ and a surface shape for its imaginary component $W_1(\vec{r}, E)$. The energy dependence of the isovector potential depths $V_1(E)$ and $W_1(E)$ has been studied in the analyses of quasielastic (p, n) scattering[15]. It has also been investigated in the comparison of neutron and proton scattering analyses from the same nuclei and through simultaneous OMP analyses of several nuclei at several energies. It is found that i) the real term $V_1(E)$ may be written as a linear E - dependence with a coefficient $\alpha_1 = -0.16 \pm 0{,}10$ MeV^{-1}, and ii) the imaginary component $W_1(E)$ also has a linear E - dependence with a coefficient $\alpha'_1 \sim \alpha_1$. These results are based on SOM studies for selected spherical nuclei[13]. For deformed nuclei in the rare earth and actinide regions, CC analyses of (n, n') inelastic scattering data below 8 MeV indicate that it is reasonable to adopt $V_1 \sim 16 - 18$ MeV and $W_1 \sim 8 - 9$ MeV[1, 9]. Despite the tremendous efforts placed on its empirical determination by various means, the isovector potential is still poorly known. Nuclar structure calculations suggest $V_1(E) = 11.5 - 0.1\ E$ (MeV) for neutrons below 80 MeV[16]

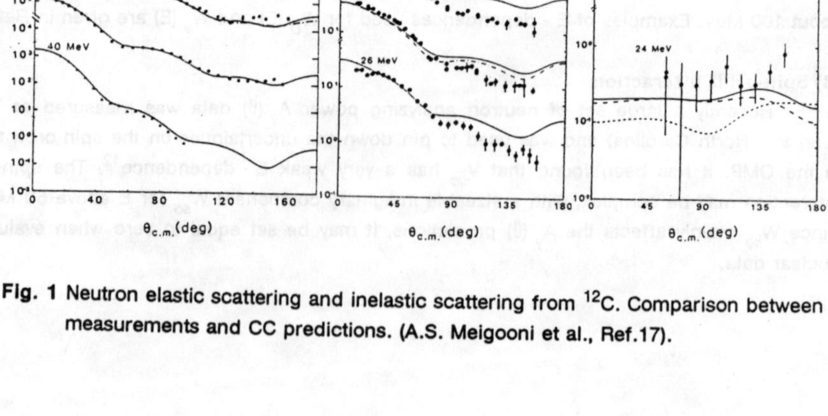

Fig. 1 Neutron elastic scattering and inelastic scattering from ^{12}C. Comparison between measurements and CC predictions. (A.S. Meigooni et al., Ref.17).

2.3 Illustrations

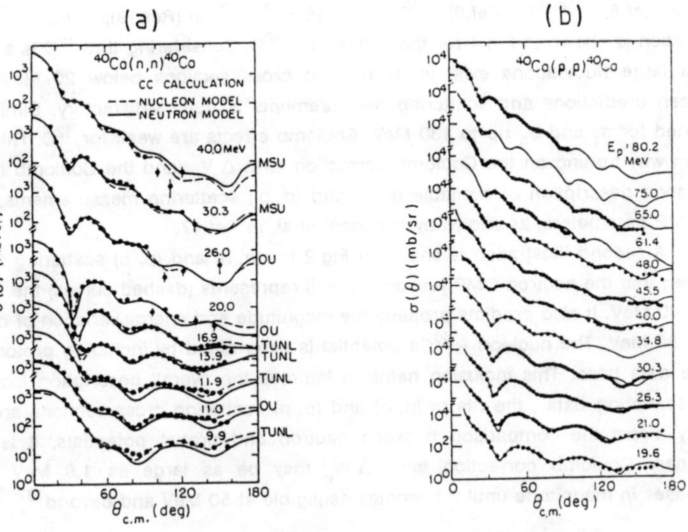

Fig. 2 Neutron (a) and proton (b) elastic scattering from ^{40}Ca. Comparison between measurements and CC predictions. (G.M. Honoré et al. Ref.12).

Much experimental information is needed to produce reliable neutron OMP parameters. In particular, total cross section (σ_T) and scattering cross section ($\sigma(\theta)$) measurements are valuable. Since $\sigma(\theta)$ data for elastic and inelastic neutron scattering rarely exist above 26 MeV, the OMP parameter optimizations heavily rely upon σ_T measurements which are often available up to ~ 100 MeV, and in a number of cases up to 600 MeV. Many careful σ_T measurements have recently been performed at ORNL and LANL. Indeed, these data are of key importance.

The neutron potentials are built in a first stage at incident energies below ~ 10 MeV, where W_v is weak. For that purpose, the well known SPRT method may be used[2]. In the second stage, the data base is gradually enlarged by incorporating data available at higher incident energies. This leads to a careful mapping of the interplay between surface and volume absorption. The final stage consists in fine tunning all the OMP parameters to reach the best overall description of all scattering and reaction data available for a nucleus.

To build proton OM potentials, $\sigma(\theta)$ and reaction cross section (σ_R) measurements may be used. The proton scattering data are often available at high energies[4] for which neutron scattering measurements are nonexistent, so that the two ensembles of experimental information complement each other in OMP analyses cast in a single framework (i.e., Lane model).

These methods have recently been tested with success for several vibrational and rotational nuclei, for instance, ^{12}C (Ref.17), ^{27}Al (Ref.18), ^{28}Si (Ref.14), ^{40}Ca (Ref.12), $^{54,56}Fe$ (Ref.8), $^{58,60}Ni$ (Ref.8), ^{59}Co (Ref.18), $^{116,120}Sn$ (Ref.19), and ^{197}Au (Ref.20). An illustration is shown in Fig.1 for the system n + ^{12}C. Considering that ^{12}C is a light nucleus for which large fluctuations exist in its neutron cross sections below 20 MeV, the agreement between predictions and scattering measurements is quite satisfactory. Similar agreement is obtained for σ_T and σ_R up to 130 MeV. Coulomb effects are weak for ^{12}C. This is probably the reason why turning on the Coulomb correction term ΔV_c (and the Coulomb interaction) leads to a good description of available (p, p) and (p, p') scattering measurements, without altering the OMP parameters, as shown by Meigooni et al. in Ref.17.

A second illustration is shown in Fig.2 for (n, n) and (p, p) scattering from ^{40}Ca. It can be seen that the neutron OMP reasonably well represents (dashed curves) the neutron $\sigma(\theta)$ data up to 40 MeV. It also predicts properly the magnitude and energy variation of σ_T measurements up to 80 MeV. The nucleon + ^{40}Ca potential is determined by including proton scattering data in the data base. This inclusion helps to improve the overall description (solid curve) of the (n,n) scattering data ; the fits to (n, n) and (p, p) scattering cross sections are of comparable quality. From the comparison between neutron and proton potentials, it is found that the imaginary Coulomb correction term ΔW_c may be as large as 1.6 MeV at 20 MeV and decreases in magnitude until it becomes negligible at 50 MeV and beyond[12].

3 . Dirac Phenomenology

There exists much evidence for the failure of the standard optical model above ~ 200 MeV[4]. These evidences are revealed in OM analyses of (p, p) scattering data, in which conventional Woods - Saxon form factors are used. Only by adopting non-standard potential form factors is it possible to achieve good fits to the data. These empirical findings together with relativistic mean-field predictions clearly point to the need for full relativistic treatments of scattering and reaction problems.

The Dirac equation meets this need provided that nucleons are considered as point - like, structureless objects. It has been employed with considerable success in (p, p) and (p, p') scattering analyses through the intermediate energy region[5, 21]. Good fits to neutron and proton scattering data below 100 MeV have also been achieved[5, 22]. In fact, this agreement between Dirac OM predictions and the data are often better than that obtained with the Schrödinger equation, i.e. in a conventional OMP analysis. Since the coupled-channel code ECIS solves both equations, it may be the proper tool to adopt in nuclear data evaluations up to 1 GeV.

4 . Conclusion

The main purpose of this paper has been to suggest methods that may be used to extend the coupled-channel calculations to incident nucleon energies higher than 20 MeV. Properties of the nucleon-nucleus OM potential below 100 MeV have been reviewed, and illustrations given. It has been shown that smooth E - dependent neutron OMP parameterizations may be established provided that σ_T data exist up to the lower end of the intermediate energy region. The validity of these parameterizations may be further tested by including proton scattering and reaction measurements in the data base. This has been successfully accomphished for T = 0 nuclei. It would be useful to repeat these tests for $T \neq 0$ nuclei in order to gain more insight into the symmetry potential at energies higher than ~ 50 MeV where U_1 (r, E) is poorly known.

Phenomenological analyses based on Schrödinger equation fail in the intermediate energy region because relativistic effects cannot be ignored. Since Dirac phenomenology has proved to be very successful for E > 100 MeV and also for E < 100 MeV, it is suggested to adopt the Dirac equation in scattering and reaction OM studies and in evaluations all the way up to 1 GeV.

5 . Acknowledgments

The author wishes to thank O. Bersillon and R.L. Walter for critical reading of the manuscript, and D. Madland for discussions on Dirac phenomenology.

6 . References

1) J. Salvy, in "*The use of the Optical Model for Calculation of neutron cross sections below 20 MeV*" (OECD/OCDE, NEANDC report N° 222'U', Paris, 1986), p.233.
2) J.P. Delaroche, Ch. Lagrange, and J. Salvy, in "*Nuclear Theory in Neutron Nuclear Data Evaluation*" (IAEA, Vienna, 1976), Vol.II, p.251.
3) A.M. Lane, *Nucl.Phys.* **35** *(1962) 676.*
4) P. Schwandt, in "*The interaction between Medium Energy Nucleon in Nuclei*", AIP Conf. Pro. N°97, ed. by H.O. MEYER (AIP, New-York, 1982), p.89.
5) B.C Clark, S. Hama, and R.L. Mercier, see Ref.4, p.260.
6) T. Tamura, *Rev. Mod. Phys.***37** *(1965) 679.*
7) J. Raynal, in "*Applied Nuclear Theory and Nuclear Model Calculations for Nuclear Technology Applications*", Trieste, Italy, 1988 (M.K Mehta and J.J Smith editors) World Scientific, Singapore, 1989, p.506.
8) R.S. Pedroni et al., *Phys. Rev.* **C38** *(1988) 2052.*
9) J.P. Delaroche et al., *Phys. Rev.* **C23** *(1981) 136.*
10) Ch. Lagrange, O. Bersillon, and D.G. Madland, *Nucl. Sci. Eng.* **83** *(1983) 396.*
11) Y. Wang and J. Rapaport, *Nucl. Phys.* **A 517** *(1990) 301*, and references therein.

12) G.M. Honoré et al., *Phys. Rev.* **C33** *(1986) 1129.*

13) J. Rapaport, *Phys. Rep.* **87** *(1982) 25.*

14) C. Howell et al.,*Phys. ReV.* **C38** *(1988) 1552.*

15) D.M. Patterson, R.R. Doering, and A. Galonsky, Nucl. Phys. **A 263** (1976) 261 ; Ch. Lagrange and A. Lejeune, *Phys. Rev.* **C25** *(1982) 2278.*

16) J.P. Jeukenne, A. Lejeune and C. Mahaux, *Phys. Rev.* **C15** *(1977) 10.*

17) A.S. Meigooni et al., *Nucl. Phys.* **A 445** *(1985) 304* ;
M. Olsson et al., .*Nucl. Phys.* **A 496** *(1989) 505.*

18) M. Nagadi, Ph.D dissertation, *Duke University (1992).*

19) P.P. Guss et al., *Phys. Rev.* **C39** *(1989) 1461.*

20) J.P. Delaroche, in "*Neutron Physics and Nuclear Data for Reactors and Other Applied Pruposes*", Harwell, UK, 1978 ed. by C. Kousnetzoff (OECD) Nuclear Energy Agency, Paris, (1978) p.366.

21) S. Hama et al., *Phys. Rev.* **C41** *(1990) 2737* ; R. Kozack and D.G. Madland, *Phys. Rev.* **C39** *(1989) 1461* ; R. DE Swiniarski, D.L. Pham and J. Raynal, *Z. Phys.* **A 343** *(1992) 179.*

22) B.C. Clark et al., in "*Neutron-Nucleus Collisions A Probe of Nuclear Structure*", AIP conf. Proc. N° 124, ed. by J. Rapaport et al. (AIP, New-York, 1985), p. 123.

REALISTIC HAMILTONIAN FOR NUCLEON SCATTERING ANALYSIS ON NON-AXIAL DEFORMABLE EVEN NUCLEI

Yu.V. PORODZINSKIJ and E.Sh. SUKHOVITSKIJ
Radiation Physics and Chemistry Problems Institute
220109, Minsk-Sosny , Republic of Belarus

ABSTRACT

An approach enabling to describe spectroscopy data as well as scattering data is suggested. The collective nucleus motion is described with the phenomenologic model of soft β- and γ- deformable rotator, the surface hexadecapole oscillation modes are also accounted for. The optical potential parameters and quadrupole as well as hexadecapole parameters for ^{238}U target nucleus are obtained.

Coupled channel method is now the basic one for analysis of experimental data on nucleon-nucleus scattering . However, the schemes of strong coupling of channels used in calculations are constructed as a rule on the basis of simplest collective nuclear Hamiltonians, which do not describe spectroscopic properties of nuclei.

In this paper it is proposed to use the phenomenologic collective model of soft to quadrupole β- and γ- vibrations rotator with effective account of hexadecapole surface modes [1,2] for the analysis of experimental data on nucleon scattering on even-even nuclei in the frame of coupled channel method.

That permits the following: (1) to analyse spectroscopic properties of nuclei (for example,energy spectrum of rotational-vibrational states, probabilities of E2 and E4 transitions, etc) from the unified theoretical positions as well as the process of nucleon scattering on them; (2) there is a chance to investigate the influence of nuclear shape parameters (for example, deformability, non-axiality) on integral and differential characteristics of nucleon scattering process; (3) it is possible to get the necessary information on parameters of nucleus shape from the analysis of experimental data on nucleon scattering (e.g. on quadrupole and hexadecapole deformations).

On Fig.1 possibilities of such a phenomenologic model are demonstrated by description of energies of excited states of positive parity for ^{238}U. On the same figure parameters of this model are presented: energy multiplier $\hbar w_0$, coefficient of deformability for longitudinal μ_β and transverse μ_γ quadrupole vibrations, parameters of non-axiality of quadrupole γ and hexadecapole γ_4 and δ_4 deformations. Parameter α is expressed with the parameters of quadrupole and hexadecapole deformations and corresponding mass parameters:

$$\alpha = (\beta_4/\beta_{20})^2 \cdot (B_4/B_2) \quad . \quad (1)$$

Model parameters were obtained from the best description of the following band-heads : ground rotational band (n_β =0, n_γ =0, k=0, τ =1), anomalous rotational band (n_β =0, n_γ =0, k=2, τ =2), β - vibrotional band (n_β =1, n_γ =0, k=0, τ =1), γ - vibrational band (n_β =0, n_γ =1, k=0, τ =1).

The model describes ^{238}U scheme of levels quite reliably (four positive parity bands) up to excitation energy ~2 MeV. Energy dependence of rotational-vibrational states on hexadecapole deformation is defined by small parameter (β_4 / β_{20})^2 and because of that it is essential for the description of levels with spins exceeding 10h. But for the analysis of data on nucleon scattering the account of hexadecapole deformation is necessary because dependence of coupling channel matrix elements on hexadecapole deformation is of the same linear order of magnitude as for the quadrupole deformation.

Good agreement of model calculations with experimental spectroscopic data permits to expect the adequate description of nuclear properties with the collective Hamiltinian and to use the obtained wave functions for the calculation of coupling channel matrix elements in coupled channel method for the task of nucleon scattering by non-axial deformable nuclei.

Potential scattering of non-relativistic particle (direct interaction) on nucleus is described by Schröedinger equation with Hamiltonian

$$\hat{H} = \hat{T} + \hat{H}_t + V(\bar{r}) \qquad ,(2)$$

where $\hat{T}$ and $V(\bar{r})$ an operator of kinetic energy of incident nucleon and potential of nucleon-nucleus interaction, correspondingly; $\hat{H}_t$ - Hamiltonian of collective motion of nucleus-target. For the description of neutron scattering by nuclei nonspherical potential is used:

$$V(\bar{r}) = -V_R \cdot f_R(\bar{r}) - 4iW_D f_D(\bar{r})\left[1 - f_D(\bar{r})\right] - \left(\frac{\hbar}{\mu_\pi \cdot c}\right)^2 \cdot V_{SO}\,\frac{1}{a_{SO} \cdot r}\, f_{SO}(\bar{r})\left[1 - f_{SO}(\bar{r})\right](\hat{\sigma} \cdot \hat{l}) \qquad ,(3)$$

with form-factors

$$f_i(\bar{r}) = \left\{1 + \exp\left[(r - \bar{R}_i)/a_i\right]\right\}^{-1}, \qquad (4)$$

radius of which in the body-fixed coordinate system is defined by expression:

$$R_i(\theta,\varphi) = R_{0i}\left\{1 + \beta_2 \cos\gamma\, Y_{20}(\theta,\varphi) + \beta_2 \sqrt{2}/2 \sin\gamma \times \left[Y_{22}(\theta,\varphi) + Y_{2-2}(\theta,\varphi)\right] + \sum a_{4\mu} Y_{4\mu}(\theta,\varphi)\right\} \qquad ,(5)$$

where β_2 and γ (quadrupole deformation and non-axiality parameters, correspondingly) are considered as dynamic variables; values $a_{4\mu}$ are expressed from deformation and non-axiality of hexadecapole vibrations [3].

Deduction of the system of coupled equations for the case, when target nucleus is considered as soft to longitudinal and transverse quadrupole vibrations rotator is similar to that described in[4].It varies from collective models, used in [5,6] only by coupling channel matrix elements. That's why suggested phenomenologic collective model with five dynamic variables could be easily realized introducing its coupling channel matrix elements into available computational codes [5,6]. That permits to spread essentially the circle of nuclei, which could be described by method of strong coupling for correct analysis of experimental data on nucleon scattering.

Parameters of potential (3) were found from the best description of the set of experimental data on total cross section in the energy region of incident neutrons 0.25-14 MeV and available experimental data on angular distributions of scattered neutrons, beginning from 3.4 MeV. At lower energies cross section contains large contribution of compound nucleus mechanism, account of which is possible, but it gives large error, which permits to use these data only for the purpose of comparison. The coupling of four first levels is assumed in our calculations, because account of three levels does not saturate coupling force. Thus accounting of fourth level changes the values of predicted low-energy strength functions, potential scattering radii and total cross sections greater than are the values of experimental errors. The optimal parameters of neutron scattering potential, which fit experimental data are obtained.

The comparison of experimental and calculated total cross sections (Fig.2) as well as differential cross sections for neutron scattering (Fig. 3,4) on ^{238}U is given.

The results of considering target nucleus as rigid axial rotator are given on the figures (Fig.3,4) by dashed curves. The same optimization search scheme for the same set of experimental data was used for the parameters of neutron scattering potential.

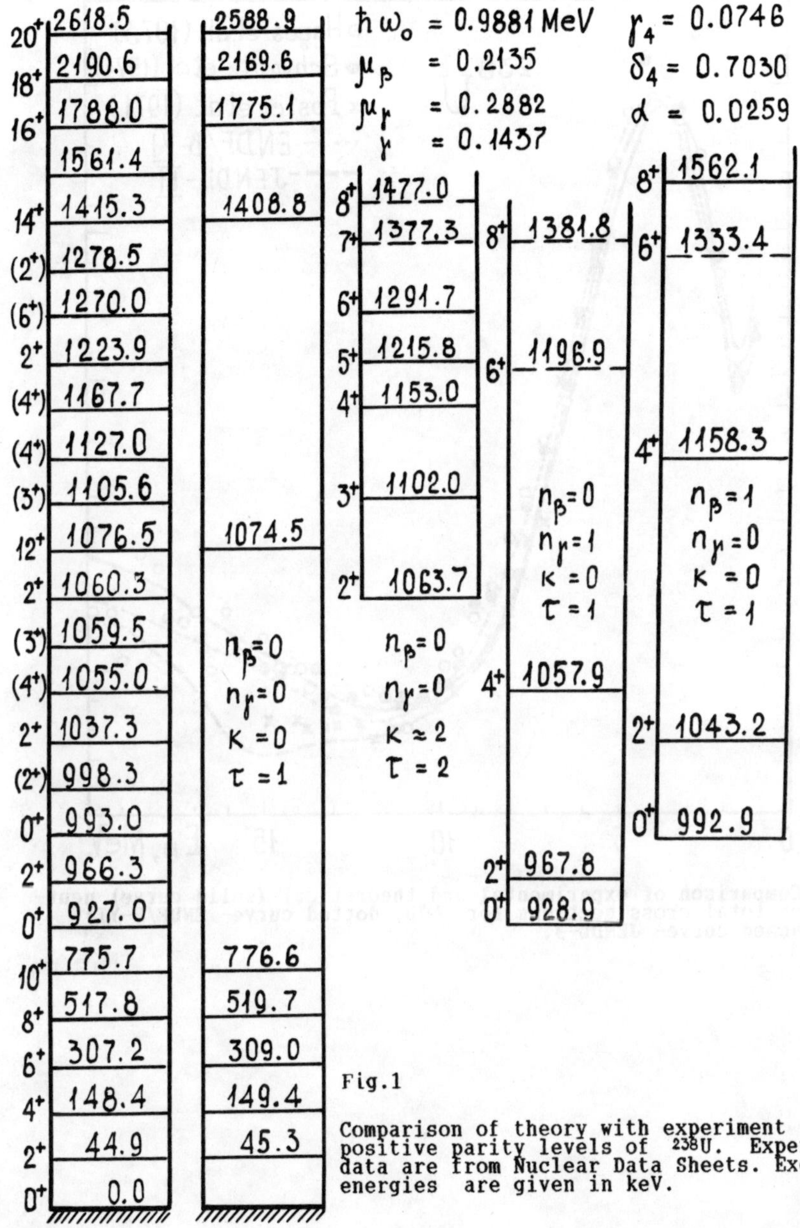

Fig.1

Comparison of theory with experiment for the positive parity levels of ^{238}U. Experimental data are from Nuclear Data Sheets. Excitation energies are given in keV.

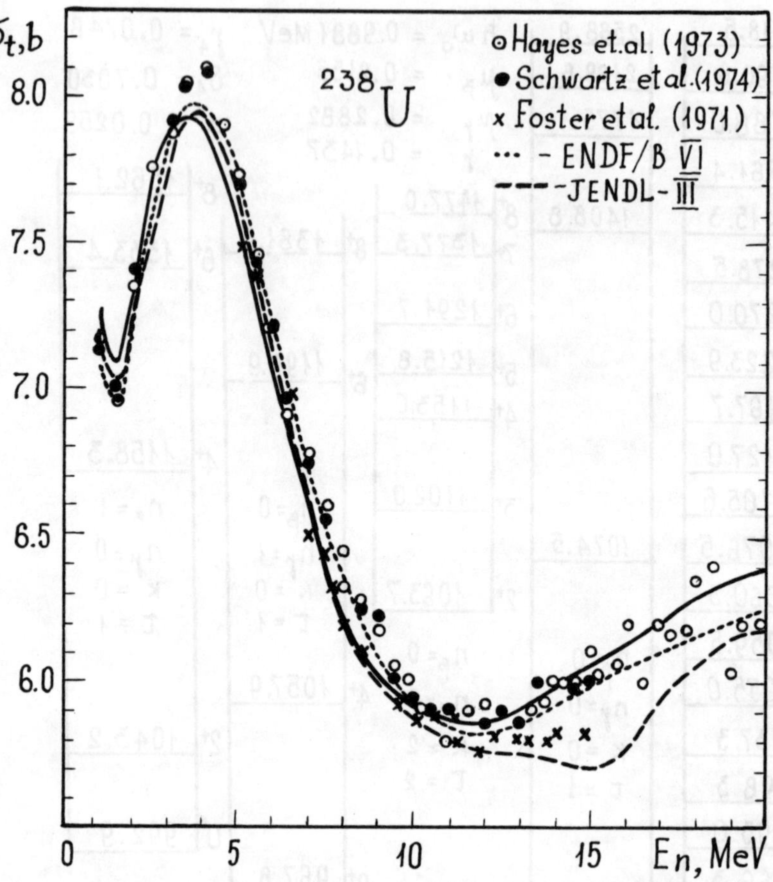

Fig.2 Comparison of experimental and theoretical (solid curve) neutron total cross sections for ^{238}U; dotted curve- ENDF/B-YI; dashed curve- JENDL-3.

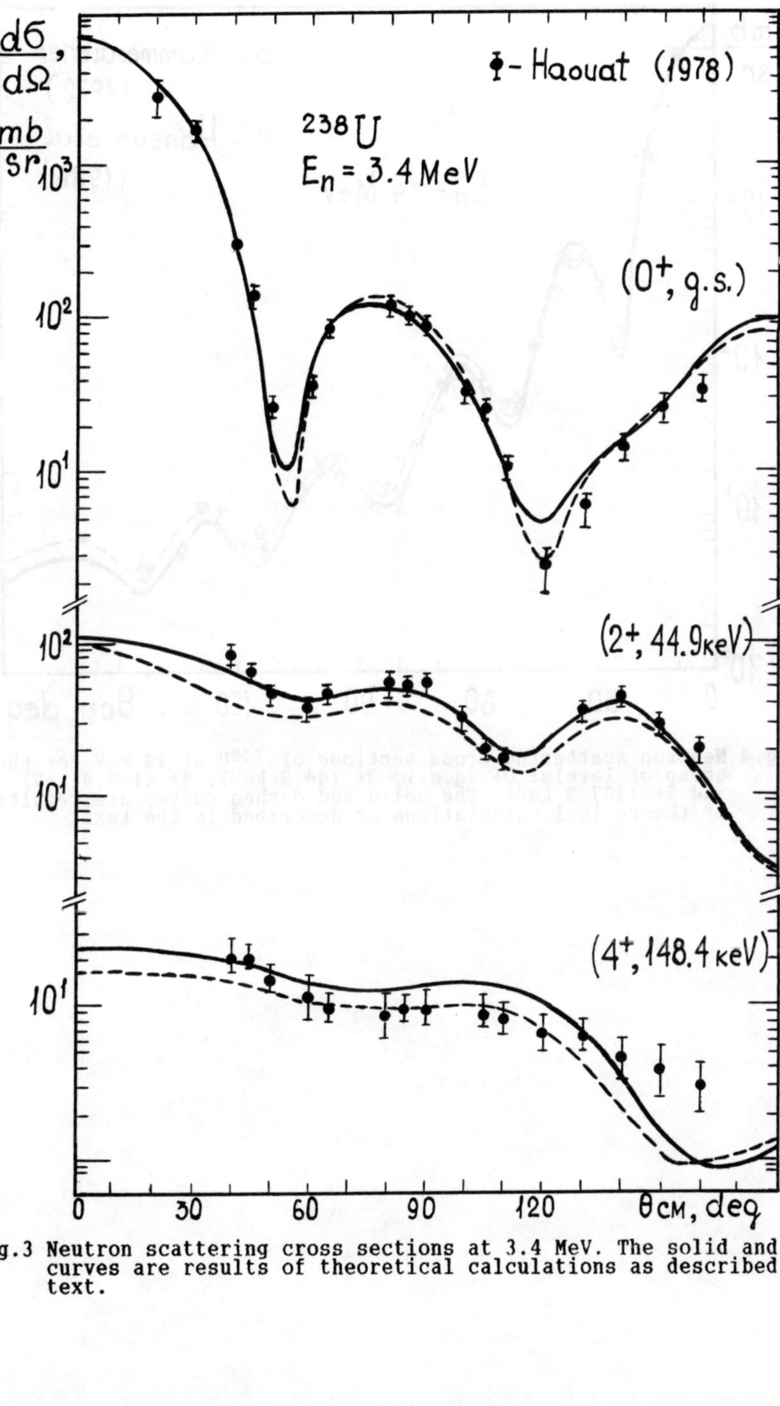

Fig.3 Neutron scattering cross sections at 3.4 MeV. The solid and dashed curves are results of theoretical calculations as described in the text.

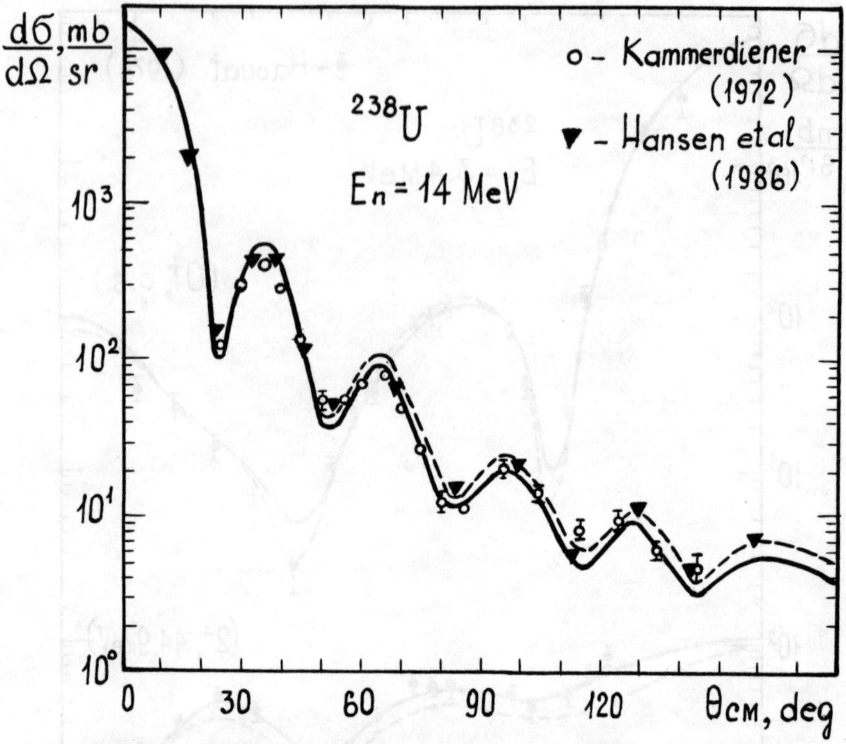

Fig.4 Neutron scattering cross sections of ^{238}U at 14 MeV for the group of levels: 0+ (g.s.), 2+ (44.9 keV), 4+ (148.4 keV) and 6+ (307.2 keV). The solid and dashed curves are results of theoretical calculations as described in the text.

Such evaluation of nucleus shape and interaction dynamics essentially improves the description of experimental data. The neutron scattering potential parameters used in the model calculations are given in Table 1.

Table 1. ^{238}U scattering potential parameters

Parameters	Rigid axial rotator	Soft non-axial rotator
V_R ,MeV	45.748-0.222xE	46.355-0.264xE
W_D ,MeV	3.477+0.333xE	2.952+0.420xE
V_{SO} ,MeV	4.963	4.838
r_R ,fm	1.257	1.245
a_R ,fm	0.654-0.006xE	0.713-0.011xE
r_D ,fm	1.258	1.249
a_D ,fm	0.448+0.016xE	0.364+0.024xE
r_{SO} ,fm	1.136	1.119
a_{SO} ,fm	0.470	0.350
	Deformation parameters	
	0.219	0.255
	0.057	0.034

The E2 and E4 transitions probabilities between low-lying rotational-vibrational states to an accuracy of second-order of magnitude were calculated using quadrupole and hexadecapole deformation parameters and formfactor radius of real potential given in Table 1.Target nucleus was considered as uniformly charged, non-axial deformable ellipse.

The comparison of the experimental and the calculated probabilities of such transitions are given in Table 2.

Table 2. The electromagnetic properties of the nucleus ^{238}U.

B(E2), $e^2 \cdot b^2$	Experiment		Theory
0+(0.0)- 2+(44.9)	11.7 ∓ 0.8	[13]	11.84
	11.70∓0.15	[14]	
6+(307.2)- 8+(517.8)	4.7 ∓ 0.6	[15]	5.52
8+(517.8)-10+(775.7)	5.2 ∓ 0.5	[15]	5.60
10+(775.7)-12+(1076.5)	5.1 ∓ 0.5	[15]	5.78
0+(0.0)- 2+(966.3)	0.017∓0.007	[16]	0.065
0+(0.0)- 2+(1060.3)	0.127∓0.009	[16]	1.01
	0.145∓0.012	[17]	
0+(0.0)- 2+(1037.3)	0.063∓0.009	[16]	0.25
	0.037∓0.015	[17]	
B(E4), $e^2 \cdot b^4$	Experiment		Theory
0+(0.0)- 4+(148.4)	0.69∓0.37	[18]	1.14
Q_{20} , e · b	Experiment		Theory
	11.12∓0.07	[18]	10.58
	10.5 ∓ 0.5	[19]	

The satisfactory agreement between experimental and calculated data is obtained. The experimental spectroscopic data and neutron scattering data may be self-consistently described in such manner.

When nucleus is supposed to be an axially symmetric rotator, there is a ~20% discrepancy in β_2 values obtained from scattering analysis and spectroscopic β_2 values. This basic inconsistency is largely removed under current approach due to non-axiality and deformability of nucleus shape. Moreover we can define the B_2/B_4 ratio. Substituting into (1) γ value derived by nuclear levels fit along with β_{20} and β_4 values obtained by scattering data fit we get $B_2/B_4 = 0.69$, in the irrotational flow model [20] $B_2/B_4 = 2$. Note also that our β_4 value is ~40% less than hexadecapole deformation parameter in case of axial-symmetric target nucleus shape. This fact is consistent with views stated above.

The diffuseness parameter of imaginary potential part grows at least twice with rising neutron energy from zero up to 14 MeV, that is consistent with the idea of heating of nucleus.

The next step in the model extension is octupole deformations incorporation, which will enable us either negative and positive parity states fitting. Then including this level into coupling scheme, we can study the door-way-states by the iterative method.

Unfortunately, the ^{238}U target nucleus is not the best example for our model application.

References

1. Yu.V.Porodzinskij, E.Sh.Sukovitskij Yadernaya Fysika 53 (1991) 64
2. Yu.V.Porodzinskij, E.Sh.Sukovitskij Yadernaya Fysika 55 (1992) 2368
3. S.G.Rohozinski, A.Sobiczewski Acta Phys. Pol. B12 (1981) 1001
4. Yu.V.Porodzinskij, E.Sh.Sukovitskij Yadernaya Fysika 54 (1991) 942
5. T.Tamura Rev.Mod.Phys. 37 (1965) 679
6. J.Raynal Computing as a Language of Physics Vienna, IAEA (1972)
7. S.H.Hayes, P.Stoler, J.M. Klement et al.Nucl. Sci.and Eng. 50 (1973) 243
8. R.B. Schwartz, R.A. Schrack, H.T. Heaton Nucl. Sci. and Eng. 54 (1974) 322
9. D.G. Foster, D.W. Glasgow Phys. Rev. 3 (1971) 576
10. G. Haouat, J. Lachkar, Ch. Lagrange et al. Rep. NEANDC(E) 196 "L"
11. J.L. Kammerdiener Rep. UCRL-51232 (1972)
12. L.F. Hansen, B.A. Pohl, C. Wong et al. Phys. Rev. C34 (1986) 2075
13. Th.W. Elze, J.R. Huizenga Nucl. Phys. A187 (1972) 545
14. F.K. McGowan, C.E. Bemis, J.L.C. Ford et al. Bull. Amer. Soc. 16 (1971) 1157
15. E. Grosse, A. Balanda, H. Emling et al. Phys. Scr. 24 (1981) 337
16. F.K. McGowan, C.E. Bemis, T.W. Milner et al. Phys.Rev. C10 (1974) 1146
17. J.G. Alessi, J.X. Saladin, C. Baktash, T. Humanic Phys. Rev. C23 (1981) 79
18. C.E. Bemis, F.K. McGowan, J.L.C. Ford et al. Phys. Tev. C8 (1973) 1466
19. G. Uefert, V. Metag, D. Habs, H.J. Specht Phys. Rev. Lett. 42 (1979) 1596
20. J.M. Eisenberg, W. Greiner Nuclear Theory V.1 Amsterdam-London, North-Holland Publishing (1970)

METHODOLOGY OF COUPLED-CHANNELS OPTICAL-MODEL ANALYSES: A STUDY OF n - ^{208}Pb SCATTERING AT LOW ENERGIES

P. G. YOUNG

Theoretical Division, Los Alamos National Laboratory, Los Alamos, NM 87545 USA

R. E. SHAMU

Western Michigan University, Kalamazoo, MI 49008 USA

We have performed a phenomenological optical model analysis of n-^{208}Pb data at neutron energies from 8.5 to 10 MeV using the coupled-channels code ECIS78. We demonstrate that central potential strengths obtained in a one-channel analysis of neutron total cross section data lead to good agreement with elastic and inelastic scattering $\sigma(\theta)$ data in coupled-channels calculations. The central potential geometry and spin-orbit potential are based on analyses of proton measurements. Excited states included in the calculations are the discrete collective states at 2.615 (3^-), 4.085 (2^+), 4.323 (4^+), 4.424 (6^+) and 4.610 (8^+) MeV; and the low-energy octupole resonance (LEOR) at 5.38 MeV (3^-). The β_L values used for the collective states are from adopted (p,p') values except for the 2.615-MeV state, which is determined from ^{208}Pb(n,n') data and is consistent with recent RPA calculations. The LEOR β_3 value was calculated from the β_3 of the 2.615-MeV state using measured energy-weighted sum rule fractions. The results of this analysis are an improved representation of the combined total, elastic $\sigma(\theta)$ and inelastic $\sigma(\theta)$ cross section data relative to spherical optical model analyses. The methodology of this analysis is described, and comparisons with spherical optical model analyses and experimental data are presented.

1. Introduction

Optical model analyses are usually an integral part of nuclear data evaluations. When Hauser-Feshbach statistical theory analyses are utilized in evaluations, optical model parameterizations are required for calculating differential elastic scattering cross sections, total cross sections, and particle transmission coefficients, which are then used to determine partial reaction cross sections. Because ^{208}Pb is doubly magic, one would expect that a conventional one-channel optical model with a spherically symmetric central potential would be especially appropriate for describing n-^{208}Pb scattering. However, results from recent one-channel phenomenological optical model analyses indicate that such a model might not be adequate for describing n-^{208}Pb scattering at low energies.[1,2] In particular, surface absorptive potential strengths determined from fits to elastic differential cross sections were found to be inconsistent with strengths deduced from total cross sections.[1] Because of the doubly magic nature of ^{208}Pb, these results are unexpected; nevertheless, they are in accord with structure calculations, several of which demonstrate that the low-lying collective states of ^{208}Pb have a significant effect on the properties of bound single particle states based on a ^{208}Pb core.[3]

The present work is a coupled-channels (CC) phenomenological optical model study of n-^{208}Pb scattering over the neutron energy range 8.5 to 10 MeV, where the influence of low-lying ^{208}Pb collective states is believed to be large.[1] Analyses over this energy range are expected to be particularly straightforward, because compound nucleus scattering, as well as contributions from higher-energy collective modes of excitation,[4] should be negligible at these energies. Results of the present CC calculations are compared with total cross sections between E_n = 8.5 and 10 MeV; with elastic scattering differential cross sections at 8.5 MeV,[5] 9.0 MeV,[6] 9.5 MeV,[1,7] and 10 MeV;[8] and with differential cross sections for inelastic scattering to the 2.615-MeV, 3^- state of ^{208}Pb at E_n = 8.5 MeV.[5] We find that a first-order vibrational model provides an excellent description of these data and demonstrate that the low-energy octupole resonance (LEOR) centered at 5.38 MeV[9] has a significant effect on n-^{208}Pb scattering at these incident energies.

2. Method of Analysis

The method of analysis was basically: (1) to adopt the optical model central potential resulting from the one-channel analysis of n-^{208}Pb total cross section data by Shamu et al.[1], and (2) to perform deformed optical model calculations progressively coupling in low-lying excited states known to be collective and comparing the results with differential scattering and total cross section data. Calculations were performed using the coupled-channels code ECIS78.[10] As in Ref. 1, the central potential geometry and spin-orbit potential used were proton-derived values,[11] and the central potential strengths over this energy range were deduced by means of one-channel calculations from neutron total cross-section data. The real potential strength employed, $V(E) = 50.0 - 0.28 E$, is identical to that of Ref. 1 and is believed to be accurate to about 1%. The surface absorptive potential strength used was $W_D(E) = 0.39 E - 0.50$. This strength was determined from neutron total cross section data from Western Michigan University[12] and Los Alamos National Laboratory[13] with the latter data multiplied by 1.0047 to account for a small normalization difference. The surface absorptive potential strength is believed to be accurate to about 10% over the energy range of interest. The volume absorptive potential strength was taken to be zero.

The collective model assumed for ^{208}Pb was a first-order vibrational model with complex coupling. Excited states included in these calculations were all those ^{208}Pb states below 10 MeV excitation energy known from various alpha-particle, proton and/or electron inelastic scattering experiments to be highly collective[4,14] as follows: discrete states at 2.615(3^-), 4.085(2^+), 4.323(4^+), 4.424(6^+), and 4.610(8^+) MeV; and the LEOR, centered at 5.38 MeV (3^-).[9] The β_L used for the discrete states were adopted proton values[15] except at 2.615 MeV, where β_3 = 0.115, a value in agreement with the ^{208}Pb(n,n') data at 8.5 MeV, was employed.* The LEOR β_3 was computed to be 0.10 using the 2.615-MeV state value and measured values of the energy-weighted sum rule (EWSR) fractions.[9,16] All collective excitations were assumed to be one-phonon states. The spin-orbit potential was not deformed.

3. Results

An overview of our results is given in Table I, where values of the reduced χ^2, χ^2_r, are listed based on comparisons with measured total cross sections (σ_t) and elastic differential cross sections [$\sigma_{el}(\theta)$]. For the $\sigma_{el}(\theta)$ data, χ^2_r are based on reported *relative* cross-section uncertainties except at 9.5 MeV,[7] where only total uncertainties were reported. For σ_t, experimental values were interpolated by means of a least-squares fit to data over the energy range 8 to 14 MeV.[2] Values of χ^2 were computed assuming a σ_t uncertainty of 1%.

---

*Agreement with β_3 = 0.115 was somewhat better than with the proton-deduced DWBA value[9], β_3 = 0.118.

TABLE I

Values of χ^2_r for various coupled-channels (CC) calculations of n-^{208}Pb scattering. For the 3$^-$ state at 2.615 MeV, $\beta_3 = 0.115$. Values in the one-channel elastic-scattering best-fit column were obtained from Ref. 1.

Energy (MeV)	Data Type	CC [All $\beta_L \equiv 0$]	CC [$3^-,2^+ \rightarrow 8^+$]	CC [$3^-,2^+ \rightarrow 8^+$, LEOR$_1$]	CC [$3^-,2^+ \rightarrow 8^+$, LEOR$_3$]	One-Chan. $\sigma_{el}(\theta)$ fit
8.5	$\sigma_{el}(\theta)$	34	6.7	3.2	3.1	2.4
	σ_t	0.0	0.4	1.8	1.1	8.4
9.0	$\sigma_{el}(\theta)$	265	58	20	20	9.0
	σ_t	0.0	0.2	0.9	0.8	7.5
9.5	$\sigma_{el}(\theta)$	62	17	6.7	7.0	3.3
	σ_t	0.0	0.1	0.5	0.3	9.2
10.0	$\sigma_{el}(\theta)$	200	42	14	15	7.3
	σ_t	0.0	0.0	0.0	0.1	11

The successive columns in Table I illustrate the improved agreement with data that occurs as the one-channel calculations (coupled-channel with all $\beta_L = 0$)$^+$ are replaced by coupled-channel results that include coupling of the ground state to collective states. Note that the label "LEOR$_3$" in Table I indicates that the LEOR near $E_x = 5.38$ MeV was represented by three fragmented states, appropriately weighted over the LEOR spreading width. Representing the LEOR by a single state at $E_x = 5.38$ MeV (LEOR$_1$ in Table I) gives results essentially equivalent to the three-state approximation. The last column in Table I gives χ^2_r values from the one-channel best fits to $\sigma_{el}(\theta)$ data from Ref. 1.

The calculated elastic scattering differential cross sections are compared directly with the measurements[5-8] between 8.5 and 10 MeV in Fig. 1. Similarly, the inelastic scattering differential cross sections to the lowest 3$^-$ level of ^{208}Pb from the coupled-channel calculations are compared with the $E_n = 8.5$ MeV data[5] in Fig. 2. The χ^2_r results in Table 1 and the cross sections in Fig. 1 show that the calculations using the potential of Ref. 1 with no collective level coupling provide a very poor fit to the elastic angular distribution data, particularly at back angles. However, Table 1 confirms that these $\beta_L \equiv 0$ calculations do reproduce experimental σ_t, which should be the case because in Ref. 1 $W_D(E)$ was deduced from a one-channel analysis of experimental σ_t data. Conversely, as shown in the far right column of Table 1, one-channel calculations which fit elastic data provide a poor fit to σ_t data. Only one parameter, W_D, differs significantly for these two

$^+$ Coupled-channels calculations performed with ECIS with zero coupling, *i.e.*, all $\beta_L \equiv 0$, are equivalent to one-channel calculations.

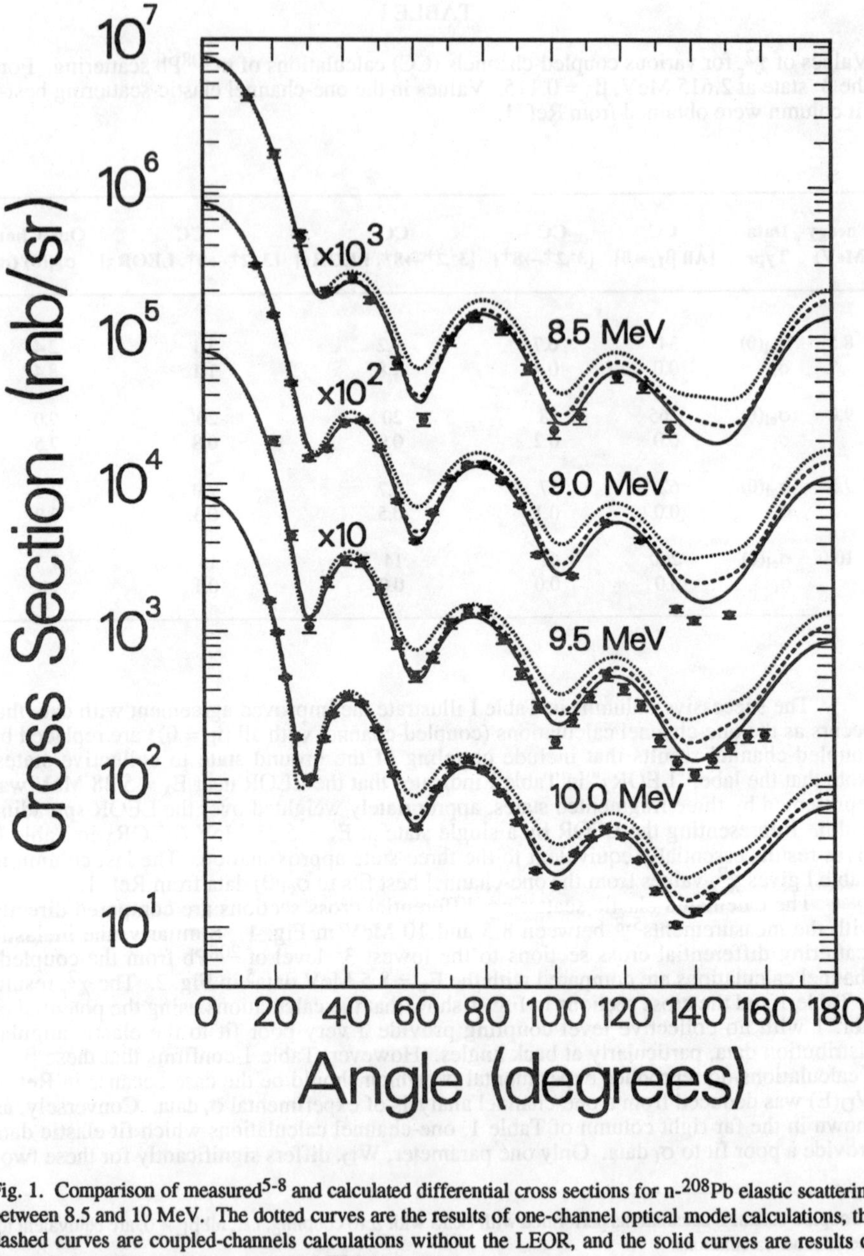

Fig. 1. Comparison of measured[5-8] and calculated differential cross sections for n-^{208}Pb elastic scattering between 8.5 and 10 MeV. The dotted curves are the results of one-channel optical model calculations, the dashed curves are coupled-channels calculations without the LEOR, and the solid curves are results of coupled-channels calculations with the LEOR included.

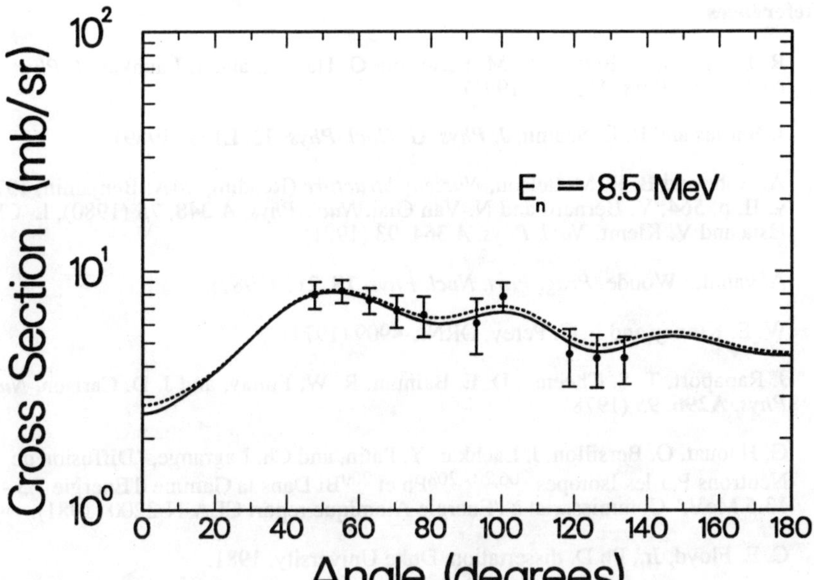

Fig. 2. Experimental[5] and calculated inelastic neutron differential scattering cross section to the 3⁻ first excited state of ^{208}Pb. The solid and dashed curves are coupled-channels optical model calculations with and without the LEOR included.

sets of calculations. The agreement with the elastic angular distributions using the potential based on σ_t data is significantly enhanced in the coupled-channels calculations, and coupling the LEOR state into the calculations further improves the results. The coupled-channels calculations are seen to reproduce the $\sigma_{el}(\theta)$ data almost as well as the *best fit* one-channel calculations and at the same time closely agree with the σ_t data.

4. Conclusions

We find that coupled-channels optical-model calculations using a first-order vibrational model with complex coupling are in good agreement with n-^{208}Pb total cross section data and elastic differential cross section data over the energy range 8.5 to 10 MeV provided all low-lying collective states of ^{208}Pb are taken into account. In particular, we find that the low energy octupole resonance centered at 5.38 MeV has a significant effect on elastic scattering at back angles and must be included in the calculation. Central potential strengths for these calculations were deduced by means of one-channel analyses of neutron total cross section data. The good agreement observed in the present work between calculated values and elastic, inelastic, and total cross section data supports the validity of this procedure.

5. References

1. R. E. Shamu, J. Barnes, S. M. Ferguson, G. Haouat, and J. Lachkar, *J. Phys. G: Nucl. Part. Phys.* **17**, 525 (1991).

2. J. Barnes and R. E. Shamu, *J. Phys. G: Nucl. Phys.* **15**, L179 (1989).

3. A. Bohr and B. R. Mottelson, *Nuclear Structure* (Reading, MA: Benjamin, 1975) v. II, p. 564; V. Bernard and N. Van Giai, *Nucl. Phys. A* **348**, 75 (1980); L. Chu-Hsia and V. Klemt, *Nucl. Phys. A* **364**, 93 (1981).

4. A. van der Woude, *Prog. Part. Nucl. Phys.* **18**, 217 (1987).

5. W. E. Kinney and F. G. Perey, ORNL-4909 (1974).

6. J. Rapaport, T. S. Cheema, D. E. Bainum, R. W. Finlay, and J. D. Carlson, *Nucl. Phys.* **A296**, 95 (1978).

7. G. Haouat, O. Bersillon, J. Lachkar, Y. Patin, and Ch. Lagrange, "Diffusion de Neutrons Par les Isotopes ^{206,207,208}Pb et ^{209}Bi Dans la Gamme d'Energie 7.5 - 13.5 MeV," Commissariat à l'Energie Atomique report CEA-N-2200 (1981).

8. C. E. Floyd, Jr., Ph.D. dissertation, Duke University, 1981.

9. Y. Fujita, M. Fujiwara, S. Morinobu, I. Katayama, T. Yamazaki, T. Itahashi, H. Ikegami, and S. I. Hayakawa, *Phys. Rev. C* **32**, 425 (1985).

10. J. Raynal, "Optical Model and Coupled-Channel Calculations in Nuclear Physics," International Atomic Energy Agency report IAEA SMR-9/8 (1972) p. 281.

11. W. van Oers, Huang Haw, N. E. Davison, A. Ingemarsson, B. Fagerström, and G. Tibell, *Phys. Rev. C* **10**, 307 (1974).

12. S. M. Ferguson and R. E. Shamu, *Bull. Am. Phys. Soc.* **21**, 985 (1976); L. J. Satkowiak, S. M. Ferguson, R. E. Shamu, and M. Soga, *Phys. Lett.* **175B**, 266 (1986).

13. P.W. Lisowski, G.F. Auchampaugh, M.S. Moore, G.L. Morgan, and R.E. Shamu, "Neutron Cross Section Measurements at WNR," Proc. Sym. on *Neutron Cross Sections from 10 to 50 MeV*, 12-14 May 1980, Brookhaven National Laboratory, BNL-NCS-51245 (1980) p. 301; R.L. Schutt, R.E. Shamu, P.W. Lisowski, M.S. Moore, and G.L. Morgan, *Phys. Lett. B* **203**, 22 (1988).

14. J. Heisenberg, J. Lichtenstadt, C. N. Papanicolas, and J.S. McCarthy, *Phys. Rev. C* **25**, 2292 (1982).

15. M. J. Martin, *Nucl. Data Sheets* **47**, 797 (1986).

16. G. R. Satchler, *Nucl. Phys.* **A195**, 1 (1972).

Equilibrium and Pre-Equilibrium Models for Calculation of Particle Emission Cross-Sections and Spectra

C. Y. Fu
Oak Ridge National Laboratory, Oak Ridge, TN 37831, USA

ABSTRACT

Improvements of equilibrium and pre-equilibrium models that have been applied to the evaluations of the recently released nuclear data libraries are reviewed. The improvements include energy-angle correlated neutron emission in the continuum, charged-particle and recoil spectra, widths-fluctuation correction for the continuum, pre-equilibrium capture, and gamma-ray strength functions. Their impacts on evaluated data are summarized.

1. Introduction

The equilibrium and pre-equilibrium nuclear reaction models had been in common use for cross section evaluations before the last symposium on nuclear data evaluation methodology[1]. Therefore, emphasis of this review is on the recent advances in these models that have been applied to the evaluation of nuclear data libraries such as EFF-2, ENDF/B-VI, and JENDL-3.

In data evaluations of general-purpose files, for which the upper incident neutron energy is 20 MeV, the equilibrium model is always of the Hauser-Feshbach (H-F) type because it is the only one that can satisfactorily account for the effects of the discrete energy levels that are present in all reaction channels. The modern H-F model includes widths-fluctuation corrections, gamma-ray strength functions, the double-humped fission barrier, and corrections for the pre-equilibrium reaction effects.

The theories of pre-equilibrium nuclear reactions, particularly those for prediction of double differential particle emission spectra, have improved rapidly since the release of ENDF/B-V in 1979. The fact that there are still many theories being used today indicates that diffculties and obscurities remain that require continued active studies. A recent (1992) book by Gadioli and Hodgson[2] details thesc theories and their respective problems. While most of these theories have their special advantages and may eventually be used for data evaluations, only those already utilized are emphasized in this paper. Some topics of an empirical nature but very useful for applications, not covered in the book, can be found in the proceedings of an NEANDC meeting[3] on pre-equilibrium reactions.

The impacts of the major theoretical improvements on recently evaluated data are summarized in the following sections.

2. Energy-Angle Correlated Neutron Emission in the Continuum

Many evaluated files in EFF-2, ENDF/B-VI, and JENDL-3 contain energy-angle correlated neutron emission spectra in the continuum for the first time. The hard part of a neutron emission spectrum is always much more forward peaked than the soft part, hence a number of angular distributions are required for each energy distribution. This fact, particularly important for shielding problems and fusion reactor design, can now be efficiently represented in the file using the new ENDF-6 formats[4].

Measured data for energy-angle correlated neutron emission are available almost exclusively for 14-MeV incident neutrons. Therefore, model codes are indispensable for fitting the available data and then used for generating the data for the incident energies from a few MeV to 20 MeV.

The theories and methods recently developed for the evaluation of double differential cross sections are semi-empirical, but they seem to serve the purpose well. One theory is the Kalbach-Mann systematics[5]. Kalbach and Mann fitted a large body of data with equilibrium and pre-equilibrium models for a wide incident energy range and a variety of incident and exit particle types by two sets of global constants, one for the equilibrium part and the other for the pre-equilibrium part. This method and its constants have been included as an option in the ENDF-6 formats. All an evaluator has to do is to determine the angle-integrated spectra, give the calculated ratios for the two parts of cross sections, and flag in the evaluated file for this option. This method has been used for some evaluations in ENDF/B-VI (^{235,238}U, ^{239}Pu) using the GNASH code[6], EFF-2 (Cr, Fe, and Ni isotopes) using MAURINA[7], and JENDL-3 Fusion File using GNASH and TNG[8].

The other method commonly used for evaluations is a unified H-F and exciton model[9]. The exciton part of this model conserves angular momentum in the same manner as the H-F theory. A forward-peaked angular distribution at high incident and outgoing energies results from partial-wave interference in the exciton part. It was found[9] that, if the incident neutron energy is not too high (less than 30 MeV), only the initial exciton states and the near-diagonal elements in the reaction matrix need to be kept for the interference. A correlation coefficient of 0.5 for these elements was found to work well for all targets considered and for incident neutron energies of 14 and 27 MeV. For all other off-diagonal elements, the random phase assumption of the H-F theory was retained. For low incident or outgoing energies for which the exciton component is small, the angular distribution approaches the usual front-back symmetry in the center-of-mass system as given by the H-F theory. This method was available in the TNG code[8] which was used for the following 21 ENDF/B-VI evaluations: ^{19}F, ^{28,29,30}Si, ^{50,52,53,54}Cr, ^{55}Mn, ^{54,56,67,58}Fe, ^{58,60,61}Ni, ^{63,65}Cu, and ^{206,207,208}Pb. The calculated results used for the ENDF/B-VI ^{208}Pb evaluation are compared with measured data[10–12] for natural Pb in Fig. 1.

A comparison with experimental data of calculated double differential neutron emission cross sections and spectra using the two methods for 27-MeV incident neutron on ^{184}W has been presented by Vonach[13]. The two methods agreed well in this case. The TNG results show rising cross sections in the back angles above 120 degrees, a consequence of angular momentum effects not accounted for by the Kalbach-Mann systematics.

3. Charged-Particle and Recoil Spectra

Similar to the neutron emission spectra, proton and alpha-particle emission spectra have been measured mostly at an incident neutron energy of 14 MeV. Equilibrium and pre-equilibrium models have again been used to extrapolate the available experimental data to other incident neutron energies. For charged particles, angular distributions are not important and hence not given in the evaluated files.

The recoil spectra for each reaction of the materials having charged-particle spectra are given for the first time. If gamma-ray production cross sections and spectra are also given separately for a given reaction, then the energy balance for that reaction can be easily achieved to any required accuracy if all of the evaluated particle spectra are generated consistently by nuclear model calculations.

Consider a reaction that emits a secondary neutron and a charged particle such as (n, np) as an example. Emission spectra should be given for each incident neutron energy for the outgoing neutrons, protons, gamma rays, and the heavy recoils. The sum of their average energies must be equal to the incident energy. For the 21 ENDF/B-VI materials listed above evaluated with the aid of TNG, the average outgoing energy for each reaction has been checked to agree with the incident energy within 1%. Note that the continuum inelastic scattering cross section is separated from the discrete ones in these files. The energy balance for exciting each discrete inelastic level can be perfectly met if the associated gamma rays are represented by a transition probability table. Then the energy balance for the inelastic continuum (MT=91) can be checked only if the gamma rays originating from the continuum are given separately. This separation can be achieved only by model calculations.

While the energy balance requirement is cumbersome for evaluators, the requirement is a necessary and stringent test for preventing large errors. Such errors were sometimes quite serious in ENDF/B-V, as shown by MacFarlane[14]. Note first that since the charged-particle and gamma-ray spectra were not given separately for each reaction in ENDF/B-V, the energy-balance checks could be made only for the sum of all reactions. For example, the difference between the incident and the outgoing energies above the Pb(n,2n) threshold in ENDF/B-V varied from -30% to +200%. It has been determined that two factors contribute to these large errors: (1) the reaction Q-value for an element having more than one isotope is always wrong, seriously so if two or more isotopes of the element have comparable natural abundances , and (2) the continuum bin widths for energy distributions were too wide just above the (n,2n) threshold to represent the average energies accurately because the (n,2n) cross section increases rapidly just above threshold. The first problem can be solved by isotopic evaluations and the second by a careful choice of energy bin widths as a function of both incident and outgoing energies. However, a serious problem remains: one cannot adjust the calculated cross sections or replace them by those from least-squares analysis without upsetting the energy balance. This problem can be solved only by balancing the energy within each reaction. This solution is one of the many reasons why equilibrium and pre-equilibrium calculations are becoming more and more important for data evaluations.

4. Widths-Fluctuation Correction for the Continuum

The H-F model does not take into account the correlation between the incident and outgoing waves in the elastic channel. The methods for its remedy, called the widths-fluctuation correction (WFC), have been reviewed and compared by Gruppelaar and Reffo[15] and more recently by Gadioli and Hodgson[2]. In these reviews, the WFC was considered only for the discrete energy region but not for the continuum. This omission results from the observation that the compound elastic cross section falls rapidly with increasing energy, becoming negligible at the threshold energy of the continuum. However, in evaluation work it frequently occurred that discrete energy levels are not experimentally known to an excitation energy high enough to neglect the WFC, resulting in a discontinuity in the calculated cross sections just above the continuum threshold. In some of the earlier evaluations, an evaluator had to smooth out such discontinuities before using the calculated results as evaluated data. A method for the WFC above the continuum threshold was not available. Two methods for including the continuum in the WFC formula of Moldauer[16] (the one commonly referred to as the classical integration formula) became available around 1980.

Reffo and Fabbri[17] suggested a channel lumping method to include the continuum bins in the same manner as the discrete levels in the WFC formula. The inelastic cross section of ^{92}Zr calculated with 18 discrete levels is shown in Fig. 2, taken from Reffo and Fabbri. The highest 14 levels were then replaced by a level-density continuum and, neglecting the WFC for the continuum, a large discontinuity at the continuum threshold shows. This abrupt change in cross section was removed if the continuum energy bins were included in the WFC formula. While this is an exagerated example because only four discrete levels were used, a small discontinuity for ^{56}Fe at the continuum threshold of 4.5 MeV with 26 discrete levels was noticed in calculations done for the earlier ENDF/B evaluations. This method has been adopted in the MAURINA code[7] used for the Cr, Fe, and Ni isotopic evaluations in EFF-2.

Another method to include the continuum in the WFC formula of Moldauer was given by Su et al.[18]. They assumed that the number of degrees of freedom in each of the continuum bins is large enough to be considered as infinity. This assumption allows the continuum transmission coefficients to enter the WFC formula in the same manner as the capture and fission transmission coefficients. While the assumption appears somewhat drastic, their example for $n + ^{235}U$ showed that the calculated compound elastic, nonelastic, inelastic, fission, and capture cross sections are self consistent and smooth varying across the inelastic continuum threshold. This method was adopted in the TNG code and was used for the 21 isotopic evaluations for ENDF/B-VI mentioned above. Some of the residual nuclides involved in the calculations for these evaluations have known levels up to rather low energies, but the evaluators (using TNG) no longer needed to worry about any discontinuities in their calculated cross sections.

5. Pre-Equilibrium Capture

A gamma ray may be emitted during equilibration in two ways: (1) an excited nucleon or hole makes a transition to a lower state, preserving the exciton number; (2) an excited nucleon annihilates a hole, reducing the exciton number by 2. As in the case for particle emission, the gamma-ray decay rates can be obtained by using the detailed balance principle. The pre-equilibrium gamma-ray emission is the dominant process in neutron capture cross sections for incident energies above a few MeV. The formula derived by Akkermans and Gruppelaar[19] has been adopted in the MAURINA and TNG codes and applied, respectively, to several EFF-2 and ENDF/B-VI evaluations. A TNG calculation for the gamma-ray production spectrum from 14-MeV neutrons on ^{93}Nb is shown in Fig. 3, taken from Shibata and Fu[20], and compared with the data and the direct-collective model calculation of Rigaud et al[21]. The calculation by Shibata and Fu did not require a normalization to fit the data. The good agreement with data, however, is probably accidental because later calculations by this author for other elements needed a scale constant to fit measured data.

6. Gamma-Ray Strength Functions

Gamma-ray strength functions are indispensable in H-F codes because they are present in all parts of the calculation. The most dominant part of the gamma-ray strength functions, that of the electric dipole (E1), is commonly modeled after a giant resonance represented by a Lorentzian shape for the E_γ dependence. Data measured for (γ,n) around the resonance peak, generally above 14 MeV, were used to determine or to globalize the parameters of the Lorentzian. While these parameters may represent the peak areas of the resonance well, the tail of the resonance below 7 MeV is highly uncertain in both absolute magnitude and shape. Unfortunately it is this tail that is the most important for applications. Two refinements, essentially for this low E_γ region, have been proposed.

Gardner and Dietrich[22] suggested replacing the Lorentzian by two Breit-Wigner resonances with E_γ-dependent widths. The separation between the peaks of the two resonances depends on deformation of the nucleus. The dependence of widths on E_γ was borrowed from the pre-equilibrium model for nucleons. Sucess of this model has been demonstrated for several technological applications by Gardner at al.[23,24].

The other improved model is the generalized Lorentzian (GLO) proposed by Kopecky and Chrien[25]. This model features (1) E_γ-dependent widths for the Lorentzian that depresses the low energy tail of the resonance, and (2) an added term that prevents the the strength function from vanishing as E_γ approaches zero.

These are the main features of the microscopic model developed by Kadmenskij et al.[26]. More detailed studies can be found in Kopecky and Uhl[27] and Uhl and Kopecky[28]. This model is being used for improving the European Activation File. If the added term is removed from the GLO, then a model called the energy-dependent Lorentzian (ELO) results. Figure 4, taken from Uhl and Kopecky, compares the GLO, ELO, and SLO (the standard Lorentian) E1 strength functions with data (see ref. 28 and references therein for details about the data).

7. Conclusions

This review has attempted to make clear that the investment in the development of nuclear model theories and codes in the early eighties has paid off soundly. It has been pointed out that there are several features in the current version of data libraries that evaluators for earlier versions simply did not have the tools to work with.

More recent developments in nuclear model theories and codes that did not impact on the EFF-2, ENDF/B-VI, or JENDL-3 evaluations are not covered in this report.

8. Acknowlegment

Research sponsored by the Office of Energy Research, Division of Nuclear Physics, U. S. Department of Energy, under contract DE-AC05-84OR21400 with Martin Marietta Energy Systems, Inc.

9. References

1. E. D. Arthur, "Calculational Methods Used to Obtain Evaluated Data Above 3 MeV," p.655, *Proc. Conf. Nuclear Data Evaluation Methods and Procedures*, eds. B. A. Magurno and S. Pearlstein, Brookhaven National Laboratory, Upton, New York, September 22-25, 1980, BNL-NCS-51563 (March 1981).
2. E. Gadioli and P. E. Hodgson, *Pre-Equilibrium Nuclear Reactions* (Clarendon Press, Oxford, 1992).
3. *Proc. Specialists' Meeting on Preequilibrium Nuclear Reactions*, ed. B. Stromaier, Semmering, Austria, February 10-12, 1988, NEANDC-245 "U".
4. *ENDF-102 Data Formats and Procedures for the Evaluated Nuclear Data File ENDF-6*, eds. P. L. Rose and C. L. Dunford, Brookhaven National Laboratory Report BNL-NCS-44945 (1990).
5. C. Kalbach and F. M. Mann, *Phys. Rev.* **C23** (1981) 112; C. Kalbach, *Phys. Rev.* **C37** (1988) 2350.
6. P. G. Young and E. D. Arthur, "GNASH: A Preequilibrium Model Code for Calculation of Cross Sections and Emission Spectra," Los Alamos National Laboratory Report LA-6947 (1977); E. D. Arthur, LA-UR-88-382 (1988).
7. M. Uhl, Institute of Radiation Research and Nuclear Physics, Vienna, Austria, Computer Code MAURINA, to be published.

8. C. Y. Fu, "A Consistent Nuclear Model for Compound and Precompound Reactions with Conservation of Angular Momentum," ORNL/TM-7042, Oak Ridge National Laboratory (1980); K. Shibata and C. Y. Fu, "Recent Improvements of the TNG Statistical Model Code," ORNL/TM-10093, Oak Ridge National Laboratory (1986).
9. C. Y. Fu, *Nucl. Sci. Eng.* **100** (1988) 61.
10. D. Hermsdorf et al., "Differential Neutron Emission Cross Sections by 14.6-MeV Neutrons," ZfK-277, Institute of Nuclear Physics, University of Dresden, Germany (1975).
11. J. L. Kammerdiener, "Neutron Spectra Emitted by Pb Irradiated by 14-MeV Neutrons," UCRL-51232, Lawrence Livermore Laboratory (1972).
12. A. Takahashi et al., "Double Differential Neutron Emission Cross Sections, Numerical Tables and Figures (1983)," Report No. A-83-01, Intense Neutron Source Facility, Osaka University, 1983.
13. H. Vonach, "Results of Recent Code Comparisons," p. 265, *Proc. Specialists' Meeting on Preequilibirum Nuclear Reactions*, ed. B. Strohmaier, Semmering, Austria, February 10-12, 1988, NEANDC-245 "U".
14. R. E. MacFarlane, "Energy Balance of ENDF/B-V," *Trans. Am. Nucl. Soc.* **33** (1979) 681.
15. H. Gruppelaar and G. Reffo, *Nucl. Sci. Eng.* **62** (1977) 756.
16. P. A. Moldauer, *Phys. Rev.* **135** (1964) 642.
17. G. Reffo and F. Fabbri, *Nucl. Sci. Eng.* **66** (1978) 251.
18. Su Zong-di, Shi Xiang-jun, Tang Xue-tian, and Tian Ye, *Phys. Energ. Fortis Phys. Nucl.* **3** (1979) 80. English translation in *Chin. Phys.* **1** (1981) 953.
19. J. M. Akkermans and H. Gruppelaar, *Phys. Lett.* **157B** (1985) 95.
20. K. Shibata and C. Y. Fu, "Recent Improvements of the TNG Statistical Model Code," ORNL/TM-10093, Oak Ridge National Laboratory (1986).
21. F. Rigaud, J. L. Irigaray, G. Y. Petit, G. Longo, and F. Saporetti, *Nucl. Phys.* **A173** (1971) 551.
22. D. G. Gardner and F. S. Dietrich, "A New Parametrization of the E1 Gamma-Ray Strength Function," p. 770, *Proc. Nuclear Cross Sections for Technology*, eds. J. L. Fowler, C. H. Johnson, and C. D. Bowman, Knoxville, Tennessee, USA, 22-26 Oct. 1979, NBS Special Publication 594 (1980).
23. D. G. Gardner, "Methods for Calculating Neutron Capture Cross Sections and Gamma-Ray Spectra," p. 62, *Neutron Radiative Capture*, ed. R. E. Chrien (Pergamon Press, Oxford, 1984).
24. D. G. Gardner and M. A. Gardner, "Some Neutron and Photon Reactions on the Ground and Isomeric States of ^{236,237}Np," p. 217, *Proc. Specialists' Meeting on Neutron Activation Cross Sections for Fission and Fusion Energy Applications*, eds. M. Wagner and H. Vonach, Argonne National Laboratory, USA, 13-15 Sep. 1989, NEANDC-259 "U".
25. J. Kopecky and R. E. Chrien, *Nucl. Phys.* **A468** (1987) 285.
26. S. G. Kadmenskij, V. P. Markushev, and V. I. Furmann, *Sov. J. Nucl. Phys.* **37** (1983) 165.

27. J. Kopecky and M. Uhl, *Phys. Rev.* **C41** (1990) 1941.
28. M. Uhl and J. Kopecky, "The Sensitivity of Statistical Model Calculations to Model Assumptions," p.977, *Proc. Nuclear Data for Science and Technology*, ed. S. M. Qaim, Jülich, Germany, 13-17 May 1991 (Springer Verlag, Berlin Heidelberg, 1992).

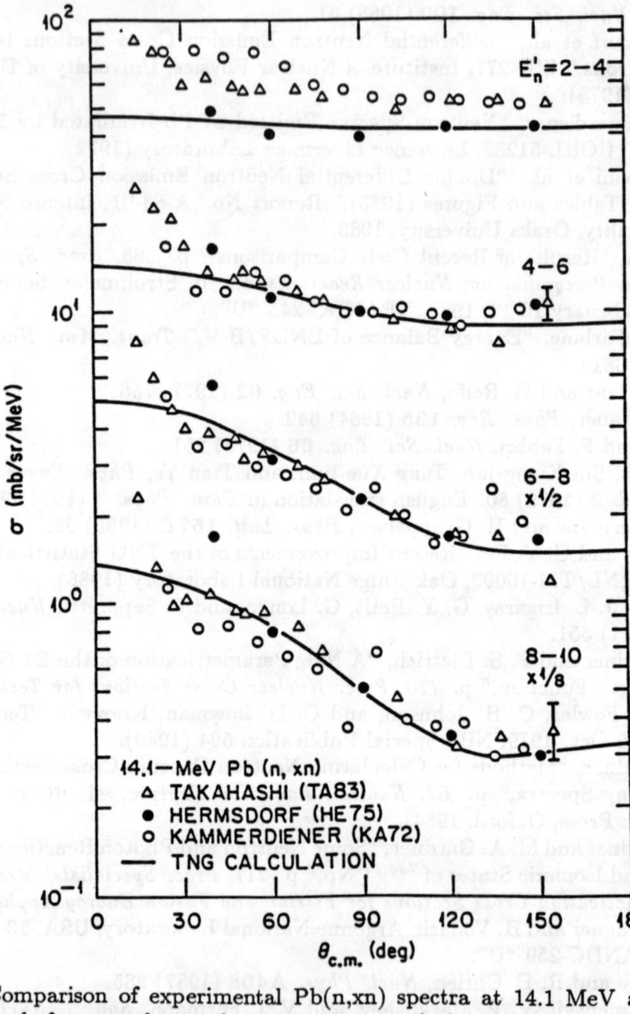

Fig. 1. Comparison of experimental Pb(n,xn) spectra at 14.1 MeV as a function of angle for several secondary energy ranges (refs. 10-12) with results of the TNG calculation[3] used for ENDF/B-VI.

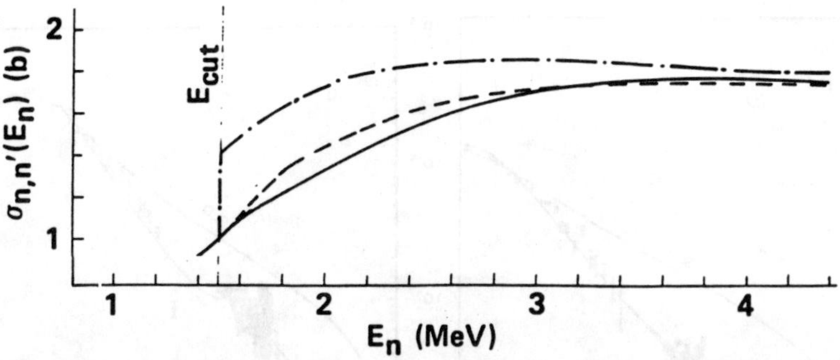

Fig. 2. Calculated neutron inelastic cross section for ^{92}Zr, taken from Reffo and Fabbri[17]. The solid line is Hauser-Feshbach calculation with widths-fluctuation correction (WFC) using 18 discrete energy levels. The dot-dashed line results if the 14 highest-energy levels were replaced by a level-density continuum for which WFC is not included. The dashed line is a case for which WFC is included for the continuum.

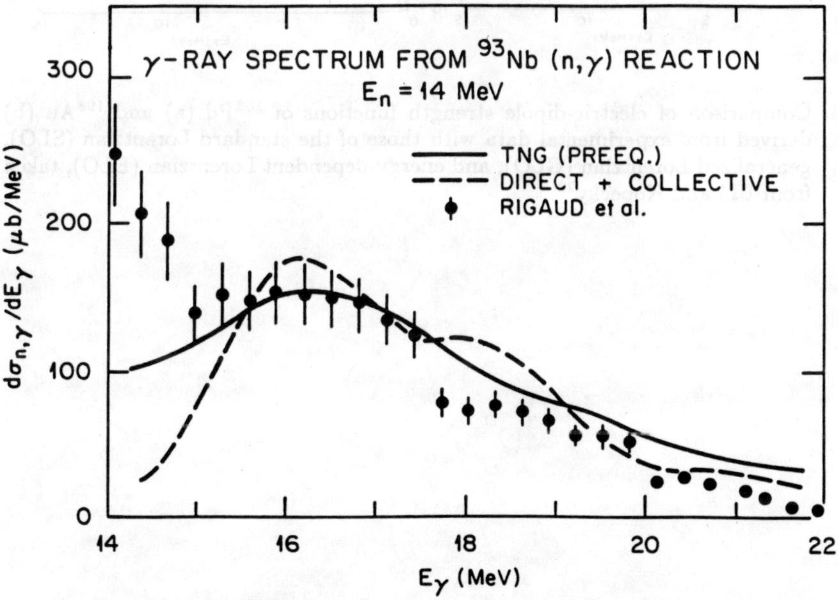

Fig. 3. The pre-equilibrium component of the capture gamma-ray spectrum for 14-MeV neutrons on ^{93}Nb, calculated by Shibata and Fu[20] using the TNG code[8]. The data and the direct-collective calculation are from Rigaud et al[21].

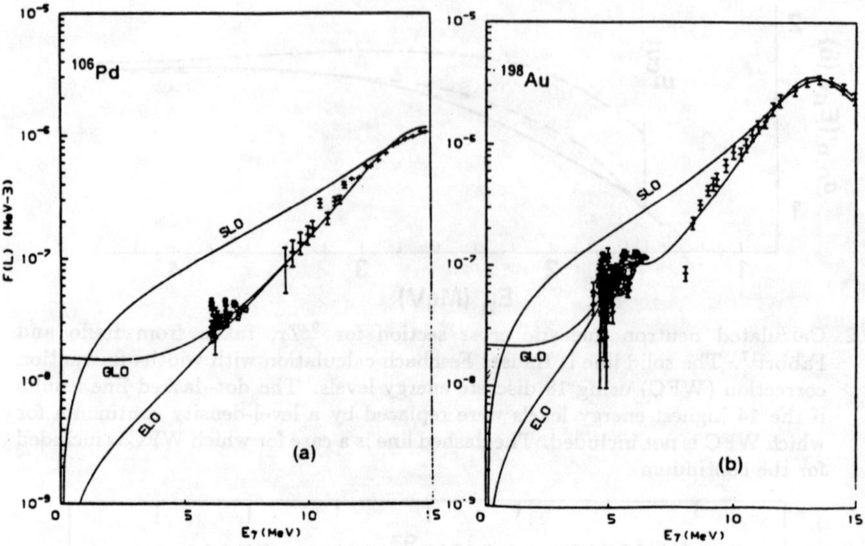

Fig. 4. Comparison of electric-dipole strength functions of ^{106}Pd (a) and ^{198}Au (b) derived from experimental data with those of the standard Lorentzian (SLO), generalized Lorentzian (GLO), and energy-dependent Lorentzian (ELO), taken from Uhl and Kopecky[28].

STATUS OF STATISTICAL MODEL CAPTURE CALCULATIONS

J. KOPECKY
Netherlands Energy Research Foundation ECN
P.O. BOX 1, 1755 ZG Petten, The Netherlands

M. UHL
Institut für Radiumforschung und Kernphysik, Universität Wien
A 1090 Wien, Boltzmanngasse 3, Austria

ABSTRACT

The most important input parameters in statistical model neutron capture calculations are γ-ray strength functions and level densities. Several models have been considered for these quantities and we have studied their influence on the calculated total radiative width, capture cross sections and γ-ray spectra. About 70 targets (90<A<250) have been used in these calculations with emphasis on the behaviour of targets from the deformed region.

1. Introduction

The statistical Hauser-Feshbach[1] model has been traditionally used to calculate radiative capture cross sections and gamma-ray spectra. Many authors have shown that the commonly used models for E1 and M1 γ-ray strength functions, namely the photoabsorption Lorentzian for E1 radiation and the single particle model for M1 radiation, often severely overpredict the available experimental data such as average radiative width or capture cross sections. We studied this overprediction[2-8] and found strong evidence for revisions of the above-mentioned models. The best fit with experimental data for E1 was obtained with a "generalized Lorentzian"[2] with a non-zero limit for ϵ_γ->0 and an energy dependent width. The spin-flip giant resonance of Lorentzian shape was recommended for M1 strength. However, by studying targets from the rare earth region[6-8] a significant underprediction of calculated data against the experimental values has been found for models, which were so successful for spherical and transitional nuclei. In this contribution we studied the impact of different auxiliary quantities on comparisons of calculated and experimental data in the light of the above mentioned discrepancies for deformed targets.

2. Auxiliary quantities in the statistical model

In the following we briefly review definitions for strength functions applied in this study. The simplest model, the <u>single particle model (SP)</u>[9], was employed for the (less important) contributions of M2, E3 and M3 radiation and occasionally also for M1 radiation with the strength (ASP) adjusted to average resonance capture (ARC) data or to systematics[4]. However, it should be noted that the SP model is at variance with a finite energy weighted sum rule.

Detailed balance theory relates the downwards strength functions to the photoabsorption. If the latter is dominated by a giant resonance of Lorentzian shape and applying the Brink hypothesis[10], the standard Lorentzian (SLO) formulation is derived

$$f_{XL}^{SLO}(\varepsilon_\gamma) = 26\times10^{-8}\left[mb^{-1}MeV^{-2}\right]\frac{1}{2L+1}\sigma_0\Gamma_0\frac{\varepsilon_\gamma^{3-2L}\Gamma_0}{(\varepsilon_\gamma^2-E_0^2)^2+\varepsilon_\gamma^2\Gamma_0^2}\,. \tag{1}$$

The parameters σ_0, E_0 and Γ_0 stand for peak cross section, energy and width of the resonance, respectively. These parameters are usually taken from experimental fits in Ref. [11]. This SLO model was employed for E1, M1 and E2 radiation. For M1 we used a spin-flip resonance as proposed by Bohr and Mottelson[12] with simple prescriptions for E_0 and Γ_0 in terms of the mass number: $E_0 = 41A^{-1/3}$ (MeV) and $\Gamma_0 = 4$ MeV[2]. The peak cross section is adjusted to reproduce available strength function data or systematics[4]. As E2 contributions are rather small, we used the global systematics[13,14] for E2 resonance parameters.

For the dominating E1 radiation some refinements have been proposed based on microscopic models. Following Dover et al.[15] the energy dependence of the width has been introduced. The theory of Fermi liquids[16] predicts an energy and temperature dependent width, $\Gamma(\varepsilon_\gamma,T) = \beta(\varepsilon_\gamma^2+4\pi T^2)$, where β is a normalization factor. The first term reflects the spreading of particle-hole states into more complex states, while the second term accounts for collisions of quasiparticles. The temperature T refers to the energy of absorbing state and can be calculated within a level density model. Kadmenskij et al.[17] proposed an expression for β such to guarantee the compatibility with the photoabsorption data, which reads as

$$\Gamma(\varepsilon_\gamma, T) = (\Gamma_0/E_0^2)\,(\varepsilon_\gamma^2+4\pi T^2)\,. \tag{2}$$

If this is included in Eq.(1) the resulting strength f_{E1} is derived from a Lorentzian with energy dependent width (ELO). In the model of Kadmenskij et al.[17] the quasi-particle fragmentation is included and the low energy limit of the dipole polarization operator considered. By adding this proper $\varepsilon_\gamma \to 0$ limit Kopecky and Chrien[2] obtained a practical expression of a generalized Lorentzian (GLO)

$$f_{E1}^{GLO}(\varepsilon_\gamma,T) = 8.68\times10^{-8}\left[mb^{-1}MeV^{-2}\right]\sigma_0\Gamma_0\left\{\frac{\varepsilon_\gamma\Gamma(\varepsilon_\gamma,T)}{(\varepsilon_\gamma^2-E_0^2)^2+\varepsilon_\gamma^2\Gamma^2(\varepsilon_\gamma,T)}+0.7\frac{\Gamma(\varepsilon_\gamma=0,T)}{E_0^3}\right\} \tag{3}$$

which for initial energies around the neutron separation energy is very close to the results of original Kadmenskij et al.[17] model.

All calculations were performed with two semi-empirical level density models, the backshifted Fermi gas model (BSGF)[18] and the model by Kataria, Ramamurthy and Kapoor (KRK)[19]. Parameters of these models were determined by fitting the average spacing of s-wave resonances D_0, adopted from Ref.[20], and low lying levels. Experience showed that D_0 should be carefully corrected for missed resonances before performing an accurate calculation. We have traced several earlier disagreements to the application of incorrect D_0 values. The influence of different optical model parameters (OMP) on the capture cross section calculations has been found to be rather weak[7], especialy if specific neutron OMP's are used. Optical potentials taken from the literature were used for particle transmission coefficients and for deformed nuclei channel coupling was accounted for.

3. Model calculations

The computations have been performed with the code MAURINA[21] for 69 target nuclei between ^{93}Nb and ^{241}Am; earlier results are presented in Refs.[3,5-8]. Total radiative widths have been calculated for all nuclei, capture excitation function and γ-ray spectra only for those targets for which the experimental information exists. Several targets have been additionally used to compare different f_{E1} and f_{M1} models with experimental information derived from ARC data. In general we did not attempt to improve the fit to the experimental data by adjusting any of model parameters (except for the revision of some D_0 values).

4. Results and discussion

For spherical targets we obtained the satisfactory simultaneous reproduction of experimental strength function (based on ARC data) and photonuclear data, total s-wave radiation width, capture cross sections and gamma-ray spectra applying the GLO(E1) and SLO(M1) models. Several examples of such agreement have been presented in Refs.[3,5-8]. In contrast the application of the SLO(E1) model clearly overpredicts all experimental quantities. The capture excitation function is compatible with both ELO and GLO models and thus insensitive to the low-energy behaviour of f_{E1}. The slope of the gamma-ray production spectrum is, however, more sensitive to the $\varepsilon_\gamma \to 0$ limit and favours the GLO model.

However, when extending these investigations to neutron capture in rare earth deformed nuclei we found contradictory results[6,7,8]. In particular the GLO(E1) model, so successful for spherical nuclei, failed to reproduce all experimental data, including the E1 strength function data. In general the experimental data were severely underpredicted by a combination of GLO(E1) and SLO(M1) prescriptions as shown in Fig. 1 for ^{156}Gd. As matter of fact the SLO model gives better results in this case. For a unified description of deformed and spherical nuclei we proposed[6] an empirical correction of the normalization factor β in Eq.(2):

$$\Gamma(\varepsilon_\gamma, T) = [k_0 + (1-k_0)(\varepsilon_\gamma - \varepsilon_0^\gamma)/(E_0 - \varepsilon_0^\gamma)](\Gamma_0/E_0^2)(\varepsilon_\gamma^2 + 4\pi^2T^2). \quad (4)$$

The constant k_0 is used to reproduce the ARC data around the arbitrarily chosen energy ε_0^γ. In order to guarantee $\Gamma(\varepsilon_\gamma \to E_0) = \Gamma_0$ an energy dependence is included. As $k_0=1$ yields again Eq.(2) we can consider Eq.(4) as a generalization. For $k_0>1$ the width is enhanced compared to the outcome of Eq.(2) and when used in Eq.(3) we refer to this model as enhanced generalized Lorentzian (EGLO). Fig.1 displays also the results for the EGLO model with $k_0=4$ and $\varepsilon_0^\gamma=4.5$ MeV and shows a reasonable simultanous reproduction of experimental data.

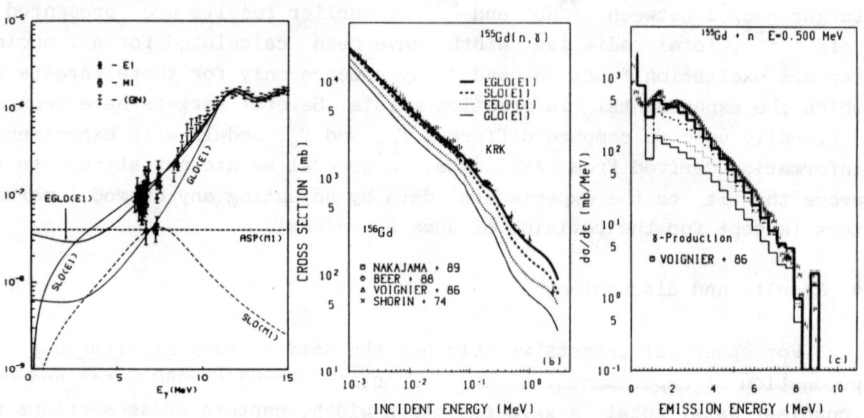

Fig. 1 (a) f_{E1} and f_{M1} models compared with ARC- and photonuclear data for Gd-156. The resulting cross sections (b) and gamma-ray spectra (c) are compared with experimental data.

The enhancement seems to be due to the increase in the spreading width of the double humped E1 giant resonance. The surprising feature of this effect is that it is common only to deformed nuclei with A<180, but disappears for several nuclei[6-8] in the deformed region above 180. The rationalization for this enhancement and the arbitrarily chosen generalization may be looked for in the dependence of the normalization constant β of the initial distribution (or density) of the single-particle p-h states (quasi-particles), which, fragmented into more complex states, determine the spreading width of the dipole resonance[16]. This density is dependent on the excitation energy and naturally also on the shell structure and thus may differ as a function of mass. The relative success of standard Lorentzian is probably due to fortuitous equality between the unphysically fixed spreading width Γ_0 and the realistic width increased by the high q-p densities for these nuclei.

In order to trace this behaviour more systematically we have extended calculations to other transitional (110<A<145) and deformed nuclei (A>170). Results for total radiative width $\langle\Gamma_\gamma\rangle$, based on the SLO model for both E1 and M1 radiation, are shown in Fig. 2 and support the earlier conclusions on spherical and transitional nuclei, namely a strong overprediction of experimental data for targets with A<145 and A>195 and a lesser one for targets with 180<A<195. If the GLO(E1)+SLO(M1) prescription is applied, a reasonable agreement is obtained, generalizing thus our earlier observations. However, for deformed targets with 150<A<180 and actinides around A=240 the SLO(E1) model agrees with the experimental data reasonably well. These conclusions are supported by a similar trend in calculated cross sections or γ-ray spectra or comparisons to ARC data.

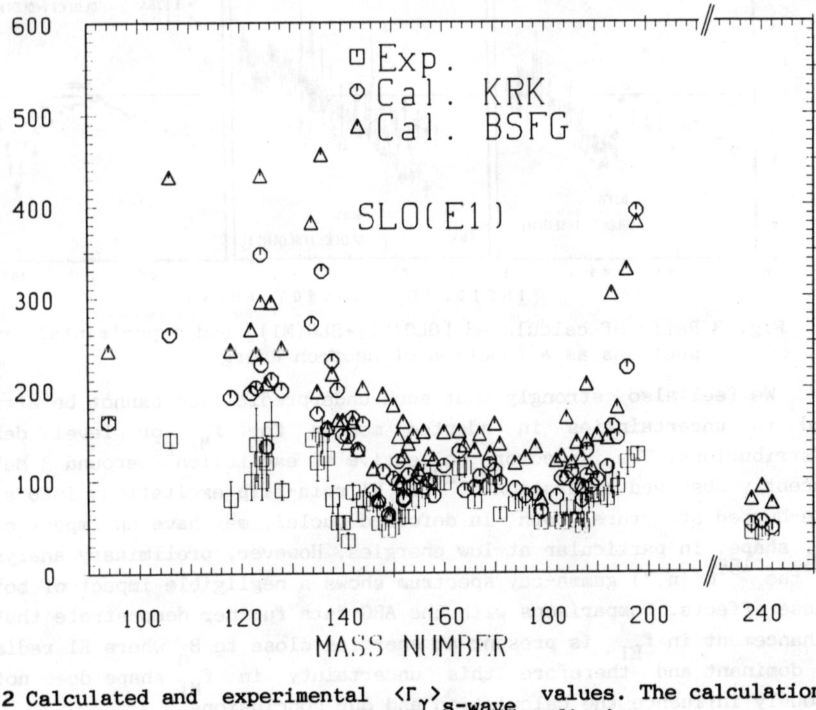

Fig. 2 Calculated and experimental $\langle\Gamma_\gamma\rangle_{s\text{-wave}}$ values. The calculations are based on the SLO model for E1 and M1 radiation.

A word of caution is, however, needed here on comparisons of statistical calculations to $\langle\Gamma_\gamma\rangle$ for s-waves. Since this quantity is based on resolved resonace data, it can carry a certain fraction of non-statistical contributions which are not accounted for by a statistical model. This may play a role e.g. around A=150, where the 4s->3p transitions may increase the experimental value of $\langle\Gamma_\gamma\rangle$. Similar observation has been made in the mass region around A=55[4,5]. This effect can be traced by plotting the $\sigma_{cal}/\sigma_{exp}$ as a function of neutron energy as is shown in Fig. 3 for three typical nuclei. The improvement of this ratio below the inelastic threshold with increasing neutron energy may support the existence of such effect. However, a sizeable underprediction of experimental data by the GLO(E1)+SLO(M1) model (see e.g. Fig. 1) at neutron energies where non-statistical effects are expected to be negligible, demonstrates that the f_{E1} enhancement has to be of some other physical origin.

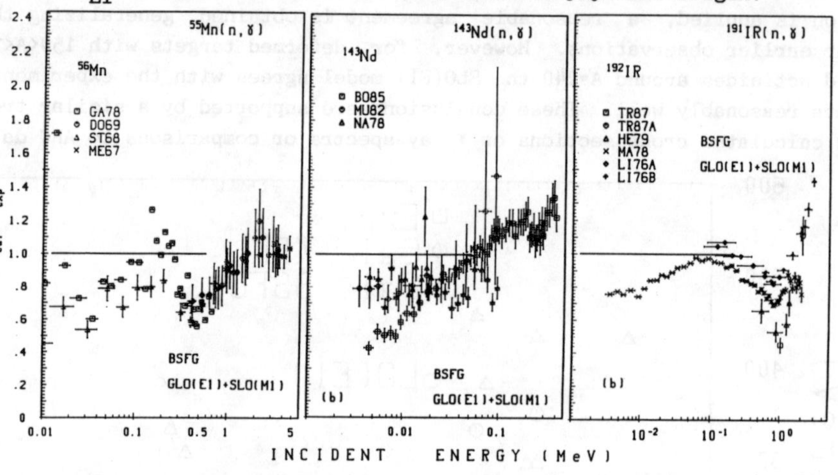

Fig. 3 Ratio of calculated [GLO(E1)+SLO(M1)] and experimental cross sections as a function of neutron energy.

We feel also strongly that such underpredicition cannot be attributed to uncertainties in adopted models for f_{M1} or level density distributions. The isovector collective M1 excitation[22] around 3 MeV and recently observed separation[23,24] of M1 spin-flip excitations into a double-humped structure, both in deformed nuclei, may have an impact on the f_{M1} shape, in particular at low energies. However, preliminary analysis[25] of the ^{155}Gd(n,γ) gamma-ray spectrum shows a negligible impact of both of these effects. Comparisons with the ARC data further demonstrate that the enhancement in f_{E1} is present at energies close to B_n where E1 radiation is dominant and therefore this uncertainty in f_{M1} shape does not seriously influence the calculation and our conclusions.

The employed level density models represent another source of uncertainty. Even if the same information on discrete levels and D_0 is used for the parametrization, both models result in appreciably different calculated quantities[7,8]. The reason is the difference in the total level densities occurring a few MeV below the neutron binding energy, where the KRK model is usually represented by the constant temperature portion. However, the sizeable underprediction of f_{E1}, in Eu and Gd isotopes by a factor of four, cannot be explained again by this difference in the total level densities, which usually influences the calculated quantities by a spread of not more than 50%.

5. Conclusion

In this contribution we have concentrated on radiative capture cross section calculations in the mass range $90<A<240$. For practical calculations we recommend, for f_{E1} the application of a EGLO model according to Eqs.(2)+(4) and for f_{M1}, a SLO prescription. The empirical parameter k_0 varies between 1 and about 4-5 for spherical and deformed nuclei, respectively, at ε_0^{γ}=4-5 MeV. These values have been estimated to reproduce f_{E1} based on several ARC data representative for these nuclei. For the transitional nuclei the situation is more complicated, values of 1-3 seem to be needed for k_0. Although for many of studied nuclei, the BSFG model was slightly superior, we suggest to keep applying both models and to use the difference between their results to assess the uncertainty due to level densities.

More theoretical studies as well as experimental data sensitive to energy distributions of f_{E1} and f_{M1} models may help to establish the origin and trend in the mass dependence of the k_0 enhancement and f_{E1+M1} behaviour for E_{γ}->0.

6. References

1. W. Hauser and H. Feshbach, Phys. Rev. **78** (1952) 366.
2. J. Kopecky and R.E. Chrien, Nucl. Phys. **A468** (1987) 285.
3. J. Kopecky and M. Uhl, Phys. Rev. **C41** (1990) 1941.
4. J. Kopecky and M. Uhl, in Proc. IAEA Specialists' Meeting on the Measurement, Calculation and Evaluation of Photon Production Cross Sections, INDC(NDS)-238 (1990), p. 103.
5. M. Uhl and J. Kopecky, idem, p. 113.
6. J. Kopecky and M. Uhl, in AIP Conderence Proceedings 234 - Capture Gamma-Ray Spectroscopy, ed. R.W. Hoff (American Insutute of Physics, New York, 1991), p. 607.
7. M. Uhl and J. Kopecky, in Nuclear Data for Science and Technology, ed. S.M. Qaim (Springer-Verlag, Berlin, 1992),p. 977.

8. M. Uhl and J. Kopecky, Contribution to Second International Symposium on Nuclear Astrophysics-Nuclei in Cosmos, Karlsruhe (BRD), July 6-10, 1992, to be published.
9. M. Blatt and V.F. Weisskopf, Theoretical Nuclear Physics, (Wiley, New York, 1952).
10. D.M. Brink, Ph. D. Thesis, (Oxford University, 1955).
11. S.S. Dietrich and B.L. Berman, At. Nucl. Data Tables, 38 (1988) 199.
12. A.G. Bohr and B.R. Mottelson, Nuclear Structure, Vol. II (Benjamin, London, 1975).
13. J. Speth and A. van der Woude, Rep. Prog. Phys. 44 (1981) 719.
14. W.V. Prestwich, M.A. Islam and T.J. Kennet, Z. Phys. A315 (1984) 103.
15. C.B. Dover, R.H. Lemmer and F.J.W Hahne, Ann. Phys. (N.Y.) 70 (1972) 458.
16. D. Pines and D. Noziers, The Theory of Quantum Liquids, (Benjamin, New York, 1966).
17. S.G. Kadmenskij, V.P. Markushev and V.I. Furman, Sov. J. Nucl. Phys. 37 (1983) 165.
18. W. Dilg, W. Schantl, H. Vonach and M. Uhl, Nucl. Phys. A217 (1973) 269.
19. S.K. Kataria, V.S. Ramamurthy and S.S. Kapoor, Phys. Rev. C18 (1978) 549.
20. S. F. Mughabghab et al., Neutron Cross Sections, Vol.1 (Academic Press, New York), Part A (1981) and Part B (1984).
21. M. Uhl, unpublished.
22. D. Boehle et al., Phys. Lett. 137B (1984) 27.
23. D. Zawischa, M. Macfarlane and J. Speth, Phys. Rev. C42 (1990) 1461.
24. A. Richter, Nucl. Phys. A522 (1991) 139.
25. J. Kopecky, M. Uhl and R.E. Chrien, ECN-RX--92-011.

SOME TOOLS OF THE TRADE WE'VE DEVELOPED FOR OUR CROSS-SECTION CALCULATIONS*

D. G. Gardner and M. A. Gardner
Lawrence Livermore National Laboratory
Nuclear Chemistry Department
Livermore, CA 94550, USA

ABSTRACT

We have modified or developed a number of computer codes, both main-frame and PC. We will mention seven codes, of which three will be discussed in some detail. The latter are: a controller-driven, double-precision version of the coupled-channel code ECIS; our latest version of STAPRE, a precompound plus Hauser-Feshbach nuclear reaction code; and NUSTART, a PC code that analyzes large sets of discrete nuclear levels and the multipole transitions among them. All main-frame codes are now being converted to the UNICOS operating system.

1. Main-Frame Codes (CRAY-YMP)

1.1. ECISPLUS

A double-precision version of ECIS[1], operated by a controller that transmits to ECIS an incident-particle energy grid plus all other required information, and saves and processes the ECIS output to produce transmission coefficients in up to five different formats.

1.2. STAPLUS

Some recent modifications to the STAPRE[2] code include: (1985) photons allowed in the entrance channel; (1989) an exciton-number and energy-dependent spin cut-off parameter for precompound evaporation; and an ad hoc, trial, collective enhancement model to modify certain level densities.

1.3. COMNUC - CASCADE

Minor changes to the original codes[3], including a new gamma-ray model and some graphics.

1.4 . ORIGAMI [4]

Modifies a calculated excitation function by folding it against a unit gaussian function that utilizes an energy-dependent resolution. This resolution function is a polynomial that can generate a variety of shapes, thus allowing a better comparison with experimental data that have a significant energy spread near thresholds.

2. IBM - PC/AT Codes

2.1. INDIGO[5]

A menu-driven, interactive, color graphics code that includes data manipulation and curve fitting.

2.2. CADISE [6]

Uses a piece-wise cubic Hermite polynomial algorithm to interpolate calculated excitation functions onto a variety of energy grids.

2.3. NUSTART[7]

Designed to handle large sets (up to 1000) of discrete nuclear levels and the multipole transitions among them when this information is in the STAPRE format. The code operates in three modes:

a) checks for physics and/or typographical errors among the branches,
b) generates STAPRE - format output from interactive input,
c) if only the levels are supplied, the code generates multipole transitions among the levels, with some options concerned with the multipole strength functions. The levels are treated as either having no collective character or as all being rotational band heads.

3. Code Performance

3.1. ECISPLUS

This "user friendly" version of ECIS is driven by a controller that reads a grid of incident particle energies, the coefficients of the energy-dependent optical model equations, and all of the other information required by ECIS such as the nuclear deformation and the energy, spin and parity of the coupled levels. At each energy, the optical model values are computed and sent to ECIS where the coupled-channel calculations are performed to the accuracy required by the user. The ECIS output is returned to the controller, where the total, the shape elastic, the direct inelastic, and the compound nucleus formation cross sections are stored. The remainder of the ECIS output is sent to the TC6 code, which computes the transmission coefficients, T_j or T_l, in one of five formats used by our Hauser-Feshbach codes. The direct reaction information is included in some formats. Figure 1 plots the log of the s-wave transmission coefficients against proton energy for the reaction ^{153}Eu + p. The ECIS results are compared with the quantum-mechanical WKB[8] coulomb barrier penetration approximation. Having a double-precision, main-frame code permits the calculation of accurate T_j values down to about 10^{-40}. Other new features include tables of Wronskian values at each energy for the ground state and each open level, and s-wave and p-wave strength functions for incident neutrons.

3.2. STAPLUS

Many modifications of the STAPRE code have been made over the years. One major modification made in 1980 was the development of the Energy-Dependent Breit-Wigner (EDBW) line shape for the E1 gamma-ray strength function[9]. It assumes that all nuclei of elements heavier than vanadium are double-peaked prolate spheroids with a separation determined by β_2. For the M1 strength function we use a function of the mass number, A, $1.44 \times 10^{-11} A^{4/3}$ MeV. Even though this does not lead to a finite sum rule value, up to 20-30 MeV no large error seems to be introduced. For the isoscalar E2 multipole, we use a single-peak resonance[10] with a peak energy of $63A^{-1/3}$ MeV, a peak width of $18.9\ A^{-1/3}$ MeV, and a peak cross section of $2.65 \times 10^{-2}\ (Z^2/A^{2/3})$ mb. The shape is either Lorentz or EDBW. Here Z is the nuclear charge number.

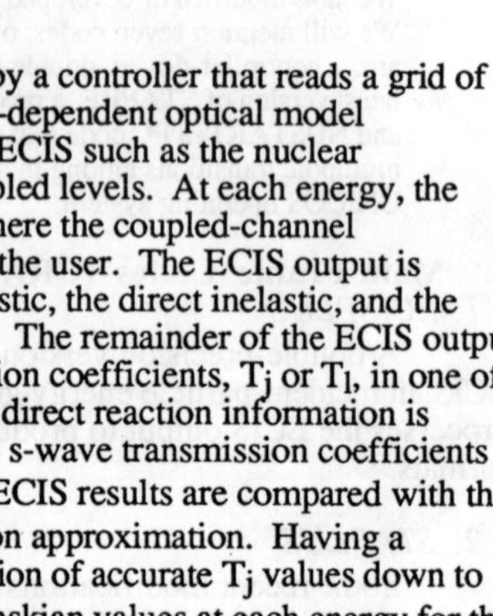

Fig. 1 Barrier penetration calculations using the coulomb radius (rc) values shown.

We have verified our systematics for many nuclei by calculating the total radiation width at the neutron separation energy (Sn), neutron capture reactions and the resulting gamma-ray spectra, measured M1/E1 ratios at low energies, average resonance capture (ARC) results with 2- and 24-MeV neutrons, and by computing the shape and magnitude of the photonuclear reactions in the 10-25 MeV range. Our most recent work involved isotopes of europium and gadolinium; figures 2a and 2b show our unadjusted results for ^{151}Eu and ^{153}Eu. Note that the splitting of the peak in ^{153}Eu is automatically predicted by our EDBW systematics. The data represent the experimenters' own fit of 2 Lorentz peaks to their data.

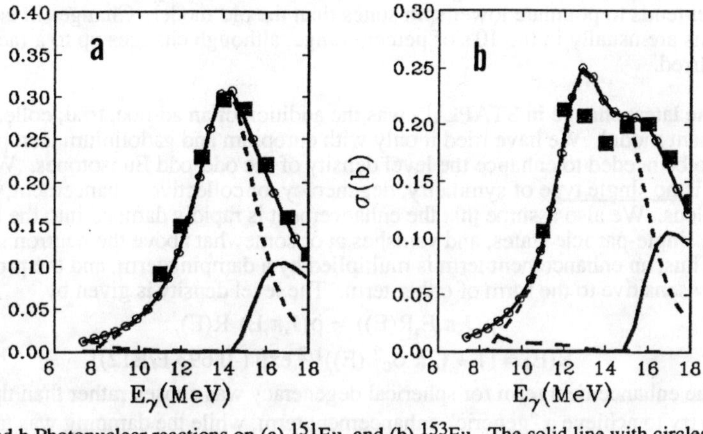

Figs. 2a and b Photonuclear reactions on (a) ^{151}Eu, and (b) ^{153}Eu. The solid line with circles is the photon absorption cross section, the long-dash line is for σ(γ,γ) the short-dash line is for σ(γ,n), the solid line is for σ(γ,2n), while the solid squares represent the experimental data.[11]

In Fig. 3 we compare the predictions of our global systematics with data from the same experimental group. In this case, it appears that the ratio of the area of the first peak to that of the second peak is not the assumed 0.5. A better fit is found with a ratio of 0.4. In our EDBW model, essentially all of the parameters can be supplied by the user, and should be done so if good data are available.

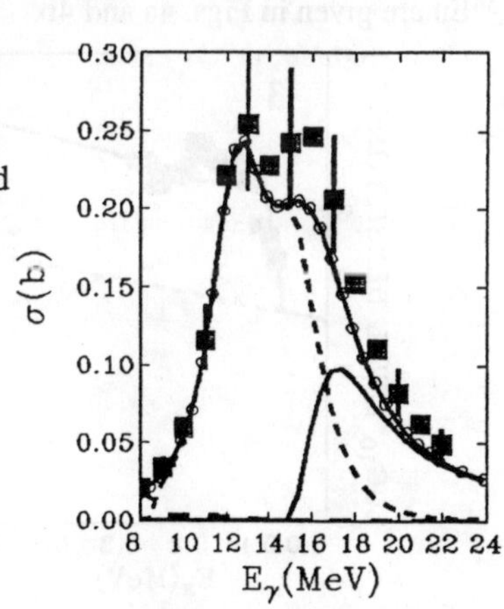

In Fig. 3 Photonuclear reactions on ^{156}Gd. Symbols and lines are as in Fig. 2.

In 1985, we modified the spin cut-off parameter, σ_p^2, used in the precompound evaporation model in a manner similar to that of Fu[12] and others. It is now a function of the number of particles and holes, N, in the particular exciton configuration ratioed to that at "equilibrium," Nc. The new spin cut-off is also dependent upon the energy, E, in the daughter nucleus by way of the compound limit, σ_c^2 (E).

$$\sigma_p^2 (N,E) = (N/Nc)\, \sigma_c^2 (E) \qquad (1)$$

Like Fu[12], we have the option of choosing for the matrix element, $|M|^2 = 262\, E^{-1} A^{-3}$. The main effects involve calculated angular distributions and isomer populations because the new model tends to populate lower spin states than the old model. Changes in isomer populations are usually in the 10's of percent range, although changes up to a factor of 2 have occurred.

The latest change in STAPLUS was the addition of an ad-hoc, trial, collective enhancement model. We have tried it only with europium and gadolinium isotopes, where it seems to be needed to enhance the level density of the odd-odd Eu isotopes. We assume that there is no single type of symmetry, degeneracy, or collective enhancement, even for a given nucleus. We also assume that the enhancement is rapidly damped into the far more numerous single-particle states, and vanishes at or somewhat above the neutron separation energy. Thus, an enhancement term is multiplied by a damping term, and the product is not greatly sensitive to the form of either term. The level density is given by

$$\rho(J,\pi,E,R(E)) = \rho(J,\pi,E)\; R(E), \qquad (2)$$

$$R(E) = (1 + (2\pi\, \sigma_c^2 (E))^{1/2} \exp(-0.693\, E/E12)). \qquad (3)$$

The enhancement term for spherical degeneracy was chosen rather than the usual $\sigma_c^2(E)$, to try to achieve a "generic" enhancement term, while the damping was taken as shown, with E12 as a free variable, to make it obvious that the term was 0.5 when E = E12. Equation 3 is strictly ad-hoc, and we are trying other formulations. The results for ^{154}Eu are given in Figs. 4a and 4b.

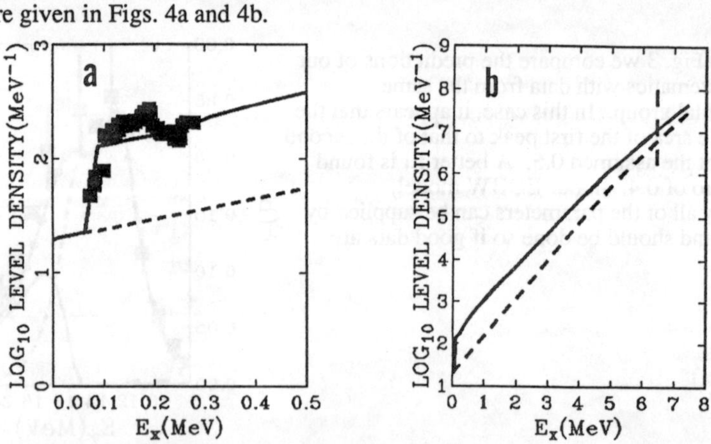

Fig. 4 Level density of ^{154}Eu (a) at low energy, and (b) up to energies beyond the neutron separation energy (vertical dashed line). The solid line is the calculation including rotational enhancement while the short-dash line is without. The solid squares are the level densities calculated from the discrete levels.

Other modifications of STAPRE include: isospin weighting factors may be applied to particle exit channels, and ARC-type calculations may be made at low-keV energies using initial spin and parity distributions calculated by COMNUC as input.

3.3. NUSTART

This PC code was initially written to examine very large sets of discrete levels and their multipole branches for typographical or physics errors. The levels and branches must be in the STAPRE format, and the code determines which multipole types can make the transition. Should the multipole type be greater than E3, and if the parent level is not an isomer, a warning is printed in the output table. A useful quantity, the branching fraction divided by (2 l +1), is also printed.

If only the discrete levels are supplied, NUSTART will compute the branching fractions under one of two assumptions: either the levels have no collective character, or they are all rotational band heads. The multipole strength functions f_{E1}, f_{M1} and f_{E2} may all be constants, or f_{E1} may be energy dependent with one or two resonance peaks. The code computes up to 20 E1 and M1 branches and 10 E2 branches for every parent level not designated as an isomer, chooses the 10 most intense, normalizes them to unity, and writes a file in the STAPRE format. The second option is used under the assumption that intraband transitions are much faster than interband transitions. The band heads are removed from the complete set for the second option, and in the branching calculation the user may choose whether to employ the K-quantum selection rules.

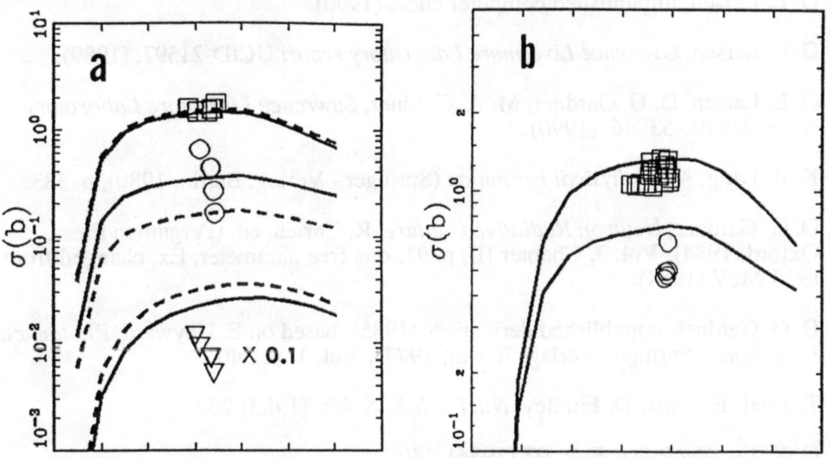

Fig. 5(a) The reaction ^{153}Eu (n, 2n) ^{152}Eu. The data[13] are squares for the 3⁻ ground state, open circles for the 0⁻ isomer, and the triangles for the 8⁻ isomer (multiplied by 0.1). The solid line used the experimental multipole branches among the levels in ^{152}Eu, while the dash line used branches from NUSTART. (b) The reaction ^{151}Eu (n, 2n) ^{150}Eu. The data are the squares for the 5⁻ ground state, and open circles for the 0⁻ isomer. No experimental branches are available, and so the solid lines now are the calculations using NUSTART branches.

Figures 5a and 5b show, respectively, the results of isomer production calculations following (n,2n) reactions of ^{153}Eu and ^{151}Eu. The lack of experimental branching information for ^{151}Eu is unfortunate, but the branchings computed by NUSTART, assuming no collective character, appear to be fairly reasonable and better than one would obtain from a partition function approach.

* This work was performed under the auspices of the U.S. Department of Energy by Lawrence Livermore National Laboratory under contract No. W-7405-Eng-48.

References

1. J. Raynal, *International Atomic Energy Agency report* IAEA-SMR-9/8, Vienna (1972).
2. M. Uhl and B. Strohmaier, *report IRK-76/1* (1976, updated 1978).
3. C. L. Dunford, *Atomics International report* AI-AEC-12931, (1970).
4. D. G. Gardner and M. A. Gardner, unpublished computer code, (1991).
5. G. L. Larsen, unpublished computer code, (1990).
6. G. L. Larsen, *Lawrence Livermore Laboratory report* UCID-21597, (1989).
7. G. L. Larsen, D. G. Gardner, M. A. Gardner, *Lawrence Livermore Laboratory report* UCRL-53946, (1990).
8. K. R. Lang, *Astrophysical Formulae* (Springer - Verlag, Berlin, 1980), p. 383.
9. D. G. Gardner, *Neutron Radiative Capture*, R. Chrien, ed. (Pergamon Press, Oxford, 1984), **Vol. 3**, Chapter III, p. 91; one free parameter, Ex, changed from 4 to 11 MeV (1984).
10. D. G. Gardner, unpublished derivation, (1985); based on E. Hayward, *Photonuclear Reactions*, (Springer-Verlag, Berlin, 1977), **Vol. 1**, p. 340.
11. T. Boal, E. Muir, D. Findley, *Nucl. Phys.* **A 406** (1983) 257.
12. C. Y. Fu, *Nucl. Sci. Eng.* **92** (1986) 440.
13. V. McLane, C. L. Dunford, P. F. Rose, *Neutron Cross Sections* (Academic Press, Boston, 1988), **Vol. 2**, p. 529.

EVALUATION OF NEUTRON-INDUCED REACTION CROSS SECTIONS ON THE NEUTRON-DEFICIENT ISOTOPES OF EUROPIUM and GADOLINIUM*

MAUREEN A. GARDNER and DONALD G. GARDNER
Lawrence Livermore National Laboratory
Livermore, California 94550 USA

ABSTRACT

We have been calculating complete sets of excitation functions for the neutron-induced reactions on the isotopes of europium(masses 145 to 154) and of gadolinium(masses 150 to 156). The approach that we are taking is to achieve best agreement between our calculated cross sections for the stable Eu and Gd isotopes and experiment, using as consistent a set as possible of all necessary calculational parameters; this set will then be used to predict all required cross sections for the unstable isotopes. A variety of experimental data in this mass region have been utilized. A potential problem for experimenters was uncovered when the spectrum fitting method is used to deduce the radiative capture cross section from prompt gamma rays.

1. Introduction

We are developing cross-section libraries that contain complete sets of excitation functions for neutron-induced reactions on the isotopes: $^{145-154}$Eu and $^{150-156}$Gd for the incident neutron energy range of 10^{-5} to 20 MeV. A large number of these isotopes are neutron-deficient and unstable and no experimental cross-section data are available. Our goal is to calculate all relevant (n,γ), (n,n'), (n,2n) and (n,3n) cross sections. The calculational modeling in this mass region is challenging for a number of reasons. The Eu and Gd isotopes lie close to the N = 82 closed neutron shell but between the closed proton shells of Z = 50 and Z = 82, and the deformation of these isotopes changes rapidly. A deformed neutron potential that works well for the heavier-mass isotopes may not be suitable for the lighter, more spherical nuclei.

Initially we had to determine if all target isotopes can be treated as rotors. The even-even Gd isotopes certainly seem to have a rotor signature. Because the deformation is changing rapidly in this mass region, we needed to find systematics that could predict values of the deformation parameter, β_2, for the light isotopes. We require β_2 values as input to coupled-channel optical model calculations and to our E1 γ-ray strength function modeling. We also hope to deduce level density systematics for the light isotopes based on the deformation parameter. It appears that the Gd optical model potential required at low energies is quite different from that required for Eu. Further, the ^{155}Gd and ^{157}Gd resonance parameters are expected to be energy sensitive because of large resonances at thermal energy; this impacts on the use of the SPRT method of optimizing the potential. The existence of low-lying energy resonances for lighter Gd isotopes is unknown. The nuclear structures for these isotopes have not been studied well experimentally, especially in the case of Eu. Many levels of the Eu isotopes are completely uncharacterized and gamma-ray branching information among the discrete levels is scant. Because of the rapid change in deformation, the Eu and Gd level densities are difficult to model. We found large disagreements between observed numbers of discrete levels and those predicted by Gilbert-Cameron systematics.

2. Some Calculational Procedures

2.1. General Approach

We hope to understand the low-energy physics in this mass region by fitting a variety of available data from Nd to Gd in a self-consistent way and then to extrapolate to the unmeasured regions. The approach that we are taking is to achieve best agreement between our calculated cross sections for the stable Eu and Gd isotopes and experiment using as consistent a set as possible of all necessary calculational parameters. This set will then be used to predict all required cross sections for the unstable isotopes. Among the experimental data utilized to determine these parameters are:

— β_2 values for some 39 nuclei
— $\Gamma\gamma$ and D_{ob} values for 7 nuclei
— s-wave and p-wave neutron strengths for 7 nuclei
— discrete level information for 16 nuclei
— (n,γ) σ's for ^{151,153}Eu and ^{152,154,155}Gd
— capture γ-ray spectrum for ^{155}Gd
— total (n,2n) σ's on ^{151}Eu and ^{152,154,155}Gd
— (n,2n) σ's to isomers for ^{151,153}Eu
— (p,n) and (p,3n) σ's on ^{151,153}Eu
— photoneutron σ's on ^{151,153}Eu and ^{156}Gd
— averaged total resonance data for ^{151,153}Eu and ^{155}Gd.

2.2. *Systematics for the Deformation Parameter, β_2*

The systematics that we used to predict β_2 values was based on the suggested[1] expression, $\beta_2{}^2$ = constant x $E\gamma^{-1}$ x $A^{-5/3}$, where $E\gamma$ is the energy difference in MeV between the ground state and the first member of the ground state rotational band. The value of the constant in this expression was determined by analyzing experimental β_2 values for some 39 nuclei from Nd to Tm; an average value of 42.5+/-3.9 was obtained for even-even nuclei and of 29.6+/-7.9 for all others. Ninety percent of the values of β_2 computed with these constants agreed with the experimental ones in this mass region to 20% or better. In Figure 1 we show the β_2 values that we computed for the specific Eu and Gd isotopes of interest. In the case of ^{153}Gd, the simple formula overestimated the deformation parameter because the ground-state band is compressed due to near-by bands; we used a β_2 value of 0.316 instead. In the case of ^{151}Eu, the low result shown prompted us to update the ground state band members we originally had adopted; this led to a new computed value of 0.188 for β_2.

2.3. Choice of Coupled-Channel Optical Model Parameters

We used our ECISPLUS version[2] of the coupled-channel code, ECIS, to obtain the neutron transmission coefficients and all required total reaction cross-sections. Because of the rapid change in the deformation parameter values as a function of mass, we found that we needed to carry out coupled-channel calculations for each isotope. A spherical-equivalent potential could not be used, especially at neutron energies below 0.1 MeV. As we mention above, we further observed that, particularly at low energies, different potentials were required for Eu and Gd. Our choice of deformed optical model potentials were those of P. Young for Eu[3] and Gd.[4] In the case of ^{155}Gd, we needed to invoke the use of a fictitious even-mass, ground-state rotational band following the approach of Ch. Lagrange, et al.[5] This "fictive" method was probably useful for one or both of the following reasons: the optical model parameters may have been optimized for the even-even isotopes and/or, because of the presence of a neutron resonance very near thermal energy, the s- and p-wave neutron strengths and the scattering radius were energy sensitive, impacting the SPRT method used to deduce the potential. In Figure 2 we show several sets of experimental data[6-8] for the ^{155}Gd(n,γ) reaction and two calculations we made, the dashed curve results when we coupled the true ground-state levels, $3/2^-$, $5/2^-$, $7/2^-$, with excitation energies of 0.060 MeV and 0.146 MeV for the $5/2^-$ and $7/2^-$ states, respectively; and the solid curve results when we coupled the "fictive" levels, 0^+,2^+,4^+, with excitation energies of 0.085 MeV and 0.282 MeV for the 2^+ and 4^+ states. However, a similar set of calculations for the ^{151}Gd(n,γ) reaction, where true and "fictive" levels were coupled, resulted in insignificant differences, perhaps due to the fact that ^{151}Gd is not very deformed.

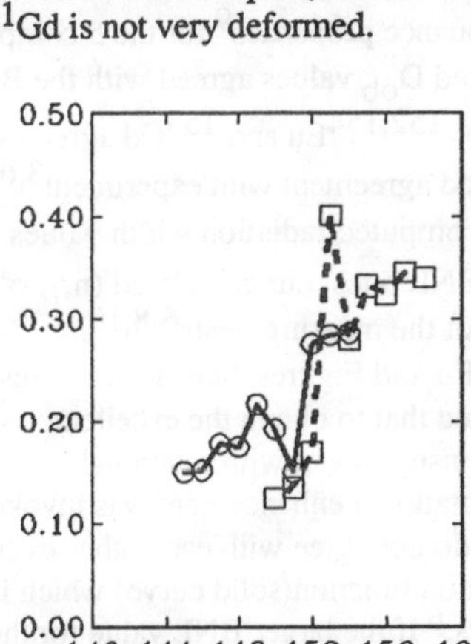

Fig. 1 Computed values for the deformation parameter, β_2, for isotopes of Eu (open circles) and of Gd (open squares).

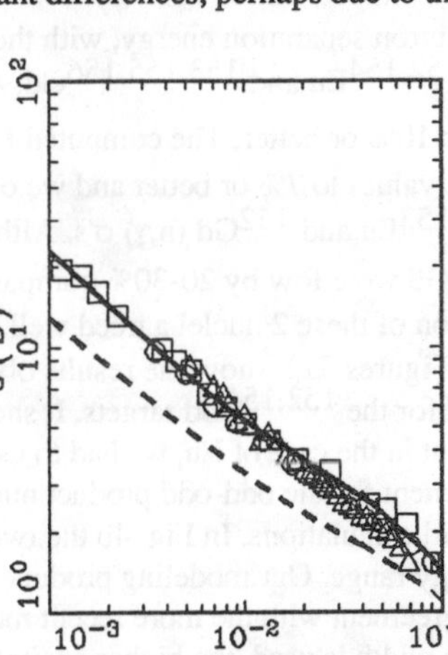

Fig. 2 Experimental data[6-8] for ^{155}Gd (n,γ) ^{156}Gd reaction compared with two calculations: dashed curve is obtained when true ground state levels are coupled; solid curve obtained when we coupled "fictive" levels.

2.4. Discrete Level Information and Level Densities

We have always tried to extract as much information as possible from complete sets of discrete level information to improve global parameterizations of level densities and to predict a variety of required calculational parameters, particulary for nuclei off the line of stability. Unfortunately, experimental discrete level information, especially for the light Eu isotopes is incomplete. Few levels are known for some isotopes, spin and parity assignments are uncertain, and multipole transition information, required for isomer population calculations, is non-existent for some nuclei. We have used our NUSTART code to provide the latter type of information.[2] We have also begun to develop systematics based on the deformation parameter, β_2, to predict some of the level density parameters, such as the pairing gap, δ. We have found that for the even-even Gd nuclei, the expression $\delta_{e,e} = 1.863 - 2.07\beta_2$, yields pairing gap values that are within 0.020-0.040 MeV of those we would deduce from the discrete level information and/or from the known values of D_{ob}, the average s-wave neutron resonance spacing. The relation, $\delta_{e,o} = \delta_{o,e} = 0.791 - 2.07\beta_2$, gives pairing gap values that are within 0.020-0.140 MeV of those we would normally deduce for the odd-mass Eu and Gd isotopes. An expression that relates $\delta_{o,o}$ to the deformation parameter is still under development.

3. Results for (n,γ) Reactions on ^{151,153}Eu and ^{152,154,155}Gd

In order to evaluate our modeling of radiation widths and level densities, particularly important to the calculation of neutron capture cross sections, we compared our calculated values of $\Gamma\gamma$, the total radiation width and of D_{ob}, the average level spacing at the neutron separation energy, with the BNL resonance properties[9] for the 5 compound nuclei, ^{152,154}Eu and ^{153,155,156}Gd. All computed D_{ob} values agreed with the BNL values to 10% or better. The computed $\Gamma\gamma$ values for ^{152,154}Eu and ^{153}Gd agreed with the BNL values to 7% or better and we obtained good agreement with experiment[3,6] for the ^{151,153}Eu and ^{152}Gd (n,γ) σ's. Although our computed radiation width values for ^{155,156}Gd were low by 20-30% compared to the BNL ones, our calculated (n,γ) σ's for production of these 2 nuclei agreed well with most of the measurements[6-8,10].

Figures 3a,b show the results obtained for Eu and Figures 4a,b show the results obtained for the ^{152,154}Gd targets. It should be noted that to obtain the excellent agreement in the case of Eu, we had to use a level density model with rotational enhancement for the odd-odd product nuclei.[2] No rotational enhancement was invoked for the Gd calculations. In Fig. 4b the two data sets do not agree with each other over the full energy range. Our modeling produces an excitation function(solid curve) which is in closer agreement with the more recent measurements.[6] If the larger BNL value for the radiation width is used, the higher excitation function (dashed curve) is obtained, in agreement with the earlier work.[8]

In the case of the ^{155}Gd(n,γ) reaction, the good agreement of our calculations with measurements was shown in Fig. 2 for incident neutron energies of 0.1 MeV and below. Above $E_n = 0.15$ MeV, the calculated and measured cross sections begin to

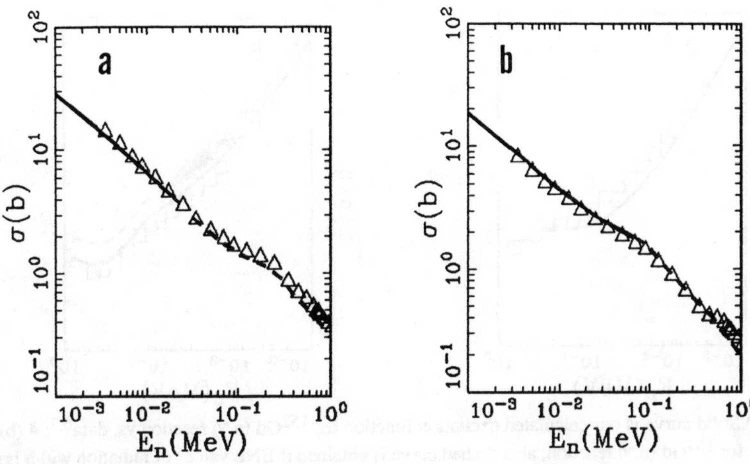

Fig. 3 (a) The solid curve is our calculated excitation function for the ^{151}Eu (n,γ) reaction vs. data;[3] (b) the same but for the ^{153}Eu (n,γ) reaction.

deviate, so that at E_n values of 0.5 to 1.0 MeV, the calculated cross sections are only about 50% of the measured ones. One might expect that the apparent undercalculation of the (n,γ) cross section could be due to an overestimation of the (n,n') reaction. However, we used 42 discrete levels up to an excitation energy of 0.664 MeV to describe the ^{155}Gd target; therefore the computed excitation function in this incident energy region is not low due to overestimating the ^{155}Gd level density.

In Figure 5 we show, for ^{155}Gd at E_n = 0.5 MeV, our calculated primary capture γ-ray spectrum(histogram) compared with the measurement of Ref. 10(open circles). As we had pointed out to the authors,[11] the original experimental spectrum must be multiplied by a factor to obtain the correct magnitude for the differential cross section values. The authors define this factor to be a value for the multiplicity. We have scaled the original measured spectrum by the multiplicity value that we compute to be most probable, 3.9. Differences between the calculated and measured spectrum are of greatest importance at gamma-ray energies below 1.5 MeV, a portion of the spectrum that the authors of Ref.10 do not measure but calculate. At E_n = 0.5 MeV, this low energy portion of the spectrum corresponds mainly to gamma-ray transitions among the discrete levels in the product nucleus, ^{156}Gd, and the inelastic levels of the target. In our calculations, this even-even nucleus is described by 42 discrete levels widely spaced up to an excitation energy of 1.909 MeV; Ref. 10 uses a level density description of the nucleus. In our calculations, this portion of the spectrum, below 1.5 Mev, corresponds to about 26% of the calculated total capture cross section. In the experimental case, this portion of the spectrum, which is calculated, is very likely a much larger percentage of the measured total capture cross section. Perhaps the level density description instead of the discrete level description of ^{156}Gd in the first 1.9 MeV leads to an overestimation of the experimental spectrum and the capture cross section deduced from it.

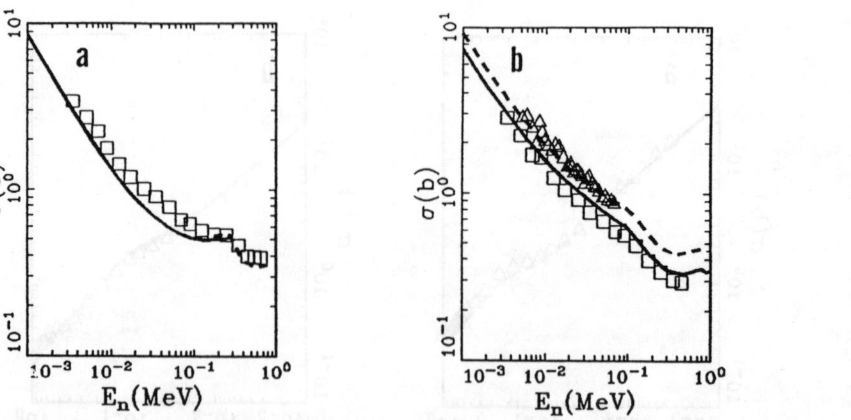

Fig. 4 (a) Solid curve is our calculated excitation function for ^{152}Gd (n,γ) reaction vs. data[6,7]; 4 (b) the same but for ^{154}Gd (n,γ) reaction, also dashed curve is obtained if BNL value[9] of radiation width is used.

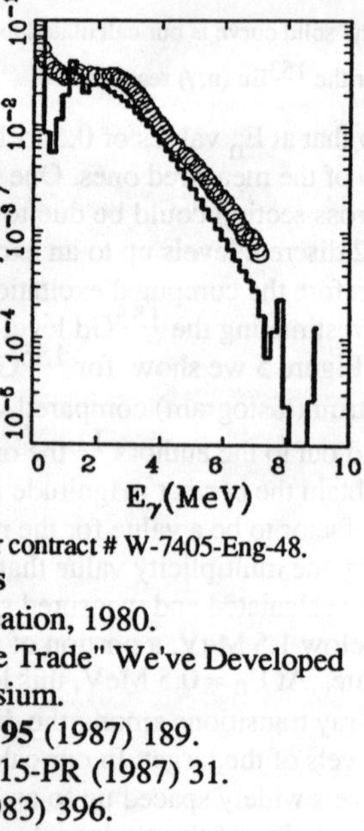

Fig. 5 Calculated primary capture γ-ray spectrum for the ^{155}Gd (n,γ) reaction at E_n = 0.5MeV (histogram) compared with measurement of Ref. 10 (open circles).

*Work performed under auspices of U.S. DOE by LLNL under contract # W-7405-Eng-48.

References

1. V. Madsen, Oregon State U., private communication, 1980.
2. D. Gardner and M. Gardner, Some "Tools of the Trade" We've Developed for Our Cross-section Calculations, this Symposium.
3. R. Macklin and P. Young, *Nucl. Sci. and Eng.* **95** (1987) 189.
4. P. Young and E. Arthur, *LANL report*, LA-10915-PR (1987) 31.
5. Ch. Lagrange, et al, *Nucl. Sci. and Eng.* **83** (1983) 396.
6. H. Beer and R. Macklin, *Astrophysical Journal* **331** (1988) 1047.
7. Y. Nakajima, et al, *NEANDC(J)-106* (1984).
8. V. Shorin, et al, *Yadernaya Fizika* **19** (1974) 5.
9. S. Mughabghab, *Neutron Cross Sections*, **Vol. 1,B** (1984).
10. J. Voignier, et al, *Nucl. Sci. and Eng.* **93** (1986) 43.
11. J. Voignier, et al., *Nucl Sci. and Eng.* **96** (1987) 343.

THEORETICAL MODELS OF NEUTRON EMISSION IN FISSION

DAVID G. MADLAND
Theoretical Division, Los Alamos National Laboratory
Los Alamos, New Mexico 87545 USA

Abstract

A brief survey of theoretical representations of two of the observables in neutron emission in fission is given, namely, the prompt fission neutron spectrum $N(E)$ and the average prompt neutron multiplicity $\bar{\nu}_p$. Early representations of the two observables are presented and their deficiencies are discussed. This is followed by summaries and examples of recent theoretical models for the calculation of these quantities. Emphasis is placed upon the predictability and accuracy of the new models. In particular, the dependencies of $N(E)$ and $\bar{\nu}_p$ upon the fissioning nucleus and its excitation energy are treated. Recent work in the calculation of the prompt fission neutron spectrum matrix $N(E, E_n)$, where E_n is the energy of the neutron inducing fission, is then discussed. Concluding remarks address the current status of our ability to calculate these observables with confidence, the direction of future theoretical efforts, and limitations to current and future calculations. Finally, recommendations are presented as to which model should be used currently and which model should be pursued in future efforts.

1. Early Representations

Two early representations of the prompt fission neutron spectrum, which are still used today, are the Maxwellian and Watt spectrum representations with parameters that are adjusted to optimally reproduce the experimental spectrum for a given fissioning system at a given excitation energy. The Maxwellian spectrum is given by

$$N(E) = (2/\pi^{1/2}T_M^{3/2})E^{1/2}exp(-E/T_M) , \tag{1}$$

where the single (temperature) parameter appearing, T_M, is related to the average energy (first moment) of the spectrum $< E >$ by

$$< E >= (3/2)T_M . \tag{2}$$

The Maxwellian spectrum is popular in applied physics because it is a closed-form expression containing a single parameter. However, this spectrum neglects the *distribution* of fission-fragment excitation energy, the *energy dependence* of the inverse process of compound nucleus formation, and the *motion* of the fission fragments emitting the neutrons. Therefore, T_M is greater than the fragment temperatures physically occurring. In practice, T_M is reduced to reproduce the tail of the experimental spectrum. To preserve normalization, this simultaneously increases $N(E)$ at lower energies in agreement with experimental measurements of the spectrum, but for the wrong physical reasons. Thus, there is *no predictive power* in this approach.

The two-parameter Watt spectrum[1] consists of a center-of-mass Maxwellian spectrum that has been transformed[2] to the laboratory system for an *average* fission fragment moving with an *average* kinetic energy per nucleon E_f. This spectrum is given by

$$N(E) = \frac{exp(-E_f/T_W)}{(\pi E_f T_W)^{1/2}} exp(-E/T_W) sinh[2(E_f E)^{1/2}/T_W] , \tag{3}$$

where E_f and the Watt temperature T_W are related to the average energy of the spectrum $< E >$ by

$$< E >= E_f + (3/2)T_W . \tag{4}$$

The Watt spectrum also neglects the ***distribution*** of fission-fragment excitation energy and the ***energy dependence*** of the inverse process of compound nucleus formation, but does account for the ***motion*** of an ***average*** fragment. However, for spontaneous fission and low-energy ($E_n < 15$ MeV) neutron-induced fission, the concept of an average fragment is generally not a good one because there are usually two average fragment masses and two average fragment kinetic energies due to the double-humped fragment mass distribution. For these reasons, the Watt spectrum, although it is more physical than the Maxwellian spectrum, has *little predictive power.* Note that if one insists on using a Watt representation, the average of the separate Watt spectra corresponding to the light and heavy fragment mass yield peaks should be taken to represent the total laboratory spectrum $N(E)$. This is then a three-parameter representation assuming statistical equilibrium exists between the nascent fragments.

At the same time that these early representations were introduced for $N(E)$, the average prompt neutron multiplicity $\bar{\nu}_p$ was modeled[3] by a simple polynomial (usually linear) in the incident neutron energy E_n , for each fissioning system considered: $\bar{\nu}_p = \nu_0 + \alpha E_n$, and again, the parameters appearing were, and are, adjusted to optimally reproduce the experimental values.

To summarize, none of the approaches described above can be used to predict $N(E)$ and/or $\bar{\nu}_p$ for a different fissioning nucleus or for a different excitation energy from what has been experimentally measured. That is, there is very little predictive power in the approaches just described.

2. Modern Approaches to the Calculation of $N(E)$ and $\bar{\nu}_p$

In recent years three new theoretical approaches have evolved for the calculation of the prompt fission neutron spectrum $N(E)$. These are the following:

- The Los Alamos approach,[4] begun in 1979, which is based upon standard nuclear evaporation theory[5] and simultaneously treats the average prompt neutron multiplicity $\bar{\nu}_p$. This approach emphasizes predictive capabilities while requiring minimal input. Refinements that treat the entire fission-fragment mass and charge distributions, instead of averages over their peak regions, have been implemented.[6,7]
- The Dresden approach,[8] begun in 1982, which is also based upon standard nuclear evaporation theory,[5] but accounts explicitly for "neutron cascade emission." This approach emphasizes a complete description, requiring a substantial amount of experimental information. The Dresden group has also employed the Los Alamos approach including the refinements mentioned above.[9,10]

- The Hauser-Feshbach statistical model[11] approach which accounts explicitly for the competition between neutron emission and gamma-ray emission in a given fission fragment. This approach, if properly applied, accounts for the influence of angular momentum and parity on neutron and gamma-ray emission whereas the Los Alamos and Dresden approaches do not.

2.1. Summary of Los Alamos Model

The original Los Alamos model[4] addresses both neutron-induced and spontaneous fission and accounts for the physical effects of

(a) the distribution of fission-fragment excitation energy,

(b) the energy dependence of the inverse process of compound nucleus formation,

(c) the center-of-mass motion of the fission fragments, and

(d) multiple-chance fission at high incident neutron energy.

To simulate the initial distribution of fission-fragment excitation energy and subsequent cooling as neutrons are emitted, a triangular approximation is used to approximate the corresponding fission-fragment residual nuclear temperature distribution. This approximation, based upon the observations of Terrell,[12] has a maximum temperature T_m that is related to the initial total average fission-fragment excitation energy $< E^* >$ by

$$T_m = (< E^* > /a)^{1/2}, \qquad (5)$$

where a is the nuclear level-density parameter. In Eq.(5), the initial total average fission-fragment excitation energy is given by

$$< E^* > = < E_r > + E_n + B_n - < E_f^{tot} >, \qquad (6)$$

where $< E_r >$ is the average energy release in fission, B_n and E_n are the separation and kinetic energies of the neutron inducing fission (set to zero for spontaneous fission), and $< E_f^{tot} >$ is the total average fission-fragment kinetic energy. These quantities are either known or can be calculated.

The energy dependence of the inverse process is treated in the center-of-mass frame by calculating the compound nucleus formation cross section $\sigma_c(\epsilon)$ using an optical-model potential with explicit isospin dependence to describe neutron-rich fission fragments more correctly. It has been found that the *shape* of $\sigma_c(\epsilon)$ with ϵ affects the *shape* of $N(E)$.

The values of the average kinetic energy per nucleon of the average light fragment A_L and average heavy fragment A_H are obtained using momentum conservation and are given by

$$E_f^{L,H} = (A_{H,L}/A_{L,H})(< E_f^{tot} > /A), \qquad (7)$$

where A is the mass number of the fissioning nucleus.

With the inclusion of these physical effects, the prompt fission neutron spectrum in the laboratory system is given by

$$N(E) = \frac{1}{2}[N(E, E_f^L, \sigma_c^L) + N(E, E_f^H, \sigma_c^H)] \ , \tag{8}$$

where

$$N(E, E_f, \sigma_c) = \frac{1}{2\sqrt{E_f T_m^2}} \int_{(\sqrt{E}-\sqrt{E_f})^2}^{(\sqrt{E}+\sqrt{E_f})^2} \sigma_c(\epsilon)\sqrt{\epsilon}\, d\epsilon \int_0^{T_m} k(T)\, T\, exp(-\epsilon/T)\, dT \ . \tag{9}$$

In this equation, ϵ is the center-of-mass neutron energy and $k(T)$ is the temperature-dependent normalization integral.[4] If $\sigma_c(\epsilon)$ is constant, Eq.(9) reduces to a closed-form expression involving the exponential integral function and the incomplete gamma function.

The average prompt neutron multiplicity $\bar{\nu}_p$ is obtained from energy conservation and is given by

$$\bar{\nu}_p = \frac{< E^* > - < E_\gamma^{tot} >}{< S_n > + < \epsilon >} \ , \tag{10}$$

where $< E_\gamma^{tot} >$ is the total average prompt gamma-ray energy, $< S_n >$ is the average fission-fragment neutron separation energy, and $< \epsilon >$ is the average center-of-mass energy of the emitted neutrons.

There are two important connections between $N(E)$ and $\bar{\nu}_p$, especially for nuclear data evaluation. The first is that the maximum temperature T_m appearing as one of three parameters in $N(E)$, also appears in the expression for $\bar{\nu}_p$ as T_m^2 through Eq. (5). The second is that the average center-of-mass neutron energy $< \epsilon >$ appearing in the expression for $\bar{\nu}_p$, is also the first moment of the center-of-mass neutron spectrum $\Phi(\epsilon)$ corresponding to the laboratory neutron spectrum $N(E)$. These two connections are very significant because they mean that if one has experimental information on either $N(E)$ or $\bar{\nu}_p$, then that information can be used as a constraint in the calculation of the other, unmeasured, observable. We note here, that there are many more measurements of $\bar{\nu}_p$ than there are of $N(E)$.

The Los Alamos model has been used to calculate $N(E)$ and $\bar{\nu}_p$ for neutron-induced and spontaneous fission, and for unmeasured as well as measured systems.

2.2. Summary of Dresden Model

The Dresden model,[8] currently known as the Complex Cascade Evaporation Model (CCEM), accounts for the physical effects of

(a) the distribution of fission-fragment excitation energy in each step of the cascade,

(b) the energy dependence of the inverse process of compound nucleus formation,

(c) the center-of-mass motion of the fission fragments,

(d) the anisotropy of the center-of-mass neutron spectrum,

(e) the complete fission-fragment mass and kinetic energy distributions, and

(f) semi-empirical fission-fragment nuclear level densities.

With information on the above physical effects given in sufficient detail, the prompt fission neutron spectrum in the laboratory system is given by

$$N(E) = \sum_A \int P(A, TKE)\, N(E, A, TKE)\, dTKE\ , \tag{11}$$

where $P(A, TKE)$ is the normalized fission-fragment mass distribution for a fixed value of the total fission-fragment kinetic energy TKE, and $N(E, A, TKE)$ is the laboratory spectrum for fixed fragment mass A and fixed TKE. The sum and integral are over all contributing fragment mass numbers and total kinetic energies, respectively. The fragment spectrum $N(E, A, TKE)$ is given by

$$N(E, A, TKE) = \int_{(\sqrt{E}-\sqrt{E_f})^2}^{(\sqrt{E}+\sqrt{E_f})^2} \frac{\phi(\epsilon, A, TKE)}{4\sqrt{\epsilon E_f}} \left\{ \frac{1 + b[(E - E_f - \epsilon)^2/4\epsilon E_f]}{[1 + (b/3)]} \right\} d\epsilon\ , \tag{12}$$

where E_f is the kinetic energy per nucleon of the fragment, b is the anisotropy coefficient, ϵ is the center-of-mass neutron energy, and $\phi(\epsilon, A, TKE)$ is the center-of-mass neutron spectrum for fixed mass and fixed TKE, given by

$$\phi(\epsilon, A, TKE) = \sum_i \int_{B_i}^{\infty} \phi_i(\epsilon, E^*, A - i)\ P_i(E^*, A, TKE)\ dE^*\ . \tag{13}$$

In this equation, the sum is over the steps i of the cascade while the integral is over the fragment excitation energy E^*, and B_i is the neutron binding energy in a fragment that has emitted i neutrons. Then, $P_i(E^*, A, TKE)$ is the excitation energy distribution *before* step i and is expressed in terms of P_{i-1} and, ultimately, P_0, which is *assumed* Gaussian. Finally, ϕ_i is the center-of-mass neutron energy spectrum given by Weisskopf.[5] The CCEM has been used extensively in studies of the ^{252}Cf(sf) reaction.

2.3. Summary of Hauser-Feshbach Approach

This approach consists of Hauser-Feshbach statistical model calculations of the de-excitation of representative nuclei of the fission-fragment mass and charge distributions, which are then weighted by the (known) yield distribution to construct the total neutron spectrum. This model accounts for the physical effects contained in the Los Alamos and Dresden models and, *in addition*, accounts for

(a) neutron and gamma-ray *competition* in the de-excitation of a given fission fragment,

(b) neutron transmission coefficients T_{lj} from an optical-model potential for *each* fragment considered,

(c) gamma-ray transmission coefficients T_γ for *each* fragment considered, and

(d) the angular momentum distribution $P(J)$ for *each* fragment considered.

Due to space limitations, a detailed description of the Hauser-Feshbach formalism for de-excitation of fission fragments is not presented here. Note, however, that crucial ingredients of these calculations include isospin-dependent neutron optical potentials for fission fragments, fragment initial excitation energy and angular momentum distributions, fragment nuclear level densities, and the partition of available excitation energy between light and heavy fragments. To date, just as in the CCEM, the Hauser-Feshbach approach has been applied primarily to studies of the ^{252}Cf(sf) reaction.

3. Representative Calculations from the Modern Approaches

Calculations using the Los Alamos model are shown in Figs. 1-5. The dependence of $N(E)$ upon the fissioning nucleus and its excitation energy is shown in Figs. 1 and 2 for Eqs. (8) and (9) evaluated in the constant cross section approximation. Figure 1 shows how the spectrum increases at high energies and decreases at low energies as the mass and charge of the fissioning nucleus increases, for thermal-neutron-induced fission. This example shows that $< E_r >$ is increasing faster with mass number A than $< E_f^{tot} >$ is increasing with charge number Z, of the fissioning nucleus. Similarly, Fig. 2 shows how the spectrum increases at high energies and decreases at low energies as the kinetic energy of the incident neutron increases, for the first-chance fission of ^{235}U. Figure 3 compares the exact [$\sigma_c(\epsilon)$ calculated from an optical potential] and approximate [σ_c = constant] versions of the Los Alamos spectrum with experimental data. Clearly, there is a preference for the exact energy-dependent cross section calculation. The Los Alamos model has also been used to calculate the neutron-induced multiple-chance fission neutron spectrum for ^{235}U up through third-chance fission.[4,6] The resulting fission neutron spectrum matrix $N(E, E_n)$, divided by the thermal spectrum $N(E, 0)$, is shown in Fig. 4. Here, the partition of the total available excitation energy into neutron emission prior to fission and neutron emission from accelerated fission fragments leads to suggestions of a *staircase effect* in the peak regions of the matrix and an *oscillatory effect* in the tail regions of the matrix. The former effect is due largely to the pre-fission evaporation neutrons while the latter effect is due largely to the occurence of cooler fission fragments following the emission of a neutron, or two neutrons, prior to fission. Figure 5 shows the application of the Los Alamos model to the calculation of the energy and angle spectrum $N(E, \Theta)$ for the ^{252}Cf(sf) reaction, by the Dresden group.[13] The calculated spectrum, from a refined Los Alamos model (their GMNM model[9]), is shown in the upper portion of the figure while the smoothed experimental data[14] is shown in the lower portion of the figure. The good overall agreement is a rather remarkable achievement by the Dresden group.

Two calculations using the Dresden CCEM are shown in Fig. 6 for the ^{252}Cf(sf) reaction. In this figure the calculations and experimental data are shown as deviations from a Maxwellian spectrum. The Dresden model(CEM), solid curves from Eqs. (11)-(13), for a center-of-mass anisotropy coefficient $b = 0.1$ ($\beta = 0.1$) agrees with experiment much better than the corresponding curve with $b = 0.0$ ($\beta = 0.0$), especially at the low energy end of the spectrum. Clearly, the anisotropy of the center-of-mass spectrum must be taken into account to obtain the most realistic description of the experimental spectrum.[15]

Two calculations using the Hauser-Feshbach approach for the ^{252}Cf(sf) reaction are shown in Figs. 7 and 8. Gerasimenko and Rubchenya[16] consider 18 representative fission fragments, and use a Fermi-gas level density and a Gaussian distribution of initial excitation energy, to obtain the total spectrum shown in good agreement with experiment in Fig. 7. They obtain even better agreement when including a center-of-mass anisotropy coefficient of $b = 0.15$, although this effect is still under study. Seeliger *et al.*[17] have shown in their Hauser-Feshbach calculations that for the right value of a "scaling factor" on the gamma-emission width, the laboratory neutron energy spectrum and neutron total angular distribution are well reproduced. This is shown in Fig. 8 for the neutron total angular distribution.

The many other calculational examples of the modern approaches that exist in the literature cannot be shown here due to space limitations.

4. Conclusions

It is concluded that prompt fission neutron spectra and average prompt neutron multiplicities can be calculated with reasonably good confidence for unmeasured as well as measured systems, and for neutron-induced as well as spontaneous fission. Current limitations to calculating $N(E)$, $\bar{\nu}_p$, $N(E, E_n)$, and $\bar{\nu}_p(E_n)$ with higher accuracy than is now possible include insufficiently accurate information on

- excitation energy partition in fission,
- fission fragment nuclear level densities,
- isospin dependence of global neutron optical-model potentials,
- fission fragment ground-state masses (for calculating $< E_r >$),
- fission fragment mass and charge distributions, and
- fission fragment initial excitation energy and initial angular momentum distributions.

It is believed that, ultimately, the Hauser-Feshbach approach will probably yield the most accurate results in the calculation of the prompt fission neutron spectrum and the average prompt neutron multiplicity. This is because the simultaneous calculation of neutron and gamma-ray emission in competition is the best way to account for the dissipation of the total fission-fragment excitation energy. A second reason is the detail inherent in the Hauser-Feshbach approach: *explicit* treatment of each fragment in a relatively large number of fission-fragment pairs.

The speakers at this symposium are to recommend methods for nuclear data evaluation. To that end, the following recommendations are made for calculating the prompt fission neutron spectrum and the average prompt neutron multiplicity:

1. Current Work: use the Los Alamos model to calculate simultaneously $N(E, E_n)$ and $\bar{\nu}_p(E_n)$ for neutron-induced fission, and $N(E)$ and $\bar{\nu}_p$ for spontaneous fission.

2. Future Work: use the Hauser-Feshbach statistical model to calculate these same quantities and, *simultaneously*, $N_\gamma(E, E_n)$ and $\bar{\nu}_{\gamma p}(E_n)$ for neutron-induced fission, and $N_\gamma(E)$ and $\bar{\nu}_{\gamma p}$ for spontaneous fission.

5. References

1. B. E. Watt, Phys.Rev. **87**, 1037 (1952).
2. N. Feather, "Emission of Neutrons from Moving Fission Fragments," BM-148, British Mission (1942).
3. F. Manero and V. A. Konshin, Atomic Energy Rev. **10**, 637 (1972).
4. D. G. Madland and J. R. Nix, Nucl. Sci. Eng. **81**, 213 (1982).
5. V. F. Weisskopf, Phys. Rev. **52**, 295 (1937).
6. D. G. Madland, R. J. LaBauve, and J. R. Nix, *Proceedings of the IAEA Consultants' Meeting on the Physics of Neutron Emission in Fission, Mito, Japan, 1988*, H. D. Lemmel, Ed. [IAEA, INDC(NDS)-220, 1989], p. 259.
7. D. G. Madland, *Proceedings of the International Conference on Nuclear Data for Science and Technology, Mito, Japan, 1988*, S. Igarasi, Ed. (Saikon Publishing, Tokyo, 1988), p. 759.
8. H. Märten and D. Seeliger, J. Phys. G **10**, 349 (1984).
9. H. Märten and D. Seeliger, Nucl. Sci. Eng. **93**, 370 (1986).
10. H. Märten, A. Ruben, and D. Seeliger, *Proceedings of the IAEA Consultants' Meeting on the Physics of Neutron Emission in Fission, Mito, Japan, 1988*, H. D. Lemmel, Ed. [IAEA, INDC(NDS)-220, 1989], p. 245.
11. W. Hauser and H. Feshbach, Phys. Rev. **87**, 366 (1952).
12. J. Terrell, Phys. Rev. **113**, 527 (1959).
13. D. Seeliger, H. Märten, and A. Ruben, *Contributions to the Theory of Fission Neutron Emission* [INDC(GDR)-057, Vienna, 1990].
14. H. Märten *et al.*, Nucl. Instr. Meth. **A264**, 375 (1988).
15. H. Märten and D. Seeliger, *Proceedings of the Advisory Group Meeting on Nuclear Standard Reference Data, Geel, Belgium, 1984*, (IAEA-TECDOC-335, Vienna, 1985), p. 255.
16. B. F. Gerasimenko and V. A. Rubchenya, *Proceedings of the Advisory Group Meeting on Properties of Neutron Sources, Leningrad, USSR, 1986*, (IAEA-TECDOC-410, Vienna, 1987), p. 208.
17. D. Seeliger, *et al.*, *Prompt Neutron Emission in Nuclear Fission*, [INDC(GDR)-056/L, Vienna, 1989], p. 8.

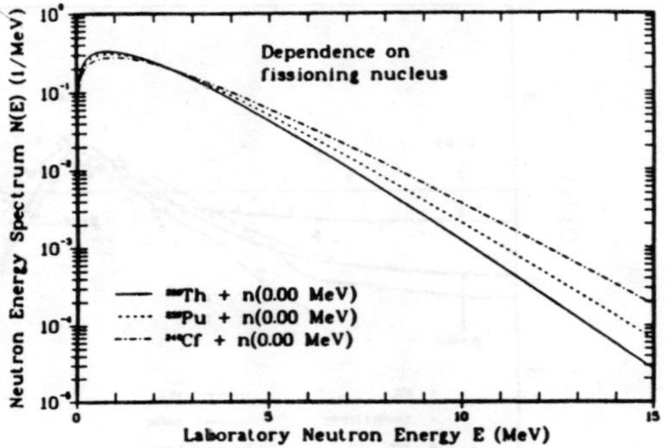

Fig. 1. Los Alamos model: dependence on fissioning nucleus.

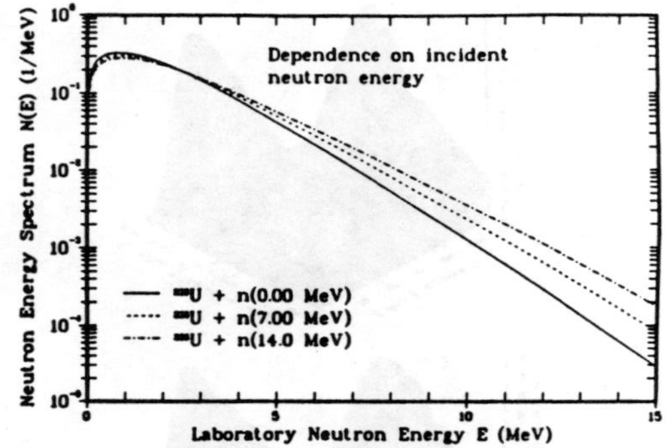

Fig. 2. Los Alamos model: dependence on excitation energy.

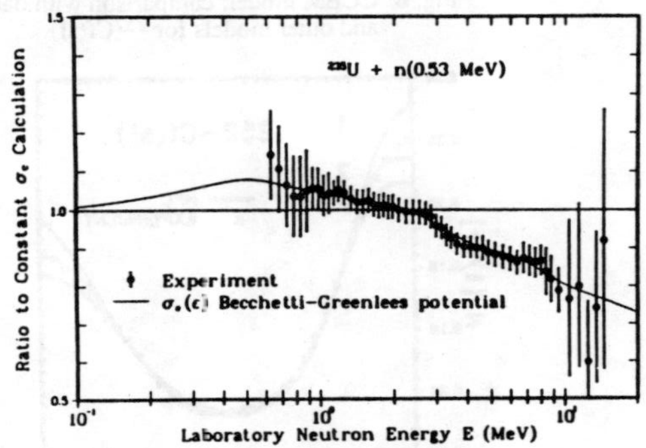

Fig. 3. Los Alamos model: comparison with experiment.

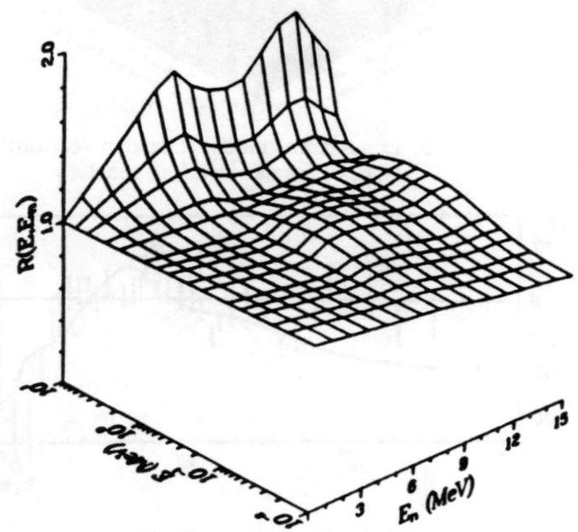

Fig. 4. Los Alamos model: $N(E, E_n)$ for the $n+^{235}U$ system.

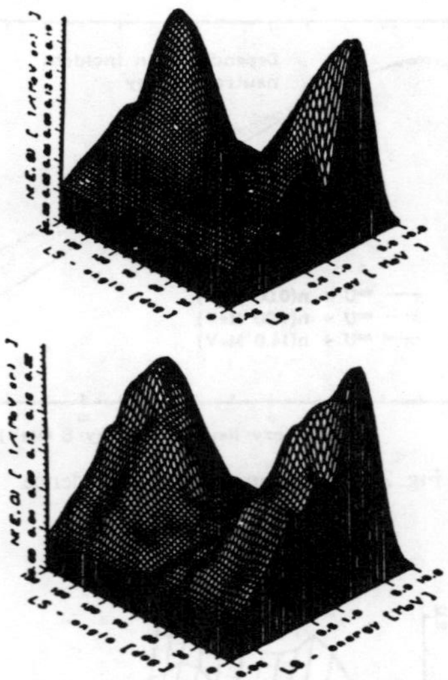

Fig. 5. Los Alamos model (Dresden version): N(E, θ) for the ^{252}Cf(sf) reaction.

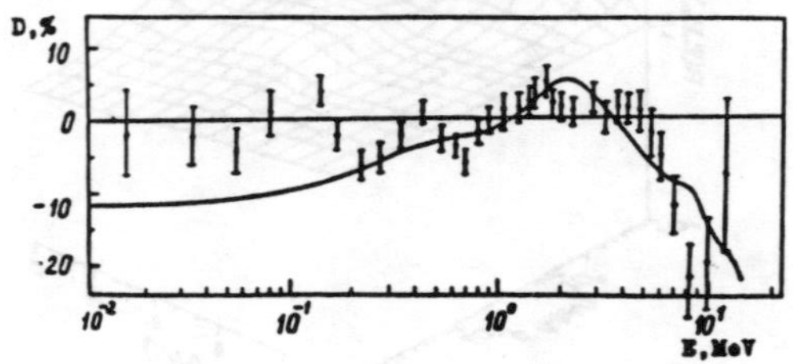

Fig. 7. Hauser-Feshbach approach for the ^{252}Cf(sf) reaction.

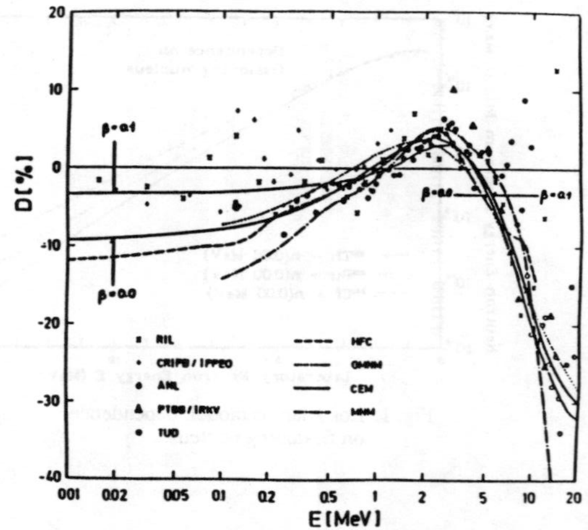

Fig. 6. CCEM model: comparison with data and other models for ^{252}Cf(sf).

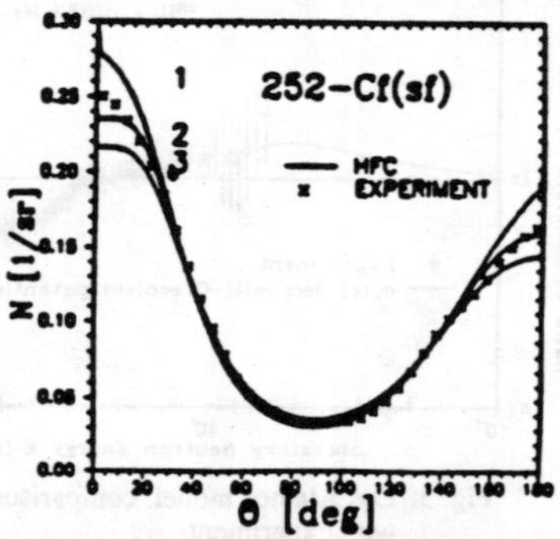

Fig. 8. Hauser-Feshbach approach for the ^{252}Cf(sf) reaction.

LEVEL DENSITY MODELS FROM PRACTICAL POINT OF VIEW

A.V.Ignatyuk

Institute of Physics and Power Engineering,
249020 Obninsk, Kaluga Region, Russia

For a practical application of the statistical models it is very important to achieve the level density description that reproduces both the average distances between neutron resonances and the observed cumulative numbers of low-lying levels. In recent years such experimental data were analyzed in various microscopic and phenomenological approaches that together with the shell and superfluid effects take into account also the collective enhancement of the level densities. The systematics of parameters obtained are reviewed and the problems of the collective enhancement damping are discussed.

1.Introduction

The statistical properties of excited nuclear levels have been a matter of concern and study for over fifty years. One of the basic statistical properties of such levels is their density. For the description of the level densities the Fermy-gas and constant temperature models are used very frequently with parameters obtained from fitting some experimental data [1]. But the physical assumptions upon which both these models are based are not sophisticated enough to allow them to account correctly for variations of level densities over the wide energy interval from the ground state to the energies much higher than the neutron separation energy. This is not surprising considering that the models discussed were first put forward over fifty years ago, when nuclear physics was in its infancy.

Some of the most important concepts, upon which current understanding of the structure of low-lying nuclear levels is based, include the shell effects, pairing correlations and collective phenomena. All these concepts have been incorporated into the Generalized Superfluid Model (GSM) developed by many authors over the last 20 years [2]. For easy fitting of experimental data the phenomenological versions of this model were developed too [2-5].

For any practical applications of the statistical models it is very important to obtain the parameters of the level density description from the reliable experimental data. The cumulative numbers of low-lying levels and the average distances between neutron resonances are usually used as such data. In this report the main problems of the corresponding parameter systematics are discussed.

2. Conformity of microscopic and phenomenological calculations

In order to obtain a consistent description of the shell, superfluid and collective effects it is necessary to use for the total level density the expression

$$\rho(U) = \rho_{qp}(U)\, k_{vibr}(U)\, k_{rot}(U), \tag{1}$$

where $\rho_{qp}(U)$ is the level density of quasi-particle excitations, k_{vibr} and k_{rot} take account of the collective enhancement of the level density due to vibrational and rotational excitations. The expressions needed to calculate $\rho_{qp}(U)$, k_{vibr} and k_{rot} are given in Refs. [2-5].

Fig.1 shows the results of the microscopic calculations of the total density of excited states for the double magic nucleus ^{208}Pb. Because the single-particle schemes are obtained usually from the analysis of the low-lying levels of the nearest odd nuclei the density calculations of all possible quasi-particle excitations do not require any fitting parameters. An addition of the most important 2^+ and 3^- collective excitations to the quasi-particle ones increases the level density only a little. We can see that the calculations both in the combinatorial and statistical approaches describe well enough the experimental data for low-lying levels and neutron resonances. The calculations reproduce also the experimental values of the spin cut-off parameter for a low energy region.

On the base of the microscopic calculations we can define the level density parameter $a = S^2/4U$, where S is the entropy and U is the excitation energy of the nucleus. The energy dependence of this parameter is shown in the upper part of the Fig.1. These changes of the level density parameter are very important for a consistent description of the shell effects in excited nuclei.

In the phenomenological approach we use almost the same equations of the superfluid model for the density of quasi-particles excitation and the collective enhancement coefficients but employ the simplified expression for the level density parameter

$$a(U,Z,A) = \begin{cases} \tilde{a}(A)\left\{1 + \dfrac{f(U-E_{cond})\delta\mathcal{E}_o(Z,A)}{U-E_{cond}}\right\} & \text{for } U \geq U_{cr}, \\ a(U_{cr}) & \text{for } U < U_{cr}. \end{cases} \tag{2}$$

Here $\tilde{a}$ is the asymptotic value of the level density parameter at high excitation energy, the dimensionless function $f(U) = 1 - \exp(-\gamma U)$ determines the energy behavior of this parameter and $\delta\mathcal{E}_0(Z,A)$ is the shell correction in the nuclear binding energy description [3]. The difference of the level density parameters obtained in the microscopical and phenomenological descrip-tions is shown in the upper part of thc Fig.1 .

The change of the energy dependence of the level density parameter (2) below and above the critical excitation energy U_{cr} is related to the phase transition of the excited nucleus from the superfluid low-energy state to the normal state in the high energy region. The critical energy is directly connected with the pairing correlation function Δ_o that is the second basic parameter of the phenomenological description. If we use the same values of the correlation functions and the collective enhancements the level densities calculated in the microscopic and phenomenological approaches will be very similar [6].

3. Ambiguities of the collective enhancement description

Using the phenomenological model described above we can extract the values of the level density enhancement coefficients from the experimental data on the densities of neutron resonances. In such analysis the asymptotic values of the level density parameters were defined as $\tilde{a} = 0.073\ A + 0.115\ A^{2/3}\ MeV^{-1}$, the shell corrections were taken from Ref.[7] and the correlation functions were approximated as $\Delta_o = 12\ A^{-1/2}$ MeV. The coefficients obtained are shown in the upper part of Fig.2. In the lower part the values of similar coefficients calculated in the adiabatic approximation are given. A correlation of both coefficients is very strong but as a rule the adiabatic evaluations give higher values of coefficients than the ones extracted from the observed density of neutron resonances. The difference of these two definitions of the level density enhancement factors demonstrates that the damping of the enhancement coefficients for high excited nuclei must be taken into account.

During the last ten years some microscopic models were developed for the consideration of the collective effects in highly excited nuclei [2,8-10]. The results of all these models demonstrate the important role of nonadiabatic effects in heated nuclei and the corresponding damping of the level density enhancement factors with the increasing of excitation energies. Various phenomenological descriptions of this damping were also proposed [4-6,8]. The temperature dependencies of the enhancement coefficients obtained in various approaches are shown in Fig.3 for typical rotational and vibrational nuclei . Big uncertainties in evaluations of the collective enhancement damping are seen and unfortunately up to now we have no experimental data that could be used to verify the predictions of different models.

The damping of the factors k_{rot} and k_{vibr} may be connected with limitations on the number of coherent bosons $\mathfrak{N}_b$ in construction of possible collective and quasi-particle nuclear excitations [8]. In some simplified consideration these coefficients may be written in the form

$$k_{rot}(t) = \mathfrak{F}_{\perp} t \left[1 - \exp\left(-2\mathfrak{N}_b^2 / \mathfrak{F}_{\perp} t\right)\right] ,$$

$$k_{vibr}^{(\lambda)}(t) = \left\{ \frac{1 - \exp\left[-(\mathfrak{N}_b + 1) w_\lambda / t\right]}{1 - \exp(-\omega_\lambda t)} \right\}^{(2\lambda+1)} ,$$

where $\mathfrak{F}_{\perp}$ is the perpendicular moment of inertia and ω_λ is the energy of collective excitations with the multipolarity λ. For $\mathfrak{N}_b \Rightarrow \infty$ these expressions give the adiabatic formulas for corresponding coefficients and the temperature dependence of the number of bosons may be approximated as $\mathfrak{N}_b(t) = \mathfrak{N}_o \exp(-t/t_b)$. Some other formulas for the enhancement factor damping were discussed in Refs. [4-6,8] . All these formulas include at least one or two parameters that can fluctuate from one nucleus to another.

4. Local and general systematics of the level density parameters

For many practical applications of the statistical models the necessity arises to get the exact agreement of the calculated level densities with definite experimental data. To account possible defects of global parameterization of the pairing correlation functions and collective enhancement damping we decided to introduce in the phenomenological GSM an additional parameter defined as the shift of the excitation energies δ_{shift} . Within the framework of such approach the set of parameters a and δ_{shift} was obtained which described simultaneously the cumulative numbers of low-lying levels and observed neutron resonance densities for all nuclei where corresponding experimental data existed [5]. These parameters are shown in Fig.4. The individual parameters fluctuate a little and these fluctuations reflect to some degree the simplifications included in the phenomenological GSM. To remove these fluctuations we need a better understanding of the collective enhancement damping for spherical, deformed and intermediate nuclei.

For any practical applications the individual parameters are preferable of course. Some examples of the experimental data description for such parameters are shown on Fig.5. Uncertainties of parameters are not very important for a prediction of the level densities in an intermediate energy region if experimental data for the neutron resonances and low-lying levels were chosen correctly.

For the study of the nuclear level densities the analysis of evaporation spectra of different particles is great interest. Fig.6 shows the level densities obtained for the (α,p), (p,n) and (α,n) reactions leading to the same residual nuclei. The energy dependencies of the level densities obtained from the spectrum analysis are in good agreement with calculations based on the individual parameters of GSM.

On the other hand for many tasks we need the level density parameters for nuclei with experimental data absent. For such goals the global parameters may be used effectively. Also some local systematics of parameters may be proposed based on extrapolations of the isotopic or isotonic changes of the individual parameters. In many cases experimental data on the cumulative number of low-lying levels might be very useful because such data permit to fit one of the individual parameters keeping the global systema-tics for others.

5. Concluding remarks

Today it seems almost obvious that in systematic analysis of the level densities of excited nuclei we must use models which are more consistent than the Fermi-gas, but inevitably more complex. The success of the Generalized Superfluid Model is attributed to the inclusion of the main well-known characteristics of nuclear structure: pairing correlations, shell effects and collective excitations. The complexities of the model are justified by the mutual consistency of the parameters obtained from the various experi-mental data and also by the close interconnection of the theoretical con-cepts used to describe the structure of low-lying nuclear levels and the statistical properties of highly excited nuclei.

References:

1) Ericson T.: Adv. Phys. 9, 425 (1960);
Gilbert A.,Cameron A.: Can. J. Phys., 43, 1446 (1965);
Dilg W., Schantl W., Vonach H., Uhl M.: Nucl. Phys.,

2) Ignatyuk A.V.: Statistical Properties of Excited Atomic Nuclei, Energoatomizdat (Moscow, 1983), translated in IAEA Report INDC-233(L), (1985).

3) Ignatyuk A.V., Istekov K.K., Smirenkin G.N.: Sov. J. Nucl. Phys., 29, 450 (1979).

4) Blochin A.I., Ignatyuk .V., Shubin Yu.N.: Sov. J. Nucl. Phys., 48, 371 (1988).

5) Grudzevich O.T., Ignatyuk A.V., Plyaskin V.I., Zelenetsky A.V.: Proc.Conf.on Nuclear Data for Science and Technology, Mito, JAERY, p.187 (1988).

6) Rastopchin E.M., Svirin M.N., Smirenkin G,N.: Sov. J. Nucl. Phys., 52, 1258 (1990).

7) Myers W.D.: Droplet Model of Atomic Nuclei. IFI/Plenum, 1977.

8) Hansen G., Jensen A.: Nucl. Phys., A406, 236 (1983).

9) Maino G., Mengoni A., Ventura A.: Phys. Rev., C42, 988 (1990).

10) Lauritzen B., Bertsch G.: Phys. Rev.,C39, 2412 (1989);
Lauritzen B., Bertsch G., Bortignon P., Broglia R.: Phys. Lett., B256, 329 (1989).

11) Grudzevich O.T., Ignatyuk A.V., Plyaskin V.I., Zelenetsky A.V.: Sov. J. Nucl. Phys., 53, 143 (1991).

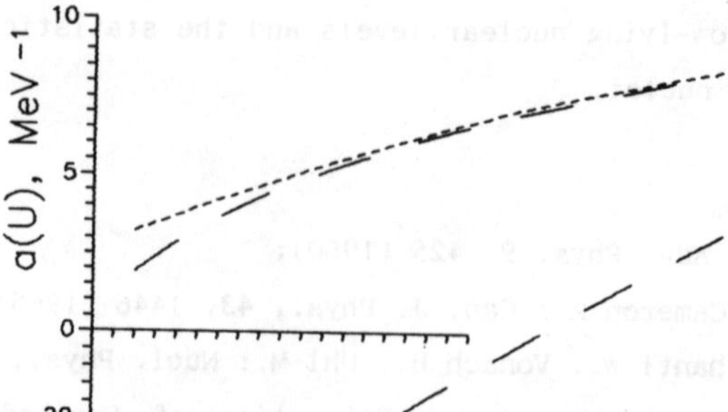

Fig.1. Level densities of ^{208}Pb calculated in various approaches: the combinatoric calculations of quasi-particle excitations (short-dash histogram), the combinatoric calculations with account of collective excitations (solid histogram) and the statistical calculations of quasi-particle and collective excitations (dot-dash curve). Experimental points are obtained from the observed low-lying levels (+) and neutron resonances (*). In the middle part the experimental and calculated values of the spin cut-off parameter are shown. In the upper part the energy dependence of the level density parameter are shown obtained in the microscopic calculations (point-dash curve) and phenomenological description (short-dash curve).

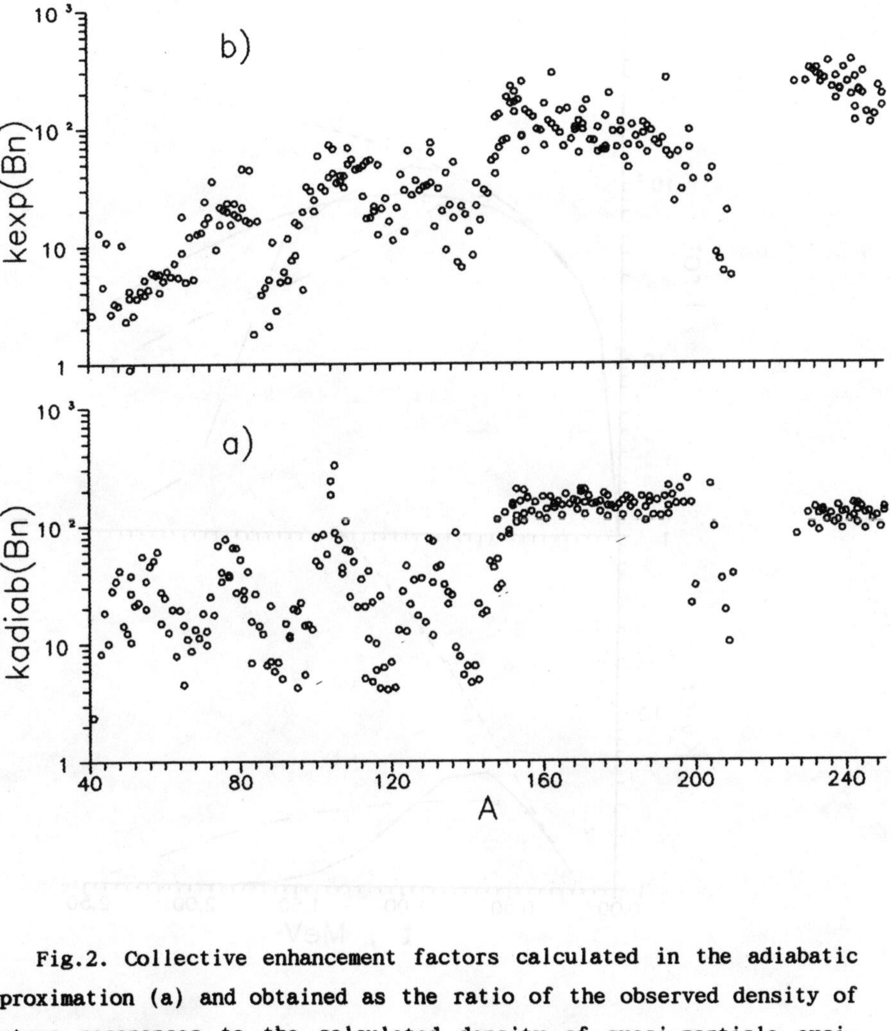

Fig.2. Collective enhancement factors calculated in the adiabatic approximation (a) and obtained as the ratio of the observed density of neutron resonances to the calculated density of quasi-particle excitations (b).

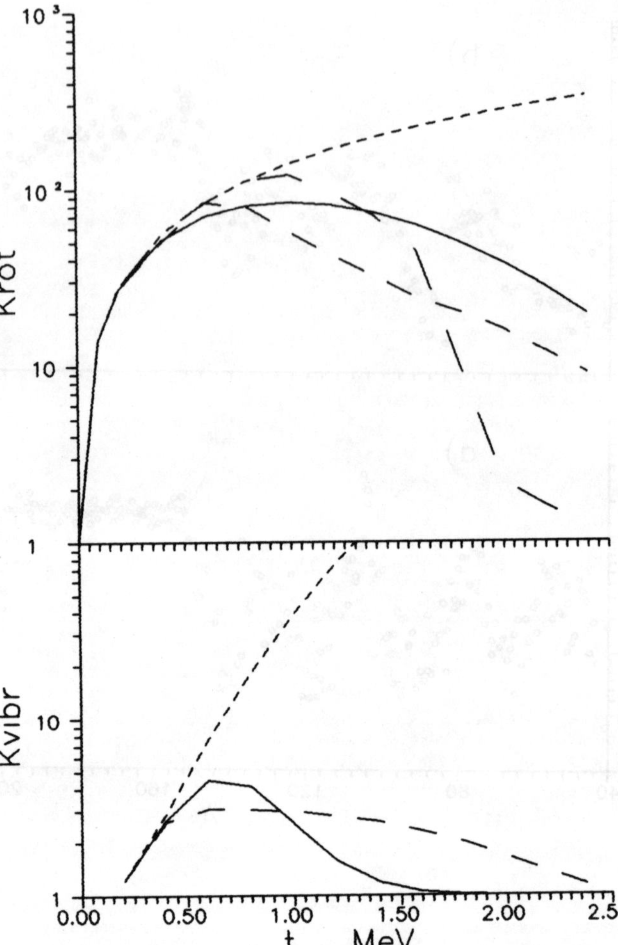

Fig.3. Temperature dependencies of the collective enhancement coefficients obtained with various models: the adiabatic approximation (short-dash curves), the interacting boson model (long-dash curves), the SU-3 model (dot-dash curves) and the phenomenological description (solid curves).

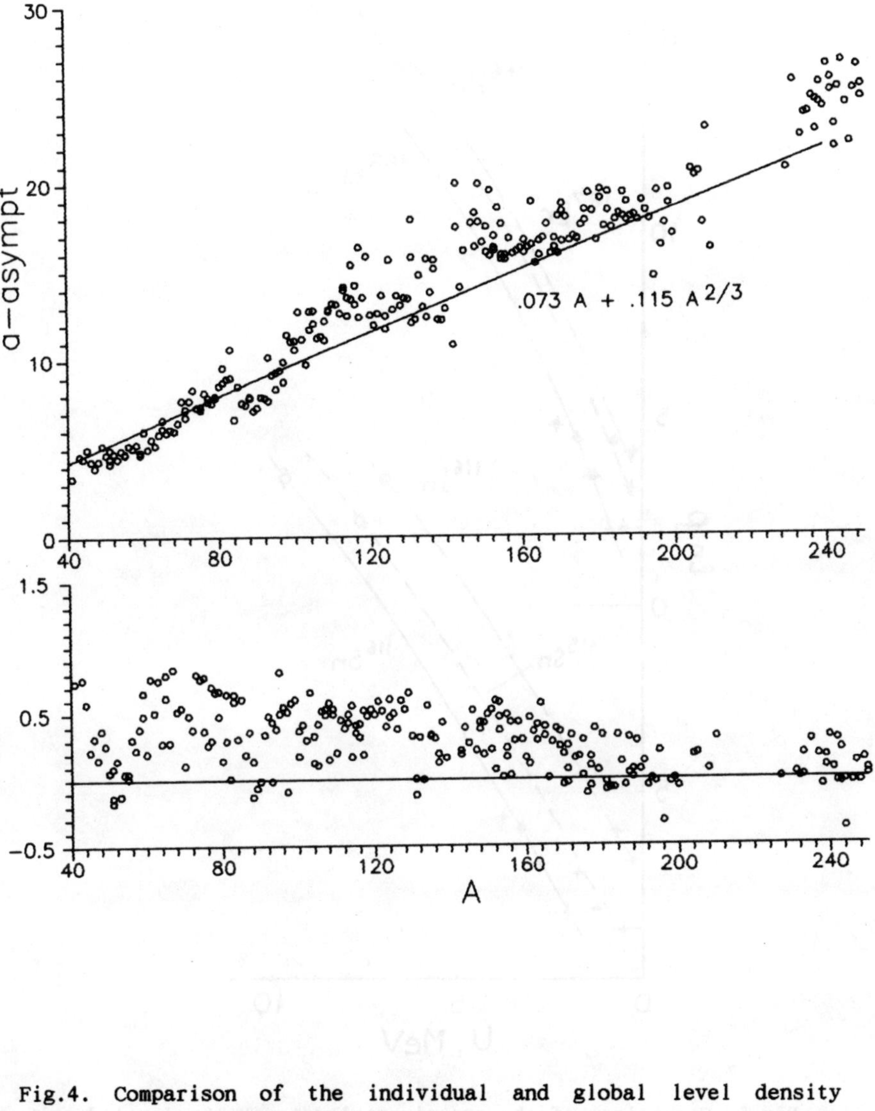

Fig.4. Comparison of the individual and global level density parameters of the phenomenological GSM.

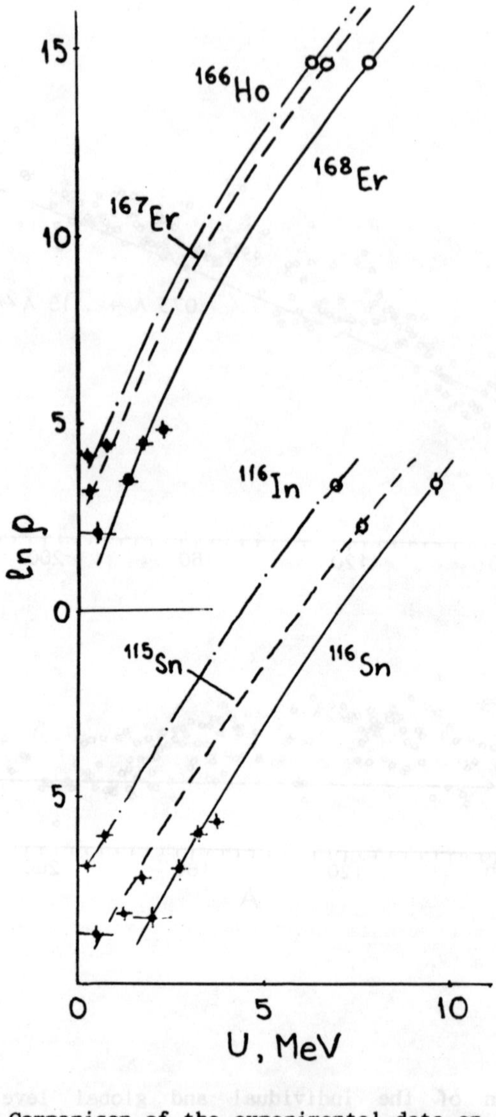

Fig.5. Comparison of the experimental data on the level density of some even-even, odd and odd-odd nuclei with the phenomenological description.

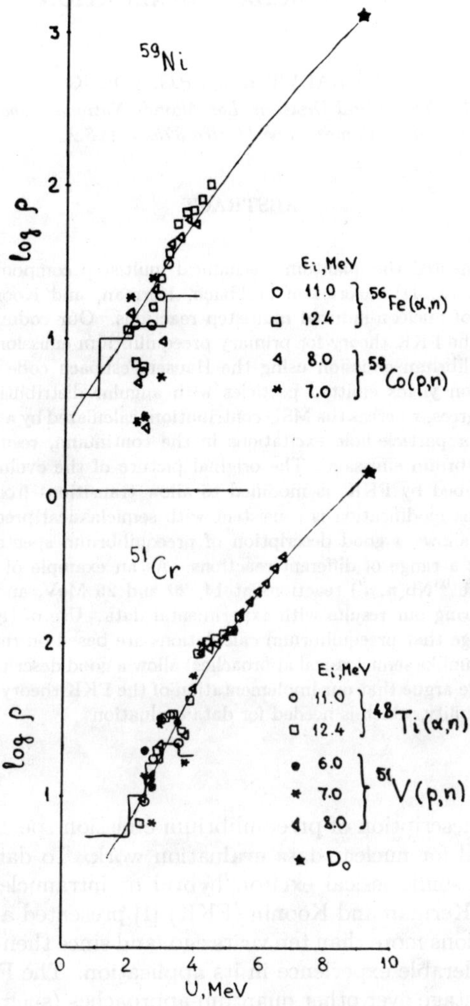

Fig.6. Comparison of the experimental data obtained from the evaporation spectra of the (p,n), (α,p) and (α,n) reactions with the phenomenological description.

APPLICATION OF MULTISTEP COMPOUND AND MULTISTEP DIRECT MODELS FOR DATA EVALUATION

M.B. CHADWICK and P.G. YOUNG
Group T2, Theoretical Division, Los Alamos National Laboratory, Los Alamos, New Mexico 87545, U.S.A.

ABSTRACT

We have implemented the quantum mechanical multistep compound (MSC) and multistep direct (MSD) theories of Feshbach, Kerman, and Koonin (FKK) for the calculation of nucleon-induced multistep reactions. Our code system, FKK-GNASH, uses the FKK theory for primary preequilibrium emission and describes subsequent equilibrium emission using the Hauser-Feshbach code GNASH. The MSC contribution yields emitted particles with angular distributions symmetric about ninety degrees, whereas the MSD contribution, calculated by averaging DWBA cross sections for particle-hole excitations in the continuum, results in forward-peaked preequilibrium emission. The original picture of the evolution of the reaction, as described by FKK, is modified to allow transitions from the MSD to MSC chain. This modification is consistent with semiclassical preequilibrium descriptions, and allows a good description of preequilibrium spectra and angular distributions for a range of different reactions. As an example of our methodology, we calculate ^{93}Nb(n,n') reactions at 14, 20 and 26 MeV, and ^{93}Nb(n,p) at 14 MeV, comparing our results with experimental data. Use of the FKK theory has the advantage that preequilibrium calculations are based on rigorous physical principles, and (unlike semiclassical approaches) allow a good description of angular distributions. We argue that our implementation of the FKK theory maximizes the level of predictability, which is needed for data evaluations.

1. Introduction

An accurate description of preequilibrium emission spectra and angular distributions is essential for nuclear data evaluation work. To date, all such descriptions have relied on semiclassical exciton/hybrid or intranuclear cascade models. However, Feshbach, Kerman and Koonin (FKK) [1] presented a quantum theory of preequilibrium reactions more than ten years ago, and since then a number of groups have obtained considerable experience in its application. The FKK preequilibrium theory has the advantage over other quantum approaches (such as that of Nishioka *et al.* [NVWY] [2]) that while it is based on fundamental physical principles, it is relatively straightforward to apply computationally.

In this paper we describe a calculational procedure for determining cross sections using our code, FKK-GNASH, which applies the FKK theory for preequilibrium emission, and treats equilibrium emission with the Hauser-Feshbach code GNASH. Our aim is to combine the fundamental physical principles embodied in the FKK theory with the predictability required for evaluation work. This approach enables angular distributions to be calculated theoretically,

In section 2 we describe how we apply the FKK theory, and discuss an important modification: inclusion of transitions from the multistep direct to multistep compound preequilibrium chain, which was ignored by FKK. In section 3, as an example of our methodology, we apply this theory to describe neutron-induced reactions on ^{93}Nb at 14, 20, and 26 MeV, considering the whole emission spectra and angular distributions. We present our conclusions in Section 4.

2. Theory

2.1. General

The FKK theory describes the evolution of a nuclear reaction as passing through a series of particle-hole ($p-h$) states of increasing complexity. In the early stages of the reaction, where the excitation energy is concentrated on a small number of particles and holes, particle emission may occur yielding the hard preequilibrium spectrum. An important development due to FKK is the distinguishing of multistep direct (MSD) and multistep compound (MSC) processes. The incident reaction flux can either populate a chain of $p-h$ states which are all bound (Q-space) giving MSC emission, or a chain of $p-h$ states in which at least one particle is unbound (P-space) giving MSD emission, see Fig. 1.

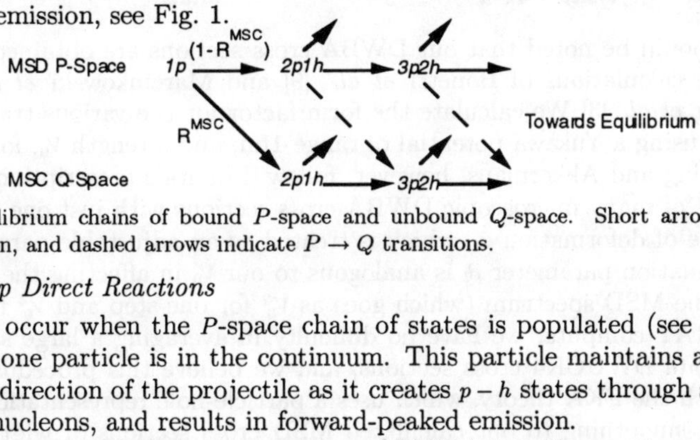

Fig.1. Preequilibrium chains of bound P-space and unbound Q-space. Short arrows represent possible emission, and dashed arrows indicate $P \rightarrow Q$ transitions.

2.2. Multistep Direct Reactions

These occur when the P-space chain of states is populated (see Fig. 1), so that at least one particle is in the continuum. This particle maintains a 'memory' of the initial direction of the projectile as it creates $p-h$ states through scatterings with bound nucleons, and results in forward-peaked emission.

MSD theory represents an extension of DWBA into the continuum, and can be derived using a Lippmann-Shwinger expansion of the transition amplitude [3]. The FKK theory differs from the theories of Tamura *et al.* [4] and Nishioka *et al.* [2] in the quantum-statistical assumptions concerning multiple scattering processes, as discussed by Koning and Akkermans [3]. It has the advantage that random matrix element assumptions are made, allowing multiple-scattering processes to be expressed in terms of a convolution of single-step scattering cross sections. For the incident energies that we consider, only one- and two-step scattering needs to be included. The one-step cross section is given by

$$\frac{d^2\sigma(E,\Omega\leftarrow E_0,\Omega_0)}{d\Omega dE}_{1-step}=\sum_l(2l+1)\rho(1p,1h,E_0-E,l)\left\langle\left[\frac{d\sigma(E,\Omega\leftarrow E_0,\Omega_0)}{d\Omega}\right]_l^{DWBA}\right\rangle, \quad (1)$$

where:-

- l is the orbital angular momentum transfer, which equals the composite spin since we follow FKK in assuming spin-zero nucleons and target nucleus.
- $\rho(1p,1h,E_0-E,l)$ is the density of $1p1h$ states with energy E_0-E and spin l. We use the Williams expression [5] with the level density parameter $g=A/13$, and pairing corrections of Dilg *et al.* [6]. A Gaussian angular momentum distribution is assumed, with spin cut-off, $\sigma_n^2=0.24nA^{2/3}$ [7], where $n=p+h$.
- $\left\langle\left[\frac{d\sigma(E,\Omega\leftarrow E_0,\Omega_0)}{d\Omega}\right]_l^{DWBA}\right\rangle$ is the average of DWBA cross sections exciting $1p1h$ states of energy E_0-E consistent with angular momentum and parity conservation. The $1p1h$ states are obtained from a spherical Nilsson model [8].

Two-step scattering is described by a convolution of one-step cross sections,

$$\frac{d^2\sigma(E,\Omega\leftarrow E_0,\Omega_0)}{d\Omega dE}_{2-step}=\frac{m}{4\pi^2\hbar^2}\int d\Omega_1\int dE_1E_1\,\frac{d^2\sigma(E,\Omega\leftarrow E_1,\Omega_1)}{d\Omega dE}\,\frac{d^2\sigma(E_1,\Omega_1\leftarrow E_0,\Omega_0)}{d\Omega dE}. \quad (2)$$

It should be noted that our DWBA cross sections are obtained in a similar way to the calculations of Bonetti *et al.* [9] and Marcinkowski *et al.* [10], and not Koning *et al.* [3] We calculate the form factor for the various transitions with DWUCK4 using a Yukawa potential of range 1fm, and strength V_0, for $1p1h$ excitations. Koning and Akkermans, however, follow Tamura's prescription of replacing the average of many microscopic DWBA cross sections with just one fictitious collective state of deformation β_l, which is treated as an adjustable parameter. Thus their deformation parameter β_l is analogous to our V_0 in affecting the overall magnitude of the MSD spectrum (which goes as V_0^2 for one-step and V_0^4 for two-step). With a CRAY computer we have no difficulty in averaging a large sample of microscopic $1p1h$ DWUCK4 cross sections, and we believe this procedure is more in keeping with the FKK theory, which uses a particle-hole representation. We apply a Gaussian smoothing to our calculated MSD cross sections of width 2 MeV, to remove artificial fluctuations which would not arise if we used deformed Nilsson single-particle states.

2.3. Multistep Compound Reactions

These occur when a chain of $p-h$ states is populated in which all the particles are bound (the Q-space of Fig. 1). The MSC processes contribution is greatest for low incident energies, becoming insignificant at high incident energies (above about 50 MeV), due to the decreasing probability of forming bound preequilibrium states with increasing energy. Matrix elements for MSC emission involving different total angular momenta, parity, and other quantum numbers, are assumed random so that no interference terms remain on averaging, yielding angular distributions that are symmetric about 90°. The MSC cross section is given by

$$\frac{d\sigma}{dE} = \pi\lambda^2 \sum_J (2J+1)\ [R^{MSC} T_J]\ \sum_{N=1}^{2} \sum_{lS\nu} \frac{\langle \Gamma_{NJ}^{\uparrow lS\nu}(E)\rangle}{\langle \Gamma_{NJ}\rangle} \prod_{M=1}^{N-1} \frac{\Gamma_{MJ}^{\downarrow}}{\Gamma_{MJ}}, \tag{3}$$

where:-

- $R^{MSC} T_J$ is the transmission coefficient for producing bound $2p1h$ states of spin J, multiplied by a reduction factor R^{MSC} (discussed below) since only a fraction of the reaction flux enters this initial state.
- N labels each preequilibrium class, e.g. $N = 1$ denotes $2p1h$. We include MSC emission from two preequilibrium stages, since the ratio of emission to total MSC widths decreases strongly with increasing N. ν labels the three different emission modes corresponding to $\Delta n = 0$, -2 and +2 [1].
- J, l, S denote the composite system spin, orbital angular momentum of the emitted particle, and residual nucleus spin respectively. Following FKK, all nucleon spins are assumed to be zero, as are the spins of the target and residual nuclei. We showed in Refs. [11, 12] that this is a reasonable approximation.
- $\langle \Gamma_{NJ}^{\uparrow lS\nu}(E)\rangle$ is the differential emission width for MSC preequilibrium decay of a nucleon with energy E, described below.
- $\langle \Gamma_{NJ}\rangle$ is the total width of the MSC stage N, J.
- $\prod_{M=1}^{N-1} \frac{\Gamma_{MJ}^{\downarrow}}{\Gamma_{MJ}}$ is the depletion factor.

The emission and damping widths for a given preequilibrium state are calculated microscopically by summing the transition matrix element over final states and averaging over initial states. This yields widths of the form

$$\langle \Gamma_{NJ}^{\uparrow lS\nu}(E)\rangle \text{ and } \langle \Gamma_{NJ}^{\downarrow}\rangle \sim X\ Y\ I^2, \tag{4}$$

where X contains all the angular momentum coupling, Y is the accessible phase space for the transition, and I is the radial overlap integral between the initial and final single-particle and -hole states. As in our MSD calculations, we use $g = A/13$, and the spin cut-off and pairing corrections from Refs. [7, 6] respectively. It is essential that Y-functions using only bound preequilibrium states are used, as described in Eqs. (21-24) of Ref. [13]. Such Y functions results in MSC emission spectra which differ considerably from those originally obtained by FKK. We calculate overlap integrals with the original FKK prescription, i.e. using constant nuclear wavefunctions, and a residual-interaction potential described by a delta-interaction of strength V_0. In previous works we determined overlap integrals using more realistic nuclear and scattering wavefunctions [14, 12]. Such an approach runs into difficulties since it is often artificially sensitive to the choice of interacting states, and also the basis set of states needed for a realistic description is often larger than practically calculable. Thus, for application purposes, the original approach of FKK is preferable. The MSC emission spectrum is independent of V_0, which cancels in the ratio of emission to total widths.

Neutron-proton distinguishability is accounted for in $N = 1$ MSC emission by reducing a two-component formalism to a one-component formalism with correction factors applied to the escape widths. We find factors of 1.037 and 1.111 for Δn=0, 2 respectively, in (n, n') or (p, p') reactions, and 0.593, 0.888 for Δn=0, 2 in (n, p) or (p, n) reactions, obtained from Ref. [14] when the flux into the Q-chain $2p1h$ stage is set equal for one and two-component approaches.

2.4. Hauser-Feshbach decay

After calculating the primary MSD and MSC preequilibrium emission, we use the Hauser-Feshbach code GNASH [15] to describe both the primary decay of compound nuclei that have survived preequilibrium decay, as well as the subsequent compound nucleus decay of residual nuclei remaining after particle emission.

The original FKK formulation calculated equilibrium emission using the "r-stage" formalism, which extended techniques used for calculating preequilibrium decay into the equilibrium regime, and FKK point out that formally it is very similar to Hauser-Feshbach theory. Since Hauser-Feshbach theory has been so successful in describing compound nucleus decay, we use it, rather than the "r-stage" approach.

2.5. Modification to FKK: $P \rightarrow Q$ Transitions

Here we present a modification to the FKK theory which, we believe, must be made in order to obtain a satisfactory description of data. Our modification leads to an increased stress on MSD processes, diminishing importance of MSC processes, and the feature of $P \rightarrow Q$ transitions, which were discounted by FKK. The need for $P \rightarrow Q$ transitions in the FKK theory was first pointed out in Ref. [11], and has since been discussed in a number of works, e.g. Refs. [16, 17]. Herman has also stated the need for this phenomena when applying the NVWY multistep theory [19]. While a theoretical formalism describing $P \rightarrow Q$ transitions has been suggested [18], we show how a detailed understanding of such mechanism is unnecessary in practical calculations, and that a requirement of unitarity is sufficient to handle these processes.

In the original FKK picture of the reaction, the incident flux splits immediately at the first interaction into the P and Q chains. Since, for example, 14 MeV reactions typically exhibit about 10% of primary emission which is forward-peaked, it was assumed that 90% of the reaction flux entered the Q chain at the initial $2p1h$ stage. However, improvements to MSC calculations (specifically, the improved Y-functions) lead to much more MSC emission for a given population of Q-states than FKK originally calculated. The consequence is that if one allows 90% of the reaction flux to enter the Q-chain at the initial $2p1h$ stage, far too much MSC preequilibrium emission occurs, overpredicting the data.

Historically, the omission of $P \rightarrow Q$ transitions also has some other origins. Firstly, it is only recently that we have developed the capability to calculate MSD as well as MSC emission, and unless one attempts to describe the whole emission spectrum (and angular distributions), it is easy to miss this effect. Most earlier analyses isolated one part of the spectrum, and assumed that only MSC, or only MSD, is important. As an example, it has been frequently assumed that MSD is insignificant at backward angles, allowing a comparison of MSC calculations with backward-angle data. Our results show this to be in error, because MSD emission, while forward peaked, still contributes relatively strongly at back angles. Secondly, we think that FKK overstated the case in claiming that "the experimental angular distributions are symmetric about 90°" when referring to Grimes' (p,n) data. At the higher emission energies this is not so, and consequently the role of MSC was probably overemphasized.

If the fraction of reaction flux entering the Q chain at the $2p1h$ stage, R^{MSC}, is reduced, and flux is allowed to cascade from $P \to Q$ states at later stages in the reaction (see dashed arrows in Fig. 1), then the amount of MSC emission is reduced, but the fraction of symmetric emission can still be maintained at the required (usually large) level. This is so because preequilibrium emission from stages more complex than $2p1h$ is relatively insignificant. In accordance with statistical ideas usually used in the exciton model, we can estimate the fraction of flux into the initial $2p1h$ Q-stage (at an energy E) as

$$R^{MSC} = \frac{\rho^B(2p, 1h, E)}{\rho(2p, 1h, E)}, \tag{5}$$

i.e. the ratio of bound to total $2p1h$ density, calculated with the restricted and unrestricted Williams expressions [12]. This statistical division is consistent with the FKK picture of a partitioning of bound and unbound states, and is expected to yield reasonable results since the Kalbach systematics [20] make a similar assumption about the MSC-MSD division, and yield very reasonable angular distribution systematics. Also, theoretical calculations [11] of the entrance strength function to the bound $2p1h$ stage indicate a small R^{MSC} compatible with (5). As shown in Fig. 1, we expect the $P \to Q$ transitions to occur predominantly via $\Delta n = +2$ processes due to the higher accessible phase space, and also phase space arguments indicate that $P \to Q$ transitions should be very likely. It would be possible to apply the same kind of reasoning as in Eq. (5), extended to more complex stages, to determine the $P \to Q$ flux for different preequilibrium stages. But since the probability of MSC emission is so insignificant for stages beyond the $2p1h$, in practice the exact details of this mechanism are not important, and it can be assumed that all the $P \to Q$ flux eventually decays by Hauser-Feshbach emission. This calculational procedure is therefore entirely well defined:-

1. Calculate MSC emission, using Eq. (5). There are no free parameters.
2. To determine the MSD spectrum, consider angle-integrated data at a high emission energy where Hauser-Feshbach decay is insignificant, but below the energies where collective effects appear. V_0, and hence the MSD spectrum, is determined by equating MSD with the difference between data and MSC.

3. Calculate collective effects using Kalka's approach [21] or coupled-channels.
4. Calculate Hauser-Feshbach emission with GNASH. Unitarity requires that the flux available for primary Hauser-Feshbach decay is $\sigma_{HF} = \sigma_R - \sigma_{MSD} - \sigma_{MSC}$, σ_R being the reaction flux. This implicitly accounts for $P \to Q$ transitions and assumes (reasonably) that preequilibrium decay following such transitions is negligible. Calculate subsequent equilibrium decay of all residual nuclei.
5. We take the residual nucleus spin distribution after preequilibrium emission from the Hauser-Feshbach spin distribution calculation, and we assume that MSC and Hauser-Feshbach emission are isotropic.

3. Results

As a test of our calculational procedure for applying the FKK theory, we have analyzed experimental data for neutron-induced reactions on ^{93}Nb at 14, 20 and 26 MeV. There have been many measurements of 14 MeV ^{93}Nb(n,p) and ^{93}Nb(n,n') reactions, which has been the focus of an IAEA Coordinated Research Program. All these sets of data exhibit clear signs of preequilibrium emission, and it represents a strong test of theory to be able to describe experiment for a range of incident energies and emission channels. The input parameters for our Hauser-Feshbach calculation were the standard ones used by GNASH in Ref. [22], with the exception that the level density parameter for ^{93}Zr was reduced to $a = 10.75$. The Wilmore-Hodgson optical potential was used for neutrons, and the Becchetti-Greenlees potential for protons. Direct collective effects were calculated using Kalka's approach [21].

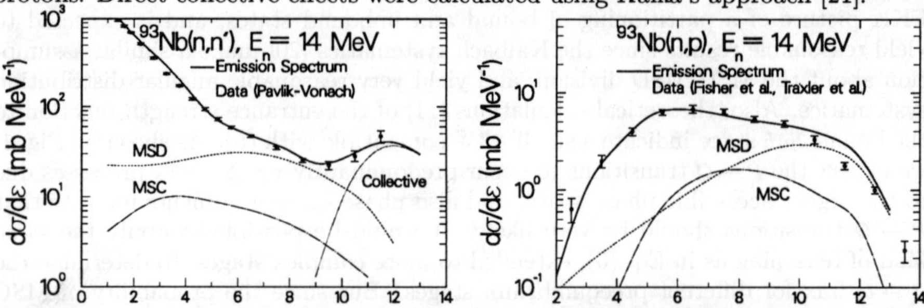

Fig. 2. FKK-GNASH calculations compared with data [23, 24] for 14 MeV ^{93}Nb(n,n') and (n,p).

Figure 2 shows our 14 MeV angle-integrated spectra compared with Vonach's evaluated ^{93}Nb(n,n') and (n,p) data. The same MSD residual interaction strength V_0=39.3 MeV was used for both the (n,n') and (n,p) channels, and it is evident that we obtain a very reasonable description of data. Perhaps even more importantly, the angular distributions for these two reactions, shown in Fig. 3, are accounted for reasonably well by our calculations. Our approach has significantly less MSC than the FKK analysis of [12] or the NVWY analysis of [25] (both of which did not include MSD and concentrated on (n,n') backward-angle data). This is because our MSD contribution, while forward-peaked, still contributes significantly at back-angles (see Fig. 3). If the amount of MSC we obtain were too small, this deficiency would be manifested in theoretical angular distributions which would be too forward-peaked. But this is not the case, and our angular distributions have about the correct forward-peaking.

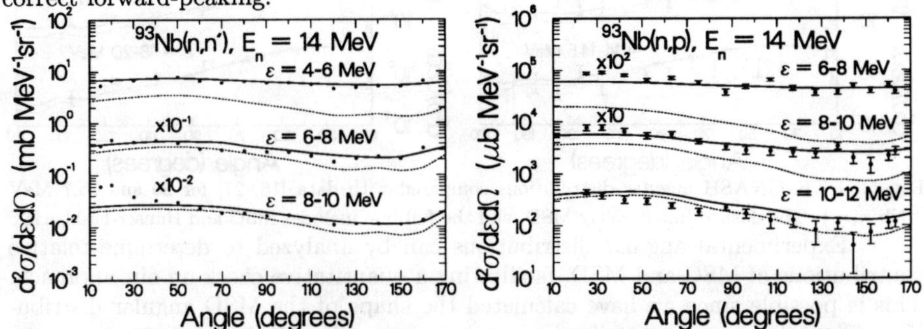

Fig. 3. FKK-GNASH angular distributions compared with data [26, 24] for 14 MeV ^{93}Nb(n, n') and (n, p). The dashed line is solely MSD, and the full line includes MSC and Hauser-Feshbach.

In Fig. 4 we compare our calculations with Marcinkowski *et al.*'s 20 and 25.7 MeV induced ^{93}Nb(n, n') data. The 25.7 MeV reaction was the focus of an intercomparison of semiclassical codes [22], and the 20 and 25.7 MeV reactions have been analyzed by Marcinkowski *et al.* [16] and Koning *et al.* [3].

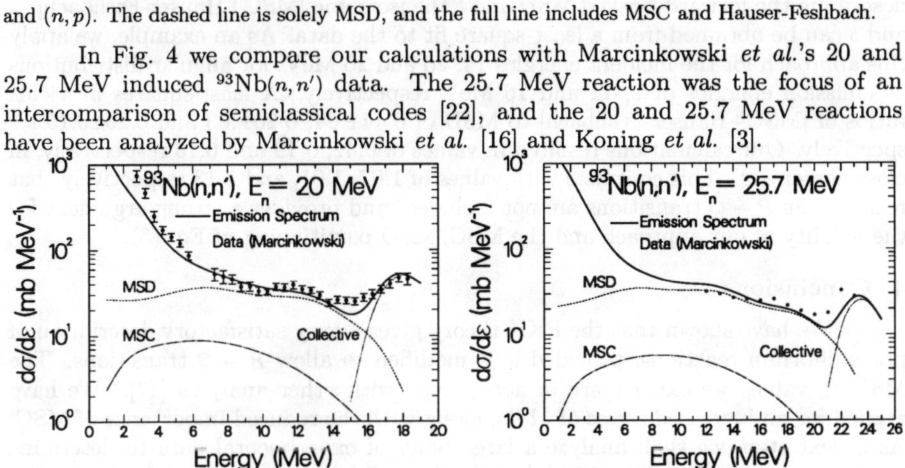

Fig. 4. FKK-GNASH spectra compared with data [16, 27] for 20 and 25.7 MeV ^{93}Nb(n, n').

We use MSD residual interaction strengths of V_0=33.6 and 27.3 MeV respectively for the 20 and 25.7 MeV reactions, and again the description of the data is quite good, though it appears that our MSD strength falls off too rapidly at the highest energies resulting in an underprediction of data at energies just below the region of collective enhancement. Preliminary calculations indicate that this deficiency is removed if we used a deformed Nilsson single-particle scheme. Figure 5 compares our predicted angular distributions with data for the 20 and 25.7 MeV reactions. Unlike semiclassical approaches, which underpredict backward-angle emission, we are able to describe the angular distributions excellently with FKK-GNASH.

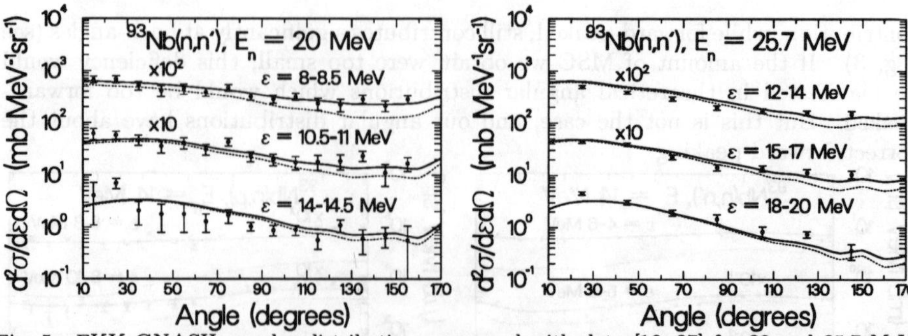

Fig. 5. FKK-GNASH angular distributions compared with data [16, 27] for 20 and 25.7 MeV ^{93}Nb(n, n'). The dashed line is solely MSD, and the full line includes MSC and Hauser-Feshbach.

Experimental angular distributions can by analyzed to determine relative contributions of MSC and MSD, facilitating a quantitative check on our approach. This is possible since we have calculated the shape of the MSD angular distribution. The total angular distribution can be expressed as $d^2\sigma/dEd\theta = a.f(\theta) + b$, $a.f(\theta)$ describing the forward-peaked MSD and b the isotropic [MSC+Hauser-Feshbach]. a and b can be obtained from a least-square fit to the data. As an example, we apply this approach for the incident energies 14, 20 and 26 MeV, for angular distributions at emission energies of 7, 11 and 16 MeV respectively. A least-squares fit yields ratios of [MSC+Hauser-Feshbach] to MSD of 0.44±.04, 0.26±.03 and 0.23±0.15 respectively. Our calculations resulted in values of 0.42, 0.19 and 0.13 respectively, in close agreement. This contrasts with values of 1.07, 1.01, and 1.13 respectively that result when $P \to Q$ transitions are not included, and provides a strong argument for the validity of our approach and the MSC/MSD partitioning of Eq. (5).

4. Conclusions

We have shown that the FKK theory gives a very satisfactory description of preequilibrium reactions, provided it is modified to allow $P \to Q$ transitions. The MSD V_0 values we extract are in accordance with other analyses [17]. We have argued for an increased stress of MSD, along with the reduced importance of MSC. As a next step, we shall analyze a large body of experimental data to determine systematic variations of V_0 with energy (and possibly mass). With such systematics we aim to use FKK-GNASH for evaluation work for reactions below 200 MeV. We shall also investigate the effects of using the FKK spin distribution correctly for subsequent Hauser-Feshbach decay, by comparing FKK-GNASH predictions with $(n, x\gamma)$ measurements at LAMPF/WNR. Finally, we think it important to undertake further statistical analyses of data, as outlined at the end of Section 3, to ensure correct MSC/MSD partitioning.

5. Acknowledgments

We would like to thank Drs. S.M. Grimes, M. Herman, P.E. Hodgson, A. Marcinkowski, D. Olaniyi and Prof. H. Vonach for helpful discussions.

5. References

1. H. Feshbach, A. Kerman, and S. Koonin, *Ann. Phys.* **125** (1980) 429.
2. H. Nishioka *et al.*, *Ann. Phys.* **172** (1986) 67.
3. A.J. Koning and J.M. Akkermans, *Ann. Phys.* **208** (1991) 216; A.J. Koning and J.M. Akkermans, submitted to *Phys. Rev.* **C**.
4. T. Tamura, T. Udagawa and H. Lenske, *Phys. Rev* **C26** (1982) 379.
5. F. C. Williams, *Nucl. Phys.* **A166** (1971) 231.
6. W. Dilg, W. Schantl, H. Vonach and M. Uhl, *Nucl. Phys* **A217** (1973) 269.
7. H. Gruppelaar, *IAEA Advisory Group Meeting on Basic and Applied Problems on Nuclear Level Densities* (Brookhaven National Laboratory 1983).
8. P.A. Seeger and R.C. Perisho *Los Alamos Report* LA-3751 (1967).
9. R. Bonetti, M. Camnasio, L. Colli Milazzo and P.E. Hodgson, *Phys. Rev.* **24** (1981) 71.
10. A. Marcinkowski, R.W. Finlay, J. Rapaport, P.E. Hodgson, and M.B. Chadwick, *Nucl. Phys.* **A501** (1989) 1.
11. M.B. Chadwick, *D. Phil thesis*, Oxford University (1989).
12. R. Bonetti, M.B. Chadwick, P.E. Hodgson, B.V. Carlson, and M.S. Hussein, *Physics Reports* **202(4)** (1991) 171.
13. P. Oblozinsky and M.B. Chadwick, *Phys. Rev.* **C42** (1989) 1652.
14. M.B. Chadwick, R. Bonetti and P.E. Hodgson, *J. Phys. G* **15** (1989) 1689.
15. P.G. Young, E.D. Arthur and M.B. Chadwick, *Los Alamos National Laboratory Report* LA-12343-MS (1992).
16. A. Marcinkowski *et al.*, *Nucl. Phys.* **A530** (1991) 75.
17. E. Gadioli and P.E. Hodgson, *Preequilibrium Nuclear Reactions* (Oxford University Press 1992).
18. H. Nishioka, H.A. Weidenmuller and S. Yoshida, *Z. Phys.* **A336** (1990) 197.
19. M.Herman, (*private communication*, September 1992).
20. C. Kalbach, *Phys. Rev.* **C37** (1988) 2350.
21. H. Kalka, M. Torjman and D. Seeliger, *Phys. Rev.* **C40** (1989) 1619. When applying Eqs. (39,40b) of this reference, we used $r_0 = 1.4$fm, and V_R=50, 69 and 55 MeV for the 14, 20 and 26 MeV reactions.
22. H. Gruppelaar and P. Nagel, *NEA Data Bank* NEANDC-204 'U' (1985).
23. A. Pavlik and H. Vonach, *Fachinformationszentrum Karlsruhe, Physics Data* **13-4** (1988).
24. R. Fischer, M. Uhl and H. Vonach, *Phys. Rev.* **C37** (1988) 578.
25. M. Herman, G. Reffo and H.A. Weidenmuller, *Nucl. Phys.* **A536** (1992) 124.
26. A. Takahashi *et al.*, Oktavian Report **A-92-01** (1992).
27. A. Marcinkowski *et al.*, *Nucl. Sci. Eng.* **83** (1983) 13.

QUANTUM-MECHANICAL MULTI-STEP DIRECT MODELS FOR NUCLEAR DATA APPLICATIONS

Arjan Koning

Nuclear Analysis Group
Netherlands Energy Research Foundation ECN
P.O. Box 1, NL-1755 ZG Petten, The Netherlands

Abstract

Various multi–step direct models have been derived and compared on a theoretical level. Subsequently, these models have been implemented in the computer code system KAPSIES, enabling a consistent comparison on the basis of the same set of nuclear parameters and the same set of numerical techniques. Continuum cross sections in the energy region between 10 and several hundreds of MeV have successfully been analysed. Both angular distributions and energy spectra can be predicted in an essentially parameter–free manner. It is demonstrated that the quantummechanical MSD models (in particular the FKK model) give an improved prediction of pre–equilibrium angular distributions as compared to the experiment–based systematics of Kalbach. This makes KAPSIES a reliable tool for nuclear data applications in the aforementioned energy region.

1 Introduction

Quantummechanical theories of pre–equilibrium reactions distinguish between multi–step compound (MSC) and multi–step direct (MSD) processes in the continuum. *Multi–step compound* reactions prevail at somewhat higher energies than those characteristic of compound nucleus decay. As with compound reactions, it is imagined that the incident particle is captured by the target nucleus. Subsequently, the stepwise reaction proceeds exclusively by the *bound* configurations of the composite system (target nucleons + incident particle). Emission takes place before the attainment of statistical equilibrium and the associated angular distribution is symmetric. If the reaction proceeds by the unbound configurations of the composite system, we speak of *multi–step direct* reactions. A crucial feature of the MSD process is that one can make a meaningful distinction between a leading fast particle and the residual nucleus. In contrast with MSC reactions, the MSD process generates the smooth forward peaked angular distributions that are characteristic for pre–equilibrium decay.

The significance of MSD reactions extends from 10 MeV to roughly 200 MeV. Since compound and multi–step compound processes are restricted to relatively low energies, MSD reactions to the continuum become very important as the incident energy increases towards several tens of MeV. This implies that the associated MSD models are relevant for a number of practical applications. At energies between 10 and 50 MeV, MSD models may provide nuclear data that are necessary for reactor–design calculations. At higher energies, MSD models can play a significant role in the analysis of the incineration of radioactive waste. Above about 150 MeV, classical approaches (such as the exciton or the intranuclear cascade model) will presumably be sufficient for the prediction of the relevant nuclear data. Below this energy however, classical models fail to reproduce the experimental data (notably the angular distributions) and quantum–mechanical models are necessary for a proper analysis of these reactions.

2 Theory

Starting from distorted wave theory, we have derived in [1] a spectrum of existing and new MSD reaction models on the basis of two different statistical postulates. The first postulate, leading-particle statistics, is based on the assumption that many states are accessible to the leading particle and that the associated DWBA matrix elements are randomly distributed. The alternative postulate is called residual-system statistics, which represents random configuration mixing due to the residual interaction within the residual nucleus. It does not make any statistical assumption about the interactions of the leading particle. Applying these physically different random postulates on first and higher order distorted wave theory leads to two distinct classes of statistical MSD theories. The models of Tamura, Udagawa and Lenske (TUL) [2] and Nishioka, Weidenmüller and Yoshida (NWY) [3] appear to be based on residual-system statistics, whereas the model of Feshbach, Kerman and Koonin (FKK) [4] can be derived using *only* leading-particle statistics. For the latter, we have shown that there is no need to invoke residual-system statistics, although this was explicitly mentioned in [5].

3 Implementation

The aforementioned multi–step direct reaction models have been implemented in the computer code system KAPSIES. In this way, we were able to perform a consistent practical comparison of the FKK, TUL and NWY model, on the basis of the same set of parameters and the same set of numerical techniques for the MSD calculations. Various experimental energy spectra and angular distributions in the pre–equilibrium region have been analyzed [6, 7]. One of the conclusions of this work was that theories based on leading-particle statistics appear to be computationally much simpler than those based on residual-system statistics, while they have essentially the same predictive power. Therefore, in this contribution we restrict ourselves to the FKK model since we consider it to be the most appropriate for routine calculations. The FKK n–step direct cross section is:

$$\frac{d^2\sigma^{(n)}(E,\Omega \leftarrow E_0,\Omega_0)}{d\Omega dE} = \frac{m}{4\pi^2\hbar^2}\int \mathrm{d}\Omega_{n-1}\int \mathrm{d}E_{n-1}E_{n-1}\frac{d^2\sigma^{(1)}(E,\Omega \leftarrow E_{n-1},\Omega_{n-1})}{d\Omega dE} \times \frac{d^2\sigma^{(n-1)}(E_{n-1},\Omega_{n-1} \leftarrow E_0,\Omega_0)}{d\Omega_{n-1}dE_{n-1}}, \quad (1)$$

where E_1, Ω_1 and E, Ω are the intermediate and outgoing energy and angle, respectively. Here, the one–step cross section is:

$$\frac{d^2\sigma^{(1)}(E,\Omega \leftarrow E_0,\Omega_0)}{d\Omega dE} = \sum_l (2l+1)R(l)\omega_{1p1h}(E_0-E)\left[\frac{d\sigma^{(1)}(E,\Omega \leftarrow E_0,\Omega_0)}{d\Omega}\right]_l^{DWBA}, \tag{2}$$

with spin distribution (with spin cutoff factor σ)

$$R(l) = \frac{2l+1}{2\sqrt{2\pi}\sigma^3}\exp\left[-\frac{(l+\frac{1}{2})^2}{2\sigma^2}\right], \tag{3}$$

and Williams particle–hole state density

$$\omega_{p,h}(E_x) = g^n \frac{(E_x - A_{p,h} - P)^{n-1}}{p!h!(n-1)!}. \tag{4}$$

where g is the single-particle state density, E_x the excitation energy, $n = p + h$ the exciton number, $A_{p,h} = [p(p-1) + h(h-1)]/4$ the Pauli correction factor and P the pairing correction. The quantities between brackets are DWBA cross sections for transferred angular momentum l. In KAPSIES, these are calculated by ECIS–88 [8].

Some results [6, 7] of the FKK calculation are displayed in fig. 1. Using standard prescriptions for the optical potential and level density parameters, a good description (simultaneously for the spectrum and the angular distribution) of the high–energy part of the pre–equilibrium region was obtained. The angular distributions are also compared with the systematics of Kalbach [9]. Figs. 1(c) and (e) clearly reveal that other reaction mechanisms (compound and MSC) dominate the reaction process at low outgoing energies.

4 Systematics

In fig. 1, the adjustments to the experimental data were achieved by the variation of only one parameter: the l–independent deformation parameter β, which is a direct measure for the magnitude of the DWBA (and also the MSD) cross section. The specific value of β completely determines the MSD cross section for all outgoing energies. For nuclear data applications, it would be interesting to predict the double–differential cross sections without the knowledge of experimental data. Therefore, we have investifgated the systematic behaviour of β in order to obtain a parameter–free model. On the basis of thirty sets of experimental (p,p') data in the energy range 29–200 MeV, we obtained the following parametrization:

$$\beta^2 = \frac{c}{A^3 E_{inc}(MeV)} \tag{5}$$

where c=2.0.10^3. With this formula, we obtained the FKK cross sections of fig. 2. The experimental data of fig. 2 were not part of the data set that led to Eq. (5). Interestingly, Eq. (5) is almost the same parametrization as Kalbach obtained [10] for the averaged squared matrix elements M^2 that appear in the semi–classical exciton model (the only difference is that, instead of the incident energy, the compound excitation energy appears in the formula for M^2). The fundamental difference with Eq. (5) is, however, that the parametrization for M^2 not only serves as a reliable method to fit the magnitude of the cross section but also represents a rather drastic physical approximation since it implies that all transitions have the same value. In the quantummechanical models all matrix elements are explicitly calculated (retaining all angular momentum rules, etc.), and their relative magnitude remains unchanged upon variation of β. This partly explains why angular distributions are described better by the MSD models than by the exciton model.

5 Conclusions

All MSD models can be derived from two alternative types of statistics, leading-particle statistics and residual-system statistics, respectively. Implementation of these models demonstrates that the problem of the underestimation of angular distributions at backward angles by the classical (exciton) models is solved by the quantum–mechanical MSD approaches. Good agreement with the experimental data is obtained, especially at the highest outgoing energies. The only free parameter of the computational model has been parametrized for (p,p') reactions, yielding satisfactory results. More extensive comparison with experimental data is required to improve the systematics and to extend it to other light–ion induced reactions.

References

[1] A. J. Koning and J. M. Akkermans, Ann. Phys. **208**, 216 (1991).

[2] T. Tamura, T. Udagawa and H. Lenske, Phys. Rev. C **26**, 379 (1982).

[3] H. Nishioka, H. A. Weidenmüller and S. Yoshida, Ann. Phys. **183**, 166 (1988).

[4] H. Feshbach, A. Kerman and S. Koonin, Ann. Phys. **125**, 429 (1980).

[5] H. Feshbach, private communication and in *Multistep Direct Reactions*, Faure, South Africa (30 Sep–2 Oct 1991), ed. R.H. Lemmer, World Scientific (1992).

[6] A.J. Koning and J.M. Akkermans, submitted to Phys. Rev. C.

[7] A.J. Koning, *Multi–step direct reactions*, PhD thesis.

[8] J. Raynal, in *Computing as a Language of Physics* (Trieste 1971), IAEA Vienna 1972. The code ECIS–88 was obtained from the NEA Data Bank, Paris.

[9] C. Kalbach, Phys. Rev. C **37**, 2350 (1988).

[10] C. Kalbach–Cline, Nuc. Phys. **A210**, 590 (1973).

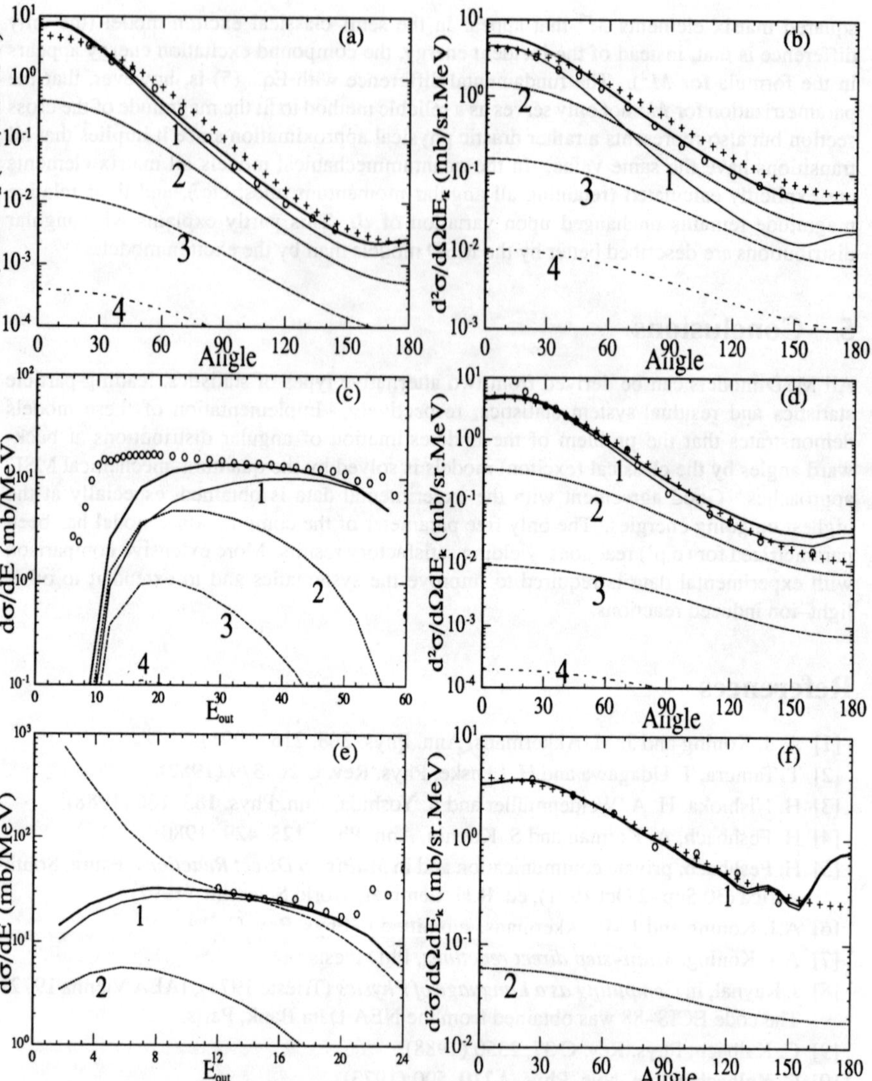

Figure 1: MSD cross-sections according to the FKK model: (a) ^{90}Zr: 80 MeV (p,p') → 60 MeV (b) ^{90}Zr: 80 MeV (p,p') → 40 MeV (c) ^{209}Bi: 62 MeV (p,p') angle–integrated (d) ^{209}Bi: 62 MeV (p,p') → 47 MeV (e) ^{93}Nb: 25.7 MeV (n,n') angle–integrated (dashed line: GNASH) (f) ^{93}Nb: 25.7 MeV (n,n') → 20 MeV. The thick solid line represents the total MSD cross section (summed over steps), the circles the experimental data and the plusses the systematics of Kalbach.

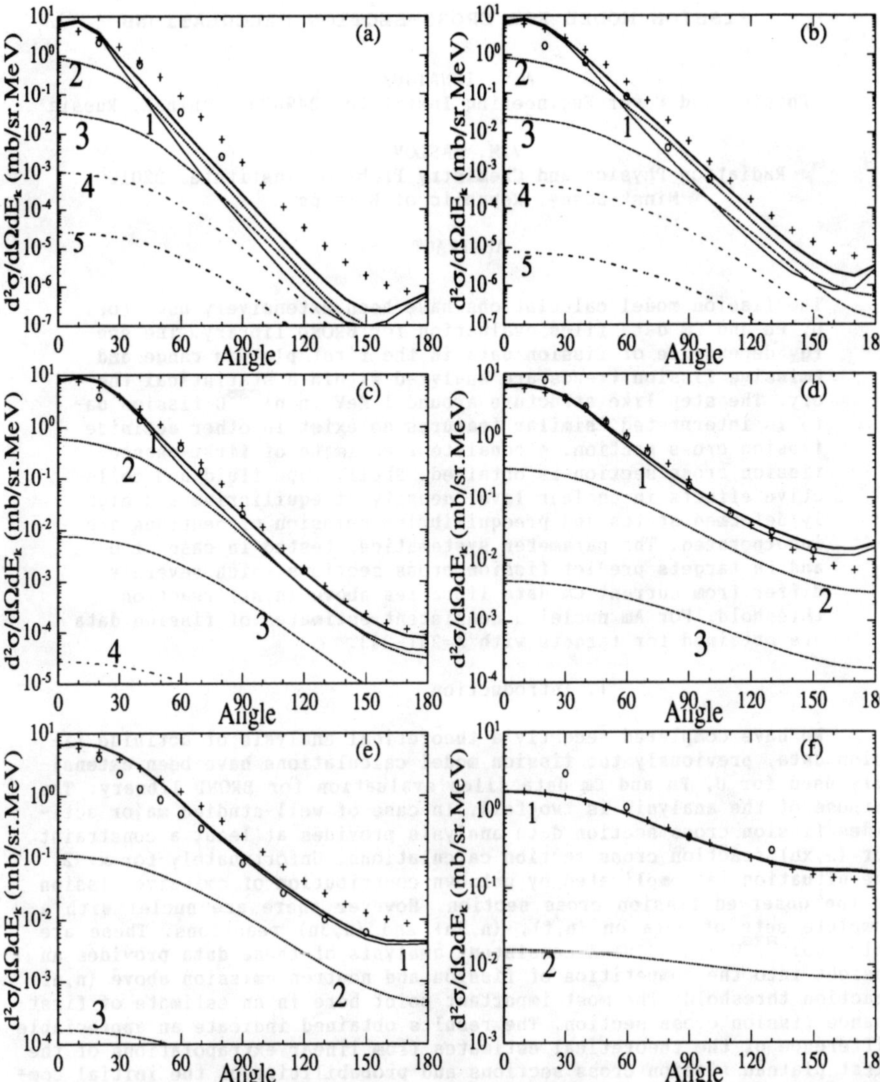

Figure 2: FKK angular distributions with the parametrization of Eq. (5): (a) ^{197}Au: 200 MeV (p,p') → 170 MeV (b) ^{58}Ni: 150 MeV (p,p') → 120 MeV (c) ^{100}Mo: 100 MeV (p,p') → 80 MeV (d) ^{165}Ho: 65 MeV (p,p') → 51 MeV (e) ^{58}Ni: 65 MeV (p,p') → 51 MeV (f) ^{197}Au: 29 MeV (p,p') → 20 MeV. The thick solid line represents the total MSD cross section (summed over steps), the circles the experimental data and the plusses the systematics of Kalbach.

FISSION MODEL FOR CROSS-SECTION CALCULATIONS

A.V. IGNATJUK
Physics and Power Engineering Institute, 249020, Obninsk, Russia

V.M. MASLOV
Radiation Physics and Chemistry Problems Institute, 220109, Minsk-Sosny, Republic of Belarus

ABSTRACT

The fission model calculations have been extensively used for U, Pu and Cm data files evaluation for BROND library. The energy dependence of fission data in the first plateau range and emissive fission trends are analyzed within a statistical theory. The step-like structure around 1 MeV in n+^{238}U fission data is interpreted, similar features do exist in other actinide fission cross section. A consistent estimate of first-chance fission cross section is obtained. Shell, superfluid and collective effects in nuclear level density at equilibrium and highly deformed states and preequilibrium emission of neutrons are incorporated. The parameter systematics, tested in case of U and Pu targets predict fission cross sections which severely differ from current Cm data libraries above (n,nf) reaction threshold. For Am nuclei a consistent estimates of fission data are obtained for targets with A=241-243.

1. Introduction

We have completed recently a theoretical analysis of actinide fission data, previously the fission model calculations have been extensively used for U, Pu and Cm data files evaluation for BROND library. The purpose of the analysis is two-fold. In case of well-studied major actinides fission cross section data analysis provides at least a constraint for (n,xn) reaction cross section calculations. Unfortunately for $x > 2$ the situation is complicated by unknown contribution of emissive fission to the observed fission cross section. However there are nuclei with a complete sets of data on (n,f),, (n,2n) and (n,3n) reactions. These are only two: ^{238}U and ^{235}U. A consistent analysis of these data provides an insight into the competition of fission and neutron emission above (n,nf) reaction threshold. The most important point here is an estimate of first-chance fission cross section. The results obtained indicate an appreciable difference of the theoretical estimates from linear extrapolations of the first plateau fission cross sections and probabilities of the initial compound nuclei beyond the (n,nf) reaction threshold. It is all the more important since the energy slope of the first plateau is different for U and Pu nuclei.Defining the first-chance fission cross section behaviour we proceed with the U and Pu fission data analysis. Here in fact the model is used to "predict" fission cross sections since calculations don't involve any free parameters. The compatibility of calculated and measured data virtually in every case makes a (n,xn) reaction cross section estimates rather reliable. The third group of nuclei consists of Am and Cm. Therefore, depending on the availability of measured data the main function of the fission model varies: from consistency analyses (^{238}U, ^{235}U and partly ^{237}Np and ^{239}Pu targets) to providing (n,xn) reaction cross sections (other U and Pu targets) and predicting neutron cross sections above (n,nf) reaction threshold (Am and Cm targets).

2. First plateau region

Starting fission cross section analysis in that region we persue two goals. First, to obtain parameter systematics for minor actinide fission cross sections. Second, to develop a model enabling extrapolation of the first-chance fission cross section beyond the threshold of the (n,nf) reaction. The plateau slope with energy $\Delta\sigma_f$ could serve as a global characteristics here. The dependence of the $\Delta\sigma_f$ on A varies from decreasing for U to increasing for Pu and Cm when A increases. So the model employed should be capable of reproducing these features.

The main ingredient of Hauser-Feshbach statistical theory calculations is level density either of fissioning nucleus and residual one. In the double-humped barrier model[1] fission in this energy range proceeds as an independent tunnelling through inner(A) and outer(B) barriers. The transmission coefficients of both barriers are governed by the level densities at the saddle deformations. These level densities were calculated with a phenomenological model[2]:

$$\rho(U,J,\pi) = K_{rot}(U,J)\ K_{vib}(U)\ \rho_{qp}(U,J,\pi), \tag{1}$$

here $\rho_{qp}(U,J,\pi)$ is the quasi-particle level density, K_{rot} and K_{vib} -factors of rotational and vibrational enchancement of level density. The quasi-particle level density $\rho(U,J,\pi)$ is defined as:

$$\rho_{qp}(U,J,\pi) = \frac{(2J+1)\ \omega_{qp}(U)}{4\sqrt{2\pi}\ \sigma_{\parallel}} \sum_{K=-J}^{J} \exp\left[-\left(\frac{K^2}{2\sigma_{\parallel}^2} - \frac{K^2}{2\sigma_{\perp}^2}\right) - \frac{J(J+1)}{2\sigma_{\perp}^2}\right] \tag{2}$$

Here $\omega_{qp}(U)$ is the intrinsic quasi-particle state density, $\sigma_{\parallel}^2$ and $\sigma_{\perp}^2$ are angular momentum distribution parameters. For details of the $\omega_{qp}(U)$, $\sigma_{\parallel}^2$ and $\sigma_{\perp}^2$ calculations see paper[2]. However to get a fair description of cumulative sums of low-lying levels the total level density

$$\rho(U) = K_{rot}K_{vib}\ \rho_{qp}(U) = K_{rot}K_{vib}\frac{\omega_{qp}}{\sqrt{2\pi}\ \sigma_{\parallel}} = T^{-1}\ \exp((U-U_o)/\ T) \tag{3}$$

was modelled by a simple constant temperature dependence in the Gilbert-Cameron approach, which here is a simple renormalization of the $\omega_{qp}(U)$

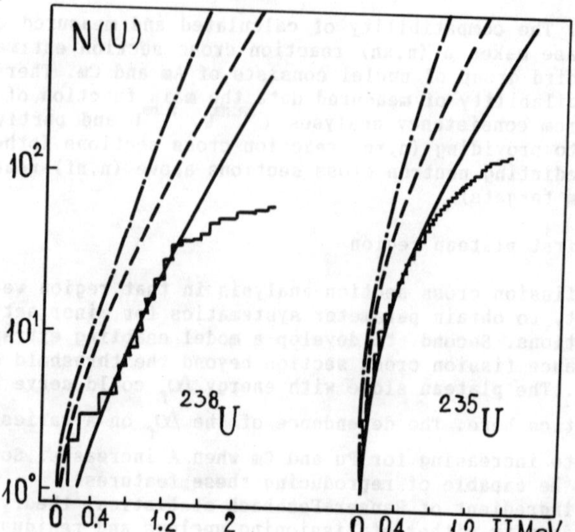

Fig.1 Cumulative number N(U) of levels. Histogram-experimental data; solid curve- equilibrium deformed state; dot-dashed and dashed curves -inner and outer saddles, respectively.

$$a(U) = \begin{cases} a\left(1 + \delta W_i \dfrac{f(U-E_{cond})}{U-E_{cond}}\right), & \text{for } U > 0.47a_{cr}\Delta_i^2 - n\Delta_i \\ a_{cr}, & \text{for } U < 0.47a_{cr}\Delta_i^2 - n\Delta_i \end{cases} \tag{4}$$

at low excitations. The instability to γ-deformations increases the total level density by a factor of $2\sqrt{2\pi}\sigma_{\parallel}$, while mass-asymmetry doubles $K_{rot} \cong \sigma_{\perp}^2$ value. This effect could be viewed on the Fig.1, showing N(U) fit for equilibrium and fissioning states. The shell effects dumping with energy was modelled by the energy dependence of the a-parameter: $f(U)=1-\exp(-0.064U)$, $E_{cond}=0.152a_{cr}\Lambda_i^2-n\Delta_i$, where n=0,1,2 for e-e,odd, o-o nuclei, respectively. For the ground state correlation function $\Lambda_o=12/\sqrt{A}$, $\delta W=M^{exp}-M^{ms}$, here M^{ms} - liquid drop mass[3]. The shell corrections $\delta W_f^{A(B)}$ for saddle points are defined relative to the liquid drop component of the fission barrier[4]. To define $\sigma_{\perp}^2$ at saddles deformed cold nuclei moments of inertia[5] were used. The only free parameter of the model is $\Lambda_f = \Delta_o + \delta$, we reproduce the plateau slope by δ variation, the δ values correlate with the $\Delta\sigma_f$

The inclusion of asymmetries at saddles results in an appreciable decrease of the theoretical barriers[6,7] and in an increase of fission barriers , derived from fitting of fission data. The axial asymmetry of the inner saddle for light U nuclei [6] appears to be the cause of rather strong A-dependence of E_f. The lacking in this case of the factor $2\sqrt{2\pi}\sigma_{\parallel}$ could be compensated by a decrease in the E_f^A by 1 MeV. Only in that case can we get a relation $E_f^A < E_f^B$ for light U nuclei with A<234[6] (236[7]). as well as qualitative agreement of theoretical and experimental barriers.

The first plateau cross section analysis gives no direct evidence of superfluid effects in $\rho(U)$. The well-known structure in $^{235}U(n,f)$ cross section around 1 MeV may be considered as a direct evidence of few-quasi-particle excitations of the fissioning even-even nucleus[8]. The intrinsic state density $\omega_{qp}(U)$ could be represented as a sum of n-quasi-particle state densities $\omega_{qpn}(U)$

$$\omega_{qp} = \sum_n \omega_{qpn}(U) = \sum_n \frac{g_n (U - U_n)^{n-1}}{((n/2)!)^2 (n-1)!}, \tag{5}$$

where $g=6a/\pi^2$. However even defining excitation threshold energies U_n correctly [9,10] one fails to fit fission data (see Fig.2). Using U_n values[10,11] we will infer $\omega_{qp}(U)$ by fitting σ_f data[11,12,13]. We suppose that the cross section magnitude is governed by the outer barrier and the E_f^J and $\hbar\omega^J$ values are fixed by fitting data for E>2 MeV. Then $E_2 = U_2 - E_f - B_n = 0.8$ MeV corresponds to two-quasi-particle state excitation threshold U_2. Four-quasi-particle excitation threshold occurs at E_4 = 2.15 MeV. The behaviour of the σ_f for E_n from 0.8 MeV up to 2.15 MeV is governed by $\omega_2(U)$ density. The $\omega_2(U)$ shape thus obtained differs from Eq(5) estimate (see Fig.3a). The residual nucleus level density was calculated by introducing o-e energy shift: $U=U+\Delta_o$ (see Fig.3b).

The ^{235}U fission data are unique since they allow to fix simultaneously the fission barrier height and the two-quasi-particle state excitation threshold. The step-like structure is most probably due to the jump-like excitation of the two-quasi-particle configuration for the outer saddle.The excitation threshold value is consistent with the fissioning nucleus pairing gap and barrier parameters, fitting the $^{235}U(n,f)$ reaction data between 0.1 and 5 MeV. Similar irregularities do exist in other actinide fission cross sections: $^{233}U(n,f)$ (0.5 MeV), $^{239}Pu(n,f)$ (1.25 MeV), $^{241}Pu(n,f)$ (1.0 MeV) (in parentheses are shown excitation threshold E_2 values).

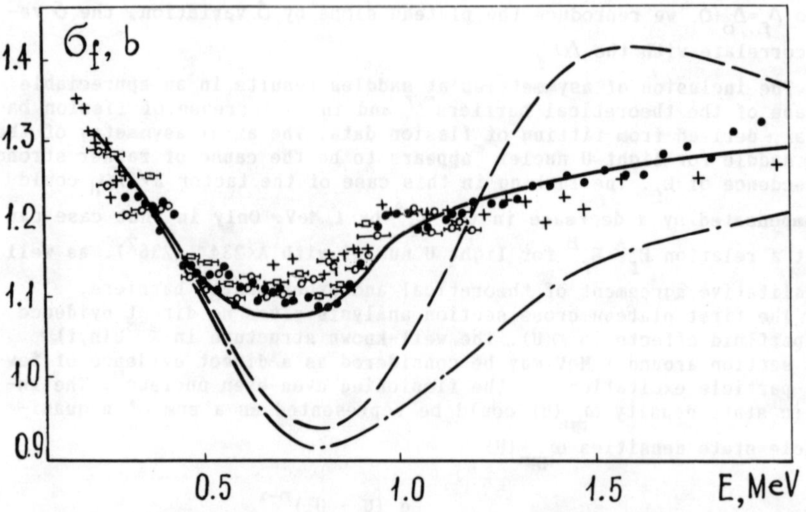

Fig. 2 Neutron-induced fission cross section for ^{235}U. Curves represent calculations with different intrinsic state densities, shown in Fig.3; data points: ○, □ -Wasson et al.[11], ● -Carlson[12], + -Poenits[13].

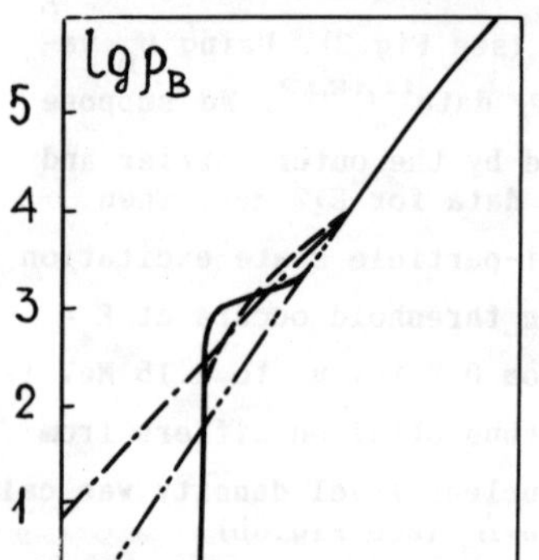

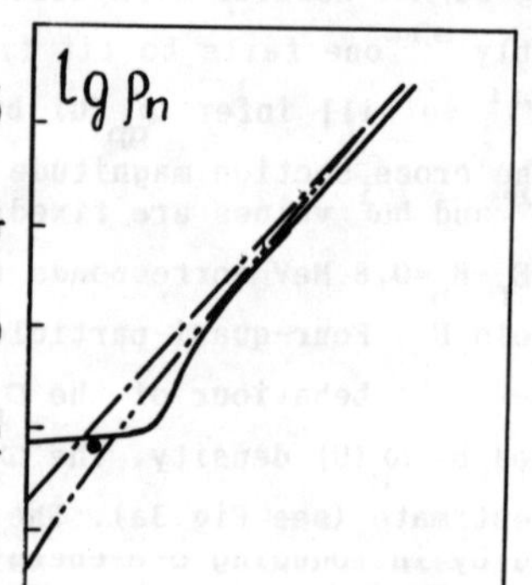

Fig.3a Level density for outer ^{236}U saddle. Solid curve corresponds to σ_f data fit; dashed curve-Eq.(5); dot-dashed curve-Eq.(3); double-dot-dashed curve-pairing model[2].
Fig. 3b Level density for residual ^{235}U nucleus. See curves on Fig. 3a; data point is defined as N(U)/U.

3. Fission cross section above (n,nf) reaction threshold

For fixed statistical model parameters of residual nuclei A, A-1... fissioning in (n,nf), (n,2nf) ... reactions, the behaviour of the first-chance fission cross section σ_f^1 should make it possible to reproduce the measured fission cross section σ_f. A consistent description of a complete set of measured data on (n,f), (n,2n) and (n,3n) reaction cross sections for the ^{235}U and ^{238}U target nuclei[14] enables one to consider the estimates of σ_f^1 and of the fission probabilities of the initial compound nuclei ^{236}U and ^{239}U as fairly realistic.

The hard component of neutron scattering spectra[15] and high energy tail of ^{238}U(n,2n) reaction cross section are interpreted as due to the pre-equilibrium evaporation of neutrons. This feature could be parametrized within a conventional exiton model[16], used in a STAPRE[17] code. By fitting the spectra for E=6-14.7 MeV we get the main parameter of the exiton model, that is the matrix element $M^2 = 10/A^3$, MeV^2. Note that when calculating exiton state density the odd-even backshift is introduced: $U = U - \Delta(2-n)$. Then the charge conservation[18] and transition

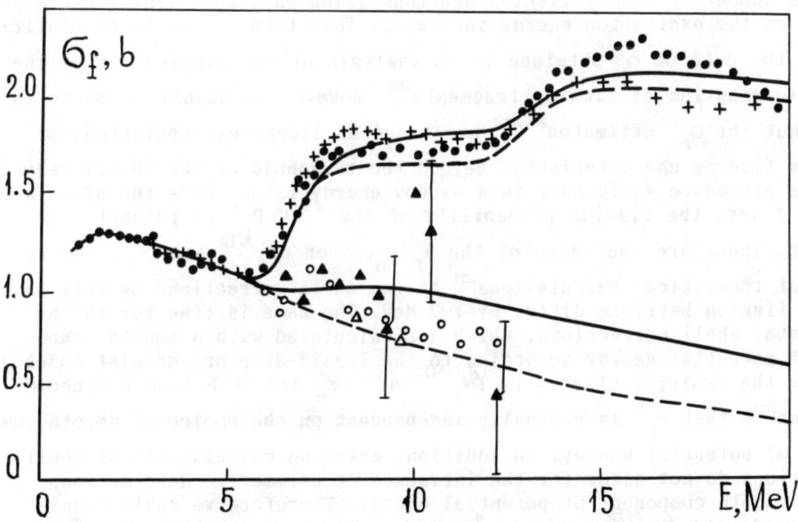

Fig.4 Neutron-induced fission cross section for ^{235}U. Solid line - calculated σ_f; dashed line - calculated with $\delta W_f^B = -1.4$ MeV σ_f; lower curves - first-chance fission cross sections; data points: ▲-Caldwell et al.[20]; ○-Frehaut[21]; Δ-Goverdovskij et al.[22]; ∇-Czirr and Sidhu[23]; ●-Leugers et al.[24].

rates renormalization[19] were employed. With all that in mind and in a STAPRE code a preequilibrium fraction emission q(E) leading to depletion of compound nucleus states population could be obtained. Largely by diminishing the odd-even dependence of q(E) we can fit the ^{235}U fission data with the same M^2 value.

The behaviour of the first-chance fission cross section is obviously related with the energy dependence of the fission probability P_f^1 of the A+1 nucleus:

$$\sigma_f^1 = \sigma_c(1-q(E))P_f^1 \qquad (6)$$

There are no direct data on the first-chance fission cross section, but for ^{235}U target nucleus there are data on the contribution α_1 of the fission of the A+1 compound nucleus to the observed cross section of the ^{235}U(n,f) reaction (Fig.4). For the neutron energies less than 9 MeV our estimate of $_f\sigma^1$ agrees with the data following from α_1 values defined by ^{236}U photofission neutron multiplicity analysis. At higher energies there is rather large scatter of the data points. The Frehaut[21] estimate of α_1, which follows from the average prompt γ-ray total energy dependence on average number of prompt fission neutrons gives an almost linear decrease of σ_f^1 as the excitation energy increases. This trend seems to be confirmed by the data on α_1 obtained in an analysis of the dispersions of the mass distribution of fission fragments[22]. However it should be borne in mind that the σ_f^1 estimates[21,22] are based on linear extrapolations of various fission characteristics beyond the threshold of the (n,nf) reaction, a procedure valid only in a narrow energy range. Once the q(E) value is fixed, the fission probability of the ^{236}U P_f^1 is parameter-dependent. These are the ratio of the a_f/a_n, then $\delta W_f^{A(B)}$ and δW_n. The results of theoretical calculations[25] of the shell corrections as well as of the fission barriers differ by 1-2 MeV. The same is true for the experimental shell corrections, which are calculated with a smooth component of potential energy according to the liquid-drop or -droplet model. However the isotopic changes of $\delta W_f^{A(B)}$ and δW_n are such (see Bjornholm and Lynn[25]) that P_f^1 is virtually independent on the choice of smooth component of potential energy. In addition, existing calculations of shell corrections do not allow for the influence of asymmetric deformations on the smooth component of potential energy. Therefore we shall consider the adopted $\delta W_f^{A(B)}$ estimates[4] to be effective, provided that δW_n are obtained with the liquid drop model. The shown on the Fig.4 trend of the first-chance fission is a manifestation of the damping of the shell effects with energy. For the σ_f^1 to be linear extrapolation from the region of first plateau, the value of δW_f^B should be reduced by $\simeq$2 MeV. The fission probability P^1_f then becomes a decreasing function of excitation energy. Unfortunately such a marked change in δW_f^B has a much weaker effect on the observed fission cross section (see Fig.4). However the quality of fit of (n,2n) and (n,3n) in that case also deteriorates. So it can be stated safely that we have got effective estimates of P_f^1 and q(E).

4. Comparison of predicted and available fission cross sections for minor actinides

As stated above the role of model calculations depends on the consistency and availability of measured data. The most extensively studied among the remaining nuclides is ^{239}Pu. The validity of the model was checked by consistent description of the (n,f) and (n,2n) data for that target[26] with much higher fissility than in case of previously studied ^{235}U and ^{238}U targets. We can state also that the adopted parameter systematics reasonably describes the competition of fission and neutron emission above (n,nf) reaction threshold. The coupled-channel optical potential parameters for ^{239}Pu calculations were obtained by fitting s- and p-wave strength functions, total cross section and angular distributions of inelastically scattered neutrons. However the quality of fit is just the same as if we use transmission coefficients for the ^{238}U target[27]. This fact justifies the using of these coefficients for the remaining nuclei. The same is true for the analysis of ^{233}U+n data[28]. The calculated σ_f curves describe fairly well the measured data for U targets (A=232-238), only in case of ^{232}U+n fission the adopted compound nucleus cross section does not permit matching the calculated curve to measured data[29] of σ_f at 14.7 MeV. The inclusion of γ-asymmetry of the inner saddle increases appreciably the density of excited levels of the fissioning nuclei and changes its energy dependence[14]. However in case of γ-symmetry of inner saddle of light U nuclei, lowering E_f^A barrier by 1 MeV one can compensate the decrease of ρ_A(U) for excitation energies above the (n,nf) reaction threshold. The observed fission cross section of the light U targets (A=232-236) could be reproduced equally well with either γ-symmetric or γ-asymmetric configurations of the inner saddles of the fissioning nuclei.

In case of Pu targets (A=238-244) the calculated curves fit the measured data up to the threshold of (n,2nf) reaction in all cases[30], except the ^{238}Pu(n,f) reaction. As in the case of U targets the deviations of the curves from the measured data at higher energies are not of a systematic character. Nonetheless in this energy range the cross sections are known with large error, which may increase due to ambiguities in ^{235}U+n fission cross section data.

The validity of the model for Z-odd targets was demonstrated in case of similar analysis of ^{237}Np+n data[31]. The analysis of current Am data files, JENDL-3 and ENDF/B-V, shows surprisingly large discrepancies in (n,f) evaluated data. This discrepancies are due to normalizing evaluated data to different measured data or pure spline fitting of various

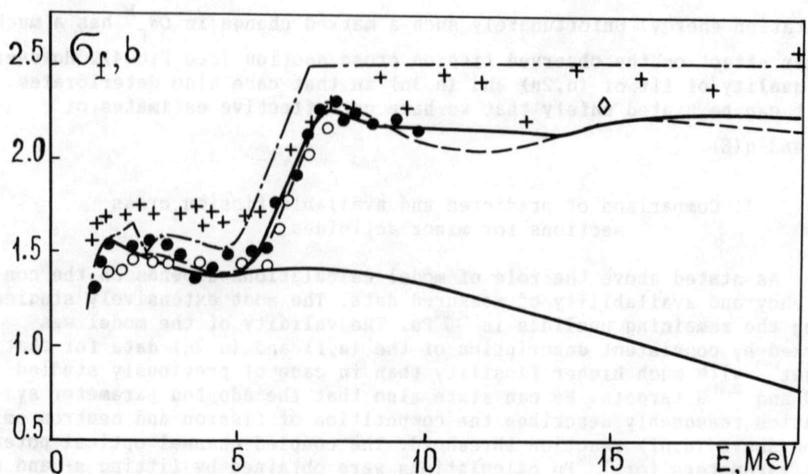

Fig.5 Neutron-induced fission cross section for ^{243}Am. Solid line-calculated σ_f; dashed line - JENDL-3; dot-dashed line - ENDF/B-V; data points: ● -Knitter et al[32].; ○ -Fursov et al[35].; + -Behrens et al[36].; ◊ -Fomushkin et al.[37]

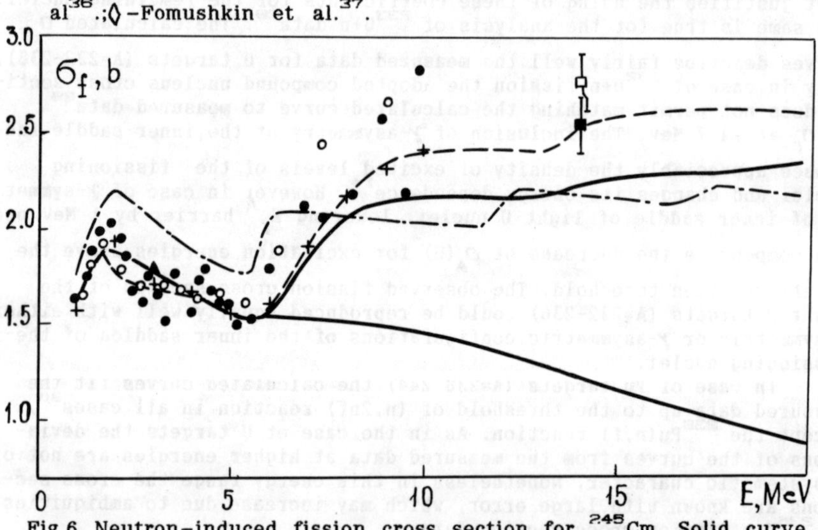

Fig.6 Neutron-induced fission cross section for ^{245}Cm. Solid curve - calculated σ_f the lower curve-σ_f^1; dashed line-JENDL-3 ;dot-dashed line-ENDF/B-V;datapoints: +,■-White et al.[38]; □ -Fomushkin et al.[40]; o,●- Khokhlov et al.[39]

data sets. A model employed heretofore could serve as a sort of a "filter" of measured data for A target, compatible with a fission data of (A-x) nuclei, fissioning in (n,xnf) reactions. In case of ^{243}Am(n,f) reaction data[32] these are ^{241}Am+n and ^{242m}Am+n data[33,34]. The simultaneous analysis of ^{241,242m,243}Am+n fission data enables one to define a consistent data sets for ^{241}Am(n,f) and ^{243}Am(n,f) and suggests a fission cross section of ^{242m}Am(n,f) above the onset of (n,nf) reaction, which is rather different from Browne et al. data[33] (see Fig.5).

The same kind of discrepancy we encounter when analyzing ^{245}Cm+n fission data [38] and recent Khokhlov et al. data[39]. Unfortunately these are the only data sets, covering the energy range higher than (n,nf) reaction threshold for Cm (A=242-248) targets. For the other targets there are data only in the first plateau region and at 14.1 MeV point. Fig. 6 compares the calculated results with data, available for ^{245}Cm+n reaction. The calculated curve fails to fit the bulk of measured data points, nonetheless one may judge good agreement with lower lying data points by Khokhlov et al.[39] (Fig.6). The most feasible way of fitting measured data at higher energies is a decreasing of the pre-equilibrium neutron emission contribution q(E) by 30%, which leads to less steep slope of σ_f^1. However we don't feel this forced fitting to measured data justified. The similar calculations were made for other Cm nuclei. The large discrepancies exist between our calculations and 14.1 MeV data[40] as well as ENDF/B-V and JENDL-3 data. That is due to inconsistencies of (n,xnf) reactions contributions with appropriate fission data in the first plateau region. From our analyses of actinide fission cross sections we have concluded that the higher target nucleus fissility , the less pronounced (n,2nf) fission reaction step in total fission cross section would be. The U, Pu and Am measured data justify that conclusion. That explains partly the origin of discrepancies of our data with JENDL-3 and ENDF/B-V data. The isotopic dependencies of measured bomb-shot[40] and calculated data at 14.1 MeV are remarkably similar. So the discrepancies virtually disappear in case of normalizing the measured data to calculated ones for ^{245}Cm+n reaction.

4. Conclusions

The simple modelling of level densities enables one to interpret step-like structures around 1 MeV netron energy and energy dependence of fission cross section in the plateau region. The former feature is a manifestation of superfluid effects. The collective effects manifest as drastic fission barrier sensitivity to the nuclear shape asymmetry. At last, shell effects damping manifests in a consistent fit of a fission cross sections for excitations higher than (n,nf) reaction threshold.

A statistical theory analysis of (n,f) and (n,xn) measured data with pre-equilibrium processes in the neutron channel included has resulted in a consistent fit of the most complete data set for ^{238}U target. In case of ^{232}Th(n,2n) reaction the parameter systematics produces consistent fit[41] of new experimental data, suggesting a possible reevaluation of that cross section in existing data files.
For Pu and U targets we get a consistency between calculated and measured fission data, the discrepancies in calculated and other evaluated

(n,xn) cross sections being notified.

For Am targets the analysis of fission data offers a possibility to define a consistent set for ^{241}Am(n,f) and ^{243}Am(n,f) reactions and predict ^{242m}Am(n,f) cross section above the onset of (n,nf) reaction. There is a certain need to revise Am data files in fast neutron region.

The analysis of existing evaluated Cm data files shows the inadequacy between JENDL-3, ENDF/B-V and statistical calculations with reasonable fission and neutron parameters. We examined the consistency of measured fission data over a mass chain A=242-248. We suppose that the reevaluation of Cm data files is highly desirable[42]. If the traditional trend of ^{245}Cm+n data above (n,nf) reaction threshold would be justified unambiguously, the reevaluation would become inevitable.

References

1. V.M. Strutinsky and S. Bjornholm, in Nuclear Structure(IAEA,Vienna, 1968),p.431.
2. A.V. Ignatjuk, K.K. Istekov, G.N. Smirenkin, Sov. J. Nucl. Phys. 29 (1979) 450.
3. W.O. Myers and W.J. Swiatecky, Ark. Fysik 36 (1967) 343.
4. V.M. Kupriyanov, G.N. Smirenkin and B.I. Fursov, Yadernaya Fyzika 39 (1984) 176.
5. A. Sobichewski and S. Bjornholm, Nucl. Phys. A202 (1973) 274.
6 W.M. Howard and P. Moller, At. Data Nucl. Tables 15 (1980) 219.
7. P. Moller, Nucl. Phys. A192 (1972) 529.
8. A.V. Ignatjuk and V.M. Maslov, in Nuclear Data for Science and Technology (Springer-Verlag,1992) p.153.
9. A.V. Ignatjuk and Yu.V. Sokolov, Yadernaya Fysika 11 (1974) 1229.
10. C.Y. Fu, Nucl. Sci. Engng. 86 (1984) 344.
1. O.A. Wasson, M.M. Maier, K.C. Duval, Nucl. Sci. Engng. 81 (1982) 196.
12. A.D. Carlson, in Proc IAEA Consult. Meeting on ^{235}U Fast Neutron Fission Cross Section, 1983, Smolenice (IAEA,Vienna,1983),p.61.
13. W.P. Poenitz, Nucl. Sci. Engng. 64 (1977) 89.
14. A.V. Ignatjuk, V.M. Maslov, A.B. Pashchenko, Sov. J. Nucl. Phys. 47 (1988) 224.
15. N.V. Kornilov, V.Ya. Baryba, O.A. Salnikov, in Neutron Physics (Tsniiatominform, Moscow,1980),vol.3, p.104.
16. C.K. Cline, Nucl. Phys. A186 (1972) 273.
17. M. Uhl and B. Strohmaier, Report IRK - 76/10 (vienna,1976).
18. E. Gadioli, E. Gadioli Erba, P.G. Sona,Nucl.Phys.A217 (1973) 589.
19. I. Ribansky, P. Oblozhinsky, E. Betak, Nucl. Phys. A205 (1973) 545.
20. J.T. Coldwell, E.J. Dowdy, R.A. Alvaretz et al., Nucl. Sci. Engng. 73 (1980) 153.
21. J. Frehaut, in Proc of the 3d Advisory Group Meeting on Transactinium Isotope Nuclear Data (Uppsala,1984),IAEATECDOC-336,1985,p.105.
22. A.A. Goverdovsky, G.A. Kudyaev, Yu. B. Ostapenko, G.N. Smirenkin, in: Proc. 18th Intern. Symp. on Physics of Fission (Gaussig,1988)
23. J.B.Czirr and J.B. Sidhu, Nucl. Sci. Engng., 57 (1975) 18.
24. B. Leugers, S, Cierjacks, P. Brotz et al., in Proc of NEANDC/NEACRP Specialists Meeting on Fast Neutron Fission Cross Sections ^{233}U, ^{235}U, ^{238}U and ^{239}Pu (Argonne,1976), ANL-79-90,1976, p.183.
25. S. Bjornholm and J.E. Lynn, Rev. Mod. Phys. 52 (1980) 725.

26. A.B. Klepatskij and V.M. Maslov, in Nuclear Data for Science and Technology (Springer-Verlag,1992), p.881.
27. Ch. Lagrange, Report INDC (FR), 56/L (NEANDC 218-L), 1982.
28. L.A. Bakhanovich, A.B. Klepatskij, V.M. Maslov et al.,in Nuclear Data for Science and Technology (Springer-Verlag,1992), p.915.
29. B.I. Fursov, E.Yu. Baranov,M.P. Klemyshev et al. Sov. J. At. Energy 61 (1986) 963.
30. A.V. Ignatjuk and V.M. Maslov, Sov. J. Nucl. Phys. 51 (1990) 781.
31. V.M. Maslov, Yadernye Konstanty 4 (1987) 18.
32. H.H. Knitter and C. Budtz-Jorgensen, Nucl. Sci. Engng. 99 (1988) 1.
33. J.C. Browne, R.M. White, R.E. Howe et al. Phys. Rev. 29 (1984) 2188.
34. J.M. Dabbs, C.H. Johnson, C.E. Bemis, Nucl. Sci. Engng. 83 (1983) 22.
35. B.I. Fursov, E.Yu. Baranov, M.P. Klemyshev et al., Rad. Effects 93 (1986) 305.
36. J.W. Behrens, J.C. Browne, E. Alles, Nucl. Sci. Engng. 77 (1981) 44.
37. E.F. Fomushkin, G.F. Novoselov, Y.I. Vinogradov, Yadernye Konstanty 3 (1984) 17.
38. R.M. White and J.C. Browne, in Nuclear Data for Science and Technology (North-Holland,1983), p.281.
39. Yu.A. Khokhlov, I.A. Ivanin, Yu.I. Vinogradov et al. Intern. Seminar on the Interactions of Neutrons with Nuclei, Dubna, 1992, p.36.
40. E.F. Fomushkin, G.F. Novoselov, G.F. Gavrilov et al.,in: Nuclear Data for Science and Technology (Springer-Verlag,1992), p.439.
41. V.M. Maslov ,Ann. Nucl. Energy, 19 (1992) 181
42. V.M. Maslov, accepted in Ann. Nucl. Energy, 1992.

EVALUATION OF TOTAL FISSION CHARACTERISTICS FOR LOW-ENERGY NEUTRON-INDUCED FISSION

UWE GOHS

Joint Institute for Nuclear Research, P. O. Box 79, 101000, Moscow
Dubna, CIS, Russia

ABSTRACT

Energy conservation consistent evaluations of total fission characteristics were performed on the basis of a phenomenological multimode model with account for the $(n,\gamma f)$ process for $J^\pi = 4^-$ neutron resonances of ^{235}U and for $J^\pi = 1^+$ neutron resonances of ^{239}Pu. Using experimental data (kinetic energy and average number of promptly emitted neutrons) as function of fragment mass number the parameters of energy partition were defined.

1. Introduction

The low-energy neutron-induced fission is described in a multiple-stage process. The first stage is the formation of the compound nucleus CN, where the excitation energy (binding energy of the incident neutron) is shared amoung the nucleus. During the next step the CN crosses a double- or triple-humped barrier. At the tops of these barriers only a few modes of motion are allowed. They give rise to the resonance structure of the fission cross section of fissile isotopes[1] and should be associated with different fission thresholds. The angular distributions of the fission fragments are assumed to be related to the channel structure at the outer barrier. Thus correlations were expected between fluctuations in fission characteristics (angular distribution, mass distribution, total kinetic energy $\overline{TKE}$, number of promptly emitted neutrons $\overline{\nu}_n$) and the fission channel configuration at the outer barrier (spin J and spin orientation quantum number K).

Bertram et al.[2] used the calculated partial fission cross section through different saddle point states in order to explain the energy dependence of the fission fragment anisotropy, $\overline{\nu}_n$, and $\overline{TKE}$ for the ^{233}U(n,f)-reaction. Hambsch et al.[3] measured the fission fragment mass distribution and the $\overline{TKE}$ for low-energy neutron-induced fission of ^{235}U. The observed fluctuations in $\overline{TKE}$ ($\Delta\overline{TKE} < 0.1\%$) and the fluctuations in $\overline{\nu}_n$ ($\Delta\overline{\nu}_n < 0.5\%$)[4] were discussed in a multimode fission model[5] and related to fluctuations in fission mode yields. Furman et al.[6] introduced a K-dependence of the mass distribution and explained the different relative frequency of fission modes by fluctuations in the relative contribution of fission channels within a two fission channel representation.

In the present work we investigate the influence of fission channels, fission modes, and the $(n,\gamma f)$ process on fission characteristics (mass yield, $\overline{TKE}$, $\overline{\nu}_n$, multiplicity of γ-rays $\overline{N}_\gamma$, and total energy of γ-rays $\overline{E}_\gamma$) within a multimode fission model including fission channels.

2. Model

In a multimode fission model[5] the mass distribution $Y_\lambda(A)$ for resonances numbered by λ (with total fission width $\Gamma_{f\lambda}$) is given by following equation

$$Y_\lambda(A) = \sum_d Y_d(A) W_{d\lambda}. \tag{1}$$

The mass yield of fission mode d $Y_d(A)$ is represented by a gaussian[5]. The population probability $W_{d\lambda}$ reflecting the relative frequency of fission mode d is related to the relative contribution $P_{k\lambda}$ of fission channel k as

$$W_{d\lambda} = \sum_k P_{k\lambda} W_d^k, \tag{2}$$

where W_d^k is the population probability of mode d for fission channel k[6]. Following this idea and including the $(n,\gamma f)$ process, the total fission characteristic $\overline{X}_\lambda$ can be expressed by

$$\overline{X}_\lambda = \frac{\overline{\Gamma}_{\gamma f}}{\Gamma_{f\lambda}} \sum_d W_d^\gamma \overline{X}_d^\gamma + \left(1 - \frac{\overline{\Gamma}_{\gamma f}}{\Gamma_{f\lambda}}\right) \sum_d \sum_k P_{k\lambda} W_d^k \overline{X}_d \tag{3}$$

with

$$\overline{X}_d^\gamma = \sum_A Y_d(A) \overline{X}_d^\gamma(A) \quad and \quad \overline{X}_d = \sum_A Y_d(A) \overline{X}_d(A). \tag{4}$$

$\overline{X}$ stands for $\overline{TKE}$, $\overline{\nu}_n$, $\overline{N}_\gamma$ or $\overline{E}_\gamma$. W_d^γ represents the population probability of fission mode d of the $(n,\gamma f)$ process. The index γ of fission characteristic $\overline{X}$ is related to this process. In previous works[7,8], the $(n,\gamma f)$ process was discussed on the basis of following relations:

$$\overline{N}_\gamma = \overline{N}_\gamma^0 + \frac{\overline{\Gamma}_{\gamma f}}{\Gamma_{f\lambda}} \overline{N}_{\gamma f} \tag{5}$$

$$\overline{E}_\gamma = \overline{E}_\gamma^0 + \frac{\overline{\Gamma}_{\gamma f}}{\Gamma_{f\lambda}} \overline{E}_{\gamma f} \tag{6}$$

$$\overline{\nu}_n = \overline{\nu}_n^0 - \frac{\overline{\Gamma}_{\gamma f}}{\Gamma_{f\lambda}} \overline{E}_{\gamma f} (\overline{\delta\nu_n/\delta E^*}) \tag{7}$$

where $\overline{N}_{\gamma f}$ and $\overline{E}_{\gamma f}$ are the average multiplicity and average energy of prefission γ-rays, respectively. $\overline{\delta\nu_n/\delta E}^*$ represents the average slope of the ν_n dependence on excitation energy E^*. In this representation, the γ-ray multiplicity $\overline{N}_\gamma^0$, the energy of γ-rays $\overline{E}_\gamma^0$, and the number of promptly emitted neutrons $\overline{\nu}_n^0$ are mean values of the (n,f) reaction concerning fluctuations in fission modes.

In the framework of the present model we assume that the excitation energy at the scission point decreases by $\overline{E}_{\gamma f}$ and the total number of γ-rays increases by $\overline{N}_{\gamma f}$. Because of unknown transition states after emission of prefission γ-rays we neglected channel effects and used the mass distribution as for the (n, f) reaction $(W_d^\gamma = W_{d\lambda})$.

The general energy balance for a given fission mode reads

$$\overline{Q}(A_L/A_H) + B_n + E_n = \overline{TKE}_d(A_L/A_H) + \overline{TXE}_d(A_L/A_H). \tag{8}$$

To keep the calculation tractable we neglected the influence of fission modes on the average energy release in fission $\overline{Q}$ for a given fragmentation A_L/A_H. B_n and E_n are the binding and the kinetic energies of the incident neutron, respectively. $\overline{TXE}_d$ denotes the total fragment excitation energy. Both terms in the rhs of Eq.(8) can be expressed with reference to scission point:

$$\overline{TKE}_d(A_L/A_H) = \overline{E}_{c,d}(A_L/A_H) + \overline{E}_{pre,d} + \overline{V}_n(A_l/A_H), \tag{9}$$

$$\overline{TXE}_d(A_L/A_H) = \overline{E}_{def,d}(A_L) + \overline{E}_{def,d}(A_H) + \overline{E}_{dis,d}. \tag{10}$$

The prescission kinetic energy $\overline{E}_{pre,d}$ and dissipative energy $\overline{E}_{dis,d}$ were calculated according to Grossmann et al.[9]. $\overline{E}_{c,d}$ and $\overline{V}_n$ represent the coulomb repulsion and the nuclear interaction, respectively. We use experimental data of kinetic energy and number of promptly emitted neutron as function of mass number in order to define the parameters of energy partition at scission point for different fission modes d and to calculate $\overline{TKE}_d$ and $\overline{TXE}_d$ for a given fragmentation.

The average excitation energy of a fragment $\overline{XE}_d(A)$ is given by

$$\overline{XE}_d(A) = \overline{\nu}_{n,d}(A)[\overline{B}_n(A) + \overline{\epsilon}_{n,d}(A)] + \overline{N}^E_{\gamma,d}(A)\overline{E}^E_{\gamma,d}(A) + \overline{N}^s_{\gamma,d}(A)\overline{E}^s_{\gamma,d}(A), \tag{11}$$

where $\overline{B}_n$ is the average neutron binding energy, and $\overline{\epsilon}_{n,d}(A)$ is the average neutron emission energy in the CMS. $\overline{N}^E_{\gamma,d}$(A) represents the average number of γ-rays due to E1/E2-transitions. $\overline{N}^s_{\gamma,d}$(A) includes statistical γ-rays and γ-rays of the $(n, \gamma f)$ process.

The most probable spin of a given fragmentation was calculated according to P. Fong[10]. Using experimental data[11,12] we determined the average spin of E1-, M1-, and E2-transition γ-rays. Assuming that all neutrons and statistical γ-rays emitted from the fragments are s-neutrons and E1 γ-rays, respectively we calculate $\overline{\nu}_n$ and $\overline{E}_\gamma$ on the basis of the statistical model[13]. The model includes the conservation relations for energy and angular momentum.

3. Application

3.1. 4^- resonances of the $^{235}U(n, f)$-reaction

Pattenden and Postma[14] measured the anisotropy of fission fragments emitted by an aligned sample of ^{235}U, irradiated by resonance neutrons. Their results described in terms of A_2 (coefficient of the second Legendre term in the angular distribution formula) suggest two open channels with K=1, 2 for $J^\pi = 4^-$ resonances. Moore et al.[15] carried out a resonance parameterization of ^{235}U within a two fission channel reduced R-matrix representation. Using these data as well as the experimental data of Hambsch et al.[3] we calculated all population values in the rhs of Eq. (2). Table 1 contents the deduced population probabilities W_d^k.

Table 1 Population probabilities W_d^k

fission mode	$W_d^{K=1}$	$W_d^{K=2}$
standard I (d=1)	0.21030±0.00573	0.18915±0.00211
standard II (d=2)	0.78935±0.00854	0.81025±0.00311
superlong (d=3)	0.00035±0.00026	0.00060±0.00010

The model parameters of energy partition at scission point were defined on the basis of experimental data of kinetic energy[3] and neutron multiplicity[16] as function of mass number. The calculated fission characteristics of the multimode fission model without account for the $(n,\gamma f)$-process (Table 2) show an anticorrelation between $\overline{TKE}$ and $\overline{\nu}_n$ as well as a correlation between $\overline{TKE}$ and $\overline{N}_\gamma$.

Table 2 Multimode fission averages without account for the $(n,\gamma f)$-process

fission mode	$\overline{TKE}^d/MeV$	$\overline{\nu}_n^d$	$\overline{N}_\gamma^d$	$\overline{E}_\gamma^d/MeV$
standard I	179.3	1.78	7.03	8.02
standard II	168.6	2.56	6.79	6.29
superlong	161.2	3.90	5.69	6.80

The comparison between experimental and calculational data[17] had shown that the observed small experimental fluctuations in total fission characteristics arise from different contributions of the two fission channels as well as the $(n,\gamma f)$-process which is characterized by $\overline{\Gamma}_{\gamma f}\overline{N}_{\gamma f}$=(0.42±0.10)meV and $\overline{\Gamma}_{\gamma f}\overline{E}_{\gamma f}$=(410±120)$eV^2$. Further, one may fairly conclude that the neutron number as function of mass depends on the contribution of fission modes. This influences the calculation of preneutron emission mass distribution from measured fragment kinetic energies and can be taken into account by an iterative way.

3.2. 1^+ resonances of the $^{239}Pu(n,f)$-reaction

In low-energy neutron-induced fission of ^{239}Pu, two separates states with $J^\pi = 1^+$ and $J^\pi = 0^+$ are available. The $J^\pi = 1^+$ strenght is represented by one fission channel[18]. Including the $(n,\gamma f)$-process as a second channel, we can evaluate the relative mass distribution and total fission characteristics of $J^\pi = 1^+$ resonances using the model described in capture 2.

Using experimental data[16,19–21] we determined the parameters of the mass distribution (Average mass number and mass variances) as well as the parameters of energy partition at scission point. The calculated total fission characteristics of different fission modes are shown in table 3.

Table 3 Multimode fission averages for thermal neutron induced fission

fission mode	$\overline{TKE}^d/MeV$	$\overline{\nu}_n^d$	$\overline{N}_\gamma^d$	$\overline{E}_\gamma^d/MeV$
standard I	184.6 ± 2.6	2.32 ± 0.28	7.01 ± 0.15	8.70 ± 0.59
standard II	174.6 ± 3.0	3.15 ± 0.30	6.80 ± 0.15	6.12 ± 0.86
standard III	163.0 ± 3.5	3.40 ± 0.35	6.25 ± 0.25	4.25 ± 0.95
superlong	171.6 ± 2.1	4.16 ± 0.47	5.59 ± 0.52	5.82 ± 2.10

To keep the calculations tractable we neglect fluctuations in fission modes standard III and superlong. Their absolute changes will be small compared to those of standard I and II. In this approximation, the difference in mass distribution for $J^\pi = 1^+$ resonances with reference to the thermal neutron induced fission arises from fluctuations in fission modes standard I and II ($\Delta W_I = -\Delta W_{II}$).

Using experimental data of $\overline{\nu}_n$, $\overline{N}_\gamma$, and $\overline{E}_\gamma$[7,8,22–24], we calculated the mass distribution as well as the parameters of the $(n,\gamma f)$ process for $J^\pi = 1^+$ resonances[25]. We reproduce the experimental differences between thermal neutron induced fission and the 0.3 eV resonance in both the number of promptly emitted neutrons and total kinetic energy (Table 4). The number of promptly emitted neutrons for $J^\pi = 1^+$ resonances in the (n,f) reaction ($\overline{\nu}_n(J^\pi = 1^+)_{(n,f)}$) and in the $(n,\gamma f)$ process ($\overline{\nu}_n(J^\pi = 1^+)_{(n,\gamma f)}$) are in agreement with the experimental data[26]. The deduced averages of the $(n,\gamma f)$ process for $J^\pi = 1^+$ resonances of ^{239}Pu are also consistent with other data. Our results indicate a larger contribution of fission mode standard I for J^π=1$^+$ resonances in comparison with the thermal neutron induced fission.

Table 4 Characteristics of $J^\pi = 1^+$ resonances

parameters	other works	this work
$\overline{\nu}_n(E_n = 0.025\ eV) - \overline{\nu}_n(E_n = 0.296\ eV)$	0.029[27]	0.019 ± 0.004
$\overline{E_k^*}(E_n = 0.025\ eV) - \overline{E_k^*}(E_n = 0.296\ eV)/keV$	-187 ± 75[20]	-160 ± 80
$\overline{\nu}_n(J^\pi = 1^+)_{(n,f)}$	2.8868 ± 0.0032[26]	2.873 ± 0.010
$\overline{\nu}_n(J^\pi = 1^+)_{(n,\gamma f)}$	2.6612[26]	2.675 ± 0.015
$\overline{E}_{\gamma f}^{1+}/MeV$	1.6[26]	1.4 ± 0.6
$\overline{\Gamma}_{\gamma f}^{1+}/meV$	3.0[26]	1.5 ± 0.5
$\overline{\Gamma}_{\gamma f}^{1+}/meV$	1.91 ± 0.81[8]	
$\overline{E}_{\gamma f}^{1+}\overline{\Gamma}_{\gamma f}^{1+}/eV^2$	4800 ± 530[7]	2150 ± 600
$\overline{E}_{\gamma f}^{1+}\overline{\Gamma}_{\gamma f}^{1+}/eV^2$	4490 ± 400[7]	
$W_1(E_n = 0.025\ eV) - W_1(E_n = 0.296\ eV)$		$(-1.1 \pm 0.7)\%$
$W_2(E_n = 0.025\ eV) - W_2(E_n = 0.296\ eV)$		$(1.1 \pm 0.7)\%$

4. Acknowledgements

Stimmulating discussions with Dr. V. I. Furman (JINR Dubna), are greatfully acknowledged.

5. References

1. A. Bohr, *Peaceful uses at. energy* (1956), p. 151.
2. W. K. Bertram, J. W. Boldeman, R. L. Walsh, and J. Caruana, *Materials of the third conference of neutron physics* (Kiev, 1975), part 5, p. 191.
3. F.-J. Hambsch, H.-H. Knitter, C. Budtz-Joergensen, and J. P. Theobald, *Nucl. Phys.* **A491** (1989) 56.
4. R. E. Howe, T. W. Phillips, and C. D. Bowman, *Phys. Rev.* **C13** (1976) 195.
5. U. Brosa, S. Grossmann, and A. Müller, *Z. Naturforsch.* **41A** (1986) 1341.

6. V. I. Furman and J. Kliman, in *XVIIth Int. Symp. on Nucl. Phys.*, ed. D. Seeliger (ZfK-646, Dresden, 1988), p.142.
7. D. Shackleton, *thesis* (Paris, 1974).
8. O. A. Scherbakov, *Phys. Elem. Part. and Atom. Nucl.* **21** (1990) 419.
9. S. Grossmann, U. Brosa, and A. Müller, *Nucl. Phys.* **A481** (1988) 340.
10. P. Fong, in *Phys. and Chem. of Fission* (Jülich, 1979) vol. 2, p. 373.
11. W. R. Phillips, I. Ahmad, H. Emling, R. Holtzmann, R. V. F. Janssens, T. L. Khoo, and M. W. Dregert, *Phys. Rev. Lett.* **57** (1986) 3257.
12. M. A. C. Hotchkis, J. L. Durell, J. B. Fitz- gerald, A. S. Mowbray, W. R. Phillips, I. Ahmad, M. P. Carpenter, R. V. F. Janssens, T. L. Khoo, E. F. Moore, L. R. Morss, Ph. Benet, and D. Ye, *Phys. Rev. Lett.* **64** (1990) 3123.
13. V. Weisskopf, *Phys. Rev.* **52** (1937) 295.
14. N. J. Pattenden and H. Postma, *Nucl. Phys.* **A167** (1971) 225.
15. M. S. Moore, G. de Saussure, and J. R. Smith, in *IAEA Cons. Meet.* (INDC(NDS)-129/GI, Vienna, 1981) p. 74.
16. V. F. Apalin, Yu. N. Gritsyuk, I. E. Kutikov, V. I. Lebedev, and L. A. Mikaelian, *Nucl. Phys.* **71** (1965) 553.
17. U. Gohs, in *Nucl. Data for Science and Technology* (Springer, Berlin, 1991), p. 98.
18. H. Derrien and G. de Saussure, Engineering Physics and Mathematics (Oak Ridge Nat. Lab., ORNL/TM-10986, Vienna, 1981), p. 74.
19. C. M. Wagemans et al., *Phys. Rev.* **C30** (1984) 218.
20. R. L. Walsh, J. W. Boldeman, and M.M. Elcombe, in *Physics and Chemistry of Fission* (Jülich, 1979), vol. 2, p. 129.
21. J. Kaufmann et al., in *Nucl. Data for Science and Technology* (Springer, Berlin, 1992), p. 131.
22. Yu. Ryabov, J. Trochon, D. Shackleton, and J. Frehaut, *Nucl. Phys.* **A216** (1973) 395.
23. L. W. Weston, J. H. Todd, *Phys. Rev.* **C10** (1974) 1402.
24. E. Dermendjiev et al., in *Nucl. Data for Science and Technology* (Springer, Berlin, 1992), p. 147.
25. U. Gohs, in *Dynamical aspects of nuclear fission* (JINR, Dubna, 1992), p. 314.
26. E. Fort, J. Frehaut, H. Tellier, And P. Long, *Nucl. Science and Engineering* **99** (1988) 375.
27. R. L. Walsh and J. W. Boldeman, *Nucl. Phys.* **A451** (1986) 113.

MODELS FOR CALCULATION OF PHOTOPRODUCTION CROSS-SECTIONS AND EMISSION SPECTRA

P. OBLOŽINSKÝ
Institute of Physics, Slovak Academy of Sciences
Dúbravská cesta 9, 842 28 Bratislava, Czechoslovakia

ABSTRACT

We review the current status and recent developments of models for calculations of photoproduction cross sections and photon emission spectra, namely the statistical model of nuclear reactions, preequilibrium models, and the unified model. An important ingrediant of all calculations is the giant dipole resonance γ ray strength function. A considerable progress was achieved in modelling the preequilibrium γ ray emission: GDR radiative processes were incorporated into phenomenological exciton and hybrid models as well as into quantum-mechanical statistical multistep models, the quasideuteron mechanism of hard photon emission was adopted for the hybrid model. From the practical point of view the most advanced seems to be the unified model where multiple particle as well multiple γ ray emission of both the preequilibrium and the statistical origin can be consistently accounted for.

1. Introduction

In this review we are concerned with nuclear reactions induced by projectiles in the energy range between 1 - 100 MeV. Of practical interest are bulk γ ray output, total γ ray spectra, discrete γ ray production cross sections, and partly also average γ ray multiplicities and angular ditributions of γ rays. Some aspects of this presentation were already discussed at two recent conferences [1, 2].

Several nuclear reaction models can calculate fully or at least partly the above quantities. A current situation is illustrated in Fig.1 where we quite schematically show total γ ray spectra induced by neutrons with 3 distinctly different incident energies $E_n = 4, 20$ and $100\ MeV$. In all three cases one observes the low energy discrete γ rays. At higher incident energy range one can see also statistical continuous component from $(n, n'\gamma)$ that extends up to about a nucleon binding energy. Furthermore, there is a relatively weak (n, γ) component in the giant dipole resonance (GDR) spectral energy range due to preequilibrium single-particle radiative mechanism. At still higher incident energy one furthermore observes hard photons arising from the two-particle (quasideuteron) preequilibrium radiative mechanism.

A basic tool in these calculations still remains the well established statistical model of nuclear reactions that includes γ emission down to realistic decay schemes of low lying levels. It should be combined with preequilibrium particle emission to account for certain important features at higher incident energies. Relatively weak components of γ spectra such as the GDR range and hard photons seem to proceed solely via preequilibrium (statistical multistep) radiative mechanisms. Of

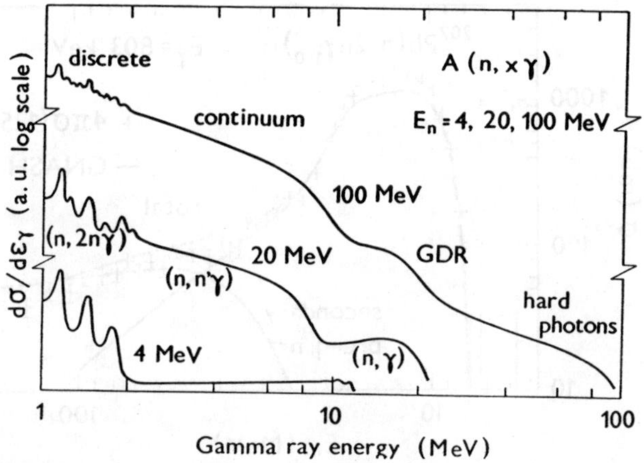

Figure 1: Schematical view of γ ray spectra for reactions with neutrons at $E_n = 4, 20, 100\ MeV$. Seen are 4 types of γ rays: discrete, continuous below nucleon binding, GDR , and hard photons.

considerable interest is the unified model that unites in a natural way preequilibrium and statistical (equilibrium) nuclear reaction concepts.

2. Statistical model

In a simple evaporation formulation of the statistical model one neglects angular-momentum effects and a book-keeping of spin and parity of states is not kept. In this case only γ transitions of the continuum-continuum type are considered. In the Hauser-Feshbach quantum-mechanical formulation such a book-keeping is essential, and continuum-continuum, continuum-discrete as well as discrete-discrete γ ray transitions can be included. Key quantities for calculating γ production are γ ray strength functions, $f_{XL}(\epsilon_\gamma)$, since they determine γ ray transmission coefficients $T_{XL}(\epsilon_\gamma)$ through the relation

$$T_{XL}(\epsilon_\gamma) = 2\pi\epsilon_\gamma^{2L+1} f_{\gamma XL}(\epsilon_\gamma), \tag{1}$$

XL being the multipole type and ϵ_γ the transition energy. The $E1$ strength function is of most importance, and the usual approach is to apply the GDR of the Lorentzian form [3] fitted to photoabsorption cross sections via relation

$$f_{\gamma E1}(\epsilon_\gamma) = \frac{1}{3\pi^2\hbar^2c^2}\frac{\sigma_{GDR}(\epsilon_\gamma)}{\epsilon_\gamma} = \frac{1}{3\pi^2\hbar^2c^2}\sigma_0\frac{\epsilon_\gamma\Gamma^2}{(\epsilon_\gamma^2 - E_0^2)^2 + \epsilon_\gamma^2\Gamma^2} \tag{2}$$

σ_0, Γ and E_0 being resonance parameters. An extension to higher excitation energies is made via the Brink-Axel hypothesis. A number of alternative functional forms can also be used [4].

There is an interesting development advocated originally in low energy neutron capture studies that suggests a violation of the Brink-Axel hypothesis. This

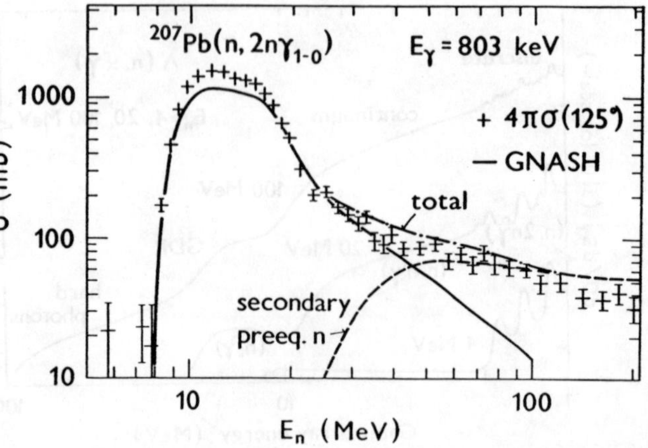

Figure 2: Excitation function for $E_\gamma = 803\ keV$ from $^{207}Pb(n, 2n\gamma)$. Shown is prediction of the statistical model with primary preequilibrium neutron emission (GNASH) and contribution due to secondary preeq neutron emission. Emission of γ rays was in all cases calculated purely statistically. (Nelson 1990 and Chadwick 1992)

shows up as a temperature-dependent GDR spreading width, and there is a non-zero limit for the strength function as $\epsilon_\gamma \to 0$. As pointed out recently [5] such γ ray strength function is in accordance with theoretical proposals and for $E1$ it reads

$$f_{\gamma_E 1}(\epsilon_\gamma, T) = \frac{8.68 \times 10^{-8}}{mbMeV^2}\sigma_0\Gamma\left(\frac{\epsilon_\gamma\Gamma(\epsilon_\gamma)}{(\epsilon_\gamma^2 - E_0^2)^2 + \epsilon_\gamma^2\Gamma^2(\epsilon_\gamma)} + \frac{2.8\pi^2\Gamma T^2}{E_0^5}\right). \tag{3}$$

The spreading width depends on the temperature T as

$$\Gamma(\epsilon_\gamma) = \Gamma\frac{\epsilon_\gamma^2 + 4\pi^2T^2}{E_0^2}. \tag{4}$$

The idea was recently applied to fast proton capture in the unified model [6].

Some other developments are of a more general nature, such as improved knowledge of level densities [7]. If one wants to calculate γ cascades down to low lying discrete levels one should use realistic decay schemes as well as branching ratios rather than theoretical predictions of these quantities. Even more important in such calculations is a proper description of nucleon emission preceding emission of γ rays. It appears that at such energies it is no more sufficient to consider statistical emission processes only, although the low energy γ deexcitation itself is fully governed by the statistical mechanism. One must consider multiple preequilibrium emission of particles (nucleons) to get proper production of discrete γ rays. This situation is illustrated in Fig. 2, where we show a case of $^{207}Pb(n, 2n\gamma)$, $E_\gamma = 803\ keV$ [8, 9].

3. Preequilibrium γ emission

Preequilibrium nuclear reactions represent a well established theoretical concept, see a recent book by Gadioli and Hodgson for more details [10]. In preequilibrium models one classifies nuclear states according to number of excited particles and holes, p and h or excitons $n = p + h$, and these states develop either into more complicated states or they decay by emitting particles or γ rays. The models are phenomenological and quantum-mechanical, and γ ray emission mechanism is the single-particle for the GDR spectral energy range ($E_\gamma < 30\ MeV$) and the two-particle (quasideuteron) for hard photons ($E_\gamma > 30\ MeV$).

3.1. Phenomenological models

The two phenomenological models, exciton and hybrid, have much similarities regarding the physics as well as the structure of basic relations to calculate spectra of emitted γ rays [11, 12, 13]. In the hybrid model, for example, one has

$$\frac{d\sigma}{d\epsilon_\gamma} = \sigma_a \sum_x \sum_n \int_0^E \frac{n p_x}{p} \frac{\omega_{n-1}(E-\epsilon) g d\epsilon}{\omega_n(E)} \frac{\lambda_\gamma(\epsilon, \epsilon_\gamma)}{\lambda_{tot}(\epsilon)} D_n \ , \tag{5}$$

where σ_a is the reaction cross section, $x = \nu, \pi$ denotes neutrons and protons, ω_n is the particle-hole state density, E is the excitation energy of the nucleus, g is the single-particle state density, ϵ is the excitation energy of the selected particle, and D_n is the depletion factor. A key quantity is the γ emission rate

$$\lambda_\gamma(\epsilon, \epsilon_\gamma) = \frac{\epsilon_\gamma^2 \sigma_{GDR}(\epsilon_\gamma)}{\pi^2 \hbar^3 c^2} \frac{g}{gn + g^2 \epsilon_\gamma} \qquad \text{for } \Delta n = 0 \tag{6}$$

and

$$\lambda_\gamma(\epsilon, \epsilon_\gamma) = \frac{\epsilon_\gamma^2 \sigma_{GDR}(\epsilon_\gamma)}{\pi^2 \hbar^3 c^2} \frac{g}{g(n-2) + g^2 \epsilon_\gamma} \frac{\omega_{n-2}(E-\epsilon_\gamma)}{\omega_{n-1}(E-\epsilon)} \quad \text{for } \Delta n = -2 \ , \tag{7}$$

where the photoabsorption cross section $\sigma_{GDR}(\epsilon_\gamma)$ is proportional to the strength function $f_{\gamma E1}$. The corresponding rate in the exciton model can be readily obtained by averaging over energies of individual excitons

$$\lambda_\gamma(\epsilon_\gamma) = \int_0^E \lambda_\gamma(\epsilon, \epsilon_\gamma) \frac{\omega_{n-1}(E-\epsilon_\gamma)}{\omega_n(E)} g d\epsilon. \tag{8}$$

As an example we show in Fig.3 [13] the γ ray spectra from the $\alpha +^{154} Sm$ reaction at the incident energy $E_\alpha = 27\ MeV$. The statistical model describes well the bulk part of the spectrum and fails completely in the high spectral energies. This high energy region is the domain of both the hybrid as well as the exciton model. It should be noted, however, that there are considerable unsolved differencies between the two models when applied to fast neutron induced reactions. The hybrid model predicts 4-8 times smaller γ ray spectra in the GDR energy range when compared to the exciton model [14]. At the moment, for neutron induced reactions we advice to apply the exciton model.

3.2. Quantum-mechanical models

Quantum-mechanical preequilibrium concept distinguishes between bound and unbound nuclear configurations giving rise to statistical multistep compound

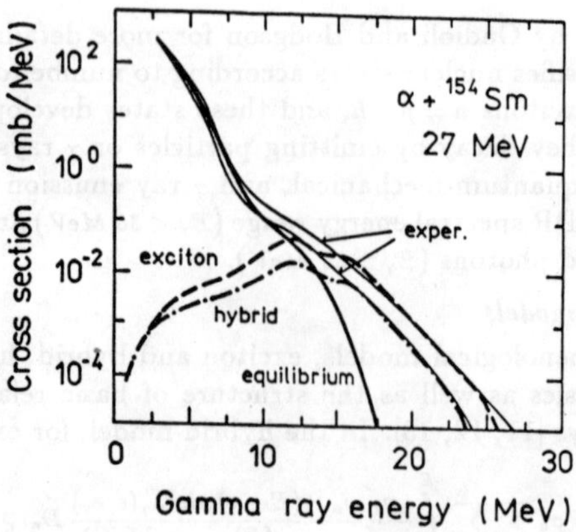

Figure 3: γ ray spectra for $\alpha + ^{154}Sm$ at $E_\alpha = 27\ MeV$. Thin full curves represent an envelope of experimental data. Shown is full spectrum predicted by the statistical model as well as (α, γ) component calculated by both the hybrid and the exciton model. (Obložinský 1988)

(SMC) and statistical multistep direct (SMD) models. These models are currently approaching a stage of practical applications. The γ emission has been first introduced into the SMC model of Feshbach-Kerman-Koonin [15] in Ref.[16] in a way close to the spirit applied in the exciton model.

A somewhat different approach was adopted by Höring et al. [18, 19]. They use the SMC scheme by Nishioka-Verbaarschot-Weidenmüller-Yoshida [17], apply the Brink-Axel hypothesis but distinguish between purely statistical $p-h$ states and collective states carying the dipole transition strength. They clasify as bound configurations all bound configurations on which the GDR is built even though the resulting state may, strictly speaking, be unbound. They furthermore introduce matrix elements for transitions between collective and statistical states. The SMC cross section for the (a, γ) reaction is

$$\sigma_{a,\gamma}^{SMC} = T_{a,Mm} \Pi_{Mm,Nn} T_{Nn,\gamma} \ , \tag{9}$$

where T_a are optical model transmission coefficients, $(\Pi^{-1})_{Mm,Nn}$ are transitional matrix elements between the states of different classes (M, N stand for the numbers of exciton pairs; $m, n = 1, 2, 3, 4$ are subclasses of which 1,2,3 denote GDR and 4 is the non-collective state), and T_γ are the γ ray transmission coefficients defined via the GDR photoabsoprtion cross section. Transitional matrix elements that involve GDR are proportional to its spreading width $(\Gamma\downarrow_{GDR} \equiv \Gamma)$

$$(\Pi^{-1})_{Mm,(M+1)n} \propto \Gamma\downarrow_{GDR} \ . \tag{10}$$

The agreement with data, as shown in Fig.4 [19], is comparable with the success of the exciton model.

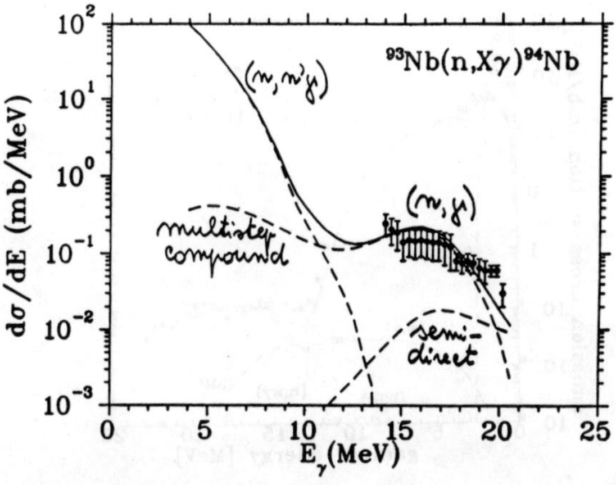

Figure 4: Multistep compound γ ray spectrum is compared with experimental data for $^{181}Ta(n,\gamma)$ at $E_n = 14.1\ MeV$. Note relatively weak semi-direct component. (Herman 1992)

A useful approach to the statistical multistep reaction concept was developed by Kalka [20, 21]. It is based on many-body theory (Green's function formalism) and random matrix physics and provides a unique description of both the statistical multistep direct and the statistical multistep compound reactions. A provision is made for $E1$ γ emission. By adopting the Brink-Axel hypothesis all quantities needed for the SMD as well as SMC calculations are related to the photoabsorption cross section. The two-step mean squared matrix element $\overline{D^2}$ is determined from

$$\sigma_{GDR}(\epsilon_\gamma) = \frac{4\pi^3}{k_\gamma^2}\rho(\epsilon_\gamma)\overline{D^2}(\epsilon_\gamma) \propto f_{\gamma E1}(\epsilon_\gamma)\,. \tag{11}$$

In the SMD part, conceptually similar to the well known direct-semidirect process, it is assumed that γ emission proceeds as a two-step process via formation and decay of the GDR. The cross section is given as

$$\frac{d\sigma_{a,\gamma}^{SMD}(E_a)}{d\epsilon_\gamma} = \frac{d\sigma_{a,\gamma}^{DSD}(E_a)}{d\epsilon_\gamma} = \frac{4\pi^3}{k_a^2}\frac{\rho(E_a)}{2}\frac{\overline{D^2}(\epsilon_\gamma)}{2L+1}\rho(\epsilon_\gamma)\,, \tag{12}$$

where $\rho(E_a)$ is the single-particle state density, $L = 1$ and $\rho(\epsilon_\gamma)$ is the state density of free γ rays. It should be noted that the above expression is quite similar to the phenomenological preequilibrium expression for the $n = 1$ term (see e.g. the exciton model). Furthermore, also the SMC part of γ emission of Kalka's approach is much the same as adopted by phenomonological models. It is thus not surprising that calculated cross sections are close to those predicted by the exciton (unified) model. An example is shown in Fig.5.

3.3. Hard photons

Considerable progress has been achieved in understanding hard photon emission ($\epsilon_\gamma > 30\ MeV$) [23, 24]. Using the concept of quasi-deuteron photo-absorption

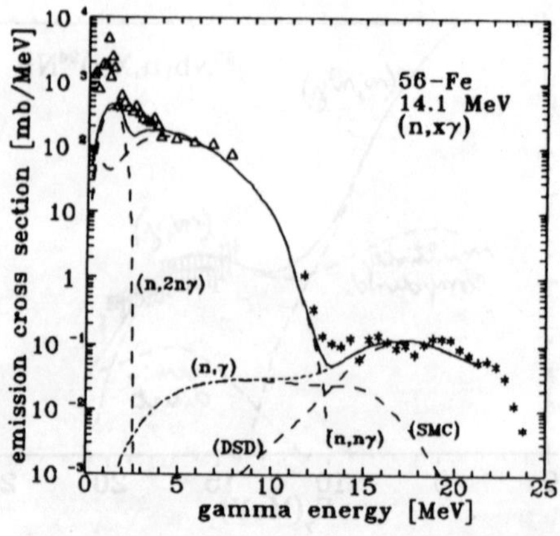

Figure 5: $(n, x\gamma)$ emission spectrum at 14.1 MeV. Note strong direct-semidirect (DSD) component and relatively weak multistep compound (SMC) component. (Kalka 1990)

the preequilibrium emission rate was proposed in the hybrid model [23]

$$\lambda_\gamma(\epsilon, \epsilon_\gamma) = \frac{\epsilon_\gamma^2 \sigma_{qd}(\epsilon_\gamma)}{\pi^2 \hbar^3 c^2} \frac{\omega_3(\epsilon - \epsilon_\gamma)}{\omega_4(\epsilon_\gamma)}, \tag{13}$$

where σ_{qd} refer to the quasi-deuteron component of the photoabsorption cross section derived in Ref.[25]. The agreement with data is surprisingly good.

4. Unified model

The unified model combines in a natural way the preequilibrium nuclear reaction concept with the equilibrium statistical decay. This is possible via the master equation approach, originally devised for the exciton model, that allows to follow time evolution of the reaction system up to its complete decay. In order to handle properly cascades of γ rays, the usual set of master equations (hundreds of equations) was enlarged to couple also different excitation energies as well as various nuclei. This gives rise typically to tens thousands of equations that can, however, be solved pretty fast even on small PC computers. Such a model has been developed over years by Běták [26, 6]. Its considerable advantage is a consistent calculation of multiple particle emission as well as multiple γ emission. A schematical picture of the initial reaction system (N, Z) together with a sequence of daughter nuclei is shown in Fig. 6. The model was quite recently extended to account for angular momenta [27] opening up a possibility to calculate also discrete γ ray production. Altogether, the model is very suitable for γ production calculations in complicated reaction chains. The set of master equations reads

$$\frac{dP(i, E, J, n, t)}{dt} = P(i, E, J, n-2, t)\, \lambda^+(i, E, J, n-2) + P(i, E, J, n+2, t)\, \lambda^-(i, E, J, n+2)$$

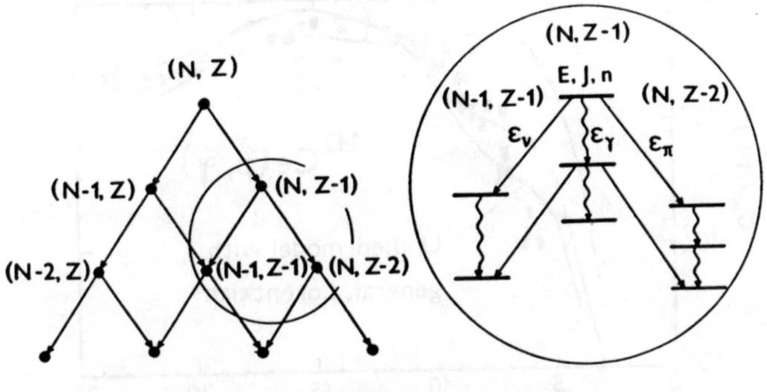

Figure 6: Schematical view of a chain of nuclei in the unified model. In a given nucleus one considers different excitation energies, spins as well as exciton numbers so that γ cascades can be calculated.

$$-P(i,E,J,n,t)\,\left[\lambda^+(i,E,J,n)+\lambda^-(i,E,J,n)+L(i,E,J,n)\right]$$
$$+\sum_{i',J',n',x}\int P(i',E',J',n',t)\,\lambda_x(\{i',E',J',n'\}\to\{i,E,J,n\};\epsilon)d\epsilon\,, \quad (14)$$

where $P(i,E,J,n,t)$ is the occupation probability of the nucleus i at the excitation energy E, spin J and exciton number n, λ^+ and λ^- are transitional rates to neighbouring states, and L is the total integrated emission rate of neutrons, protons and γ rays

$$L(i,E,J,n)=\sum_{x=\nu,\pi,\gamma}\int\lambda_x(i,E,J,n;\epsilon)d\epsilon\,. \quad (15)$$

The last term in (14) ensures the coupling of different nuclei and various excitation energies. Emission rates λ_x consider spin conservation so that for γ rays one uses the angular momentum coupling terms of the type

$$x^{nJ}_{n'J'}=\frac{3(2J+1)}{R'_n(J')}\sum_{j_1j_2j_3}(2j_1+1)R_1(j_1)(2j_2+1)R_1(j_2)R_{n-1}(j_3)$$
$$\times\begin{pmatrix} j_2 & 1 & j_1\\ \frac{1}{2} & 0 & -\frac{1}{2}\end{pmatrix}^2\begin{Bmatrix} j_2 & j_3 & J'\\ J & 1 & j_1\end{Bmatrix}^2, \quad (16)$$

where j_1,j_2,j_3 are spins and R is the spin part of the state density. The structure of the γ emission rate is similar to that discussed above for the hybrid/exciton model. The quantities of interest are time integrals

$$\tau(i,E,J,n)=\int_0^\infty P(i,E,J,n,t)dt \quad (17)$$

and the cross section for the reaction $(a,bc\gamma)$ is

$$\frac{d\sigma_{a,bc\gamma}}{d\epsilon_\gamma}=\sigma_a\sum_{E,J,n}\tau(i=bc,E,J,n)\,\lambda_\gamma(i=bc,E,J,n;\epsilon_\gamma)\,. \quad (18)$$

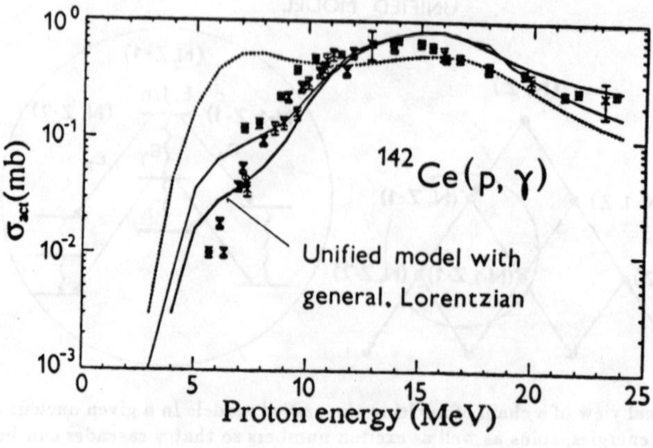

Figure 7: Cross sections of the $^{142}Ce(p,\gamma)$ reaction calculated by the unified model with the generalized Lorentzian of Eq.(3). (Běták 1992)

The above formalism is encoded into the program PEGAS (Preeq-Eq-Gamma-And-Spin) [27]. An example can be seen in Fig.7 [6].

5. Conclusions

Current theoretical models allows one to calculate bulk γ production, discrete γ production as well as γ emission spectra including their high energy parts. The firm basis for these calculations is the statistical model of nuclear reactions combined with preequilibrium particle emission. Preequilibrium γ emission is a useful tool to calculate high energy parts of γ ray spectra (GDR range and hard photons). Approaches are available within the phenomenological as well as statistical multistep compound/direct models to calculate such spectra. A very consistent treatment of γ emission provides the unified model. The above concepts were transfered into several useful computer codes. Their typical representatives are:

- Statistical γ with preeq particles: STAPRE (Uhl), TNG (Fu), GNASH (Young)
- Statistical multistep γ: EXIFON (Kalka)
- Unified γ: PEGAS (Běták and Obložinský).

This work was supported in part by the International Atomic Energy Agency (contract No. 5148/RB) and the Slovak Academy of Sciences (grant No. 042/92).

6. References

1. P. Obložinský, in *Capture Gamma-Ray Spectroscopy*, Pacific Grove, California 1990, ed. R.W. Hoff (AIP Conf. Proc. 238, AIP, New York, 1990) p.714.

2. P. Obložinský and M.B. Chadwick, in *Nuclear Data for Science and Technology*, Jülich, Germany 1991, ed. S.M. Qaim (Springer Verlag, Berlin, 1992), p. 842.
3. S.S. Dietrich and B.L. Berman, *At. Nucl. Data Tables* **38** (1989) 199.
4. D.G. Gardner, in *Neutron Radiative Capture*, ed. R.E. Chrien (Pergamon Press, Oxford, 1984), p.62.
5. J. Kopecky and M. Uhl, *Phys. Rev.* **C41** (1990) 1941.
6. E. Běták, F. Cvelbar and J.Kopecky, *Report* ECN-RX 92-012 (Petten, 1992); to appear in *Phys. Rev. C.*
7. Eds. G. Reffo, M. Herman and G. Maino, *Nuclear Level Densities*, Proc. of the OECD Meeting, Bologna, Italy 1989 (World Scientific, Singapore, 1992).
8. R.O. Nelson et al., as Ref.[1], p.697.
9. M.B. Chadwick and P.G. Young, *Private Communication* (1992).
10. E. Gadioli and P.E. Hodgson, *Pre-Equilibrium Nuclear Reactions* (Clarendon Press, Oxford, 1992).
11. J.M. Akkermans and H. Gruppelaar, *Phys. Lett.* **B157** (1985) 95.
12. P. Obložinský, *Phys. Rev.* **C35** (1987) 407.
13. P. Obložinský, *Phys Lett.* **B215** (1988) 597.
14. G. Reffo, M. Blann and Remington, *Phys. Rev.* **C41** (1990) 403.
15. H. Feshbach, A. Kerman and S. Koonin, *Ann. Phys. (NY)* **125** (1980) 429.
16. P. Obložinský and M.B. Chadwick, *Phys. Rev.* **C41**(1990) 1652.
17. H. Nishioka, J.M. Verbaarschot, H.A. Weidenmüller and S. Yoshida *Ann. Phys. (NY)* **172** (1986) 67.
18. A. Höring and H.A. Weidenmüller, *Report* DOE/ER/40561-069-INT92-00-20 (1992) unpublished.
19. M. Herman, A. Höring and G. Reffo, *Report* DOE/ER/40561-070-INT92-00-21 (1992) unpublished.
20. H. Kalka, M. Torjman and D. Seeliger *Phys. Rev.* **C40** (1989) 1619.
21. H. Kalka, *Code EXIFON*, Report INDC(GDR)-060/L (IAEA, Vienna, 1990).
22. H. Kalka, as Ref.[2], p.897.
23. P. Obložinský, *Phys. Rev.* **C44** (1989) 1591.
24. M.B. Chadwick and G. Reffo, *Phys.Rev.* **C43** (1991) 919.
25. M.B. Chadwick, P. Obložinský, P.E. Hodgson and G. Reffo, Phys. Rev. **C45** (1991) 814.
26. E. Běták, *Code PEQAG*, Report INDC(CSR)-016/LJ (IAEA, Vienna, 1989).
27. E. Běták and P. Obložinský, *Code PEGAS* (1992), under preparation.

MEDIUM ENERGY DATA

METHODS USED TO PRODUCE NUCLEAR DATA FILES FOR 0 - 10 GEV INCIDENT NEUTRONS AND PROTONS

SOL PEARLSTEIN

Department of Nuclear Energy, Brookhaven National Laboratory
Upton, New York 11973, U.S.A.

ABSTRACT

Nuclear models for calculating medium energy data are in need of a unified theory. For energies above 100 MeV the incident nucleon wavelength is smaller than its collision mean-free-path within the target nucleus so that an intra-nuclear-cascade model, for example, is appropriate. For energies below 100 MeV the incident nucleon wavelength is larger than the dimensions of the target nucleus so that a partial wave nuclear model, for example, is appropriate. Comparisons between calculations and experiments show discrepancies even when using appropriate models. Experimental data alone is rarely convergent or sufficient to define an evaluation over the complete range of interest. The evaluator is therefore left to selectively choose from a diverse arsenal of experimental data, nuclear models and nuclear systematics those tools that may best help him to complete an evaluation. These points are illustrated by the use of calculation and experiment in the evaluation of nucleon data for ^{12}C.

1. Introduction

1.1. Evaluation Tools

There is an overlap of the nuclear data requirements for applications that include space exploration, astrophysics, accelerator shield design, spallation neutron sources, radiation therapy and medical isotope production. The total energy range of interest is 0 - 10 GeV. There is no one nuclear theory or model code validated over this energy range. For energies above 100 MeV the incident nucleon wavelength is smaller than its collision mean-free-path within the target nucleus so that an intra-nuclear-cascade model, for example, is appropriate. For energies below 100 MeV the incident nucleon wavelength is larger than the dimensions of the target nucleus so that a partial wave nuclear model, for example, is appropriate. These characteristics can be seen by looking at the total cross section for neutrons on Pb as shown in Fig. 1. Below about 200 MeV the total cross section is oscillatory mainly due to the Ramsauer effect of interference between the incident neutron wave traversing the target nucleus and the neutron wave going around. Above 200 MeV the total cross section is relatively flat as expected for nucleon-nucleon collisions.

In the case of the well measured total cross section for neutrons on Pb even the most indecisive evaluators could agree on an evaluated cross section but the experimental data alone is rarely convergent or sufficient to define an evaluation over the complete phase space of interest. Some nuclear model codes are a marvel in the sense that they provide answers to all of the questions asked by the evaluator. To determine if the answers are correct comparisons between calculations and

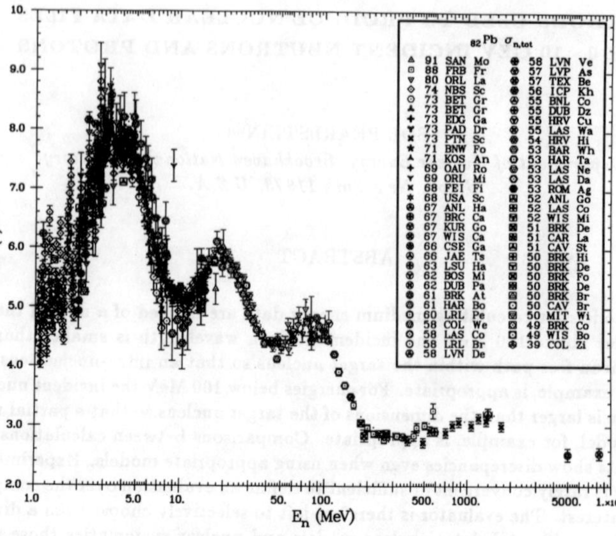

Figure 1: Total cross section for neutrons on lead.

experiments are made when possible. This approach is not fool proof because some experiments can systematically be wrong in both magnitude and shape.

The evaluator is therefore left to cautiously choose those methods that may best help him to complete an evaluation. They are the following:

1. Experimental data bases. Data contained in this article or references to such data are typical of that obtainable from the National Nuclear Data Center[1].
2. Nuclear model codes. There are many codes useful to evaluators of medium energy data. Some information about medium energy codes have been collected[2,3,4].
3. Nuclear systematics. These are semi-empirical formulas that can be useful in estimating data where measurements have not been made. Formulas based on medium energy systematics are readily available[4,5].
4. Arithmetic strategies. Statistical averaging, spline fitting and smoothing, interpolation, extrapolation, etc., may be useful to filling gaps and arriving at recommended values for the cross section.

1.2. Evaluated Data for ^{12}C

As an example of the methods used to produce data files for 0 - 10 GeV incident neutrons and protons the data for a ^{12}C target is discussed. The ^{12}C target is an intermediate situation where the number of target nucleons may be too many for

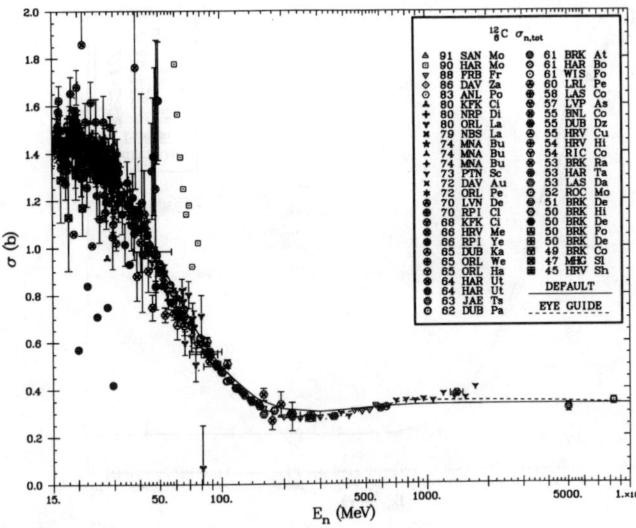

Figure 2: Total cross section for neutrons on carbon. See text for explanation of curves.

few-body analysis and too few for statistical modelling. Codes, admittedly suspect in terms of their validity for ^{12}C analysis, are used with extensive comparisons with experimental data to help recommend the evaluated data.

For evaluated data files having a low energy upper limit, e.g. 20 MeV, reactions involve the emission of only a few products and the reaction path is generally unique, e.g. (n,a) predominates over (n,2d) or (n,n^{3}He). For medium energy data files the nucleon total, elastic scattering, and nonelastic cross sections; multiple particle production cross sections, e.g. (n,xn), and residual nucleus formation cross sections, e.g. $^{12}C(p,X)^7Be$, are the practical objectives.

2. ^{12}C Total, Elastic and Nonelastic Data

Curves using nuclear model codes and systematics similar to those developed earlier[5] will be used as a trial curve (labeled DEFAULT) to compare calculation with experiment. The total cross section in the energy range 15 to 10000 MeV is shown in Fig. 2. Because natural carbon is 98.9% ^{12}C elemental carbon data is interchangeable for ^{12}C data for such cross sections. The experimental data will not be individually referenced in this report but expanded reference titles for those symbols appearing in the legend box are available from the author and the data are obtainable in computerized form[1]. The DEFAULT curve for the total cross section is quite satisfactory. The agreement can be improved by lowering the curve a bit

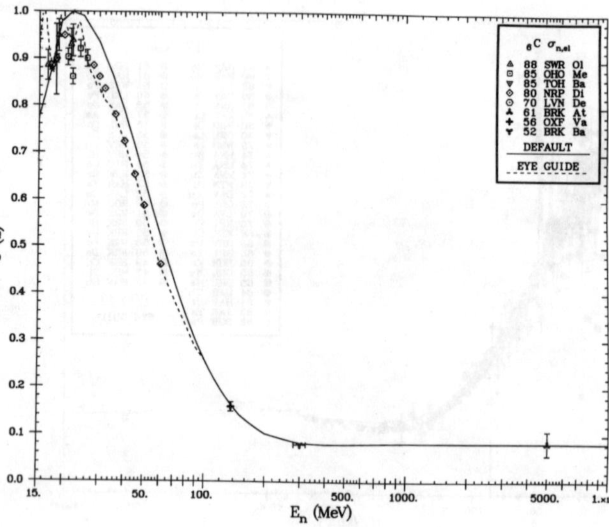

Figure 3: Elastic scattering cross section for neutrons on carbon. See text for explanation of curves.

about 200 MeV and raising the curve a bit about 1000 MeV. Below 32 MeV data from ENDF/B-VI[6] for elemental carbon[7] is used. At 32 MeV the ENDF/B-VI total cross section is 1.306 b.

The elastic scattering cross section over the same energy range is shown in Fig. 3. The agreement between calculation and experiment is excellent for the high energy points available but the curve should be revised downward about 10% below 60 MeV to join ENDF/B-VI smoothly at 32 MeV where the elastic scattering cross section is 0.810 b.

The nonelastic cross section is shown in Fig. 4. Again, the agreement between calculation and experiment is excellent for the high energy points available but the curve should be revised upward about 10% below 60 MeV to join ENDF/B-VI smoothly at 32 MeV where the nonelastic cross section is 0.496 b.

The DEFAULT curve for protons on ^{12}C nonelastic cross section shown in Fig. 5 is obtained from optical model calculations using global parameters[5]. Good agreement with an experimental data compilation[8] can be obtained at high energies by about a 5% lowering of the curve above 100 MeV. Below 30 MeV single-particle effects are present which are not modeled in the optical model calculations. The evaluation would follow the eye-guide shown in the figure.

3. Residual Nucleus Formation Data

In the work described here for the calculation of residual nucleus formation

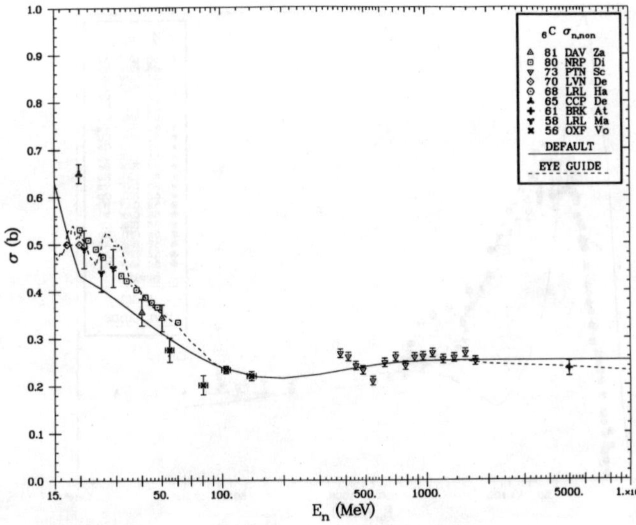

Figure 4: Nonelastic cross section for neutrons on carbon. See text for explanation of curves.

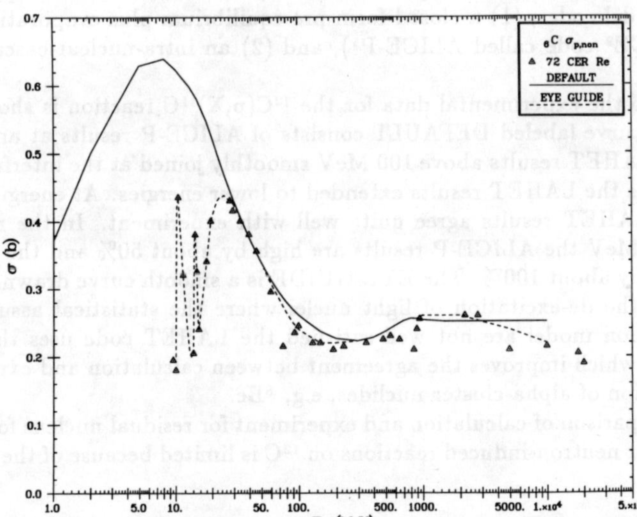

Figure 5: Nonelastic cross section for protons on carbon. See text for explanation of curves.

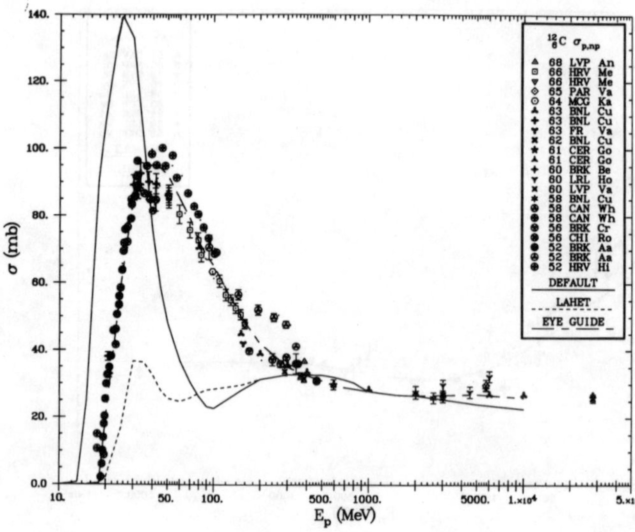

Figure 6: Cross section for the $^{12}C(p,X)^{11}C$ reaction. See text for explanation of curves.

cross sections from neutron- and proton-induced reactions, reliance is placed on two nuclear model codes; (1) a closed form pre-equilibrium plus evaporation code, (modified ALICE[9] code called ALICE-P[5]), and (2) an intra-nuclear-cascade code (LAHET[10]).

In Fig. 6 the experimental data for the $^{12}C(p,X)^{11}C$ reaction is shown with 3 curves. The curve labeled DEFAULT consists of ALICE-P results at and below 100 MeV and LAHET results above 100 MeV smoothly joined at the interface. The LAHET curve is the LAHET results extended to lower energies. At energies above 300 MeV the LAHET results agree quite well with experiment. In the region of the peak at 50 MeV the ALICE-P results are high by about 50% and the LAHET results are low by about 100%. The EYE-GUIDE is a smooth curve drawn through the data. For the de-excitation of light nuclei where the statistical assumptions of the evaporation model are not well satisfied the LAHET code uses the Fermi breakup model which improves the agreement between calculation and experiment for the production of alpha-cluster nuclides, e.g. ^{8}Be.

The comparison of calculation and experiment for residual nucleus formation cross sections for neutron-induced reactions on ^{12}C is limited because of the scarcity

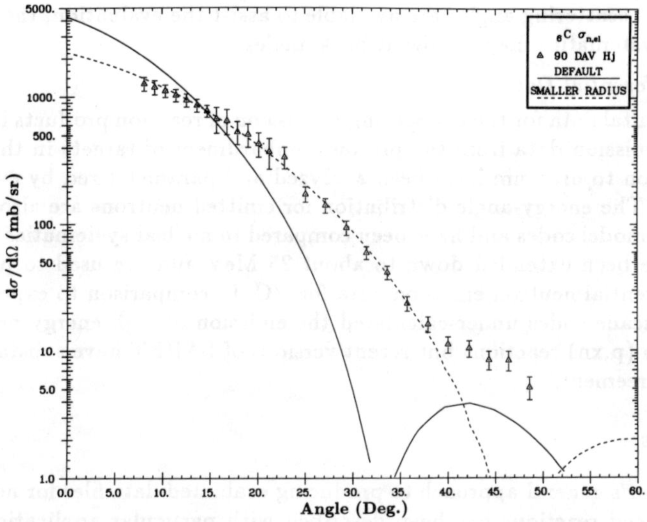

Figure 7: Elastic scattering angular distribution for 65 MeV neutrons on carbon. See text for explanation of curves.

of medium energy neutron reaction data.

4. Single- and Double-Differential Data

1. Elastic Scattering

Elastic scattering data for 65 MeV neutrons on carbon in the zero-linear-momentum system are shown in Fig. 7. The DEFAULT curve is based on nuclear systematics[3,11] that is generally valid for small angle scattering. Here, the modeled data is too forward peaked. The second diffraction peak occurs at about the same angle in both the measured and modeled data but the modeled peak magnitude is lower. Reducing the radius in the calculation brings the measured and calculated curves into agreement at forward angles but increases the disagreement at larger angles. These conclusions also hold true in comparisons with measurements of forward peaked elastic neutron scattering from carbon at higher energies.

Elastic scattering of charged particles includes components from Coulomb scattering, nuclear scattering, and the interference between them. For elastic scattering of protons the nuclear elastic scattering is assumed to be similar to neutron elastic scattering.

Data for elastic scattering of nucleons in the forward angles, which are usually the most important ones in transport calculations, can be parameterized by a simple diffraction model. Unless experimental data covering a wide range of inci-

dent energies and scattering angles are available to assist the evaluation, the curves obtained from systematics may be low at back angles.

1.1. Double-Differential Data

Experimental data for the energy-angle emission of reaction products is available. Neutron emission data from the proton bombardment of targets in the mass range from carbon to uranium have been analyzed and parameterized by evaporation formulae[12]. The energy-angle distribution for emitted neutrons are also calculated by nuclear model codes and have been compared to nuclear systematics[13]. The systematics have been extended down to about 25 MeV and are used to produce the double-differential neutron emission data for ^{12}C. In comparison to experiment intra-nuclear-cascade codes under-calculated the emission of high energy neutrons at back angles in (p,xn) reactions but recent versions of LAHET have substantially improved the agreement.

5. Conclusions

The author's general approach to producing evaluated data files for neutron- and proton-induced reactions has been described with particular application to a ^{12}C target. For a light mass target such as ^{12}C the nuclear model calculations can be deficient. The calculations may be of the right order of magnitude but revisions are frequently required to obtain agreement with experimental data. The completed ^{12}C evaluations have been given to the National Nuclear Data Center.

5. Acknowledgements

This work was performed at Brookhaven National Laboratory and supported by the U.S. Department of Energy.

6. References

1. National Nuclear Data Center, Brookhaven National Laboratory, Upton, New York 11973. Tel:(516)282-2901, FAX:-2806, E-Mail addresses are Bitnet:NNDC@BNL, Internet:NNDC@BNL.GOV, use Internet:NNDC@BNLND2.DNE.BNL.GOV for on-line access to neutron, charged particle, nuclear structure and radioactive decay data and indices to bibliography.
2. S. Pearlstein, *Summary of the Meeting of the Medium Energy Nuclear Data Working Group*, May 21-22, 1986 Brookhaven Nat. Lab. (BNL-NCS-38404, 1986).
3. S. Pearlstein, *Summary of the Meeting of the Medium Energy Nuclear Data Working Group*, May 27-29, 1987 NASA Langley Res. Ctr. (BNL-NCS-40070, 1987).

4. S. Pearlstein, in *Proc. Int. Conf. on Nuclear Data for Science and Technology*, May 30-June 3, 1988, Mito, p. 1115 (Saikon Pub. 1988).
5. S. Pearlstein, *Astrophysics J.* **346** (1989) 1049.
6. ENDF/B, *Evaluated Nuclear Data File/ Version B.* Distributed by the National Nuclear Data Center.
7. C. Y. Fu, E. J. Axton and F. G. Perey, *ENDF/B-VI, Evaluated Data for 0-32 MeV Neutrons incident on Elemental Carbon*, Completed August 1989. See Refs. 1 and 6.
8. P. U. Renberg et. al., *Nucl. Phys.* **A183** (1972) 81.
9. M. Blann, *Code ALICE/85/300* and succeeding versions. Lawrence Livermore Nat. Lab. (UCID-20169 1985).
10. R. E. Prael and H. Lichtenstein, in *User Guide to LCS: The LAHET Code System*, Las Alamos Nat. Lab. (LA-UR-89-3014 1989).
11. S. Pearlstein, *Nucl. Sci. Eng.* **49** (1972) 162.
12. S. Pearlstein, *Nucl. Sci. Eng.* **95** (1987) 116.
13. S. Pearlstein, *An Intercomparison of Medium Energy Cross Section Codes* , Brookhaven Nat. Lab. (BNL-NCS-52155, 1988).

USE OF THE NUCLEAR MODEL CODE GNASH TO CALCULATE CROSS-SECTION DATA AT ENERGIES UP TO 100 MEV

P. G. YOUNG, M. B. CHADWICK, and M. BOZOIAN

Group T-2, Theoretical Division, Los Alamos National Laboratory
Los Alamos, NM 87545, USA

ABSTRACT

The nuclear theory code GNASH has been used to calculate nuclear data for incident neutrons, protons, and deuterons at energies up to 100 MeV. Several nuclear models and theories are important in the 10 - 100 MeV energy range, including Hauser-Feshbach statistical theory, spherical and deformed optical models, preequilibrium theory, nuclear level densities, fission theory, and direct reaction theory. In this paper we summarize general features of the models in GNASH and describe the methodology utilized to determine relevant model parameters. We illustrate the significance of several of the models and include comparisons with experimental data for certain target materials that are important in applications.

1. Introduction

The first version of the GNASH nuclear theory code was completed in 1974,[1] and it has been developed continually since that time. Several versions of the code are in use, including one adapted for use in generating activation cross sections for the Japanese fusion program,[2] and two versions with modifications and approximations suitable mainly for higher energy calculations.[3] The code discussed here is the base version and has been used for calculations at energies as low as 1 keV and as high as 100 MeV. It is best summarized as the statistical Hauser-Feshbach plus preequilibrium version with full angular momentum conservation.[4]

GNASH has been used to support many data evaluations at incident energies below 20 MeV, including several of the ENDF/B-VI evaluations. In recent years the code has been extended for work at higher energies and has been utilized in evaluations up to incident energies of 100 MeV. In particular, it was used for comprehensive analyses of neutron-induced reactions to 40 MeV for ^{56}Fe,[5] to 50 MeV for ^{59}Co,[6] and to 100 MeV with initial calculations for ^{208}Pb.[7] Additionally, a set of neutron- and proton-induced transport data libraries were generated with GNASH to an incident energy of 100 MeV for ^{27}Al, ^{28}Si, ^{56}Fe, ^{184}W, and ^{238}U,[8] and similar libraries are presently under development for deuteron-induced reactions to 100 MeV on ^{27}Al and ^{208}Pb.

In this paper we summarize the methods and procedures used in performing calculations with GNASH in the energy range of 10 - 100 MeV. Because the basic methodology is to utilize physically meaningful nuclear model parameters that result in reasonable descriptions of experimental data, we summarize in Sec. 2 the major models that are used in GNASH. In Sec. 3 we describe some aspects of choosing model parameters for higher energies, and we illustrate features of the nuclear models by comparing calculations with experimental data. Finally, Sec. 4 includes some conclusions and summary remarks concerning the use of GNASH at higher energies.

2. Nuclear Models Used in GNASH

A block diagram summarizing the various input and output quantities for a GNASH calculation is included as Fig. 1 The input information includes separate files of discrete level decay data for all significant residual nuclei that are formed in a given problem, optical model transmission coefficients, direct reaction cross sections, and ground-state mass, spin and parity tables. The remaining model parameters, reaction sequences, calculational energy range, etc., are provided in an INPUT file.

GNASH implements Hauser-Feshbach theory in an open-ended sequence of reaction chains, with full conservation of angular momentum. Reaction chains are defined in the code input, either by direct specification of the sequence or by setting a flag to indicate a particular decay chain option. A schematic illustration of the decay processes that occur for each compound nucleus in a decay sequence is given in Fig. 2. Note in the simplified expressions at the bottom of the figure that the change in population of states in the residual nucleus (K') from transitions from the initial (K) nucleus is proportional to the transmission coefficient, $T(U\text{-}U'\text{-}B')$, for the radiation and to the level density, $\rho(U')$, of the final nucleus. Transmission coefficients for particles are obtained from optical model calculations and for gamma rays from a giant dipole resonance model. In addition to being important for providing transmission coefficients, optical model potentials are used to obtain the initial compound nucleus formation cross section, which then determines the overall normalization of all calculated emission cross sections. Continuum level densities are obtained from phenomenological level density functions,[9,10] which are matched at lower excitation energies to the available discrete level data. Both the discrete and continuum structure data are utilized in GNASH calculations.

Preequilibrium corrections become increasingly important at energies above 20 MeV and are made in GNASH using the exciton model of Kalbach.[11] We demonstrate below that for certain applications it is also important to include multiple preequilibrium effects for secondary neutrons and protons as well as the usual corrections in the primary decay channels. For actinide studies, the code contains a detailed fission model, allowing use of up to three uncoupled fission barriers. The fission model and its use at higher energies is described elsewhere[12] and is not repeated here. The various models in GNASH are expected to be applicable over a wide energy range, and the version of the code described here has been utilized from incident energies below 1 keV to above 100 MeV.

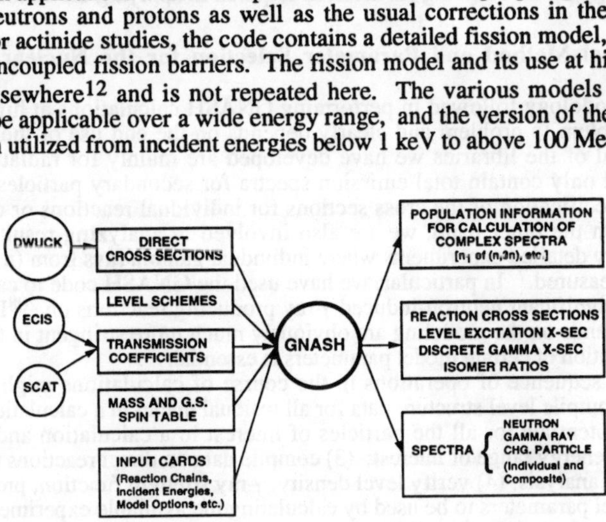

Fig. 1. Schematic representation of the input and output capabilities of the GNASH code.

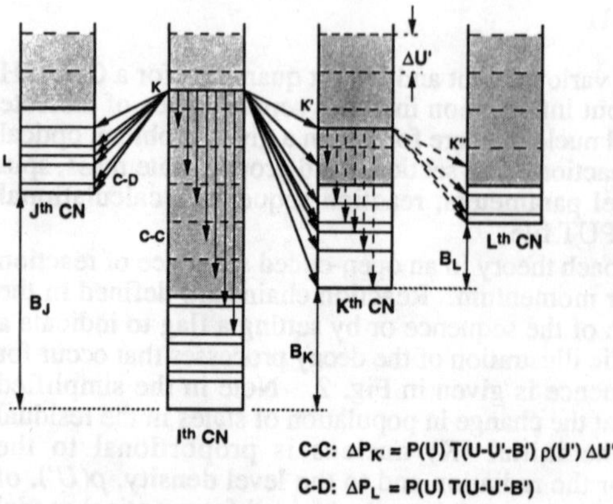

Fig. 2. Illustration of the decay of a continuum energy bin in the Ith compound nucleus to continuum (C - C) and discrete (C - D) states in the Jth and K^{th} compound systems. Simplified equations depicting the continuum to continuum and continuum to discrete population increments are given.

Referring to Fig. 1, the primary output from GNASH are absolute angle-integrated particle and gamma-ray spectra, and excitation and deexcitation cross sections of discrete states. The calculated spectra are integrated and summed to provide absolute reaction cross sections. The present configuration of the code permits the incident particle type to be neutrons, protons, deuterons, tritons, ^{3}He, or ^{4}He. In addition to gamma rays, these same particles are permitted in the decay channels. Angular effects are not included in the calculations, as the results are normally used in combination with the systematics-based parameterization of Kalbach[13] to determine angular distributions. For this purpose, the code provides tables of ratios of preequilibrium to total emission cross sections as functions of emission energy for all outgoing particles, as required in applying the Kalbach relations.

3. Calculational Method and Parameter Selection for the Nuclear Models

The methodology followed in performing GNASH calculations at higher energies varies from problem to problem and clearly depends on the end use of the results. For example, several of the libraries we have developed are mainly for radiation transport calculations and only contain total emission spectra for secondary particles and gamma rays, without specification of the cross sections for individual reactions or discrete level transitions.[8] On the other hand, we are also involved in analyzing results from measurements in very detailed experiments where individual gamma rays from (n,xγ) reactions (x≈ 1-10) are measured.[7] In particular, we have used the GNASH code to calculate cross sections for all significant neutron-induced γ-ray producing reactions on ^{208}Pb up to E_n = 100 MeV. Demands on the modeling are obviously much more stringent in this case, and careful determination of certain model parameters is essential.

A typical sequence of operations in the course of calculations might include the following: (1) compile level structure data for all residual nuclei in a calculation; (2) obtain optical model potentials for all the particles of interest in a calculation and verify their validity over the energy range of interest; (3) compile data on direct reactions that might be of interest for an analysis; (4) verify level density, γ-ray strength function, preequilibrium, and fission model parameters to be used by calculating the available experimental data; (5) verify by sample calculations that all important decay channels are included in the problem. Elaboration of some of these steps with examples is given below. In particular, we compare the effects of different model or parameter choices in calculations of different reactions, with emphasis on ^{208}Pb(n,nxγ) cross sections for specific γ-rays.

3.1 *Optical Potentials*

There are a large number of global optical model potentials available for incident nucleons, and these are frequently used in GNASH calculations where accuracy requirements are only modest. However, a basic problem exists in carrying analyses to energies of 100 MeV or higher, because transmission coefficients are required in the Hauser-Feshbach calculations to low particle energies and global potentials usually do not cover such a large energy range. One method of handling this problem is to combine two potentials that span the energy region of the calculation, with particular attention to selecting a matching energy where the reaction cross sections from the two potentials are consistent. We have frequently combined the Madland[14] higher energy potential for protons and neutrons with one of the common lower energy potentials.[15]

Best results are usually obtained if analyses have been performed leading to potentials for the specific target nucleus and energy range under investigation. We are currently using GNASH to analyze high-resolution neutron-induced gamma-ray production measurements on Pb isotopes from the WNR/PSR facility at LAMPF. Three of the potentials we are studying are compared to neutron total cross section measurements[16] on ^{208}Pb in the top half of Fig. 3, and to the nonelastic (reaction) cross section in the bottom half. Of the three optical models illustrated, only the Schutt et al.[17] analysis was directed specifically at n + ^{208}Pb data over the entire energy range shown. In particular, the Schutt et al. potential results from an analysis of n + ^{208}Pb elastic scattering data to 40 MeV and total cross section data between 2 and 250 MeV. The Finlay et al.[18] potential results from a similar analysis of data between 7 and 50 MeV. For the calculations in Fig. 3, the Finlay potential was utilized to about 60 MeV and the Madland potential[14] at higher energies. The coupled-channel potential is a simple extrapolation of the low-energy Young and Shamu potential described at this symposium,[19] adjusted to reasonably reproduce differential

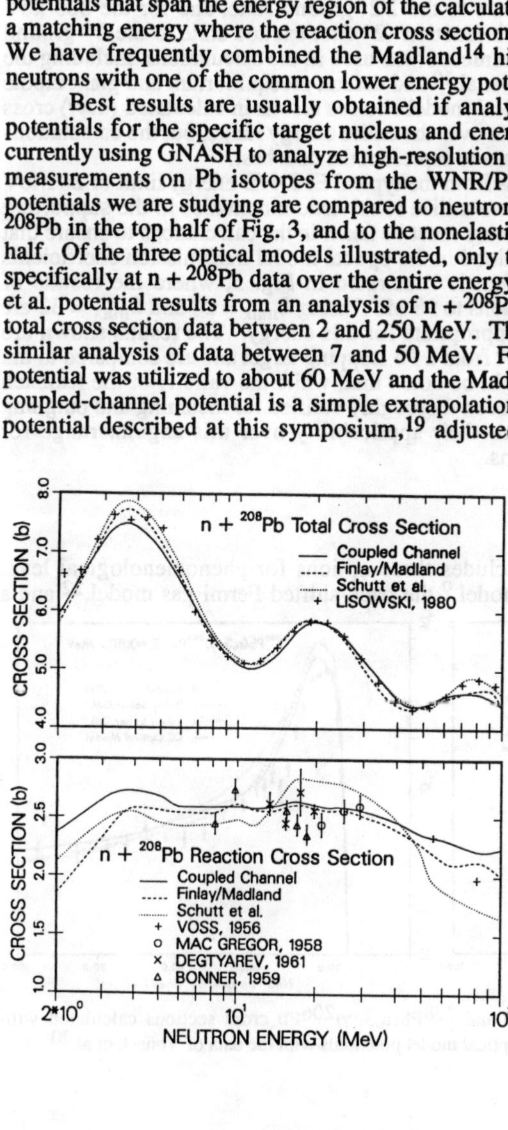

Fig. 3. Total and reaction cross sections for n + ^{208}Pb reactions between 2 and 100 MeV.

elastic scattering data near 20, 30, and 40 MeV. All three potentials agree with the σ_T measurement to within ≈ 2% at energies below 60 MeV, but the coupled-channel and Madland potentials lie ≈5-7% below the data in the 70-100 MeV range. The three potentials result in reaction cross sections that are reasonably consistent with data, although the Schutt et al. curve has structure near 20 MeV that is not indicated by the measurements.

Results from the n + ^{208}Pb calculations using the above three optical potentials are compared in Fig. 4 to experimental results from Vonach et al.[20] The two cases shown are for the 570-keV gamma ray from the ^{208}Pb(n,2nγ)^{207}Pb reaction, corresponding to the transition from the 5/2$^-$ first excited state to the 1/2$^-$ ground state, and for the 803-keV gamma ray from the ^{208}Pb(n,3nγ)^{206}Pb reaction (2$^+$ first excited state to the 0$^+$ ground state). A standard set of models and parameters was used in the calculations, including the level density representation by Ignatyuk et al.[10] and default preequilibrium and giant dipole resonance parameters. None of the optical model choices results in calculated (n,xγ) cross sections that agree perfectly with the data over the entire energy range, but the calculations with the coupled-channel potential appears to represent the data best.

In extending GNASH calculations from the typical ENDF/B energy limit of 20 MeV to a range of 100 MeV, it is preferable to utilize relativistic kinematics in the supporting optical model calculations. In addition, care must be taken in the calculations to ensure that appropriate angular momentum is permitted in the optical model transmission coefficients provided to GNASH. This latter effect is illustrated in Fig. 5, where the results of GNASH calculations of d + ^{208}Pb reactions to 98 MeV using ℓ_{max} = 20 and ℓ_{max} = 32 are compared, with the latter limit being appropriate for this energy. The results shown are neutron emission spectra at 30° and 105° from a Pb stopping target and are compared to the measurements of Martin et al.[21] It might be noted that, while the validity of the Kalbach angular distribution systematics[13] have been thoroughly verified for incoming and outgoing nucleons, the results in Fig. 4 confirm their applicability over this angular range for incident deuterons and emission neutrons.

3.2 *Level Densities*

The GNASH code presently includes three options for phenomenological level densities: the Gilbert and Cameron model,[9] the back-shifted Fermi gas model,[22] and a

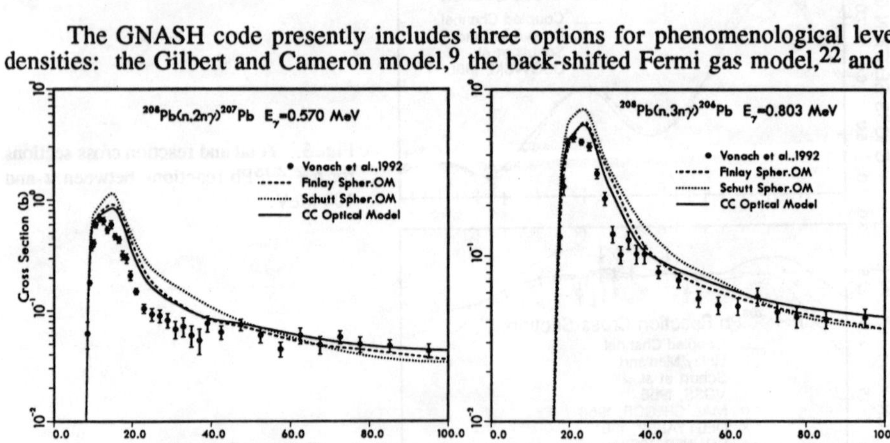

Fig. 4. Comparison of ^{208}Pb(n,2nγ)^{207}Pb and ^{208}Pb(n,3nγ)^{206}Pb cross sections calculated with transmission coefficients from three different optical model potentials with the data of Vonach et al.[20]

form of Fermi gas model by Ignatyuk et al.[10] that includes an energy-dependent level density parameter and the damping out of shell effects at higher excitation energies. The Ignatyuk model is in better agreement with microscopic Fermi gas models, and its more realistic treatment of shell effects is especially important at higher excitation energies for calculations on target nuclei such as ^{208}Pb that lie at or near shell closures. The computed level density of ^{208}Pb using the Gilbert and Cameron formulation and that of Ignatyuk et al. differ at excitation energies above 50 MeV by many orders of magnitude.[8]

The results of GNASH calculations of ^{208}Pb(n,xnγ) cross sections using the Gilbert-Cameron and Ignatyuk formulations are compared in Fig. 6 with the measurements of Vonach et al.[20] The cases shown are for the 570-keV gamma ray from the ^{208}Pb(,2nγ) reaction, similar to Fig. 4, and the 961-keV gamma ray from the ^{208}Pb(n,7nγ) reaction

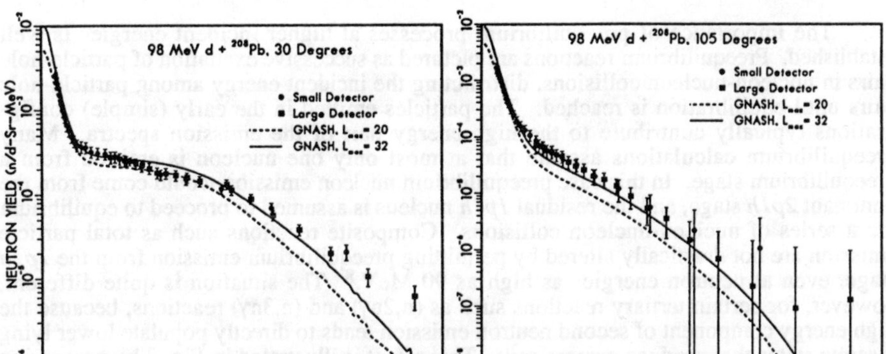

Fig. 5. Neutron emission spectra calculated with two different angular momentum cutoffs from a stopping Pb target bombarded with 98 MeV deuterons at 30° and 105°. The measurements are from Martin et al.[21]

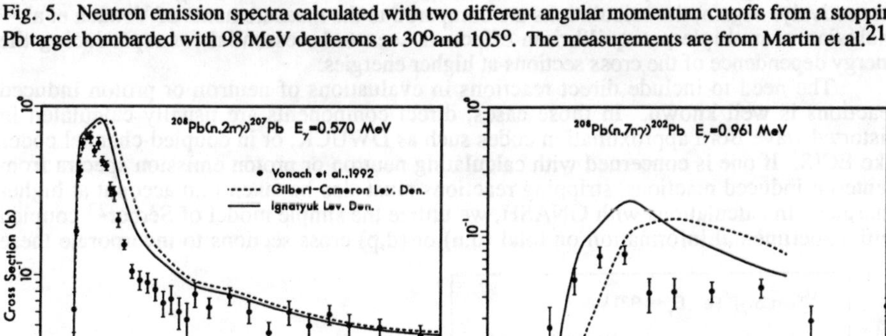

Fig. 6. Comparison of ^{208}Pb(n,2nγ)^{207}Pb and ^{208}Pb(n,7nγ)^{202}Pb cross sections calculated with two level density models[9,10] with the data of Vonach et al.[20]

from the 2+ first excited state to the ground state of ^{202}Pb. The main effect of using the Gilbert and Cameron model is to successively shift the shape of (n,xnγ) reactions to higher energies, with the effect becoming larger with increasing x. The Ignatyuk formulation clearly agrees better with the experimental data.

For nuclei more removed from shell closures, the effect seen in Fig 6 is much less pronounced. In Fig. 7 a calculation of the ^{56}Fe(n,2nγ)^{55}Fe cross section for the 931-keV gamma ray is shown, together with the measurements of Larson.[22] The calculations with both the Gilbert-Cameron and Ignatyuk level density models result in essentially equivalent agreement with the experimental data in Fig. 7.

3.3 *Preequilibrium and Direct Reactions*

The importance of preequilibrium processes at higher incident energies is well established. Preequilibrium reactions are pictured as successive excitation of particle-hole pairs in nucleon-nucleon collisions, distributing the incident energy among particle-hole pairs until equilibration is reached. The particles emitted in the early (simple) configurations typically contribute to the high-energy part of the emission spectra. Many preequilibrium calculations assume that at most only one nucleon is emitted from a preequilibrium stage. In this case preequilibrium nucleon emission would come from the dominant *2p1h* stage, and the residual *1p1h* nucleus is assumed to proceed to equilibrium via a series of nucleon-nucleon collisions. Composite reactions such as total particle emission are not drastically altered by permitting preequilibrium emission from the *1p1h* stage, even at neutron energies as high as 90 MeV.[8] The situation is quite different, however, for certain tertiary reactions such as (n,2nγ) and (n,3nγ) reactions, because the high energy component of second neutron emission tends to directly populate lower lying discrete states that produce gamma rays. This effect is illustrated in Fig. 8 by comparing GNASH calculations with and without multiple preequilibrium reactions for the 570-keV and 803-keV gamma rays from ^{208}Pb(n,2nγ)^{207}Pb and ^{208}Pb(n,3nγ)^{206}Pb reactions, respectively. Again, the calculations are compared to the measurements of Vonach et al.[20] Inclusion of multiple preequilibrium effects is seen to be essential for reproducing the energy dependence of the cross sections at higher energies.

The need to include direct reactions in evaluations of neutron or proton induced reactions is well known. In those cases, direct components are usually calculated in distorted wave Born approximation codes such as DWUCK, or in coupled-channel codes like ECIS. If one is concerned with calculating neutron or proton emission spectra from deuteron-induced reactions, stripping reactions must also be taken into account at higher energies. In calculations with GNASH, we utilize the simple model of Serber[23] coupled with experimental information on total (d,n) or (d,p) cross sections to incorporate these

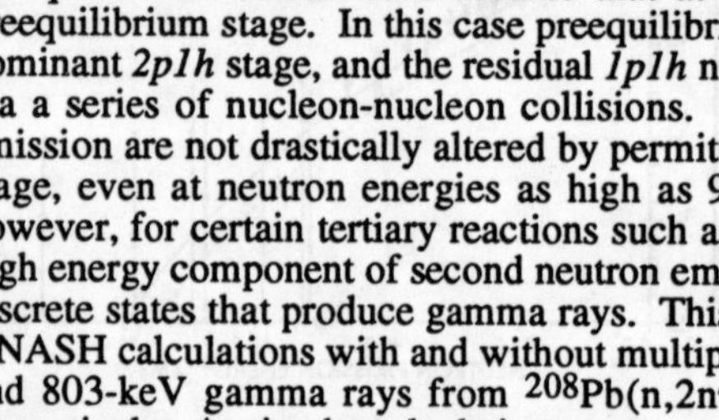

Fig. 7. Comparison of calculated and measured[22] values of the ^{56}Fe(n,2nγ)^{55}Fe cross section for the 0.931-MeV gamma ray with the Gilbert and Cameron[9] and Ignatyuk et al.[10] level density models.

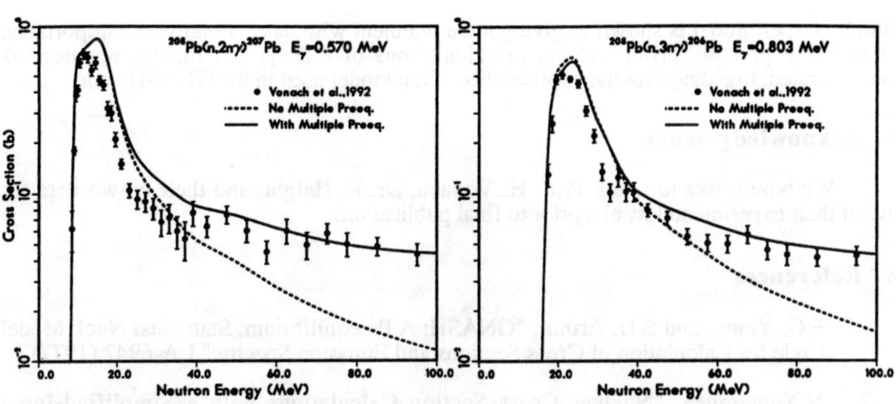

Fig. 8. Comparison of $^{208}Pb(n,2n\gamma)^{207}Pb$ and $^{208}Pb(n,3n\gamma)^{206}Pb$ cross sections calculated with and without multiple preequilibrium effects with the data of Vonach et al.[20]

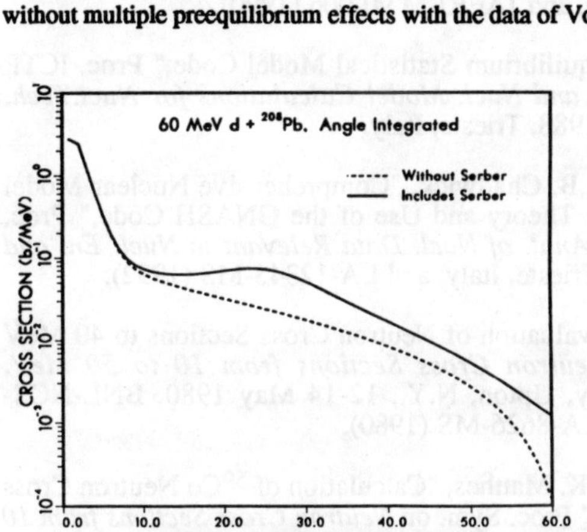

Fig. 9. Calculated neutron emission spectra from 60-MeV d + ^{208}Pb reactions with and without direct components from a Serber model.[23]

direct effects. This method was used in the calculations of neutron emission from 98-MeV deuteron reactions with ^{208}Pb shown in Fig. 5, which agree well with stopping target measurements. The extent of the Serber contribution is shown in the angle-integrated neutron emission spectrum of Fig. 9 for 60-MeV deuterons on ^{208}Pb. The contribution is a maximum near 1/2 of the incident deuteron energy and increases the cross section at emission energies above $\approx$ 15 MeV by a factor of 2 or more.

4. Conclusions

The use of different models and parameter choices in calculations with the GNASH code confirms the importance of sound optical model potentials and level density models that include reasonable representations of shell effects. The importance of including direct reactions in deuteron-induced nucleon-emission calculations is confirmed, and use of a simple Serber model is shown to give good agreement with data. Finally, the importance of multiple preequilibrium effects in calculations of (n,2nγ) and (n,3nγ) reactions is demonstrated, together with the applicability of the model used in the GNASH code.

5. Acknowledgments

We would like to thank Prof. H. Vonach, Dr. R. Haight, and their co-workers for use of their experimental results prior to final publication.

6. References

1. P.G. Young and E.D. Arthur, "GNASH: A Preequilibrium, Statistical Nucl. Model Code for Calculation of Cross Sections and Emission Spectra," LA-6947 (1977).

2. N.Yamamuro, "Nuclear Cross Section Calculations with a Simplified-Input Version of ELIESE-GNASH Joint Program," Proc. Int. Conf. on *Nucl. Data for Sci. and Tech.*, Mito (1988) p.489, and JAERI-M 90-006 (1990).

3. E.D. Arthur, "The GNASH Preequilibrium Statistical Model Code," Proc. ICTP Wksp. on *Applied Nucl. Theory and Nucl. Model Calculations for Nucl.Tech. Applications*, 15 Feb. - 18 Mar., 1988, Trieste, Italy.

4. P.G. Young, E.D. Arthur, and M.B. Chadwick, "Comprehensive Nuclear Model Calculations: Introduction to the Theory and Use of the GNASH Code," Proc. ICTP Workshop on *Comp. and Anal. of Nucl. Data Relevant to Nucl. En. and Safety*, 10 Feb. - 13 Mar., 1992, Trieste, Italy, and LA-12343-MS (1992).

5. E.D. Arthur and P.G. Young, "Evaluation of Neutron Cross Sections to 40 MeV for ^{54,56}Fe," Proc. Sym. on *Neutron Cross Sections from 10 to 50 MeV*, Brookhaven National Laboratory, Upton, N.Y., 12-14 May 1980, BNL-NCS-51245 (1980), v. II, p. 731, and LA-8626-MS (1980).

6. E.D. Arthur, P.G. Young, and W.K. Matthes, "Calculation of ^{59}Co Neutron Cross Sections between 3 and 50 MeV," Proc. Sym. on *Neutron Cross Sections from 10 to 50 MeV*, Brookhaven National Laboratory, 12-14 May 1980, BNL-NCS-51245 (1980), v. II, p. 751.

7. P. G. Young, R. C. Haight, R. O. Nelson, S. A. Wender, C. M. Laymon, G. L. Morgan, D. M. Drake, M. Drosg, H. Vonach, A. Pavlik, S. Tagesen, D. C. Larson, and D. S. Dale, "Calculation of (n,xγ) Cross Sections between Threshold and 100 MeV for Fe and Pb Isotopes: Comparisons with Experimental Data," IAEA CRP Meeting on *Methods for the Calculation of Neutron Nuclear Data for Structural Materials of Fast and Fusion Reactors*, Vienna, Austria, 20-22 June 1990, INDC(NDS)-247 (1991) p. 239.

8. P. G. Young, E. D. Arthur, M. Bozoian, T. R. England, G. M. Hale, R. J. LaBauve, R. C. Little, R. E. MacFarlane, D. G. Madland, R. T. Perry, and W. B. Wilson, *Trans. Amer. Nucl. Soc.* **60**, 271 (1989), and LA-11753-MS (1990).

9. A. Gilbert and A. G. W. Cameron, *Can. J. Phys.* **43**, 1446 (1965).

10. A. V. Ignatyuk, G. N. Smirenkin, and A. S. Tishin, *Sov. J. Nucl. Phys.* **21**, 255 (1975).

11. C. Kalbach, *Z. Phys. A* **283**, 401 (1977).

12. E.D. Arthur and P.G. Young, "Calculated Medium-Energy Fission Cross Sections," Proc. Conf. *50 Years with Nuclear Fission*, National Institute of Standards and Tech., Gaithersburg, Md., 25-28 April 1989, p. 931; M. Bozoian, E.D. Arthur, D.C. George, D.G. Madland, and P.G. Young, "Preliminary Calculations of Medium-Energy Fission Cross Sections and Spectra," *ibid*, p. 624.

13. C. Kalbach,*Phys. Rev. C* **37**, 2350 (1988); see also C. Kalbach and F. M. Mann, *Phys. Rev. C* **23**, 112 (1981).

14. D.G. Madland, "A Preliminary Medium-Energy Nucleon-Nucleus Phenomenological Optical Model Potential," Proc. IAEA Advisory Group on *Nuclear Theory for Fast Neutron Data Evaluation*, Beijing, PRC, 12-16 Oct. 1987, IAEA-TECDOC-483 (1988) p.80.

15. For example, D. Wilmore and P.E. Hodgson, *Nucl. Phys.* **55**, 673 (1964); F. D. Becchetti, Jr., and G. W. Greenlees, *Phys. Rev.* **182**, 1190 (1969).

16. P.W. Lisowski, G.F. Auchampaugh, M.S. Moore, G.L. Morgan, and R.E. Shamu, "Neutron Cross Section Measurements at WNR," Proc. Sym. on *Neutron Cross Sections from 10 to 50 MeV*, 12-14 May 1980, Brookhaven National Laboratory, BNL-NCS-51245 (1980) p. 301.

17. R.L. Schutt, R.E. Shamu, P.W. Lisowski, M.S. Moore, and G.L. Morgan, *Phys. Lett. B* **203**, 22 (1988).

18. R. W. Finlay, J. R.M. Annand, T. S. Cheema, and J. Rapaport, *Phys. Rev. C* **30**, 796 (1984).

19. P.G. Young and R.E. Shamu, Proc. Sym. on *Nuclear Data Evaluation Methodology*, 12-16 October 1992, Brookhaven National Laboratory, to be issued.

20. H. Vonach and R. Haight, personal communication; see also R. C. Haight et al., *Bull. Am. Phys. Soc.* **35**, 1038 (1990).

21. J. Martin, R. Kiziah, D. Ek, N. Miller, B. Johnston, D. Rutherford, and W. Wilson, "Neutron and Gamma-Ray Production from Thick Targets Bombarded with 98-MeV Deuterons," AFWL, Albuquerque, NM, personal comm. (1988).

22. D. C. Larson, "High-Resolution Structural Material (n,xγ) Production Cross Sections for $0.2 < E_n \leq 40$ MeV," Proc. Int. Conf. on *Nuclear Data for Basic and Applied Science,* May 13-17, 1985, Santa Fe, N. M., V.1, p. 71.

23. W. Dilg, W. Schantl, H. Vonach, and M. Uhl, *Nucl. Phys.* **A217**, 269 (1973).

24. R. Serber, *Phys. Rev.* **72**, 1008 (1947).

Evaluation of the Cross-Sections of (n,xn) and (p,xn) Reactions for Intermediate Energies

Lunyov V.P., Masterov V.S., Shubin Yu.N.
Institute of Physical and Power Engineering.
249020, Obninsk, Kaluga Region, Russia.

Abstract

The methods of the excitation function calculations, using preequilibrium and evaporation theory for the the neutron and proton induced reactions are analyzed in energy range up to 100 MeV. The calculated (n,xn) reaction cross sections for the fission product nuclei are in satisfactory agreement with the experimental data available and are close to the JENDL-3 evaluations. The preequilibrium emission contribution to the reaction cross sections in α-particle channel is estimated. The cross sections for the formation of both ground and isomer states of lead isotopes are calculated and arising problems are discussed for the multiple particle emission reactions. The sensitivity of the cross section calculation results to the relation between level density parameters in competing channels for the magic nuclei is shown. The necessity of the consistent parameter choice in the adjacent energy regions is noted.

1. Introduction.

The importance of the nuclear data at intermediate energies for many applications explains the growing interest to and the activity in this area [1-7]. Large volume of experimental data on neutron and proton induced reaction cross sections is accumulated [8-14]. However the experimental data available can not satisfy all needs in intermediate energies region. The energy interval (1-1000 MeV) is large, so various approaches were developed which differ one from another in the important physical assumptions used for their realization. Their predictive abilities are dependent on the applicability limits, details of the elaboration, the experimental test possibility and the formulation of the problem for which the method was worked out. To get the data needed various methods are used: both systematics with the parameters selected from comparison with available experimental data and the models based on various assumptions and the combination of both [1-7].

The intermediate energies region has specific features. The models and methods developed for high energies can lead to significant errors if applied at intermediate energies below 100 MeV [1], because the nuclei structure is not taken into account as a rule. On the other hand the extrapolation of the methods developed for low energies into the energy region high above neutron binding energy requires special care. There are strong indications to this effect in energy dependence of some physical phenomena connected with nuclei structure.

In some laboratories the computer codes have been developed based on intranuclear cascade model combined with the evaporation model and radiation transport [15-18]. The agreement is satisfactory as a rule for integral values. However there are cases when the similar codes used for the same experimental cases give different results.

Another group of codes based on the evaporation and preequilibrium theories, for example ALICE [19], GNASH [20], STAPRE [21] have been extensively used to calculate cross sections and secondary particle energy spectra in the tens of MeV range. The codes using similar physical ideas vary in detalization degree of input information and consequently in the features that can be described. Most widely used is ALICE code which with minimum input information allows to calculate charged particle induced reaction cross sections up to 200 MeV. Such analysis has been performed by Michel et. al. [6]. STAPRE code requires more detailed information for every competition channel and all decay steps, but it allows to calculate the other characteristics, in particular, isomer ratios. It should be noted that increasing of the incident energy and consequently the number of open channels makes getting detailed input data rather difficult. As a result the accuracy of description decreases and advantages of more complicated codes vanish. To improve the calculation methods one needs to investigate the nucleon-nucleus interaction on the base of microscopic theory and to pay attention to the parameters consistency.

In this paper we would draw attention to the energy region 20-100 MeV where the calculation methods developed to evaluate nuclear data for reactors (E≤20 MeV) meet methods developed for higher energies. It is this energy region that is important for many problems, in particular for setting up nuclear waste management concept, for the choice of optimum ways of medical radionuclides production and others. The neutron induced reaction excitation functions are calculated and analyzed for the fission product nuclei. The proton induced reaction cross sections are calculated for the lead region nuclei with the formation of the radioisotope ^{201}Tl and lead isomers states. The importance of taking into account of α-particle preequilibrium emission is stressed for the (n,α) and (p,α) reaction cross section evaluations. The dependence of the calculation results on the relations between the level density parameters for the channels with nucleon and α-particle emission is shown.

2. Excitation functions of neutron induced reactions on fission product nuclei.

To work out the nuclear waste management concept and to study a feasibility of principal technology in accelerator based transmutation problem it is necessary to know reaction cross sections both for structural materials, long-lived fission product nuclei and minor actinides in a very broad energy range. The results of neutron induced reaction excitation function evaluations for long-lived fission-product nuclei ^{135}Cs, ^{129}I, ^{90}Sr are compared with experimental data on ^{133}Cs(n,2n), ^{127}I(n,2n) and ^{88}Sr(n,p) reactions in Figures 1-3. Such comparison has some sense, because of the weak dependence of the reaction cross sections on mass number in this energy region. One can say the results of the cross section calculations with ALICE code are in a reasonable agreement with the experimental data. The agreement with the experiment for (n,p) reaction on ^{135}Cs and ^{129}I is not so good. It should be noted that the experimental data are scarce. They have been taken from papers [22-36].

JENDL-3 [37] includes the evaluated data on neutron induced reactions for some fission product nuclei in energy range up to 20 MeV. The (n,xn) reaction cross section calculation results for ^{96}Zr and ^{148}Sm are compared with JENDL-3 data on (n,2n) and (n,3n) reactions in Figures 4,5. The agreement between calculated and evaluated results is satisfactory, although some shift of the threshold energies for $^{96}Zr(n,3n)$ reaction is observed. The calculated (n,2n) and (n,3n) reaction cross sections for ^{147}Sm and ^{151}Sm isotopes coincide with the evaluated cross sections within 40%. There is a marked difference between the calculated results and evaluations for the (n,pxn) reactions with one proton emission. The calculation results of $^{96}Zr(n,2pxn)$ reactions with two protons emission are shown in Figure 6. One can see the considerable difference between calculated and evaluated (n,α) cross sections. Such discrepancy is probably due to the fact that in the applied version of ALICE the preequilibrium α-emission was not taken into account. The results of (n,α), (n,2n) and (n,p) reaction excitation functions calculations performed with modified version ALICE/ASH for ^{133}Cs are shown in Figure 7. The preequilibrium α-emission contribution is calculated by ALICE/ASH in accordance with Iwamoto and Harada approach [38]. The results of (n,α) reaction cross section calculations without the preequilibrium α-emission process differ significantly from the experimental data [39-41].

The libraries of experimental and evaluated data, accumulated in energy range up to 20 MeV play a significant role in testing of the methods used in the intermediate energy range. The consistent approach for the whole energy region should have physical model parameters that are close one to another in adjacent energy regions and vary in a consistent way. ALICE code uses as a rule quasiclassical linear dependence of level density parameter on the mass number. In paper [3] where such dependence has been used the conclusion was made that the calculation results are insensitive to the absolute value of the level density parameters. The level density parameter systematics based on the direct measurement of the neutron resonances density at neutron binding energy and the spectroscopical information at low energies are well known. They show that there exists the shell structure that is manifested in a strong difference of level density parameters of nuclei with magic neutron and proton numbers from neighboring ones. A large majority of nuclear data evaluations below 20 MeV is based on those systematics to a certain degree. Therefore the consistent approach should describe the transition from shell values to quasiclassical ones. Such possibility is given the by Ignatyuk et.al. approach [42], although in some mass number regions the model parameters may require more precise definition. The analysis performed has shown that a weak sensitivity of calculation results to the level density parameters absolute values takes place only for the reactions with nucleon emission. For the decay channels when the residual nucleus is far from double magic one and consequently the decay channel level density parameters differ considerably this dependence is rather strong. The results of ^{208}Pb(n,2pxn) reaction cross section calculations with α-particle emission for the various values of level density parameters in the competing channels using systematics by Dilg et.al. [43], Ignatyuk et. al. [42] and quasiclassical dependence a=A/9 are shown in Figure 8. One can see the large difference in calculation results for the level density parameters used. Such strong dependence is explained by large difference of the level density parameters in α-particle and nucleon decay channels.

As result the α-emission probability increases as compared to the case when the level density parameters are approximately equal.

3. The (p,xn) reaction excitation functions.

The proton induced reaction excitation functions in medium energy region have been analyzed by many authors [6,44]. It has been shown in paper [44] that for the medium mass nuclei the results of calculations with ALICE code give the reasonable description for the reaction cross sections for the most cases. However there exists the region for the neutron deficient nuclei when the assumptions used are not valid. Therefore it is worthwhile to analyze the data for the other nuclei, in particular for ones with A∾200. This mass region is important for many applications, in particular for the choice of optimum ways to produce the medical radioisotopes and for neutron spallation source activation evaluation in accelerator based waste transmutation projects.

The results of ^{203}Tl(p,xn) reaction excitation functions calculations are shown in Figure 9 together with the experimental data available. For the (p,2n) reaction the isomer state formation cross section only has been measured. The calculations by ALICE (solid curves) are in satisfactory agreement with the experimental data on (p,3n) and (p,4n) reaction cross sections when ^{201}Pb and ^{200}Pb isotopes are produced, but do not describe (p,2n) reaction cross section with isomer ^{202m}Pb production.

The isomer ratios description problem for the reactions with multiple particle emission is far from being solved. There exist the isomer ratios systematics [42,43]. However they are rather limited and comprise the data on (n,p), (n,α) and (n,2n) reactions in 14 MeV energy region only. For the some cases the data can be calculated on the base of the statistical theory. However the large discrepancy between the theory and experimental data exists particularly for small (<0.1) isomer ratio values, connected with the uncertainty in the level schemes, the γ-emission probabilities and the level density spin dependence [46].

One can evaluate the isomer and ground state formation cross sections by STAPRE code [21]. The calculations were performed using the optical model parameters from systematics by Blann et.al. [19]. The matrix element of intranuclear nucleon-nucleon interaction has been chosen so that the preequilibrium processes contributions calculated by STAPRE and ALICE are approximately equal. In so far as the preequilibrium model formulations used in those codes differ the contributions were chosen equal within 20%. The discrete excited level schemes and the electromagnetic transition probabilities have been taken from compilation [47]. It should be noted that the calculated excitation functions depend on the level scheme used, on the number of discrete levels coupled both with isomer and ground states. Besides that STAPRE code needs input data on level density and γ-width for every competing channel. The level density parameters have been chosen from [43]. The results of the reaction excitation function calculations by STAPRE for ^{200,201,202}Pb ground states and for isomer ^{202m}Pb formation in ^{203}Tl(p,xn) reaction are shown in Figure 9 by dashed lines. It should be noted that the calculated ground state formation cross sections are similar for both codes but STAPRE results differ somewhat from ALICE calculations in maximum values and some energy shift. The calculated isomer state formation cross sections are in agreement with the experimental data for ^{203}Tl(p,2n)^{202m}Pb reaction. The excitation functions for ^{205}Tl(p,xn) reactions have been measured by Lagunas-Solar et.al. [48,49]. The calculated cross sections are in good agreement with the experimental data [48] for ^{205}Tl(p,3n) ^{203}Pb and ^{205}Tl(p,5n)^{201}Pb reactions (Figures 10-11). The agreement with later measurements performed by the same authors [49] is somewhat worse. The experimental data on isomer formation reaction cross sections are compared with calculated ones for $^{205}Tl(p,2n)^{204m}Pb$ and $^{205}Tl(p,4n)^{202m}Pb$ reactions in Figure 10. For these reactions the agreement is worse than in the case of $^{203}Tl(p,2n)$ reaction. Apparently it is connected with the assumption used that ^{202m}Pb isomer is formed through ^{203}Pb nucleus decay to the lowest isomer state and is not formed through radiative transitions from higher states. It shows the necessity of more thorough input data analysis for every competing channel. The difference between STAPRE and ALICE codes in the preequilibrium model formalism and the level density parameters of the statistical model leads to the larger difference of the calculation results when the decay chain considered is longer.

The results of the proton induced reaction cross section calculations on lead isotopes $^{204,205,207,208}Pb$ with the ^{201}Pb and ^{201}Bi formation is shown in Figure 12. These reactions determine main contribution to the medical radioisotope ^{201}Tl production due to decay chain ^{201}Bi-^{201}Pb-^{201}Tl. For the $^{206}Pb(p,6n)^{201}Bi$ reaction the calculation are in reasonable agreement with experimental data on both excitation functions and the total yield in 51-66 Mev energy interval [45].

Figure 13 shows the calculated curves and experimental data on $^{88}Sr(p,2p3n)$ and $^{88}Sr(p,5n)$ reactions. The calculations were performed both with account of the preequilibrium α-emission (solid curve) and without it. One can see the importance of preequilibrium mechanism in α-emission description.

As conclusion one can say that the results of calculations by ALICE code can serve as a reasonable evaluation for the (n,xn) and (p,xn) reactions except the isomer states formation reactions. Less satisfactory results are obtained for the (n,pxn) and (p,pxn) reactions with proton emission. For the (n,2pxn) and (p,2pxn) reactions with α-emission it is necessary to take into account the preequilibrium α-decay. Often the description of the experimental data can be achieved by different methods. The important criterion to choose the method is the consistency of the model parameters.

REFERENCES

1. Pearlstein S., J.Astrophys., v.346 (1989) p.1049.
2. Pearlstein S., Nucl.Sci.and Eng., v.95 (1987) p.116.
3. Fukahori T. and Pearlstein S., Presented at IAEA Advisory Group Meeting on Intermediate Nuclear Data Applications, Vienna, Austria, 9-12 october 1990. INDC(NDS)-245 (1991) p.93.
4. Blann M., ibid p.63.
5. Nakahara Y., ibid p.77.
6. Michel R., Reiffer F. and Stuck R., Nucl.Phys., v.A441 (1985) p.617.
   Michel R. and Galas M., Nucl.Phys., v.A404 (1983) p.77.
   Michel R. and Galas M., Int.Jour.Appl.Radiat.Isot., v.34 (1983) p.1325.
   Michel R., Brinkman G., Nucl.Phys., v.A338 (1980) p.167.
   Michel R., Brinkman G., Weigel H., Herr W., Nucl.Phys., v.A322 (1979) p.40.
7. JAERI-M.92-039, INDC(JPN)-158/L, Proced. of the Specialists' Meeting on High Energy Nuclear Data. October 3-4 (1991) JAERI Tokai. ed. Tokio Fukahori.
8. Bauhoff W., Atom.Data and Nucl.Data Tables., v.35 (1986) p.429.
9. McCamis R.H., Davison N.E., van Oers W.T.H., Carlson R.E. and Cox A.J.,Can.J.Phys., v.64 (1986) p.685.
10. McLane V., Dunford C.L. and Rose P.E., Neutron Cross Sections, BNL-325,v.2, Academic Press (1988).
11. Pearlstein S., Nuclear Data for Science and Technology, Proc.Int.Conf., 1988, Mito, Japan, p.1115.
12. Shutt R.H., Shamu R.E., Lisowski P.W., Moore M.S. and Morgan G.L., Phys.Lett., v.B203 (1988) p.22.
13. Franz J., Grotz H.P., Lehman L., Rossle E., Shmitt H. and Morgan G.L., Nucl.Phys., v.A490 (1988) p.667.

14. Arthur E.D., Madland D.G. and McClellan D.M., Bibliographic Survey of Medium Inclusive Reaction Data., LA-10699-MS (1986).
15. Bertini H.W.,Phys.Rev. v.188 (1969) p.1711.
16. Barashenkov V.S., Toneev V.D., Chigrinov C.A., Atomic Energy v.37 (1974) p.480.
17. HETC. Code Package CCC-178, ORNL-4744., Radiation Shielding Information Center, Oak Ridge National Laboratory.
18. Filges D., et.al., 1984, Validation of the Intranuclear-Cascade Evaporation Model for Particle Production. Kernforshungsanlage, Julich, JUL-1960.
19. Blann M. and Vonach H.K., Phys.Rev., v.C28 (1983) 1475. Blann M. Code ALICE/85/300, Lawrence Livermore National Laboratory (1984) UCID 20169.
20. Young P.G. and Arthur E.D. GNASH: A Preequilibrium Statistical Nuclear Model Code for Calculation of Cross-Sections and Emission Spectra., (1977), LANL LA-6947.
21. Uhl V., Stromaier B. STAPRE-code. Preprint IRK-76/01 (1976) Vienna.
22. Lu W.D., Ranakumar N. Phys.Rev., v.C1 (1970) p.350.
23. Leich D.A., Borg R.J., Lanier V.B. Paper presented at 16th Lunar and Planetary Conference, Houston, TX.
24. Bormann M. et.al. Z.Phys., v.A227 (1976) p.203.
25. Martin H.C., Tashek R.F. Phys.Rev., v.89 (1953) p.1302.
26. Barral R.C. et.al. AFWL-TR-68-134 (1969).
27. Barral R.C., Silbergeld M. Nucl.Phys., v.A138 (1969) p. 387.
28. Santry D.C. EANDC(CAN)-46/L (1979) p.4.
29. Husain et.al. Phys.Rev., v.C1 (1970) p.1233.
30. Strohal F. Nucl.Phys., v.30 (1962) p.49.
31. Dresler J. Report INR-1464 (1973) p.12.
32. Levkovski V.N. et.al. Sov.Jour.Nucl.Phys., v.8(1) (1968) p.7.
33. Cohen A.V., White P.H. Nucl.Phys., v.1 (1956) p.73.
34. Library for Fission Products, Version 2., JAERI-M89-204, NEANDC (J)-145-U (1989).

35. Iwamoto A., Harada K. Phys.Rev., v.C26 (1982) p.1821. Sato N., Iwamoto A., Harada K. Phys.Rev., v.C28 (1983) p.1527.
36. Bychkov V.M., Manokhin V.N., Paschenko A.B., Plyaskin V.I. The cross sections of neutron induced threshold reactions., Energoizdat, Moscow (1982).
37. Fischler B., Traxler G., Uhl M., Vonach H. Phys.Rev., v. C30 (1984) p.72.
38. Avrigeanu M., Ivascu M., Avrigeanu V. Z.Phys., v.A329 (1988) p.177.
39. Ignatyuk A.V., Istekov K.K. and Smirenkin G.N. Yad.Fiz., v.29 (1979) p.875.
40. Dilg W. et.al. Nucl.Phys., v.A217, (1973) p.269.
41. Shubin Yu.N. Proc. of Advisory Group Meeting on Intermediate Nuclear Data Applications, Vienna, Austria, 9-12 october 1990. INDC(NDS)-245 (1991) p.151.
42. Kopecky J., Gruppelaar H. Systematics of Neutron Induced Isomeric Cross-Section Ratios at 14.5 MeV. Report ECN-200 (1987).
43. Grudzevich O.T., Ignatyuk A.V. et.al. Evaluation of Isomeric Excitation Functions in Neutron Induced Reactions. Report IAEA CRP Meeting, Vienna (1991).
44. Table of Isotopes. Seventh Edition., ed. C.M. Lederer., Wiley Interscience., NY (1978).
45. Lagunas-Solar M.C. et.al. Int.Jour.Appl.Rad.Isot., v.29 (1978) p.159.
46. Lagunas-Solar M.C. et.al. Int.Jour.Appl.Rad.Isot., v.31 (1980) p.117.
47. Blue J.W., Liu D.C., Smathers J.B., Med.Phys., v.5 (1978) p.532.
48. Bell R.E., Skargard H.M., Can.J.Phys., v.34 (1956) p.745.
49. Gadioli E., Gadioli-Erba E., Hogan J.J., Phys.Rev., v. C16 (1976) p.1404.

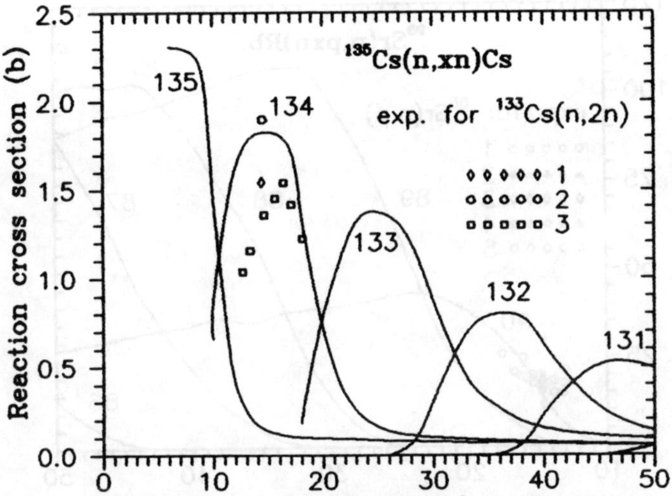

Fig.1. Excitation functions for ^{135}Cs(n,xn) reaction. Experimental data for reaction ^{133}Cs(n,2n): 1 – [22], 2 – [23], 3 – [24].

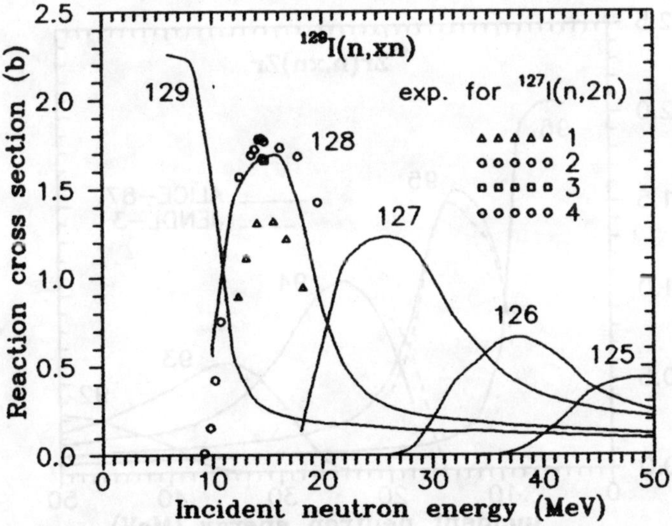

Fig.2. Excitation functions for ^{129}I(n,xn) reaction. Experimental data for reaction ^{127}I(n,2n): 1 – [25], 2 – [26], 3 – [27], 4 – [28].

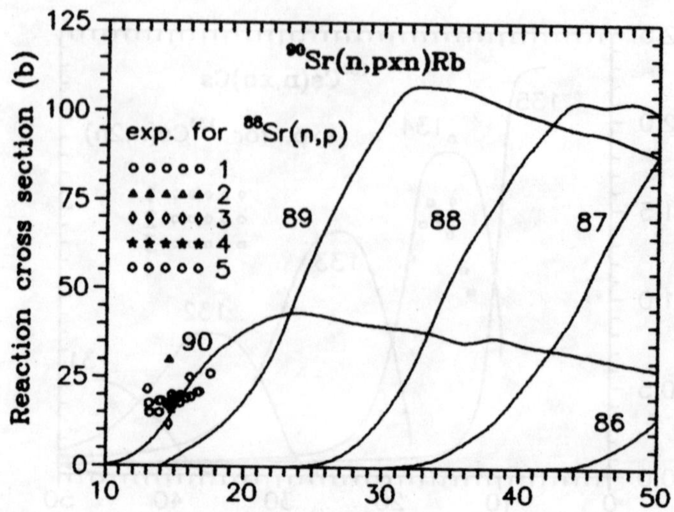

Fig.3. Excitation functions for ^{90}Sr(n,pxn) reaction.
Experimental data for reaction ^{88}Sr(n,p):
1– [29], 2 – [30], 3 – [31], 4 – [32], 5 – [33].

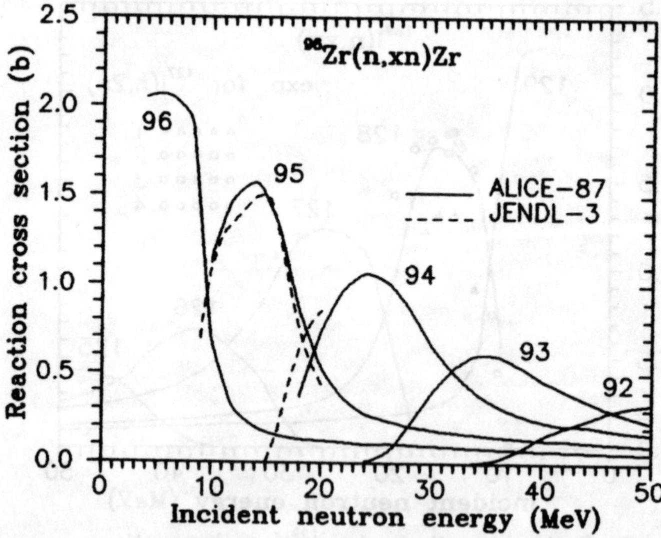

Fig.4. Excitation functions for ^{96}Zr(n,xn) reaction.
JENDL-3 data for (n,2n) and (n,3n) reactions.

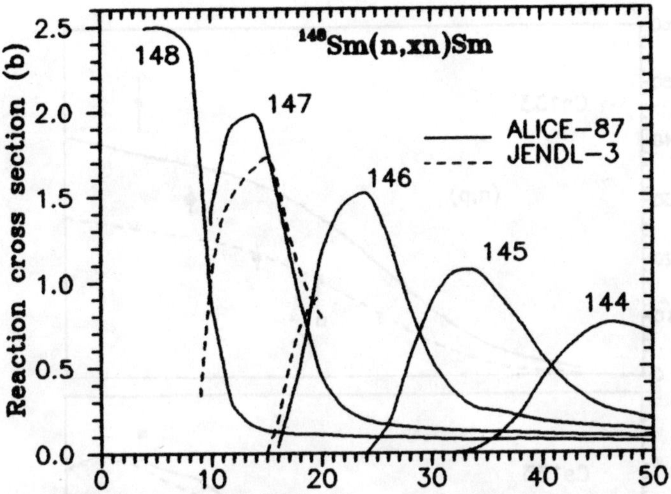

Fig.5. Excitation functions for ^{148}Sm(n,xn) reaction. JENDL-3 data for (n,2n) and (n,3n) reactions.

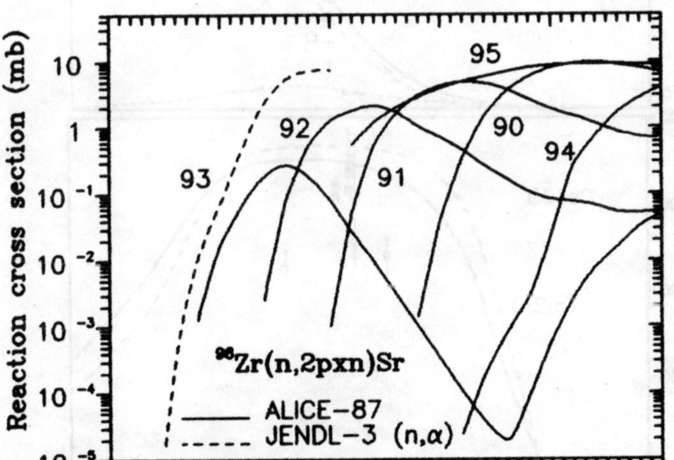

Fig.6. Excitation functions for ^{96}Zr(n,2pxn) reaction. JENDL-3 data for (n,α) reaction.

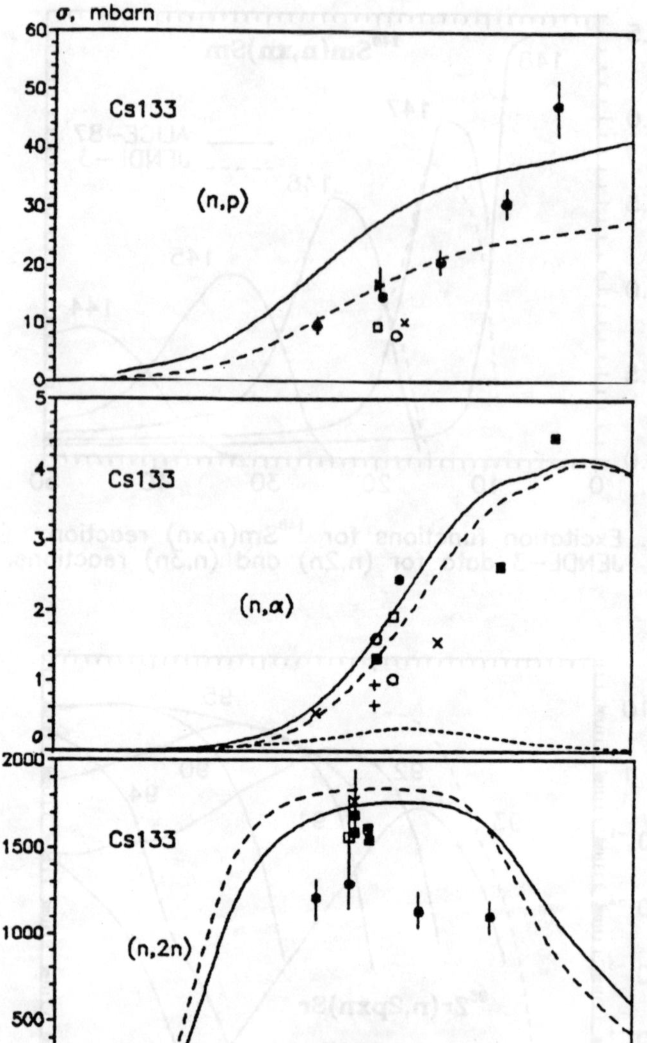

Fig.7. Excitation functions for (n,2n),(n,p) and (n,α) reactions on ^{133}Cs. Solid curve - -ALICE/ASH with parameter a equals A/9, dashed-ALICE/ASH with parameter a taken according to [39], dotted-ALICE-87 calculations. Experimental data from [35-38].

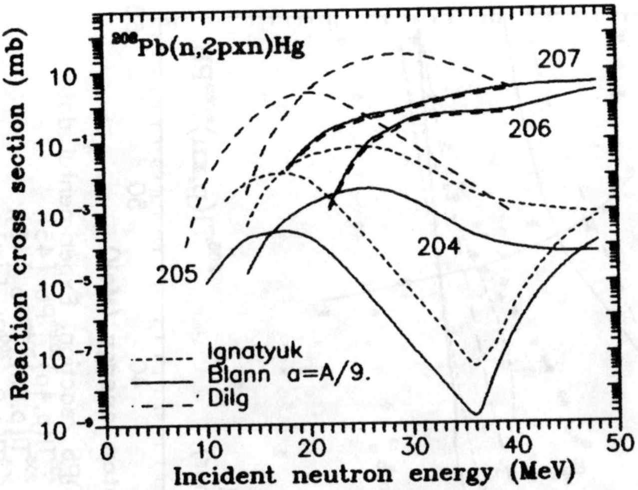

Fig.8. Excitation functions for ^{208}Pb(n,2pxn) reaction calculated with different level density models parameters.

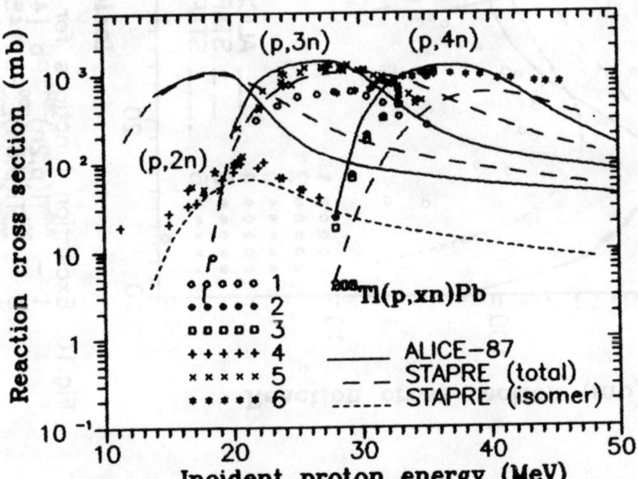

Fig.9. Excitation functions for ^{203}Tl(p,xn)Pb reaction. Experimental data: 1 – ^{203}Tl(p,3n)^{201}Pb [47], 2 – ^{203}Tl(p,4n)^{200}Pb [47], 3 – ^{203}Tl(p,4n)^{200}Pb [47], 4 – ^{203}Tl(p,2n)^{202m}Pb [45], 5 – ^{203}Tl(p,2n)^{201}Pb [45], 6 – ^{203}Tl(p,4n)^{200}Pb [45].

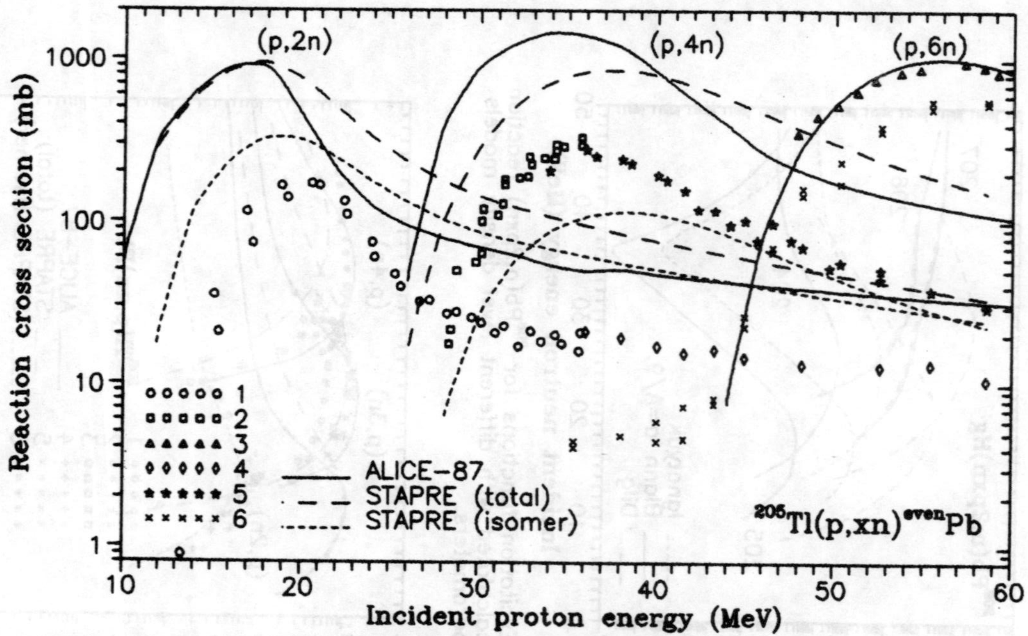

Fig.10. Excitation functions for ^{205}Tl(p,xn)Pb reaction. Experimental data:
1 – ^{205}Tl(p,2n)^{204m}Pb [45], 2 – ^{205}Tl(p,4n)^{202m}Pb [45],
3 – ^{205}Tl(p,6n)^{200}Pb [45], 4 – ^{205}Tl(p,2n)^{204m}Pb [46],
5 – ^{205}Tl(p,4n)^{202m}Pb [46], 6 – ^{205}Tl(p,6n)^{200}Pb [46].

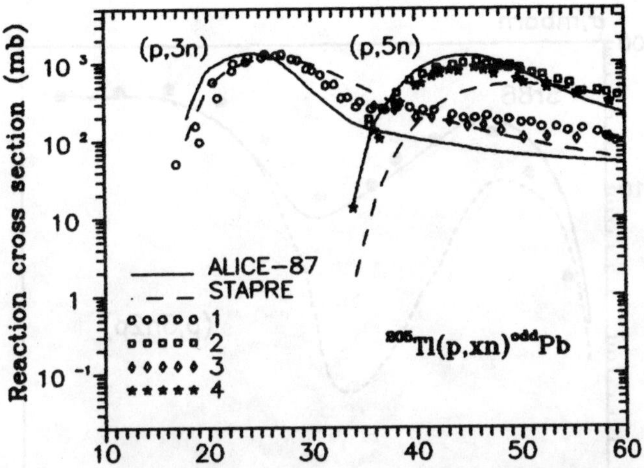

Fig.11. Excitation functions for ^{205}Tl(p,xn)Pb reaction. Experimental data: 1 – ^{205}Tl(p,3n)^{203}Pb [45], 2 – ^{205}Tl(p,5n)^{201}Pb [45], 3 – ^{205}Tl(p,3n)^{203}Pb [45], 4 – ^{205}Tl(p,5n)^{201}Pb [46].

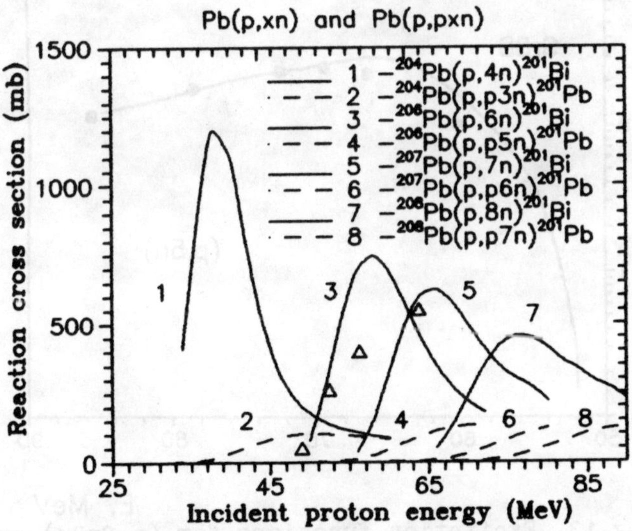

Fig.12. Excitation functions for Pb(p,xn)^{201}Bi and Pb(p,pxn)^{201}Pb reaction. Experimental data from [48].

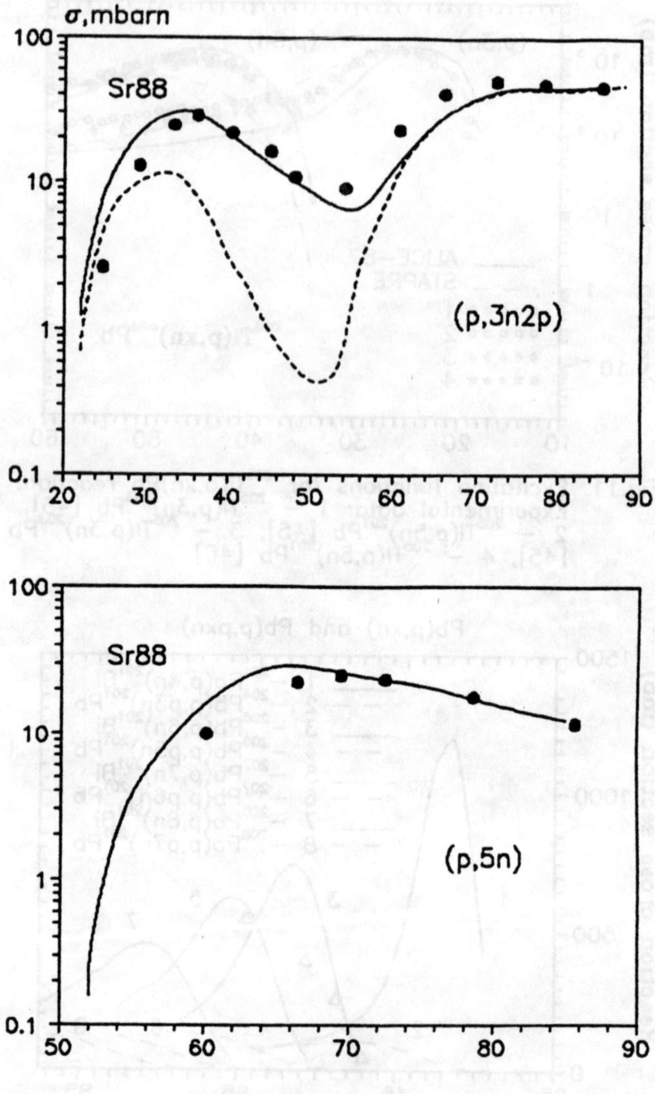

Fig.13. Excitation functions for (p,2p3n) and (p,p5n) reactions for ^{88}Sr. Experimental data from [49].

STATISTICAL MULTISTEP REACTIONS NEAR 100 MeV

H. KALKA
Technische Universität Dresden, Mommsenstr. 13,
O-8036 Dresden, Germany

ABSTRACT

The effective interaction is treated as a random matrix from the parameter-free GOE. In this way, the whole information about transition rates and cross sections is deduced from the single-particle state density, or dimension of the model space. The latter can be determined from the shell model. SMD calculations for (p,n)-reactions near 100 MeV are presented.

1. Introduction

In reaction theory we distinguish between statistical multistep compound (SMC) and statistical multistep direct (SMD) processes[1,2]. Some typical diagrams of the SMC process are

(1)

Here, the first graph denotes the bound-to-bound transition; the second graph symbolizes the unbound-bound transition, i.e., the formation of a 2p1h-doorway state. (The double line indicates unbound states.) The inverse process of the latter describes (one possible mode of) the decay of the composite system into the continuum. For simplicity, in these diagrams the spectator particles are omitted.

Otherwise, the SMD processes are based on transitions between continuum states,

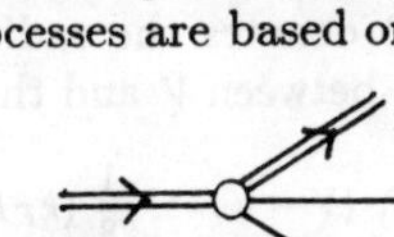

(2)

To calculate transition strengths or widths, $2\pi\overline{V^2_{\mu\nu}}\rho$, in previous approaches two informations were necessary: the effective 2-body interaction V and the final state density ρ. In this paper, however, $2\pi\overline{V^2_{\mu\nu}}\rho$ will be directly related to the parameter-free Gaussian Orthogonal Ensemble (GOE). This will be done for the diagrams in Eq. (1) at low energies (Sec. 4 and 5) and for the diagram in Eq. (2) at high energies (Sec. 6).

2. GOE and Shell Model

In random matrix theory[3] the effective interaction V is replaced by the GOE. The first and second moments are defined by

$$\overline{V_{\mu\nu}} = 0 \qquad \text{and} \qquad \overline{V^2_{\mu\nu}} = (1+\delta_{\mu\nu})\, v^2, \tag{3}$$

whereas the variance v^2 is related to the mean level-spacing D by

$$\sqrt{2\pi v^2} = D. \tag{4}$$

Note that, the GOE defines an ensemble where the particle rank k of the interaction is equal to the number N of (quasi-)particles in the system, $k = N$. Thus, for a two-body force we have to consider ($N = 2$)-particle spaces with the corresponding mean level-distance $D = D_2$. (The general case $k \leq N$ is described by the Embedded GOE. It follows by "propagation" of the GOE from k- to N-particle space using "geometrical" coefficients.)

Effective forces act in truncated (finite) shell-model spaces[4]. Usually, the model space is restricted to one oscillator shell near the Fermi energy E_F. In this way, single-particle (sp) states in the interval $\Delta = 40A^{-1/3}$ MeV around $E_F = 40$ MeV are involved in the excitation process. The total number of sp states is then $d_1 = g\Delta$, where $g = A/13$ MeV^{-1} represents the sp state density at E_F. Further, the mean spacing between sp states is $D_1 = 4\Delta/d_1 = 4/g$, where the factor 4 considers the spin-isospin degeneracy.

The dimension of the $N = 2$ particle space is $d_2 = \binom{d_1}{2}$; the corresponding mean level spacing is $D_2 = 4(2\Delta)/d_2$. Thus, from Eq. (4) we get the simple relation,

$$2\pi v^2 \left(\frac{g}{4}\right)^4 \Delta^2 = 1. \tag{5}$$

3. Strength of the Effective Interaction

Using the above results, the strength F_0 of the surface-delta interaction[5,6]

$$V(1,2) = \frac{F_0}{A}\frac{4\pi R^2}{u^4(R)}\,\delta(\mathbf{r}_1 - \mathbf{r}_2)\delta(r_1 - R), \tag{6}$$

can be determined. (Here, $u(R)$ denotes the radial wave function at the nuclear surface $R = r_0 A^{1/3}$.) The relation between V and the variance v^2 of the GOE is

$$v^2 \equiv \frac{1}{d_2(d_2+1)}\,\mathrm{Tr}\,(V - \overline{V}) \;\simeq\; \frac{1}{8}\,(k_F R)^{-2}\left(\frac{F_0}{A}\right)^2. \tag{7}$$

The last equality can be calculated[7] if the sum over model states is simulated by a sum over angular momenta. From Eq. (5) we then obtain

$$F_0 \;=\; 32\,(\pi g^3\Delta)^{-1/2}A \;=\; 134\,A^{-1/3}\;\text{ in MeV}. \tag{8}$$

For $A > 60$ this value coincide with $F_0 = 27.5 \pm 6.5$ MeV found in nuclear structure calculations[5].

4. Damping of sp Bound States

The first diagram in Eq. (1) describes the decay of a bound sp state into a 2p1h-configuration. The corresponding damping width is given by

$$\Gamma_1\!\downarrow\; = 2\pi v^2\rho_1^{(+)} = \frac{64}{g}\left(\frac{E}{\Delta}\right)^2 \tag{9}$$

(for $E \leq \Delta$), where the final state density $\rho_1^{(+)} = \rho_3 = g^3E^2/4$ and Eq. (5) was used. The energy dependence, $\Gamma_1\!\downarrow \propto E^2$, agrees with the predictions of Fermi-liquid theory[8]. Within a factor smaller 2 the theoretical values of Eq. (9) for $A > 60$ coincide with the experimental data[8] inferred from transfer reactions.

4. Damping of sp Unbound States

The neutron strength function Γ_n/D_n describes the decay of an unbound sp state with kinetic energy E_c into a 2p1h-configuration of energy $E = E_c + B$, where B is the binding energy. It is depicted in Eq. (1) and can be expressed by

$$\frac{\Gamma_n}{D_n} = \frac{\Gamma_1\!\downarrow \mathcal{P}_l}{D_{sp}}, \tag{10}$$

where $D_{sp} = \pi\hbar K/(MR) \simeq 4\Delta\sqrt{1 + E/E_F}$. Further, $\mathcal{P}_l = 4k_cK(K + k_c)^{-1}$ represents the penetration coefficient with $\hbar^2k_c^2 = 2ME_c$ and $\hbar^2K^2 = 2M(E_F + E)$. In the limit of small kinetic energy, $E_c \to 0$, and for $l = 0$ we get

$$\frac{\Gamma_n}{D_n} = \frac{64}{g\Delta^3}\frac{B^2E_F^{1/2}}{|V_0|}E_c^{1/2}. \tag{11}$$

For $V_0 = -48$ MeV, $B \simeq -(E_F + V_0) = 8$ MeV, and the reference energy $E_c = 1$ eV, we obtain from Eq. (11) the value $1.1\cdot10^{-4}$ independent of A, which should be compared with the black nucleus predictions[9] of $1.0\cdot10^{-4}$. The agreement is surprisingly well.

6. SMD Processes

In a straigthforward way, Eq. (5) can be generalized to the case of transitions between two unbound states as depicted in Eq. (2). For this reason, the sp state density at Fermi energy, $g/4 = \rho_1(E_F)$, should be replaced by $\rho_c \equiv \rho_1(E_c)$ with $c = a, b$. After partial-wave expansion we then obtain,

$$2\pi v^2\rho_a\rho_b\left(\frac{g}{4}\right)^2\Delta^2 = \sum_{l_a}(2l_a + 1) \;\simeq\; (k_aR)^2. \tag{12}$$

The validity of this estimate will be tested in SMD calculations at high energies (where SMC processes are negligible). Therefore, the analytical SMD model[10,11] which guarantees the unitarity of the S-matrix (probability conservation) will be used.

The cross section is given by

$$\frac{d\sigma_a^{\rm SMD}}{dE_b} = y_{ab}\;\pi R^2\sum_s\frac{4\;(dx^{(s)}/dE_b)}{(1 + x_a)\;(1 + x_b)^2}. \tag{13}$$

Restricting to neutrons and protons the "branching ratio" is simply assumed as $y_{ab} = \frac{1}{2}$. The radius parameter for R is taken from Ref. 12. In this calculations the sum runs over all s-step contributions. Further, we have $x_c \equiv \sum_s\int_0^{E_c}dE(dx^{(s)}/dE)$, and

$$\frac{dx^{(s)}}{dE_b} = 4\pi\ (k_a R)^{-2} C^s f_s(E_a - E_b) \sqrt{\frac{E_b}{E_a}} \tag{14}$$

with

$$f_s(U) = \sum_{i,j=1}^{s} (-)^{i+j} \binom{s}{i} \binom{s}{j} \frac{(U - iB - jE_F)^{2s-1}}{(2s-1)!} \Theta(U - iB - jE_F). \tag{15}$$

Due to the denominator in Eq. (14), which results from the optical-model wave functions, probability conservation is guarenteed.

The transition strength, i.e., the whole dynamics of the process, is contained in the energy-independent quantity

$$C = \frac{\pi}{4}\ (k_a R)^{-2} v^2 \rho_a^2\, 4 \left(\frac{g}{4}\right)^2 = \frac{1}{2\Delta^2}, \tag{16}$$

where the last equality follows from Eq. (5).

The magnitute and A-dependence of C was proved in (p,n)-reactions at 90-Zr and 208-Pb. Agreement with experimental data[13] shown in Figs. 1 to 3 was obtained by scaling C with a constant factor of 0.7 for all reactions. This additional factor results from the uncertainty in the estimation of D_2 or Δ, for example.

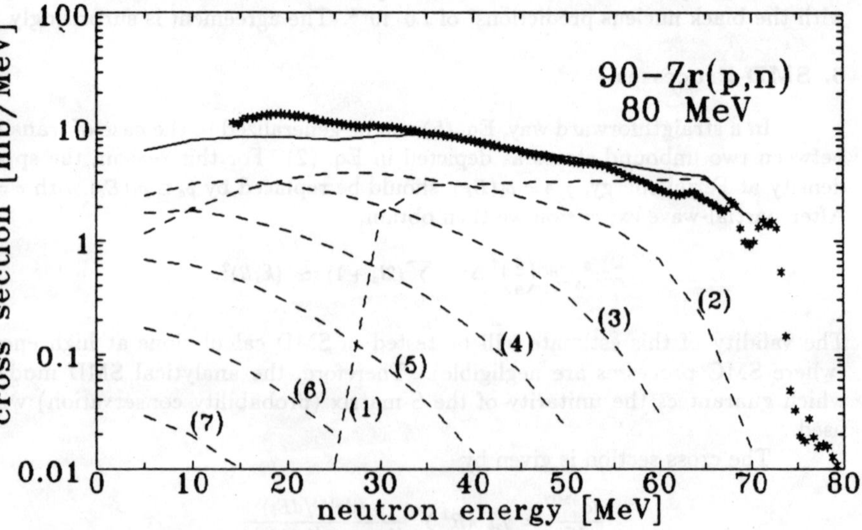

Fig. 1: Calculational and experimental (p,n)-cross section for 90-Zr at 80 MeV incident energy. The numbers in brackets denote the individual s-step contributions. Experimental data are taken from Ref. 13.

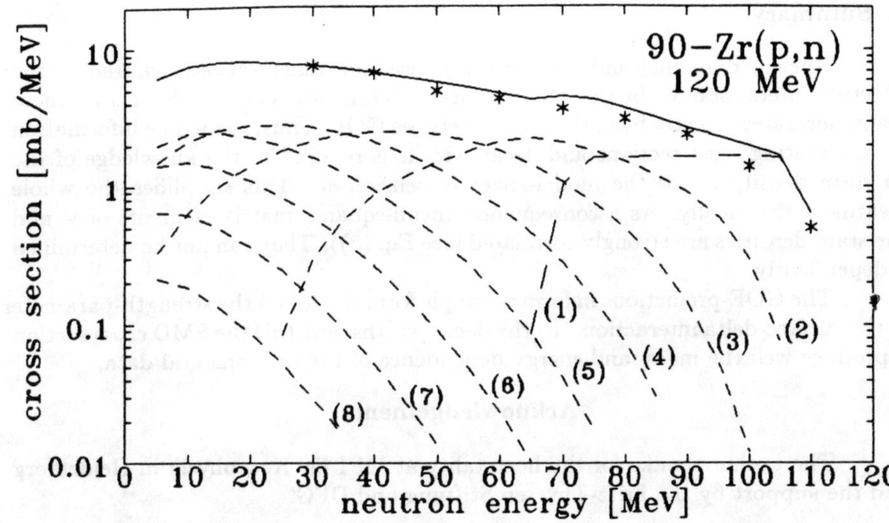

Fig. 2: Calculational and experimental (p,n)-cross section for 90-Zr at 120 MeV incident energy. The numbers in brackets denote the individual s-step contributions. Experimental data are taken from Ref. 13.

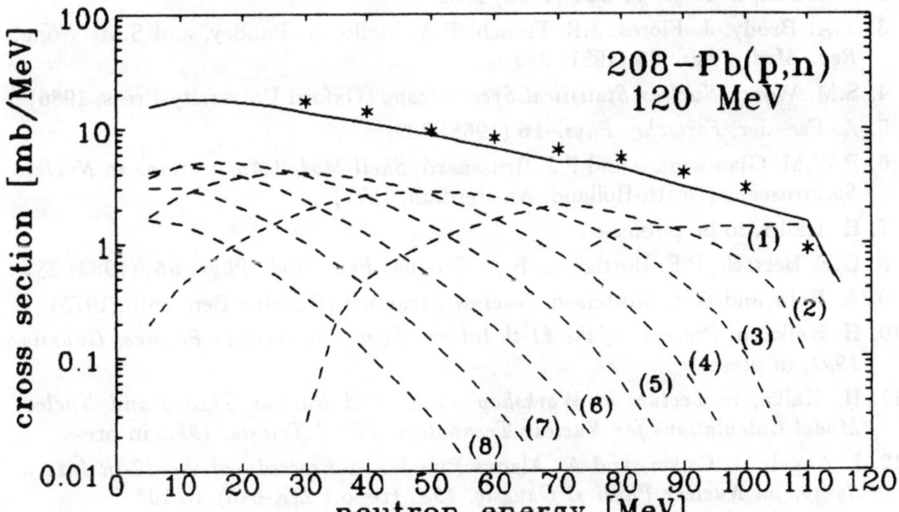

Fig. 3: Same as Fig. 2 but for 208-Pb.

7. Summary

Nuclear dynamics and reactions are based on transition rates caused by the effective interaction V. In this paper a first attempt was done to determine these transition rates directly from the parameter-free GOE. Thus, the whole information for calculating cross sections and decay widths is reduced to the knowledge of the sp state density ρ_1, or the model-space dimension d_1. This simplifies the whole treatment drastically. As a consequence, mean-squared matrix elements of V and the state densities are strongly correlated (see Eq. (5)). They can not be determined independently.

The GOE-predictions in form of simple formulas for (i) the strength-parameter of the surface-delta interaction, (ii) the decay widths, and (iii) the SMD cross section reproduce well the mass- and energy dependence of the experimental data.

Acknowledgement

The author thanks for the hospitality at MPI für Kernphysik in Heidelberg and the support by the Fritz-Thyssen Stiftung and DFG.

References

1. E. Gadioli and P.E. Hodgson, *Pre-Equilibrium Nuclear Reactions* (Clarendon Press, Oxford, 1992).
2. H. Kalka, *Z. Phys.* **A 341** (1992) 289.
3. T.A. Brody, J. Flores, J.B. French, P.A. Mello, A. Pandey, and S.M. Wong, *Rev. Mod. Phys.* **53** (1981) 385.
4. S.M. Wong, *Nuclear Statistical Spectroscopy* (Oxford University Press, 1986).
5. A. Faessler, *Fortschr. Phys.* **16** (1968) 309.
6. P.W.M. Glaudemans and P.J. Brussaard, *Shell-Modell Applications in Nuclear Spectroscopy* (North-Holland, Amsterdam, 1977).
7. H. Kalka, to be published.
8. G.F. Bertsch, P.F. Bortignon, R.A. Broglia, *Rev. Mod. Phys.* **55**, (1983) 287.
9. A. Bohr and B.R. Mottelson, *Nuclear Structure* (Reading Benjamin, 1975).
10. H. Kalka, in *Proceed. of the 21-th Intern. Symp. on Nuclear Physics, Gaussig, 1991*, in press.
11. H. Kalka, in *Lecture at Workshop on Applied Nuclear Theory and Nuclear Model Calculations for Nuclear Technology, ICTP Trieste, 1992*, in press.
12. I. Angeli, J. Csikai, and A. Algora Pineda, in *Proceed. of the 17-th Intern. Symp. on Nuclear Physics, Gaussig, 1987* (report ZfK-646), p. 103.
13. M. Trabandt, W. Scobel, H.M. Blann, R. Pohl, C. Byrd, C.C. Foster, and R. Bonetti, *Phys. Rev.* **C39**, (1989) 452.

METHODS OF HEAVY ION CROSS-SECTION MODELING FOR GALACTIC COSMIC RAY SHIELDING APPLICATIONS

L. W. TOWNSEND
NASA Langley Research Center
Mail Stop 493
Hampton, Virginia 23681-0001, USA

ABSTRACT

Heavy ion cross section data for use in galactic cosmic ray shielding applications typically consist of total reaction cross sections and nuclear fragmentation (breakup) cross sections. While the data base for total reaction cross sections is adequate, and accurate theoretical models are in-hand, the same is not true for the nuclear fragmentation cross section data base. Relevant experimental fragmentation data are sparse, individual data sets are often incomplete, and differences between measurements for the same collision pairs by different experimenters often disagree by more than the quoted uncertainties. Theoretical models under development are hampered by the paucity of experimental data. Semiempirical parameterizations exist; however, their accuracies are limited by experimental data uncertainties and the inconsistencies between different data sets. These issues and methods used to reconcile differences between calculations and measurements will be discussed. Recommendations for improving future results will be made.

1. Introduction

Whenever high-energy galactic cosmic rays (GCRs) traverse bulk matter, such as a spacecraft or an astronaut's body, their radiation fields undergo compositional changes resulting from atomic and nuclear interactions with the materials in their paths. These changes in the field composition depend upon the thickness, material composition, and geometric configuration of the

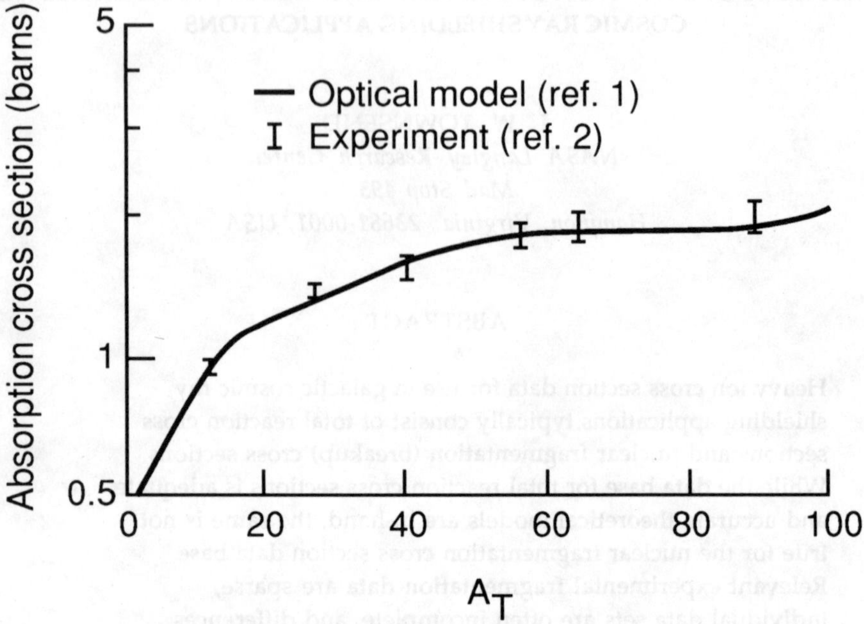

Fig. 1 - Absorption cross sections as a function of target mass number, for carbon beams at 83 A MeV.

target. They are usually described by transport models which relate the transmitted fluxes to the incident fluxes. At the atomic level, a propagating ion loses most of its energy to electronic excitation and ionization of the target material. These energy losses per collision are small (on the order of ev), but they occur very frequently (collision mean-free path $\lambda \sim 10^{-6}$ cm). Collisions at the nuclear level are infrequent ($\lambda \sim 1$ - 10 cm), but violent and usually result in the breakup of the incident and/or target nucleus (fragmentation). Since spacecraft and body organ self-shielding thicknesses are comparable to nuclear collision mean-free paths, most ions will undergo one or more nuclear interactions before exiting the spacecraft or coming to rest.

The attenuation of galactic cosmic rays by nuclear collisions is usually represented by total reaction (absorption) cross sections. At present the data base for these cross sections is adequate, and accurate theoretical models based upon quantum-mechanical scattering theory are available.[1] Typical results are shown in figure 1. The experimental data are taken from reference 2.

Nominal differences between theory and experiment for absorption cross sections are only a few percent.

Although the attenuation of galactic cosmic rays by nuclear collision is well understood, the nuclear fragmentation cross section data base, which describes the production of lighter particles by the breakup of heavier ones, is not. At present there is no suitably accurate theory for predicting nuclear fragmentation cross sections for all collision pairs at all energies of interest. Because relevant experimental data are sparse, individual data sets are often incomplete, and differences between measurements for the same collision pairs by different experimenters often disagree by more than the quoted uncertainties, existing theoretical models, and parameterizations are not yet accurate enough for precise transport applications.

In this work, inconsistencies and shortcomings of existing models and data sets are discussed. Methods used to evaluate and reconcile differences between theory and experiment are described. Recommendations for improving future results are made.

2. Fragmentation Models

Nuclear fragmentation cross sections currently used in GCR shielding studies are obtained from semiempirical formulations based either upon an extension of the Rudstam formulas for hydrogen targets,[3] or upon a geometric abrasion-ablation collision model.[4,5] Typical differences between calculated and measured cross sections using these models are approximately 25 - 50 percent. The larger differences appear to be associated with isotopic rather than elemental production cross sections. Neither model treats light ion production adequately. Details of these models are found in the references.

In addition to semiempirical methods, fragmentation models based upon formal quantum scattering theory are also under concurrent development at NASA Langley Research Center. For helium ion fragmentation, an effective, three-body, multiple-scattering approach is used to describe two-body dissociation on composite targets.[6] For heavier projectile ions, a quantum-mechanical, abrasion-ablation theory using an optical potential is being developed.[7] It is anticipated that these more formal fragmentation models will be incorporated into the Langley GCR transport codes[8] in the future. For the present, however, their development is severely hindered by a paucity of experimental data.

3. Nuclear Fragmentation Data Sets

For GCR shielding purposes, fragmentation cross sections for nuclei with charge numbers Z = 2 (helium) through Z = 28 (nickel), having kinetic energies in the range of ~ 100 A MeV to ~ 5 A GeV, colliding with targets representative of spacecraft structural and body tissue materials are of primary interest.

3.1 *Shortcomings*

To date, very few of the necessary cross sections have been measured. Of those which have been measured, most are for hydrogen targets which are of interest mainly for astrophysics studies. Other shortcomings include:

(1) incomplete data sets (many/most fragment production cross sections not measured);
(2) neutron production cross sections are generally not measured;
(3) cross sections for most projectile-target combinations at a given beam energy are typically measured only once;
(4) inconsistencies between data sets when measurements are (infrequently) repeated by different groups at similar energies;
(5) "cross sections" which are not really cross sections because they are either not corrected or only partially corrected for such things as detector acceptance, target thickness, etc.

As an example of these shortcomings, figure 2 displays elemental production cross sections for iron beams fragmenting in copper targets at similar beam energies of 1.88 A GeV[9] and 1.55 A GeV.[10] Theoretically, the reported cross-section differences cannot be accounted for by the slight difference in beam energies.

3.2 *Experimental Data Needs*

From a theoretical perspective, the experimental data base needs to be significantly expanded. For each projectile-target-energy combination, complete data sets must be obtained. This includes, as a minimum, elemental and isotopic (if possible) production cross sections for all secondaries with Z ≥ 1. In addition, secondary neutron spectra and cross

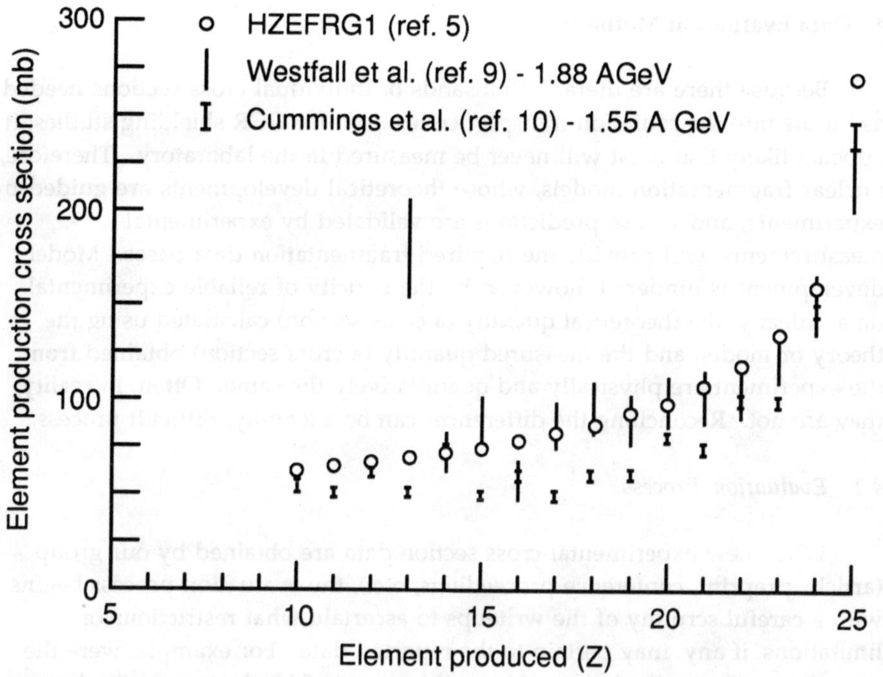

Fig. 2 - Element production cross sections for iron beams colliding with copper targets.

sections must also be measured. The incident ions of most interest are carbon, oxygen, neon, magnesium, silicon, and iron (not ordered by relative importance) because these ions and their secondary reaction products are expected to contribute most of the GCR risk to astronauts on deep-space missions.[11] For consistency and reliability in the data sets, some of the experiments should be independently repeated to ensure that unknown or unrecognized errors are not present. Scientific issues which require resolution include clearly establishing the dependence, if any, of fragment production cross sections on (1) the incident kinetic energy of the fragmenting nucleus; and (2) the mass and charge numbers of the target nucleus.

4. Data Evaluation Methods

Because there are literally thousands of individual cross sections needed as inputs into the radiation transport codes used for GCR shielding studies, it appears likely that most will never be measured in the laboratory. Therefore, nuclear fragmentation models, whose theoretical developments are guided by experiments, and whose predictions are validated by experimental measurements, will provide the required fragmentation data bases. Model development is hindered, however, by the paucity of reliable experimental data. Ideally, the theoretical quantity (a cross section) calculated using the theory or model, and the measured quantity (a cross section) obtained from the experiment are physically and quantitatively the same. Often, in reality, they are not. Reconciling the differences can be a lengthy, difficult process.

4.1 Evaluation Process

When new experimental cross section data are obtained by our group (article, preprint, conference proceedings, etc.), the evaluation process begins with a careful scrutiny of the writeups to ascertain what restrictions or limitations, if any, may pertain to the reported data. For example, were the measurements obtained using thin or thick targets? If the targets used were thick, have the results been corrected? Have detector acceptance corrections been applied? Are the data actual total cross sections, or are they values obtained by integrating over limited ranges of angles and energies? Are the reported uncertainties statistical? Do they include any systematic error contributions?

The next step in the evaluation process is to compare the theoretical predictions with the experimental measurements. The comparisons are either quantitative or qualitative. Quantitative comparisons are absolute and usually involve directly comparing calculated to measured cross sections. Although this is the method of choice, it is only useful if the measured and calculated cross sections represent the same physical quantities. If the calculation gives cross sections representing specific secondary reaction products produced at all angles and energies, a direct comparison to measured cross sections which include contributions only over limited ranges of energies and/or angles may not be quantitatively meaningful. Instead, qualitative comparisons such as, cross section ratios or trends may be all that are possible.

Quantitative agreement between theory and experiment is evaluated by investigating the distribution of cross section differences. If theory and experiment generally agree (subjective evaluation), are some modifications to the theory indicated? If theory and experiment disagree, is the problem the theory or the experiment? The latter may be indicated if the theory generally agrees with other data for similar reactions. An illustration of this can be found in the hadronic dissociation analyses for 14.6A GeV silicon beams presented in ref. 7.

Using the results in figure 2 as a quantitative evaluation example, we find for 1.88 A GeV iron fragmentations[9] that 69 percent of the calculated cross sections are within the quoted experimental errors, 85 percent are within 25 percent of the experimental values, and over 92 percent are within 50 percent of the experimental values. The agreement appears to be excellent. On the other hand, for 1.55 A GeV iron fragmentations[10] only 6 percent of the calculated cross sections are within the quoted experimental errors, 25 percent are within 10 percent of the experimental values, 44 percent are within 25 percent of the experimental values, and over 87 percent are within 50 percent of the experimental values. Overall, the agreement between theory and the 1.55 A GeV data set is not as impressive as for the 1.88 A GeV data set. These apparent inconsistencies appear to be experimental rather than theoretical. Therefore, the initial focus is on trying to resolve the apparent differences between data sets, rather than modifying the theoretical model. Qualitatively comparing the two experimental data sets and the model calculations, we note that the odd-even variation in cross section as a function of fragment charge Z is less pronounced for the 1.88A GeV experimental data than for the 1.55A GeV data, and is absent altogether in the calculated cross sections. We also note that the cross sections generally decrease as Z decreases (for $Z \leq 16$). Whether this trend continues for smaller Z values is unclear since both data sets only measure limited portions of the secondary particle production spectra. Finally, we note that the 1.88A GeV cross sections are typically larger than the 1.55A GeV values for the same element produced. Although some systematic differences might arise from the slight differences in energies of the two experiments, the magnitudes of the these cross section differences (up to ~ 40 mb) are greater than expected and cannot be accounted for by theory. Similar systematic differences are noted for other targets common to both experiments.

4.2 *Data Systematics*

Studies of early heavy ion fragmentation data[12] suggest that the fragment production cross sections are independent of beam energy above a certain threshold (limiting fragmentation hypothesis) and that the cross section $\sigma_{BT}{}^{F}$ for producing a particular fragment (F) in a beam (B) -target (T) interaction can be factored

$$\sigma_{BT}^{F} = \sigma_{B}{}^{F} \gamma_{T} \tag{1}$$

where $\sigma_B{}^F$ depends only the beam and fragment and γ_T depends only on the target (strong factorization). Eq. (1) is sometimes modified so that γ_T becomes γ_{BT}, thereby incorporating a beam dependence into the second factor (weak factorization). The validity of these hypotheses is unclear. If they are valid, there would be an enormous simplification in the GCR shielding problem because the large number of required fragmentation cross section measurements mentioned in section 3 would be drastically reduced. Recent studies[10] suggest that these hypotheses may only be valid to about 20-30 percent.

5. Recommendations

Communication between theorists and experimentalists is essential to minimizing misconceptions and misunderstandings. Limitations of experimental data sets must be clearly indicated by the experimentalist, and their ramifications appreciated by the theorist. Sometimes the model calculations and experimental data cannot be compared directly because corrections to the theoretical calculations, such as detector acceptance in angle and energy or cutoffs for various thresholds, must be applied before comparisons become meaningful. As a general rule, if the limitations of a theory or experiment are not explicitly indicated, users will assume that there are none.

For thick shields, GCR particle fluences and doses are eventually dominated by light ions and nucleons.[8] Neutrons, in particular, may pose a significant health risk for future astronauts. Therefore, experimental data sets must be as complete as possible. If practical, all secondary production cross sections (including nucleons) should be measured. In addition to expanding the overall size of the experimental data base, and repeating some measurements to verify their correctness, the heavy ion fragmentation data base needs to be extended to include measurements of spectral and angular distributions of the fragments. This will facilitate the development of theoretical models for use in three-dimensional heavy ion transport/shielding codes. Exclusive channel measurements of fragmentation cross sections should also be considered in order to enhance our understanding of the fragmentation process.

Finally, a realistic evaluation of the existing data base needs to be undertaken. The limitations, shortcomings, and restrictions of current models and experimental data sets must be readily available to current and future users.

6. Concluding Remarks

The present status of heavy ion cross section data for use in galactic cosmic ray shielding applications is a mixture of the adequate and the inadequate. For total reaction cross sections, the data base is adequate and accurate theoretical formulations are in-hand. For nuclear fragmentation reactions, the opposite is true. The existing models are not yet sufficiently accurate[1] and their further development is hindered by an inadequate experimental data base. Specific recommendations concerning the breadth and depth of the required data were offered. The need for improved communication between theorists and experimentalists was discussed. Finally, it was recommended that realistic evaluations of current models and existing data sets be carried out.

7. Acknowledgements

The author wishes to thank Dr. Ram Tripathi and Dr. Ferdous Khan for aid with the analyses and for useful discussions on the content of the manuscript.

8. References

1. L. W. Townsend and J. W. Wilson, NASA Reference Publication No. RP-1134, May 1985.
2. S. Kox, A. Gamp, R. Cherkaoui, A. J. Cole, N. Longequene, J. Menet, C. Perrin, and J. B. Viano, *Nucl. Phys.* **A420** (1984) 162.

3. R. Silberberg, C. H. Tsao, and J. R. Letaw, *Ap. J. Suppl.* **58** (1985) 873.
4. J. W. Wilson, L. W. Townsend, and F. F. Badavi, *Nucl. Inst. & Methods* **B18** (1987) 225.
5. L. W. Townsend, J. W. Wilson, R. K. Tripathi, J. W. Norbury, F. F. Badavi, and F. Khan, NASA Technical Paper (to be published).
6. F. A. Cucinotta, L. W. Townsend, and J. W. Wilson, *Phys. Lett. B* **282** (1992) 1.
7. L. W. Townsend, J. W. Norbury, and F. Khan, *Phys. Rev.* C **43** (1991) R2045.
8. J. W. Wilson, L. W. Townsend, W. Schimmerling, J. E. Nealy, G. S. Khandelwal, F. A. Cucinotta, L. C. Simonsen, F. Khan, J. L. Shinn, and J. W. Norbury, NASA Reference Publication No. RP-1257, December 1991.
9. G. D. Westfall, L. W. Wilson, H. J. Crawford, D. E. Greiner, and H. H. Heckman, *Phys. Rev. C* **19** (1979) 1309.
10. J. R. Cummings, W. R. Binns, T. L. Garrard, M. H. Israel, J. Klarmann, E. C. Stone, and C. J. Waddington, *Phys. Rev.* C **42** (1990) 2508; *ibid* 2530.
11. L. W. Townsend, J. W. Wilson, F. A. Cucinotta, and J. L. Shinn, *Nucl. Tracks and Radiat. Meas.* **20** (1992) 65.
12. E. M. Friedlander and H. H. Heckman, in *Treatise on Heavy Ion Science*, ed. D. A. Bromley (Plenum, New York, 1985), Vol. 4, p. 403.

MODEL CROSS-SECTION CALCULATIONS USING LAHET

R. E. PRAEL
Los Alamos National Laboratory
Los Alamos, NM, USA

ABSTRACT

The current status of LAHET is discussed. The effect of a multistage preequilibrium exciton model following the INC is examined for neutron emission benchmark calculations, as is the use of a Fermi breakup model for light nuclei rather than an evaporation model. Comparisons are made also for recent fission cross section experiments, and a discussion of helium production cross sections is presented.

1. Introduction

LAHET is a Monte Carlo code for the transport and interaction of nucleons, pions, muons, light ions, and antinucleons in complex geometry[1]; it may also be used without particle transport to generate particle production cross sections. LAHET is the result of a major effort at Los Alamos National Laboratory to develop a code system based on the LANL version of the HETC Monte Carlo code for the transport of nucleons, pions, and muons, which was originally developed at Oak Ridge National Laboratory[2]. Due to the many new features added at LANL, the present code has been renamed LAHET, and the system of codes based on LAHET designated as the LAHET Code System (LCS).

Within the LCS, cross section output is generated from LAHET history file output by the XSEX utility code[1]. For the calculation of cross sections, LAHET is run in a non-transport mode, with the interaction physics invoked at the source particle energy. The source energy may be described by a tabulated distribution, providing cross sections averaged over an energy distribution. The XSEX code is then used to process a history file, generating double-differential and energy- and angle-integrated cross sections or yields.

2. Physics Models in LAHET

As an alternative to the Bertini intranuclear cascade model[6], LAHET contains the INC routines from the ISABEL code. The ISABEL INC model is an extension by Yariv and Fraenkel[3] of the VEGAS code[7] It has the capability of treating nucleus-nucleus interactions as well as particle-nucleus interactions. It allows for interactions ("CAS-CAS") between particles both of which are excited above the Fermi sea[8]. The nuclear density is represented by up to 16 density steps, rather than three as in the Bertini INC. It also allows antiproton annihilation[9], with emission of kaons and pions. As presently implemented in LAHET, only projectiles with $A \leq 4$ are allowed, and antiproton annihilation is not presently allowed in particle transport problems.

LAHET includes as user options two models for fission induced by high energy interactions: the ORNL model by Alsmiller and others[10], and the Rutherford Appelton Laboratory model by Atchison[11]; the fission models are employed with the evaporation model of Dresner[12].

In LAHET, the Fermi breakup model[4] (FBM) has replaced the evaporation model for the disintegration of light nuclei; it treats the deexcitation process as a sequence of simultaneous breakups of the excited nucleus into two or more products, each of which may be a stable or unstable nucleus or a nucleon. Any unstable product nucleus is subject to subsequent breakup. The probability for a given breakup channel is primarily determined from the available phase space, with probabilities for two-body channels modified by Coulomb barrier, angular momentum, and isospin factors. The model is applied only for residual nuclei with $A \leq 17$, replacing the evaporation model for these nuclei. In the LAHET implementation, only two- and three-body breakup channels are considered; it is an abbreviated form of a more extensive implementation of the Fermi breakup model, with up to 7-body simultaneous breakup, used previously for cross section calculations on light nuclei[13].

A significant addition to LAHET is the multistage preequilibrium exciton model (MPM) as described in reference 5. The MPM is invoked at the completion of the INC, with an initial particle-hole configuration and excitation energy determined by the outcome of the cascade. At each stage in the MPM, the excited nucleus may emit a neutron, proton, deuteron, triton, ^{3}He, or alpha; alternatively, the nuclear configuration may evolve toward an equilibrium exciton number by increasing the exciton number by one particle-hole pair. The MPM terminates upon reaching the equilibrium exciton number; the evaporation model (or the Fermi breakup model) is then applied to the residual nucleus with the remaining excitation energy. The linkage of the MPM to each of the INC models is discussed in reference 14.

In the implementation of the MPM, the inverse reaction cross sections are represented by the parameterization of Chatterjee[15]. The potentials from which the inverse reaction cross sections are obtained are those selected by Kalmbach[16] for the PRECO-D2 code.

Three optional level density formulations have been implemented. In the default case, the level density parameter a, is obtained from the energy dependent formulation of Ignatyuk[17] as implemented in GNASH[18]. The other options in LAHET are the model originally used in the evaporation model of HETC[12] and a mass dependent alternative more recently developed for use in HETC[19].

LAHET differs from HETC in the use of cutoff energies for particles escaping from the nucleus during the intranuclear cascade. For either INC model, the neutron cutoff energy is uniformly distributed between zero and *twice* the mean binding energy.

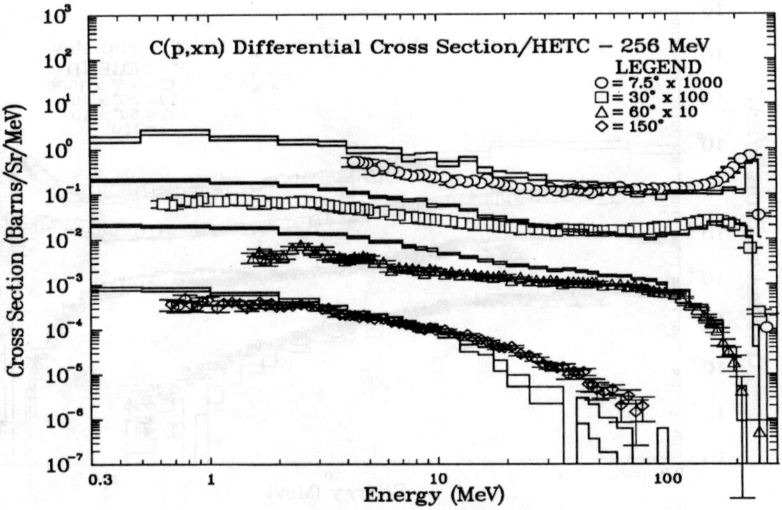

Figure 1: HETC calculation for 256 MeV protons on natural carbon.

The Coulomb barrier is randomly distributed in a form simulating a Coulomb barrier transmission probability; the maximum of the Coulomb barrier and the neutron cutoff is then used as the proton cutoff. The sampling for the cutoff energies is performed once for each projectile-target interaction; the barriers thus defined are then applied to every particle emission in the resulting cascade. This procedure, admittedly artificial, has the effect of preventing a discontinuity in the particle emission spectrum while preserving quite well the mean particle emission rates.

Another small addition to the intranuclear cascade procedure is applied to (p,n) and (n,p) INC reactions only. In this case, the outgoing particle energy is corrected by the binding energy difference in the entrance and exit channels. The modification greatly improves the realism in the high energy emission spectrum and significantly improves the overall energy balance in the INC.

3. Neutron Emission Benchmark Calculations

A full set of benchmark calculations for experiments[20,21] with 113 MeV and 256 MeV protons on a variety of targets are currently available for both thin[14] and thick targets[22]. Release of benchmark calculations for 596 MeV and 800 MeV protons awaits final publication of the experimental data.

The three examples shown here illustrate the use of the new physics options. All three cases, shown in figures 1, 2, and 3, use the Bertini INC for 256 MeV protons incident on natural carbon. The data points shown are from reference 21.

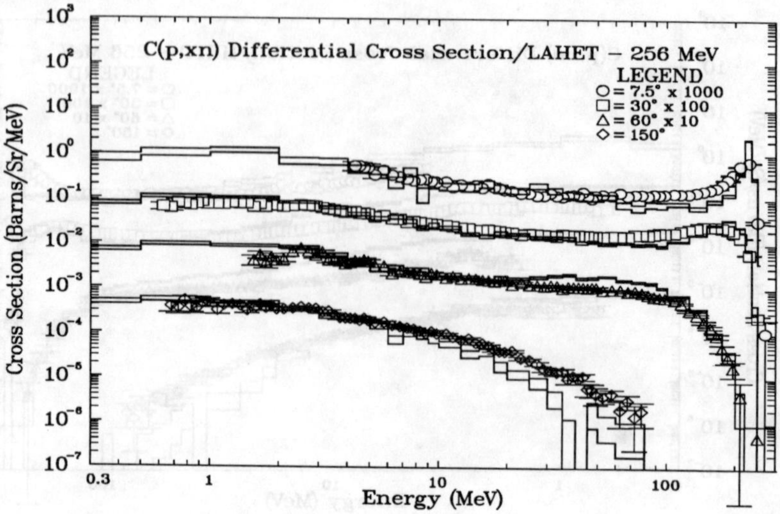

Figure 2: LAHET calculation for 256 MeV protons on natural carbon, without preequilibrium model.

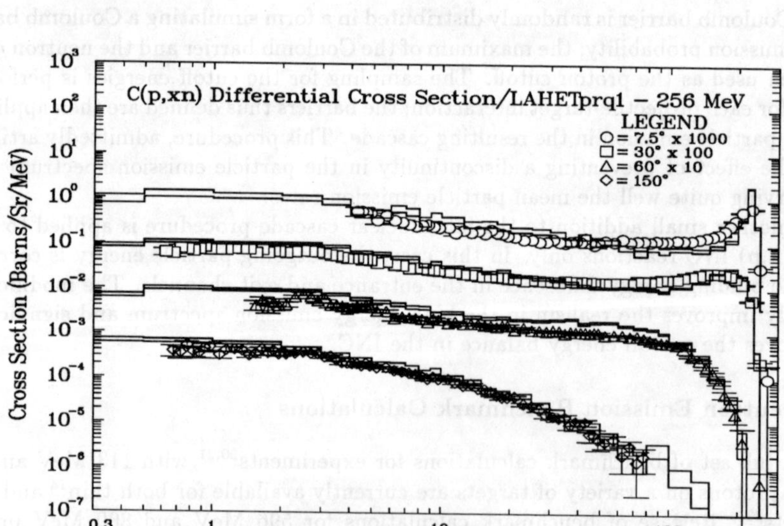

Figure 3: LAHET calculation for 256 MeV protons on natural carbon, with preequilibrium model.

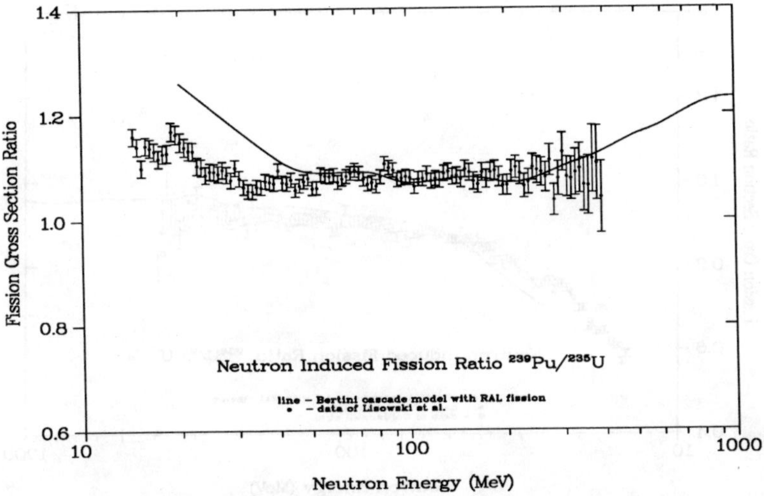

Figure 4: Neutron induced fission ratio ^{239}Pu/^{235}U.

The calculation in figure 1 was made with the LANL version of HETC and uses neither the Fermi breakup model nor the MPM. The results shown in figure 2 were made with LAHET without the use of the MPM. The significant features of the difference in the calculations are 1) the low energy neutron emission, reduced by the FBM, and 2) the 3MeV to 20 MeV spectrum, modified by the random cutoffs in the INC.

In passing to figure 3, the effect of the MPM is observed. The back angle neutron emission is much enhanced, producing a good fit to the data; at forward angles, the emission now becomes somewhat too high from 5 MeV to 20 MeV. In general, the effect of the MPM is more significant at lower energies and with heavier nuclei. Above 500 MeV, the impact is relatively minor. In general, the improvement in the back angle spectrum is frequently accompanied by some loss of agreement in the forward direction at energies below 20 MeV.

4. Neutron-induced Fission Cross Section Benchmarks

The data from neutron-induced fission experiments performed at Los Alamos National Laboratory over the energy range 15 MeV to 400 MeV is nearing publication[23]. The fission cross section ratio data (relative to ^{235}U) provide excellent benchmarking opportunities for the fission models used in LAHET and other codes. In figures 4 through 7, four cases are shown with the final ratio data[23] and calculations made with HETC.

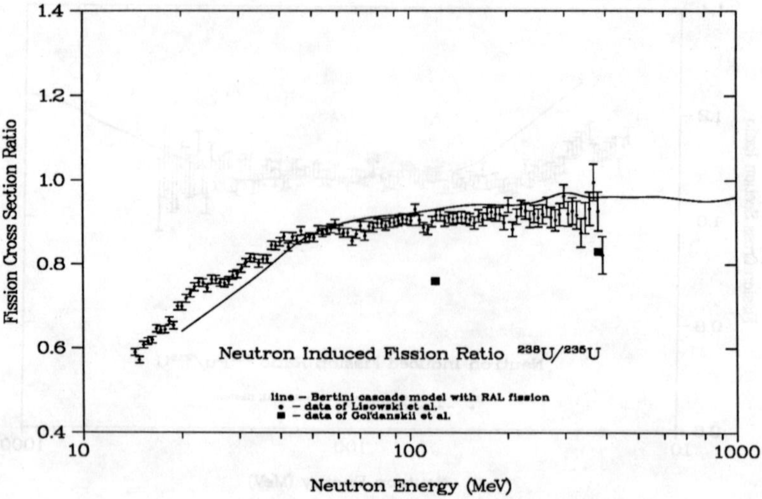

Figure 5: Neutron induced fission ratio $^{238}U/^{235}U$.

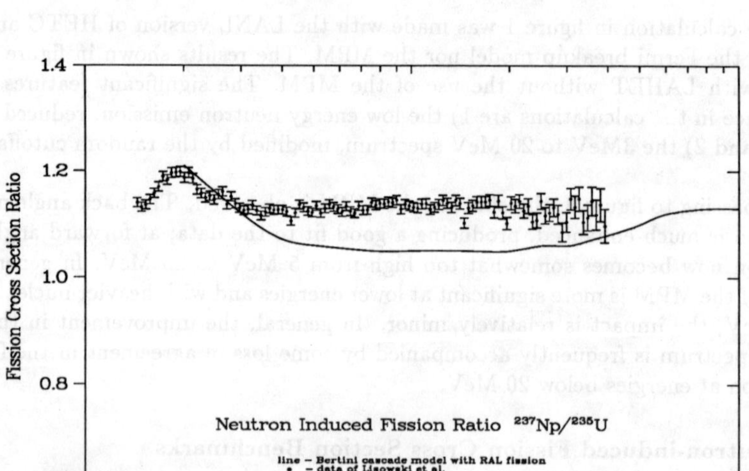

Figure 6: Neutron induced fission ratio $^{237}Np/^{235}U$.

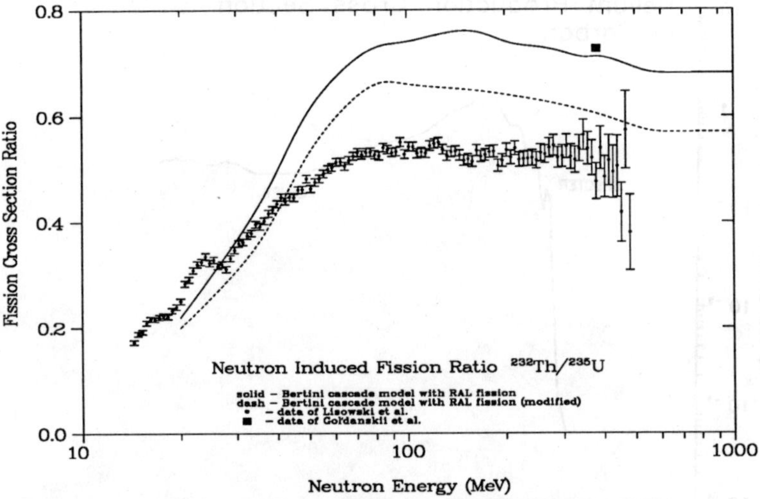

Figure 7: Neutron induced fission ratio $^{232}Th/^{235}U$.

The calculations shown represent the zero point in the benchmarking effort, having been made with the LANL version of HETC, using the Bertini INC and the RAL fission model, without the MPM or the newer level density options. Although the agreement for ^{239}Pu and ^{238}U is good over a wide energy range, the agreement for the other two cases has worsened with reevaluation of the data. Evaluation of the effects of the newer code features may allow some parameter fitting in the fission models. The modification shown for ^{232}Th involved adjusting the parameters for $Z = 89$ in the RAL fission model; the effect on higher Z targets is negligible, and the modification is currently used in LAHET.

5. A Note on Helium Production Cross Sections

M. S. Wechsler has reported an examination of the helium production cross sections calculated by LAHET for neutrons on a variety of targets[24]. He has noted that the calculations are generally very low compared to available experimental data and significantly discontinuous with (and generally lower, by factors of 4 or 5, than) ENDF cross sections at 20 MeV. Older VNMTC calculations, using DFF as an evaporation code rather than the EVAP model, produced much higher and more consistent helium production cross sections.

However one case stands out, as shown in figure 8, where the "SPECTER" cross section is derived from ENDF data. For the case of neutrons on carbon, the Fermi breakup model is used exclusively and the evaporation model is never invoked. The

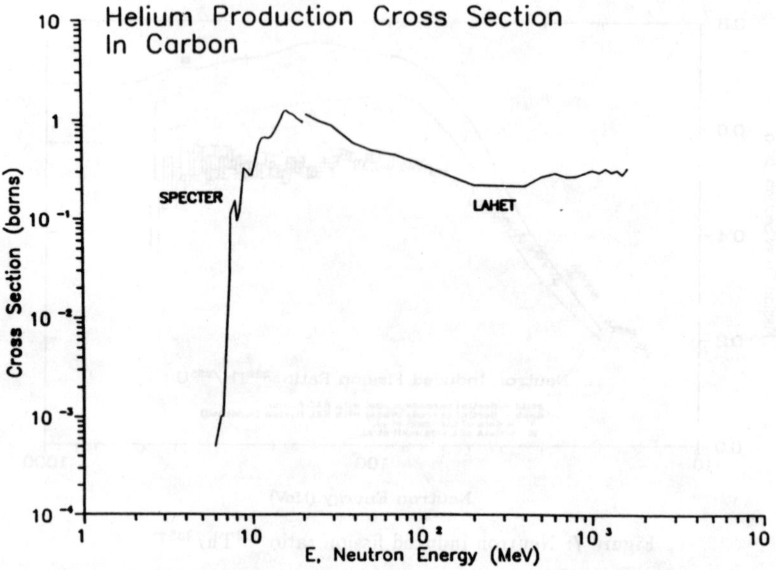

Figure 8: Neutron-induced helium production cross sections for carbon

FBM strongly enhances the multiple α particle disintegration modes. The results shown in figure 8 are very consistent.

6. Summary

The neutron emission benchmark calculations have shown the utility of the MPM and the FBM for the calculation of double-differential neutron production cross sections. The benchmark series will be completed with the calculations for the 596 MeV and 800 MeV data.

The publication of the data for neutron-induced fission to 400 MeV provides an excellent benchmarking opportunity for the fission models used in LAHET and similar codes. The sensitivity to level density parameterization may also be defined with benchmark calculations. It is suggested that adjustment of fission model parameters in view of the extensive data be made a model development priority.

Recent interest in gas production in targets suggests that benchmarking of the various formulations of the evaporation model be undertaken in this area. A potential direction for development is to reformulate the evaporation model in terms of the parameterized optical model inverse reaction cross sections now used with the MPM.

7. References

1. Richard E. Prael and Henry Lichtenstein, "User Guide to LCS: The LAHET Code System", LA-UR-89-3014, Los Alamos National Laboratory (September 1989).
2. Radiation Shielding Information Center, " HETC Monte Carlo High-Energy Nucleon-Meson Transport Code", Report CCC-178, Oak Ridge National Laboratory (August 1977).
3. Y. Yariv and Z. Fraenkel, *Phys. Rev. C, 20* (1979), p. 2227.
4. D. J. Brenner, R. E. Prael, J. F. Dicello, and M. Zaider, "Improved Calculations of Energy Deposition from Fast Neutrons", in *Proceedings Fourth Symposium on Neutron Dosimetry,* EUR-7448, Munich-Neuherberg (1981).
5. R. E. Prael and Michael Bozoian, "Adaptation of the Multistage Preequilibrium Model for the Monte Carlo Method (I)", LA-UR-88-3238, Los Alamos National Laboratory (September 1988).
6. H. W. Bertini, *Phys. Rev. 188* (1969), p. 1711.
7. K. Chen et al., *Phys. Rev., 166* (1968), p. 949.
8. Y. Yariv and Z. Fraenkel, *Phys. Rev. C, 24* (1981), p. 488.
9. M. R. Clover, R. M. DeVries, N. J. DiGiacomo, and Y. Yariv, *Phys. Rev. C, 26* (1982), p. 2138.
10. J. Barish et al., "HETFIS High-Energy Nucleon-Meson Transport Code with Fission", ORNL/TM-7882, Oak Ridge National Laboratory (July 1981).
11. F. Atchison, "Spallation and Fission in Heavy Metal Nuclei under Medium Energy Proton Bombardment", in *Targets for Neutron Beam Spallation Sources,* Jül-Conf-34, Kernforschungsanlage Jülich GmbH (January 1980).
12. L. Dresner,"EVAP – A Fortran Program for Calculating the Evaporation of Various Particles from Excited Compound Nuclei", ORNL-TM-196, Oak Ridge National Laboratory (April 1962).
13. D. J. Brenner and R. E. Prael," Calculated Differential Secondary-particle Production Cross Sections after Nonelastic Neutron Interactions with Carbon and Oxygen between 10 and 60 MeV", *Atomic and Nuclear Data Tables* **41**,71-130 (1989).
14. R. E. Prael, "LAHET Benchmark Calculations of Differential Neutron Production Cross Sections for 113 MeV and 256 MeV Protons", LA-UR-89-3347, Los Alamos National Laboratory (September 1989).
15. A. Chatterjee, K. H. N. Murthy, and S. K. Gupta, *Pramāna 16* (1981), p. 391.
16. C. Kalbach, "PRECO-D2: Program for Calculating Preequilibrium and Direct Reaction Double Differential Cross Sections", LA-10248-MS, Los Alamos National Laboratory (1985).
17. A. V. Ignatyuk, G. N. Smirenkin, and A. S. Tishin, *Sov. J. Nucl. Phys. 21* (1975), p. 256.
18. E. D. Arthur,"The GNASH Preequilibrium-Statistical Model Code", LA-UR-88-382, Los Alamos National Laboratory (February 1988).

19. P. Cloth et al., "The KFA-Version of the High-Energy Transport Code HETC and the Generalized Evaluation Code SIMPEL", Jül-Spez-196, Kernforschungsanlage Jülich GmbH (March 1983).
20. M. M. Meier et al., "Differential Neutron Production Cross Sections and Neutron Yields from Stopping-Length Targets for 113-MeV Protons", LA-11518-MS, Los Alamos National Laboratory (March 1989).
21. M. M. Meier et al., "Differential Neutron Production Cross Sections and Neutron Yields from Stopping-Length Targets for 256-MeV Protons", LA-11656-MS, Los Alamos National Laboratory (August 1989).
22. R. E. Prael, "LAHET Benchmark Calculations of Neutron Yields from from Stopping-Length Targets for 113 MeV and 256 MeV Protons", LA-UR-90-1620, Los Alamos National Laboratory (May 1990).
23. P. W. Lisowski, Los Alamos National Laboratory (unpublished).
24. M. S. Wechsler, Iowa State University, private communication.

IMPROVEMENTS AND PERSPECTIVES OF INC-MODELS IN THE MEDIUM- AND HIGH-ENERGY RANGE

DETLEF FILGES, PETER CLOTH
Institut für Kernphysik, Forschungszentrum Jülich, Postfach 1913
D-5170 Jülich, Germany

and

DIETMAR THEIS
Institut für Strahlen- und Kernphysik, Universität Bonn, Nußallee 14-16,
D-5300 Bonn, Germany

Abstract

In the medium energy range the original version of the HETC-Bertini INC-model simulates inelastic collisions between the projectile and a nucleon inside the target nucleus via formation and decay of a Δ_{33}-resonance. Therefore, only single and double pion production can occure. The INC-Bertini model was extended by using the socalled HADRIN model for the elementary collisions inside the nucleus, which takes into account all resonances contributing more than 2% to the total cross section upto kinetic energies of 5 GeV. To obtain a precise description of the cascade, angular distributions were used from phase shift calculations for elastic nucleon-nucleon, elastic and charge exchange pion-nucleon scattering distributions. For pion and nucleon projectiles also the total cross section is calculated from those phase shifts. At higher energies - above several GeV - the Bertini model of HETC is not adequate for nuclear event generation, whereas the modelling of higher energy hadron nucleus collisions is found to be adequately realized in the FLUKA code. The mergence of the corresponding codes, Bertini, INC of the HETC system and EVENTQ of FLUKA, into a single system by INC-coupling is described.

1. Introduction

The Monte Carlo simulation of medium and high energy hadron events and cascades in matter requires nuclear event generators, i.e. computer codes treating hadron nucleus collisions in all ranges of incident particle momentum. Normally the shower induced by a high energy projectile comprises also particles in the medium energy range. The physical modelling in these two regimes of momentum is, however, considerably different and adequate codes do not cover the whole energy scale in question. As a consequence in simulation, the complete shower usually has to be computed subsequently by at least two different codes. Best results in hadron nucleus collision simulations are obtained by computer codes, which generate secondary particles by simulating the intranuclear cascade as far as possible from first principles. With such codes e.g. strict energy and momentum conservation in the final state can be achieved.

2. The Intranuclear Cascade Codes

A widely used code, which meets the above conditions in the medium energy range is the INC (intranuclear cascade code) of Bertini[1]. Usually this code is run within the shower Monte Carlo program HETC[2,3,4]. A good idea of the capability of the Bertini INC code may be achieved in Figs. 1 and 2 (see also Ref. 5).

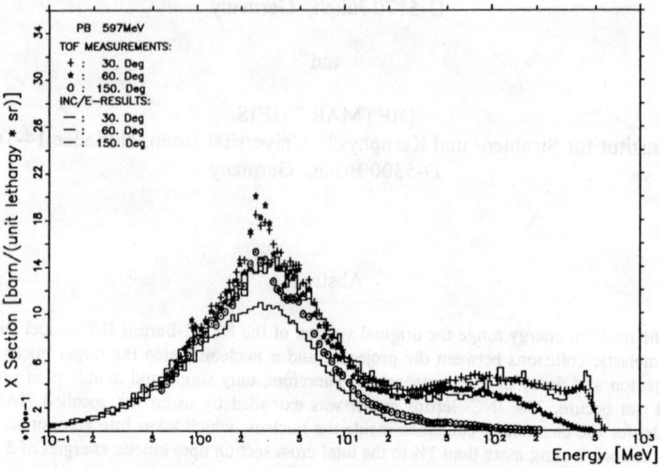

Fig. 1: Double differential neutron production cross section for incident proton on lead at 597 MeV between 30° and 150° in the laboratory system. Full line Bertini-HETC, symbols measured data (details Ref.[5])

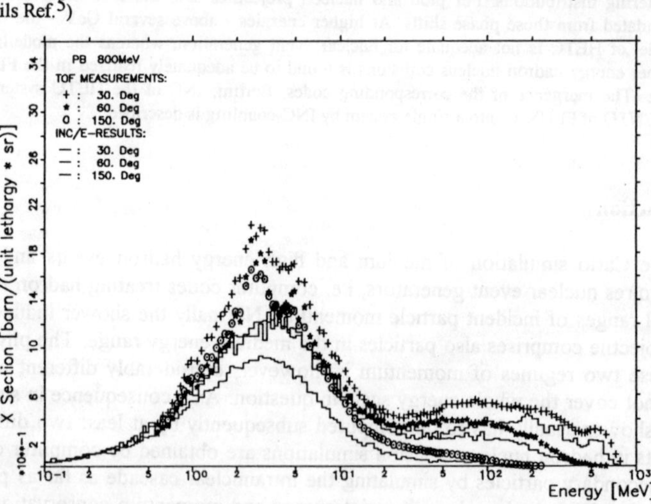

Fig. 2: Double differential neutron production cross section for incident proton on lead at 800 MeV between 30° and 150° in the laboratory system. Full line Bertini-HETC, symbols measured data (details Ref.[5])

All comparisons with INC simulations show good agreement for E_n>20 MeV and justify the usage of the code in a shower Monte Carlo program for the medium energy range. The part of the cross section below 20 MeV is mainly due to evaporation of the residual target nuclei after the high energy reaction. An adequate high energy code for shower simulation is the EVENTQ of the shower Monte Carlo FLUKA[6], well known in high energy physics. For reasons given above it was strongly desired to have the features of the two event generators, INC and EVENTQ, unified in one code, i.e. coupling of the two models on the base of the intranuclear cascade. Coupling via the secondary particles was found to be not satisfying and adequate. INC is an intranuclear particle transport Monte Carlo code considering a nucleus model with varying nucleon density and nuclear potential represented by 3 concentric spherical shells. The nucleons are assumed to have a Fermi gas momentum distribution. Elastic and inelastic nucleon-nucleon and pion-nucleon cross sections are used to find types and sites of the collisions inside the nucleus. Double differential cross sections are used to find the final states of these collision. Complete energy and momentum balance is observed in each of them. No other hadrons than nucleons and pions are allowed in INC. The main features of the Bertini INC model, in combination with the evaporation code EVAP, referred to as in the INCE model, are illustrated in Fig. 3.

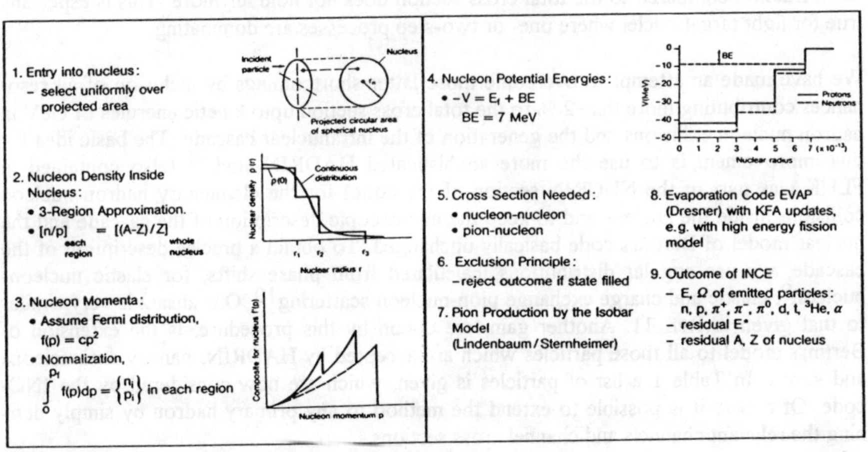

Fig. 3: The Bertini intranuclear cascade model and the excited nucleus evaporation model of the HETC code

EVENTQ performs nuclear collisions in 3 phases: The high energy collisions are treated by the "Dual Parton Multi-Chain Fragmentation Model"[7]. The "soft cascade" called part of the hadron nucleus collision is treated as breaking up the available energy for this part into portions taken away by nucleons. The excitation energy of the residual nucleus is also set aside for later use (e.g. evaporation). The fractions of primary energy going into the 3 phases are pre-estimated using inelasticity functions for the projectile nucleus system. In this model in each of the interacting hadrons the valence quarks are separated into two coloured systems (quarks, diquarks), which share the particle momentum. These systems then are combined to form two colour-neutral chains, which finally hadronize into a multiparticle final state. In addition to this non-diffractive component of particle production described so far, the single diffractive component is included. In single diffractive events the chain generation takes place between an exchange meson and one of the interacting hadrons. Hadronization is performed via the Monte Carlo chain decay fragmentation model[7] contained in EVENTQ. The intranuclear cascade is simulated as a number of hadron nucleon collisions inside the nucleus, without considering the spatial distribution, i.e. the collision sites remain uncertain.

3. Improvements of the Bertini-INC in the Medium Energy Range

One of the reasons of the limitations of the INC-model of Bertini to relatively low energies is the fact, that only the formation of the Δ_{33} resonance is included. This must not be a limitation for more inclusive problems, e.g. to calculate the energy deposition in a thick absorber, but may lead to discrepancies when using the code for simulations, where the correct multiplicity in each event is essential or in calculations of double differential cross sections, e.g. applying trigger conditions may suppress those final states which are making up the major part of the total cross section but not effect other channels which have a relatively small contribution. In this case the argument that the latter can be neglected because of their small fraction compared to the total cross section does not hold anymore. This is especially true for light target nuclei where one- or two-step processes are dominating.

We have made an attempt to overcome those latter shortcomings by inclusion of all resonances contributing more than 2 % to the total cross section upto kinetic energies of GeV in hadron nucleon collisions and the generation of the intranuclear cascade. The basic idea for this improvement is to use the more sophisticated HADRIN-model[8] (also contained in FLUKA as part of the NUCRIN section of the code) for the elementary hadron nucleon collisions inside the nucleus and to keep the microscopic description of the cascade and the nuclear model of Bertini's code basically unchanged. To obtain a precise description of the cascade, we use angular distributions, calculated from phase shifts, for elastic nucleon-nucleon[9], elastic and charge exchange pion-nucleon scattering[10]. Our ansatz is very similar to that given in Ref. 11. Another gain, we obtain by this procedure, is the extension of Bertini's model to all those particles which are accepted by HADRIN, namely antinucleons and kaons. In Table 1 a list of particles is given, which are now considered by the INC-code. Of course it is possible to extend the method to any primary hadron by simply defining the relevant channels and channel cross sections.

Table 1: Particles considered in the extended Bertini INC version

Primary hadrons	Secondary particles*
proton	proton
anti-proton	anti-proton
neutron	neutron
anti-neutron	anti-neutron
π^+, π^-	π^+, π^-, π^0
K^+, K^-	K^+, K^-, e^+, e^-
	μ^+, μ^- etc.

*Secondary particles are hyperons, leptons, and photons resulting also from decay of hadrons (s. also Ref. 8)

In Figures 4-6 some examples of calculations of double differential production cross sections are shown in comparison with measured data and some other calculations with other codes.

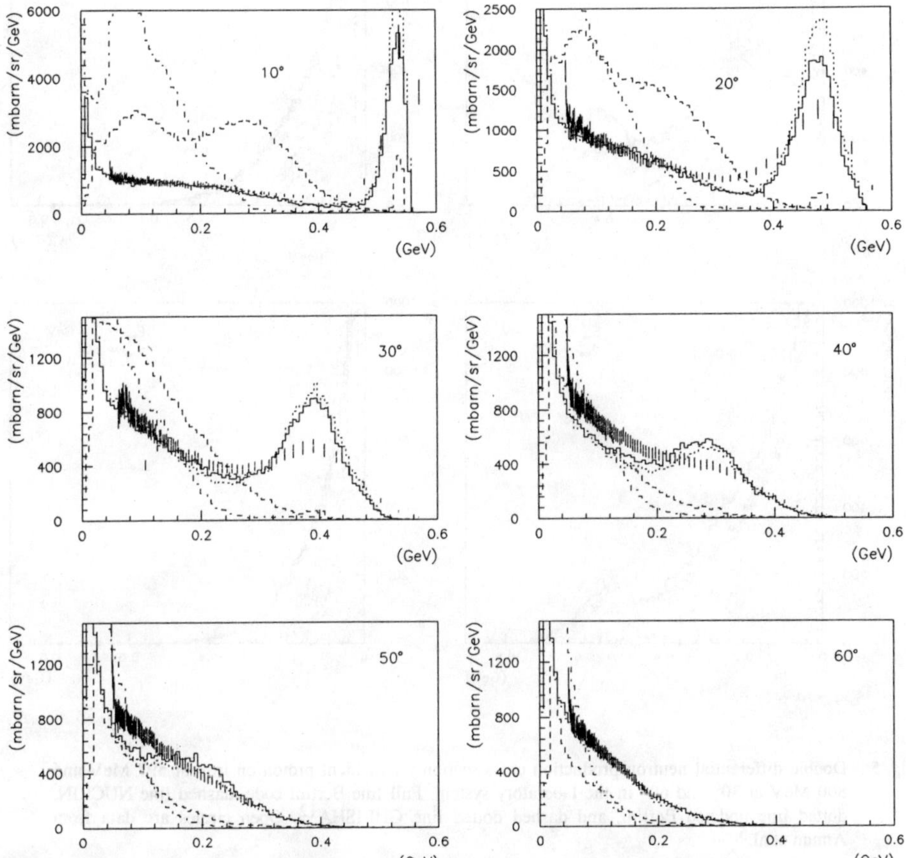

Fig. 4: Double differential proton production cross section for incident proton on iron at 558 MeV between 10° and 60° in the laboratory system. Full line Bertini code, dashed line NUCRIN, dotted line updated Bertini, and dashed dotted line GHEISHA[12]. Data from Beck et al.[13]

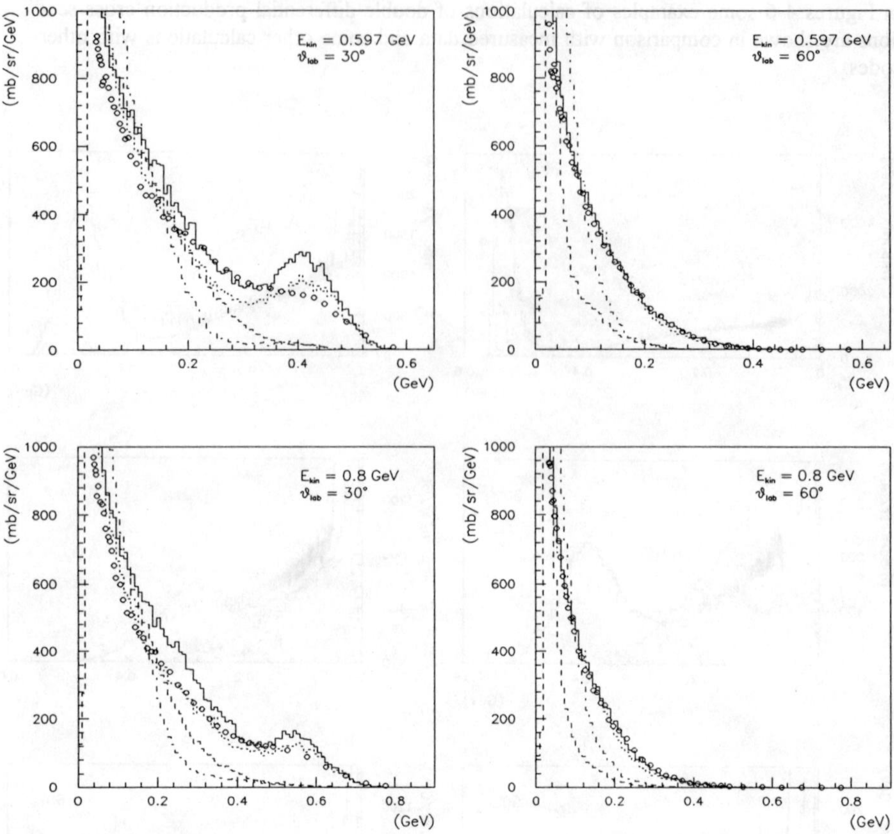

Fig. 5: Double differential neutron production cross section for incident proton on iron at 597 MeV and 800 MeV at 30° and 60° in the laboratory system. Full line Bertini code, dashed line NUCRIN, dotted line updated Bertini, and dashed dotted line GHEISHA[12]. Open circles are data from Amian et al.[5]

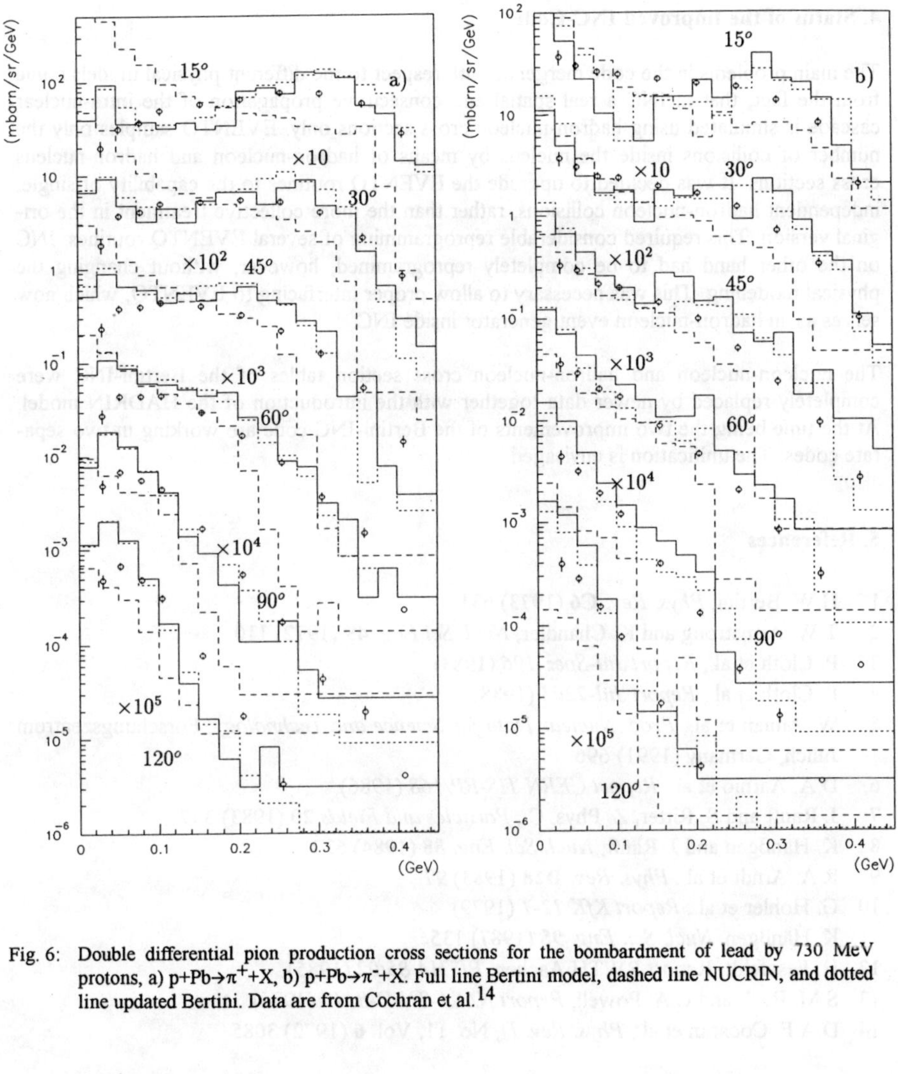

Fig. 6: Double differential pion production cross sections for the bombardment of lead by 730 MeV protons, a) p+Pb→π^++X, b) p+Pb→π^-+X. Full line Bertini model, dashed line NUCRIN, and dotted line updated Bertini. Data are from Cochran et al.[14]

4. Status of the Improved INC-Code

The main problems in the code mergence with respect to the different physical models come from the fact, that in INC a real spatial and consecutive propagation of the intra-nuclear cascade is simulated using hadron-nucleon cross sections only. EVENTQ samples only the number of collisions inside the nucleus by means of hadron-nucleon and hadron-nucleus cross sections. It was decided to upgrade the EVENTQ routines to the capability of single, independent hadron-nucleon collisions, rather than the more collective treatment in the original version. This required considerable reprogramming of several EVENTQ routines. INC on the other hand had to be completely reprogrammed, however, without changing the physical modelling. This was necessary to allow proper interfacing to EVENTQ, which now serves as an hadron-nucleon event generator inside INC.

The nucleon-nucleon and hadron-nucleon cross section tables of the Bertini-INC were completely replaced by newer data together with the introduction of the HADRIN model. At the time being the two improvements of the Bertini-INC code are working in two separate codes. The unification is envisaged.

5. References

1. H.W. Bertini, *Phys, Rev.* **C6** (1973) 631
2. T.W. Armstrong and K. Chandler, *Nucl. Sci.Eng.* **49** (1972) 110
3. P. Cloth et al., *Report Jül-Spez-196* (1983)
4. P. Cloth et al., *Report Jül-2203* (1988)
5. W. Amian et al., *Proc. Nuclear Data for Science and Technology*, Forschungszentrum Jülich, Germany (1991) 696
6. D.A. Aarnio et al., *Report CERN TIS-RP/168* (1986)
7. J. Ranft and S. Ritter, Z. Phys. C., *Particles and Fields* **20** (1983) 347
8. K. Hänßgen and J. Ranft, *Nucl. Sci. Eng.* **88** (1984) 537
9. R.A. Arndt et al., *Phys. Rev.* **D28** (1983) 97
10. G. Höhler et al., *Report KfK 12-1* (1979)
11. K. Hänßgen, *Nucl. Sci. Eng.* **95** (1987) 135
12. H. Fesefeldt, Report RWTH Aachen, *PITHA 85/02* (1985)
13. S.M. Beck and C.A. Powell, *Report NASA-TN-D-8119* (1976)
14. D.A.F. Cochran et al., *Phys. Rev. D*, No. 11, Vol. **6** (1972) 3085

MOVING SOURCE MODEL FOR EVALUATING THE DOUBLE DIFFERENTIAL CROSS-SECTIONS ON NEUTRON-EMISSION AND FRAGMENTATION

K. ISHIBASHI, S. SAKAGUCHI,* K. HIGO,** Y. WAKUTA
Department of Nuclear Engineering, Kyushu University
Hakozaki, Higashi-ku, Fukuoka 812, Japan
and

H. TAKADA, T. NISHIDA, Y. NAKAHARA
Japan Atomic Energy Research Institute
Tokai-mura, Naka-gun, Ibaraki 319-11, Japan

ABSTRACT

Moving source models are utilized to reproduce the neutron spectra in spallation reactions. At incident proton energies below 400 MeV, a new idea of a Watt moving source model is introduced. The combination of the standard and Watt moving source models enables us to parameterize the double differential cross sections of neutron emission for incident protons of 80 to 800 MeV. For the systematics of fragmentation reactions induced by 0.5-3GeV protons, a liquid-gas phase-transition model is adopted in addition to the moving source model. Both the mass yield and double differential cross sections are well reproduced.

1. Introduction

The spallation reaction is caused by bombarding targets with particles having an energy above a few hundred MeV. The reaction produces a number of neutrons. This reaction is then usable for providing an intense spallation neutron source or transmuting long-lived radioactive wastes. In spite of the usefulness for engineering purposes, the accumulation of the measured data is poor for the proton induced spallation reaction. Therefore, it is important to find the systematic behavior in the neutron data[1] as well as the fragmentation cross sections. We parameterize these cross sections to extend to different energy regions or target nuclei.

A moving source (MS) model has been used for describing protons[2] emitted by several-GeV proton-induced spallation reactions. The model expresses the angular distribution of emitted particles by the use of a single parameter. At present, the experimental neutron spectra in the spallation reaction are available at only a few incident energies and emission angles. Hence, we will apply this model to study the systematics of the neutron data. At incident energies below 400 MeV, in addition, we will introduce a Watt moving

* Science and Technology Agency, Kasumigaseki, Chiyoda-ku, Tokyo 100, Japan
**Kyushu Matsushita Electric Co., Minoshima, Hakata-ku, Fukuoka 812, Japan

source (WMS) model in accordance with the change of the collision phenomena.

As is well known, up to proton energies of several hundred MeV, the spallation reaction consists of reaction processes such as cascade, preequilibrium and evaporation. Furthermore, the fragmentation process is induced above these energies. The fragmentation process emits intermediate- mass fragments such as Na and Mg, which are never released by the above processes. The fragmentation reaction has been analyzed by either the liquid-gas phase transition model[3] or the quantum molecular dynamics (QMD)[4] approach. In this study, the former model is chosen for parameterization from the viewpoint of nuclear data evaluation. We attempt to investigate the systematics of the fragmentation mass yield by the liquid-gas phase transition model, and to find those of the double differential cross section by employing the moving source models.

2. Neutron Data Analysis

2.1 Moving Source (MS) Model

The situation of the MS model is shown[5] in Fig. 1, where a projectile interacts with a nucleus that is indicated by a circle. The illustrations on the left and the right hand sides show the collision phenomenon in the laboratory and the moving frames, respectively. A cascade collision is seen in the left hand side. When the velocity of the moving frame β is properly chosen, the collision is observed to be isotropically developed as shown in the right hand side. In this moving frame, neutrons are assumed to be emitted isotropically at a nuclear temperature T (MeV) with a Maxwellian energy distribution. The invariant cross section, therefore, is written as

$$E_{tot}(d^3\sigma / dp^3) = A_m \exp(-E^*_{kin} / T) \tag{1}$$

where E_{tot} (MeV) is the total neutron energy including rest mass, A_m(mb/sr(MeV/c)2c) a variable parameter, and E_{kin}^* (MeV) the emitted neutron energy in the moving frame. The Lorentz transformation is written as

$$E^*_{kin} + m = \frac{E_{kin} + m - p\beta\cos\theta}{\sqrt{1-\beta^2}} \tag{2}$$

where m is the neutron mass (MeV), Ekin (MeV) the kinetic energy, p(MeV/c) the momentum, and θ the emission angle. These are quantities in the laboratory frame. Then, Eq. (1) is converted into the equation in the laboratory frame as

$$\begin{aligned} E_{tot}(d^3\sigma / dp^3) &= \frac{1}{p}(d^2\sigma / d\Omega dE_{kin}) \\ &= A_m \exp\left[-\left\{\frac{E_{kin} + m - p\beta\cos\theta}{\sqrt{1-\beta^2}} - m\right\} / T\right] \end{aligned} \tag{3}$$

The parameters A_m, T, and β are adjustable in fitting the equation with the experimental differential cross sections. We call Am amplitude parameter, T temperature parameter, and β velocity parameter. The angular distribution in the laboratory frame is dominantly influenced by the value of β. To the spallation reaction, the MS model is applied in a form of summation of three components as

$$E_{tot}(d^3\sigma / dp^3) = \frac{1}{p}(d^2\sigma / d\Omega dE_{kin})$$
$$= \sum_{i=1}^{3} A_{mi} \exp\left[-\left\{\frac{E_{kin}+m-p\beta_i\cos\theta}{\sqrt{1-\beta_i^2}} - m\right\} / T_i\right] \tag{4}$$

The subscripts i=1, 2 and 3 stand for the intranuclear-cascade, the preequilibrium and nuclear-evaporation (equilibrium) processes, respectively.

Figure 2 shows the results of the parameter fitting for the data[6] of 585 MeV proton incident on a lead target. The experimental neutron spectra are plotted by marks, and the fitting results by solid curves. The agreement between the fitting and the experimental data is excellent. The solid curve at 150 deg is decomposed into the three processes, as plotted by the dashed lines. The parameters obtained for targets from carbon to uranium are listed in Table 1.

2.2 Watt Moving Source (WMS) Model

For the measurement data of 113 MeV proton incidence on iron[7], the results of fitting by the MS model are shown by the dashed curves in Fig.3. In the forward direction of 7.5 and 30 deg, the dashed curves are in poor agreement with the data at neutron energies above several ten MeV. The neutrons in this energy range are emitted mainly from the intranuclear- cascade process. The MS model thus reproduced the experimental data

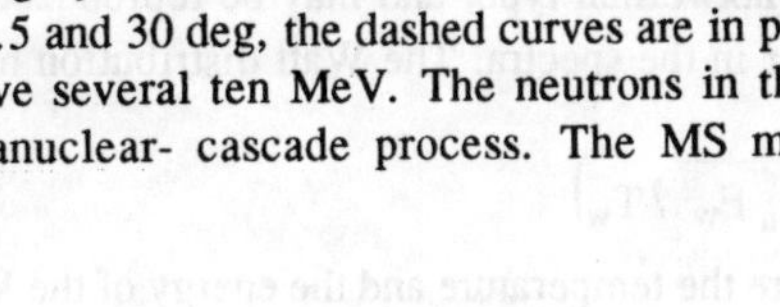

Fig. 1. Moving source model.

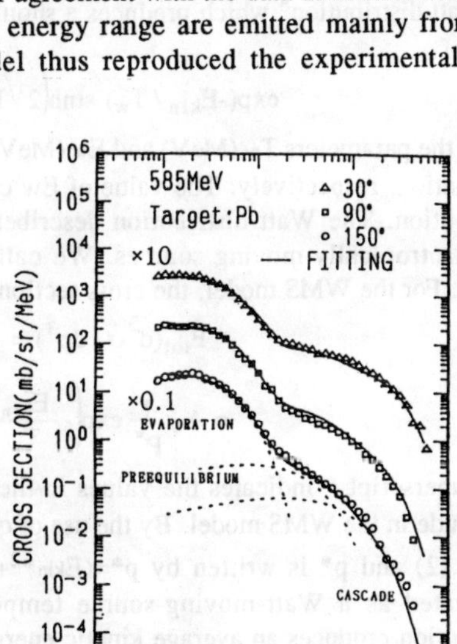

Fig. 2. Experimental neutron spectra[6] and the results by the moving source model.

Table 1. Parameters obtained for the 585 MeV proton-induced reactions.

	Amplitude mb/sr $(\mathrm{MeV}/c)^2c$			Temperature MeV			Velocity		
	A_{m1} 10^{-4}	A_{m2} 10^{-3}	A_{m3} 10^{-2}	T_1	T_2	T_3	β_1 10^{-1}	β_2 10^{-2}	β_3 10^{-3}
C	6.77	2.20	8.20	82.0	16.7	2.23	2.89	8.83	15.8
Al	22.6	8.59	33.3	79.8	14.6	2.19	2.49	6.11	18.4
Fe	48.6	15.4	77.3	73.9	17.7	2.67	2.54	6.42	11.2
Nb	69.8	24.4	205.	72.8	19.3	2.61	2.39	6.07	7.24
In	89.5	35.7	325.	70.5	18.1	2.55	2.34	5.11	5.04
Ta	109.	54.3	705.	76.1	19.9	2.53	2.31	6.34	4.33
Pb	99.4	50.9	788.	75.9	22.9	2.57	2.26	6.02	3.70
U	127.	72.1	1370	78.8	20.4	2.40	2.17	6.95	3.49

unsuccessfully in the incident proton energy region below a few hundred MeV. The disagreement in Fig. 3 suggests that the energy spectra in the moving frame are different from the simple exponential function like the Maxwellian type, and may be reproduced by the Watt distribution[8] which produces a shoulder in the spectra. The Watt distribution has a form of

$$\exp(-E_{kin}/T_w)\sinh\left(2\sqrt{E^*_{kin}\,E_w}\;/T_w\right) \tag{5}$$

where the parameters T_w (MeV) and E_w (MeV) are the temperature and the energy of the Watt distribution, respectively. The value of E_w corresponds to the characteristic velocity of the distribution. The Watt distribution describes the situation where neutrons are evaporated from isotropically-moving sources. We call this model the Watt moving source (WMS) model. For the WMS model, the cross section is written in the laboratory frame as

$$E_{tot}(d^3\sigma/dp^3) = \frac{1}{p}(d^2\sigma/d\Omega dE_{kin})$$

$$= A_w\frac{1}{P^*}\exp\left(-\frac{E^*_{kin}}{T_w}\right)\sinh\left(-\frac{2\sqrt{E^*_{kin}\,E_w}}{T_w}\right) \tag{6}$$

The superscript * indicates the values in the moving frame. The parameter A_w shows an amplitude in the WMS model. By the use of quantities in the laboratory frame, $E_{kin}{}^*$ is given by Eq.(2) and p* is written by $p^*=(E_{kin}{}^{*2}+2mE_{kin}{}^*)^{1/2}$. The parameters T_w and E_w are designated as a Watt-moving-source temperature and energy, respectively. The Watt distribution produces an average kinetic energy $<E>= 3/2T_w+E_w$ in the first approximation, while the Maxwell distribution gives $<E>=3/2T$ in the moving frame. The two parameters in the WMS model are then converted into a temperature T' which corresponds to the MS model.

$$T'=T_w+2/3E_w \tag{7}$$

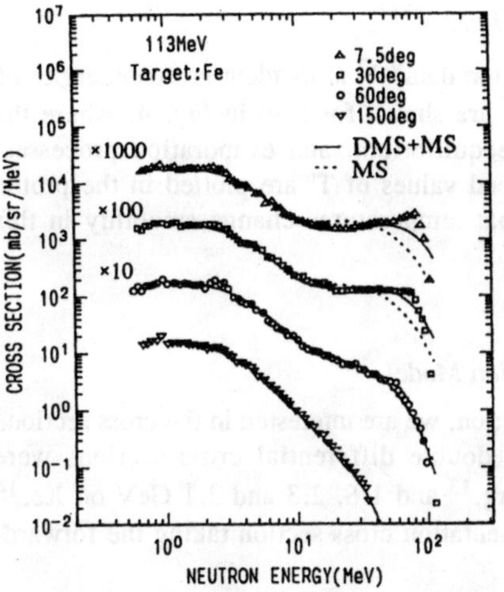

Fig. 3. Experimental neutron spectra[7] and the results by Watt moving source model.

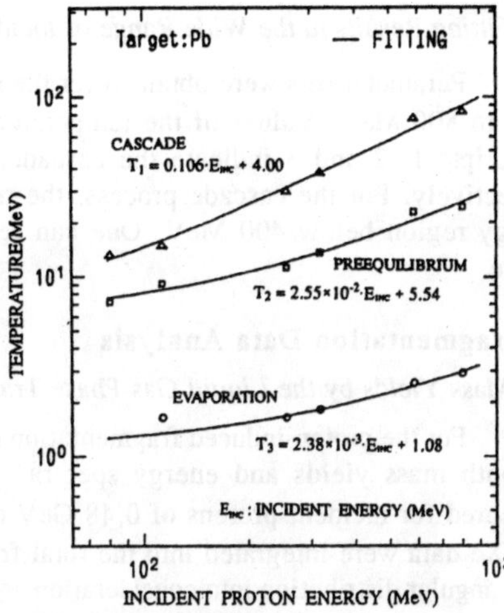

Fig. 4. Temperature parameter T for 80-800 MeV proton incident on lead.

The WMS model is applicable only to the initial process that is directly induced by the incident protons. The other remaining processes are treated by the MS model. The solid curves in Fig. 3 show the results of fitting by the combination of WMS and MS models. The experimental data are expressed better by the solid than the dashed curves.The parameters obtained are listed in Table 2 for 113 MeV proton incidence on targets of carbon to uranium.

Table 2. Parameters obtained for the 113 MeV proton-induced reactions.

	Amplitude mb/sr $(MeV/c)^2c$			Temperature MeV				Velocity		
	A_w 10^{-2}	A_{m2} 10^{-3}	A_{m3} 10^{-2}	T_w	E_w	T_2	T_3	β_1 10^{-1}	β_2 10^{-3}	β_3 10^{-3}
C	9.38	3.47	1.85	7.64	12.7	6.60	1.72	2.53	5.59	16.8
O	7.81	5.37	2.63	5.98	13.3	6.41	2.39	2.42	5.59	17.4
Al	13.3	9.02	14.7	7.11	13.5	7.78	2.12	2.26	2.50	5.44
Fe	27.0	17.1	62.9	7.25	11.7	7.05	1.89	2.04	5.48	4.92
W	101.	136.	1500	7.38	12.2	6.67	1.41	1.74	6.04	2.29
Pb	154.	70.8	1160	8.89	9.12	9.14	1.35	1.85	2.59	2.00
U	94.9	63.6	1530	8.34	11.2	9.64	1.65	1.81	10.5	2.25

2.3 Fitting Results in the Wide Range of Incident Energy

Parameter sets were obtained for the neutron data[9-12] at incident proton energies of 80.5 to 800 MeV. Values of the temperature T are shown for lead in Fig. 4, where the subscripts 1, 2 and 3 indicate the cascade, preequilibrium, and evaporation processes, respectively. For the cascade process, the reduced values of T' are plotted in the proton energy region below 400 MeV. One can see that temperatures change smoothly in this figure.

3. Fragmentation Data Analysis

3.1 Mass Yields by the Liquid-Gas-Phase Transition Model

For the proton-induced fragmentation reaction, we are interested in the cross sections on both mass yields and energy spectra. The double differential cross sections were measured for incident protons of 0.48 GeV on Ag,[13] and 1.6, 2.3 and 3.1 GeV on Xe.[14] The Xe data were integrated into the total fragmentation cross section taking the forward-peak angular distribution into consideration.

For the reaction analysis, we adopt the model that the liquid-gas-phase transition takes place in highly-excited nuclei. In this model, a nucleus stays in a liquid phase in the initial state. When an incident proton with high energy makes the nucleus in a highly excited state, the phase in some region of the nucleus is supposed to change from the liquid to the gaseous phase. The region of the gaseous phase may elongate across the nucleus, and lead to break-up of the nucleus, resulting in production of fragments.

Panagioutou[3] has proposed a simple form of the fragment formation cross section Y(A) on the basis on the liquid-gas phase-transition model:

$$Y(A)= C\, A^{-2.0}\exp\,[\{-F_A^S(T)\cdot A^{2/3} - F_A^B(T)\cdot A + \mu\,(T)\cdot A\}/T], \tag{8}$$

where A is the mass number, T the nuclear temperature(MeV), $F_A^S(T)$ the surface free energy(MeV), $F_A^B(T)$ the volume free energy(MeV), and $\mu\,(T)$ the chemical potential. The nuclear temperature at fragmentation T is the characteristic parameter that expresses the overall shape of the cross section vs. mass number. The expressions for $F_A^S(T)$, and $F_A^B(T)-\mu\,(T)$ are given in reference 3. When T is smaller than the critical temperature T_c=12MeV, $F_A^B(T)$ equals $\mu\,(T)$: For $T<T_c$, Eq.(8) is written by[3]

$$Y(A)=CA^{-2.0}\exp\,[-18.4\,(1-T/T_c)^2A^{2/3}\,/\,T] \tag{9}$$

For $T>T_c$, $F_A^s(T)$ is negledible, and the equation becomes

$$Y(A)=CA^{-2.0}\exp\,[-9.0\,(1-T/T_c)^2A\,/\,T] \tag{10}$$

For the fitting to the experimental data, the quantities of the temperature T and the amplitude C are treated as adjustable parameters. The experimental data[13-15] are shown in Figs. 5 and 6 with the fitting results plotted by the Monte Carlo sampling. The parameters obtained are listed in Table 3.

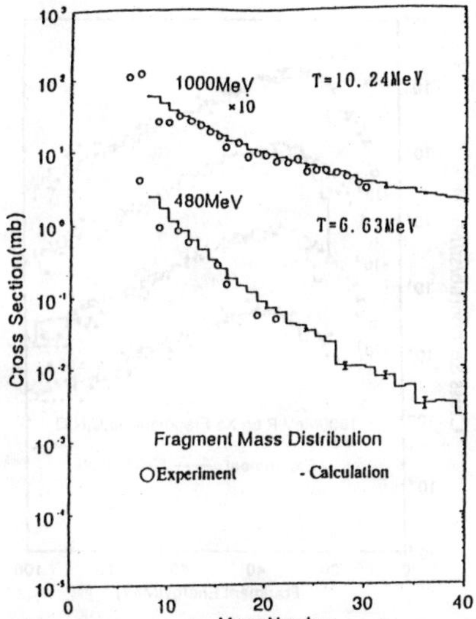

Fig. 5. Fragment mass yields for silver target.[13,15] Lines are sampled by Eq. (8).

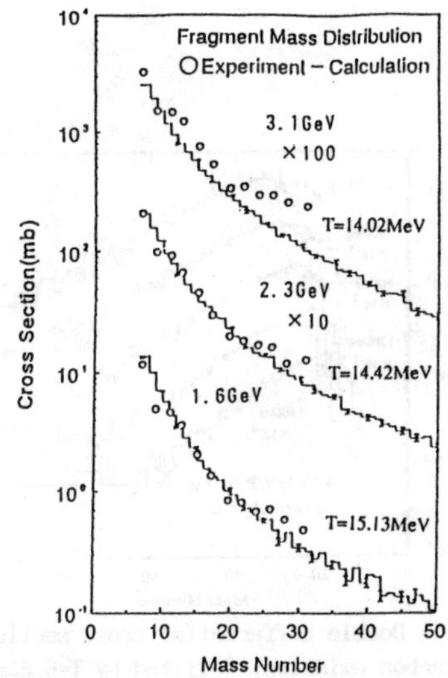

Fig. 6. Fragment mass yields for xenon target.[14] Lines are sampled by Eq. (8).

3.2 Double Differential Cross Sections by the Moving Source Model

To reproduce the angular distribution of the fragment energy spectra, the moving source model is assumed to hold good, where fragments are considered to be isotropically emitted with a Maxwellian temperature T in a moving frame. The same value as in the mass yield evaluation is taken as the temperature T. According to the experimental indication, the moving source velocity βs is considered as a function of average fragment kinetic energy <E_{kin}>:[13]

Table 3. Parameters obtained for fragment mass yields for proton reactions.

Target nucleus	proton energy, GeV	Temperature T, MeV	Amplitude C, mb
Ag	0.48	6.63	94.8
Ag	1.0	10.3	394
Ni	1.0	12.0	105
Xe	1.6	14.0	457
Xe	2.3	14.4	929
Xe	3.1	15.1	1.14×10^{3}

Table 4. Coulomb Barrier height with the simple classical estimation and the experimental indication.

Target nucleus	Fragment nucleus	Classical value, MeV	Experimen. indic., MeV
Ag	^{7}Li	24.1	26.0
	^{12}C	43.0	38.0
	^{15}N	48.0	39.0
	^{24}Na	69.3	42.0
Xe	^{9}Be	34.4	28.0
	^{12}C	47.9	32.0
	^{15}N	53.7	35.0
	^{16}O	60.0	38.0

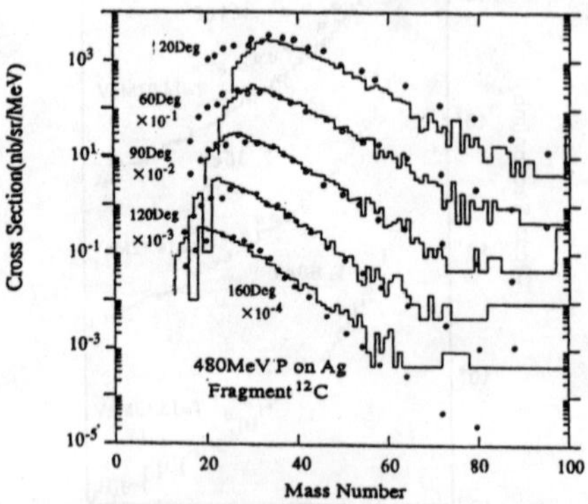

Fig. 7. Double differential cross section for carbon emission,[13] fitted by T=6.63MeV.

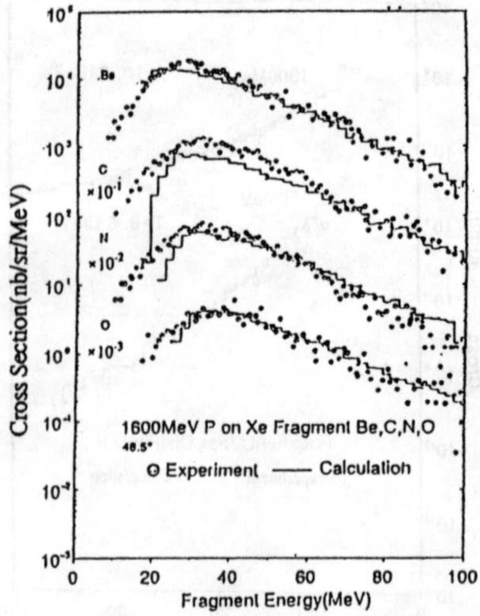

Fig. 8. Double differential cross section for fragments,[14] fitted by T=15.1MeV.

$$\beta_s = a+b\sqrt{1 - \frac{m_f^2}{(\langle E_{kin}\rangle+m_f)^2}} \tag{11}$$

where m_f is the fragment mass (MeV) and Ekin the fragment kinetic energy(MeV). The parameters are evaluated as $a=1.38E_{kin}^{-3}$ and $b=1.34E_{kin}^{-1}$. The Coulomb barrier height was checked as listed in Table 4. The expansion of the nuclear volume at the high excitation results in a decrease of Coulomb barrier height in comparison to the simple classical estimation that is based on the normal values of nuclear densities and radii,-i.e. the ground states. For simplicity in this study, the coulomb barrier height is taken to be half of the simple estimation, and to have a standard deviation of 3 and 6 MeV for silver and xenon data, respectively. The barrier penetrability was calculated by the WKB approximation. The double differential cross sections of fragment emission[13,14] are shown in Figs. 7 and 8, where the experimental data are reproduced in an acceptable manner.

4. Conclusion

The use of the moving source model with three reaction processes reproduced the experimental neutron spectra in the incident proton energy region of 600 to 800 MeV. At energies below 400 MeV, the introduction of the Watt moving source model for the cascade process was found to lead to good agreement with the experimental spectra. These moving source models were found to be applicable to the nuclear data evaluation for a wide range of target masses and incident energies. For the fragmentation cross section, the liquid-gas phase-transition model based on the nuclear temperature was useful in reproducing the mass yield distribution. This temperature is applicable to representing the double differential cross sections by the use of the moving source model.

5. Acknowledgements

The authors gratefully acknowledge Dr. H. Yasuda of Japan Atomic Energy Research Institute for his continued encouragements.

6. References

1. S.Pearlstein, *Nucl.Sci.Eng.* **95** (1987) 116.
2. T.-A.Shibata et al., *Nucl.Phys.* **A408** (1983) 525.
3. A.D.Panagioutou et al., *Phys.Rev.Lett.* **52** (1984) 496, and *Phys.Rev.* **C31**(1985) 55.
4. J.Aichelin et al., *Phys.Rev.* **C37** (1988) 2451.
5. K.Ishibashi et al., *J.Nucl.Sci.Tech.* **29** (1992) 499
6. S.Cierjacks et al., *Phys.Rev.* **C36** (1987) 1976.
7. M.M.Meier et al., *Nucl.Sci.Eng.* **102** (1989) 310.
8. J.Terrell, *Phys.Rev.* **113** (1957) 527.
9. M.Trabandt et al., *Phys.Rev.* **C39** (1989) 452.
10. M.M.Meier et al., *LA-11656-MS (1989)*, and *Nucl.Sci.Eng.* **110** (1992) 289.
11. M.M.Meier et al., *Rad.Effect.* **96** (1986) 73.
12. S.D. Howe et al., *LA-UR-85-3660* (1985).
13. R.E.L. Green et al., *Phys.Rev.* **C29** (1984) 1806.
14. N.T. Porile et al., *Phys.Rev.* **C39** (1989) 1914.
15. L.N. Andronenko et al., *Phys.Lett.* **B174** (1986) 18.

CHARGED-PARTICLE AND PHOTO-NUCLEAR DATA

COMPUTATION TECHNIQUE USED TO SEARCH FOR NUCLEAR PHYSICAL CONSTANTS FOR THERMONUCLEAR REACTIONS

A.G. Zvenigorodskii, S.N. Abramovich, B.Ja. Guzhovskii, V.A. Zherebcov, L.M. Lazarev

All-Russian Scientifican Research Institute of Experimental Physics, 607200, Arzamas-16, Nizhni-Novgorod region, Russia

1. Introduction

The works in evaluation of nuclear physical constants in light nuclei ($z \leq 5$) interactions are justified by the fundamental role the latter play in the nucleosynthes in the Universe and by the role which is not less important for humanity and which they should play in the development of new thermonuclear energetics. The astrophysical aspects of nuclear data applications being vastly reported about in reviews and monographs[1] the "Earth" aspects of their applications in thermonuclear fusion problems and in thermonuclear device computations are not completely generalized. Data on nuclear-physical constants (NPC) for light nuclei interactions are published in numerous specialized scientific journals and issuies which are not easy accessible for the majority of potential users; moreover, the experimental data on interaction cross-sections are often unfit for direct applications in computations as they are presented as separate points of incident particle energies or of nuclei-products angles. In the majority of cases the NPC's may be recommended for applications only after estimation of all experimental data, including the computed cross sections, obtained with the known models of nuclear reactions.

2. Compilation and Pretreatment of Nuclear Data

At the first step of compilation and estimation of nuclear-physical data for thermonuclear applications we limited ourselves by data on reaction cross-sections: total cross-sections, reaction velocities for equilibrium plasma; data on elastic and and non-elastic scattering.

The bibliographic searching was based on F.Ajzenberg issues[2], charged particle bibliography by Brookhaven National Laboratory[3]. In contrast to neutron data for which the international cooperation in compilation and estimation exists for a long time and a vast library in EXFOR format is accumulated[4], the works in charged particle data compilation and estimation were done by separate and scarcely connecting groups.

In the field of interest by now there are two substantial compilations. The Evaluated Charge Particle Library of Livermore Lawrence National Laboratory[5], (ECPL), contains the estimated data for (p,d,t,τ,α)-particles interacting with targets beginning with hydrogen to ^{16}O, and DATLIB library[6] dealing with the same nuclei and particles but concentrating more on the compilation step than on estimation.

Table 1

Brief descrption of information stored in Experimental Data Library

Reaction	Energy MeV	Number /Date	Reaction	Energy MeV	Number /Date
^{2}H(p,γ)	.02-45	9/62-84	^{6}Li(t,2n)	4.8-12	2/60-62
^{2}H(p,n+p)	.08-19	10/61-78	^{6}Li(τ,n)	3.-25	3/61-75
^{2}H(p,EL)	.8-24	24/51-88	^{6}Li(τ,p)	.5-17	9/56-90
^{2}H(d,γ)	.03-19	15/69-91	^{6}Li(τ,d)	.4-18	5/68-81
^{2}H(d,n+p)	Eth-20	3/56-62	^{6}Li(τ,n)	.7-2	1/ 80
^{2}H(d,n)	.01-25	44/54-90	^{7}Li(p,γ)	.3-15	18/54-91
^{2}H(d,p)	.01-14	31/51-90	^{7}Li(p,n)	Eth-26	24/55-82
^{2}H(d,EL)	.1-13	15/52-85	^{7}Li(p,p)	.2-12	17/51-82
^{3}H(p,γ)	.1-30	8/53-83	^{7}Li(p,α)	.02-12	21/54-89
^{3}H(p,n)	Eth-32	23/53-82	^{7}Li(d,p)	.28-19	20/52-86
^{3}H(p,d)	Eth-14	2/71-74	^{7}Li(d,n)	.13-19	11/52-74
^{3}H(p,EL)	.03-14	15/51-71	^{7}Li(d,t)	1.4-15	5/55-90
^{3}H(d,γ)	.03-20	6/59-86	^{7}Li(d,d)	.4-15	9/55-76
^{3}H(d,n)	.02-19	27/52-87	^{7}Li(d,α)	.15-3	8/63-86
^{3}H(d,EL)	.08-14	7/52-71	^{7}Li(d,n)	2.8-11	1/ 84
^{3}H(t,2n)	.02-3	8/51-85	^{7}Li(t,γ)	.5-1	1/ 78
^{3}H(t,γ)	.4-24	4/68-73	^{7}Li(t,n)	.12-2	3/60-91
^{3}He(p,EL)	.3-12	10/57-80	^{7}Li(t,p)	3.3-15	3/71-91
^{3}He(d,γ)	.03-17	4/54-85	^{7}Li(t,α)	.04-22	8/54-83
^{3}He(d,p)	.04-20	19/51-90	^{7}Li(t,n)	.3-12	5/51-84
^{3}He(d,EL)	.3-20	11/54-81	^{7}Li(τ,n)	.4-15	4/62-66
^{3}He(t,n)^{5}Li	.4-26	5/63-75	^{7}Li(τ,p)	1.0-16	9/66-90
^{3}He(t,d)	.1-1.5	6/53-73	^{7}Li(τ,t)	1.7-6	3/63-78
^{3}He(t,EL)	4.2-21	3/68-71	^{7}Li(τ,α)	.8-12	5/65-70
^{3}He(t,x)^{1}N	.1-1.9	2/60-67	^{7}Li(α,n)	4.1-11	8/56-84
^{3}He(τ,γ)	.9-27	2/67-74	^{9}Be(p,γ)	.1-2	5/57-73
^{3}He(τ,2p)	.03-9	6/66-87	^{9}Be(p,n)	Eth-29	15/50-76
^{3}He(τ,EL)	3.-19	4/67-68	^{9}Be(p,p)	1.7-15	28/56-83
^{3}He(α,γ)	.1-4	11/59-88	^{9}Be(p,d)	.3-12	18/56-81
^{4}He(p,EL)	.5-18	21/54-77	^{9}Be(p,α)	.1-12	12/51-77
^{4}He(d,EL)	.9-27	20/54-83	^{9}Be(d,γ)	.4-12	7/63-84
^{4}He(t,γ)	.1-2	5/59-87	^{9}Be(d,n)	.5-16	19/57-88
^{6}Li(p,γ)	8.-18	4/79-87	^{9}Be(d,p)	.1-23	37/52-89
^{6}Li(p,p)	2.-14	11/63-74	^{9}Be(d,d)	.2-14	14/56-83
^{6}Li(p,τ)	.05-18	28/56-89	^{9}Be(d,t)	.1-18	28/52-89
^{6}Li(d,n)	.02-17	21/52-84	^{9}Be(d,α)	.1-14	17/57-80
^{6}Li(d,p)	.01-12	35/52-84	^{9}Be(t,n)	.6-9	5/61-77
^{6}Li(d,t)	.3-15	3/55-67	^{9}Be(t,α)	.5-6	4/68-85
^{6}Li(d,α)	.03-15	33/54-89	^{9}Be(τ,γ)	1.0-20	4/64-74
^{6}Li(d,d)	1.8-15	10/55-76	^{9}Be(τ,n)	.9-27	11/62-84
^{6}Li(t,d)	1.5-10	2/61-84	^{9}Be(τ,p)	1.8-14	9/56-77
^{6}Li(t,p)	.3-19	2/80-84	^{9}Be(τ,d)	3.6-10	3/60-67
^{6}Li(t,n)	.1-2	6/61-86	^{9}Be(τ,α)	1.3-18	7/63-78

* - Number/Date is number of publication and date of publications.

We decided to gather together the experimental data from numerous publications and to translate them, presented as rule in the graphic form, into the computer media in the format suitable for the subsequent works. As an initial presentation, EXFOR format was chosen as for latter there were the language concepts and decent software by that time. To accomplish graphic information translation into computer representation, the corresponding hardware-software complex was developed[7]. Table 1 presents brief characteristics of the experimental information stored in our library.

3.Approximation Functions for Evaluated Data

To chose the method of data estimation we proceed from the following assumption. This method must:

1, provide a possibility of using the statistical procedure which takes into account random and systematic errors in common process;

2, give opportunity to obtain evaluated curve in a form which is convenient for machine representation;

3, permit in essential manner to join the different parts of estimated curve (for reactions with charghed particles it is especially important as low-energy part of the cross section excitation curvies is usually described by analytical function);

4, simplify to obtain integral characteristics,for example,reaction raties.

As for the majority of interaction cross sections of the charged particles with light nuclei the excitation functions are characterized by gradual dependence of cross section on energy, it seems natural to chose a spline-polynome as an approximation function.

The method of approximation spline-curve presentation is based on the model described in ref.[8]. The parameters of the approximation curve are found out from the condition of minimum of a weighted sum of the squares for experimental points deviation from the approximation curve:

$$MIN\,\chi^2=\sum_{i=1}^{N} W_i(y_i^{\exp}-y_i^{apr})^2 \qquad (3.1)$$

where y_i^{exp}, y_i^{apr} - experimental and aproximate values in x_i-points.

It is supposed that x_i is exactly known and y_i^{exp} is known, but with a certain error σ_i (standart deviation) and $w_i=1/\sigma_i^2$.

In our case as it was earlier mentioned a polynomial spline of degree (p=0,1,2,3) is used. Usually splines of 1 defect are used when in grid nodes all derivatives of to p-1 order are contineous. However, one has frequently enough to use less smooth functions (e.g. due to abrupt leaps or steep bendings in the evaluation curve). Our technique developed allows to employ the splines of any defect, not necessarily of one and the same in all nodes, and the need of enlargement is revealed mostly automatically when the nodes are being fitted.

As it became evident from the practice the most suitable are the cubic splines (p=3) but in some specific problems other degrees are also required. The technigue applied allows in particular to approximate by broken lines (p=1) and step functions (p=0).

$$y^{apr}(x)=\sum_{1}^{N} a_l \cdot Q_l^{(k)}(x), \qquad (3.2)$$

where N - space dimensions, where the spline is defined; $Q_l^{(k)}(x)$- a basis spline with l number and k defect. Basis spline $Q_l^{(k)}(x)$ with k defect in x_l node is defined by the expression:

$$Q_l^{(k)}(x)=\sum_{j}^{p+2-k} \frac{(x_{l+j}-x)_+^p}{(x_{l+j}-x_l)^{k-1}\cdot \prod\limits_{m=0,m\neq j}^{p+2-k}(x_{l+j}-x_{l-m})} \text{ for } x>0,$$

$$Q_l^{(k)}(x)=0 \text{ for } x<0,$$

$$(x_{l+j}-x)_+^p=(x_{l+j}-x)^p \text{ for } x_{l+j}>x,$$

$$(x_{l+j}-x)_+^p=0 \text{ for } x_{l+j}<x \qquad (3.3)$$

To define (3.2) spline, coefficients a_l and nodes x_l will be unknown. With the nodes fixed the spline coefficients can be easily determined: the problem lies only in solution of the system of linear equations. In ref.[8] the statistic model oriented onto nuclear experiment data treatment is described in detail. This model gives the possibility to obtain the evaluation curve according to the experiment results of different authors with account of accidental and systematic errors. In this case it was assumed that accidental errors of each author and systematic errors of a group of authors are distributed according to the normal Law with a zero mathematic expectation. As the systematic errors are unknown, as a rule, we were to determine them in the expertise of each experimental work. This systematic error was taken as the first approximation. The final value of the systematic error for the data of the individual author was determined considering the data set of various authors.

In the known approximation methods by spline usually it makes the limits leaving the subjective considerations for the node selection. To exclude the subjectiveness in x_l node grid determination, one must define x_l values also out of the condition of minimum (3.1). In this case the problem becomes substantially non-linear. We solved the non-linear problem by Newton method to find the numerical minimum. Only x_l-values are found approximately, the coefficients a_l at the specified nodes are always determined accurately, i.e. in fact Newton method adequately serves to solve the minimization problem for $\chi^2(x)=\min \chi^2(a,x)$ function. The partial derivatives needed for this method are calculated analytically. Their calculation occurs to be rather cheap due to the limits of basis spline carriers. Dramatically bigger expenses of computer time for node optimization are justified by substantial increasing of fitting quality. In this case the description accuracy becomes higher with less nodes, the estimation objectiveness increases due to reduction of the subjective factor in node position selection. The node number decrease allows to avoid physically ungrounded curve oscillations in many cases.

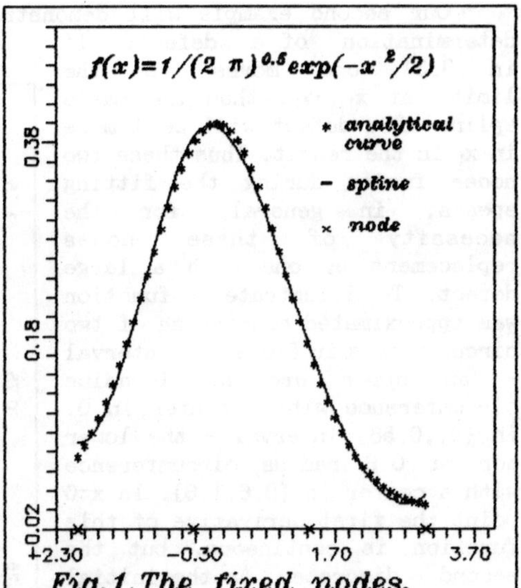

Fig.1.The fixed nodes.

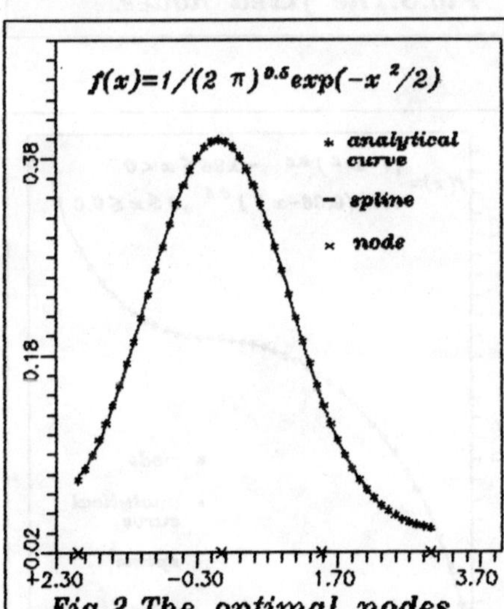

Fig.2.The optimal nodes.

To illustrate the effect on node fitting on the approximation quality we shall give a simple example. Truelly speaking there can´t be an objective criterion for comparison of "fixed" and "optimal" nodes, as any optimal position can be announced fixed. So, let us choose a rather neutral function without vividly expressed peculiarities and compare the uniformed placed nodes and optimal ones. The Gauss peak part is approximated, the result with the fixed nodes is shown in fig.1, and with optimal - in fig.2.Spline nodes are marked by crosses in x-axis. These illustration speak for themselves. One must also pay attention that the optimal position does not dramatically differ from the uniform one, so one can state that it might have been hardly foreseen rather than a inform one.

Our second example will demonstrate the possibility of automatic determination of a defect. If in (3.3) one moves to the limit at $x_{l+1} \to x_l$, than the basis spline of a defect will be 1 more in x_l in the result. Thus these two nodes fusion during the fitting speaks, in general, for the necessity of these nodes replacement by one with a large defect. To illustrate a function was approximated consisting of two circular arcs in {-0.94,0} interval - an upper arc of 1-radius circumference with a center in 0. in {0.,0.58} interval - the lower acr of 0.6 radius circumference with a center in {0.6,1.6}. In x=0 point the first derivative of this function is contineous, but the second - discrete. As the initial approximation the uniform node position was chosen. Fig.3 shows the spline with these nodes. During fitting the third and the fourth nodes fused. Then they were replaced by one node where the spline defect was increased for 1 and the fitting was continued. The result is in fig.4. This node appeared in 0.000103 point, i.e. the break point for the second derivative being practically invisible according to the starting data was determined rather precisely. Here, one can also state that without the method described one could hardly manage to identify such node positions and conclude on the degree on function smoothness. The examples considered allow to conclude that the expenses on the nodes fitting are fully justified. They are incomparably smaller than the expenses on the achievement of the experimental information and drastically increase the estimation quality.

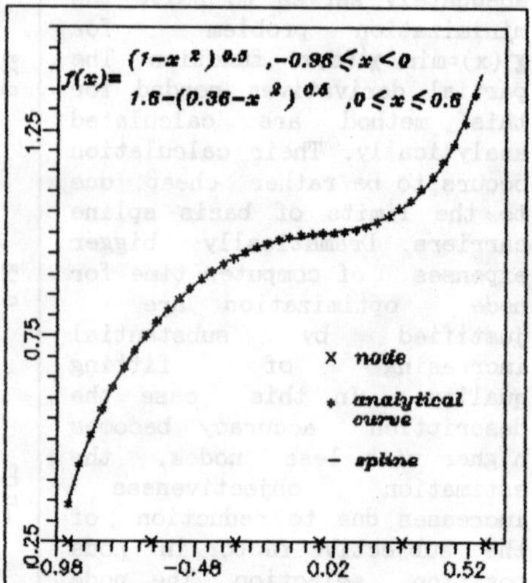

Fig.3.The fixed nodes.

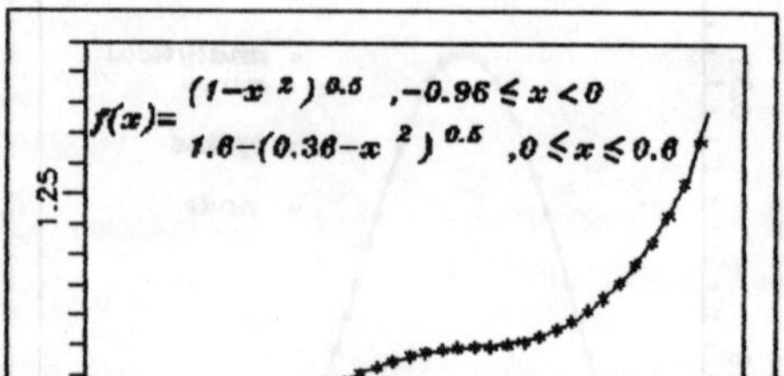

Fig.4.The optimal nodes. The nodes 3 and 4 fused.

To obtain a spline-curve for practical applications the spline presentation by formula (3.2) isn't quite suitable due to the subsequent huge calculations with a_l-coefficient. Far more convenient to obtain the coefficients via simple polynome of p-degree in each $[x_l, x_{l+1}]$ grid interval according to $(x-x_l)$ degrees.

A certain interest is in thermonuclear reaction velocities. The attemps to present the reaction velocities by analytical or partial-analytical functions have been done not once[9-11], however the description precision is substantially lower than the presition of the obtained <σv> values in a wide temperature (0.0001-1 MeV) range even for the most successful work[9]. Thus, we have searched for the approximation suited for computer application and allowing to preserve the precision spline-approximation for Maxwell
S-factor, <S>, bound with <σv> relationship[12] was used:

$$\langle\sigma v\rangle = 8\hbar(3^{5/2}\pi Z_1 Z_2 e^2 M)\langle S\rangle[1+(5/12)\tau]\tau^2\exp(-\tau)$$

$$\tau = 3(2kT)^{-1/3}(\pi Z_1 Z_2 e^2 M^{1/2}\hbar^{-1})^{2/3} \qquad (3.4)$$

where Z_1, Z_2 - the atomic numbers of colliding nuclei.
M=m1*m2/(m1+m2) - their reduced mass in carbon units.

The calculated according to this relations <S> values for the reactions of interest were approximated[13] with the third-order spline-functions in the following way. First, for each reaction the search for optimal node positions was determined by this method minding their minimal number at the specified maximal spline-description deviation from the tabular presentation which does not exceed 1%. Table 3 shows this search results for thermonuclear reaction $^3H(d,n)^4He$. One can calculate <S> values out of table (3) according to the expression (3.5):

Table 3

Spline-Coefficients of S-factor Descriptions for $^3H(d,n)^4He$ reaction.

Node j Number	Energy kT_j, MeV	Spline Coefficients B_0	B_1	B_2	B_3
1.	0.0001	12.119	0.46066	-0.47387	0.20145
2.	0.00522	18.994	6.1672	1.9166	-5.0665
3.	0.01158	22.562	-0.41778	-10.186	3.1819
4.	0.05001	10.111	-9.7910	3.7801	-0.54457
5.	0.31210	1.5133	-1.4253	0.78847	-0.18110
6.	1.	0.6368	0.0	0.0	0.0

where B_m - spline-function coefficients corresponding to j-node;
kT^j - temperatures for j-node, MeV;
kT - current temperature corresponding to the following condition: $kT_j \le kT \le kT_{j+1}$

$$\langle S\rangle = \sum_{m=0}^{3} B_m \cdot \ln(kT/kT_j), \quad kT_j < kT < kT_{j+1}. \quad (3.5)$$

Then one node set for all main reactions (this is the requirement of computation program) was searched for under the condition that the maximal deviation from the tabular values would not exceed 1%. For the considered reaction range we managed to achieve this at 10 nodes.

4. Interaction Cross Section Extrapolation in Zero Energy Region.

In the problem of astrophysics and controlled thermonuclear fusion the computation of average velocities of thermonuclear reactions is based on the knowledge of light nuclei reaction cross section in the range from zero to some MeV's.

The cross sections can be experimentally measured from 10 keV and more. It seems impossible to go down to the zero energy region due to the exponential cross section reduction caused by Coulomb repulsive forces. In the energy region inaccesible for measurements the reaction cross sections can be obtained via theoretical extrapolation of the experimental data. The extrapolation of cross sections for scattering or reactions demands the account of channelling of nuclear processes and resonance states for a compound nucleus. R-matrix theory[14] can not help in solution of this task as it suits for solution of one-channel and one-resonance problems. Last 20 years Feshbach theory[15] is extensively applied in the calculations of light nuclei reaction cross sections and states. The resonating group technique (RGT) being an individual case of Feshbach theory made the calculations of light nuclei cross sections in zero energy region possible. The RGT calculation results depend on functionbasis choice (on the nucleus model choice) over which the wave function system is expanded into series[16].

The modelless theory seems to be one which will give a more reliable extrapolation. In the report presented at this conference one of the possible approaches of the modelless theory is suggested based on Feshbach theory. It consists of the cross section presentation in kR,1 energy region in the form of analytical energy function with a constant parameter set having a definite physical sense[17]. The parameter values are estimated by the leastsquare method (LSM) with a theoretical approximation of the experimental data in the energy region where the analytical function cross section presentation is just. The cross section is extrapolated into E=0 region with the theoretical function having parameters calculated by LSM. The validity of theoretical presentation in the form of a series according energy degrees grows when the energy reaches zero. Thus, the extrapolation accuracy is determined, mainly, by the accuracy of theory parameter calculations. When the compound nucleus resonances are present the problem of parameter calculations with LSM is not linear. This fact and the presence of the experimental errors lead to parameter correlation and problem complication. As a measure for the correctness of parameter determination the limitations on the parameters imposed by the unitarity of reaction matrix can serve.

The noticeable deviations from them will testify for insufficient number of parameters in the theoretical approximating function (insufficient number of decomposition terms into series according to energy degrees).

As an example the calculation of differential S-factor[17] is given for $^7Li(t,\alpha)^6He$ reaction done on the theoretical approximation of the differential cross section measured in ref.[18] at 90 angle in energy range E_t=80-500 keV. In this energy region two resonances of the compound nucleus take place: 1^- with parameters E_r=80.7 keV, Γ=22.7 keV and 2^+ with parameters E_r=181.0 keV, Γ =37.4 keV. The energy dependence of differential S-factor showes on the fig. 5. The extrapolation into $E=10^{-7}$ MeV gives: $S(E,90^0)$=6.428 mb*MeV/sr. This value is correct at $E\to0$ if the resonance and threshold states are absent in zero energy region.

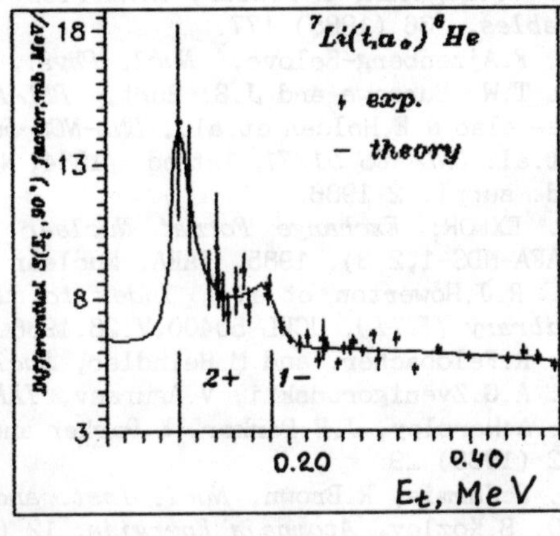

Figure 5.

5. Conclusion

Today, there are independent libraries of experimental and evaluated data for cross-sections of reactions and scattering which are of interest for thermonuclear applications. They are the library of Livermore National Laboratory ECPL-86 and nuclear data library of Russian Nuclear Center EDLF-90. The first comparison of the evaluated data obtained by russian and american researchers has shown that a good agreement is evident for the cross-sections of the main thermonuclear reactions but it is not the same for the reaction velocities and the cause seems to be only in the mathematical treatment of original data. The greater discrepancy ought to be expected in the reaction velocity presentation for non-equilibrium plasma. Today, we are inclining to the idea that to approximate the reaction velocities of the non-equilibrium plasma it is convenient to apply a two-dimensional spline obtained at the expense of uni-dimensional section unification. A convenient presentation of evaluated cross section dependences for charged particles in the form of the tables of the corresponding spline coefficients makes desirable to introduce this presentation into the accepted in the world format for the evaluated data ENDF-VI

6.References

1. G.Caughlan, A.Powler, M.Harries et.al., *Atom. Data and Nucl. Data Tables,* 36 (1985) 177.
2. F.Ajzenberg-Selove, *Nucl. Phys.*, A490 (1988) 1.
3. T.W. Burrows and J.S. Burt, *BNL-NCS-50640*,4th ed, suppl. 1980; see also N.E.Holden et.al., *BNL-NCS-50640*, 4th ed. sup. 2 1982; N.E.Holden et.al. *BNL-NCS-51771*, 1st ed. 1984; N.E.Holden et.al., *BNL-NCS-51771*, 1st ed. suppl. 2 1986.
4. EXFOR: *Exchange Format Nuclear Data Library* (Documentation series IAEA-NDS-1,2,3), 1985, IAEA, Nuclear Data Section, Vienna.
5. R.J.Howerton et.al, *Index to the LLNL Evaluated Charged Particle Library (ECPL)*, UCRL-50400,V.28,1986.
6. R.Feldbacher, and M.Heindler, *Nucl. Inst. and Meth.*, A271 (1988) 55.
7. A.G.Zvenigorodskii, V.Agureev, *IAEA-INDC(CCP)* 222/L (1984) 31-41.
8. A.Horsley, J.B.Parker, K.Parker and J.A.Price, *Nucl Inst. and Methods,* 62 (1968) 29.
9. N.Jarmie, R.Brown, *Nucl. Inst. and Meth.*, B10/11 (1985) 405-410.
10. B.Kozlov, *Atomnaja Energija,* 12 (1962) 238.
11. L.Hively, *Nucl. Fusion,* 17 (1977) 873.
12. N.Jarmie, R.Brown, R.Hardekopf, *Phys. Rev.*, C29 (1984) 2031.
13. S.Abramovich, B.Guzhovskii, S.Dunaeva, *in Nuclear Data for Science and Technology.* Proc. of Intern. Conference, ed S.M.Qaim (Julich, 1992)
14. A.M.Lane and P.G.Thomas, *Rev. Mod. Phys.*, 30 (1958) 257.
15. H.Feshbach, *Ann. Phys.*, 5(1958)357.
16. S.Abramovich,B.Guzhovskii,L.Lazarev,*Particles and Nuclei*, 23(1992)305.
17. L.Lazarev, *Ukrainskij Fizicheskij Zhurnal*, 36 (1989) 661-667.
18. S.Abramovich, L.A.Morkin, V.I.Serov, *Izv. ANSSSR ser. fiz.*,51(1987)930.

CHARGED-PARTICLE EVALUATION METHODOLOGY AT LLNL

DAVID A. RESLER and ROGER M. WHITE
Nuclear Data Group, Physics Department
Lawrence Livermore National Laboratory
P.O. Box 808, L-298
Livermore, CA 94550, USA

ABSTRACT

We recently completed charged-particle evaluations for reactions of hydrogen and helium important to fusion applications. In this paper we emphasize the methods developed which are unique to the evaluation of low-energy charged-particle data. In addition, we discuss general methods we developed to eliminate some of the tedious aspects of data evaluation such as the extraction of numerical values from graphs when tabulated experimental data are unavailable. We also discuss software tools developed for visualization of experimental data and to construct curves using a computer mouse or a variety of least-squares techniques ranging from simple polynomial and cubic-spline fits to R-matrix analyses.

1. Introduction

In this paper we discuss the methods developed to evaluate the fusion cross sections for the ^{2}H(d,n)^{3}He, ^{2}H(d,p)^{3}H, ^{3}H(t,2n)^{4}He, ^{3}H(d,n)^{4}He, and ^{3}He(d,p)^{4}He reactions.[1] Evaluation of these charged-particle reactions requires techniques not typically encountered in performing evaluations of neutron-induced reactions. We also discuss some techniques developed for these evaluations which have relevance to neutron-induced reactions as well.

For low-energy charged-particle reactions, the cross section decreases many orders of magnitude per order of magnitude decrease in energy. In stark contrast to the behavior of neutron-induced reactions near threshold, these low-energy charged-particle cross sections are of utmost importance to fusion applications. In particular, to calculate the Maxwellian-averaged reaction rate for the ^{2}H(d,p)^{3}H reaction at a plasma temperature of 0.1 keV, the cross section must be known only for $E_{d(lab)}$ between 1 and 10 keV, energies lower than the available data (see Fig. 1). Therefore, an accurate extrapolation must be made to low energies where the cross section is changing by orders of magnitude over a small energy interval.

To make this extrapolation manageable, the cross section is rewritten in a different form. The low-energy charged-particle cross section for an exothermic reaction can be written as

$$\sigma(E) = S(E)\frac{1}{E}exp(-C/\sqrt{E}) , \qquad (1)$$

where E is the incident projectile energy (lab) and C is a constant which depends on the charges and masses of the incident and target charged particles. This constant can be written

$$C = \frac{2\sqrt{2}\pi^2 e^2}{h} \cdot Z_1 Z_2 \cdot \sqrt{\frac{m_1 m_2}{m_1 + m_2}} = 0.9895139 \cdot Z_1 Z_2 \cdot \sqrt{\frac{m_1 m_2}{m_1 + m_2}} \qquad (2)$$

where the masses are given in mass units, u. The function $S(E)$ is called the astrophysical S-factor. It represents the nuclear part of the reaction cross section and is, in general, a slowly varying function of energy. In Fig. 2, we show the astrophysical S-factor for the same reaction cross section data and evaluation shown in Fig. 1. It is apparent that by converting the cross section to the S-factor, the extrapolation to low-energy can be made much more accurately.

2. Data Compilation and Standardization

2.1. Scanning and Digitization Tools

The references used for our charged-particle evaluations date back to 1946.[1] Some authors gave tabulated results for their experimental information while others provided only graphical information. For many works containing only plots, it was not possible to obtain the tabulated data.

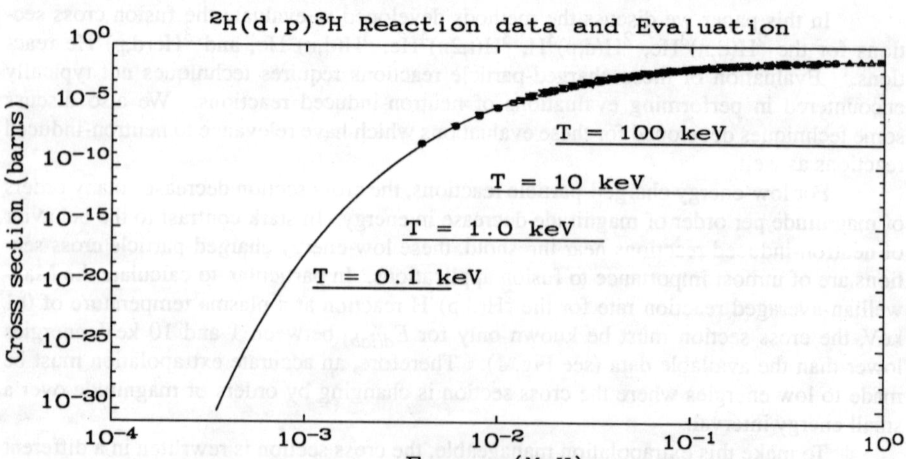

Figure 1: Total reaction cross section for the $^2H(d,p)^3H$ reaction. The experimental values are plotted as symbols and our evaluation is the curved solid line. The horizontal lines indicate the energy range over which the cross section must be known for calculating Maxwellian-averaged reaction rates for plasma temperatures of 0.1, 1.0, 10, and 100 keV.

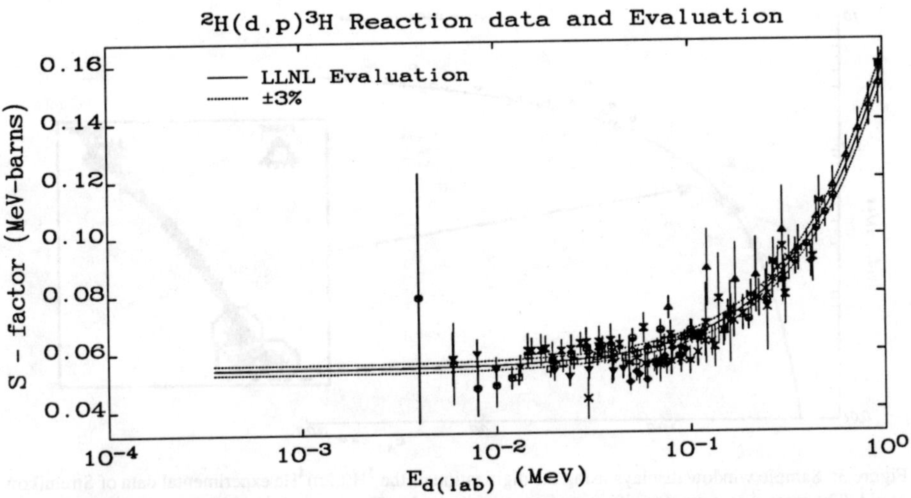

Figure 2: Data and evaluation of the $^2H(d,p)^3H$ cross section shown in Fig. 1 converted to astrophysical S-factor using Eqs. (1) and (2). Notice that the S-factor is nearly constant at the lower energies, whereas the cross section changes by orders of magnitude.

Using a commercially available scanner and software package, we scanned these graphical images into our computer. We developed a digitization software package[2] that takes a bitmap image of the graph and places it into an X-windows display. Using the computer mouse and keyboard, the user can define and calibrate the axes of the graph. The software package corrects for rotations of the graph and for any nonorthogonality of the axes. As shown in Fig. 3, once the axes are set, the code gives a real-time display of the present position of the mouse cursor in the units of the graph. By placing the cursor over a point and clicking the mouse button, the (x,y) values of that point are recorded in a file. The digitization code allows the user to magnify any part of the graph so that the selection of points becomes as accurate as the original placement of the points on the graph.

2.2. *Common Data Format*

All charged-particle evaluation work was performed on UNIX workstations. We adopted a simple two-(x,y) or three-(x,y,Δy) column, space-delimited format for the total integrated reaction cross section files. In our compilations of charged-particle cross sections, we found the experimental data presented in many different forms. In all cases, we transformed the experimental data to common units, i.e., energies in MeV and cross sections in barns. Some experimental data were given as differential cross sections which we integrated. Depending on the quality and number of differential cross sections, the fitting of an angular distribution to a Legendre polynomial expansion can be sometimes difficult.

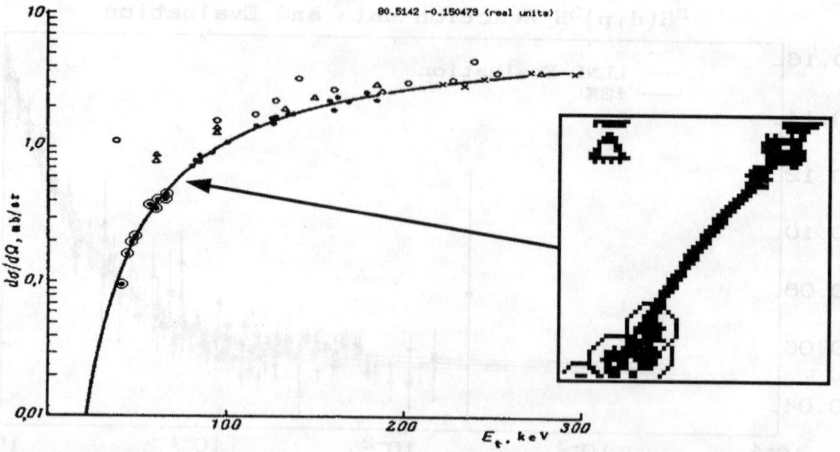

Figure 3: Sample window displays using the digitization of the $^3H(t,2n)^4He$ experimental data of Strelnikov *et al.*[1] The inset shows the magnification feature of the code. The data points which have been circled indicate those points which have already been digitized. Note the live time display of the coordinates of the mouse cursor at the top of the plot. The Y value displayed is $\log_{10}$ of the actual Y value.

For data sets with large errors and/or limited ranges in angular coverage, particular care must be used to pick the correct number of Legendre polynomials in the fit. The primary guide to determining the optimum number of Legendre polynomials to use is the overall chi-square value. In addition, we observe whether or not the value of the B_0 coefficient (where $\sigma_{int} = 4\pi B_0$) changes significantly with the addition of more polynomials. Graphically, we check the resulting fit to make sure there are no pathologies, i.e, no negative cross sections nor extremely large back angle cross sections. Finally, we check that the maximum number of polynomials used is a smoothly varying function of energy as are the coefficients of the fit.

3. Evaluation Tools

3.1. Graphics

For evaluation work, we have developed a highly-interactive, general-purpose graphics code, QPX, for making quick plots in an X-windows environment. QPX is written mostly in FORTRAN-77 with the graphics interface written in C, making the code easily transportable to other workstation environments. For output, PostScript files are produced and sent directly to a laser printer or imported into documents such as this.

Data sets are easily loaded into QPX (currently dimensioned for up to 150 data sets of 50,000 points each). The user can make plots showing all or only subsets of the data, change the limits of the axes, change from an (x,y) plot to a (log x, y) plot, or chose from many other options. Using the computer mouse to move a graphics cursor on the display screen, the evaluator can define knots of a cubic-spline curve. In this manner, it is fast and easy to draw a smoothed curve through multiple data sets. The left mouse button places a knot at the location of the mouse cursor and the middle mouse button removes the knot closest to the location of the mouse cursor. Each time, the resulting cubic-spline curve is redrawn on the screen without removing the previous curves which allows the user to see exactly how the curve changes each time a knot is added or removed. The right mouse button is used to refresh the screen to show only the last curve drawn. After determining the "best" cubic-spline curve through the data sets, the user can enter a command to output the knots of the cubic-spline to a file.

3.2. Data Manipulation Tools

The utility code CSPLINE was written to expand the knots back into a cubic-spline curve and the user can specify the number of (x,y) points to place between each of the original knots. Typically many more points are included than are actually necessary. The utility code THINER is used to thin the (x,y) points of the curve under whatever assumption of interpolation scheme (usually linear-linear) and to whatever tolerance the evaluator chooses. To transform the representation of data between cross sections and S-factors, we developed the utility codes CSGEN and SFGEN. These codes require only the names of the input and output data files and the names of the incident projectile and target. The codes determine the charges and the mass values from built-in mass tables thus reducing the potential for errors introduced by the user. The importance of these procedures is described in the next section.

The utility code RATIO can be used to take the ratio between any two data files. The code determines automatically whether the data files contain two or three columns and if the third column is present in either or both files, the errors are propagated to the output. The user can also specify the interpolation scheme to be used between the tabulated points of each input file. The output file will contain the superset of X values contained in the two input files. If the same interpolation scheme is used with both input files, then the output file can also be used with the same interpolation scheme. For example, this utility is important in trying to identify slight shape differences between a data set and an evaluation.

3.3. Importance of These Tools to Charged-Particle Evaluations

For the nonresonance charged-particle evaluations, the experimental data were transformed from cross sections to S-factors using SFGEN. The evaluation was performed by drawing (or fitting) a cubic-spline curve through the S-factor data using QPX. The utility code CSPLINE was used to fill in the points between the knots (in the S-factor representation) resulting in several thousand (x,y) values. CSGEN was then used to transform the S-factors back into cross sections. THINER was used to thin the cross section results for linear-linear interpolation to a tolerance of 0.1%. THINER required that, at a minimum, at least every other point in the cross section file be discarded to insure that the linear-linear interpolation was good to the specified tolerance. In cases where this requirement was not met, we went back to the S-factor file, included more points with CSPLINE, carried out the conversion to cross section with CSGEN, and repeated the thinning procedure. In this way, we could guarantee that the resulting cross section file could be linear-linear interpolated to the tolerance specified by the evaluator.

For all of the charged-particle reactions we evaluated, the integrated cross section files contained ~750 points and were cut off at energies below which the cross section was less than 10^{-31} barns, the limit of single-precision on many 32-bit machines. This lower energy limit is more than sufficient to calculate Maxwellian-averaged reaction rates for plasma temperatures as low as 100 eV (see Fig. 1).

4. Least-Squares Fitting Tools

For some charged-particle reactions, more analytic methods than 'by eye' were required for the evaluations. This was particularly true for the ^{3}H(d,n)^{4}He and ^{3}He(d,p)^{4}He reactions where the low-energy cross sections are dominated by resonance behavior. We developed automatic curve fitting utilities based on the least-squares technique. However, several difficulties are encountered in this approach. First, it is not always possible to determine the experimental errors for each of the measured data sets. In these cases, we made what we felt were reasonable estimates of the errors. Second, when errors were given, they were not always stated precisely as to whether they were absolute or relative. This information is important when normalizing one measurement relative to another. Third, some experimenters make mistakes and the quoted errors are not a true reflection of the actual errors. This is particularly evident in the highly discrepant data sets for the ^{3}He(d,p)^{4}He reaction shown in Fig. 4. For all these reasons, we used the total error (as stated or estimated) in our fitting of integrated cross sections. (The exception was in the fitting of angular distributions where we estimated the split of the total error into relative and absolute parts in all cases.)

To automate the fitting procedure and to keep track of the large amount of experimental data, we chose a simple format (see Fig. 5) where each data set was placed into a single master file separated by header information identifying the data sets. We could easily select either individual data points or entire data sets to be excluded from the master file for a given fitting procedure. General features of our fitting utilities include the usage of this single input file with the ability to interactively pick the energy range over which the fit would be made and whether or not to let the overall normalizations of the data sets to vary.

4.1. *Simple Functions*

The utility POLYFIT fits a polynomial of specified order to a set of data. The low-energy regions of both the ^{2}H(d,n)^{3}He and ^{2}H(d,p)^{3}H reactions are nearly linear when plotted as S-factor vs. energy. Using POLYFIT, we performed linear and quadratic fits to these data using many different energy ranges and including/excluding various experimental data sets. The final least-squares results were virtually identical to the cubic-spline used in the final evaluation.

The low-energy portion of the ^{3}H(t,2n)^{4}He reaction has a minimum near $E_{t(lab)}$ =

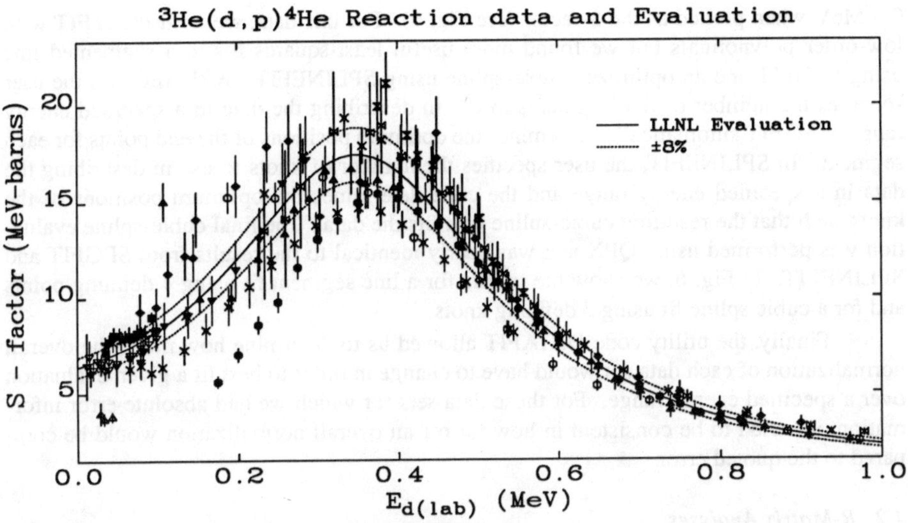

Figure 4: Plot of all experimental data for the ^{3}He(d,p)^{4}He reaction[1] in terms of the astrophysical S-factor vs. $E_{d(lab)}$ from 0 to 1.0 MeV. Notice that the data are more discrepant than the quoted errors.

```
# DATA SET: jarmie85cs
#
     2.97773E-02    5.39167E-04    2.74207E-05
     4.46661E-02    2.28516E-03    1.07321E-04
     5.98972E-02    5.44618E-03    2.78479E-04
     7.46148E-02    8.92136E-03    4.75513E-04
     8.95034E-02    1.29377E-02    6.68555E-04
     1.04050E-01    1.65896E-02    8.49873E-04
     1.13805E-01    1.85874E-02    9.30630E-04
#
# DATA SET: serovcs
#
     3.12500E-02    7.07221E-04    9.83037E-05
     3.70000E-02    1.42651E-03    1.56916E-04
     4.28000E-02    2.28614E-03    2.26328E-04
     4.68000E-02    3.00535E-03    2.01358E-04
     5.03000E-02    3.51758E-03    2.88441E-04
     5.84000E-02    4.50223E-03    3.42169E-04
     8.26000E-02    1.03608E-02    6.63088E-04
     1.09850E-01    1.61181E-02    9.99324E-04
     1.55400E-01    2.30496E-02    1.38298E-03
```

Figure 5: Part of the master data file for the ^{3}H(t,2n)^{4}He reaction.

0.3 MeV when plotted as the S-factor (see Fig. 6). For this data, we used POLYFIT with low-order polynomials but we found more useful least-squares fits to a segmented line using SEGFIT and an optimized cubic-spline using SPLINEFIT. With SEGFIT, the user specifies the number of line segments to use in describing the data in a specified energy range. SEGFIT automatically determines the optimum positions of the end points for each segment. In SPLINEFIT, the user specifies the number of knots to use in describing the data in a specified energy range and the code determines the optimum positions of the knots such that the resulting cubic-spline best fits the data. The final cubic-spline evaluation was performed using QPX and was nearly identical to the results from SEGFIT and SPLINEFIT. In Fig. 6, we show the results for a line segment fit using 4 defining points and for a cubic spline fit using 3 defining knots.

Finally, the utility code DATAFIT allowed us to determine how much the overall normalization of each data set would have to change in order to best fit a given evaluation over a specified energy range. For those data sets for which we had absolute error information, we tried to be consistent in how far off an overall normalization would be compared to the quoted error.

4.2. R-Matrix Analyses

For the resonance reactions $^3H(d,n)^4He$ and $^3He(d,p)^4He$, we performed least-squares fitting using the R-matrix formalism. Most calculations used one-level fits employing three parameters—the level energy and the primary entrance and exit channel

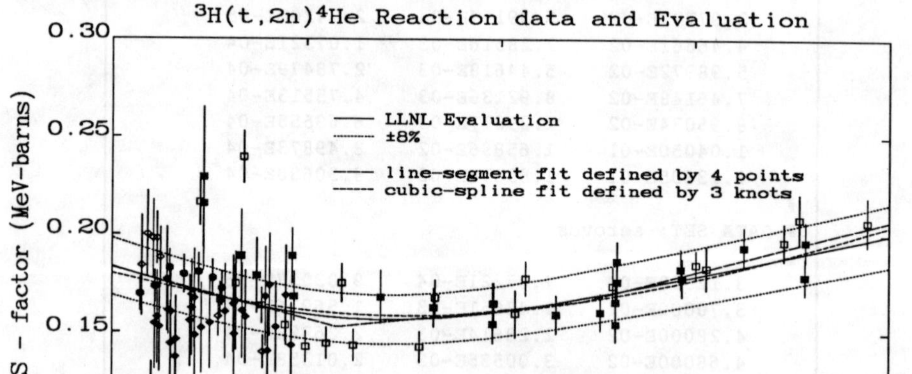

Figure 6: Plot of all experimental data and evaluation[1] of the $^3H(t,2n)^4He$ reaction for $E_{t(lab)}$ from 0 to 0.8 MeV. Also shown are the results from an optimized line-segment fit defined by 4 points and an optimized cubic-spline fit defined by 3 knots.

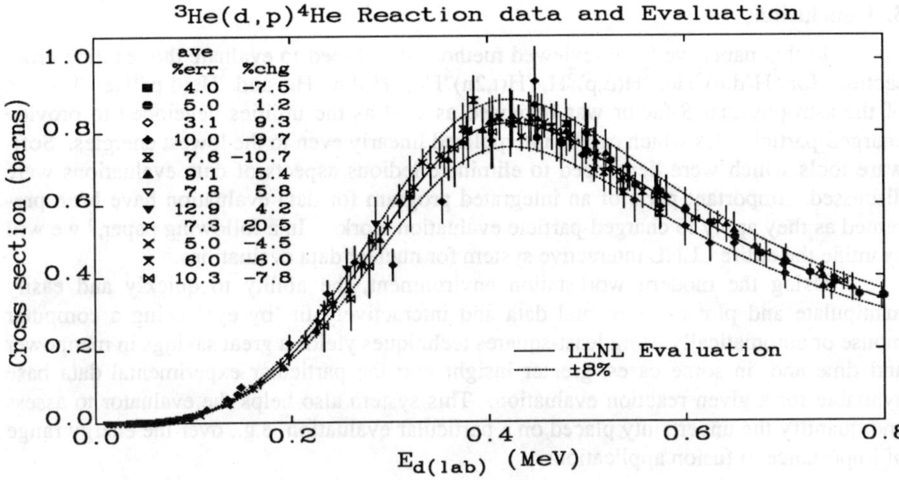

Figure 7: Experimental data and evaluation[1] for the $^3He(d,p)^4He$ reaction in terms of cross section vs. E_d. The *ave %err* is the average total error of each data set. The *%chg* is the renormalization of each data set for the final R-matrix fit.

widths. Because of the differences observed in fitting selected subsets of the low-energy data and different energy ranges, we concluded that the best low-energy fit could be obtained from a one-level analysis and that the quality of the data base does not allow much improvement with a more complicated multilevel analysis.

The large discrepancies in the experimental data sets for the $^3He(d,p)^4He$ reaction shown in Fig. 4 caused special problems. In general, the *shapes* of the data sets are in good agreement. However, without renormalizing some or all data sets, the resulting least-squares R-matrix fit produces a result which is too broad in shape and too low in magnitude and therefore does not match the shape of any of the data sets. The problem is then what to do about the normalization of the discrepant data. It is obvious from Fig. 4 that the differences in the various data sets are larger than the experimenters' quoted errors but it is not obvious which experiments are correct. To deal with this problem, we allowed the individual normalization of each set of data to vary simultaneously with the determination of the R-matrix parameters subject to the constraint that the average overall normalization of all the data sets remain unchanged. Ideally, one would constrain the changes in normalization by the absolute errors, but because of not knowing the absolute errors for all of the measurements, this was not possible. With this procedure, we determined the shape of the $^3He(d,p)^4He$ resonance. To validate the resonance shape determined in this way, we calculated R-matrix fits under identical conditions except that we fixed the average overall normalization of all the data to be +4% and −4% from the original value. The shape of the integrated cross section did not change. In Fig. 7 we show the final shape and normalization of the resonance evaluation for the $^3He(d,p)^4He$ reaction derived from this procedure.

5. Conclusions

In this paper we have reviewed methods developed to evaluate the reaction cross sections for $^2H(d,n)^3He$, $^2H(d,p)^3H$, $^3H(t,2n)^4He$, $^3H(d,n)^4He$, and $^3He(d,p)^4He$. The use of the astrophysical S-factor was reviewed as well as the utilities developed to provide charged-particle files which can be interpolated linearly even to the lowest energies. Software tools which were developed to eliminate tedious aspects of data evaluations were discussed. Important parts of an integrated program for data evaluation have been presented as they apply to charged-particle evaluation work. In a following paper,[3] we will examine the entire LLNL interactive system for nuclear data evaluations.

Using the modern workstation environment, the ability to quickly and easily manipulate and plot experimental data and interactively fit 'by eye' using a computer mouse or automatically using least-squares techniques yields a great savings in manpower and time and, in some cases, greater insight into the particular experimental data base available for a given reaction evaluation. This system also helps the evaluator to assess and quantify the uncertainty placed on a particular evaluation, e.g., over the energy range of importance to fusion applications.

6. Acknowledgments

The authors wish to thank Robert J. Howerton for many useful discussions during all aspects of this work. Work performed under the auspices of the U.S. Department of Energy by the Lawrence Livermore National Laboratory under contract number W-7405-ENG-48.

7. References

1. R. M. White, D. A. Resler, and S. I. Warshaw, *Evaluation of Charged-Particle Reactions for Fusion Applications*, in the proceedings of *Nuclear Data for Science and Technology*, ed. S. M. Qaim (Springer-Verlag, Berlin, 1992), and references therein.
2. J. D. McGowan, *Digitization Program XDIGITIZE* (unpublished).
3. D. A. Resler and R. M. White, *The LLNL Interactive System for Nuclear Data Evaluation*, paper to appear later in these proceedings.

TECHNIQUES USED FOR CHARGED PARTICLE NUCLEAR DATA EVALUATIONS AT CNDC

ZHUANG YOUXIANG SUN ZUXUN
Chinese Nuclear Data Center, China Institute of Atomic Energy
P. O. Box 275 (41), Beijing 102413, P. R. China

ABSTRACT

The methods and techniques used for Charged Particle Nuclear Data (CPND) evaluation at Chinese Nuclear Data Center (CNDC) are summarized in this report, including compilation and evaluation of experimental data, nuclear reaction theory and model calculation, systemstics research and comprehensive recommendation etc..

1. Introduction

Charged Particle Nuclear Data (CPND) have been widely used in basic scientific researches on nuclear force, nuclear structure, nuclear reaction mechanism and nuclear astrophysics etc.; also used in engineering and technology such as space radiation effects, neutron sources, biomedical isotopes production, radiation therapy, activation analyses, radiation processing, accelerator shielding and fusion reactor designs.

Due to Coulomb repulsion between incident charged particle and target nucleus, the cross sections of charged particle nuclear reaction are much smaller than those of neutron, they are in about millibarn region; besides, there is a Coulomb threshold, which is dependent upon Coulomb barrier, when incident charged particle energy is less than the Coulomb threshold, the nuclear reaction can not be induced, even if reaction energy $Q>0$. In general, the threshold for a reaction of $Q>0$ equals to zero. These are the two characters of charged particle nuclear reaction.

Compared with neutron experimental data, CPND are scarce and existing data are scattered, usually the measured data are not enough to do an evaluation. Therefore, Chinese Nuclear Data Center (CNDC) puts the stress on the establishment of nuclear reaction theory and model calculation codes for CPND.

By the end of fifties, the measurements of CPND with light particle incident were started in China Institute of Atomic Energy. About ten years later, the studies of nuclear reactions induced by heavy ion have been performed. From 1975 on, the evaluations[1] of

CPND on light, intermediate, heavy and fissionable nuclides have been carried out at CNDC in the energy region from threshold to 20 MeV or above.

In general, data are generated from experiment, theory and systematics. The experiment is the first and direct source, and the base of both the latters.

The main task of evaluators consists in the collection, comparison, critical assessment and selection of experimental data and associated covariances; nuclear model calculations and systematics must be used to fill gaps and remove inconsistencies in the available experimental data; followed by the derivation of prefered values by appropriate combination procedures, or self–consistent sets of those if necessary.

2. Evaluation of Experimental Data

2.1. Compilation and Collection of CPND

The measured data need to be compiled in EXFOR format for preserving, exchanging and using. CNDC has compiled the CPND measured in China, and sent to NDS / IAEA.

It is high important to make full use of all the available information on nuclear data. This is one of cardinal principles of nuclear data evaluataion; because the more information is utilized, the more reliable and accurate recommendation is obtained. Therefore the related experimental data must be collected as complete as possible.

The bibliographies and indexes to CPND used at CNDC are as follows :

Nuclear Science Abstracts; Nuclear Data Table; Atomic Data and Nuclear Data Table; INIS Atomindex; USRL–50400, BNL–NCS–50640, BNL–NCS–51771; CPND EXFOR Master–File Index.

The numeral data can be got from EXFOR entries, some papers and their authors sometimes; however, it is necessary for this purpose to read graph, because parts of the experimental data have not been compiled in EXFOR entries yet. A part of numeral data used for CPND evaluation at CNDC were taken from the figures of related articles and reports. The error of reading graph is about $1\sim5\%$.

2.2. Comparison and Evaluation of CPND

The first step, experimental data are analysed on the basis of measuring method and instrument, correction factors and covariance matrixes etc.

The main methods used in measurements of CPND are as follows :

(1) Measurement of outgoing particles : "E / Delta E" and "E / t" measurements, magnetic deflection, mass–separator, pulse–height discrimination;

(2) Measurement of production nucleus : activation, chemical separation, collection of recoil nucleus, and their combination.

In a complex reaction, the method with particle identification is better than the others, and the more the instrumental resolution is high, the more the measured result is accurate.

In absolute determination, the incident charged particle flux is often measured by beam current integration. The unstability of beam current integrator is one of important sources for data difference sometimes, because there are usually some interferences in experiment, if they are not noticed and eliminated. In addition, the background due to big elastic scattering is also a factor not to be ignored.

In relative measurement, if the standard data of monitor are renewed, the result should be renormalized.

When any correction and covariance in evaluated experiments are unreasonable or omitted, we adjust or supplement them, if possible; otherwise, such experimental data shouldn't be adopted.

After this process, available experimental data have been selected.

The second step, all chosen data are plotted in different figures according to data categories respectively, in order to compare each other. If the discrepancy among them is large, a further and refined analysis is needed to decide which should be accepted or rejected. If it is difficult to do this, the basic principle of that the minority is subordinate to the majority, or equal–weight average can be adopted.

Currently, simultaneous evaluation is proposed and applied to estimate cross sections, especially for important ones. It is advanced in comparison with individual evaluation usually adopted in many works because of the following reasons. In the former method, it can be expected to reduce contribution of systematic errors in experiments to uncertainties of evaluated results, and original values of experimental data are used without renormalizing even ratio data.

Simultaneous evaluation method has been used in neutron data evaluation at CNDC. It will be utilized for evaluating CPND.

2.3. Data Processing

The data fit programs with Legendre polynomials[2] for angular distribution, orthogonal polynomials[3] and spline functions[4] with knot optimization are used at CNDC. Both the latters can estimate not only the evaluated cross sections but also the covariances associated with them. The both quantities can be estimated taking account of correlation on the cross section measurements.

Model theory is also used to fit experimental data[5], as shown in Fig. 1. Sometimes, eye–guide is convenient and useful for this purpose.

Because of large systematic errors and discrepancies between different experimental results which in many cases are very difficult to disclose and eliminate, or lack of experiment data, the further and more precise measurement is necessary, if possible.

If evaluated experimental data are not quite enough for applied requirement, theoretical model calculation, systematics research or comprehensive recommendation are required.

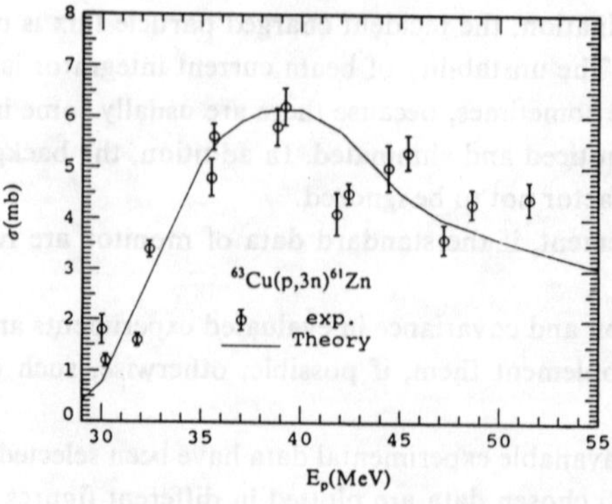

Fig. 1 The excitation function of $^{63}Cu(p,3n)^{61}Zn$ reaction

3. Nuclear Reaction Theory and Model Calculation

Many CPND needed for application are either very difficult to measure or have to wait for being measured for unpredictable time delays. To a large extent due to the practical requirements, nuclear reaction theories and models have been developed to the extent, that with appropriate parameters based on related available experimental data the required limited accuracies for some of those "unmeasurable" data can approximately be met. These developments concern particularly the theoretical description of fission process, improvement in the statistical and optical models of nuclear reaction, the introduction of pre–equilibrium and multistep processes in the description of nuclear reaction and a further understanding of the physical foundations of nuclear model parameters. Parallel to these theoretical developments powerful computer codes have been developed which allow very detailed calculations of nuclear reaction data for structural and fissionable materials.

The present power of appropriately parameterised nuclear reaction theories and nuclear model calculation can be described as follows :

(1) They can be employed for inter– and extra–polation of experimental results[5], see Fig. 2;

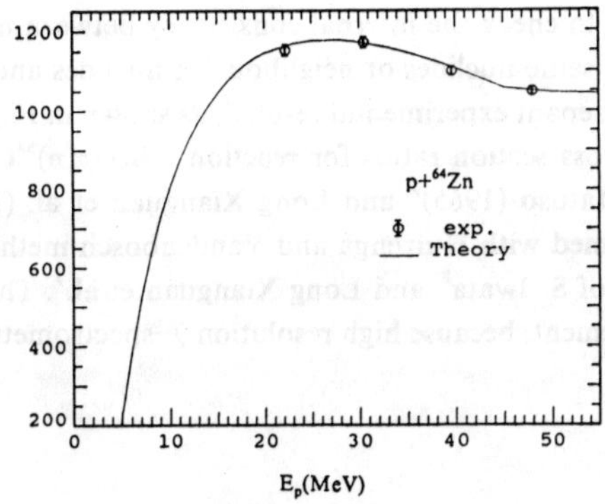

Fig. 2 The reaction cross sections of p+^{64}Zn — CMUP2 calculation

(2) They can be used for the prediction of unknown nuclear data such as reaction cross section, angular and energy distributions as well as double differential cross section etc.[6], see Fig. 3;

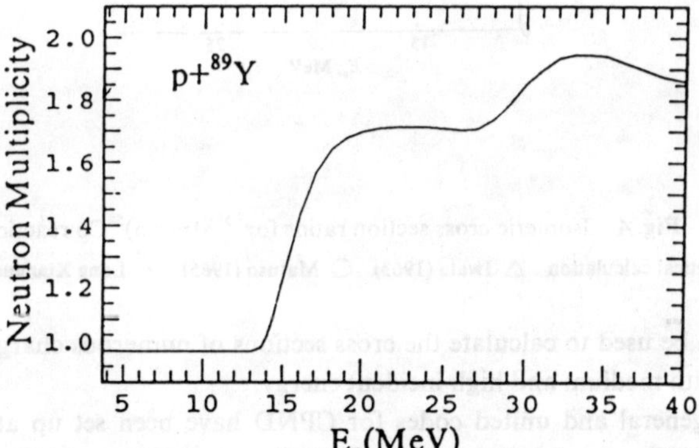

Fig. 3 The calculated neutron multiplicity of ^{89}Y induced by proton below 40 MeV

(3) They can be used to check the internal consistency between measurements of different nuclear data of the same nuclides or neighbouring nuclides and can occasionally help in deciding among discrepant experimental results[7], as shown in Fig. 4.

The isomeric cross section ratios for reaction $^{55}Mn(\alpha,n)^{58}Co$ were determined by S. Iwata (1962)[8], T. Matuso (1965)[9] and Long Xianguan et al. (1989)[7]. The theoretical calculation was performed with Huizenga and Vandenbosch method[10], it trends towards the experimental data of S. Iwata[8] and Long Xianguan et al.[7]. The latter is a newer and more accurate measurement, because high resolution γ–spectrometry technique was used.

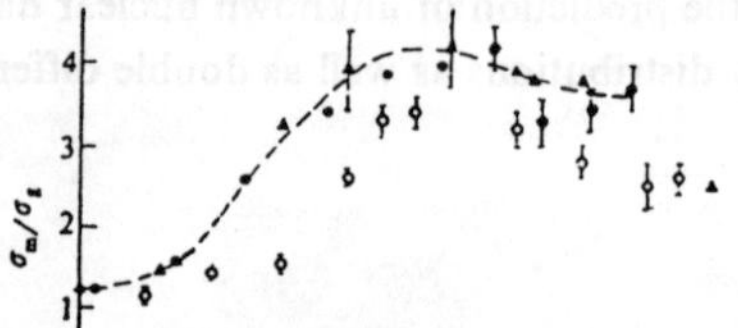

Fig. 4 Isomeric cross section ratios for $^{55}Mn(\alpha,n)^{58}Co$ reaction

— Theoretical calculation △ Iwata (1963) ○ Matuso (1965) • Long Xianguan et al. (1989)

(4) They can be used to calculate the cross sections of numerous charged particle nuclear reactions with medium and high incident energy.

Some general and united codes for CPND have been set up at CNDC. For instance, a comprehensive R–matrix analysis RCA code[11] based on the multichannel and multilevel R–matrix theory[12] is for light nuclei; model calculation codes CMUP2[13] based on the optical model and the unified treatment of exciton model and evaporation model is for medium and heavy weight nuclei and CFUP1[14] based on MUP2[15] code for fissile ones, they can be used to calculate the related CPND in the energy region from threshold to 50 MeV, or even to 100 MeV according to the plan of 1992. See Fig. 5[16].

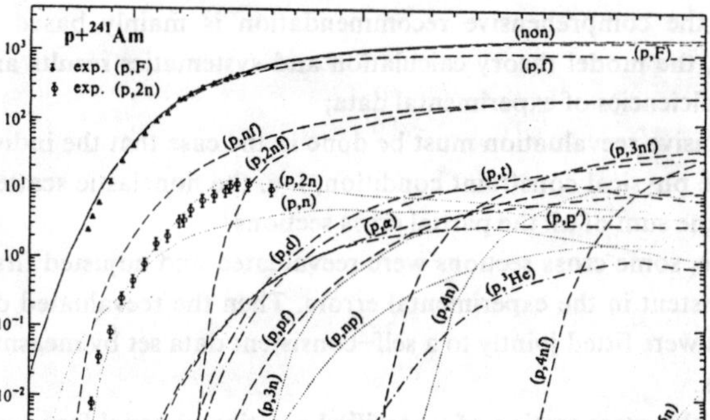

Fig. 5 The calculated particle emission cross sections in reaction p+^{241}Am below 35 MeV

— CFUP1 calculation

4. Systematics Research

The CPND of the nuclides which have a short life time or very low abundance are either scarce or scattered, some of these data can complemented by systematics.

The method settled a problem by from individuality to generality such as systematics has been applied in many sciences.

If systematical behavior can be found out from existing experimental data of other nuclei, unknown physical quantities can be derived from known ones with the aid of systemataics research.

A series of systematics research on neutron data have been done at CNDC, such as average total radiative widths[17], thermal fission cross sections of actinides[18], (n,2n)[19], (n,charged–particle) and (n,γ) cross sections[20~23].

The systematicses of (p,n), (p,2n) and (d,n), (d,2n) reaction cross sections are bening studied by means of simplified model formulas and empirical parameters at CNDC. The results will be published in "Communication of Nuclear Data Progress".

5. Comprehensive Recommendation

A final comprehensive recommendation is usually necessary after the evaluations for various channels accomplished individually, when a complete set of all the partial reaction data is required.

(1) Generally, the comprehensive recommendation is mainly based on the evaluated measured data, the model theory calculation and systematics results are usually used to make up the deficiencies of experimental data;

(2) A comprehnsive reevaluation must be done in the case that the individual evaluations do not meet the physical constraint condition, i. e. the nonelastic scattering cross section is not equal to the sum of all the partial cross sections.

In this case, some cross sections were reevaluated and adjusted firstly to make them to be self–consistent in the experimental errors. Then the reevaluated data of various reaction channels were fitted jointly to a self–consistent data set by measns of one of the following methods.

(1) To obtain the cross section of a specified reaction channel from other varions channels according to certain physical relation. For example, a reaction cross section (usually they are deficient in measured data or with poor accuracy) can be deduced from the nonelastic scattering cross section minus the sum of all the other partial reaction cross sections;

(2) To make the excitation functions of various reaction channels self–consistent by the aid of simulteneously fitting of the various individual evaluations with the B–spline function or polynomial.

If there is a need for the data of a specific reaction channel but not for a complete set of all the partial reaction channels, the basic principles of comprehensive recommendation can be given as follows :

(1) There are more accurate and enough experimental data, the recommended values are taken from evaluated ones;

(2) If experimental data are scarce, theoretical model calculation or systematics can represent them well, model or systematics calculation can be accepted, see Fig. 6[5].

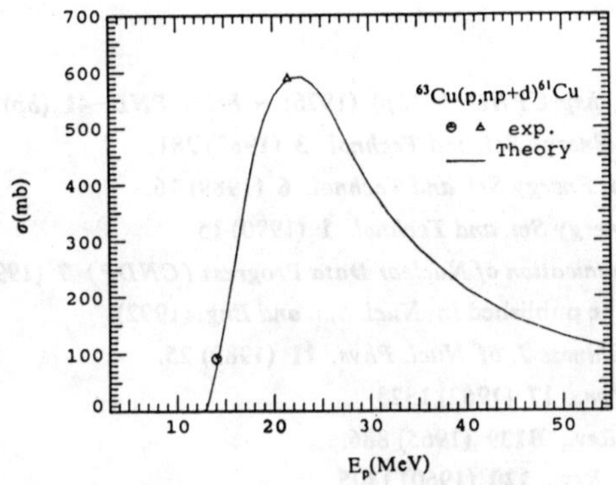

Fig. 6 The excitation function of ^{63}Cu(p,np+d)^{61}Cu reaction

6. Summary

Using mentioned–above charged–particle evaluation techniques, CNDC has evaluated thermal nuclear reaction cross sections for fusion application and activation cross sections of intermediate and heavy nuclei for isotopes production and activation analysis, such as ^{2}H(d,n)^{3}He, ^{2}H(d,p)^{3}H, ^{3}H(d,n)^{4}He, ^{3}He(d,p)^{4}He, d+^{6,7}Li, t+^{6,7}Li, ^{3}He+^{6,7}Li and ^{56}Fe(p,γ)^{57}Co, ^{57}Fe(p,n)^{57}Ci, ^{235}U(d,2n)^{235}Np, ^{235}U(p,n)^{235}Np. The evaluated results have satisfied the needs of the users.

However, the model calculation program for light nuclides still need to be set up at CNDC. The existing codes CMUP2 and CFUP1 wil be developed in the energy region from throshold to 100 MeV or more. It is necessary for CNDC to make a further research on the systematics and covariance of CPND. A problem awaiting solution in the world is how to establish a set of standard data of monitor for CPND measurements.

Authors are very grateful to Drs. C. L. Dunford and R. M. White of U. S. A. for giving a chance to summarize the CPND work at CNDC, and Dr. Wang Dahai of IAEA for his helpful advices.

References

1. Youxiang Zhuang et al., *hsj–CPND–1–(bp)* (1976) ~ *hsj–CPND–41 (bp)* (1992).
2. Baosheng Yu et al., *At. Energy Sci. and Technol.* **3** (1983) 281.
3. Zhixiang Zhao et al., *At. Energy Sci. and Technol.* **6** (1989) 16.
4. Tingjin Liu et al., *At. Energy Sci. and Technol.* **1** (1990) 15.
5. Yinlu Han et al., *Communication of Nuclear Data Progress (CNDP)* **7** (1992) 45.
6. Qingbiao Shen et al., to be published in *Nucl. Sci. and Eng.* (1992).
7. Xianguan Long et al., *Chinese J. of Nucl. Phys.* **11** (1989) 25.
8. S. Iwata, *J. Phys. Soc. Jpn.*, **17** (1962) 1323.
9. T. Matuso et al., *Phys. Rev.*, **B139** (1965) 886.
10. J. Huizenga et al., *Phys. Rev.*, **120** (1960) 1305.
11. Zhengpeng Chen et al., *CNDP* **4** (1990) 20.
12. A. M. Lane et al., Rev. *Mod. Phys.*, **30** (1958) 257.
13. Chonghai Cai and Qingbiao Shen, *CNDP* **3** (1990) 29, *CNDP* **4**, (1990) 18.
14. Qingbiao Shen, *CNDP* **5** (1991) 36.
15. Chonghai Cai et al., *MUP2(IAEA–0907/01), News from the NEA Data Bank,* No. **1** (1986) 7.
16. Youxiang Zhuang and Qingbiao Shen, *CNDP* **8** (1992).
17. Youxiang Zhuang et al., *At. Energy Sci. and Technol.* **1** (1985) 34.
18. Fuhua Cu et al., *At. Energy Sci. and Technol.* **1** (1985) 39.
19. Jin Zhang et al., *Proc. of Inter. Conf. on Nuclear Data for Basic and Applied Science, Santa Fe* 1985, p. 1611; *IAEA–TECDOC–457* (1989) p. 159.
20. Zhixiang Zhao et al., *LNucl. Sci. and Eng.* **99** (1988) 367.
21. Zhixiang Zhao et al., *Proc. of Inter. Conf. on Nuclera Data for Science and Technology*, Mito, Japan, 1988, p. 513.
22. Zhixiang Zhao et al., *Chines J. of Nucl. Phys.* **11** (1989) 71.
23. Zhixiang Zhao et al., *Chines J. of Nucl. Phys.* **13** (1991) 139.

THE METHOD OF REDUCTION AS PHOTONUCLEAR DATA EVALUATION TOOL

N.G.EFIMKIN AND V.V.VARLAMOV

Centre for Photonuclear Experiments Data
Institute of Nuclear Physics, Moscow State University
119899 Moscow, Russia

ABSTRACT

An approach based on the Method of Reduction (MR) is proposed for the evaluation of the photonuclear reaction cross sections obtained using different apparatus functions, calibrations and normalizations. This approach consists of using the real apparatus function (photon spectrum) of each individual experiment to reduce the data to a representation generated by an apparatus function of better quality. The task is to find the most reasonably achievable monoenergetic representation (MRAMR) of the information about the cross section contained in the experimental observables. The MR is used to obtain the MRAMR of the reaction cross sections for several target nuclei. The results are compared with tagged photon data where it is possible.

Introduction

The absence of sufficiently intensive beams of monoenergetic photons during whole period of investigations of the γ-quanta interaction with atomic nuclei from the beginning (in 50-ies) till now is one of the main problems of these studies. This makes experimenters to use various methods for creation of conditions in which the effective photon energy spectrum in any approach can be interpreted similar to the monoenergetic one.

Today considerable number of the methods for such kind of condition creation can be listed, for example:

- carrying out the experiments with bremsstrahlung (further BR) or other kind of γ-quanta with continuous energy spectrum and solving afterwards the appropriate problem of unfolding for which a large number of mathematical methods are used;
- application of the difference methods for continuous beams of bremsstrahlung with neighboring end-point energies;
- application of the beams of quasimonoenergetic (further QME) photons, with energy spectrum which looks like Gaussian or Lorentzian function and have finite width:
  - annihilation in flight of relativistic positrons;
  - Compton scattering of γ-quanta;
  - inverse reactions of radiative capture;
- application of tagged bremsstrahlung photons (further TP).

The apparatus functions for several methods of "monoenergetic" spectrum condition creation are shown in Fig. 1.

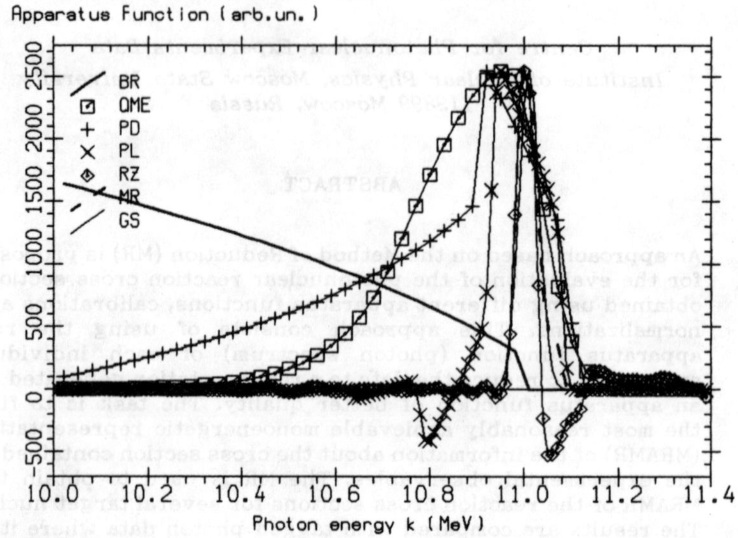

Fig. 1. The experimental apparatus functions: BR - bremsstrahlung beam; QME - quasiminoenergetic beam; and the apparatus functions of methods for the bremsstrahlung data analysis: PD - photon difference methods; PL - Penfold-Leiss method (150 keV analysis bin); RZ - regularization method; MR - method of reduction (resolution 50 keV). The 50-keV width Gaussian (GS) is shown for comparison.

This simple enumeration confirms all the complexities which arise in comparative analysis of the data obtained in various experiments, because of very different conditions of them, first of all - differences in apparatus functions. These differences result in certain systematic disagreements between the experimental results, which sometimes are very significant. The systematic disagreements in value and shape of photonuclear reaction cross sections obtained using BR and QME γ-quanta beams are widely known[1]. The main of these disagreements are that the QME cross sections are much smother than BR ones and QME cross sections have amplitudes (absolute values) smaller than BR ones.

The character of disagreements mentioned above is illustrated in Fig. 2. This is one of fairly obvious illustration of the discrepancies being discussed for the detailed comparison of the results of both BR and QME measurements of $^{18}O((\gamma,n) + 2(\gamma,2n))$ reaction cross section performed in joint work[2] of Melbourne and Livermore groups. The problem of principal possibility to obtain in BR experiment the result being in agreement with the result of QME experiment was investigated in this work. The data of[2] are presented in Fig. 2 and give the possibility to make three obvious conclusions:

- in spite of similarity in general shape of both cross sections, practically all QME resonances have in the average the smaller amplitudes than BR resonances: $\langle A_{BR}\rangle/\langle A_{QME}\rangle = 1.17$;
- integrated cross sections for energy range 8-28 MeV differ also: 187 (BR) and 177 (QME) MeVmb;
- under all the formality of resonance width determination in the cross section having very complicated structure the majority of QME resonances have the average width larger than BR resonances: $\langle \Gamma_{QME}\rangle/\langle \Gamma_{BR}\rangle = 1.25$.

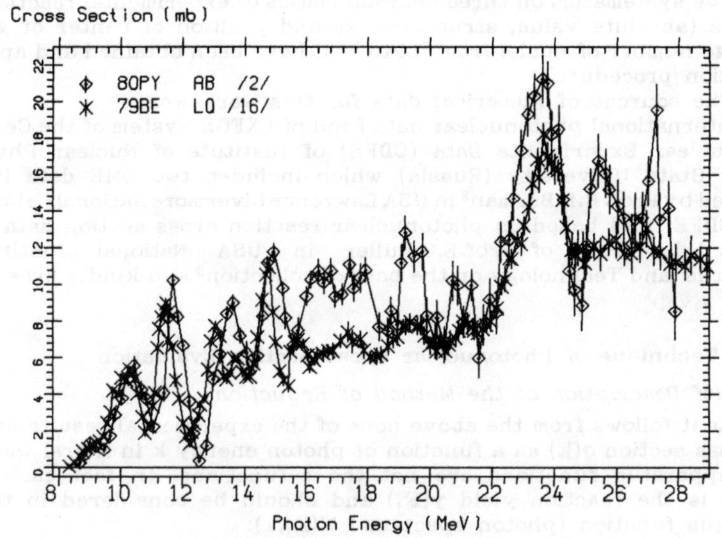

Fig. 2. Comparison of BR[2] and QME[16] results for ^{18}O.

It should be underlined that experimental results differ (and sometimes significantly) not only in general shape and absolute value, but also in energy positions of structure resonances (and cross section at all). Moreover the differences in energy positions of resonances and structures in the cross sections depend on the photon energies[3].

It is obvious that if the cross sections under comparison are displaced in energy against each other, the joint evaluation procedure which does not take into account this circumstance and do not use appropriate correction will lead to very smoothed evaluated cross section. It will mean that very important information about the character structures with the widths smaller than this displacement of initial cross sections will be lost in the evaluation result.

It is obvious also that joint analysis and evaluation of photonuclear reaction cross sections in conditions described above are possible only for the data which are in the same representation. This representation means that the data are transformed to the view as being obtained for some unique apparatus function and for identical or very close cross section and energy scales.

It is clear that for correct evaluation the study of character of real errors of the data should be done on the basis of both experimental method uncertainties and the uncertainties which can be obtained from joint analysis of all set of the data.

This work is devoted namely to create the method of photonuclear reaction cross section evaluation using the mathematical method of reduction which is one in family of image restoration theories.

The systematics on three characteristics of experimental reaction cross sections (absolute value, structureness and position of center of gravity) for large number of nuclei from ^{6}Li to ^{238}U have been obtained and applied in evaluation procedure.

The sources of numerical data for this work were:
- the international photonuclear data fund of EXFOR System of the Centre for Photonuclear Experiments Data (CDFE) of Institute of Nuclear Physics of Moscow State University (Russia) which includes two QME data libraries prepared by Prof. B.L.Berman[8] in USA Lawrence Livermore National Laboratory;
- the CDFE fund based on photonuclear reaction cross section data library prepared by group of Prof.E.G.Fuller in USA National Institute of Standards and Technology on the base of collection[9] and kindly presented to CDFE.

1. The Technique of Photonuclear Cross Section Evaluation

1.1. Brief Description of the Method of Reduction

As it follows from the above none of the experimental results estimates the cross section $\sigma(k)$ as a function of photon energy k in literal way, since their apparatus functions are not the δ-functions. In fact each of the results is the reaction yield $\mathbf{y}(E_j)$ and should be considered in terms of apparatus function (photon spectrum) $W(E_j,k)$:

$$\mathbf{y}(E_j) = = \alpha \int_{E_{threshold}}^{E_j} W(E_j,k)\ \sigma(k)\ dk + \nu_j\ , \qquad (1)$$

where ν_j is statistical uncertainty (additive noise), described by means of the covariance matrix Σ.

To solve the problem of cross section interpretation the method of reduction[11,12,13] was applied. The essence of the method follows.

The experimentally measured reaction yield (in expression (1) for which $W(E_j,k)$ means as well any of effective photon spectra) can be written in operator form:

$$\mathbf{y} = \mathbf{A}\ \sigma + \nu \qquad (2)$$

In the method of reduction (MR) the experimental yield $\mathbf{y}$ is transformed to the vector

$$\hat{\sigma} = \mathbf{U}\ \sigma + \hat{\nu}, \qquad (3)$$

which is interpreted as a result of experiment, the apparatus function (photon spectrum) of which $u(E_i,k)$ described by means of operator **U** has the desired properties, e.g. it might be ordered to be as monoenergetic as possible.

To solve this problem the linear operator **R** should be found so that the vector

$$\hat{\sigma} = \mathbf{R}\mathbf{y}$$

could be interpreted as a least square estimator of vector $\mathbf{U}\sigma$. The condition of solution existence should be satisfied:

$$\mathbf{R}\mathbf{A} = \mathbf{U}, \tag{4}$$

since only in this case the vector $\hat{\sigma}$ and its uncertainty can be found without a priori information about the unknown σ:

$$\hat{\sigma} = \mathbf{R}\mathbf{y} = \mathbf{U}\sigma + (\mathbf{R}\mathbf{A} - \mathbf{U})\sigma + \mathbf{R}\nu \tag{5}$$

The total uncertainty $M\,\|\mathbf{R}\nu\|^2$ of the vector $\hat{\sigma}$ is minimum[11,12], if:

$$\mathbf{R} = \mathbf{U}(\Sigma^{-1/2}\,\mathbf{A})^{-}\,\Sigma^{-1/2} \tag{6}$$

where $\mathbf{X}^{-}$ is pseudoinversion of the operator **X**,
$\|\mathbf{x}\|$ is norm of vector **x**,
M is assembly averaging.

The covariance matrix **G**, characterizing the uncertainty

$$\hat{\nu} = \mathbf{R}\nu$$

of the vector $\hat{\sigma}$, equals

$$\mathbf{G} = \mathbf{R}\,\Sigma\,\mathbf{R}^{*}. \tag{7}$$

It is clear that not all possible **U** operators can satisfy the solution condition (4), and of the **U** operators that do provide a solution, not all will meet the researcher's needs.

Therefore, the next step is to construct the operator **U**, which satisfies the condition (4) of solution existence, and at the same time agrees with the researchers' concept of a monoenergetic source spectrum. One of the most acceptable line shapes to use for this purpose is Gaussian.

Let us take the operator **Uo** to be the desired apparatus function having, e.g., a Gaussian line shape. Then we must find an operator **U**, closest to **Uo**, as the product (**KA**) of an arbitrary operator **K** with operator **A** (to satisfy condition (4)), and use **K** to minimize the difference between **U** and **Uo** as given by the following form:

$$\|\mathbf{U}-\mathbf{Uo}\|_2{}^2 = \mathrm{tr}((\mathbf{U}-\mathbf{Uo})(\mathbf{U}-\mathbf{Uo})^{*}), \tag{8}$$

where $\|\mathbf{X}\|_2{}^2$ means the Hilbert-Schmidt norm of the operator **X**. Then

$$\mathbf{U} = \mathbf{Uo}\mathbf{A}^{-}\mathbf{A} \tag{9}$$

and **U** is determined only by the operator **A** of the experiment and the operator **Uo**. The discrepancy between **U** and **Uo** is characterized by the value of $\|\mathbf{U}-\mathbf{Uo}\|_2{}^2$

Meanwhile it is very important to point out that the method described is not the method of "unfolding" or reconstruction of the cross section $\sigma(k)$ from experimental yield **y**. The method just transforms the data obtained with some experimental apparatus function **A** to the form they would have being measured by means of apparatus function **U**.

Consequently, there is no need to solve the ill-posed problem (1).

In the frame of the MR it is possible to find the most reasonably achievable monoenergetic representation (MRAMR) for the cross section.

The advantages of the MR in application to the problems discussed can be demonstrated with the analysis of the model cross section built of two 10 keV width resonances at the energies 11.0 and 11.2 MeV.

The yield **y** of QME experiment was simulated (Fig. 3a (stars)) with the energy step 100 keV for the resolution 330 keV with the apparatus function A presented in Fig. 3b (stars). The reduction result **Ry** (Fig 3a (dashes)) with apparatus function U (Fig. 3b (dashes)) having resolution 150 keV is compared in Fig. 3a with exact model cross section Uo σ with the resolution 150 keV (solid line).

1.2. Recalibration-Renormalization Procedure

The MR allows one to remove the most complicated source of data disagreement, caused by the difference of apparatus functions. The reduction of the data obtained in different experiments to the same representation (the same apparatus functions) provides the principal opportunity for their joint analysis and simultaneous evaluation.

The other two kinds of disagreements were pointed out above: the systematic uncertainties in normalization and energy calibration. They are smooth functions of energy and their possible values are known.

To make the systematic errors clearly seen it is necessary to reduce all of the data to so poor resolution that the statistical errors were negligible After that the smooth curves should be moved toward each other using a priori information about systematic error distributions.

There are two kinds of poor resolution cross section shapes.

1. In the first case researcher can easily identify the shape of the curve and "guess" which regions of different curves should coincide. A typical example - the cross sections at Oxygen measured in MSU INP[13] and LLNL[14]. These curves contain several resonances which are easily "recognizable".

In this case ("manual" procedure) it is possible to order the limits of corresponding regions and to fit the parameters a_i(additive adjustment), b_i (normalization adjustment), c_i(energy shift):

$$a + (1 + b)\ \sigma\ (E + c), \qquad (10)$$

where $i = 2 \div N$, and N - number of curves, $a_1 = b_1 = c_1 = 0$, to minimize the sum of square differences between each curve and one of them chosen arbitrarily and marked by index 1. Afterwards the weighted average values <a>, <b>, <c> are computed and subtracted from the parameters obtained:

$$\hat{a}_i = a_i - \langle a\rangle \quad \text{etc.}$$

Therefore the adjustments $\hat{a}_i$, $\hat{b}_i$, $\hat{c}_i$, corresponding to energies c_i of curve number i are found for each marked region. These adjustments should be applied to achieve the coincidence of marked regions of the curves.

2. In the second case the curves have only one or two broad maxima, e.g.[14,15] cross section for Praseodymium. As against the first case the decision on the peak identification is very complicated. To reconcile the curves the special "objective" procedure has been worked out.

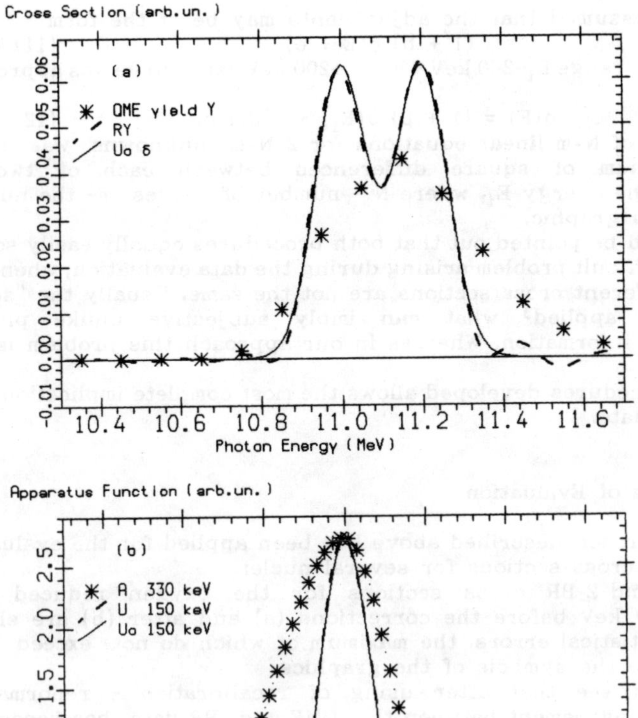

Fig. 3. The method of reduction capabilities:
a) - cross sections (stars QME yield, simulated with the apparatus function A, (Fig.3b - stars - resolution 330 keV); dashed line - reduction result Ry corresponding to apparatus function U (Fig.3b - dashed line) of resolution 150 keV; solid line - U0σ - model cross section folded with 150 keV-width Gaussian U0 (Fig.3b - solid line));
b) - apparatus functions A, U and U0, normalized for comparison.

It was assumed that the adjustments may be of the form

$$(1 + b)\ \sigma(E + c). \qquad (11)$$

For the energy range E_i−200 keV < E < E_i+200 keV each curve was approximated by the line

$$\sigma(E) = (1 + b)\ \sigma(E_i) + \ (E_i)\ c. \qquad (12)$$

The set of N·m linear equations for 2·N·m unknowns was built to minimize the sum of square differences between each of two cross sections for the energy E_i, where N - number of curves, m– the number of points in each graphic.

It should be pointed out that both procedures equally easily solve one of the most difficult problem arising during the data evaluation, when energy ranges of different cross sections are not the same. Usually the "sewings" together are applied[3] what can imply subjective unlike physically substantiated information. Whereas in our approach this problem is solved automatically.

The procedures developed allows the most complete implication of both QME and BR data.

2. The Results of Evaluation

The approach described above has been applied for the evaluation of photoneutron cross sections for several nuclei.

5 QME and 2 BR cross sections for the Oxygen reduced to the resolution 600 keV before the corrections (a) and after (b) are shown in Fig.4. The statistical errors, the maximum of which do not exceed 0.1 mb, are well inside the symbols of the graphics.

One can see that after using of recalibration - renormalization procedure the agreement between the QME and BR data has become much better.

The adjustments obtained were applied to correct the QME yields. Afterwards all 5 corrected yields were put to the simultaneous procedure of reduction without a priori information. The result of reduction to the resolution of 240 keV - the final result of photoneutron cross section evaluation at Oxygen - is shown in Fig.5b). All experimental cross sections from the database are shown in Fig.5a for comparison.

The initial experimental and final evaluated data on total photoneutron cross sections for the ^{208}Pb is presented in Fig.6.

The evaluation cross sections have been obtained also for ^{6,7}Li, ^{12}C, ^{27}Al, ^{28}Si, ^{40}Ca, ^{63,65}Cu, ^{90}Zr, ^{141}Pr, ^{208}Pb.

Conclusion

The systematical analysis of the data on the total photoneutron reaction cross section obtained in the experiments of different kind has been carried out for different nuclei. It was found that the reasons of the well known discrepancies in the data are the differences of effective apparatus functions (photon spectra) used, calibration and normalization. It was pointed out that these three types of the systematic errors are energy dependent, that essentially complicates the use of traditional methods of data evaluation.

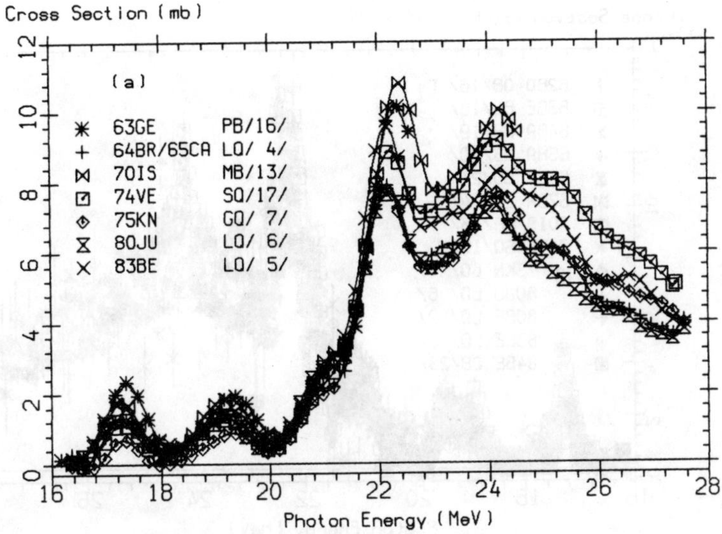

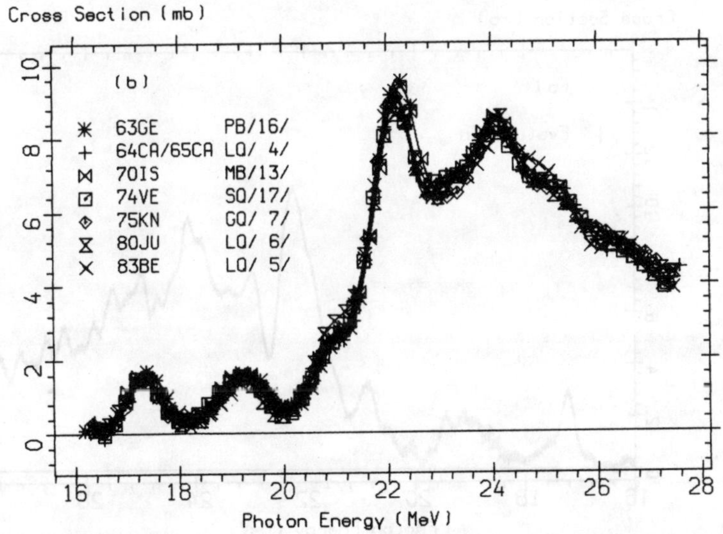

Fig.4. The results of photoneutron cross section processing for Oxygen:
a) - 5 QME and 2 BR data reduced to 600 keV resolution;
b) - the same data after the recalibration - renormalization.

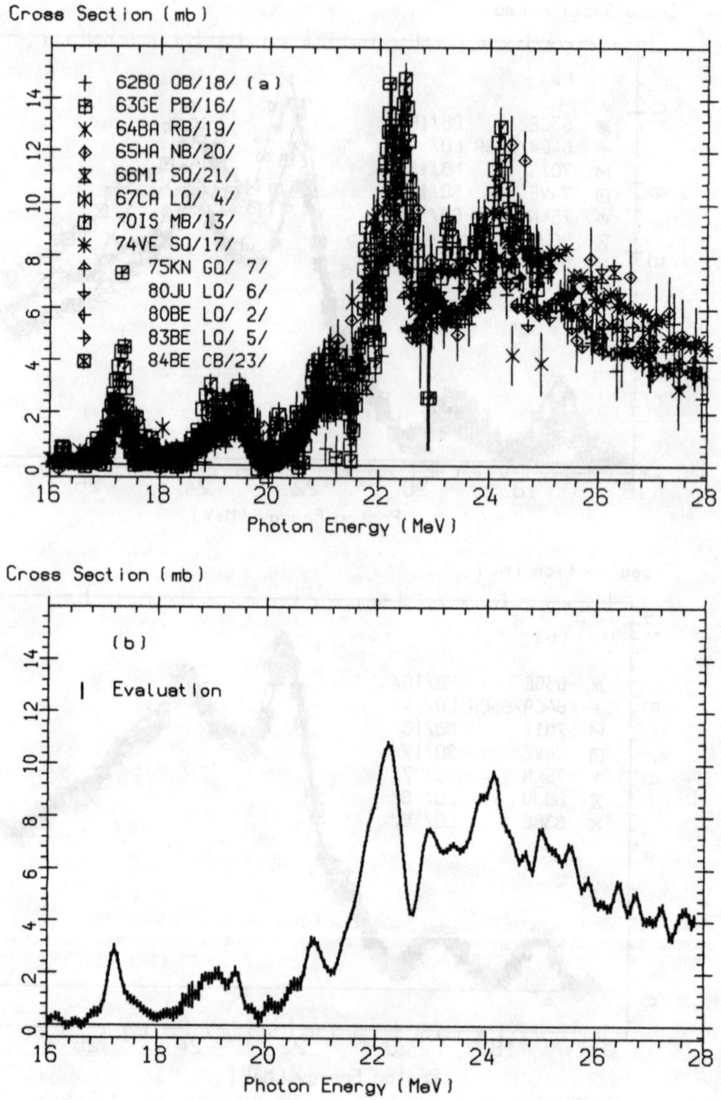

Fig.5. The data base of initial photoneutron reaction cposs section for Oxygen (a) and the result of evaluation with the resolution 240 keV(b).

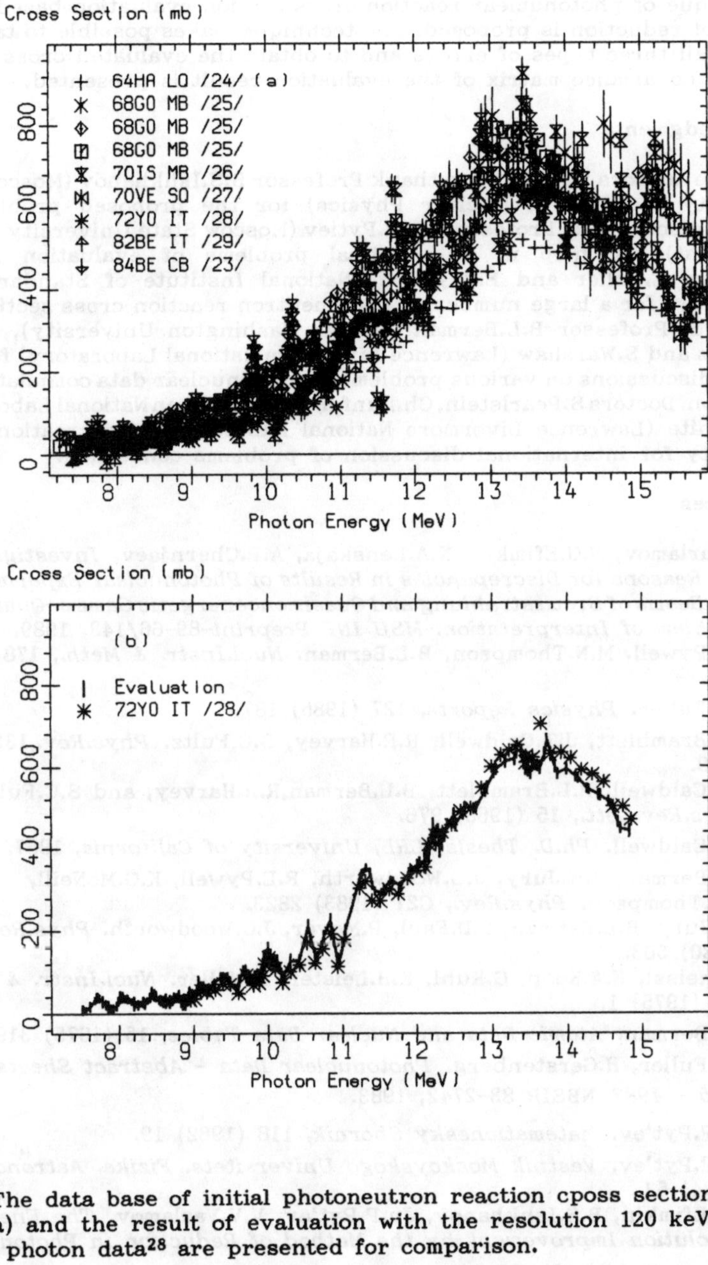

Fig.6. The data base of initial photoneutron reaction cposs section for Lead (a) and the result of evaluation with the resolution 120 keV(b). The tagged photon data[28] are presented for comparison.

A technique of photonuclear reaction cross section evaluation based on the method of reduction is proposed. The technique makes possible to take into account all three types of errors and to obtain the evaluated cross section with the covariance matrix of the evaluation result is presented.

Acknowledgments

The authors would like to thank Professor B.S.Ishkhanov (Moscow State University Institute of Nuclear Physics) for the proposed problem and valuable discussions, Professor Yu.P.Pytiev (Moscow State University Physics Department) for help in mathematical problems of evaluation method, Professors E.Fuller and E.Hayward (National Institute of Standards and Technology) for a large number of photoneutron reaction cross section data presented, Professor B.L.Berman (George Washington University), Doctors R.Alvarez and S.Warshaw (Lawrence Livermore National Laboratory) for very fruitful discussions on various problems of photonuclear data compilation and evaluation, Doctors S.Pearlstein, Ch.Dunford (Brookhaven National Laboratory) and R.White (Lawrence Livermore National Laboratory) for creation of the possibility for international discussion of problems concerned.

References

1. V.V.Varlamov, N.G.Efimkin, N.A.Lenskaja, A.P.Chernjaev, *Investigation of the Reasons for Discrepancies in Results of Photonuclear Experiments at the Beams of Bremsstrahlung and Quasimonoenergetic Gamma-Quanta. The Problem of Interpretation. MSU INP Preprint*-89-66/143, 1989.
2. R.E.Pywell, M.N.Thompson, B.L.Berman. *Nucl.Instr. & Meth.*, **178** (1980) 149.
3. E.G.Fuller. *Physics Reports*, **127** (1985) 187.
4. R.L.Bramblett, J.T.Caldwell, R.R.Harvey, S.C.Fultz. *Phys.Rev.*,**133** (1964) B869.
   J.T.Caldwell, R.L.Bramblett, B.L.Berman,R.R.Harvey, and S.C.Fultz. *Phys.Rev.Lett.*, **15** (1965) 976.
   J.T.Caldwell. *Ph.D. Thesis, LRL, University of California*, 1967.
5. B.L.Berman, J.W.Jury, J.G.Woodworth, R.E.Pywell, K.G.McNeill, M.N.Thompson. *Phys.Rev.*, **C27** (1983) 2823.
6. J.W.Jury, B.L.Berman, D.D.Faul, P.Meyer, J.G.Woodworth. *Phys.Rev.*, **C21** (1980) 503.
7. U.Kneissl, E.A.Koop, G.Kuhl, K.H.Leister, A.Weller. *Nucl.Instr. & Meth.*, **127** (1975) 1.
8. B.L.Berman. *Atomic Data and Nuclear Data Tables*, **15** (1975) 319.
9. E.G.Fuller, H.Gerstenberg. *Photonuclear Data - Abstract Sheets. 1955 - 1982.* NBSIR 83-2742, 1983.
10. Yu.P.Pyt'ev. *Matematichesky Sbornik*, **118** (1982) 19.
11. Yu.P.Pyt'ev. *Vestnik Moskovskogo Universiteta. Fizika. Astronomia*, **25** (1984) 53.
12. N.G.Efimkin, B.S.Ishkhanov, Ju.P.Pyt'ev, V.V.Varlamov. *The Energy Resolution Improvement by the Method of Reduction in Photonuclear*

Experiments. MSU INP Preprint-91-35/239, 1991.
N.G.Efimkin, B.S.Ishkhanov, Yu.P.Pyt'ev. *Vestnik Moskovskogo Universiteta. Fizika. Astronomia*, **32** (1991) 48.
13. B.S.Ishkhanov, I.M.Kapitonov, E.V.Lazutin, I.M.Piskarev, V.S.Sopov, V.G.Shevchenko. *Yadernaya Fizika*, **12** (1970) 892.
14. H.Beil, R.Bergere, P.Carlos, A.Lepretre, A.Veyssiere, A.Parlag. *Nucl.Phys.*, **A172** (1971) 426.
15. T.K.Deague, R.J.J.Stewart. *Nucl.Phys.*, **A191** (1972) 305.
16. J.G.Woodworth, K.G.McNeil, J.W.Jury, R.A.Alvarez, B.L.Berman, D.D.Faul, P.Meyer. *Phys.Rev.*, **C19** (1979) 1667.
17. A.Veyssiere, H.Beil, R.Bergere, P.Carlos, A.Lepretre, A.De Miniac. *Nucl.Phys.*, **A227** (1974) 513.
18. L.N.Bolen, W.D.Whitehead. *Phys.Rev.Lett.*, **9** (1962) 458.
19. G.Baciu. *Rev.Rom.Phys.*, **9** (1964) 783.
20. E.Hayward, T.Stovall. *Nucl.Phys.*, **69** (1965) 241.
21. J.Miller, C.Schuhl, G.Tamas, C.Tzara. *J. de Phys.*, **27** (1966) 8.
22. B.L.Berman, D.D.Faul, P.Meyer, D.L.Olson. *Phys.Rev.*, **C22** (1980) 2273.
23. S.N.Belyaev, O.V.Vasil'ev, A.B.Kozin, A.A.Nechkin,V.A.Semenov. *Izvestiya AN SSSR, Seriya Fizicheskaya*, **48** (1984) 1940.
24. R.R.Harvey, J.T.Caldwell, R.L.Bramblett, S.C.Fultz. *Phys.Rev.*, **136** (1964) B126-31.
25. B.I.Goryachev, B.S.Ishkhanov, I.M.Kapitonov, V.G.Shevchenko. *JETP Lett.*, **7** (1968) 161.
26. B.S.Ishkhanov, I.M.Kapitonov, E.V.Lazutin, I.M.Piskarev, O.P.Shevchenko. *Sov.J.Nucl.Phys.*, **12** (1971) 370.
27. A.Veyssiere, H.Beil, R.Bergere, P.Carlos, A.Lepretre. *Nucl.Phys.*, **A159** (1970) 561.
28. L.M.Young. *Ph.D. Thesis, University of Illinois*, 1972.
29. Z.W.Bell, L.S.Cardman, and P.Axel. *Phys.Rev.*, **C25** (1982) 791.

METHODS USED IN PHOTONUCLEAR DATA EVALUATION AT JNDC

NORIO KISHIDA
Advanced Technology Dep't, Technology Development Div., CRC Research Institute, Inc. 1–3D–16, Nakase, Mihama–ku, Chiba–shi, Chiba 261–01, Japan

ABSTRACT

The present status of evaluation work on photonuclear data and evaluation methods used at JNDC are described. The evaluation work began in 1989. The photonuclear data for all natural isotopes will have been evaluated to the end of 1990's. Theoretical evaluation as well as experimental data evaluation are employed. Nuclear model calculation based on pre-equilibrium and evaporation model is carried out and reproduce experimental data quite well. Evaluation results for the ^{12}C, ^{27}Al, ^{100}Mo and ^{209}Bi nuclei are shown as examples.

1. Introduction

A Photo-Reaction Data Working Group was organized as one of working groups of the Japanese Nuclear Data Committee (JNDC) in 1989. Various kinds of photonuclear reaction cross sections have been being evaluated by each member of the group. We, the author and the other group members, purpose to evaluate photonuclear data for all natural isotopes, several transuranic nuclides and some fission products. From a survey of many literatures, we reached the conclusion that it is difficult to construct the photonuclear data file if we adhere to evaluation method to use measured cross sections only. Because there are not sufficient experimental data necessary for the evaluation. For instance, there scarcely exist energy spectra and DDX for emitted particles which were measured using quasi monoenergetic gamma-rays. We are therefore proceeding the evaluation work with the help of theoretical calculation based on statistical nuclear reaction models.

2. Evaluation Plan

2.1. Objective Nuclides

Now we are evaluating the photonuclear data for C, N, O, Na, Mg, Al, Ca, Ti, Cr, Mn, Fe, Co, Ni, Cu, Zr, Mo, Ta, W, Pb, Bi and U. We intend to begin the next evaluation work for H, D, Be, Si, Cl, Ar, K, Sr, Sn, Cs, Pt, Au, Th, Pu and Np from 1994 and to finish it by 1996. The remaining nuclides will have been evaluated by the end of the 1990's.

2.2. Evaluated Cross Sections

The photonuclear cross sections that are to be contained in the forthcoming data file are as follows: photonuclear absorption cross sections not to include elastic cross sections, photoneutron cross sections, photoproton cross sections, yield cross sections for photoneutrons, photoprotons, photodeuterons, phototritons, photo-^{3}He-particles and photo-alpha-particles, single neutron emission cross sections, double neutron emission cross sections, energy spectra, angular distributions and DDX for photoneutrons and photoprotons, and nuclide production cross sections. The maximum energy of incident photons is 140 MeV because the evaluation work becomes very hard once the pion production channel opens. The photonuclear data file is planned to be constructed using ENDF−6 format.

3. Experimental Data

The EXFOR file for photonuclear reaction data was obtained from NEA Data Bank and includes the monoenergetic photoneutron cross sections that Dietrich and Berman[1] have compiled. In evaluation for the nuclides that we can not obtain sufficient quantities of the cross sections from the file, we read values of the cross sections from the figures in literatures. In order to seek the literatures, the abstract sheets[2] edited at NIST are utilized. In addition, Computer Index for Photonuclear Reaction Data[3] like CINDA is being made by one of the group members to aid our evaluation work. The literatures which were published between 1950 and 1990 and treat the photonuclear reactions are registered on the index file.

4. Evaluation of Photoabsorption Cross Sections

4.1. Data Evaluation for Photoabsorption Cross Sections

It is thought that nuclear photoabsorption process up to 140 MeV mainly takes place via excitation to the giant electric dipole resonance (GDR) or via disintegration of a proton-neutron pair (quasideuteron) in a nucleus. The latter process has been originally proposed by Levinger[4] and is called the quasideuteron model (QDM) at present. The former process is dominant for incident photon energies in the range up to about 40 MeV, while the latter process is dominant in the energy range from 40 MeV to 140 MeV.

It is widely recognized that experimental excitation functions of photoabsorption cross sections for GDR region can be well reproduced by superposition of Lorentz resonance curves:

$$\sigma_{abs}^{GDR}(E) = \sum_{i=1}^{n} \frac{\sigma_i^{abs}}{1 + \left[(E^2 - E_i^2)^2 / E^2\Gamma_i^2\right]}, \tag{1}$$

where the Lorentz parameters E_i, σ_i^{abs} and Γ_i are the i-th resonance energy, its peak cross section and its full width at half-maximum, respectively, E is the incident photon energy, and n is the number of resonances. Strictly speaking, σ_i^{abs} and Γ_i are not constant, but energy dependent. Furthermore, σ_i^{abs} is divided into many factors: the ground state spin, that of GDR, its total width, its partial width for de-excitation to the ground state, the wave length of the incident photon and some other physical constants.

The QDM photoabsorption cross section $\sigma_{abs}^{QDM}(E)$ is expressed in terms of the free deuteron photodisintegration cross section $\sigma_d(E)$,

$$\sigma_{abs}^{QDM}(E) = \frac{L}{A} NZ\sigma_d(E)f(E), \tag{2}$$

where L is the Levinger parameter, $f(E)$ is the Pauli-blocking function. The meanings of A, N and Z seem to be obvious. Recently, Chadwick *et al.*[5] have proposed a new derivation of the Pauli-blocking function. As a result, it become possible to calculate the QDM cross sections without free parameters. They also showed that the calculated photoabsorption cross sections are in good agreement with experimental data.[5]

On the basis of the facts mentioned above, we have decided to express the photoabsorption cross section $\sigma_{abs}(E)$ as the sum of $\sigma_{abs}^{GDR}(E)$ and $\sigma_{abs}^{QDM}(E)$,

$$\sigma_{abs}(E) = \sigma_{abs}^{GDR}(E) + \sigma_{abs}^{QDM}(E). \tag{3}$$

In case of existence of the measured photoabsorption cross sections with high accuracy, a least-squares fitting procedure, in which Eq. (3) is the fitted expression and E_i, σ_i^{abs}, Γ_i and L are the fitted parameters, is employed to determine the evaluated photoabsorption cross sections. Although Chadwick *et al.*[5] derived the value $L = 6.5$ without theoretical ambiguity and use of its value indeed reproduced very well many experimental data for nuclides heavier than Sn, we found that this value failed to reproduce the measured values for nuclides lighter than Al and always underestimated those values in the present evaluation. We therefore treat L as a free parameter at present.

Evaluation methods for no experimental photoabsorption cross sections are described in the following two subsections.

4.2. The Case with Photoneutron Data

In a large number of nuclides, it is usual that photoabsorption cross sections have not been measured, but those of photoneutrons have been done. Thus we need the evaluation methods to estimate the photoabsorption cross sections $\sigma_{abs}(E)$ from the photoneutron cross sections $\sigma_{(\gamma,nx)}(E)$. In fact we can carry out such evaluation if we are able to obtain the branching ratios for the photoneutron reactions. Then the absorption cross sections are derived from

$$\sigma_{abs}(E) = \frac{\sigma_{(\gamma,nx)}(E)}{R_{(\gamma,nx)}(E)}, \tag{4}$$

where $R_{(\gamma,nx)}(E)$ represents the branching ratio.

In this evaluation method the evaluated photoneutron cross sections are indispensable. The photoneutron cross sections are expressed as

$$\sigma_{(\gamma,nx)}(E) = \sigma_{(\gamma,nx)}^{GDR}(E) + R_{(\gamma,nx)}(E)\sigma_{abs}^{QDM}(E), \tag{5}$$

where $\sigma_{(\gamma,nx)}^{GDR}(E)$ represents the photoneutron cross sections with excitation to GDR and it is written by the expression that σ_i^{abs} of Eq. (1) is replaced with $\sigma_i^{(\gamma,nx)}$. We have been obtaining the evaluated photoneutron cross sections by fitting Eq. (5) to those obtained by experiment. In addition, since the branching ratios must be known before the fitting, we calculate those ratios using evaporation and pre-equilibrium nuclear model codes, ALICE–F[6] and MCPHOTO, which is modified versions of ALICE–P[7] and MCEXCITON,[8] respectively. The reason why these nuclear models can apply to calculation of the branching ratios is to be described in Sec. 5.

4.3. The Case of No Experimental Data

In the case of no experimental data, it is difficult to obtain accurate evaluated values of the photoabsorption cross sections in the GDR energy region (less than about 40 MeV). Because it is rather hard to calculate precise E1 excitation strength of nuclei purely theoretically. On the other hand, it is well known that GDR of spherical nuclei heavier than ^{51}V is reproduced quite nicely by single-line Lorentz curve.[1] Fortunately, a systematics to give the resonance energy E_R and the resonance width Γ_{tot} have been found via analyses of many experimental data: Berman and Fultz[9] have obtained a systematics to E_R,

$$E_R = 77.9A^{-1/3}\left(1 - e^{-A/238}\right) + 34.5A^{-1/6}e^{-A/238} \quad \text{MeV}, \tag{6}$$

and Carlos *et al.*[10] have gotten that to Γ_{tot},

$$\Gamma_{tot} = (0.026 \pm 0.005)\, E^{1.91\pm0.1} \quad \text{MeV}. \tag{7}$$

On the other hand, the resonance peak strength σ_R can be obtained from a classical sum rule[11] for the strength of electric dipole photon absorption and is given by

$$\sigma_R = 60\frac{NZ}{A}\frac{2}{\pi\Gamma_{tot}} \quad \text{mb}. \tag{8}$$

However, since use of this value always results in underestimation to the experimental photoabsorption cross sections, we use the modified value that is multiplied by 1.2 according to the suggestion by Berman and Fultz.[9]

There is completely no systematics to calculate the photoabsorption cross sections for nuclei lighter than Ti, because shape of their excitation functions reveals quite complicated structure and very different shape every nucleus. This fact results from the excitation process into unbound individual E1 states taking place. Hence, we believe that it is necessary to perform quantum mechanical calculation of nuclear structure in order to obtain the evaluated values for such light nuclei.

5. Theoretical Evaluation

Since it is evident not to construct an evaluated photonuclear data file via experimental data evaluation only, we need to perform theoretical evaluation as well as the data evaluation.

Wu and Chang[12] have first applied a pre-equilibrium reaction model to analysis of the photonuclear reactions. They showed that the model can reproduce neutron and proton energy spectra resulting from irradiation by bremsstrahlung gamma-rays. In 1983, Blann *et al.*[13] have revealed that the model can also reproduce excitation functions of various kinds of photoneutron cross sections, which had been measured using quasi-monoenergetic photons. Then it was made clear to be able to apply the model to the photonuclear reactions.

From the viewpoint of a microscopic nuclear theory, the excitation process of GDR is considered as production process of a particle–hole excited state. On the other hand, QDM assumes that an incident photon excites one neutron and one proton above the Fermi level and leaves two holes in the Fermi see. It is therefore thought that pre-equilibrium, especially exciton models can describe the de-excitation process of photonuclear reactions. Consequently, reaction branching ratios, energy spectra, nuclide production cross sections and other quantities can be obtained from model calculations by ALICE–F and MCPHOTO.

We make the exciton model calculation initiate from a one particle and one hole (1p–1h) configuration in excitation to GDR and from a 2p–2h or a 2p-1h configuration in the quasi-deuteron process. Use of the 2p–1h initial configuration was originally proposed by Blann *et al.*[13] The reason is because the two holes produced from a quasi-deuteron are correlated so that they jointly provide one degree of freedom rather than two. In fact, results of two types of calculations do not present large difference. A level density parameter necessary in evaporation calculation following the pre-equilibrium stage is adjusted in such a way to reproduce the ratio of single neutron emission cross sections and those of double neutron emission if those cross sections have been measured.

6. Some Evaluation Results

6.1. Result for ^{12}C Nucleus

In Fig. 1 the evaluated results[14] for ^{12}C are shown together with the experimental data on photoabsorption,[15] photoneutron[16] and photoproton[17] cross sections. The measured cross sections were simultaneously fitted into by Eq. (3), Eq. (5) and two equations similar to Eq. (5) corresponding to photoprotons and photo-alpha-particles. In this fitting the evaluator certainly considered the condition that the absorption cross sections are sum of the remaining three types of cross sections and those of gamma-emission, and he determined the evaluated cross sections so as to obtain well reproduction to the experimental photoabsorption and photoneutron cross sections. Because he judged that these cross sections have high accuracy in comparison with the photoproton and photo-alpha-particle[18] cross sections not to be measured using 4π-detectors. Then all branching ratios greater than 40 MeV were fixed to certain values which do not vary with the incident photon energies.

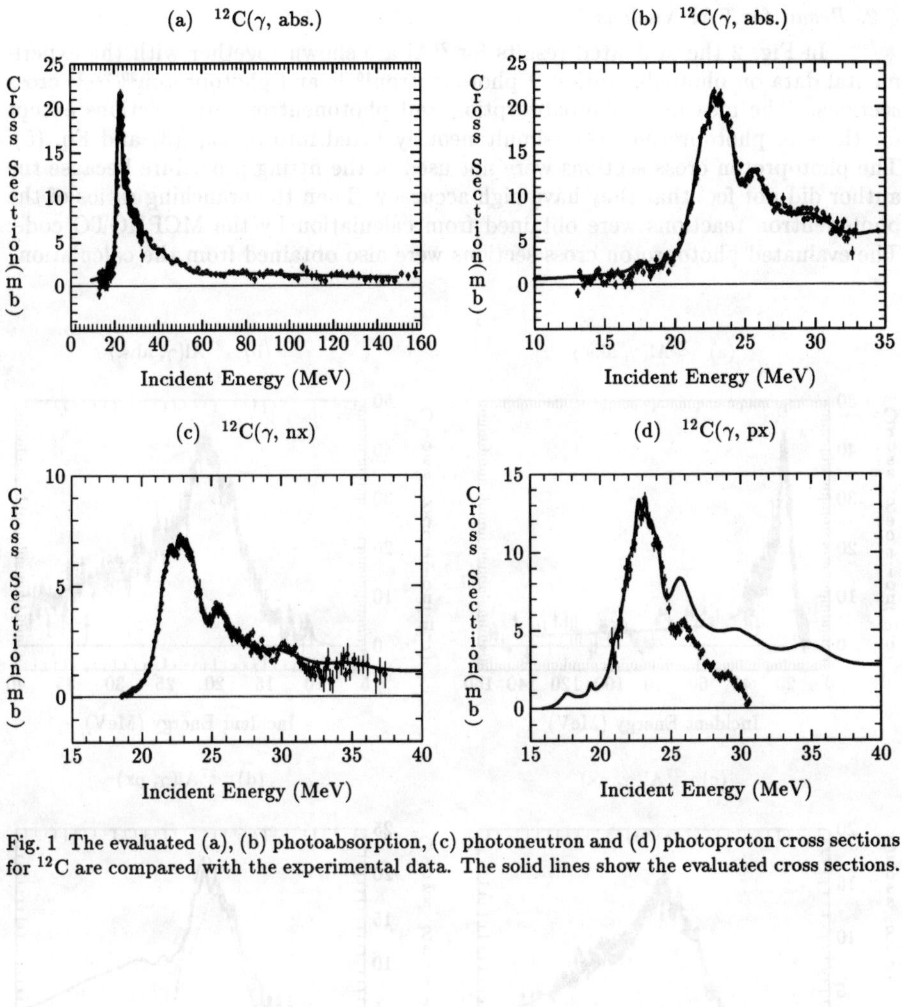

Fig. 1 The evaluated (a), (b) photoabsorption, (c) photoneutron and (d) photoproton cross sections for ^{12}C are compared with the experimental data. The solid lines show the evaluated cross sections.

6.2. Result for ^{27}Al *Nucleus*

In Fig. 2 the evaluated results for ^{27}Al are shown together with the experimental data on photoabsorption,[15] photoneutron[16, 19] and photoproton[20, 21, 22] cross sections. The measured photoabsorption and photoneutron cross sections except for those of photoprotons were simultaneously fitted into by Eq. (3) and Eq. (5). The photoproton cross sections were not used in the fitting procedure because the author did not feel that they have high accuracy. Then the branching ratios of the photoneutron reactions were obtained from calculation by the MCPHOTO code. The evaluated photoproton cross sections were also obtained from the calculation.

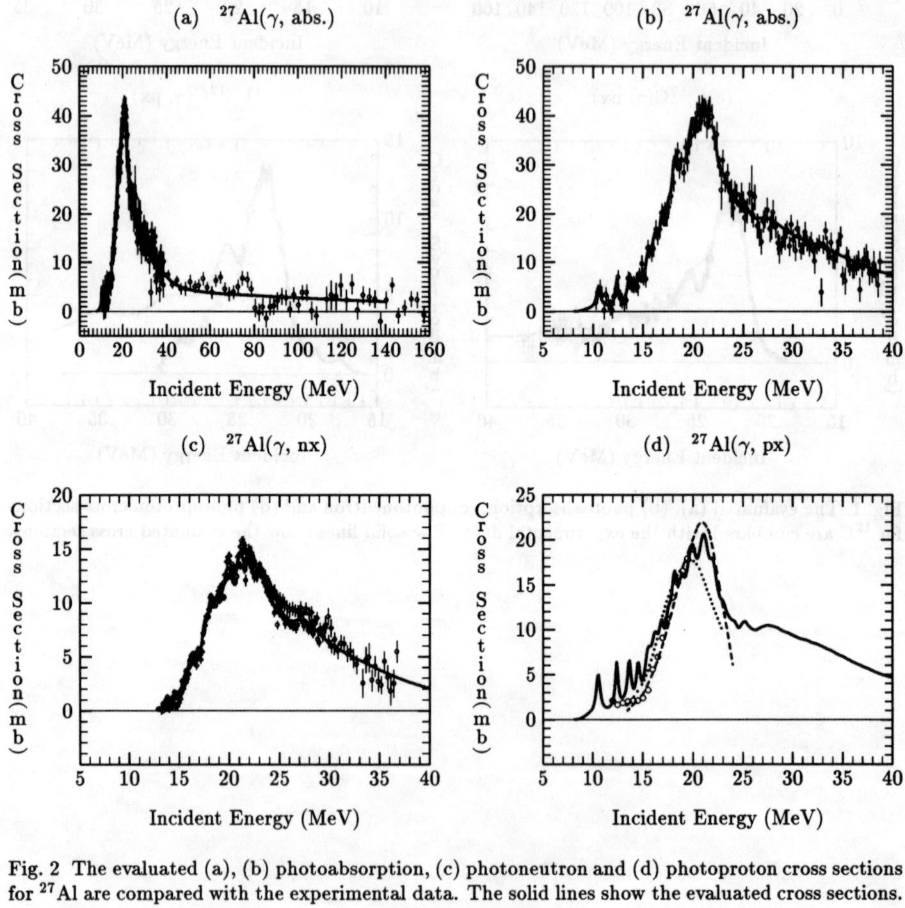

Fig. 2 The evaluated (a), (b) photoabsorption, (c) photoneutron and (d) photoproton cross sections for ^{27}Al are compared with the experimental data. The solid lines show the evaluated cross sections.

6.3. Result for ^{100}Mo *Nucleus*

In Fig. 3 the evaluated photoneutron cross sections for ^{100}Mo are shown together with the experimental data.[23] The measured photoneutron cross sections were fitted into by Eq. (5). Then the branching ratios of the photoneutron reactions were obtained from calculation by MCPHOTO. The evaluated single, double and triple neutron emission cross sections were also obtained by the calculation. It is clearly seen that the calculated cross sections except for those of triple emission reproduce the experimental data rather well. The reason why the calculation fails to reproduce those of triple neutron emission is because the employed value of the level density parameter was probably inadequate for ^{97}Mo.

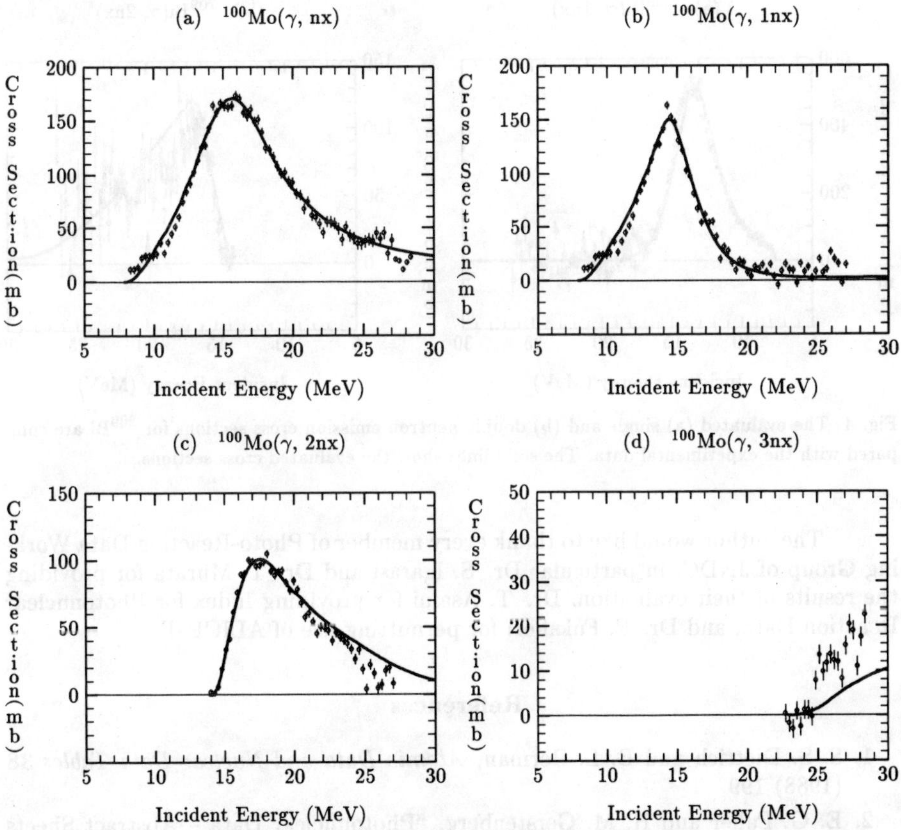

Fig. 3 The evaluated (a) photoneutron, (b) single neutron emission, (c) double neutron emision and (d) triple neutron emission cross sections for ^{100}Mo are compared with the experimental data. The solid lines show the evaluated cross sections.

6.4. Result for ^{209}Bi *Nucleus*

In Fig. 4 the evaluated single and double neutron emission cross sections[24] for ^{209}Bi are shown together with the experimental data.[25] In this evaluation the measured photoneutron cross sections were considered as photoabsorption cross sections, so that those were fitted into by Eq. (3). This is because photoproton cross sections are negligibly small owing to large Coulomb barrier. In addition the evaluator did not vary the Levinger parameter. The evaluated single and double neutron emission cross sections were calculated by ALICE–F and reproduce the experimental data quite well.

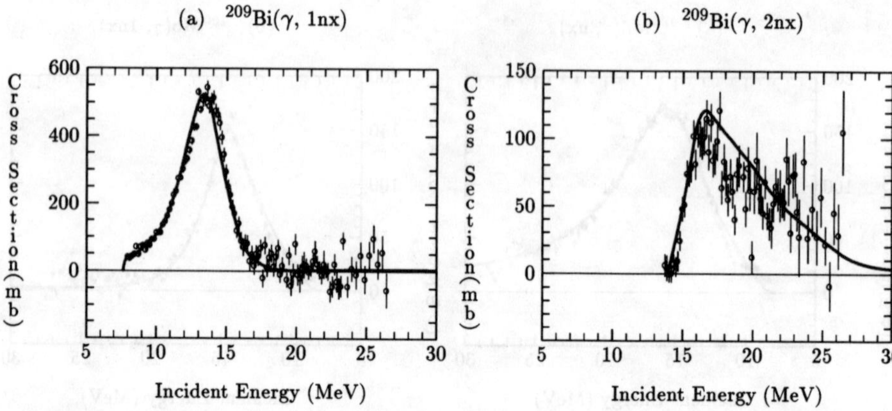

Fig. 4 The evaluated (a) single and (b) double neutron emission cross sections for ^{209}Bi are compared with the experimental data. The solid lines show the evaluated cross sections.

The author would like to thank every member of Photo-Reaction Data Working Group of JNDC; in particular Dr. S. Igarasi and Dr. T. Murata for providing the results of their evaluation, Dr. T. Asami for providing Index for Photonuclear Reaction Data, and Dr. F. Fukahori for permitting use of ALICE–F.

References

1. S. S. Dietrich and B. L. Berman, *Atomic Data and Nuclear Data Tables* **38** (1988) 199.
2. E. G. Fuller and H. M. Gerstenberg, "Photonuclear Data – Abstract Sheets 1955 – 1982," *NBSIR 83–2742*, National Bureau of Standards(1983–1985).
3. T. Asami, private communication.
4. J. S. Levinger, *Phy. Rev.* **84** (1951) 43.

5. M. B. Chadwick, P. Obložinský, P. E. Hodgson and G. Reffo, *Phy. Rev.* **C37** (1991) 814.
6. T. Fukahori, private communication.
7. S. Pearlstein, *Astrophysi. J.* **346** (1989) 1049.
8. N. Kishida and H. Kadotani, in *Proc. Int. Conf. on Nuclear Data for Science and Technology*, ed. S. Igarasi (Saikon, Tokyo, May/June 1988), p. 1143.
9. B. L. Berman and S. C. Fultz, *Rev. Mod. Phys.* **47** (1975) 713.
10. P. Carlos, R. Bergère, H. Beil, A. Leprêtre and A. Veyssière, *Nucl. Phys.* **A219** (1974) 61.
11. J. S. Levinger, *Ann. Rev. Nucl. Sci.* **4** (1954) 13.
12. J. R. Wu and C. C. Chang, *Phys. Rev.* **C16** (1977) 1812.
13. M. Blann, B. L. Berman and T. T. Komoto, *Phys. Rev.* **C28** (1983) 2286.
14. T. Murata, private communication.
15. J. Ahrens, H. Borchert, K. H. Czock, H. B. Eppler, H. Gimm, H. Gundrum, M. Kröning, P. Riehn, G. Sita Ram, A. Zieger and B. Ziegler, *Nucl. Phys.* **A251** (1975) 479.
16. S. C. Fultz, J. T. Caldwell, B. L. Berman, R. L. Bramblett and R. R. Harvey, *Phys. Rev.* **143** (1966) 790.
17. R. Carchon, R. Van de Vyver, H. Ferdinande, J. Devos and E. Van Camp, *Phys. Rev.* **C14** (1976) 456.
18. F. K. Goward and J. J. Wilkins, *Proc. Roy. Soc(London)* **A217** (1953) 357.
19. A. Veyssière, H. Beil, R. Bergère, P. Carlos, A. Leprêtre and A. de Miniac *Nucl. Phys.* **A227** (1974) 513.
20. R. E. Chrien and A. H. Benade, *Phys. Rev.* **119** (1960) 748.
21. J. Halpern and A. K. Mann, *Phys. Rev.* **83** (1951) 370.
22. K. Shoda, K. Abe, T. Ishizuka, N. Kawamura and M. Kimura *J. Phys. Soc. Japan* **17** (1962) 735.
23. H. Beil, R. Bergère, P. Carlos, A. Leprêtre, A. de Miniac and A. Veyssière, *Nucl. Phys.* **A227** (1974) 427.
24. S. Igarasi, private communication.
25. R. R. Harvey, J. T. Caldwell, R. L. Bramblett and S. C. Fultz, *Phys. Rev.* **133** (1964) B1149.

DECAY DATA EVALUATION

DECAY DATA EVALUATION FOR ENDF/B-VI

T. R. England
Theoretical Division, Los Alamos National Laboratory
Los Alamos, NM 87545, USA

J. Katakura
Japan Atomic Energy Research Institute,
Tokai-mura, Naka-gun, Ibaraki 319-11, Japan

F. M. Mann
Westinghouse Hanford Company
P.O. Box 1970, Richland, WA 99352, USA

C. W. Reich
Idaho National Engineering Laboratory
P.O. Box 1625, Idaho Falls, ID 83415, USA

R. E. Schenter
Westinghouse Hanford Company
P.O. Box 1970, Richland, WA 99352, USA

W. B. Wilson
Theoretical Division, Los Alamos National Laboratory
Los Alamos, NM 87545, USA

ABSTRACT

ENDF/B-VI decay files depart from earlier versions in that model data are now used as needed to provide a complete set of average decay energies and spectra. Delayed neutron spectra are also included for the first time. There are 979 radioactive nuclides in the file, and all but 8 have decay spectra. The use of model data has eliminated the "Pandemonium" problem in aggregate fission product calculations, and it greatly improves decay heat calculations. This report describes the data base with an emphasis on the need for models used in combining measurement and model data. Extraction of only the evaluated measurement data is simple, should it be desired by users. We present here an abbreviated chronology of the development of ENDF/B-VI.

1. Introduction

The primary motivation for decay data in ENDF/B-VI is its use in applications. This requires reasonably complete data, including model calculations where there are no measured data. Because of the need for completeness, the Evaluated Nuclear Structure Data file (ENSDF) serves primarily as a starting point for modifications. At the time we started forming the new ENDF file we found that ENSDF would be adequate for only about half of the nuclides needed, and it could supply partial information for about 120 more. These 120 and an additional 300 would require model calculations if there were to be any significant improvements over earlier versions of ENDF for applications.

The reader should understand that ENDF/B-IV and -V files enjoyed many widely reported successes that will not be discussed here. Rather, we emphasize only the problems in order to explain the basis of decisions to incorporate model data. The earlier versions did have some model data, as described in Ref. 1, but not for beta, gamma, and delayed neutron spectra.

Table I summarizes the content of the nuclide decay files in terms of a grouping into classes of activation, actinide and fission products. The actual decay file combines these classes. Some of the fission product nuclides (50) are also activation products, and any common nuclides are removed in the combined ENDF/B-VI file after the modifications discussed in this paper. As a class, more than half of the fission products have incomplete measurements that must be supplemented by model calculations.

The activation and actinide nuclides have measured spectroscopic data converted to ENDF/B formats at Westinghouse Hanford Co. At Los Alamos we tested these data for consistency, Q-values, etc., and found some obvious errors that are not discussed in this paper but were reported in the CSEWG minutes of 1989-90.

Table I indicates 979 unstable nuclides, all but eight having essentially complete spectra, including delayed neutron spectra. Our intent is to explain how the current files depart significantly from earlier evaluations. This is best accomplished by showing you some aspects of the chronological past, covered in part by Refs. 1-31. A description of earlier evaluations and some aspects of the current evaluation are found in Refs. 1-15.

Table I

ENDF/B-VI Decay Data in Activation, Actinide, and Fission-Product Files*

Data Type	File Activation	Actinide	Fission Product
Total Nuclides	158	108	891
Stable	0	0	127
Isomeric States	22	5	168
Delayed Neutron Precursors	0	0	171
Nuclides with Spectra	158	108	755
Gamma and/or X-Ray	119	98	736
Beta and/or Discrete Elec.	119	96	677
Positron and/or e.c.	61	0	40
Alpha	12	71	3

*50 of the activation nuclides are also included in the fission product count. The ENDF/B-VI final decay file is combined into a set that includes 979 unstable nuclides.

2. Delayed Neutrons

There were no delayed neutron spectra in ENDF/B-IV and -V decay files. There are 271 delayed neutron precursors in the current files, all with spectra. Just before ENDF/B-V was issued, we realized that the 6-group representation really only covered the delayed neutron spectra from 70 keV to 1.2 MeV, usually for only four of the time groups and for only six of the fissioning nuclides. An extensive effort was made to provide a new data base for the individual precursors and their subsequent use in generating new six-group data for all fissioning nuclides that have evaluated yields. This was an extensive effort that succeeded only after several years of work. Figure 1, for example, shows aggregate delayed neutron spectra calculated with -VI spectra data compared with results of measurements at the University of Lowell.

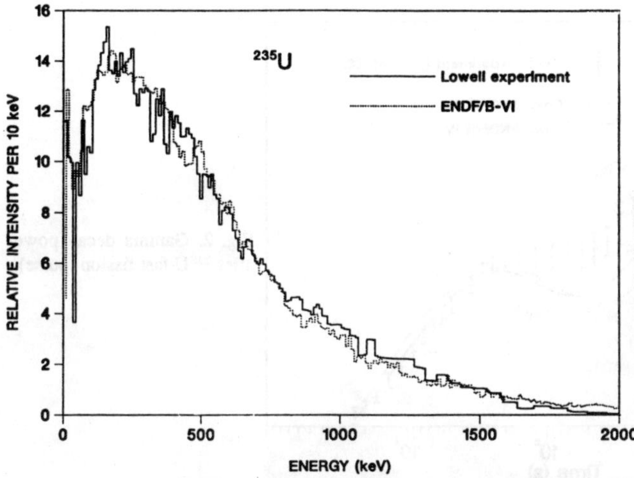

Fig. 1. Comparisons with Lowell delayed neutron spectra for delay interval (0.79-125 s).

The spectral data are in several hundred 10-keV energy bins (up to the maximum emission energy per precursor). The six-group data, derived from the decay and fission product yield files, are in 300 bins, each bin being 10-keV wide (ENDF/B-V used 28 variable energy bins, as did ENDF/B-IV). Some of the spectra are derived from simple models, and some are augmentations of measured spectra with model values. This work is described in over 20 Los Alamos reports and publications, and it will not be disucssed further here. References 11-16 describe the data base.

3. ENDF/B-IV and ENDF/B-V Comparisons

Table II shows the decay and cross-section content in versions -IV and -V. Version -IV had been reasonably successful, and we expected that the large increase in measured data for -V would clear up most of the remaining discrepancies in comparisons with measurements. As we reported in 1981,[7] the -V data showed poorer agreement with decay heat measurements than did -IV for some fuels and certainly for gamma and beta components. This is emphasized by the comparisons with ^{238}U and ^{235}U in Figs. 2-3.

Table II

ENDF/B-IV and -V Decay Decay and Cross-Section Content (Fission Products)

Quantity	ENDF/B-IV (1974-75)	ENDF/B-V (1979-80)
Total nuclides	824	877
No. with cross sections	181	196
Unstable nuclides	711	750
No. with beta or gamma average	711	750
No. with beta or gamma spectra	180	264
No. with delayed neutron Pn	55	105
No. with delayed neutron spectra	0	0
No. with some model spectra	0	0

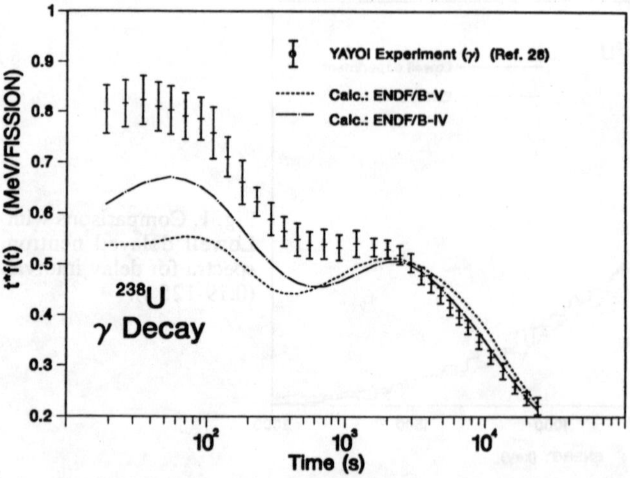

Fig. 2. Gamma decay power after ^{238}U fast fission (pulse).

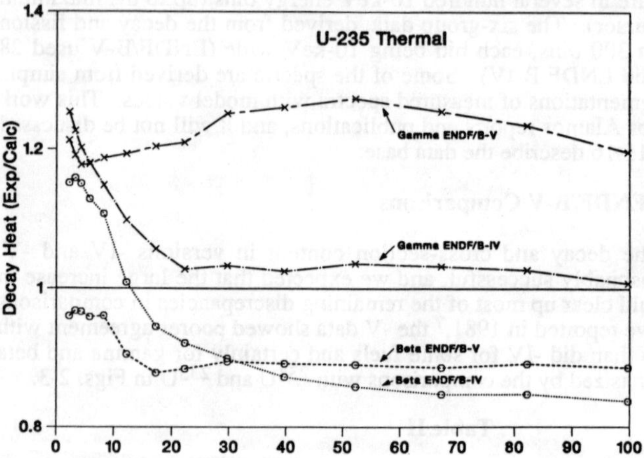

Fig. 3. Ratio of experiment to calculated decay heat for a 1 -s irradiation at constant flux (J.K. Dickens).[31]

Figure 4 is a comparison of gamma spectra using -V data and ORNL measurements. At about 2.5 MeV, the aggregate spectrum formed from the -V data is an order of magnitude low, and it becomes worse at higher energies. Only 264 of the products had gamma spectra, so this situation was qualitatively expected, but such large differences were *not* anticipated. (The same result occurs for other fissionable nuclides.)

Our Japanese colleagues found identical deficiencies using an independent data compilation (Refs. 16-23, 29). They started using the gross theory of beta decay to correct high-Q (>5-MeV) nuclides. These generally appear to suffer from incomplete gamma spectra, causing the total gamma energy to be low and the beta to be high--the so called "Pandemonium" problem. Later, the gross theory was applied to other nuclides that appeared to have deficient, incomplete, or no measured spectra. At the time (1980-85) this effort was directed to the average energies. Later, a gamma cascade model was added for use in estimating spectra.

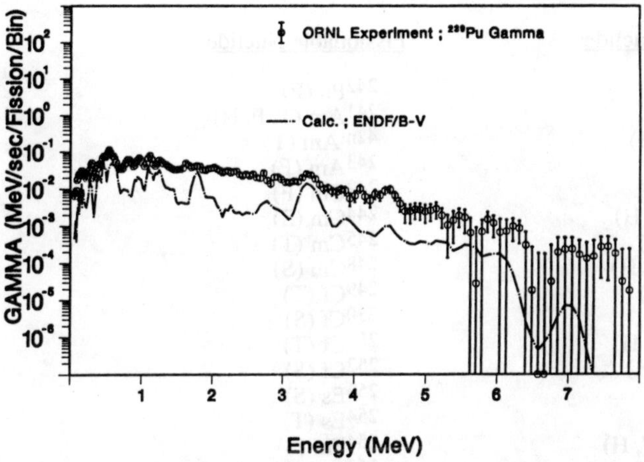

Fig. 4. Gamma spectrum after ^{239}Pu thermal neutron fission (T_{irr} = 1.0 s, $T_{cool.}$ = 2.2 s).

For applications, we clearly could not use ENDF/B-V decay energies. We formed a library consisting in part of the gross theory results and entirely of ENDF/B-V parameters other than decay energies. This effort is summarized in an appendix of the *Summary of ENDF/B-V* report.[8]

So, while we continued working on delayed neutron data and fission product yield evaluations, we stopped using the ENDF/B-V decay energies, in about 1981. There were several libraries formed for average decay energies; for aggregate spectra, we formed exponential fits that incorporated measured results. Nothing was done to improve the individual nuclide spectra in ENDF/B-V before about 1987 (however, we did create updates for programmatic use at LANL).

During 1987 we completed the delayed neutron data along with a preliminary set of fission-product yields for 34 nuclides (later reevaluated in 1989, and ultimately released as mod 0 for ENDF/B-VI.

Table III lists the yield sets. The 1989 yield data have all been reevaluated[30] and 10 yield sets added; we hope to have these released as mod 1 soon. This table lists the 36 nuclides that have one or more yield sets--a total of 60 sets of yields, both independent and cumulative. The yield evaluations were discussed at the JAERI meeting in May-June of this year (1992).

3. ENDF/B-VI Decay Data

In 1987 we received the gross theory code and began using it to fill in the missing energies. It wasn't clear at that time how adequate the new spectroscopic data would be. In December 1988 we decided to use the gross theory code with a gamma cascade option to produce unmeasured spectral data and decay energies, should the new spectroscopic data prove to be inadequate.[24-27]

Table III

Evaluated Yield Sets for ENDF/B (Mod O)

Fissionable Nuclide	Fissionable Nuclide
^{227}Th (T)	^{242}Pu (F)
^{229}Th (T)	^{241}Am (T, F, H)
^{232}Th (F, H)	^{42m}Am (T)
^{231}Pa (F)	^{243}Am (F)
^{232}U (T)	^{242}Cm (F)
^{233}U (T, F, H)	^{244}Cm (S)
^{234}U (F, H)	^{245}Cm (T)
^{235}U (T, F, H)	^{248}Cm (S)
^{236}U (F, H)	^{249}Cf (T)
^{237}U (F)	^{250}Cf (S)
^{238}U (S, F, H)	^{251}Cf (T)
^{237}Np (F, H)	^{252}Cf (S)
^{238}Np (F)	^{253}Es (S)
^{238}Pu (F)	^{254}Es (T)
^{239}Pu (T, F, H)	^{254}Fm (S)
^{240}Pu (F, H)	^{255}Fm (T)
^{241}Pu (T, F)	^{256}Fm (S)

S = Spontaneous, T = Thermal, F = Pooled Fast, and H = 14 MeV

Ten new sets completed: ^{243}Cm(T,F), ^{246}Cm(S,F), ^{244}Cm(F), ^{248}Cm(F), ^{242}Pu(T,H), ^{237}Np(T), and ^{240}Pu(T). These will be included in Mod 1; all other sets have also been reevaluated for Mod 1.

Early in 1989 we replaced the spectroscopic data with new evaluations. Approximately 120 of the 510 nuclides having such line data either had obviously incomplete spectra or were suspected of having the "Pandemonium" problem. For average energies they were replaced with gross theory calculations, and all others used ENDF/B-V data. Figure 5 illustrates the disappointing result. Due to this and similar results for other fissioning nuclides, it seemed clearly necessary to use theory to improve agreement with such aggregate measurements. *Such integral comparisons became a necessary part of the evaluation process.*.

We formed a total of six libraries, checking for agreement with the aggregate decay heat measurements in each case. Each library was based on temporal comparisons, identifying the probable nuclide(s) that caused a lack of agreement; also we examined in detail each nuclide to determine those not showing agreement between line data calculations and Q-values, etc. Admittedly, the process was not exact; as a result, anyone can argue with some (but not all) of the particular choices we made (*e.g.*, ^{102}T$_c$ has been questioned in Ref. 26).

Figure 2 compared ENDF/B-IV and -V with gamma measurements for ^{238}U. Figure 6 is the same comparison using the energies in ENDF/B-VI. Figures 7a-7c show comparisons for ^{235}U thermal fission. For other fissioning nuclides the results are almost as good.

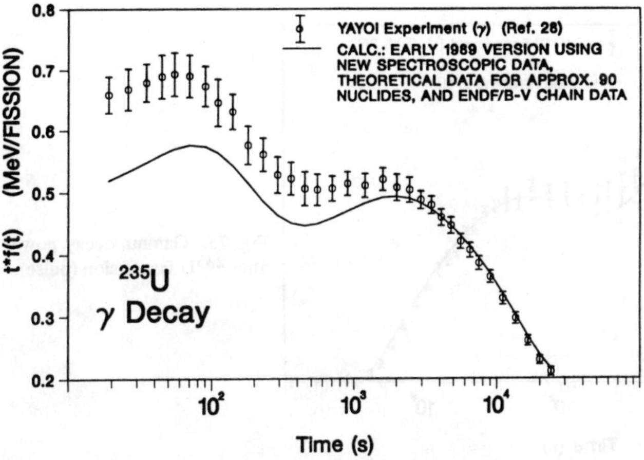

Fig. 5. Gamma decay power after ^{235}U fast fission (pulse).

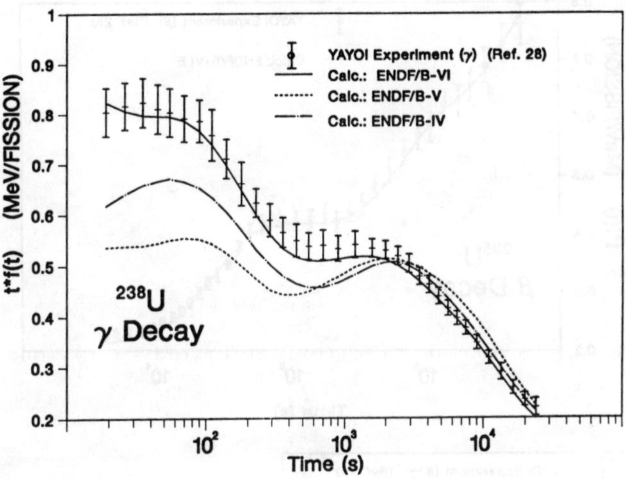

Fig. 6. Gamma decay power after ^{238}U fast fission (pulse).

The spectra present another problem; here we made a decision that cost us much work: we would not only use gross theory to generate spectra for unmeasured nuclides, but, for those having some measurements, we would augment the spectra with theory, including augmentation of any spectra that failed to reproduce the average gamma energies within ~ 5%. In the process, we did not want to lose the spectroscopic data.

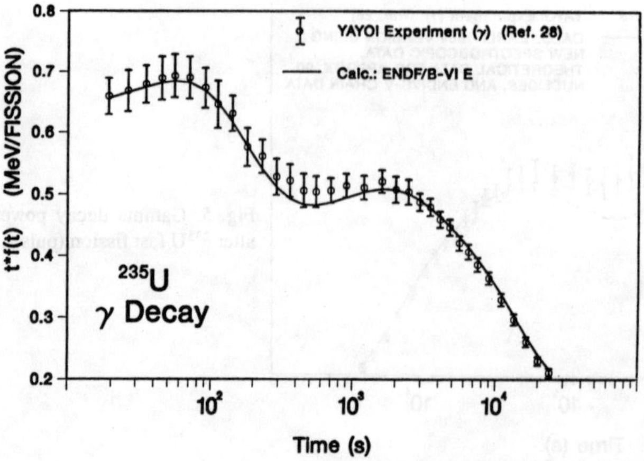

Fig. 7a. Gamma decay power after ^{235}U fast fission (pulse).

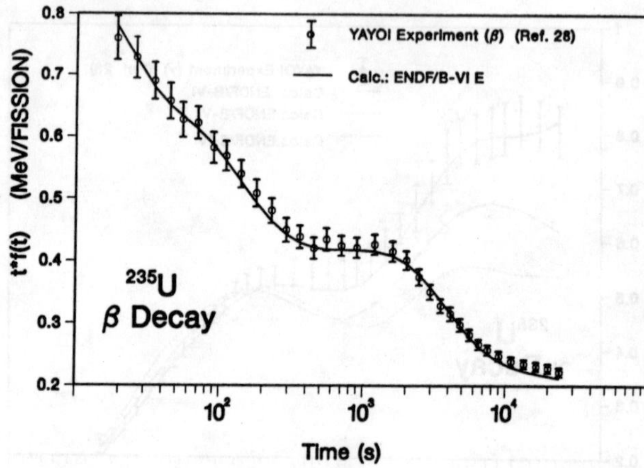

Fig. 7b. Beta decay power after ^{235}U fast fission (pulse).

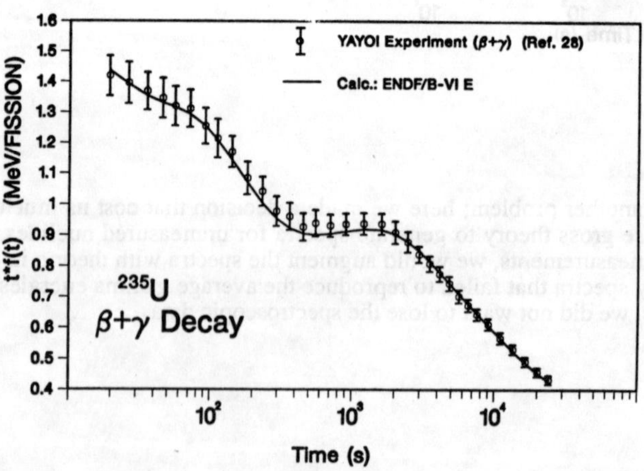

Fig. 7c. Total decay power after ^{235}U fast fission (pulse).

All added spectra occur in the class of nuclides labeled as fission products, and all are continuous spectra, including the extensive delayed neutron spectra. The files contain a simple key (starting with "LANL Modifications") that identify the 420 nuclides having some added theoretical values; also their total energies are listed. The documentation report (in progress) will contain a table with explicit identification of these nuclides and modifications. Anyone who wants to eliminate the theoretical values of spectra, can simply ignore the continuous spectra and change the normalization factor for beta spectra.

Earlier we discussed the comparison of ENDF-V spectra with Oak Ridge measurements. Figure 8 presents a spectral comparison using -V and -VI data. We have compared over 400 such spectra and included these in Ref. 27. The particular manner in which we used gross theory calculations is also discussed in Ref. 27.

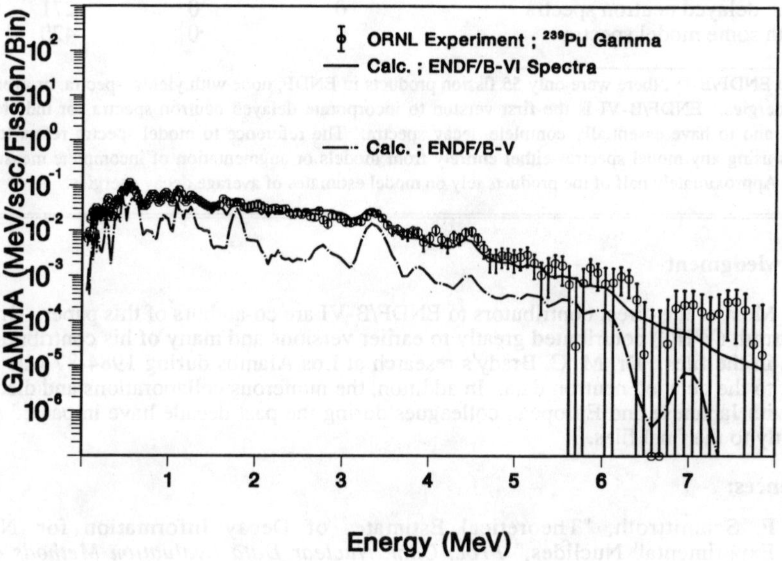

Fig. 8. Gamma spectrum after ^{239}Pu thermal neutron fission ($T_{irr} = 1.0$ s, $T_{cool} = 2.2$ s).

Conclusion

Only a few comparisons have been described in order to validate the necessity of using theory. Obviously, many more comparisons were needed as part of the evaluation process.

Table IV is a comparison of the content of the fission products in -IV, -V, and -VI. With the exception of 8 of the products, all contain essentially complete spectra, including 271 delayed neutron spectra. 420 products have either theory or theory augmenting the measured data. The files contain an additional 215 actinides and activation products having only measured spectra. The decay files contain a total of 979 nuclides.

No measured data are lost; on the other hand, for applications, much has been gained.

Table IV

Decay and Cross Section Evolution[a] (Fission Products)

Quantity	ENDF/B-IV (1974-75)	ENDF/B-V (1979-80)	ENDF/B-VI (1991-92)
Total Nuclides	824	877	891
No. with Cross Sections	181	196	196
Unstable Nuclides	711	750	764
No. with beta or gamma average	711	750	764
No. with beta or gamma spectra	180	264	755
Nol with delayed Neutron Pn	55	105	271
No. with delayed neutron spectra	0	0	271
No. with some model spectra	0	0	420

[a]Prior to ENDF/B-IV, there were only 55 fission products in ENDF, none with yields, spectra, or average decay energies. ENDF/B-VI is the first version to incorporate delayed neutron spectra for individual nuclides and to have essentially complete decay spectra. The reference to model spectra refers to all products using any model spectra--either entirely from models or augmentation of incomplete measured spectra. Approximately half of the products rely on model estimates of average decay energies.

Acknowledgment

Most of the direct contributors to ENDF/B-VI are co-authors of this paper. Dr. F. Schmittroth (WHC) contributed greatly to earlier versions and many of his contributions remain in the files. Dr. M. C. Brady's research at Los Alamos during 1984-87 was important to the delayed neutron data. In addition, the numerous collaborations and discussions with Japanese and European colleagues during the past decade have impacted significantly to the final files.

References:

1. F. Schmittroth, "Theoretical Estimates of Decay Information for 'Non-Experimental' Nuclides," *Proc. Conf. Nuclear Data Evaluation Methods and Procedures,"* Brookhaven National Laboratory, Upton, New York, 22-25 September 1980, BNL-NCS-51361 (March 1981), Vol. I, p. 185.

2. R. L. Bunting and C. W. Reich, "Evaluation Procedures for Experimental Decay Data," *Proc. Conf. Nuclear Data Evaluation Methods and Procedures,"* Brookhaven National Laboratory, Upton, New York, 22-25 September 1980, BNL-NCS-51361 (March 1981), Vol. I, p. 163.

3. C. W. Reich, R. L. Bunting, *Nucl. Sci. Eng.* **82**, 132 (1982).

4. C. W. Reich, R. G. Helmer, M. H. Putnam, "Radioactive-Nuclide Decay Data for ENDF/B," Idaho National Engineering Laboratory report ANCR-1157 (ENDF-210) (August 1974).

5. "Decay Data of the Transactinium Nuclides," IAEA Technical Report Series No. 261, International Atomic Energy Agency, Vienna (1986).

6. T. R. England, R. E. Schenter, and F. Schmittroth, "Integral Decay-Heat Measurements and Comparison to ENDF/B-IV, and V," Los Alamos Scientific Laboratory report LA-7422-MS [NUREG/CR-0305] (August 1978). Also published in *Proc. Seminar Nuclear Data Problems for Thermal Reactor Applications*, Brookhaven National Laboratory, 22-24 May 1978. [EPRI-NP-1098, BNL-NC-25047, ENDF-270, June 1979.]

7. W. B. Wilson, T. R. England, and D. E. Wessol, "ENDF/B-V Actinide and Fission Product Decay Power," *Trans. Am. Nucl. Soc.* **38**, 651 (June 1981).

8. T. R. England, W. B. Wilson, R. E. Schenter, and F. M. Mann, "ENDF/B-V Summary Data for Fission Products and Actinides," Electric Power Research Institute report NP-3787 (ENDF-332) (December 1984).

9. F. M. Mann, "Status of Dosimetry and Activation Data," *Proc. Int. Conf. Nuclear Data For Science and Technology*, Mito, Japan, 30 May-3 June 1988, p. 1013.

10. C. W. Reich, "Review of Nuclear Data of Relevance for the Decay Heat Problem," *Proc. Specialists' Mtg. on Data for Decay Heat Predictions*, Studsvik, Sweden, 7-10 September 1987, NEANDC-245 'U" (1987).

11. F. M. Mann, M. Schreiber, R. E. Schenter, and T. R. England, "Evaluation of Delayed Neutron Emission Probabilities," *Nucl. Sci. Eng.* **87**, pp. 418-431 (August 1984).

12. T. R. England, W. B. Wilson, R. E. Schenter, and F. M. Mann, "Aggregate Delayed Neutron Intensities and Spectra Using Augmented ENDF/B-V Precursor Data," Nucl. Sci. Eng. **62**, pp. 139-155 (October 1983).

13. M. C. Brady and T. R. England, "Delayed Neutron Data and Group Parameters for 43 Fissioning Systems," *Nucl. Sci. Eng.* **103**, No. 2, 129-149 (October 1989).

14. G. Couchell, P. Bennett, M. Haghighi, E. Jacobs, D. Pullen, W. Schier, Q. Sharfuddin, R. Tanczyn, M. Villani, "Measurements of Delayed Neutron Energy Spectra of ^{235}U, ^{238}U, and ^{239}Pu," *Proc. 1988 Int. Reactor Physics Conf.*, Jackson Hole, Wyoming, September 1988.

15. *Proc. Specialists' Mtg. on Delayed Neutrons*, University of Birmingham, Birmingham, England, 15-19 September 1986, proceedings issued June 1988 (see papers by England *et al.*, Mann, Goulding, Reeder, Couchell, Weaver).

16. J. Katakura and R. Nakasima, "Average Decay Energies in JNDC file Version 2," *Proc. Specialists' Mtg. on Data for Decay Heat Prediction,* Studsvik, Sweden, 7-10 September 1987, NEACRP-302 'L', NEANDC-245 'U' (1987).

17. M. Akiyama and J. Katakura, "Measured Data of Delayed Gamma-Ray Spectra from Fissions of ^{232}Th, ^{233}U, ^{235}U, ^{238}U, and ^{239}Pu by Fast Neutrons: Tabular Data," Japan Atomic Energy Research Inst. report JAERI-M 88-252 (1988).

18. T. Yoshida and R. Nakasma, *J. Nucl. Sci. Tech.* **18**, No. 6, 393-407 (June 1981).

19. T. Yoshida and J. Katakura, "Calculation of the Delayed Gamma-Ray Energy Spectra from Aggregate Fission Product Nuclides," *Nucl. Sci. Eng.* **93** (1986).

20. J. Katakura and T. Yoshida, "Gamma-Ray Spectrum Data Library of Fission Product Nuclides and Its Assessment," Japan Atomic Energy Research Inst. report JAERI 1311 [NEANDC(J)-124/U, INDC(JPN-110/L)] (1988).

21. T. Yoshida, "Theoretical Calculation of Decay Data of Short-Lived Nuclides for JNDC Fission Product Decay File, Japan Atomic Energy Research Inst. report JAERI-M-83-127 [NEANDC(J)-92/U, INDC(JAPO)-79/L] (August 1983).

22. K. Tasaka, H. Ihara, M. Akiyama, T. Yoshida, Z. Matumot, and R. Nakasima, "JNDC Nuclear Data Library of Fission Products," Japan Atomic Energy Research Inst. report JAERI 1287 [NEANDC(J) 91/U, INDC(JAPO 78/L)] (1983).

23. K. Takahashi, M. Yamada, and T. Kondoh, "Beta-Decay Half-Lives Calculated on the Gross Theory," Atom. Data and Nucl. Data Tables **12**, 101 (1973) (see also references therein).

24. T. R. England, F. M. Mann, C. W. Reich, R. E. Schenter, "ENDF/B-VI Radioactive Decay and Yield Libraries, *Trans. Am. Nucl. Soc.*, **60**, 614 (1989).

25. T. R. England *et al.*, in *Nuclear Theory and Applications Progress Report, Jan. 1, 1989-April 1, 1990*, Los Alamos National Laboratory report LA-11972-PR, pp. 44, 96 (December 1990).

26. C. W. Reich and T. R. England, "The File of Evaluated Decay Data in ENDF/B," Trans. Am. Nucl. Soc., Annual Mtg., 2-6 June 1991, Orlando, Florida, **63**, 163.

27. J. Katakura and T. R. England, "Augmentation of ENDF/B Fission Product Gamma-Ray Spectra by Calculated Spectra, Los Alamos National Laboratory report LA-12125-MS (ENDF-352) (October 1991).

28. M. Akiyama, S. An, "Measurements of Fission-Product Decay Heat for Fast Reactors," *Proc. Int. Conf. on Nuclear Data for Science and Technology*, Antwerp, 6 September 1982 (Reidel Pub., Dordrecht, 1983), p. 237. (See also M. Akiyama, J. Katakura, "Measured Data of Delayed Gamma-Ray Spectra from Fission of ^{232}Th, ^{233}U, ^{235}U, ^{238}U, and ^{239}Pu by Fast Neutrons: Tabular Data," Japan Atomic Energy Research Inst. report JAERI-M-88-252 (December 1988).

29. K. Takahashi, M. Yamada, *Prog. Theor. Phys.* **41**, 1470 (1969).

30. T. R. England and B. F. Rider, "Evaluation and Compilation of Fission Product Yields, 1989," (ENDF-349), in progress, to be published as a Los Alamos National Laboratory report as the primary documentation for ENDF/B-VI.

31. J. K. Dickens *et al.*, Nucl. Sci. Eng. **74**, 107 (1980).

SYMPOSIUM ON NUCLEAR DATA EVALUATION METHODOLOGY

Decay Data Evaluation For JEF-2

Blachot Jean
CEA Grenoble, France

Nordborg Claes
OECD/NEA Data Bank, France

Abstract

The JEF-2.2 radioactive decay data library contain information for 2344 radioactive nuclides and has been compiled at the NEAData Bank. The library results from a cooperation of physicists from UK, Sweden, Germany, and France.The data originated from different sources. A selected combination of the UK and the French data files, supplemented with data from ENSDF, constituted the basis for the library. Different evaluations and measurements have been used to complement this starter file. A number of average beta and gamma energies obtained from experiments by Rudstam et.al. have for example been included, as well as some theoretically calculated average energies and half lives by Klapdor-Kleingrothaus et.al.. Delayed neutron branching ratios of precursors (Pn) have been taken from a recent measurement and an evaluation by Rudstam et.al.. The atomic mass adjustment of Audi and Wapstra of 1988 have been used to update most of the decay energies. A small data base (TBRA) was created to follow recent publications not yet entered in the ENSDF data base.Great care has been taken to verify the consistency of the data. In particular the ratio between the available energy and the total energies in different spectra were tested and adjusted as close as possible to 1.Decay heat calculation will be presented as part of the validation of this new library.The JEF-2.2 radioactive decay data library with its documentation (JEF Report 13 and the program for PC computers developed at Univ. of Birmingham, UK) will be extensively validated and work on the improvement of the data will continue for subsequent versions of the library.

1. Introduction

1.1 Joint Evaluated File

The Joint Evaluated File (JEF) project was started in 1981 to bring together, in the OECD/NEA Data Bank member countries, the evaluation efforts in order to create a comprehensive reference data base for fission reactor application.

The special work on Radioactive Decay Data started somewhat later. This part of the library resulted from a scientific collaboration between laboratories in France, Germany, Sweden, and the United Kingdom.

The first version of the library, JEF-1, was completed in 1985 but the JEF Scientific Coordination Group decided to produce a second improved version of the file. The starting point for this new file was the JEF-1 data.

1.2 Origin of the JEF-1 decay data library

The JEF-1 file was mainly built by merging data from the CEA and the UK libraries.

The French file has been used for a long time in France. The data were available through a local French Atomic Energy (CEA) network (CISI). Many programs were written[1] to update the file, to retrieve the information and to present the data. This file was also used as library input for decay heat calculation with the PEPIN[2] and the MECCYCO[2] codes. The CEA file are in a special data format. A computer code[3] was written to convert the data into the ENDF[4] format. Summary and test results were written into the comment part of the file. An example is given in Table 1.

Table 1

Sample of CEA library

```
35-BR- 85 NEADB    RCOM-JUL83 JEF DECAY DATA WORKING GROUP
----JEF-2          MATERIAL 3552
 *********************************************************
     DATA WERE TAKEN FROM THE FRENCH LIBRARY
 *********************************************************
***  35BR   85
 **DECAY DATA EVALUATION CEA OCT. 1984
***J.BLACHOT-B.DUCHEMIN-C.FICHE-B.NIMAL ***

   MEAN BETA- ENERGY                        1042.15    KEV
   MEAN GAMMA ENERGY                          66.03     KEV
   EFFECTIVE Q VALUE                       2566.17    KEV
   CALCULATED Q VALUE                      2577.82    KEV
      % DEVIATION                                0.45
                        ***COMMENTS***
* 07/08/80 * ENSDF: 85BR B- DECAY
              75NU03,75HU02,71ER15  80NDS
```

The UK library were already in ENDF format. The evaluators in UK used the COGEND[5] program. This program output the data in the correct format but also performs a few tests and calculate some parameters not directly measured. The Table 2 shows of the UK Data.

Table 2

Sample of UK library

```
35-BR- 80MNEADB       RCOM-JUL83 JEF DECAY DATA WORKING GROUP
*****************************************************************
 JEF-2
   DATA WERE TAKEN FROM THE UK LIBRARY
*****************************************************************

 35-BR- 80 M      DECAY SCHEME EVALUATION  A.TOBIAS   SEP-80
DECAY DATA FROM ENSDF-EXTRACTION PROGRAM BY B.S.J.DAVIES

       MEAN GAMMA ENERGY               =       14.6203  KEV
       MEAN X-RAY ENERGY               =        9.5408  KEV
       MEAN AUGER ELECTRON ENERGY      =        8.0464  KEV
       MEAN CONVERSION ELECTRON ENERGY=        53.9239  KEV
       EFFECTIVE Q VALUE               =       85.9499  KEV
       UNCERTAINTY IN EFFECTIVE Q (%) =         0.4654
      CALCULATED Q VALUE               =      86.1314  KEV
      % DEVIATION                      =      -0.2111

    ENDF/B- V DATA PRODUCED BY COGEND (A.TOBIAS JAN-1984)
```

2. The compilation of the JEF-2 decay data file

2.1 JEF Working Group

A working group was created to coordinate the work. The summary records of the working group are reported in internal JEF documents. The policy of the group was to recommend as correct and recent data as possible, but also to be careful on the consistency of the recommended data. In particuliar the ratio between the effective Q-value and the calculated Q-value was the main criterium.

(a) effective Q-value $=\sum$ Qi BRi

where Qi and BRi are the Q-value and branching ratio of the i-th decay mode

(b) calculated Q-value $=\sum$ Eai Pai

where Eai and Pai are the energies and emission probabilities of all spectra

2.1.1 ENSDF

Bhat presented, at the last Nuclear Data meeting in Julich[6] the Evaluated Nuclear Structure Data File (ENSDF), the description of the network and the content of the file. The description of the file is also given in a BNL report[7].

This file is mainly oriented to nuclear structure.
The decay data are present in the file , but ordered following the final nuclide. To combine data from a parent with different decay modes, different data sets have to be merged. For instance, if a nuclide decays via internal transition and beta- decay , one data set (IT decay) will be found in the A,Z part and the other (beta-) in the A,Z+1 part of the file. Often the data sets are from different evaluators and common parameters, such as half-life, branching etc., might differ. Only the Q values are to a certain extend standardized as they are always taken from one of the Wapstra-Audi mass evaluations.
The ENSDF network has recently been relatively active in updating the mass-chains as can be seen from figure 1.

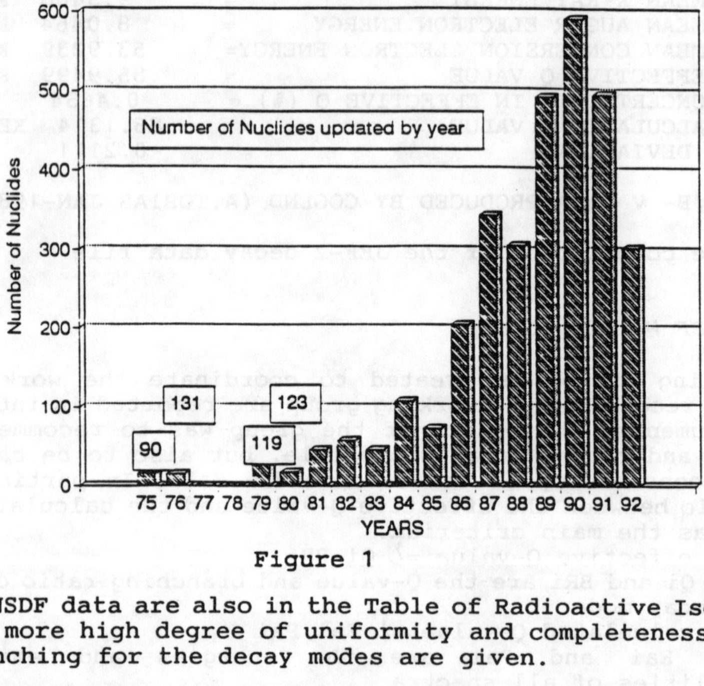

Figure 1

The ENSDF data are also in the Table of Radioactive Isotopes with a more high degree of uniformity and completeness[8]. But no branching for the decay modes are given.

2.1.2 RADLST

The computer code RADLST[9] is designed to calculate the nuclear and atomic radiations associated with the radioactive decay of nuclei. It uses, as its primary input, decay data in the ENSDF format and presents the calculations in a variety of forms. One of the output formats is the ENDF format.

The file contains the decay modes, the branching energies, the intensities for various nuclear radiations, conversion electrons, electron-positron pairs from internal pair formation. The energies and intensities of the associated atomic radiations (X-rays and Auger electrons) are also calculated. The program does not output prompt or delayed neutrons nor decay protons.

2.2 Compilation program

The actual compilation of the JEF-2 decay data file is performed by a semiautomatic computer code, originally written by T. Nakagawa. The program reads the basic input files stored on the computer and uses an index file, where information on all corrections and additions are stored. It makes the necessary modification to the basic data, performs a number of internal consistency checks and documents all what has been done or found in the comment part of the final data set. Among the items considered for internal checking, and sometimes automatic correction, could be mentioned: Sum of branchings, energy consistencies, Q-values, and FT-values.

2.3 Statistics on the JEF-2 file

Some statistics on the JEF-2 file are given in Table 3

Table 3

Status of JEF2-2

Total No. of Nuclides =2597			
Radioactive Nuclides	2344	Total No. of Gamma	69835
Ground State	1840	Total No. of Beta-	7851
1st Excited State	470	Total No. of Beta+	7876
2nd Excited State	34	Total No. of Alpha	1733
Stable Nuclides	253	Total No. of Electron	37227
Nuclides with Spectra	=2139	Total No. of X-Ray Lines	6249

3. Modification to the basic file

3.1 Mean Ebeta Egamma energies.

Rudstam et.al.[10] have carried out extensive investigations on beta and gamma spectra of individual fission products. The description of the gamma and the beta measurement and the analysis of the data have been published . Some of these data can be compared with mean energies derived from level schemes found in ENSDF. When the level scheme is well known, the agreement in the two data sets is generally good as can be seen from the figure 2a[9].

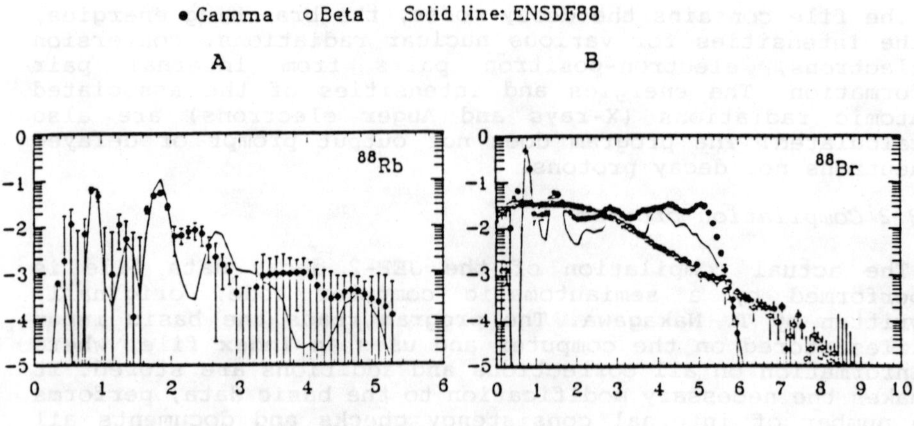

Figure 2

For many nuclides with incomplete level scheme, in particular those with high Q-beta, the high energy gammas were often not measured. See figure 2b. Some nuclides have not enough data to derive a complete level scheme. Mean beta and gamma energies of 55 nuclides of Rudstam were included in the JEF-2 file.

3.2 The P(n)

A number of compilations and evaluations of delayed neutron branchings have been reported during recent years (see for example England[11]). The values in the ENSDF data base often differs from the data in these evaluations, which sometimes use unpublished data or internal reports, not yet referenced in the Nuclear Structure Reference (NSR[12]) file.
A recent evaluation of Rudstam[13] presents an updated list of delayed-neutron branchings. In order to fill the gaps, where the existing data are meager or even lacking, an extensive series of measurements have been carried out at the OSIRIS isotope on-line separator facility at Studsvik. As a result of this effort delayed-neutron branchings have been measured for 52 precursors, in many cases with a considerable improvement in accuracy. It is always difficult to judge the quality of the different experimental results; a value with a small uncertainty can still be erroneous. Rudstam discard values which are far outside of the bulk of determinations, even if their error estimates are small. If the data are widely discrepant, Rudstam uses an unweighted average. The P(n) values in the JEF-2 file have been tested[14,15] using a summation calculation to derive the nu bar for the 39 fissile systems.

The results are in general agreement with the recommended values[16] for Nu-bar. Only the 238U system needs more work but the problem seems to come from the JEF-2 fission yield library.

3.3 Theoretical Calculation for un-measured parameters

Most of the calculations of the parameters associated with the beta decay was done by the gross theory[17] during recent years. To improve this calculation which essentially ignores the existence of the nuclear shell model Klapdor et.al. have calculated microscopically the beta decay with a Tamm-Dancoff approximation (TDA) approach. A description of the model can be found in a paper of the group[18] .The JEF Working Group has decided to use the calculated values ofKlapdor-Kleingrothaus, only when no experimental data were available. 17 values adopted in JEF for half-lives, 101 mean beta and gamma energies, and 32 P(n) values.

3.4 Nuclear binding energies and Q values

For decay heat calculation the uncertainty in decay energy have to be reduced. The progress in Nuclear Physics during these last decade could be easily illustred by comparison with data in two masses evaluation
The Table 4 shows the difference for 4 nuclides of A=142.

Table 4

Comparison between 1964 and 1991

Nuclide	QB- ([19])		QB- ([20])	
142XE			5040	100
142Cs			7303	10
142BA	2200	100	2213	5
142LA	4510	30	4512	4

Using the recent data of Audi, we have compared with JEF2 as shown in Table 5

Table 5
Comparison between Audi et al[20] and JEF2(partial)

Nuclide	:RTYP	: Q(Audi)	: Q(JEF2A)(kev)	: DIFF.(%)
12Mg 21	: 2.0	: 13093.00	: 13098.00	: 0.0
16S 29	: 2.0	: 13793.00	: 13790.00	: 0.0
17Cl 32	: 2.0	: 12686.00	: 12687.00	: 0.0
40Zr 83	: 2.0	: 5868.00	: 6020.00	: -2.5
32Ge 85	: 1.0	: 10355.00	: 10080.00	: 2.7
32Ge 85	: 1.5	: 4621.00	: 5033.49	: -8.9

3.5 Half-lives.

A small data base, TBRA[21] have been created. It contains data for radioactive nuclides (in the case of metastable isotopes only half-lives greater than one millisecond) for the level energy of the isomer, the spin and parity, the half-life and its uncertainty, the different decay modes and their branchings, the date of the last update and a flag indicating if the data were taken from another source than ENSDF. An example is given in Table 6 for the mass 79.

Table 6

The 'TBRA' file

NUCLIDE	LEV		SPIN	T		DT	BRA	REFE
79CU	0	0.0		0.188	S	25	%B-=100$%B-N=55 17	F 91KR
79ZN	0	0.0		1.0	S	1	%B-=100 $%B-N=1.3 4	F 86EK
79GA	0	0.0	(3/2-)	3.00	S	9	%B-=100$%B-N=0.055 12	T 82
79GE	0	0.0	(1/2-)	19.1	S	3	%B-=100	T 82
79GE	2	185.	(7/2+)	39.0	S	10	%B-=96 1$ %IT=4 1	T 82
79AS	0	0.0	3/2-	9.01	M	15	%B-=100	T 82
79SE	0	0.0	7/2+	6.5E+4	Y	LT	%B-=100	T 82
79SE	2	95.7	1/2-	3.91	M	5	%IT=100	T 82
79KR	0	0.0	1/2-	35.04	H	10	%EC+%B+=100	T 82
79KR	2	129.	7/2+	50	S	3	%IT=100	T 82
79RB	0	0.0	5/2+	22.9	M	5	%EC+%B+=100	T 82
79SR	0	0.0	(3/2-)	2.25	M	10	%EC+%B+=100	T 82
79 Y	0	0.0	5/2+	14.4	S	15	%EC+%B+=100	F 92GR

136 nuclides have new references in TBRA. These new references are only for papers which contain half-life information mainly on newly identified nuclides.
Using TBRA, it has recently been possible to update many nuclides in JEF-2. Only half-lives, decay modes and branchings have been updated using TBRA. The comparison between the two files have been made with a small FORTRAN program.

Table 7
Comparison between TBRA and JEF2 (partial)

Nuclide	T (TBRA)		T (JEF2A)		DIFF.(%)
14Si 32	0.1720E+03	Y	0.3296E+03	y	-91.6
29Cu 76	0.3500E+00	S	0.6100E+00	s	-74.2
33As 88	0.8400E-01	S	0.5600E+00	s	-566.6
41Nb 89 m	0.1180E+01	H	0.2030E+01	h	-72.0
34Se 90	0.3480E+00	S	0.1800E+00	s	48.2
41Nb 91	0.6800E+03	Y	0.9990E+04	y	-1369.1
38Sr102	0.3550E+03	MS	0.6800E-01	s	80.8

400 nuclides are not yet in JEF-2 but are present in TBRA. These nuclides may be added in the next update.

3.6 The contribution of A. Nichols

A re-evaluation of the decay data for 126 heavy elements and actinides (Hg206 to Es253) has been made by A. Nichols[22] This re-evaluations include lists of the references used to produce the proposed decay schemes and comments identifying their deficiency. A particular care was applied in testing the consistency of the data.

4. Documentation of the JEF-2 file

The paper documentation of the JEF-2 decay data library (JEF Report 13) is under preparation and will contain the following information:

a. A table containing,for each isotope, the most important decay parameters, such as average energies, half-lives, Q-values, spin and parity, branching, and number of spectral lines.

b. A list of the origin of all isotopes and the reference for any modifications to half-lives, average energies, P(n) values, and the eventual inclusion of continuum spectra.

c. A table showing graphically the decay of the mass-chains from A=71 to A=170, including cumulative and chain yields for the most important fissionable systems.

d. A comparison of the main decay parameters from the JEF-2, ENDF/B-VI, and JNDC-2.0 libraries.

A computer code, for personal computers (PC's), containing general information such as half-life, Q-values, average decay energies, as well as detailed gamma and alpha spectra for all isotopes in the JEF-2 decay data file, was developed in Birmingham, and will be part of the official documentation of the library.

5. Testing

5.1 Individual testing.

In November 1990 the ENDF/B-VI and JEF-2 libraries (both in a preliminary stage of development) were compared for fission product nuclides. A report giving the list of nuclides missing in the 2 libraries and the difference in decay modes and branching and also the number of transitions in the different spectra, was published[23]

5.2 Decay Heat

Storrer[24] has used the JEF-2.2 data to calculate decay heat as shown in figure 4.

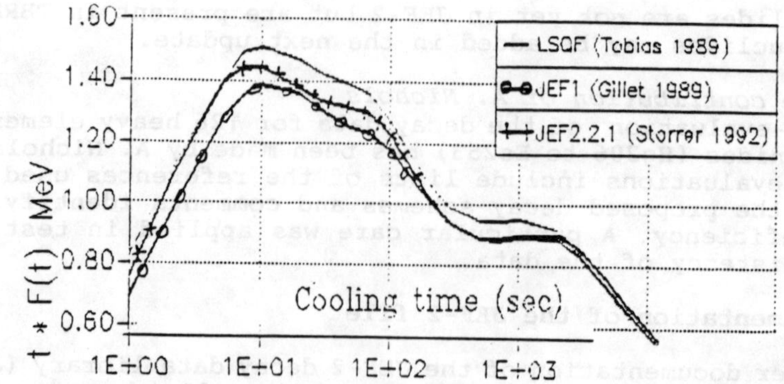

Figure 4

6. Suggested improvements

The creation of a small file, such as TBRA, that summarizes parameters from decay data sets could be useful to compare the future updates of the JEF library with data in ENSDF.

7.Acknowledgements

One of us J.B. wants to thanks O. Bersillon ,C. Phillis B and J. P. Nimal, B. Duchemin.

8. References

1. J. Blachot, C. Fiche ,*At. Nucl. Data Tables* **20** (1977) 241
2. B. Duchemin and C. Nordborg in *Nuclear Data for Science and Technology* ed.S.M. Qaim (Springer-Verlag,1991), p556
3. B. Nimal, B. Duchemin SERMA/LEPF/86/744 (1986)
4. P. F. Rose and C.L. Dunford *ENDF-102* **BNL-NCS-44945**
5. A. Tobias *COGEND* **RD/B/N4147 (1977) RD/B/N4309 (1978)**
6. M.R. Bhat ENSDF, idem to 2, p. 891
7. J.K. Tuli, **BNL-NCS-51655-Rev.87**
8. T. W. Burrows, *The program Radlst* **BNL-NCS-52142** (1988)
9. G. Rudstam et al, *At. Nuc. Data Tables.* **45** (1990) 239
10. E. Browne,R. B. Firestone, Table of Radioactive Isotopes A Wiley-Interscience Publication (1986)
11. T. R. England et al, **LA-UR-86-2693**
12. S. Ramavataram, C.L. Dunford, **BNL-NCS-51800**
13. G. Rudstam submitted to *At. Nucl. Data Tables*
14. J. Blachot, idem to 2 ,p. 89
15. A. Mills idem to 2 ,p. 86
16. J. Blachot et al, NEANDC-299"U"
17. K. Takahashi *Prog. Theor. Phys.* **41** (1969) 1470
18. A. Staudt et al. , *At. Nuc. Data Tables.* **44** (1990) 79
19. J.H. Mattauch and A.H. Wapstra, *Nucl Phys.* **67** (1964) 1
20. G. Audi, *private communication* , Nov. 1991
21. O. Bersillon and J. Blachot, *CEA N 2643* (1990)
22. A.L. Nichols , *CNDC(91)* p61
23. J. Blachot and T.R. England Unpublished (1990)
24. F. Storrer, Private communication,and Thesis (1992)

COVARIANCE DATA GENERATION

COVARIANCES OF EVALUATED NUCLEAR DATA – PRESENT STATUS AND PROSPECTS FOR THE FUTURE

H. Vonach

Institut für Radiumforschung und Kernphysik, Universität Wien
Boltzmanngasse 3
1090–Vienna, Austria

ABSTRACT

The paper reviews the needs for covariance information both with respect to amount and the quality of the data needed for the various applications, the different methods for obtaining such informations and the present status of covariance files in existing evaluated nuclear data files. Finally the paper discusses the priorities for future work in this field and makes some proposals for improvement of the present still rather unsatisfactory situation.

1. Introduction:

At the 1986 IAEA specialist Meeting [1] R.W. Peelle pointed out "The need for variance/covariance data of evaluated nuclear data rests on the simple but powerful proposition that there would be little sense in evaluating a million cross section quantities if no corresponding information were recorded on how well the numbers are known". This statement leads us to the central problem, the necessity to create and store an enormous amount of information; in principle for every evaluated quantity information about its uncertainty is needed. Up to now this goal has only been achieved to a very limited extent and thus I will try to address the question how the present situation could be improved in the most efficient way.

For this purpose I will first discuss the needs for covariances in some detail both with respect to the quantity and the quality of information needed for different applications. Then (as this is the subject of a number of subsequent papers) I will shortly review the various methods for creating uncertainty information at different levels of sophistication and summarize the status of the covariance information in the present evaluated nuclear data files. Finally I will try to address the question how the present limited resources could be used most efficiently to improve the existing rather unsatisfactory situation.

2. Needs for covariances of evaluated nuclear data

Our need for covariance information is basically twofold:

1) Rough uncertainty information for the whole body of evaluated nuclear data, that is for all materials and all types of data, especially cross sections, angular distributions and secondary energy distributions.

2) Detailed, rather precise covariance data for a rather limited number of important materials and cross section types.

Uncertainty information of type 1 is needed for a quick estimate of the influence of the uncertainties of nuclear data on any nuclear response that might become of technological interest. Such estimates will in many cases allow the decision that nuclear data uncertainties are not important or at least to reduce the problem from many materials and cross section types to a few important ones. For this purpose it will be sufficient to have what I want to call "qualitative" covariance information that is essentially reasonably safe estimates of the maximum cross section uncertainties in a very crude bin structure (e.g. a very short MF 33 LB 1 sub–subsection); however this information should be as complete as the corresponding cross section files.

On the other hand there exists a rather small number of materials and cross section types for which high accuracy is needed e.g. cross section standards, fission cross sections of major actinides, cross sections for neutron dosimetry reactions, cross sections for structural materials and cross sections for neutron multipliers to be used in fusion reactor technology. In these cases it is essential that the evaluations really provide the best information on these cross sections which can be extracted from the existing experimental data base and theoretical knowledge and to provide accurate estimates of the uncertainties of the evaluated "precise" cross sections. In such cases it is necessary to combine all available information quantitatively by generalized least–squares techniques (as for example successfully completed in the ENDF/B VI standards evaluation [2]) and covariances are then obtained automatically as a by–product of such evaluations based on advanced statistical methods.

This needed qualitative and quantitative covariance information is still lacking to a large extent. For the majority of the evaluated cross sections there exists no covariance information at all and for many important materials we have only qualitative instead of quantitative covariance information. For example for ^{56}Fe, one of the most important materials for fusion reactor shielding total cross section uncertainties of the most recent evaluated files (derived by rough qualitative methods) are estimated to be about 3% [3,4], whereas quantitative evaluation of the existing data base can reduce these uncertainties to well below 1% as shown in a contribution to last week's NSC Specialist Meeting [5]. Thus in order to satisfy the needs also in the future both qualitative and quantitative methods for generation of covariance information will be needed.

3. Methods for generating covariance information

3.1. *Qualitative methods*

For getting rough estimates of cross section uncertainties a number of qualitative methods have been developped [3,4]. If at least some experimental data are available the uncertainties are mostly based on the scatter of these data. If no experimental data are available qualitative estimates have been either based on the scatter of results between different evaluations [3] or even more approximate on the observation that in general the relative cross section uncertainty is negatively correlated with the absolute cross section values and thus some empirical relation between these quantities is used to get rough estimates for the cross section variances [4].

Although these methods may appear rather crude, they have been remarkably successful in estimating the uncertainties of evaluated cross sections from nuclear model calculations even in the absence of experimental data as demonstrated by the comparison of evaluations using such qualitative covariances with later evaluations using an improved data base and quantitative evaluation methods [5]. Examples for this are given in figures 1 and 2. On average it seems that the uncertainties tend to be somewhat overestimated by the evaluators using qualitative methods especially if the uncertainties are estimated intuitively for the scatter of experimental data, because many authors tend to choose "uncertainty bands" which include most of the data points, although in case of several measurements the uncertainty of the evaluated (averaged) cross sections is expected to be considerably better than those of the individual measurements.

3.2. *Quantitative methods:*

In case of the discussed qualitative methods the evaluation itself and the uncertainty estimates are two different successive steps which actually have been performed by different scientists [3]. Generation of quantitative covariance information on the other hand is an integral part of the evaluation process by means of generalized least–squares or Bayesian methods. These methods are well–developped [6] and subject of the next lectures, thus I will only devote a few remarks to this subject:

1) Evaluation purely based on experimental data: Here the main problems are inconsistent experimental data. To my experience [7] this problem can be handled adequately by both rejection of obviously wrong data and accounting for smaller inconsistencies by increasing all data uncertainties so as to achieve a value of chi–square per degree of freedom of about unity [7,8]. Evaluations based on one data set only (essentially using one single measurement as evaluated cross section) is always dangerous. Especially if such measurement contradicts theoretical expectations, a reliable evaluation is only possible after further measurements [8,9].

2) Evaluations based on joint use of experiment and theory: This is one of the most important problems. For materials important in fission or fusion technology e.g. iron we need consistent and complete high quality evaluations. Although usually a large number of accurate cross section measurements exist for such materials, the demand for completeness makes it necessary to use cross sections from model calculations at least for those cross sections where the measurements cover only a part of the necessary energy range. In general these calculations depend on a considerable number of parameters (optical potential parameters, level–density parameters), thus in principle one could determine these parameters and their covariance matrix from a least–squares fit to all available experimental data as recently proposed and discussed by D. Muir [9] in considerable detail. In practice attempts to do this have frequently run into the difficulty that the model calculations even with variation of all adjustable parameters were not able to simultaneously reproduce the full set of experimental cross sections especially if some of these had been determined to accuracies of only a few percent. Thus it appears that at the few percent level deficiencies of the present theoretical models for cross section calculations show up, which limit the accuracy of model calculations even with optimum adjustment of parameters to values of about 5 – 10%, which is much worse than obtainable in carefully designed experiments. Thus at present it is not quite clear how to combine the results of precision measurement and of model calculations in a consistent way which retains the accuracy of the experimental data base. One way might be to add some additional flexibility to the theoretical descriptions by including effects like statistical fluctuations in level densities or intermediate structures which might permit better fits to the experimental data. Alternatively one could rely completely on the experimental data for the cross section types and neutron energy ranges covered by accurate measurements and use theory only as a means for extra– and interpolation to fill the gaps.

3) A new approach towards solving this problem has recently been presented at last weeks meeting on covariances at Oak Ridge [5]. Basically the following procedure was used:

a) As a starting a recent evaluation (EFF–2) based on nuclear model calculations was selected.

b) Qualitative covariance matrices for these calculated cross sections were estimated from comparison of the selected evaluation with other recent evaluations. A Gaussian type correlation function with energy dependent width was assumed for calculation of the off–diagonal elements of the covariance matrices of all cross sections.

c) Results of step 1 and 2 served as prior data vector and covariance matrix for the Bayesian Code GLUCS which was used to improve this prior by successive addition of all available relevant data so as to obtain an improved set of evaluated cross sections and an improved covariance matrix. This final result retains the advantage of the calculation–based evaluation (completeness and consistency); in addition it makes full use of the experimental data base resulting in much reduced uncertainties for those energy ranges and cross section types for which precision measurements exist.

The improvements in accuracy achievable in this way are considerable as shown in table 1. Of course these improvements depend strongly on the quality of the experimental data base, however just for the materials most important for applications precision measurements of many cross sections have been performed, which have not been fully utilized in present evaluations.

TABLE 1 Cross section uncertainties of the new theory + experiment evaluation of ref. 5 for ^{56}Fe compared to other recent evaluations

Type of cross section	Cross section uncertainties within range of evaluation (.85 – 20 MeV (%))		
	Ref. 5	EFF–2	ENDF–B/VI
σ_t	.50 – 1.55	3.0	3.33 – 3.96
σ_{non}	0.78 – 2.40	3.0 – 15	4.45 – 6.46
$\sigma_{n,n1}$	1.33 – 4.6	15 – 60	4.9 – 95.3
$\sigma_{n,n2}$	3.14 – 61.2	20 – 60	11.2 – 257.4
$\sigma_{n,n3}$	4.86 – 93.8	15 – 90	10.5 – 36.2
$\sigma_{n,cont}$	3.15 – 29.5	15 – 35	10.0 – 37.4
$\sigma_{n,p}$	0.44 – 6.0	3 – 20	1.51 – 18.75
σ_{nnp}	16.1 – 47.2	30 – 50	20 – 52.1
$\sigma_{n,2n}$	3.0 – 7.80	20 – 30	10.0 – 83.9
$\sigma_{n,\alpha}$	4.81 – 81.4	20 – 80	19.5 – 40.6

4. Status of covariance information in the recent evaluated data files

The status of the covariance information in the important recent evaluated nuclear data files is shown in table 2.

TABLE 2 Status of covariance information in important recent evaluated nuclear data files

File	Covariance Information
ENDF/B VI Standards	excellent quantitative complete covariance information
ENDF/B VI General Purpose File	covariance mostly derived by qualitative methods for cross sections in the continuum range for isotopes of Li, B, C, N,O, V, Cr, Mn, Fe, Co, Ni, Cn, Y, Nb, An, Pb, Bi, V and Pn
JENDL–3	none
JEF–2	none
BROND	none
EFF–2	qualitative covariance information for continuum cross sections and elastic scattering angular distributions for ^{52}Cr, ^{56}Fe, ^{58}Ni and ^{60}Ni
EAF (European Activation File	qualitative covariance information to be added in the near future
IRDF 90 (International Dosimetry File)	satisfactory quantitative covariance information for about 75% of the reactions, for the remaining reactions an update of the existing evaluations and their covariances is desirable

The situation shown in this table may be summarized as follows:

a) Satisfactory covariance information exists only for the neutron cross section standards and for the majority of the neutron dosimetry reactions.

b) For the most important materials the ENDF/B VI general purpose file (and in case of the structural materials also EFF) provide covariance information for continuum cross sections. This information, however, was mostly derived by qualitative methods only.

c) No covariances exist for a large number of materials of lesser importance.

d) With very few exceptions no covariances exist for cross sections in the resonance range.

e) Except for the structural materials no covariances exist for the angular distributions of elastic scattering.

f) There is no covariance information on any material for the energy and angular distributions of secondary neutrons nor of total and energy differential gamma–

production cross sections. These quantities can at present not even be described in the ENDF/B VI format.

5. Needs and priorities for future work

Basically future work on covariances should focus on two problems of about equal importance.

1) Creation of rather complete covariance information by means of rough qualitative methods. This information should include not only cross sections, but all neutron nuclear data of interest in fusion and fission technology especially elastic angular distribution, energy and angular distributions of secondary neutrons (needed for neutron transport calculations) and γ–production cross sections (needed for calculation of γ–heating).

2) Accurate quantitative covariance information for a relatively small number of most important materials selected by sensitivity calculations within the various user communities.

Successful work towards these two goals will also require some further improvements in our methodology. In detail I would like to mention just three points.

1) There should be new formats for specifying the uncertainties of coupled energy–angular distributions of secondary neutrons and for total and energy differential γ–production cross sections. Suggestions for such formats have been made at the recent NSC specialist meeting [10].

2) It would be highly desirable if the "intuitive" methods for generation of qualitative covariance information could be formalized in a way as to allow "mass–production" of such covariances with limited resources. The procedures chosen for creation of uncertainties for the European Activation File (EAF) are certainly a very important step in this direction.

3) More work seems necessary on the problem how to perform accurate quantitative evaluations, which involve combining our knowledge from both accurate measurements and model calculations.

LITERATURE

[1] W.Peelle,
Report INDC(NDC)–192/L, Vienna 1988

[2] A.D. Carlson et al.,
Proc.Int.Conf.on Nuclear Data for Science and Technology, Santa Fe, 13.–17.5.1985, Gordon and Breach, N.Y.

[3] S. Tagesen and H. Vonach,
Proc.Int.Conf.on Nuclear Data for Science and Technology, Jülich, 13.–17.5.1992, Springer Berlin, ed. S.M.Qaim, p. 871

[4] D.M. Hetrick, D.C. Larson and C.Y. Fu,
Report ORNL/TM 11763, Feb. 1991

[5] V. Pronyaev, H. Vonach, S. Tagesen and M. Wagner,
Contribution to the NSC Specialist Meeting on Evaluation and Processing of covariance data, Oak Ridge, Oct. 1992 (in press)

[6] D.L. Smith,
Probability Statistics and Data Uncertainties in Nuclear Science and Technology, Am.Nucl.Soc., 1991

[7] H. Vonach,
Proc.Int.Conf. on Nuclear Data Evaluation Methods and Procedures, Brookhaven, 22.–25. Sept. 1980, Report BNL–NSC–51363 (1981), p. 621

[8] M. Wagner et al.,
Physics Data 13–5, Fachinformationszentrum Karlsruhe (1990)

[9] D.D. Muir,
Proc.of Workshop on Computation and Analysis of Nuclear Data relevant to Nuclear Energy and Safety, Triest, 10.2.–13.3.1992 (in press)

[10] H. Vonach
Contribution to the NSC Specialist Meeting on Evaluation and Processing of covariance data, Oak Ridge, Oct. 1992 (in press)

GENERAL PROCEDURES AND COMPUTATIONAL METHODS FOR GENERATING COVARIANCE MATRICES

C.BASTIAN

CEC-JRC-Central Bureau for Nuclear Measurements

B-2440 Geel, Belgium

ABSTRACT

The covariance matrix of a set of n observations may be broken down in an uncorrelated and a correlated part. The latter may in turn be expressed as the product of a rectangular matrix with its own transpose, in as far as common sources of uncertainties are known. This covariance representation scheme may be applied to the analysis of neutron induced reaction measurements, starting from raw multichannel counting spectra, going through calibration, crunching, correction, etc.. steps and delivering a spectrum of measured point values of the cross-section. To perform and record the analysis of cross-section measurements on a PC or a workstation, we have devised a file format AGS in which intermediate and final results are stored as spectra including the current covariance information in the form introduced before. Every step of the analysis is designed as a program reading one or more spectra from the AGS file, taking specific sources of uncertainty into account, and appending one or more new spectra to the file.

Introduction

The debate within the nuclear community is still open as to wether experimenters should provide explicit covariance matrices along with the measured values, or detailed tables of errors with correlation information[1]. In the case of large sets of cross-section values like those measured using accelerator facilities, the second approach is not practical : on-site data reduction and analysis involves many steps of computation, which complicates the survey of the uncertainties. The amount of work involved in putting up a posteriori a complete table of errors and correlations, later on in re-assembling these data to a covariance matrix is indeed formidable. Nevertheless, there is an understandable request from the side of the evaluators to know exactly how the covariance matrix -if provided with the data- was generated.

In the following we propose a solution in the form of a file format capable of documenting step by step the on-site reduction and analysis of a multichannel cross-section measurement. The file format includes a record of the covariance matrix propagation through every step of the analysis.

1 Compact covariance representations

1.1 The case of raw multichannel counting spectra

A counting histogram as produced by a multichannel acquisition system consists of a vector of **n** count numbers $c_1, c_2, ...c_n$. On these values, the uncertainties due to counting statistics are not correlated. Their individual variance i.e. $<\Delta c_i \times \Delta c_i>$. is generally estimated by c_i , the count number itself. Accordingly, the covariance matrix of a raw multichannel counting spectrum is diagonal, with diagonal values identical to the count numbers. There is no need for covariance storage in that very special case.

1.2 Building the covariance of data from the literature

In order to be used in an evaluation, a set $m_1, m_2, ...m_n$ of **n** measurements of related physical quantities must be complemented with their covariance, a $n \times n$ matrix of the average uncertainty product estimators $<\Delta m_i \times \Delta m_j>$.

When the measured values are only available in the literature as single points from different authors and/or distinct experimental runs, most of the work of the evaluator consists in extracting and weighing the "paper uncertainties" and in identifying the sources of uncertainties which are specific or common to some of the measurements[2,3]. Having eventually identified the source of common uncertainties as a set of **k** parameters $p_1, p_2, ...p_k$, one may express the uncertainty on any measurement m_i as:

$$\Delta m_i = \delta m_i + (\partial m_i/\partial p_1) \times \Delta p_1 + ... + (\partial m_i/\partial p_k) \times \Delta p_k \quad (1)$$

where

δm_i	is the amount of uncertainty which is specific to m_i
$\partial m_i/\partial p_k$	is the sensitivity of m_i towards the source parameter p_k
Δp_k	is the uncertainty on p_k

By definition, the δm_i's are independent from each other and from the Δp_k's

$$<\delta m_i \times \delta m_j> = 0 \qquad \text{if } i \neq j \quad (2)$$

$$<\delta m_i \times \Delta p_k> = 0 \qquad \text{any } i,k \quad (3)$$

We will assume in this section that the Δp_k's are independent from each other.

$$<\Delta p_k \times \Delta p_m> = 0 \qquad \text{if } k \neq m \quad (4)$$

Going now to vector-matrix notation we express the covariance as

$$\mathbf{C = (\delta M + S \times \Delta P) \otimes (\delta M^T + \Delta P^T \times S^T)} \quad (5)$$

where

$\mathbf{\delta M, \Delta P}$	**are vectors containing the δm_i's and Δp_k's**
$\mathbf{S}$	**is the sensitivity matrix of the $\partial m_i/\partial p_k$'s**
$\mathbf{S^T}$	**is the transpose of $\mathbf{S}$**
$\otimes$	**denotes the average tensor product of two vectors**
$\times$	**denotes the usual matrix product.**

Eqs(2,4) imply that $\mathbf{\delta M \otimes \delta M^T}$ and $\mathbf{\Delta P \otimes \Delta P^T}$ are diagonal matrices: We therefore set

$\mathbf{\delta M \otimes \delta M^T = U}$ **uncorrelated part of the covariance matrix**

and factorize

$$\mathbf{\Delta P \otimes \Delta P^T = T \times T^T} \quad (6)$$

T is diagonal : its terms are the square roots of the variances of the $\mathbf{p_k}$'s

Setting now

$$\mathbf{ST = S \times T} \quad (7)$$

and noting that, due to Eq (4)

$$\mathbf{\delta M \otimes \Delta P^T = \Delta P \otimes \Delta M^T = 0}$$

we may reduce Eq (5) to:

$$\mathbf{C = U + ST \times ST^T} \quad (8)$$

where

U is purely diagonal and may be stored as a vector of n values

ST is rectangular and has the same dimensions $\mathbf{n \times k}$ as S

1.3 Covariance breakdown in the general case

The covariance breakdown of Eq(8) may be applied in all cases where the sources of uncertainty are identified and the sensitivity matrix $\mathbf{S}$ is known. In general, the uncertainties on the source parameters $\mathbf{p_1}, \mathbf{p_2}, \ldots\mathbf{p_k}$ may be correlated, so that their covariance matrix $\mathbf{\Delta P \otimes \Delta P^T}$ has non-diagonal terms. This happens typically with background corrected spectra produced by a least-squares fit operation.

Nevertheless the covariance matrix $\mathbf{\Delta P \otimes \Delta P^T}$ is by definition symmetric and positive definite: it may be factorized to satisfy Eq(6). In this case, the standard Cholesky method of factorisation produces $\mathbf{T}$ as a lower triangular matrix of $\mathbf{k \times k}$ so that $\mathbf{ST}$ has the dimensions $\mathbf{n \times k}$, and the breakdown scheme of Eq(8) applies.

Let us remark that the uncorrelated term $\mathbf{U}$ has the same dimension as for raw counting spectra (Sec 1.1), which may be considered as a special case of Eq(8) without the correlated term.

The covariance representation in Eq(8) obviously saves storage space when the number $\mathbf{n}$ of measured points is much larger than the number $\mathbf{k}$ of causes of uncertainties. The classical matrix representation uses $\mathbf{n \times n}$ numbers. A smarter symmetric storage scheme would use $\mathbf{n \times (n+1)/2}$ numbers. The representation scheme of Eq (8) needs only $\mathbf{n \times (k+1)}$ numbers to convey the same amount of information.

<u>**N.B.**</u> :Established ways of representation of the covariance of <u>evaluated</u> cross-section data points e.g. the LB rules of ENDF[4] involve a certain amount of approximation. The representation scheme introduced here avoids approximations and reduces redundance. However our purpose is the presentation of measured data in a reliable form to be used as the <u>input</u> of an evaluation program. We do not support here the idea of presenting the <u>results</u> of an evaluation in another form than the ENDF format.

2. Covariance propagation through the steps of an analysis

2.1. Reduction and analysis as a sequence of operations on spectra

The computing process starting from raw multichannel counting spectra and producing a set of measured cross-section values may be seen as a sequence of steps e.g condensation, normalization, background correction, etc... Every of these steps may be devised as a computer program reading the results of the preceding steps as spectra and producing spectra to be used as the input of the following steps.

We thereby define a <u>spectrum</u> as a data set including at least a vector of $\mathbf{n}$ data points with its associated covariance matrix in the representation of Eq(8), i.e. a vector $\mathbf{U}$ of $\mathbf{n}$ values plus a rectangular matrix $\mathbf{ST}$ of $\mathbf{n \times k}$ values. We shall see further on that a spectrum may also contain supplementary information e.g. on the channel bounds or on experimental parameters of the measurement.

2.2 .Elementary operations on spectra

The simplest operation on a spectrum **Y** is probably the multiplication of all values with an experimental constant α known to the uncertainty $\Delta\alpha$, i.e. $\mathbf{Y_1} = \alpha.\mathbf{Y}$

The covariance of the result in the form of Eq(8) will be

$$\mathbf{C_1} = \mathbf{U_1} + \mathbf{ST_1} \times \mathbf{ST_1}^T$$

where $\mathbf{U_1} = \alpha^2.\mathbf{U}$ and $\mathbf{ST_1}$ is equal to **ST** right-laminated (extended) with a column containing $\mathbf{Y}.\Delta\alpha$ in order to account for the new common source of uncertainty.

In general, a unary operation on a spectrum as symbolized by

$$\mathbf{Y_1} = \mathbf{F}\ (\mathbf{Y},\alpha)$$

will be accompanied by

$$\mathbf{U_1} = (\partial\mathbf{F}/\partial\mathbf{Y})^2.\mathbf{U}$$

$$\mathbf{ST_1} = \mathbf{ST},(\partial\mathbf{F}/\partial\alpha).\Delta\alpha$$

where the comma is a matrix lamination and the dot a term-by term multiplication.

The most obvious binary operation is the addition of two spectra of same size. As well known, the variance vectors **U** add simply in this case. The **ST** matrices are to be laminated if their source uncertainties are independent from each other, else added if they are identical. In general, a binary operation expressed by:

$$\mathbf{Y_3} = \mathbf{F}\ (\mathbf{Y_1},\mathbf{Y_2})$$

will be accompanied by

$$\mathbf{U_3} = (\partial\mathbf{F}/\partial\mathbf{Y_1})^2.\mathbf{U_1} + (\partial\mathbf{F}/\partial\mathbf{Y_2})^2.\mathbf{U_2}$$

$$\mathbf{ST_3} = ((\partial\mathbf{F}/\partial\mathbf{Y_1}).\mathbf{ST_1}),\ ((\partial\mathbf{F}/\partial\mathbf{Y_2}).\mathbf{ST_2})$$

(assuming the independence of the sources of uncertainty)

2.3 Selection in and catenation of spectra

Extracting a subset of values out of a spectrum produces a spectrum of smaller size. Catenating two spectra produces a spectrum of larger size. Both types of operation are currently used in analysis. The concurrent operations on the variance vector **U** and the **ST** matrix are obvious, because their dimensions must coincide with that of the vector of values before and after the operation. However, these operations make only sense if the spectra also contains a **n×2** matrix of bounds of the counting channels or a vector of **n** neutron energies corresponding to the measured cross-sections.

2.4 Least-squares fit operations

Fitting a measured spectrum with an analytical function **F(E)** of the neutron energy also requires the availability of an energy value for each measured point. In that case, the least squares-fit procedure uses the covariance matrix constructed from **U** and **ST** to determine the **k** parameters of the function and their covariance matrix. The fitted spectrum has variance $\mathbf{U_1 = 0}$. Its $\mathbf{ST_1}$ matrix is **k** columns wide and **n** lines high

3. The file format AGS (Analysis of Geel Spectra)

3.1 Components of a spectrum

In order to record the proceedings of an analysis, we now introduce a flexible binary file format to which the results of operations on spectra may be appended step by step. From the preceding, we know that all the data concerning a spectrum and its covariance may be contained in

a matrix of channel bounds or a vector of neutron energies (B)
a vector of measured values (Y)
a vector of variances (U)
a matrix of correlated uncertainties (ST)

All these data subsets may be seen as matrices, possibly with only 1 column, and stored all in the same way. Besides, some of these matrices may consist of integer values (e.g raw counting spectra) and some of double floating point data. Finally, we associate 3 vectors of experimental constants to every spectrum:

Global constants of the measurement
Local constants, specific to a spectrum
Scaler values, result of single channel counting

3.2. Matrix tags and matrix filing

Matrices are stored"flat"in the AGS file , i.e. one line catenated after another
To identify and retrieve a matrix, we use a matrix tag, i.e. a data structure with
The type of data (4 for long integer, 8 for double floating point)
The number of lines of the matrix (usually equal to the size of the spectrum)
The number of columns (1 for values or variances, 2 for ch. bounds, ? for ST)
The offset i.e. the position in bytes where the data begins in the file
(In the C language, we define the data type MAT_TAG for that purpose)

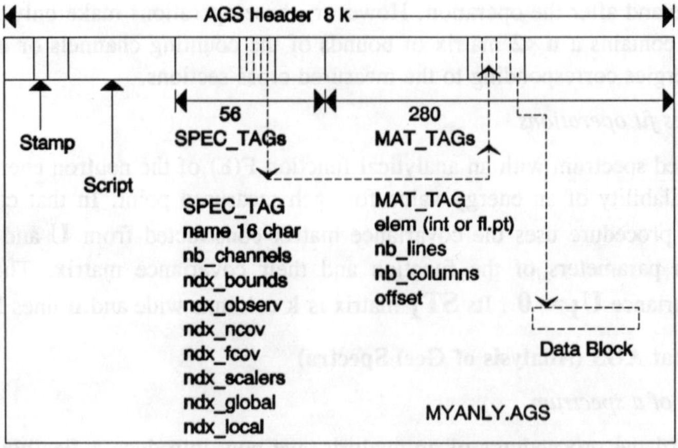

Fig.1:The AGS file structure

3.3 .The AGS file header

The book-keeping of an AGS file is concentrated in a header structure 8192 bytes long.(See Fig.1) In particular, it contains an array of 280 matrix tags as described in Sec 3.2.. Any matrix tag may thus be accessed by its index in this array

Further, a spectrum consists of a set of matrices, as explained in Sec 3.1.
The data of any spectrum are accessible via a spectrum tag, a data structure containing

The <u>name</u> of the spectrum (16 characters)
MAT_TAG <u>indexes</u> for the <u>bounds</u>, the <u>values</u>, the <u>U</u> and the <u>ST</u> matrices
MAT_TAG <u>indexes</u> for the <u>global</u>, <u>local</u> and <u>scaler</u> constants

In the C language, we define the data type SPEC_TAG for that purpose.

The AGS header contains an array of 56 SPEC_TAG's to register new spectra as they are appended to the file. If a spectrum does not include all matrix terms, the missing indexes are set to -1.Finally the header has room for a script of the operations which were applied to build-up the file, plus different counters and check values.

3.4 Available software

We have developed a software package for practical applications of the AGS file format. The development was done in the C-language concurrently under the MS/DOS operating system on the Personal Computer and under the VMS system on the microVAX/II. In both cases the C-language compiler was compliant with the ANSI standard. A large part of the package consist in dedicated function libraries in the spirit of the C standard libraries.

Our present purpose is to make AGS software tools available on a wide range of small and medium size systems: we will be soon implementing this AGS software on a UNIX workstation. In the following we shall review what is already available in this package (status October '92).

3.4.1 Service programs

ags_scan.c	lists the spectrum contents of an AGS file
ags_scri.c	lists the script i.e the past history of the generation of an AGS file
ags_mpty.c	generates an empty AGS file
ags_xfr.c	transfers(copies) spectra from one AGS file to another
ags_list.c	displays the contents of a spectrum
ags_edit.c	edits the experimental constants of a spectrum
ags_xplt.c	prepares the ASCII output for a plot program

3.4.2 Function libraries

ags_defs.h	definition of AGS structures and global variables
ags_lib.h	basic manipulation of AGS data
ags_iob.h	parallel stream I/O on spectrum components
histo.h	fluctuation analysis and crunching of counting histograms
mtx_oper.h	matrix operations (factorisation, inversion, etc..)

3.4.3 Application programs

The analysis of a transmission measurement was developed as a test case of the AGS approach. It includes the import of counting spectra (program ags_gene.c), the condensation of fluctuations (ags_catn.c), the dead-time and background corrections (ags_dtco.c, ags_bkco.c), the calculation of the transmission and of the cross-section (ags_trxs.c). Although some of these programs may be reusable, new programs will have to be written (preferably in C) to analyse other types of measurement.

References

1. D.L. Smith, *Probability, statistics and data uncertainties in nuclear science* Neutron Physics and Nuclear Data in Science and technology OECD Nuclear Energy Agency NDC series (1991)

2. E.J. Axton *CBNM Internal Report GE/PH/01/86*, (June 1986)

3. C.Bastian *Proc. of IAEA Specialist's meeting on Covariance Methods Rome, Italy 17-19 Nov 1986.* INDC(NDS)-192/L PP. 89-115.

4. P.F.Rose, C.L.Dunford *Data formats and procedures for ENDF-6* ENDF-102, BNL-NCS-44945, (July 1990)

COVARIANCE METHODS FOR STANDARDS EVALUATIONS

Robert Peelle
Consultant, Nuclear Data Measurement and Evaluation Section
Oak Ridge National Laboratory, P. O. Box 2008
Oak Ridge, Tennessee, USA 37831-6356

ABSTRACT

Neutron cross-section standards evaluations differ from most others in that more accurate results are expected and in that a greater number and variety of interdependent measurements should be included. Therefore, cross sections and covariances for several reactions need to be evaluated in the same procedure. The constraints of reaction theory should be included. The multiphase work of the Standards Subcommittee of the Cross Section Evaluation Working Group provides examples of problems encountered in such covariance analyses, and in some cases approaches available to solve them. The small output uncertainties that are characteristic of such data combination projects are discussed to help reconcile them with the qualitative impression of experts that larger uncertainties actually prevail. For those cross sections constrained by reaction theory, improved techniques are needed to treat energy regions where existing data are inconsistent with plausible resonance structures. Suggestions are included for covariance methods that might be employed in future standards evaluations.

1. Introduction

The idea behind neutron standard cross sections is to allow concentrated effort for very accurate measurements, effort that does not have to be repeated for each new nuclide studied.[1] Other measurements can be done relative to such a standard and then normalized appropriately. Uncertainties in such standard cross sections have to be sufficiently small that they do not make the overall uncertainties in the final results too large. Indeed, no other nuclide is apt to have a cross section known better than a standard based on the same reaction type. Neutron cross section standards help determine the incident particle flux, a quantity that is more easily obtained from the beam current in an experiment on reactions induced by charged particles. While the existence of a well-examined set of standard cross sections has proven to be very useful, many important experiments do not utilize standards as such.

To utilize a standard properly it is essential that variance-covariance quantities representing the standard be available both to aid the decision whether to use it and to permit quantification of covariance data from the resulting experiment. (Terms like "covariance data" are used here to refer to elements of an appropriately defined variance-covariance matrix.) Moreover, the term standard reference cross section almost begs for use of an objective evaluation process that will produce covariance data along with a consistent set of standard cross sections.

The reactions chosen for standard cross sections have been defined to achieve smoothness, availability of instruments to allow determination and utilization of the standard, and the desire for standards based on specific reaction types. Consensus reactions for standards development are shown in the table below. Note that most of these reactions are

Reaction Used as Standard	Energy Range as Standard
H(n,n)	10 eV to about 50 MeV
^{6}Li(n,T)	subthermal to ~ 0.5 MeV
^{10}B(n,α_1) and ^{10}B(n,α_T)	subthermal to ~0.2 MeV
^{197}Au(n,γ)	~ 0.2 MeV to 1-2 MeV
^{235}U(n,f)	0.1 or 0.2 MeV to 20 MeV

important to nuclear technology without considering their use as standards, though not necessarily in the same energy regions. Inclusion of other cross section as standards or in a simultaneous evaluation process would affect only the dimensions of the problem.

Most of the examples below arose in the work on the CSEWG Standards Subcommittee cooperative evaluation for ENDF/B-VI;[2,3] however, this paper does not document that work and indeed the development of the ENDF/B-VI covariance data is not complete. While it is clear that the other authors of ref. 2 were involved in the ideas discussed below, the development of ideas in the present paper reflects the present author's viewpoints. Most of the considerations below will impact any new effort to evaluate the set of standard cross sections.

In the following we identify those aspects of the standards data base that affect the treatment of covariances, indicate ways expected problems might be approached, and finally discuss how the covariance matrix of the resulting standard might be presented for use.

2. Data Base and Reaction Theory Determine Strategy

There is a large data base for the standards that includes almost every possible type of absolute and relative measurement. Past efforts have determined that it is impossible to handle the complexity well using graphs and a "light table" to obtain a consensus set. The task is made more difficult because the quantities reported by experimenters are often not the quantities observed, for example cross section shape measurements are normalized to prior data from the same author, or an earlier standard.[4] This complexity demands use of least-squares or related data combination procedures that can take into account uncertainties and correlations in the data.

The evaluation complexity caused by the interdependence of the standards is effectively reduced if a hierarchical scheme can be used that recognizes, for example, that the H(n,n) cross section is known to a far smaller uncertainty than others, so that its cross section values are not influenced by the other standards. The small uncertainty depends on the smoothness of the energy dependence of this cross section derived from the theoretical relation satisfied by the scattering cross section and angular distribution. Some detectors are sensitive only to the protons scattered into particular angular ranges, and the uncertainty in the evaluated angular distribution for this case remains to be worked out.

The shapes of the (n,α) reactions in ^{6}Li and ^{10}B are constrained by the R-matrix reaction theory,[5] but valid fits depend on proper identification of the pertinent resonances. Again, the advantage of smoothness leads to higher utility as well as small nominal cross section uncertainties if the neutron and other relevant data can be fit with the theory. Since these standards are defined over a neutron energy region that overlaps the standards based on fission and capture cross sections, it is not clear that a hierarchical relationship can be assumed.

In principle, theory can also be helpful in constraining evaluation of fission and particularly capture standards,[6] though up to now not much use has been made of such techniques. The evaluation of ^{238}U data to obtain consensus neutron capture cross sections is an important exception.[7]

Evaluation strategy, or at least complexity, is affected by use of "standard instruments" to determine the neutron flux in the underlying measurements. While a variety of such instruments has been used over the years, the presently most popular ones capable of event timing have been "black detectors." A black detector is an organic scintillation counter, usually mounted in a shield with an axial collimator, arranged so that nearly all fast neutrons that reach the scintillator scatter a few times before escape, and thereby produce a pulse above a reasonable bias level.[8] The energy-dependent efficiency of such devices could enter into enough measurements that it has a status similar to another standard cross section. Correlations generated among measurements dependent on the counter must be considered unless the uncertainties associated with its use are shown to be quite inconsequential.

Overall, circumstances lead to a data fitting process that may be able to treat n-p scattering separately, but should include other data in a global fitting process that includes all relevant data and covariance information and produces both the standard cross sections and the related covariance data. The fitting process needs to take advantage of reaction theory for the light-element standards, but might well utilize average cross sections rather than resonance parameters as the fitting parameters for the capture and fission reactions. Though many details could be handled differently, this logic motivated the procedure used in the work of Ref 2.

3. Dealing with Inherent Tactical Problems

If at least some cross sections will be represented by discrete values rather than parameters from model fits, an energy grid must be defined. The technique chosen will affect the ease of the covariance representation. A clear definition of the grid is essential because from some measurements several points may occur in each interval, and some "pointwise" data may have broad resolution functions that affect more than one discrete interval. Perhaps the most convenient choice overall is the ENDF-6 pointwise representation as used in GLUCS.[9] To fix ideas, assume linear-linear interpolation of the cross section between points on the energy grid. This choice suffers the difficulty that ENDF-6 formats for File 33 do not include an option that represents the covariance of a cross section at an energy grid point with such an interpolation law.[10] (Rather, for processing ease, with one exception they provide energy intervals within which the cross section is fully correlated.) Possibly, File 30 could be used for the covariances of the "model parameters" that are just the cross sections at the energy grid points.[11] (Such a procedure might be attractive here since other parameters in the same global fit are R-matrix parameters.) In any case, the measurement covariance matrices need to be reduced to an energy grid consistent with that used for cross sections. This is a complex process since systematic uncertainties usually dominate the measurement uncertainties and since many experimental papers do not allow direct extraction of a credible data covariance matrix. If the off-diagonal covariances are neglected, the effect will be to reduce too much the uncertainties in cross sections averaged over broader intervals. If the off-diagonal elements are too large the value of the data set is degraded and may impose too much shape stiffness when combined with other data sets. While correlation matrices are interesting to review, primary data is rarely in that form. To avoid too-large off-diagonal elements, the author recommends building up a measurement covariance matrix (relative or absolute) directly from the contributions of identified components.[12]

Once input data sets have been separated according to the measurements actually performed, it is possible to look at overlapping data sets. Perhaps some data set has such large uncertainties that it can effectively have no influence on the final result; if so there may be no need to retain that set.

At this stage the recognition of data inconsistencies should begin. The author believes that delay in dealing with discrepancies leads only to unanswerable questions at later stages in the analysis when tools for correction become cruder and cruder. (The reason is that the interpretation of results from the least-squares technique depends on input data consistency.) Alas, little effective general guidance can be given except to follow traditional steps,[13] perhaps elaborated to deal with data sets rather than individual points. Early recognition is helpful because community consensus may be important in finding acceptance of results obtained after deleting or de-weighting some well-respected measurements.

The size of the covariance matrix resulting from a standards evaluation of the type discussed here is an obstacle. Tens of thousands of data points may have been considered, perhaps 600 cross section values and resonance parameters are produced and at the output there are a few hundred thousand covariance elements that likely must be stored and used in double precision. Considering advances in computer storage technology and the likelihood that efforts of this magnitude will not often be repeated, one must consider the difficulties in this area as merely technical.

4. Presenting the Covariance Matrix

The most convenient method of presentation and inclusion of covariance data in an evaluation will usually be a collapsed matrix, specifically the covariance matrix of average cross sections on a grid considerably more coarse (by a factor of at least four) than that utilized for the evaluation. It could then be considered a reasonable approximation to represent the matrix using a selected File 33 format. With the reduced matrix dimension, the number of digits available in the 11-column format may be adequate to assure that the matrix remains positive. Alternatively, formatted files could be abandoned and a standard 32 or 64 bit binary file structure agreed to. The portion of the matrix pertinent to one reaction (e.g., ^{6}Li(n,a)) might then fit on a present-day 3.5 in. diskette.

If the collapsed matrix approach is followed, there are the problems that the cross section would be represented as fully correlated over each energy interval of the collapsed structure, and correspondingly the diagonal elements (squared cross section uncertainties) would be even smaller than with the narrow bins. A possible compromise might be to retain the elements from the full matrix near the diagonal, but use a collapsed matrix for covariances more distant from the diagonal. Generating such a file would be a bit complicated.

Under ENDF-6 covariance formats a special sub-subsection is recommended (LB=8) to take account of fluctuations in the average cross sections caused by the underlying resonance structure.[14] This requirement would apply to the ^{197}Au(n,γ) and ^{235}U(n,f) standards, and would meet the qualitatively required criterion that the average cross section is less well known if averaged over very small intervals. Macklin has analyzed this fine structure effect for ^{197}Au(n,γ).[15]

5. Are the Evaluated Standards Uncertainties Too Small?

It is expected that any global evaluation of the neutron standards performed now would find fractional-percent uncertainties prevalent over the selected energy regions. When this occurred following the ENDF/B-VI evaluation, there was comment that the uncertainties must be overly optimistic. Possibly so, but the point of this section is to suggest that, with exceptions, small uncertainties on the evaluated cross sections are to be expected because of the general wealth of available data.

First, the uncertainties apply to averages rather than point cross sections, while experimenters may think of the uncertainties as representing a region no larger than the available energy resolution. As indicated above, for the heavy element standards the uncertainties must be larger for cross sections averaged over narrower intervals. Second, the cross sections are not independent, so the uncertainty on the average over 16 of the grid points will be significantly larger than 1/4 that for a single point. Put another way, there might be adequate acceptance of an evaluated result with 100 points over energy, each with 2% relative uncertainty and zero correlation specified. Yet for that case the spectrum averaged cross section could have an uncertainty as small as 0.2%. An actual evaluation with uncertainties of about 0.5% and appropriate covariance terms could well have an uncertainty for the spectrum-averaged cross section larger than 0.2%. This logic does not hold well for the light element standards, where the cross sections are so small that those for nearby energy points are more highly correlated.

There are cases where one should be wary of least-squares uncertainties that seem small. Unless some "fix" is automatically included, a least squares program just propagates input uncertainties and correlations to the output. That is, the scatter among input data does not affect the output covariance quantities. Overall chi-square tests are used to uncover discrepancies, but such overall tests do not identify the portion of the input data responsible for the discrepancy. If a region of unresolved data discrepancy is known, one can be rather sure that the corresponding regions of the output are flawed. A goal would be to devise a relatively rigorous way to expand output uncertainties in just those regions. One might also be concerned about broad energy regions where there is no substantial data supporting an R-matrix analysis, if there is any possibility that resonance structure could be hidden there.

Often, a correction for overall lack of fit is performed by multiplying all covariances by the overall chi-square sum divided by the number of degrees of freedom. According to the above discussion this is appropriate only when there is no specific region of discrepancy, and the need or desire to make such a correction suggests that adequate evaluation and correction of the input data has not been possible. Segmenting the overall chi-square value is rigorously possible only for values or blocks of input data uncorrelated to the others. Whenever this condition pertains, the data combination program used should make the chi-square values for the segmented problem accessible to provide aid in pinpointing problems.

This discussion of how to understand the small values of the nominal uncertainties in the standards does not imply that measurements should cease. If most of the measurements in a particular region have been performed similarly, nominal uncertainties could be small but misleading because of common unrecognized systematic uncertainties. For example, an unrecognized type of background could make all measurements too large. Therefore, well-done diverse experiments are welcomed just because they may turn up problems. Experiments are also needed where evaluators have had to make relatively arbitrary choices among disparate data sets.

6. Suggestions for the Future

Within a decade a new evaluation of the standards should be started. While the work of Ref. 2 used covariance techniques as well as it could, I envision a future evaluation being as thorough, but avoiding a number of difficulties.

As indicated above, the greatest effort is involved in study of the input data, evaluation of appropriate covariance information, and the handling of discrepancies. More constructive ways need to be developed to deal with discrepancies when data sets rather than individual points are involved.

If the judgement prevails that knowledge of the capture and fission standards can influence the best values of some of the light element standards in their important domain, then the idea makes sense of a joint analysis based on R-matrix fits to the light-element cross sections and discrete cross section values with an interpolation law for fission and capture. The whole process should be brought to a single computing machine so that iteration can be performed. (Iteration is performed so that the derivative matrix can be computed at the solution point.) By using the "Bayes method" as in GLUCS (known as the adjustment equations in the study of integral experiments), it is possible to avoid including all the experimental data at one time. However, if this is done the iteration process and the values obtained from the chi-square consistency tests must be treated very carefully to avoid confusion.

As a near-term step, the covariance data from the ENDF/B-VI standards need to be made available, certainly including those for the n-p scattering cross sections. This case may be a good opportunity to use the File 30 technique developed by Muir[16].

7. Acknowledgement

The author is indebted to W. P. Poenitz, G. M. Hale, and A. D. Carlson, as well as others unnamed, for many conversations that must have produced many of the ideas in this paper. L. W. Weston and D. C. Larson provided helpful reviews. Research sponsored by the Office of Energy Research, Division of Nuclear Physics, U. S. Department of Energy, under contract DE-AC05-84OR21400 with Martin Marietta Energy Systems, Inc.

8. References

1. R. F. Taschek, in *Neutron Standards and Flux Normalization* (U.S. Atomic Energy Commission Division of Technical Information, AEC Symposium Series 23, CONF-701002, 1971) p. 3.

2. A. D. Carlson, W. P. Poenitz, G. M. Hale, and R. W. Peelle, in *Nuclear Standard Reference Data* (International Atomic Energy Agency IAEA-TECDOC-335, 1985) p. 77. A.D. Carlson et al. , in *Nuclear Data for Basic and Applied Science*, eds. P.G. Young et al. (Gordon and Breach, Science Publishers, Inc. 1986) p. 1429. R. Peelle and H. Condé, in *Nuclear Data for Science and Technology*, ed. S. Igarasi (Saikon Publishing Co., Tokyo 102, 1988) p. 1005.
3. A. D. Carlson et al., *The ENDF/B-VI Neutron Cross Section Measurement Standards*, (draft of final report, private communication, National Institute of Standards and Technology, 1992).
4. W. P. Poenitz, in *Proceedings of the Conference on Nuclear Data Evaluation Methods and Procedures*, eds. B. A. Magurno and S. Pearlstein, (Brookhaven National Laboratory Report BNL-NCS-51363, 1981) p. 254.
5. G. M. Hale, in *Proceedings of the Conference on Nuclear Data Evaluation Methods and Procedures*, eds. B. A. Magurno and S. Pearlstein, (Brookhaven National Laboratory Report BNL-NCS-51363, 1981) p. 509.
6. W. P. Poenitz and R. W. Peelle, in *Covariance Methods and Practices in the Field of Nuclear Data*, (International Atomic Energy Agency Report INDC(NDS)-192/L, 1988) p. 72.
7. F. H. Froehner, *Nuclear Science and Engineering* 103 (1989) 119.
8. W. P. Poenitz, *The Black Neutron Detector*, (Argonne National Laboratory Report ANL-7915, 1972). G. P. Lamaze et al., in *Nuclear Cross Sections and Technology*, eds. R. A. Schrack and C.D. Bowman (National Bureau of Standards Special Publication 425, 1975) p. 73.
9. D. M. Hetrick and C. Y. Fu, *GLUCS: A Generalized Least-Squares Program for Updating Cross-Sections with Correlated Data Sets*, (Oak Ridge National Laboratory Report ORNL/TM-7341) 1980.
10. *Data Formats and Procedures for the Evaluated Nuclear Data File ENDF-6*, eds. P. F. Rose and C. L. Dunford (Brookhaven National Laboratory Report BNL-NCS 44945, ENDF-102, 1991) p. 33.1ff.
11. R. W. Peelle and D. W. Muir, in *Application of Critical Experiments and Operating Data to Core Design via Formal Methods of Cross Section Data Adjustment*, Ed. D. Wade (Nuclear Energy Agency document NEACRP-L-307, 1988) p. 311.
12. J. K. Dickens et al., *Nucl. Sci. Engineering* 74 (1980) p. 126ff. D.C. Larson et al., *Application of New Techniques To ORELA Neutron Transmission Measurements And Their Uncertainty Analysis: The Case Of Natural Nickel From 2 keV to 20 MeV*, (Oak Ridge National Laboratory Report ORNL/TM-8203, 1983) p. 37ff.
13. R. Peelle, paper 3.1 in these proceedings.
14. R. Peelle, in *Data Formats and Procedures for the Evaluated Nuclear Data File ENDF-6*, eds. P.F. Rose and C. L. Dunford , op. cit. pp. 33.17f, 33.26f.
15. R. L. Macklin, J. Halperin, and R.R. Winters, *Phys. Rev.* C11 (1975) 1270.
16. D. Muir, in *Data Formats and Procedures for the Evaluated Nuclear Data File ENDF-6*, eds. P.F. Rose and C. L. Dunford, op.cit. p. 30.7ff.

GENERATION OF COVARIANCE MATRIX FOR DERIVED QUANTITIES AND THEIR LEAST SQUARE FITTING

Zhixiang Zhao
Chinese Nuclear Data Center, P.O. Box 275 (41), Beijing, China

F. G. Perey
ORNL, TN 37831–6356, Oak Ridge, USA

I. Introduction

Most data, such as nuclear cross sections, that are reported as having been measured were, in fact, not directly measured. Instead,their values were derived using some quantities that were directly observed in other experiments. Whenever this occurs, the so–called "Law of Error Propagation (LEP)" is used to generate the covariance matrix of these derived quantities from the uncertainties associated with the directly measured quantities. Even though this LEP is well known to all experimenters and evaluators, there appears to be some circumstances where different of opinion seem to exist today concerning its proper application. We became aware of this through informal discussions of a problem proposed by R. W. Peelle[1] which has been referred to as "Peelle's Pertinent Puzzle (PPP)". In fact, PPP phenomena happen not only in the situation of single parameter combinition but also in least squares fitting for a measured curve.

We will show that PPP arose from an incorrect application of the LEP, where different estimates for the true values were used in computing the approximate covariance matrix associated with the derived quantities. In this report we analyze the general problem of generating the covariance matrix associated with derived quantities and the use of such matrices in data evaluations using the least squares method.

II. The Law of Error Propagation

Let us denote by **D**

$$\mathbf{D} = (a(x_1) \cdots a(x_N)\ b_1 \cdots b_n)^t \tag{2.1}$$

the true values of physical quantities. We assume that these true values are not known. Let us denote by $\mathbf{D}^*$

$$\mathbf{D}^* = (a^*(x_1) \cdots a^*(x_N)\ b_1^* \cdots b_n^*)^t \tag{2.2}$$

the directly measured data for **D** with covariance matrix $\mathbf{V_D}$.

We denote by $\delta\mathbf{D}$ the deviation of the measurement $\mathbf{D}^*$ from the true values **D**

$$\delta\mathbf{D} = (\mathbf{D}^* - \mathbf{D}) \tag{2.3}$$

and, using angle brackets to denote expectation values, we finally have

$$V_D = \langle \delta\mathbf{D} \cdot \delta\mathbf{D}^t \rangle \tag{2.4}$$

We now introduce physical quantities, whose true values we denote y(x), that is related to **D** as follows :

$$y(x_i) = \psi_i(a(x_i), \mathbf{b}) \tag{2.5}$$

Let us now introduce a vector **G** whose components are **y** and **b**

$$\mathbf{G} = (\mathbf{y}\ \mathbf{b})' \tag{2.6}$$

where

$$\mathbf{y} = (y(x_1) \cdots y(x_N))' \ ; \ \mathbf{b} = (b_1 \cdots b_n)'$$

and we refer to G as derived data vector. The so called LEP deals with generating the covariance matrix V_G associated with G^* which is derived from directly measured data D^* and V_D associated with D^* through deriving relation (2.5).

When function $\psi_i(a(x_i), \mathbf{b})$ are non–linear function, truncated Taylor expansions about the true values $a(x_i)$ and **b** are used to linearize them. With such truncated expansions the deviations of the derived values are linear functions of the directly measured values and one can then readily calculate the covariance matrix of derived values. Proceeding formally we have

$$y^*(x_i) = \sum_{m=0}^{\infty} \frac{1}{m!}[(a_i^* - a_i)\frac{\partial}{\partial a_i} + \sum_{j=1}^{n}(b_j^* - b_j)\frac{\partial}{\partial b_j}]^m \psi_i$$

$$\approx \psi_i + \psi_a(i,i)\delta a_i + \sum_{j=1}^{n} \psi_b(i,j)\delta b_j \tag{2.7}$$

where $\psi_i = \psi_i(a_i, \mathbf{b})$ and

$$\psi_a(i,i) = \frac{\partial \psi_i}{\partial a_i} \qquad \psi_a(i,j) = 0 (i \neq j) \tag{2.8}$$

$$\psi_b(i,j) = \frac{\partial \psi_i}{\partial b_j} \tag{2.9}$$

It should be emphasized that what appear in the partial derivetives terms, ψ_a and ψ_b are true values a and **b**, not a^* and $\mathbf{b}^*$. In practice, the true values a and **b** are not known and ψ_a, ψ_b in the expansion (2.7) must be calculated at some estimates of a and **b**. In matrix notation, we have

$$\delta\mathbf{G} = \mathbf{G}^* - \mathbf{G} \approx \hat{S}\delta\mathbf{D} \tag{2.10}$$

where

$$\hat{S} = \begin{bmatrix} \hat{\psi}_a(N \times N) & \hat{\psi}_b(N \times n) \\ 0\ (n \times N) & I\ (n \times n) \end{bmatrix} \tag{2.11}$$

where I is an unite matrix and **0** a null matrix.

At finally we have

$$V_G = \langle \delta\mathbf{G} \cdot \delta\mathbf{G}' \rangle \approx \hat{S} V_D \hat{S}' \tag{2.12}$$

When the derived quantities are non–linear functions of the measured quantities, one can obtain only an estimate for the covariance matrix associated with the derived quantities, this estimated covariance matrix is based upon some specific estimates for the true values of the measured quantities. Consequently, some consistency problem may arise if such an estimated covariance matrix is subsequently used in conjunction with other results that are themselves based upon different estimates for the true values of measured quantities.

III. Least Squares Fitting to Derived Quantities

Let us denote by

$$y(x) = f(x;\mathbf{C}) \tag{3.1}$$

the theoretical expression being fitted to the derived data, where x is an independent variable and

$$\mathbf{C} = (C_1 \cdots C_m)^t \tag{3.2}$$

is the vector of the parameter being adjusted to fit the data. We now analyze two different ways of obtaining a least square estimates for the parameters **C** and show that these two different methods yield identical results. In the first method we fit directly measured quantities. In the second method we calculate the derived quantities and their associated covariance matrix,which are then fitted.

From Fitting Directly Measured Data

We assume that

$$a(x_i) = \Phi_i(y(x_i),\mathbf{b}) \tag{3.3}$$

is an inverse function of (2.5). By using (3.1), we have

$$a(x_i) = \Phi_i(f(x_i;C),\mathbf{b}) \tag{3.4}$$

Based on the principle of least square, we can obtain the least square estimates for **C** and **b** by minimizing with respect to $\mathbf{P} = (\mathbf{C}\ \mathbf{b})^t$ the following quadratic expression

$$Q(\mathbf{D}) = (\mathbf{D}^* - \mathbf{D})^t V_{\mathbf{D}}^{-1}(D^* - \mathbf{D}) \tag{3.5}$$

As usual this non-linear least square problem can be solved by iteration. Performing a Taylor expansion of **D** about some estimates for **P**, we have

$$\mathbf{D} \approx \mathbf{D}^{(k)} + \hat{U}_k(\mathbf{P} - \mathbf{P}^{(k)}) \tag{3.6}$$

with

$$D^{(k)}(i) = \Phi_i(f(x_i;C^{(k)};b^{(k)}) \quad i \leqslant N \tag{3.7}$$

$$D^{(k)}(i+N) = b_i^{(k)} \qquad i = 1, \cdots, n \tag{3.8}$$

$$\hat{U}_k = \begin{bmatrix} \hat{\Phi}_c(N \times m) & \hat{\Phi}_b(N \times n) \\ 0\ (n \times m) & I\ (n \times n) \end{bmatrix} \tag{3.9}$$

where

$$\Phi_i^{(k)} = \Phi_i(f(x_i;C^{(k)}),\mathbf{b}^{(k)}) \tag{3.10}$$

$$\hat{\Phi}_c(i,j) = \frac{\partial \Phi_i^{(k)}}{\partial C_j^{(k)}} ; \hat{\Phi}_b(i,j) = \frac{\partial \Phi_i^{(k)}}{\partial b_j^{(k)}} \tag{3.11}$$

Substituting the Taylor expansion (3.6) into (3.5), we obtain

$$Q(\mathbf{D}) \approx (\mathbf{D}^* - \mathbf{D}^{(k)} - \hat{U}_k(\mathbf{P} - \mathbf{P}^{(k)}))^t V_D^{-1}(\mathbf{D}^* - \mathbf{D}^{(k)} - \hat{U}_k(\mathbf{P} - \mathbf{P}^{(k)})) \tag{3.12}$$

Denoting by $\mathbf{P}^{(k+1)}$ the values of **P** which minimizes (3.12) and by $\mathbf{V}^{(k+1)}$ the covariance matrix of $\mathbf{P}^{(K+1)}$, we have

$$\mathbf{P}^{(k+1)} = \mathbf{P}^{(k)} + (\hat{U}_k^t V_D^{-1} \hat{\mathbf{U}}_k)^{-1} \hat{\mathbf{U}}_k^t V_D^{-1}(D^* - D^{(k)}) \tag{3.13}$$

$$V^{(k+1)} = (\hat{U}_k^t V_D^{-1} \hat{U}_k)^{-1} \tag{3.14}$$

We call (3.13) and (3.14) as the Solution in Data Space.

From Fitting Derived Quantities

Let us denote by $\mathbf{G}^*$ the derived vector

$$\mathbf{G}^* = (y_1^* \cdots y_N^* \; b^*)^t \tag{3.15}$$

and by $\mathbf{G}$ the true values

$$\mathbf{G} = (y_1 \cdots y_N \; b)^t \tag{3.16}$$

An alternate way of obtaining a least square estimates for $\mathbf{P}$ is to minimize, also with respect to $\mathbf{P}$, a quadratic expression in which derived values $\mathbf{G}$ and $\mathbf{G}^*$ appear explicitly. We now need to express the deviation $\delta\mathbf{G}$ in terms of deviation $\delta\mathbf{D}$ in order to generate the covariance matrix, $\mathbf{V_G}$, associated with the derived quantities $\mathbf{G}^*$. As discussed in section 2, the elements of the transformation matrix that transforms $\delta\mathbf{D}$ into $\delta\mathbf{G}$ are the partial derivatives of $\psi_i(a_i^*, \mathbf{b}^*)$ evaluated at some estimates of the true values $\mathbf{a}$ and $\mathbf{b}$. Let us denote by $\mathbf{a}^{(k)}$ and $\mathbf{b}^{(k)}$ these estimates. Performing a Taylor expansion of $\mathbf{G}^*$ about these estimates, we have

$$\mathbf{G}^* \approx \mathbf{G}_k^* = \hat{\mathbf{G}}_k + \hat{S}_k(\mathbf{D}^* - \mathbf{D}_k) \tag{3.17}$$

where

$$\hat{\mathbf{G}}_k = (\psi_1^{(k)} \cdots \psi_N^{(k)} \; b^{(k)})^t \tag{3.18}$$

$$\hat{S}_k = \begin{bmatrix} \hat{\psi}_a(N \times N) & \hat{\psi}_b(N \times n) \\ 0\ (n \times N) & I\ (n \times n) \end{bmatrix} \tag{3.19}$$

$$\hat{\psi}_a(i,i) = \frac{\partial \psi_i^{(k)}}{\partial a_i^{(k)}} \quad \hat{\psi}_a(i,j) = 0 \quad (i \neq j) \tag{3.20}$$

$$\hat{\psi}_b(i,j) = \frac{\partial \psi_i^{(k)}}{\partial b_j^{(k)}}; \; \psi_i^{(k)} = \psi_i(a_i^{(k)}, \mathbf{b}^{(k)}); \; a_i^{(k)} = \Phi_i(f(x_i; \mathbf{C}^{(k)}), \mathbf{b}^{(k)}) \tag{3.21}$$

and we have

$$\delta\mathbf{G}_k = \hat{S}_k \delta\mathbf{D} \tag{3.22}$$

$$V_{Gk} = \langle \delta\mathbf{G}_k \cdot \delta G_k^t \rangle = \hat{S}_k V_D \hat{S}_k^t \tag{3.23}$$

On the other hand,

$$\mathbf{G} \approx \mathbf{G}_k = \hat{\mathbf{G}}_k + \hat{T}_k(\mathbf{P} - \mathbf{P}^{(k)}) \tag{3.24}$$

where

$$\hat{T}_k = \begin{bmatrix} \hat{F}_c(N \times m) & 0\ (N \times n) \\ 0\ (n \times m) & I\ (n \times n) \end{bmatrix} \tag{3.25}$$

$$\hat{F}_c(i,j) = \frac{\partial}{\partial C_j^{(k)}} f(x_i; C^{(k)}) \tag{3.26}$$

Minimizing with respect to $\mathbf{P} = (\mathbf{C}\ \mathbf{b})^t$ the following quadratic expression

$$Q(G) \approx (\mathbf{G}_k^* - \hat{\mathbf{G}}_k - \hat{T}_k(\mathbf{P} - \mathbf{P}^{(k)}))^t V_{Gk}^{-1} (\mathbf{G}_k^* - \hat{\mathbf{G}}_k - \hat{T}_k(\mathbf{P} - \mathbf{P}^{(k)})) \tag{3.27}$$

we have

$$P^{(k+1)} = P^{(k)} + (\hat{T}_k' V_{Gk}^{-1} \hat{T}_k)^{-1} \hat{T}_k' V_{Gk}^{-1} (\mathbf{G}_k^* - \hat{\mathbf{G}}_k) \quad (3.28)$$

$$V^{(k+1)} = (\hat{T}_k' V_{Gk}^{-1} \hat{T}_k)^{-1} \quad (3.29)$$

We call (3.28) and (3.29) as the Solution in Derived Space.

It can be prooved that the Solution in Derived Space is identical to one in Data space[2].

It should be noted that if F(x;C) is a linear function of m components of C then (3.24) is exact, and the notation T_k is somewhat misleading since the elements of T_k are constants. However, even if F(x;C) is a linear function of the m components of C, one must iterate (3.28) because we have assumed that the function $\psi_i(a_i,b)$ was nonlinear and consequently the covariance matrix of the derived data V_{Gk} is an approximation which will be a function of $P^{(k)}$. To our knowledge, very few if any of least square fitting codes that allow for an iterated solution of the type represented by (3.28) where one recalculates the covariance matrix of the data being fitted based upon an updated estimates for the parameters being determined.However, it is not necessary that we directly fit the derived data $\mathbf{G}^*$. As we noted, the same end results can be achieved by fitting the directly measured data $\mathbf{D}^*$ with its covariance matrix V_D.

IV. Computational Example

The directly measured data are given as follows:

X_i = 0.8 1.0 2.3 3.4 4.5 7.4 8.8 9.7

a_i^* = 19. 30. 27. 41. 52. 53. 63. 78.

We assume that all of the data a_i^* are independent and have 10% standard error. Furthermore,we have the independent measurement of the quantity b_1, b_1^* = 1.0± 0.2.We are interested in the derived quantity

$$y_i = \psi_i(a(x_i),\mathbf{b}) = a(x_i) b_1 \quad (4.1)$$

By using formalism given by this work, (3.1) to (3.29), the the least square estimate for C_1 and C_2 from fitting directly measured data or from fitting the derived values are obtained through an iteration and given as follows:$\hat{C}_1 = 17.12 \pm 3.819$;$\hat{C}_2 = 5.689 \pm 1.244$;$\rho(\hat{C}_1,\hat{C}_2) = 0.69$. The fitted curve is given by the one labelled A in Figure 1.

It should be noted that a number of evaluator might attempt to fit the derived data in the following manner. The derived data vector would be taken as follows

$$\mathbf{G}^* = (a_1^* \, b_1^* \, \cdots \, a_8^* \, b_8^*)' \quad (4.2)$$

and the elements of its associated covariance matrix taken as

$$V_G(i,i) = (b_1^*)^2 Var(a_i^*) + (a_i^*)^2 Var(b_1^*) \quad (4.3)$$

$$V_G(i,j) = a_i^* a_j^* Var(b_1^*) \quad i \neq j \quad (4.4)$$

The derived data vector and its covariance matrix would be treated as if they were directly measured data and the following quadratic expression would be minimized with respect to parameter vector $\mathbf{P}$

$$Q(G) = (G^* - F\mathbf{P})' V_G^{-1} (G^* - F\mathbf{P}) \quad (4.5)$$

where F(1,j) = 1, F(2,j) = X_j and $\mathbf{P} = (C_1, C_2)^t$.

If we were indeed to solve this problem in that manner the results would be as follows: $\hat{C}_1 = 10.47 \pm 3.167$; $\hat{C}_2 = 3.478 \pm 1.002$; $\rho(\hat{C}_1, \hat{C}_2) = 0.55$, and the curve labelled B in Figure 1, rather than those given in the curve labelled A.

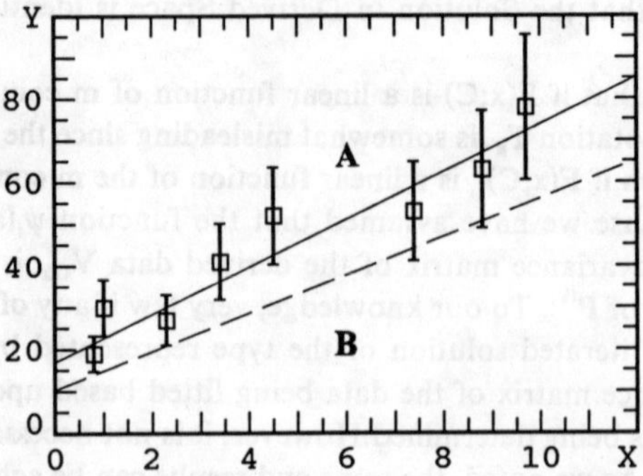

Fig.1 Fitting of Derived Curve

Acknowledgments

The authors are grateful to many individuals, among them R. W. Peelle, F. H. Froehner, S. Chiba, H. Vonach, D. L. Smith, and D. W. Muir, with whom the authors have engaged in fruitful discussions on this subject.

References

1 R. W. Peelle, "Peelle's Pertinent Puzzle", private communication, Oct. 13, 1987, Revised Oct. 28(1988)

2 Zhixiang Zhao and F. G. Perey, "The covariance matrix of derived quantities and their combination", ORNL/TM−12106, (1992)

APPROXIMATE METHODS FOR DERIVATION OF COVARIANCE DATA

S. TAGESEN
Institut für Radiumforschung und Kernphysik
University of Vienna, Boltzmanngasse 3, A–1090 Vienna, Austria

and

D. C. LARSON
Engineering Physics and Mathematics Division
Oak Ridge National Laboratory, Oak Ridge, Tennessee 37831, USA

ABSTRACT

Several approaches for the derivation of covariance information for evaluated nuclear data files (EFF2 and ENDF/B–VI) have been developed and used at IRK and ORNL respectively. Considerations governing the choice of a distinct method depending on the quantity and quality of available data are presented, advantages/disadvantages are discussed and examples of results are given.

1. Introduction

The predominant part of the data contained in evaluated nuclear data files has been derived by nuclear model calculations with parameters adjusted to fit existing experimental data as closely as possible. This is due to the requirement of completeness (many of the quantities needed have not been measured at all) and consistency (the data files have numerous required constraints, e.g., partial cross sections have to sum up exactly to totals). As all model calculations have to admit some approximations, there is no overall good fit to all experimental data and we have thus to accept model– and situationspecific uncertainties, which are not generally provided by the model codes.

Nevertheless there is an urgent requirement from applications, to have reliable uncertainty and covariance information available. Thus evaluators have to develop some means of estimating these numbers from other sources of information and/or evidence.

2. General Approaches to Derivation of Uncertainty Components

2.1. The Rigourous Method

This method is mentioned here only for completeness: If a sufficient number of experimental data exists, which all have a complete table of all relevant uncertainty contributions with them, separated into components which are fully correlated each but mutually independent, one may construct the full covariance matrix by applying a suitable uncertainty propagating code, e.g. GLUCS[1]. Unfortunately this is possible only for some "standard cross sections", which have been subject to extensive analysis due to their importance.

2.2. Few Experimental Data Available

The most frequently encountered situation is, that several data(sets) exist which have served also for fitting the nuclear model parameters, but which do not provide adequate information on uncertainty and mostly nothing at all on correlations. The problem for the evaluator then is, to deduce from this incomplete information sufficient evidence to construct the full covariance matrix in a way which does certainly not underestimate, but also not largely overestimate uncertainties and correlations.

Up to the 1980th authors rarely supply much more than uncertainties from counting statistics. Sometimes some information on uncertainties of some corrections, e.g. for dead–time and background is given in addition. The former are certainly short–range correlated (SR) contributions, the latter may give rise to medium range (MR) contributions, which will be discussed later. As a first approach plotting of the data will already reveal much of the extractable information.

When reaction cross sections are grouped into rather narrow "energy–bins", scatter of the data inside each of the groups will reflect mainly SR correlated uncertainty components. Inconsistencies within energy groups will indicate the presence of additional uncertainty components, not taken into account in the uncertainty ascribed by the authors.

If we calculate mean cross sections over such energy groups and plot these together with the data, we can get more information. Most obvious are long–range (LR) correlations, caused by e.g. normalization errors, uncertainties in the determination of sample properties and irradiation– or measurement–geometry. They manifest themselves in a regular shift of the excitation curve towards one side of the mean, i.e., the sum of all residua deviates more than the given uncertainty from zero

$$| \sum_i (\sigma_{mean,i} - \sigma_i) | > \text{unc} \qquad \text{(i...group index)} \qquad (1)$$

for a certain data set, σ representing the cross section σ(E) at a given incident particle energy.

If energy groups are chosen to span a few MeV each, elements of a partial covariance matrix representing MR correlations may be estimated for data sets:

$$var_i = ((\sigma_{mean,all} - \sigma_{mean,set}) / \sigma_{mean,all})^2 \qquad (2)$$

i is a group index like in (1). For calculating the off–diagonal elements of this partial covariance matrix an assumption has to be made on a function reflecting the decrease of correlation with increasing difference in incident particle energy. A rough estimate of the half–width of such a function may be derived from the changes of sign of the residua as a function of energy.

As has been pointed out by Peelle[2], with some precautions and under favourable conditions (several nondiscrepant data sets of equal quality over the whole energy range) one might even deduce a reasonable full covariance matrix from the scatter of the data by applying the formula

$$cov_{ij} = (1/W_{ij}) \sum_{sets} w_{set} (\sigma_{i,set} - \sigma_{i,all}) (\sigma_{j,set} - \sigma_{j,all}) \qquad (3)$$

where the w_{set} and W_{ij} are set weights and combined weights which have to be assigned properly by the evaluator, i and j indicate energy groups.

2.3. Data from Model Calculations Only

If for an excitation function there are (almost) only data from model calculations available, the evaluator may not have recourse to any statistical assistance. He is left in fact with two options:

– decide from insight into the model calculation (assumptions, approximations, sensitivity to parameter choices, etc.) on the expected uncertainty of results, i.e. make an "advanced guess", or

– try to find at least some evidence on predictive power and uncertainty of various types of model calculations. This may be possible, if solutions for the same data representation problem by different (fairly) independent models are available. Differences in the model–calculated results, all assumed to be derived with the best parameters for the respective model, are then likely to reflect the attainable precision of prediction. A practical application of this assumption will be shown in section 5.

3. Requirements for Processing and Future Update of Covariance Information

While generating covariance matrices for evaluated nuclear data files one always has to bear in mind the destination and future evolution of these data. Approximate methods, which we are talking about, will create a startup and first approximation to uncertainties to be supplemented by more and more experimental data later on. Two aspects are of main concern in this context:

3.1. Keeping Sufficient Freedom to Allow for Modifications by Future High Precision Measurements

Cross sections for a certain reaction as a function of incident particle energy present two types of information which are somewhat different in nature. One is "shape–information", derived in great detail especially in high (energy–)resolution measurements, the other is "absolute value" information, related to normalization, flux determination, reference cross sections etc.. To allow for improvement of the former one must avoid strong MR or LR correlations, because these make an excitation function rigid against twist. Experimental evidence, however, very often suggests contrary. Due to computer controlled experiments and data acquisition data can be collected in long stable runs with statistical uncertainty being only a small fraction of other (highly correlated) uncertainty contributions. Basing an uncertainty estimate solely on one such data set – though tempting – would create serious problems in future updates.

3.2. The Data Handling Problem Related to the Size of Matrices

The mathematical procedures for propagation of covariances, in application calculations as well as in update codes involve at least matrix multiplications, to include new information also requires matrix inversion. It is therefore highly desirable to keep the size of the matrices limited to prevent from problems of numerical precision (rounding errors) and excessive consumption of computer time.

The ENDF/B format specification[3] allows for different types of presentation of the covariance information distinguished by the "LB–flag" and different "sub-subsections". One might thus break down the whole matrix into SR, MR and LR components and store those efficiently in separate sub–subsections, e.g. with LB=0 or LB=1, with different energy grids. This would result in a considerable reduction of matrix dimensions for applications, where it is not required to restore the full covariance matrix. Evaluation codes, however, have up to now no provision for decomposition of the covariance matrix after processing of new data and would end up in any case with a huge matrix based on the union energy grid. One might therefore be forced to compromise already at the estimate stage and design a full covariance matrix (LB=5 flag) with a reduced (with respect to cross section data) number of energy groups.

4. Methods Used at ORNL

Forced by application requirements – uncertainty assessment in design parameters of fusion reactors – procedures were developed for the generation of covariances for all reactions given in file 3 of ENDF/B–VI for the isotopes of Cr, Fe, Ni, Cu and Pb. Four approaches have been used, depending upon the quantity and quality of available experimental data.

4.1. Rigourous Method

For a total of 6 reactions, where sufficient experimental data with complete uncertainty information were available, a Bayesian analysis using the GLUCS code was done using ENDF/B–V as a prior. This procedure provides updated cross sections and complete covariances (LB=5 in ENDF/B terms) on an energy grid with a moderate number of groups.

4.2. Sparse or Incomplete Experimental Data

SR and LR correlations have been assembled using separate sub–subsections of file 33 with LB=1 flag. The fractional components were assigned using either the procedures explained in section 2.2. or available information on measurement– and normalization uncertainties etc.. An absolute component for the whole energy range from threshold to 20 MeV was included using a LB=0 sub–subsection if the reaction has cross section values which differ by orders of magnitude, as e.g. at total cross section minima, near threshold or at high incident energies. This absolute component was typically taken as a percentage of the smallest cross section value in the file, varying between 5% for large cross section minima and over 100% for reactions with minima << 1 mb.

4.3. Cross Section Data from Model Calculations

Several of the evaluated excitation functions depend predominantly on TNG calculations[4]. In this case SR uncertainties have been assigned according to table 1:

Table 1. Estimated uncertainties as a function of cross section

cross section X (mb)	standard deviation
$500 < X$	10 %
$100 < X < 500$	15 %
$30 < X < 100$	20 %
$X < 30$	≥ 30 %

A LR uncertainty component (one energy interval from threshold to 20 MeV) was then assigned by using one half of the minimum value of the SR component. Subsequently the SR uncertainties were divided by two to compensate for the effective removal of the LR component.

The ENDF/B specifications require a sub-subsection with LB=8 to ensure that processed covariance matrices are positive-definite. This was done by taking a fraction F from the SR covariance component and multiplying this value by the square of the cross section for the appropriate energy interval from file 3. The SR values were then finalized by taking (1–F) times the initial SR components. The fraction F chosen was 0.01 for the total, elastic, nonelastic and capture cross section, 0.1 for all other reactions. The choice of a small fraction is important in cases, where the energies in file 3 are close together, since the processed group variance from an LB=8 sub-subsection is inversely proportional to the group width.

$$VAR_{jj} = F_k * dE_k / dE_j \qquad (4)$$

where dE_k is the width on the file 33 grid, while dE_j is on the file 3 grid. This would lead to unreasonably high standard deviations if e.g. file 33 uses MeV–groups and file 3 point distances of keV or less.

4.4. "Derived" Covariances for MT = 1, 2, 3 and 4

"Derived" in ENDF/B context means that a cross section may be determined by summing up other (given) cross sections. Consequently its uncertainties may be calculated from the uncertainties of the other cross sections. This is indicated in file 33 by giving "NC type" sub–subsections instead of "NI type" sub–subsections which give covariances or correlations explicitly. The method of giving "NC type" sub–subsections may thus take advantage of any existing constraints among the various cross sections. In the case of this ORNL study the following considerations have lead to a "mixed" representation with "NC type" and "NI type" sub–subsections

a. The total cross section (MT=1) uncertainties are generally well known at thermal energies, in the resonance region, and up to 20 MeV.

b. The elastic scattering cross section (MT=2) uncertainties are generally well known at thermal energies and in the resolved resonance region, but not from the end of the resolved resonance region to several MeV due to structure which occurs as a result of obtaining (file)3/(MT)2 from 3/1 – 3/3.

c. The nonelastic cross section (MT=3) uncertainties up to the threshold of the first reaction are given by the capture cross section uncertainties, and above the threshold are defined by data and model uncertainties.

d. The total inelastic cross section (MT=4) uncertainties may be estimated more accurately from the uncertainties for the nonelastic and other partial components of the nonelastic than obtaining simply by summing uncertainties given for MT = 51–91, which may be large.

With these considerations in mind the uncertainties for MT = 3 are given as derived (33/102) from 1.E–5 eV to the threshold of the first reaction and explicitly, based on uncertainties estimated from data and optical model, for the rest of the energy range up to 20 MeV.

The uncertainties for MT = 4 are derived from threshold to 20 MeV as (33/3 – 33/16 – 33/22 – 33/28 – 33/102 – 33/103 – ...)

For the elastic cross section MT = 2 uncertainties from 1.E–5 eV to close to the first resonance are given explicitly to ensure the correct thermal uncertainty, then to the first reaction threshold derived as (33/1 – 33/102), and from threshold to 20 MeV are derived as (33/1 – 33/3).

Finally, uncertainties for the total cross section MT = 1 are specified derived as 33/2 + 33/102 for the region where it is explicitly in MT = 2. For the rest of the energy range uncertainties are given explicitly, based on experimental data.

5. Methods Used at IRK

At the IRK a research project supported by the Commission of the European Communitys' Fusion Technology Programme, with the goal of deriving covariance information for the most important reactions of the main isotopes of the structural materials, Cr–52, Fe–56, Ni–58 and Ni–60 as given in the European Fusion File (EFF[5]), has been carried out. The method chosen – comparison of different independent evaluations – had been suggested by H. Vonach[6] 1987 and has been mentioned in section 2.3.

For all four isotopes four evaluations of equal quality have been performed in Japan (JENDL–3)[7], U.S. (ENDF/B–VI)[3], U.S.S.R. (BROND)[8] and Europe (EFF) approximately at the same time. All these evaluations are based on about the same experimental data base and the same knowledge of nuclear reaction theory. Thus their results represent the cross section predictions of four independent experienced experts based on our present experimental and theoretical knowledge. It can be expected that all important uncertainties in the evaluation process, such as different judgement of the quality of different experimental data sets, use of different parametrizations in the nuclear reaction models, uncertainties in the preferred input parameters, e.g. level densities, and use of slightly different approaches concerning the description of direct and preequilibrium processes will manifest themselves as differences between the four evaluations. Fig. 1 may serve to illustrate this situation.

We have therefore adopted as our working hypothesis, that these evaluations constitute a quasi–statistical ensemble and the uncertainty of the EFF cross sections in question is approximately given by the maximum difference to any of the remaining three evaluations. As we define our uncertainties in the usual way as effective standard deviations, this is a quite conservative assumption since we admit a chance of at least 35%, that the "true cross section value" may be found outside the "band" spanned by the four evaluations. In absence of more precise methods for the uncertainty estimates we decided however, that this safety margin was necessary.

In order to actually implement the method plots were made comparing the four evaluations for each of the four isotopes and the reactions characterized by the following ENDF/B "MT" numbers: MT1(total), MT3(non), MT4(inel), MT16(n,2n), MT28(n,np), MT51(n,n1), MT102(n,γ), MT103(n,p) and MT107(n,α).

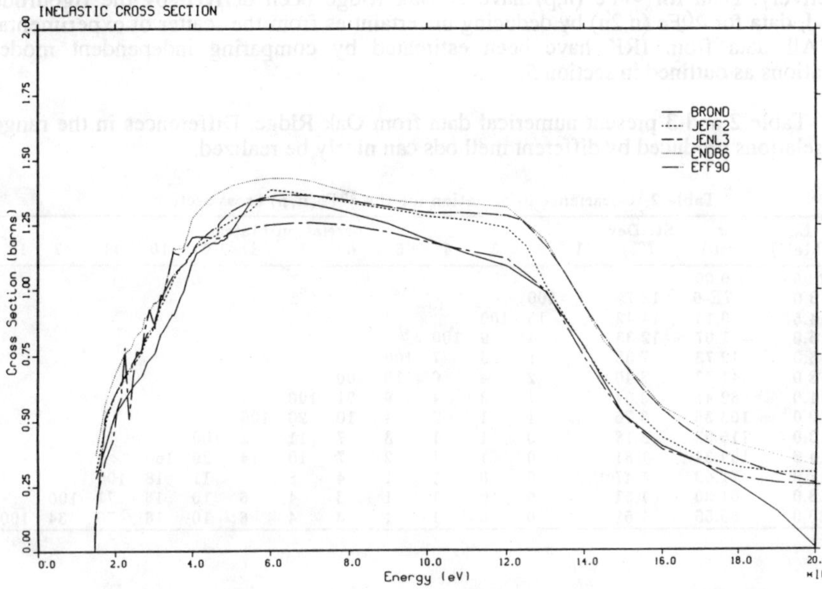

Fig. 1: Comparison of ^{52}Cr inelastic cross sections in various evaluated data files.

By looking at fig. 1, we can also state, that there is certainly some correlation, but generally not extending over energy ranges of more than 5 MeV. After close inspection of a total of 45 sets of excitation functions we decided to describe the correlations between the cross section uncertainties by a gaussian distribution function with a half–width of 4 MeV. The choice of the correlation function determined in turn to a large extent the energy grid for the representation of the covariance matrices. Adjacent energy bins of 1 MeV width are 84% correlated, thus intervals smaller than .5 MeV are certainly of no practical use. For future improvement by adding new information it appeared, however, favourable to use not too crude a structure. We therefore decided to use an energy grid of 28 bins, identical for all excitation functions, with .5 MeV bins from the resonance region up to 6 MeV, then 1 MeV groups up to 13 MeV, again .5 MeV groups for the important energy range up to 15 MeV and 1 MeV bins beyond.

Lead by experience from updating such an evaluation by experimental data[9] we have recently slightly modified this decision. It appeared more appropriate to introduce a "sliding width" for the correlation distribution function, starting with 1 MeV at 1 MeV incident particle energy, to leave more freedom for shape improvements, as discussed in section 3.1. Consequently also the energy grid was refined to 0.2 MeV bins up to 3 MeV, leading to a total of 41 energy groups now.

6. Examples of Results

A few examples may demonstrate the results produced at ORNL and IRK respectively. Data for ^{56}Fe (n,p) have at Oak Ridge been derived by the rigourous method, data for ^{56}Fe (n,2n) by deducing uncertainties from the scatter of experimental data. All data from IRK have been estimated by comparing independent model calculations as outlined in section 5.

Table 2 and 3 present numerical data from Oak Ridge. Differences in the range of correlations produced by different methods can nicely be realized.

Table 2. Covariance information for the $^{56}Fe(n,p)$ cross section

	E_n (MeV)	σ (mb)	St. Dev. (%)	Correlation Matrix 1	2	3	4	5	6	7	8	9	10	11	12	13
1	2.9655	0.00														
2	3.0	7E-6	18.75		100											
3	4.5	0.14	14.42		13	100										
4	5.0	1.07	12.33		4	9	100									
5	6.0	12.73	7.38		1	3	7	100								
6	8.0	41.73	3.40		2	4	6	16	100							
7	11.0	82.41	3.64		1	3	4	9	21	100						
8	12.0	103.35	3.76		1	1	2	4	10	20	100					
9	13.0	115.23	3.18		0	1	1	3	7	11	22	100				
10	14.6	109.34	1.81		0	1	1	2	7	10	14	29	100			
11	17.0	71.63	1.47		0	0	1	1	4	5	7	11	18	100		
12	18.0	61.30	1.51		0	0	1	1	3	4	6	10	16	71	100	
13	20.0	50.55	1.51		0	0	1	1	3	4	6	10	16	70	34	100

Table 3. Covariance information for the $^{56}Fe(n,2n)$ cross section

	E_n (MeV)	σ (mb)	St. Dev. (%)	Correlation Matrix 1	2	3	4	5	6	7	8	9	10	11	12	13
1	11.402	0.00														
2	11.5	2.24	282.52		100											
3	11.6	7.59	83.87		100	100										
4	11.8	25.04	27.08		94	97	100									
5	12.0	49.93	17.48		73	75	78	100								
6	12.4	111.59	11.71		50	53	61	81	100							
7	13.5	317.68	9.97		22	26	37	69	91	100						
8	14.0	402.04	10.84		16	20	31	37	46	49	100					
9	15.5	556.52	10.33		13	17	28	36	47	51	86	100				
10	16.0	582.71	10.78		12	16	27	34	44	49	44	46	100			
11	17.5	611.45	10.69		11	15	26	34	45	49	45	46	83	100		
12	18.0	609.92	10.77		11	15	26	34	44	48	44	46	44	44	100	
13	20.0	575.40	10.89		12	15	26	34	44	48	44	46	44	44	82	100

Fig. 2 compares graphically the respective uncertainty assignments, fig. 3 is a cut through the covariance matrix at E_n = 14.1 MeV. The IRK estimate for the width of the uncertainty distribution function is somewhat larger than with the ORNL GLUCS result, while in fig. 4 – estimate from scatter – only LR correlations have been assigned differently.

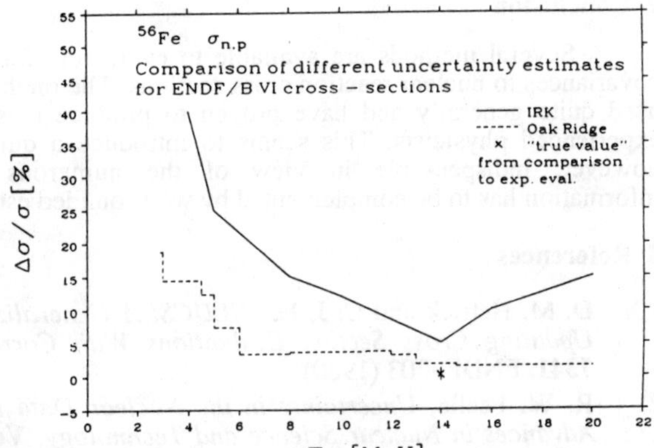

Fig. 2: Comparison of uncertainty estimates from IRK and ORNL. The "true value" indicates the uncertainty at 14 MeV, derived by a rigourous evaluation of all available experimental data up to 1991[10].

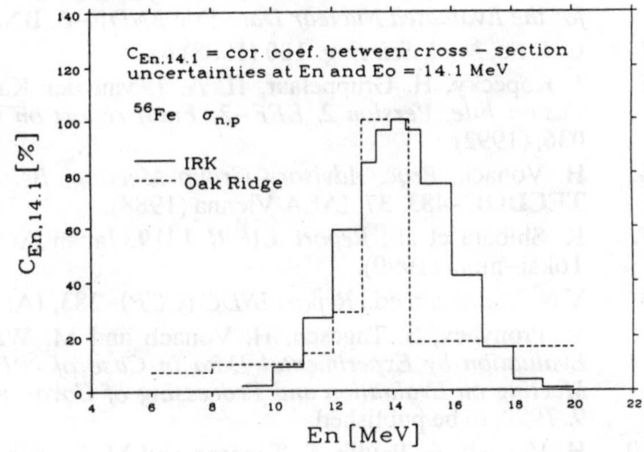

Fig. 3: Comparison of correlation distribution function at 14 MeV between IRK and GLUCS result.

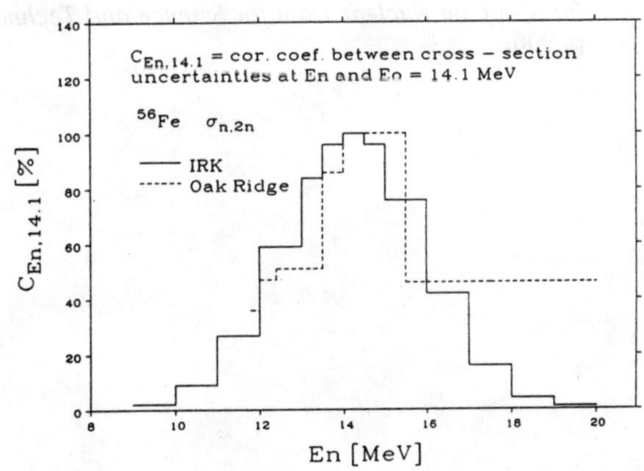

Fig. 4: Comparison of correlation distribution function at 14 MeV between IRK and ORNL result from scatter of experimental data.

7. Conclusion

Several methods are available to evaluators for a "preliminary" assignment of covariances to nuclear reaction cross sections. The methods are sufficiently simple to be used quite generally and have proven to produce reasonable results in the hands of experienced physicists. This seems to introduce a quite subjective component. It is, however, indispensable in view of the numerous situations where incomplete information has to be complemented by well founded estimates.

8. References

1. D. M. Hetrick and C. J. Fu, *GLUCS: A Generalized Least Squares Program for Updating Cross Section Evaluations With Correlated Data Sets*, **ORNL/TM 7341**, ENDF-303 (1980)
2. R. W. Peelle, *Uncertainty in the Nuclear Data used for Reactor Calculations, Advances in Nuclear Science and Technology*, **Vol. 14**, Lewis and Becker, eds., Plenum Press, New York (1982) p.11 ff.
3. P. F. Rose and C. L. Dunford, editors, *ENDF−102 Data Formats and Procedures for the Evaluated Nuclear Data File ENDF−6*, **BNL−NCS−44945** (1990)
4. C. J. Fu, *Nucl. Sci. Eng.* **100** (1988) 61
5. J. Kopecky, H. Gruppelaar, H. A. J. van der Kamp and D. Nierop, *European Fusion File, Version 2, EFF−2; Final report on basic data file*, **ECN−C−−92−036**, (1992).
6. H. Vonach, *Proc. Advisory Group Meeting, Beijing, Oct. 12−16, 1987*, **IAEA TECDOC−483**, 37, IAEA Vienna (1988).
7. K. Shibata et al., *Report JAERI* **1319**, Japan Atomic Energy Research Institut, Tokai-mura (1990).
8. V.N. Manokhin ed., *Report INDC (CCP)*−**283**, IAEA, Vienna (1988).
9. V. Pronyaev, S. Tagesen, H. Vonach and M. Wagner, *Updating a Theoretical Evaluation by Experimental Data in Case of ^{56}Fe, Proc. NEANSC Specialists' Meeting on Evaluation and Processing of Covariance Data, ORNL, October 7−9, 1992*, to be published.
10. H. Vonach, A. Pavlik, S. Tagesen and M. Wagner, *Evaluation of 14 MeV Cross Sections for the Main Isotopes of the Structural Materials Cr, Fe and Ni, Proc. Int. Conf. on Nuclear Data for Science and Technology, Jülich, 13−17 May 1991*, p. 906.

DETERMINATION OF COVARIANCE OF CALCULATED CROSS SECTIONS

YUKINORI KANDA
Department of Energy Conversion Engineering,
Kyushu University,
Kasuga, Fukuoka, 816 / Japan

ABSTRACT

The covariance of calculated cross sections have been generated with the method which are adopted by individual evaluators in order to express their understanding for the covariance. Three methods are briefly reviewed and compared quoting the results for Fe. The feasture of the three are discussed connecting the result to the mutual differences are revealed to some extent.

A method to visualize a correlation matrix, called a graphical correlation matrix, is developed to compare easily each other. It is made definite that the method is considerably valid to clarify differences between correlation matrices. By using it, the characteristics of the correlation matrices depending on the parameters used in cross section calculation are examined to reach the method of covariance generation for calculated cross sections.

1. Introduction

Nuclear data files are required not only to supply reliable evaluated values but also to contain comprehensively consistent data that is the total cross section is equal to the sum of the partial ones and all the kinds of the partial cross sections are included except negligibly small ones. This is absolutely necessary whether the available experimental data exist or not. There is, however, no guarantee that the requirement mentioned above is always met sufficiently with measurements. Sometimes, a measurement can never be performed even if experimenters make full use of a latest technology. Experiments always imply proper uncertainty even when they are technically possible. When available measurements for an interested cross section are nothing or insufficient to evaluate it from experiments, we must calculate it by using theoretical formulae. At present, while the potentiality producing experimental data base is decreasing in the world, the requirement of the evaluated cross sections, for example in fusion application, is increasing in quality and quantity. In addition, the measurement of the required cross section is technically difficult. No matter what there are enough available measurements to produce the cross section curve, the quantitative presentation of the reliability on the evaluated data is currently demanded in the nuclear data file[1]. It is expressed with a covariance. Naturally the calculated data should also have the covariance. Methods valid for these issues have been developing in the nuclear data evaluation field. The subject in this work is to review methodology to determine the covariances of the calculated cross sections.

In the following section, three methods to generate the covariance of calculated cross sections are reviewed. In section 3, the features of the three methods are discussed comparing graphical presentations of the deduced covariances. In the last section, conclusions are given.

2. Methods of Covariance Generation for Calculated Cross Sections

2.1 Analogy with Experimental Uncertainty

In this section, a method of covariance generation for calculated cross sections without experimental information concerning is presented. This method is interesting from a viewpoint that the uncertainty for the calculated cross section is substituted with the one inferred analogically from experimental uncertainties. This work was performed to estimate the covariance of some cross sections of medium nuclei for ENDF/B-VI by Hetrick et al. [2] in Oak Ridge National Laboratory.

The cross section was calculated with the TNG code[3] which is an advanced multistep Hauser-Feshbach code including precompound and compound contributions to cross sections. The uncertainties for the calculated cross sections are assigned as shown in Table1, based on the magnitude of the cross section. The authors of the original paper[2] said "This assumes that nuclear model codes are more reliable in their predictions for reactions which have large cross sections than for reactions which have small cross sections". The assumption seems to be an idea from a statistical error in an experiment. The author's opinion is that the statement in the original paper is not acceptable and it is reasonable to be analogy with experimental uncertainty. Therefore, the title of this sub-section is named. The covariance file for the calculated cross sections is constructed with the procedures in which short-range (i.e.small energy intervals, typically 2 to 4MeV) correlations and for some cross sections, long-rang (i.e. large energy intervals, typically 1 to 3 intervals from threshold to 20MeV) correlations are assembled in order that the covariance matrices match with the ENDF-6 format and, in particular, to ensure that they are positive-definite.

2.2 Distribution of Recent Evaluations

The method of covariance generation proposed by Vonach[4] was applied to the isotopes Cr, Fe, and Ni for the evaluation of the EFF-2 file by Vonach et al. [4] in Institute fuer Radiumforschung und Kernphysik der Universitaet Wien (IRK). In this section, its summary is presented. The basic hypothesis assumed in the method is that the uncertainty of the evaluated cross section is approximately given by the maximum difference to any of the results of the other evaluations. Namely, the covariance is generated from a distribution of available evaluations.

Three recent evaluations, JENDL-3, ENDF/B-VI and BROND correspond to the evaluations described above. All these evaluations are based on about the same experimental and theoretical data base. These results, therefore, can be regarded as the products of four independent expert groups based on the present experimental and theoretical knowledge in our field. It can be expected that all main uncertainties in the evaluation manifest themselves as mutual differences in the four evaluations. Thus, as their basic hypothesis, they assume that the true cross section values are expected to fall within the bands defined by the maximum and minimum of the four evaluated cross sections. This assumption is approximately equivalent to the statement that the true value of some quantity will in general be found within the range of results of four independent measurements. If this analogy between experiments and evaluations can be acceptable it is reasonable according to a conventional error theory that the four evaluations are the events sampled from the population of the evaluation in a form of Gaussian distribution and the true cross section value is the most probable value in the distribution. After the visual inspection of the distribution of the evaluations, they decided to describe the correlations between the cross section uncertainties at E_1 and E_2 respectively by a Gaussian with half-width of 4MeV :

$$\exp\left[-\frac{(E_1 - E_2)^2}{4}\cdot \ln 2\right]. \tag{1}$$

They mentioned that this is admittedly a crude approximation, however, in absence of evidence for the contrary in the present situation it seemed reasonable to assume the most simple form for the shape of the correlation function. The correlation matrices can be easily imagined from eq. (1) : Their off-diagonal components are equally systematic. So, they are not presented as tables.

The standard deviation for the evaluated cross sections was derived by visual inspection of the figures of the four evaluation curves.

2.3 Uncertainty for Parameters Used in Calculation

The method described in this section has been developeted in the author's group in Kyushu University[6,7,8]. Reaction cross sections are calculated with Hauser-Feshbach model and the covariances for them are estimated from the uncertainties of the level density parameters which are one of the leading parameters in the Hauser-Feshbach model formulae.

A detailed description of our procedures for covariance generation is not presented in this report since they can be found elsewhere[9] and a brief description is given here to some extent in order to be compatible with the two methods in the preceding sections.

Cross section calculation is conducted with the GNASH code[10] equivalent to the TNG code[3]. The covariances for calculated cross sections are estimated in the two ways in which a difference is in the choice of the level density parameters in the Hauser-Feshbach formulae. One is unchanged values of Gilbert-Cameron[11] : Alternative is adjusted values with Bayesian analysis so as to reproduce simultaneously the experimental data for the interesting and competing reactions. These are processing with the KALMAN code[12] which has been developed and applied to the works[6,7,8] in the author's group. It has been recently named. In both calculations, all conditions except for the level density parameters are quite equal. In the present work, the calculated covariances depend on only the uncertainties of the level density parameters because the uncertainties for the other parameters such as optical model parameters are not taken into account. This is equivalent to the assumption that the formulae in the GNASH code and the used parameters except the level density parameters are completely accurate. Actually it is evident that they are not accurate. As a result, therefore, all the uncertainties concerning charge the level density parameters and they result in the generated covariances. There is no grounds to claim that this is correct but no means to make correct at present.

3. Comparison of the Methods and Generated Covariances

3.1 Comparison of the Covariances determined with the Three Methods

In the methods of ORNL and IRK, the standard deviations of the evaluated cross sections are estimated from the analogy to experimental values and the distribution of the evaluated values in different files, respectively, and the covariances are constructed from them. In the method of Kyushu Univ., the covariances are calculated from those uncertainties of those parameters in the theoretical formulae used in the cross section calculation which are estimated on the basis of the uncertainties of experiments. Therefore, the last method can connect between experimental uncertainties and generated covariances even though the propagating process from the former to the latter is extremely complicated.

It is difficult to search the differences among the deduced covariances produced by the three methods because it is a two-dimensional array of numbers. To simplify a comparison, the respective row at 14MeV of the covariance matrix for ^{56}Fe(n,p) in the three works is shown together in Fig.1. The respective correlation matrices generated with the methods of ORNL and Kyushu Univ. do not have in principle equal components in the own but in the IRK method the matrix components at every row is always same according to the assumptions.

Figure 2 shows the comparison of those standard deviations for the evaluated cross sections of ^{56}Fe(n,p) in the three evaluations which correspond to the diagonal elements of the evaluated covariance. The ORNL curve in Fig.2 represents the primarily assumed values themselves for the calculated cross sections. (In the ORNL report of Hetrick et al.[2], the author can not identify whether the criterion for deciding of standard deviation is applied to ^{56}Fe(n,p) or not. The ORNL curve in Fig.2 is qualitatively suitable for this statement.) The Kyushu Univ. curve is almost along the zero line, because it is computed from so many experimental data that the standard deviations are markedly decreased. Simply considering, the method is equivalent to averaging experimental data and then the more number of data is the smaller standard deviation is resulted. Though the extremely small value is right in the algorithm of the method, it is a problem whether it is acceptable or not. One feels that it must be too small to agree. There, however, is not the decisive solution to make ones understand. In contrast to it, the IRK curve seems to be reasonable comparing the figure plotting available experiments, Fig.3. It was evaluated from the difference of the four evaluations, which reflect the distribution of the existing measurements plotted in Fig.3. If the experimental data exist in a narrow band, the evaluated cross section in this region has a small standard deviation. This similarly results in a small difference of the four evaluations. Accordingly, the IRK curve involves to be reasonable. Combining the above two discussions, a typical problem in nuclear data evaluation appears as the one to be solved in future : Namely, the result deduced through correct algorithm is not always acceptable in a physical sense.

3.2 *Comparison of the correlation calculated on various conditions*

The status of the existing experiments for ^{56}Fe(n,p) and ^{54}Fe(n,p) is shown in Figs.3 and 4, respectively. The excitation functions computed with the original level density parameters, called prior one, and the adjusted values, posterior one, respectively. In figs.5 and 6, are shown the graphical correlation which are plotted individual rows in a correlation matrix by moving the peaks, that is, the diagonal points to a common position. A sharp curve means strong short range correlation, and a wide curve does dull changing correlation. A typical dull changing curve is shown Fig. 7, which is the correlation for the prior excitation function of ^{56}Fe(n,p). As examples, the correlation matrices of ^{56}Fe(n,p) are show in Tables 2 and 3 corresponding to Figs. 5 and 7, respectively.

From a comparison between Figs. 5 and 7, the influence of experiments to the correlation can be seen the prior correlation curves generally are dull and are changed to the sharp curves. The difference of both expresses the effects of experiments. Similarly, the prior curve for ^{54}Fe(n,p) is in Fig. 8. The curve of the 14MeV row shown in Fig. 1 is not representative of the ^{56}Fe(n,p) correlation matrix and is strongly influenced from experiments.

To make clear the effects of existing experiments, the curves for the individual rows are plotted around the home position of respective peak in Fig. 9 for ^{56}Fe(n,p). Though the comparison of these figures is expected to give new knowledge for determination of covariances of calculated cross sections, it is too complicate to reach the goal.

4. Conclusion

Three methods to generate the covariances of calculated cross sections, that is, the methods of ORNL, IRK and Kyushu Univ., are briefly reviewed. The correlation matrices and standard deviations are compared and discussed on their features. A conclusion from these discussions is that the result deduced through some correct algorithm is not always acceptable in a physical sense. To develop the method for determination of covariances of calculated cross section, it should be solved in future.

A method to visualize a correlation matrix has been developed and is applied to search the covariances for Fe. The resultant graphys are used to define the effects of existing experiments.

The author are indebted to Dr. T. Kawano, K. Kamitsubo, T. Iwamoto, and N. Fujikawa for their discussion and for their help to preparing of the report.

5. References

1. Y. Kanda, "New Perspective in Covariance Evaluation for Nuclear Data" in *Proc. International Conference on Nuclear Data for Science and Technology*, 13-17 May 1991, Juelich, Germany, p. 176 (Springer-Verlag, Berlin, 1992).
2. D. M. Hetrick, D. C. Larson, and C. Y. Fu, "Generation of Covariance Files for the Isotopes of Cr, Fe, Ni, Cu, and Pb in ENDF/B-VI", OR-NL /TM-11763, ENDF-350 (1991).
3. C. Y. Fu, *Nucl. Sci. Eng.* **100**, (1988) 61.

4. H. Vonach, in *Proc. Advisory Group Meeting on the Use of Nuclear theory for Fast Neutron Nuclear Data Evaluation*, Oct. 1987, Beijing, IAEA TECDOC-483, p.37 (IAEA, Vienna, 1988).
5. H. Vonach, S. Tangesen, M. Wagner, and A. Pavlik, "Uncertainty Estimates for the Fast Neutron Cross-Sections of the European Fusion File EFF for 52Cr, 56Fe, 58Ni, and 60Ni and Evaluation of the 14MeV Cross-Sections of these Isotopes from the existing experimental data base", EFF-DOC-85 (Institute fuer Radiumforshung und Kernphysik der Universitaet Wien, 1990).
6. Y. Kanda and Y. Uenohara, "A method to Evaluate Covariances for Correlated Nuclear Data", in *Proc. of the IAEA Specialists' Meeting on Covariance Methods and Practices in the Field of Nuclear Data*, 17-19 Nov. 1986, Rome, Italy, INDC(NDS) 192/U, (IAEA, Vienna, 1988) p.81.
7. Y. Kanda and Y. Uenohara, "Covariances of Nuclear Model Parameters Generated from Experimental Information" ibid. p.98.
8. Y. Kanda and Y. Uenohara, "Covariance Matrices Evaluated by Different Methods for Some Neutron-Dosimeter Reactions", in *Proc. of the International Conference on Nuclear Data for Science and Technology*, Mito, Japan, 30 May-3June, p.1041 (Saikon Publishing Co. Tokyo, 1989).
9. Y. Kanda, "Generation of Covariance Data from Combined Experimental and Theoretical Data", *NEANSC Specialists' Meeting on Evaluation and Processing of Covariance Data*, 7-9 Oct. 1992, Oak Ridge, USA.
10. E. D. Arthur, "The GNASH Preequilibrium-Statistical Nuclear Model Code", in *Applied Nuclear Theory and Nuclear Model Calculations for Nuclear Technology Applications : Proceedings of a Workshop at trieste*, Italy, 15 February-19 March 1988, (World Scientific Publishing Co. Ltd., Singapore) pp.676-725 (1989).
11. A. Gilbert and A. C. W. Cameron, *Can. J. Phys.*, **43** (1965)1466.
12. T. Kawano, Y. Uenohara and Y. Kanda, "Program KALMAN, A General Purpose Computer Program for Parameter Estimation, User's Guide" (Private Communication, 1992, To be published elsewhere).

Table 1. Assigned uncertainties for calculated cross sections.

Cross Section X (mb)	Standard Deviation
500 < X	10%
100 < X < 500	15%
30 < X < 100	20%
X < 30	> 30%

Table 2 Covariance for $^{56}Fe(n,p)$ with posterior level density parameters.

ENERGY [MeV]	ERROR [%]	CORRELATION [x1000]															
5.0	3.97E-02	1000															
6.0	2.41E-01	164	1000														
7.0	5.48E-01	73	995	1000													
8.0	6.65E-01	56	991	997	1000												
9.0	6.15E-01	52	972	979	991	1000											
10.0	4.97E-01	52	906	914	940	977	1000										
11.0	3.83E-01	48	732	740	785	859	947	1000									
12.0	3.13E-01	34	354	361	423	534	698	889	1000								
13.0	3.04E-01	12	-70	-65	-2	118	315	594	894	1000							
14.0	2.90E-01	0	-191	-187	-134	-28	150	429	771	962	1000						
15.0	2.56E-01	-1	-52	-48	-13	55	178	381	647	817	925	1000					
16.0	2.51E-01	1	168	170	178	193	225	298	406	499	658	894	1000				
17.0	2.62E-01	4	270	272	259	234	199	174	170	212	394	712	951	1000			
18.0	2.75E-01	8	361	362	336	285	206	111	17	4	184	541	860	975	1000		
19.0	2.56E-01	7	359	359	326	262	162	38	-86	-107	77	448	800	946	993	1000	
20.0	2.12E-01	8	294	293	255	184	75	-55	-171	-167	31	406	765	924	978	993	1000

Table 3 Covariance for $^{56}Fe(n,p)$ with prior level density parameters.

ENERGY [MeV]	ERROR [%]	CORRELATION [x1000]															
5.0	4.04E-02	1000															
6.0	1.91E+00	40	1000														
7.0	4.53E+00	32	999	1000													
8.0	7.46E+00	51	979	984	1000												
9.0	1.06E+01	70	932	941	986	1000											
10.0	1.37E+01	82	883	894	959	992	1000										
11.0	1.67E+01	90	843	856	934	980	996	1000									
12.0	1.93E+01	95	812	826	913	967	990	998	1000								
13.0	2.14E+01	98	792	806	898	958	985	995	999	1000							
14.0	2.19E+01	99	786	800	894	955	983	995	999	999	1000						
15.0	2.03E+01	99	790	804	897	957	985	995	999	999	999	1000					
16.0	1.75E+01	98	797	812	902	961	986	996	999	999	999	999	1000				
17.0	1.41E+01	97	803	817	906	963	988	997	999	999	999	999	999	1000			
18.0	1.10E+01	96	812	826	912	967	990	997	999	998	998	999	999	999	1000		
19.0	8.18E+00	95	817	830	915	969	990	997	999	998	998	998	999	999	999	1000	
20.0	5.82E+00	96	813	827	913	967	989	997	998	998	998	998	999	999	999	999	1000

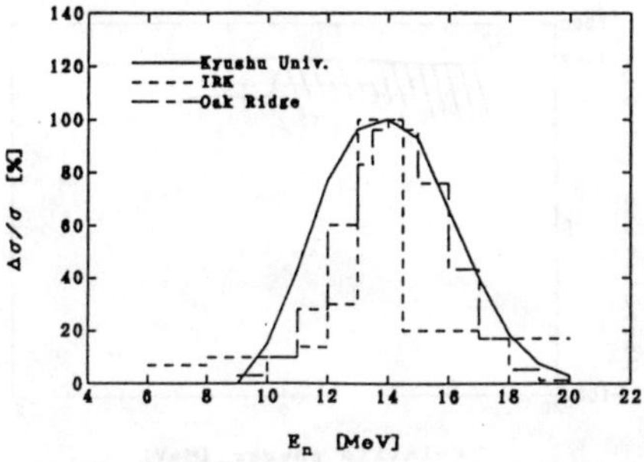

Fig. 1 Comparison of the respective row at 14MeV of the correlation matrix for 56Fe (n, p) in Kyushu Univ., IRK and ORNL.

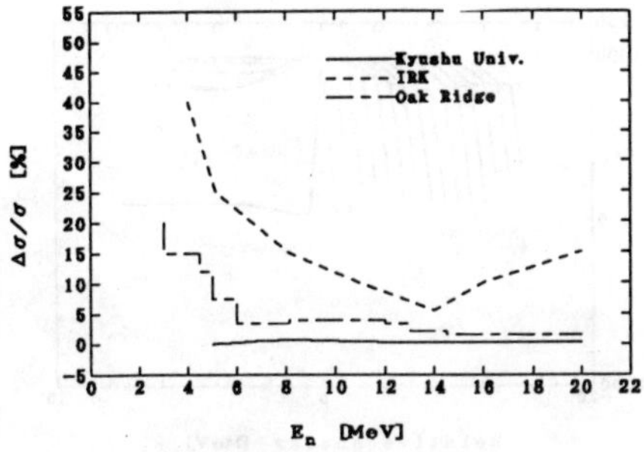

Fig. 2 Comparison of the respective standard deviation for the evaluated cross-sections of 56 (n, p) in Kyushu Univ., IRK and ORNL.

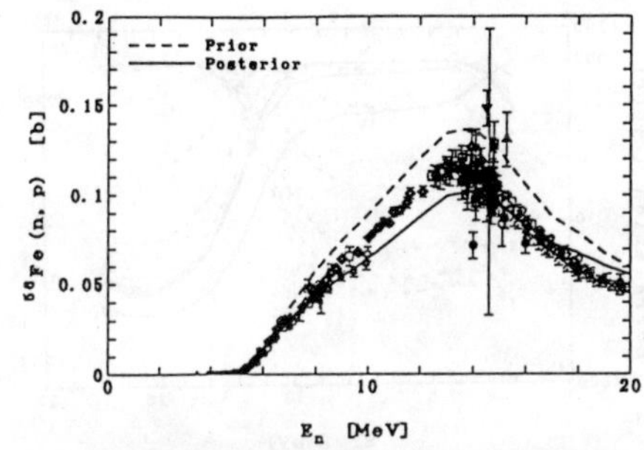

Fig. 3 Excitation functions of 56Fe (n, p). The dashed line is the thoretical calculation with the original level density parameters (prior) and the solid line with the adjusted ones (posterior).

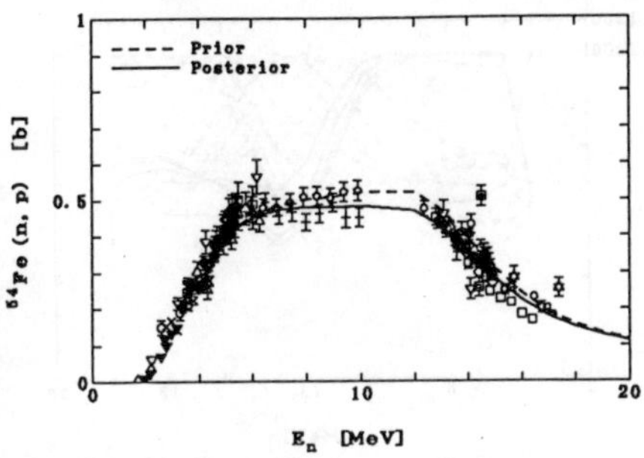

Fig. 4 Excitation functions of 54Fe (n, n). See the caption of Fig. 3.

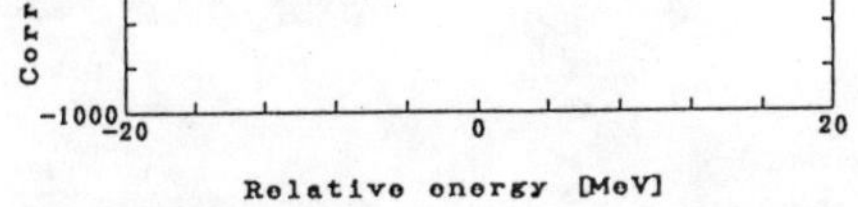

Fig. 5 The graphical peak-moved correlation matrix of 56Fe (n, p) computed with the adjusted level density parameters (posterior) corresponding to the solid line in Fig. 3. A line is graphically visualized a row in the correlation matrix. The diagonal value of the respective rows is at the origin of the abscissa.

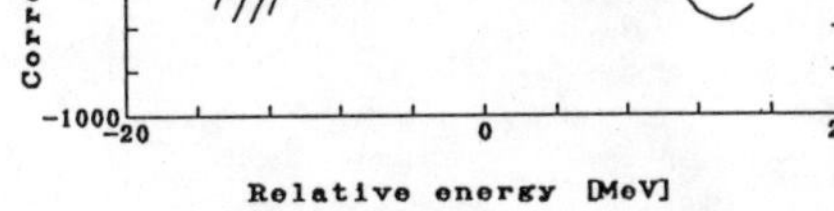

Fig. 6 The graphical peak-moved correlation matrix of 54Fe (n, p) computed with the adjusted level density parameters (posterior) corresponding to the solid line in Fig. 4. Seethe caption of Fig. 5.

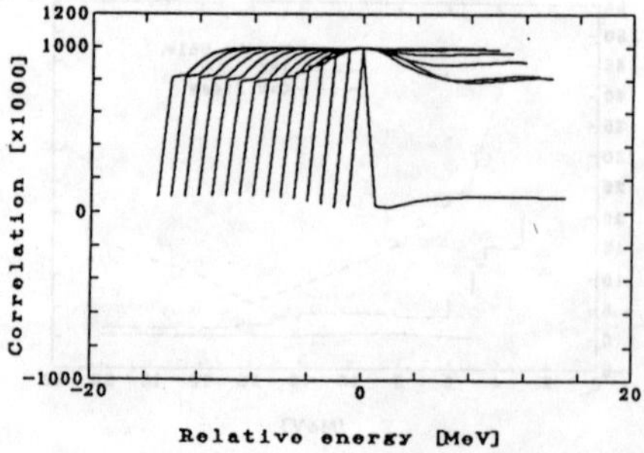

Fig. 7 The graphical peak-moved correlation matrix of 56Fe (n, p) computed with the original level density parameters (prior) corresponding to the dashed line in Fig. 3. See the caption of Fig. 5.

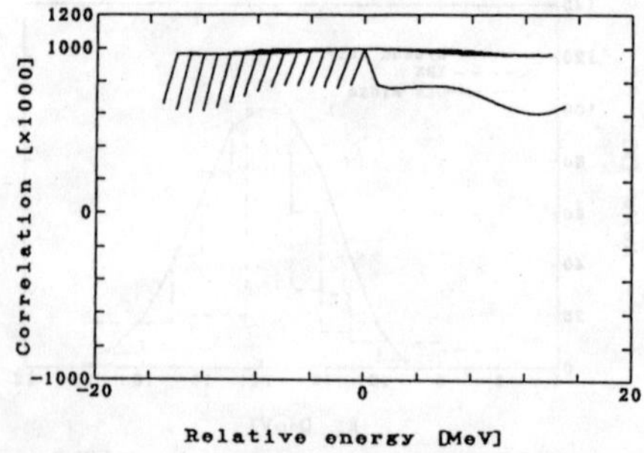

Fig. 8 The graphical peak-moved correlation matrix of 54Fe (n, p) computed with the original level density parameters (prior) corresponding to the dashed line in Fig. 4. See the caption of Fig. 5.

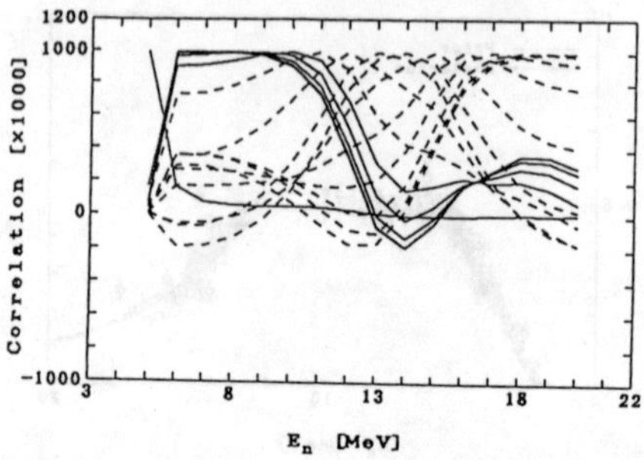

Fig. 9 The graphical correlation matrix of 56Fe (n, p) computed with the adjusted level density parameters (posterior) corresponding to the solid line in Fig. 3. The solid lines express the rows at 5-10MeV, respectively, and the dashed lines do at 11-20MeV. The data base of these curves are identical to the one of Fig. 5, and in this figure the individual rows are plotted as they are.

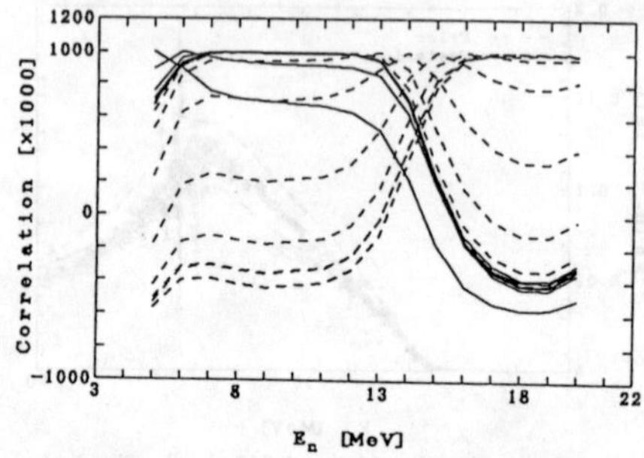

Fig. 10 The graphical correlation matrix of 54Fe (n, p) computed with the adjusted level density parameters (posterior) corresponding to the solid line in Fig. 4. This data base is identical to the oneof Fig. 6. See the caption of Fig. 9.

STATUS AND COVARIANCE FILES OF THE SCATTERING MATRIX OF GRAPHITE

F. C. DIFILIPPO, J. P. RENIER, B. A. WORLEY
Oak Ridge National Laboratory
P.O. Box 2008, Bldg. 6025
Oak Ridge, Tennessee 37831-6363 USA

ABSTRACT

Calculations related to the temperature coefficient of reactivity of enriched gas-cooled reactors show the high sensitivity of this parameter to the proper description of thermalization effects in the moderator; additionally, calculation of the temperature dependence of the inelastic scattering cross section correlates the errors of the cross sections as functions of the temperature. The ENDF/B evaluation of $S(\alpha,\beta,\mathrm{T})$ is still based on the approximations (incoherent component only) and phonon spectras of the early 60's. Subsequent measurements showed that the structure observed in $S(\alpha,\beta,\mathrm{T})$ cannot be described using the incoherent approximation, and soon after the availability of highly intense neutron beams and large specimens of pyrolitic graphite better calculation of the phonon spectra were possible. Our calculations of the probability distributions of the momentum and energy transfer, α and β, allow one to define a range of α and β for which comparison of experimental and theoretical data is of most interest for reactor analysis, and to point out regions of deficient resolution or excessive details in the present α,β mesh used in the ENDF/B files. Because the phonon spectra defines $S(\alpha,\beta,\mathrm{T})$ we have found mathematical formulas that allow the calculation of the covariance matrix of S by propagating the errors of the phonon spectra.

1. Introduction

Some results related to the neutronic performance of highly enriched Modular High Temperature Gas-Cooled Reactors (MHTGR) poisoned with highly absorbing targets are shown in Fig. 1. Because of the competition between high production and destruction terms, the reactivity balance of this configuration is particularly sensitive to details of the thermal neutron spectra which in turn depends on the inelastic scattering of the neutrons with the graphite moderator. Three of the four sets of values for the temperature coefficient of reactivity shown in Fig. 1 were computed with standard ENDF/B files for this process available in early 1991. The unexpected peak and valley that appear in Fig. 1 were related to the lack of continuity in the shape of the scattering matrices $S(\alpha,\beta,\mathrm{T})$ tabulated in ENDF/B. Further analysis revealed that this behavior was related to numerical problems with the code GASKET[1] used to generate $S(\alpha,\beta,\mathrm{T})$; once the problem was solved, a smooth temperature coefficient of reactivity, shown in Fig. 1 was found. This sensitivity and the need to compute the error in the calculations of the temperature coefficient of reactivity with the proper accounting for the correlation in temperature drove our efforts to characterize the present status of the inelastic cross section of graphite and to calculate its covariance file. Section 2 defines the ranges of α and β relevant to reactor physics, Section 3 deals with the present status of our knowledge of the inelastic scattering of neutrons by graphite and Section 4 describes our methodology to compute the covariance of the scattering matrix.

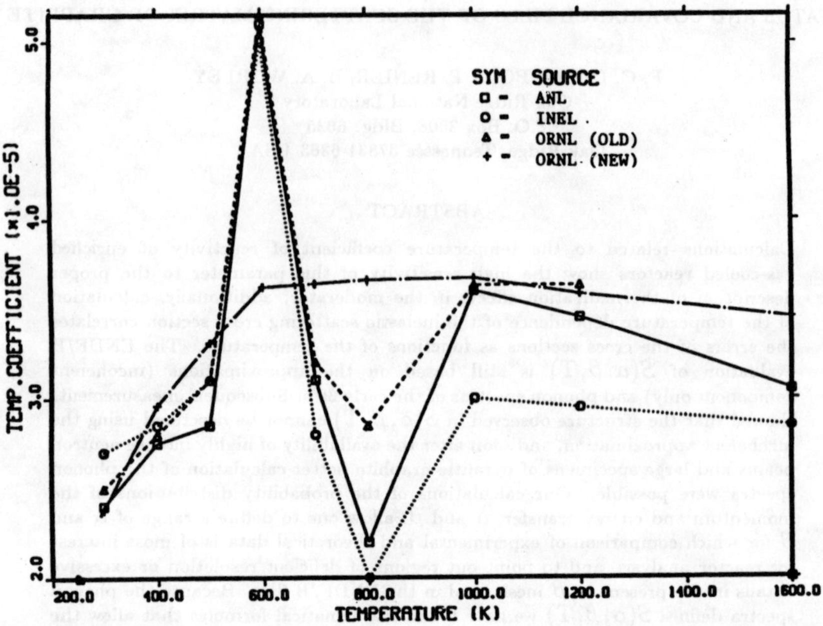

Fig. 1. Temperature coefficient of reactivity (absolute value) calculated with new and old scattering matrices for graphite.

2. Probability Distributions for the Momentum and Energy Transfers

The double differential inelastic scattering cross section for neutrons in a medium with temperature T is customarily written as:

$$\sigma_{si}(\varepsilon_o, \varepsilon, \theta, \phi) = \frac{d^2\sigma}{d\varepsilon d\Omega} = \frac{\sigma_b}{4\pi}\left(\frac{\varepsilon}{\varepsilon_o}\right)^{1/2} e^{-\beta/2} \quad S(\alpha, \beta, \mathrm{T}) \qquad (1)$$

where (1) the initial and final energies of the neutron, ε_o and ε, are in k_BT units; (2) σ_b is the bound scattering cross section of the scattering nucleus; (3) $S(\alpha, \beta, \mathrm{T})$ is the dimensionless scattering matrix; (4) $\beta = \varepsilon - \varepsilon_o$ is the energy transfer; and (5) $\alpha = \left(\varepsilon_o + \varepsilon - 2\sqrt{\varepsilon\varepsilon_o}\,\mu\right)/A$ (where A is the mass of the scattering nucleus in neutron mass units and $\mu = \cos\theta$), is proportional to the square of the momentum transfer. Integrating Eq. (1) with respect to the azimuthal variable, ϕ and passing from $d\mu$ to $d\alpha$ we obtain the differential cross section for the inelastic scattering of a neutron with energy ε_o resulting in a transference of energy and momentum β and α to the neutron. In order to obtain the probability distributions for α and β, we have to integrate over ε_o with the proper neutron spectrum; for fixed α and β the limits of ε_o are $(\alpha A - \beta)^2/4\alpha A$ and infinite. For the particular case of a Maxwellian spectra, where $M(\varepsilon_o) = \varepsilon_o\, e^{-\varepsilon_o}$, we obtain:

$$P(\alpha,\beta) = \frac{\sigma_b\ A}{4\ \overline{\sigma}_{si}}\ e^{-\left(\frac{\alpha A}{4}+\frac{\beta^2}{4\alpha A}\right)}\ S(\alpha,\beta,\mathrm{T}) \tag{2}$$

where $\overline{\sigma}_{si} = \overline{\sigma}_{si}(\mathrm{T})$ is the total inelastic cross section averaged over the Maxwellian spectra; numerical integrations over β and α give the probability distributions of, respectively, α and β. The analysis of Eq. (2) addresses two important issues: (1) Relevant ranges of α and β to compare experimental and theoretical values of $S(\alpha,\beta,\mathrm{T})$ and (2) proper choice of α and β grids to tabulate $S(\alpha,\beta,\mathrm{T})$. For example, present ENDF/B files use a fixed α,β grid [in units of 0.0253 ev (T = 293.6°K)]; there are 40 values for α, from 0.252 to 13.386, and 80 values for β, from 0. to 40.46. The first value of α, 0.252, seems to be large for describing low momentum transfer at T ~ 293.6°K, while the last value of β, 40.46, corresponds to extremely high and unlikely energy transfers.

3. Comparison of Experimental and Calculated Scattering Matrices

Present ENDF/B evaluations of the scattering matrices of graphite are based on the incoherent approximation implemented in the GASKET code.[1] For the particular case of a monoatomic polycrystal the scattering matrix in the incoherent approximation depends entirely on the frequency distribution of the phonon spectra, $\rho(\omega)$.

Three phonon spectras for graphite are available in the literature. The simplest and oldest one is the spectra proposed by Egelstaff[2] which is quadratic up to a frequency w_1, linear up to w_2, flat up to w_3, and zero beyond w_3; this spectra was found to be consistent with early measurements of $S(\alpha,\beta,\mathrm{T})$. A more basic approach corresponds to the spectra calculated by Young and Koppel[3] who found $\rho(w)$ by solving the dynamic of the lattice; four elastic constants describe the four forces considered.

The availability of high intensity neutron beams and large specimens of pyrolytic graphite allowed the direct measurement of the elastic forces of relevance for a better calculation of $\rho(w)$. Thus, the ORNL phonon spectra[4] was calculated by solving the dynamic of the lattice by including tensor forces up to the fourth neighbor with a total of eight elastic constants.

Direct measurements of $S(\alpha,\beta,\mathrm{T})$ are scarce, some of the documents are difficult to obtain or do not contain sufficient details for evaluation of the experiments. Of four known authors, Refs. 5 to 8, only one had published in the open literature and produced a document with enough information to be considered.

The experimental data of Carvalho[8] is compared in Fig. 2 with GASKET calculations corresponding to the three different phonon spectras. Neither of the spectras reproduce the considerable structure of the experimental data which seems to be related, as Carvalho showed, to coherent effects not included in the GASKET calculations. Despite the large differences between the phonon spectra, none of them seem to produce better results than the others; e.g., at $\beta = 0.15$, as Fig. 2 shows, the ORNL spectra produces results that agree quite well with the average of the experimental data while for $\beta = 0.60$ this is the case for the Egelstaff spectra.

Following the nomenclature of Ref. 1, $\mathcal{S} \equiv e^{-\beta/2} S(\alpha, \beta)$ (the nonsymmetric scattering matrix) depends on:

$$\gamma(\tau, \mathrm{T}) = \int_o^\infty \frac{P(\eta)}{\eta} \coth\left(\frac{\eta}{2}\right) \cos \eta\tau \, d\eta \tag{3}$$

and

$$\gamma_1(\tau, \mathrm{T}) = \int_o^\infty \frac{P(\eta)}{\eta} \sin(\eta\tau) d\eta \tag{4}$$

where $P(\eta) = (k_B \mathrm{T}/\hbar)\, \rho(w = k_B \mathrm{T}\eta/\hbar)$.

The functionals of the phonon distribution, γ and γ_1, depend on the parameters of the elastic forces of the lattice; their fluctuations are then function of the partial derivatives of ρ which can be obtained numerically by looking at how the number of states in the first Brillouin zone change when the constants of the elastic forces change.

The previous section shows that the Egelstaff spectra produces reasonable results for S. Additionally its simplicity considerably helps the implementation of the above described procedure. For this spectra the covariance of the scattering matrix [calling $\overrightarrow{x} = (\alpha, \beta, \mathrm{T})$ and $\overrightarrow{y} = (\alpha', \beta', \mathrm{T}')$] is

$$\begin{aligned} C\left(\overrightarrow{x}, \overrightarrow{y}\right) = {} & P_2\left(\overrightarrow{x}\right) P_2\left(\overrightarrow{y}\right) \left(\frac{w_3 - w_2}{w_o^2}\right)^2 \overline{(\Delta w_2)^2} \\ & + P_3\left(\overrightarrow{x}\right) P_3\left(\overrightarrow{y}\right) \left(\frac{w_2}{w_o^2}\right)^2 \overline{(\Delta w_3)^2} \end{aligned} \tag{5}$$

where w_2, w_3, and w_o are parameters of the spectra and P_2 and P_3 are defined by the following set of equations (p_1 and p_3 are the relative weights of the translational and vibrational degrees of freedom):

$$P_2(\alpha, \beta, \mathrm{T}) = \alpha p_3 \gamma(0, \mathrm{T}) \mathcal{S} + \frac{D}{\pi}\, \alpha p_3 \int_o^\infty d\tau \left[Q_1(\tau) \cos \lambda\tau + R_1(\tau) \sin \lambda\tau\right] \tag{6}$$

$$P_3(\alpha, \beta, \mathrm{T}) = P_2 - p_3 \mathcal{S} \alpha h(0, \mathrm{T}) + \frac{D}{\pi}\, \alpha p_3 \int_o^\infty d\tau \left[Q_2(\tau) \cos \lambda\tau - R_2(\tau) \sin \lambda\tau\right] \tag{7}$$

$$Q_1 = e^{\alpha p_3 \gamma(\tau, \mathrm{T}) - \alpha p_1 \tau^2} \left[\gamma_1 \sin(p_3 \alpha \gamma_1) - \gamma(\tau, \mathrm{T}) \cos(p_3 \alpha \gamma_1)\right] \tag{8}$$

$$R_1 = e^{\alpha p_3 \gamma(\tau, \mathrm{T}) - \alpha p_1 \tau^2} \left[\gamma_1 \cos(p_3 \alpha \gamma_1) + \gamma(\tau, \mathrm{T}) \sin(p_3 \alpha \gamma_1)\right] \tag{9}$$

$$h(\tau, \mathrm{T}) = \frac{1}{\eta_3} \coth\left(\frac{\eta_3}{2}\right) \cos \eta_3 \tau \tag{10}$$

$$\eta_3 = w_3 \hbar / k_B\, T \tag{11}$$

$$Q_2 = e^{\alpha p_3 \gamma(\tau, \mathrm{T}) - \alpha p_1 \tau^2} \left[h(\tau, \mathrm{T}) \cos(p_3 \alpha \gamma_1) - g(\tau, \mathrm{T}) \sin(p_3 \alpha \gamma_1)\right] \tag{12}$$

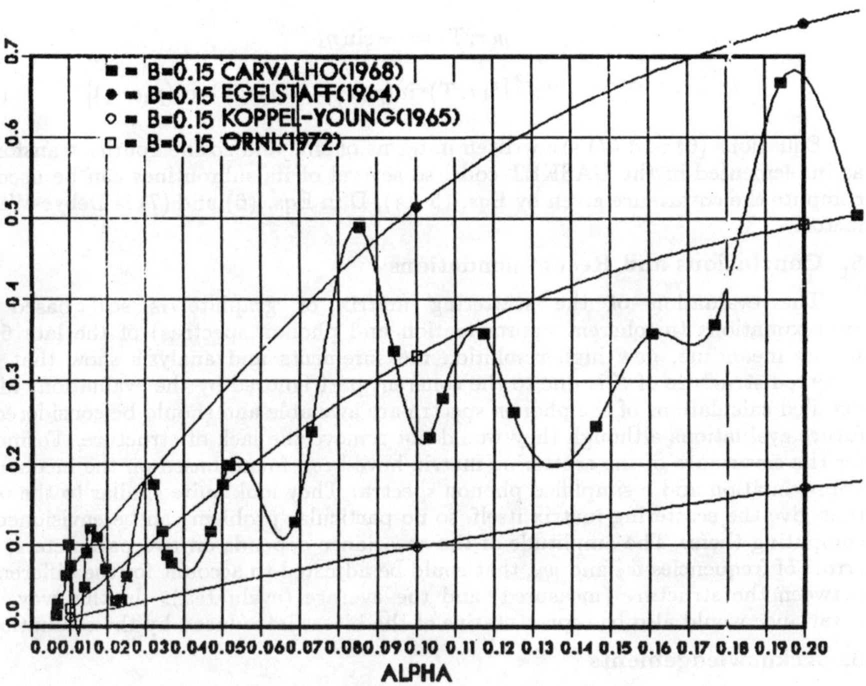

Fig. 2. The experimental results of Carvalho, graphite at 533°K, $\beta = 0.15$, are compared with GASKET calculations using three different phonon spectras.

4. Covariance of the Scattering Matrix

The discussions of the previous sections show that the incoherent approximation used in the evaluation of $S(\alpha, \beta, \mathrm{T})$ has a bias and that there are not enough experimental data to evaluate the covariance file "externally" (that is by comparing the whole experimental data with the evaluation).

The evaluated scattering matrix is anyway a component of the real one; if its covariance matrix is found, an amplification factor has to be applied to include the effects of the biases. The covariance of the real scattering matrix can then be approximated by choosing an amplification factor compatible with the fluctuations shown in Fig. 2.

The first step is then to find the covariance matrix of $S(\alpha, \beta, \mathrm{T})$ in the incoherent approximation. Because under these conditions S is mainly determined by the phonon spectra, the problem is reduced to the calculations of the propagation of the errors of $\rho(w)$ and its covariance. Since the phonon spectra is determined by a set of a few elastic constants, the error of any functional involving $\rho(w)$ has to be found by sampling the distribution of the errors of the elastic constants. More explicitly and

$$g(\tau, \mathrm{T}) = \frac{1}{\eta_3} \sin \eta_3 \tag{13}$$

$$R_2 = e^{\alpha p_3 \gamma(\tau,\mathrm{T}) - \alpha p_1 \tau^2} \left[h(\tau, \mathrm{T}) \sin(p_3 \alpha \gamma_1) + g(\tau, \mathrm{T}) \cos(p_3 \alpha \gamma_1) \right] \tag{14}$$

Equations (6) and (7) are written in terms of *sine* and *cosine* Fourier transforms as implemented in the GASKET code, so several of its subroutines can be used to compute the covariance given by Eqs. (5–14), D in Eqs. (6) and (7) is Debye-Waller factor.

5. Conclusions and Recommendations

The evaluation of the scattering matrix of graphite is still based on approximations (incoherent approximation and phonon spectras) of the late 60's. In the meantime, new high resolution measurements and analysis show that the observed structure of S is due to the coherent part ignored by the evaluation. More detailed calculations of the phonon spectra are available and should be considered in future evaluations although they would not remove the lack of structure. Formulas for the covariance of the scattering matrix have been found based on the incoherent approximation and a simplified phonon spectra. They look quite similar to the ones that give the scattering matrix itself, so no particular problem can be envisioned in computing them. The amplitude of the covariance depends on two parameters, the errors of frequencies w_2 and w_3, that could be adjusted to account for the differences between the structure (measured) and the average (evaluated). In this way, the covariance would also be representative of the biases introduced by the evaluation.

6. Acknowledgements

This research was sponsored by the U.S. Department of Energy, under contract DE-AC05-84OR21400, with Martin Marietta Energy Systems, Inc.

7. References

1. J. V. Koppel, J. R. Triplett and Y. D. Naliboff, "GASKET: A Unified Code for Thermal Neutron Scattering," GA-7417, September 1966.
2. N. F. Wikner, G. D. Joanou and D. E. Parks, "Neutron Thermalization in Graphite," *Nucl. Sci. and Eng.*, **19** (1964), p. 108.
3. J. A. Young and J. V. Koppel, "Phonon Spectra of Graphite," *The Journal of Chemical Physics*, **42** (1965), p. 357.
4. R. Nicklow, N. Wakabayashi and H. G. Smith, "Lattice Dynamics of Pyrolytic Graphite," *Phy. Rev. B*, **5** (1972), p. 4951.
5. P. A. Egelstaff, "Compilation of Early Scattering Data," *AERE-R 3931*, (1962).
6. B. C. Haywood, "A Compilation of Scattering Law for Graphite, Be and BeO at 22°C," *AERE-R 4732* (1964).
7. W. Wittemore, "Neutron Scattering by Reactor Grade Graphite," GA-8284 (1967).
8. F. Carvalho, "Inelastic Scattering of Thermal Neutrons in Graphite," *Nucl. Sci. and Eng.*, **34** (1968), p. 34.

Description of Covariance Data in ENDF-6 Format

D. W. Muir
Nuclear Theory and Applications Group
Theoretical Division
Los Alamos National Laboratory
Los Alamos, NM 87545
U. S. A.

ABSTRACT

In this paper we discuss the strengths and weaknesses of the current ENDF-6 covariance formats and explore the interaction between covariance formats and evaluation methodology, with the aim of encouraging more effective use of formatted covariances by data evaluators in communicating the quality of their evaluations to data users. Topics include the purpose of data covariance files, quality requirements, file-size limitations, the use of computer calculus to obtain covariances for theoretical evaluations, status of covariance processing codes, and future needs and plans.

1. Introduction

Formatted data covariances are numerical data that are included in a data evaluation in order to describe the accuracy of the data. Such covariances are a property of both the experimental data selected by the evaluator and the analytical methods employed to interpret or fit the experimental data.

1.1. Relation of Data Covariances to Least Squares

Evaluators should make routine use of the generalized method of least squares [1]. It is a practical method of evaluation, and it produces objective evaluations of both the data and their covariances simultaneously. Although the generalized method of least squares is often incorrectly called "Bayes' Method", generalized least squares can be derived without any reference whatever to Bayes' theorem. All of the commonly-used mathematical expressions of generalized least squares can be obtained more simply, yet with complete rigor, by taking the evaluated data to be equal to those unbiased linear combinations of the measured data having the least uncertainty. The minimization of uncertainty is, in fact, exactly what a data evaluator is trying to achieve. In this view, evaluation of data covariances is an integral part of the process of data evaluation, not a separate topic involving new concepts and new analyses.

2. User Requirements for Covariances

Very often, data users need to estimate uncertainties in their computed results due to the uncertainty in their input nuclear data. These analysts need reasonably accurate uncertainties in order to avoid the costly waste associated with setting overly conservative design margins. It has been estimated [2], for example, that adding 1.0 cm to the ITER blanket would cost an additional $5 million. This suggests that formatted data covariances may also be of value in demonstrating the economic benefits to be expected from new nuclear-data research and development.

A second major use of covariances is in "data combination" studies, as discussed in several papers at this symposium. Here data covariances are needed to give a proper weight to the basic nuclear data, relative to other available information (such as measured spectra or reaction rates), in analyses such as: (a) neutron-energy-spectrum unfolding by the thin-foil activation method, (b) statistical adjustment of multigroup data taking into account relevant integral experiments, (c) remote sensing of material data (oil well logging, space exploration, etc.). In these fields, covariance data occupies "center stage" in a manner analogous to the role of (n,xγ) data in radiation shielding and double-differential neutron-emission data in fusion neutronics.

3. ENDF Covariance Formats

In both of the major application areas mentioned above, one needs to estimate the uncertainties in various sums of the basic data. To propagate the uncertainties of individual data values σ_i forward into the uncertainties of sums, is necessary to know both the standard deviations $\Delta\sigma_i$ and the correlation coefficients ρ_{ij}. These familiar quantities from statistics combine to determine the covariances,

$$\mathrm{cov}(\sigma_i, \sigma_j) \equiv \langle \delta\sigma_i \, \delta\sigma_j \rangle = \Delta\sigma_i \, \Delta\sigma_j \, \rho_{ij}$$

The covariance *matrix* $D(\sigma)$ is a square matrix having $\mathrm{cov}(\sigma_i, \sigma_j)$ as its ij-th element.

There is a fundamental problem in storing data covariances in a data file of practical size. In a modern nuclear-data evaluation, there are typically 10^6 data points. There would be no problem associated with storing a sufficient number of standard deviations, if that were the only requirement. The fundamental problem is that, associated with 10^6 data quantities, there are 5×10^{11} (!) potentially non-trivial correlation coefficients. In the ENDF-6 format [3], this fundamental size problem is attacked in one of three different ways: (a) to claim (or hope) that the data being evaluated will not be summed, so correlations can safely be ignored; (b) to claim that the correlation coefficients are constant (typically 1.0 or 0.0) over broad energy ranges, so that the data storage problem reduces to specifying just the boundaries of these ranges of constant correlation; or (c) to identify a smaller number (perhaps 10 to 50) of underlying parameters $\{\alpha_i, i=1,n\}$ that represent the main features of the uncertainty of the data, so that the data covariances can be stored in factored form, $D(\sigma) = R\,D(\alpha)\,R^T$, where the sensitivity matrix R is used to store, within the file, explicit derivatives of the data with respect to the parameters, $\delta\sigma_i = R_{ij}\,\delta\alpha_j$.

Choice (b) above corresponds to the standard approach, used in ENDF-6 Files (MF-values) 31, 33, 35, and the proposed MF 36 [4]. The main advantage of this approach is that is extremely compact. It is especially suitable for use in evaluations that are primarily the result of the direct averaging of experimental data in energy bins. The sharp boundaries of the correlation regions are, in most cases, unrealistic, and this has led to problems in application and interpretation of some existing data covariance files. In using this approach, it is also difficult for the evaluator to be sure that the very approximate treatment of correlations preserves the uncertainty in computed quantities in all of the various areas in which the file may be applied.

Choice (c) above, is the approach followed in the ENDF-6 File-30 format [5]. This format is quite new, having been approved only in 1988 (after the completion of most of the data evaluations in the ENDF/B-VI library). This format has no sharp correlation boundaries, and it meets a wide variety of future covariance formatting needs simultaneously. Examples include (a) uncertainties of nuclear-model based evaluations; (b) uncertainties from the fitting of smooth curves (e.g., cubic splines or Padé approximants [6]) to experimental data; (c) uncertainties in MF-6 data (double-differential emission data, radionuclide production, multiplicities, etc.); and (d) uncertainties in fission-product yield and decay data. In addition, File 30 has a capability to represent "cross-file" type correlations (e.g., between MF-33 and MF-35 data) that is absent in the older formats.

As when using any other method of data compression, evaluators who choose to use File 30 should be careful to retain a sufficient number of parameters to avoid loss of uncertainty information needed in the intended applications.

The structure of data stored in MF-30 format is shown in Table I below. Covariances stored in this format should be reasonably easy to process, using a code such as NJOY [7], into accurate covariances of multigoup data in arbitrary user-supplied group structures. Work has begun on the addition of a File-30 processing capability to NJOY.

Table I. Organization of Data in the File-30 Format of ENDF-6

"data" block	"$\partial/\partial\alpha_1$" block	"$\partial/\partial\alpha_2$" block	etc.
$\sigma(E)$	$\alpha_1\,\partial/\partial\alpha_1[\sigma(E)]$	$\alpha_2\,\partial/\partial\alpha_2[\sigma(E)]$	etc.
$df/d\Omega$	$\alpha_1\,\partial/\partial\alpha_1[df/d\Omega]$	$\alpha_2\,\partial/\partial\alpha_2[df/d\Omega]$	etc.
dg/dE'	$\alpha_1\,\partial/\partial\alpha_1[dg/dE']$	$\alpha_2\,\partial/\partial\alpha_2[dg/dE']$	etc.
etc.	etc.	etc.	etc.

4. Remarks Concerning Theoretical Evaluations

With the availability of ever faster and less expensive computers, in the next few years one can expect a noticeable shift in the way that nuclear data evaluations are performed, away from methods that place a large burden on the intuition, interpretive skills, and time of the human evaluator and toward methods that make heavier use of computer-intensive nuclear theory and statistical-inference methods. This is not to say that theory will make experiment obsolete. Continuing emphasis will have to be placed on the gathering of quality experimental data, especially where particular data are needed to normalize the theories. However, it can be expected that theory will be used increasingly as a routine tool in the evaluation of the data. In addition to making efficient use of the evaluator's time, nuclear theory is well-suited to this task because it (a) automatically enforces conservation laws (e.g., energy balance), and (b) facilitates the production of complete evaluations when there are gaps in the energy, angle, and isotope coverage of the available experimental data.

To these important and well-known advantages, one can now add the additional advantage that theory can be very helpful in evaluating the data covariances needed in applications. The connection between nuclear theory and data covariances is not new. Optical model analyses have for many years provided fits to experimental data accompanied by covariances of the optical model parameters, and hence the fitted cross sections. Also, the R-matrix evaluation codes EDA [8] and SAMMY [9] likewise have been used routinely to evaluate not only data but also data covariances for the ENDF/B library and similar techniques also have been employed for some time at Karlsruhe [10] and elsewhere.

What *is* new is the technology of "computer calculus" [11], just now reaching maturity. With this tool, an arbitrarily complex nuclear-theory FORTRAN program (for example, Hauser-Feshbach or coupled-channels programs) now can be converted more-or-less automatically to a program that computes both the nuclear data and the sensitivities of the data to the model input parameters.

By coupling sensitivity versions of the most important nuclear-model codes to a nonlinear generalized least squares program, one can construct a general-purpose evaluation system that produces evaluations of both nuclear data and nuclear-model input parameters, both accompanied by covariances. The covariances of the computed nuclear data will be produced in exactly the factored form, $D(\sigma) = R\, D(\alpha)\, R^T$, required by the File-30 format, as described in Section 3 above.

Since the technique of computer calculus has not been applied previously in nuclear data evaluation studies, it is useful to review this subject before proceeding.

5. Notes on Computer Calculus

A nuclear model calculation can be characterized as the computation of a set of dependent variables y_i (the computed results), as functions $y_i(\mathbf{c})$ of several independent variables c_j (the parameters). We are mainly interested in those particular input parameters that effect the output of the program in a smooth way, so that one can actually define the partial derivatives $R_{ij} = \partial y_i/\partial c_j$ over some range of parameter values about the nominal ones $\mathbf{c}_0$. In the computer calculus approach, a pre-compiler program searches the user's FORTRAN program for arithmetic statements and constructs from them all additional statements required for the ultimate calculation of numerical values of the derivatives R_{ij} at the reference point $\mathbf{c}_0$. The calculation of ratios of finite differences is not involved in this approach, so the computed sensitivities are generally as precise as any other computed quantity.

The basic ideas of computer calculus are best illustrated by considering how it would work in an idealized computer program. One can imagine that all variables involved in the calculation reside in the computer memory for the entire duration of the calculation. Every storage location can be assigned to one of two arrays, represented by the columns vectors **y** and **c**. To **c** we assign all storage locations that remain unchanged from the beginning to the end of the calculation, which includes all of the nuclear-model parameters and other input. To **y** we assign all storage locations that are modified by one or more program statements. Thus **y** contains all intermediate and final results of the calculation, including the computed nuclear data. We indicate by $\mathbf{y}_0$ the initial state of the "result" vector, by $\mathbf{y}_1$ the result vector after the execution of the first program statement, and by $\mathbf{y}_n$ the result vector after n statements, and so on. (Note the distinction we draw between subscripted scalars y_i and subscripted vectors $\mathbf{y}_n$.) We can represent symbolically the evolution of the result vector, responding to the effect of successive statements, as follows:

$$\mathbf{y}_1 = \mathbf{f}_1(\mathbf{y}_0, \mathbf{c})$$

$$\mathbf{y}_2 = \mathbf{f}_2(\mathbf{y}_1, \mathbf{c})$$

$$\cdots$$

$$\mathbf{y}_n = \mathbf{f}_n(\mathbf{y}_{n-1}, \mathbf{c})$$

The function $\mathbf{f}_n$ is just the right-hand-side of the n-th program statement. This schematization of the process of computing applies equally well to scalar and to vector processors. If there are a total of N program statements, then the functions $y_i(\mathbf{c})$ are just $y_i(\mathbf{c}) = [\mathbf{y}_N(\mathbf{c})]_i$. We introduce the notation $\mathbf{R}_n$ to denote the sensitivity of $\mathbf{y}_n$ to **c**,

$$\mathbf{R}_n \equiv \left.\frac{\partial \mathbf{y}_n}{\partial \mathbf{c}}\right|_{\mathbf{c}_0},$$

where the symbol on the right denotes a matrix having, as its ij-th element, the derivative of $[\mathbf{y}_n(\mathbf{c})]_i$ with respect to c_j. The desired (final) sensitivities are then given by $R_{ij} = [\mathbf{R}_N]_{ij}$.

It is not obvious, but the *derivatives* of computed results can be computed in the same stepwise manner as that used in computing the actual data. As illustrated below, this can be achieved through repeated application of the chain rule of calculus:

$$\frac{\partial \mathbf{y}_1}{\partial \mathbf{c}} = \frac{\partial \mathbf{f}_1}{\partial \mathbf{y}_0}\frac{\partial \mathbf{y}_0}{\partial \mathbf{c}} + \frac{\partial \mathbf{f}_1}{\partial \mathbf{c}} = \frac{\partial \mathbf{f}_1}{\partial \mathbf{y}_0}\,\mathbf{0} + \frac{\partial \mathbf{f}_1}{\partial \mathbf{c}} = \mathbf{R}_1$$

$$\frac{\partial \mathbf{y}_2}{\partial \mathbf{c}} = \frac{\partial \mathbf{f}_2}{\partial \mathbf{y}_1}\frac{\partial \mathbf{y}_1}{\partial \mathbf{c}} + \frac{\partial \mathbf{f}_2}{\partial \mathbf{c}} = \frac{\partial \mathbf{f}_2}{\partial \mathbf{y}_1}\,\mathbf{R}_1 + \frac{\partial \mathbf{f}_2}{\partial \mathbf{c}} = \mathbf{R}_2$$

$$\cdots$$

$$\frac{\partial \mathbf{y}_n}{\partial \mathbf{c}} = \frac{\partial \mathbf{f}_n}{\partial \mathbf{y}_{n-1}}\frac{\partial \mathbf{y}_{n-1}}{\partial \mathbf{c}} + \frac{\partial \mathbf{f}_n}{\partial \mathbf{c}} = \frac{\partial \mathbf{f}_n}{\partial \mathbf{y}_{n-1}}\,\mathbf{R}_{n-1} + \frac{\partial \mathbf{f}_n}{\partial \mathbf{c}} = \mathbf{R}_n$$

By inserting these derivative lines into locations just prior to the corresponding lines of the original program, the calculation of the derivatives of the results of an arbitrarily complex computer calculation can be achieved by a long sequence of relatively simple calculations, each involving only the computation of the numerical values of the derivatives of the right-hand side of the next program statement with respect to the static (**c**) and dynamic (**y**) variables, followed by a matrix multiplication and matrix addition.

To make this approach fully practical, it is necessary that the complete operation of reading a line of the original code, adding the required lines of coding to compute the derivatives $\mathbf{R}_n$ and writing a sensitivity-enhanced code must all be done automatically by another computer program. Such a program must necessarily have much of the complexity and generality of a computer-language compiler, plus knowledge of the laws of calculus.

At least two modern software packages are now available for performing such automated sensitivity conversions of existing FORTRAN programs. One, named GRESS (an acronym for Gradient Enhanced Software System), has been under development at Oak Ridge National Laboratory for several years [12]. The latest version, GRESS 2.0, was released in November 1991. GRESS 2.0 is machine dependent, but versions are supported on VAX (both the VMS and ULTRIX operating systems), IBM RS/6000, Sun, and Hewlett-Packard 9000 computers. In the following Section we present an example giving preliminary results of applying GRESS to the statistical-preequilibrium nuclear reaction theory code GNASH [13]. A second package for automated sensitivity conversions, ADIFOR [14], is being developed at Argonnne National Laboratory. Although not yet tested in the GNASH type environment, ADIFOR appears to offer some interesting new features.

6. Example of Sensitivity and Uncertainty Analysis using GNASH

As an example of the application of the ideas developed in the previous section, we present here a short progress report on an ongoing project to develop a sensitivity-analysis version of the GNASH preequilibrium-statistical nuclear reaction theory code. This sensitivity code version will make it possible to do automated searches on the model input parameters in order to improve consistency of code output with measured nuclear data and, at the same time, provide quantitative estimates of the uncertainties (covariances) of the computed cross sections and spectra.

The strategy we chose in pursuing this objective was to implement the August 1991 version of GNASH on a VAX 8700 and then to utilize the VAX/VMS version of GRESS to operate on the resulting VAX-compatible GNASH source code to produce GNASH-91S, the desired sensitivity computing version. The first step, to convert GNASH-91 from CRAY- to VAX-compatible FORTRAN was successfully completed. Nine separate test problems involving the reactions of fast neutrons, protons and alpha particles with ^{55}Mn, ^{56}Fe, ^{93}Nb, ^{235}U, ^{238}U, respectively, gave results on the VAX that are essentially identical to results from analogous runs on a CRAY XMP-48. There was a factor of 5 running time penalty on the VAX relative to the CRAY. The VAX version of GNASH, together with the results of the nine test problems (both VAX and CRAY output) have been released to the Radiation Shielding Information Center at Oak Ridge.

The second phase of the work, the application of GRESS, Version 2.0, to develop the sensitivity version of GNASH-91S, required some interaction with the authors of GRESS and a few minor corrections to their code. It also required some minor rearrangements in GNASH, none of which changed the computed results. The largest single change required was to convert all READs and WRITEs to binary scratch files to in-core READs and WRITEs. This is necessary because allowing data to exit from the computer core memory breaks the logical chain necessary for propagation of the derivatives. Many derivatives of various results with respect to various code input parameters have been obtained, and, in all cases examined to date, very satisfactory agreement has been obtained with pseudo-derivatives obtained by direct recalculation with altered input.

As an example of the application of GNASH-91S, we have used it to calculate the neutron emission spectrum from the bombardment of ^{55}Mn with 10-MeV alpha particles. The GNASH input for this calculation was adopted from a sample problem used by P. G. Young in his computer exercises on GNASH-91 [13] at the joint IAEA/ICTP Workshop on Computation and Analysis of Nuclear Data, Trieste, Italy, February-March 1992.

In Fig. 1 are shown the logarithmic sensitivities of the resulting neutron emission energy spectrum ($\alpha_i\ \partial/\partial\alpha_i$[dg/dE']) to F2 (the normalization factor in the Kalbach pre-equilibrium model), GG (the composite nucleus state density in the preequilibrium model), the normalization factors [GGDNORM(E1), GGDNORM(E2), and GGDNORM(M1)] used with the gamma-ray strength functions, and to the level-density parameters A2 used in the Ignatyuk level density model for ^{55}Mn, ^{58}Fe, and ^{58,59}Co. All but the level-density parameters are GNASH input parameters. The level-density parameters are defined within the code as part of the Ignatyuk model.

The sensitivity and uncertainty analysis capabilities of GNASH-91S come with a definite running-time penalty. In the case of 10-MeV alpha particles incident on ^{55}Mn, the running-time penalty amounted to a factor of 10 for a 2-parameter problem, a factor of 20 for a 17-parameter problem, and a factor of 28 for a 37-parameter problem.

The incremental cost of adding an additional parameter to a sensitivity calculation is therefore similar in magnitude to the cost of performing one additional recalculation with altered input to the original code to achieve somewhat the same result. Even though the GRESS path does not appear likely to save substantial central-processor running time, the real advantage of the approach is that it is suitable for the full automation of large-scale parameter optimization studies. The central-processor times associated with the approach, while not negligible, are certainly in the acceptable range.

In addition to being a useful tool for generating nuclear data with defendable covariances, the GRESS+model code+least-squares approach discussed above appears to be applicable in the context of the project under way in the IAEA Nuclear Data Section to assemble a comprehensive Reference Nuclear Parameter Library [15]. There are also clear connections with the interest reported at this symposium in the further development of large scale nuclear-data evaluation systems, such as the IDA system at Bologna and the INDES system at Tokai.

7. Model Deficiency

When the parameters of a given model are optimized to fit a particular experiment, it can easily happen that even the best fit is still not very good, as evidenced by a high value of

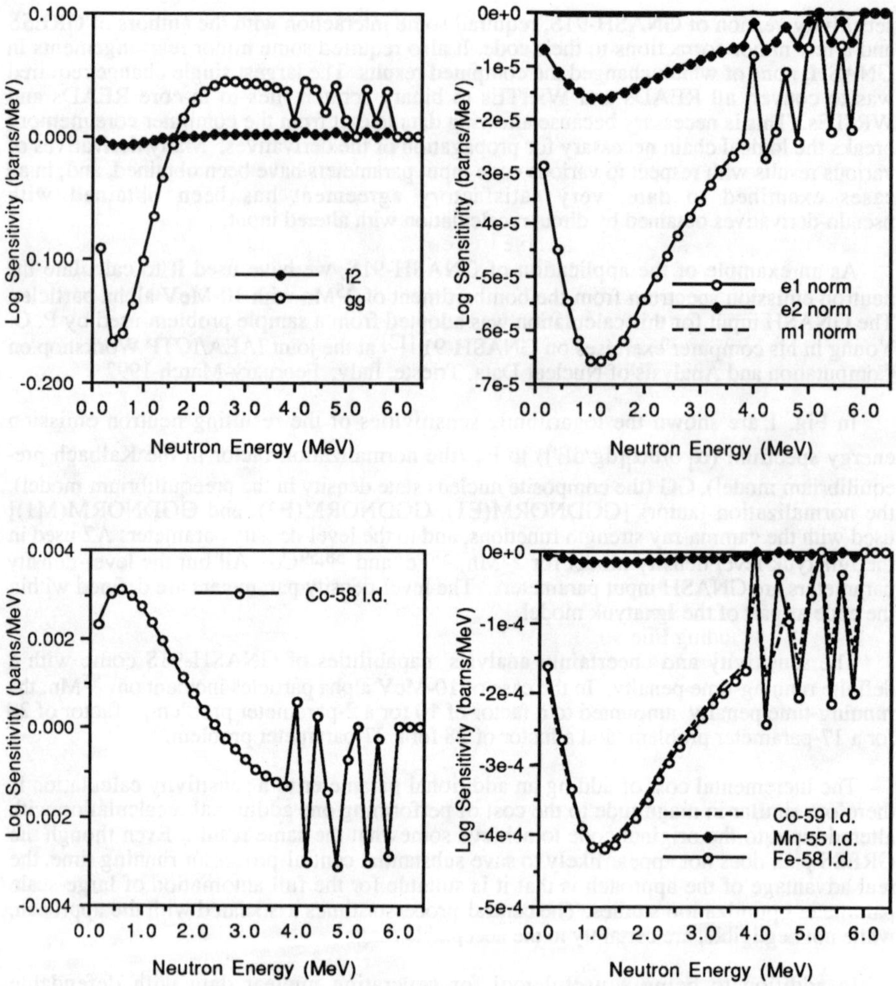

Fig. 1. Parameter sensitivities of neutron emission spectrum from α + ^{55}Mn at 10 MeV.

chi-squared. This is, first of all, extremely valuable information in itself, because it is *only* when a large chi-squared is obtained from a least squares analysis that the theorist has a clear signal from the laboratory that something is missing from the nuclear model. (Without performing such a least-squares analysis, information on model adequacy or deficiency is obscured by the large pre-fit uncertainty of the parameters.)

For users of a nuclear model code other than the author, the most reasonable response to a poor least-squares fit is to apply some *ad hoc* correction to the computed data to obtain better agreement with experiment. The question arises as to what to say about the data covariances after such a correction has been applied to the data.

First of all, it is clearly incorrect to treat such model deficiency as an additional random error. The best corrective action may, in fact, be no action at all. That is, the final *covariances* (as opposed to the computed data themselves) obtained by fitting a deficient model to experiment may be usable in many cases without any post-fit modification. In support of this conjecture, one can note that evaluated uncertainties should be reasonable, but extreme accuracy in their numerical values is not needed. In addition, any deviations of the data from the final model fit will be treated by the least squares code as statistical fluctuations (even if they are very unlikely ones), so that the uncertainties output by the code will similar in magnitude for both a poor fit and a good one, being primarily a reflection of the uncertainties in the input experimental data. Thus going through the hypothetical exercise of correcting all of the deficiencies in the model and then repeating the least-squares analysis would certainly improve the fit, but might not change the final uncertainties very much.

8. Future Needs and Plans

The ENDF-6 covariancc formats are reasonably complete and offer sufficient flexibility to meet most anticipated near-term needs. The situation in the processing codes is not as good. The NJOY code, for example, now processes about 90 percent of the ENDF-6 covariance formats. Unfortunately, the features not yet treated include many important new capabilities, including File 30. Progress in remedying the situation by updating the NJOY covariance modules has been slowed by funding limitations. The Los Alamos Nuclear Theory and Applications Group welcomes collaborative work, exchange visits, etc., as a means of moving forward with this necessary work.

9. Acknowledgments

The author wishes to thank Prof. H. Vonach of the Institut für Radiumforschung und Kernphysik of the University of Vienna for providing time on the University's new VAX 4000 for collaborative work on GRESS and GNASH. Thanks also go to Dr. J. Horwedel of Oak Ridge National Laboratory for assistance and advice on the GRESS code.

10. Refcrences

1. D. W. Muir, "Evaluation of Correlated Data Using Partitioned Least Squares: A Minimum-Variance Derivation," *Nucl. Sci. and Eng.* **101**, 88-93 (January 1989).

2. M. Z. Youssef, "Comparison of PCA vs. Tungsten in the TIBER/ITER Inboard Shield and Impact of Nuclear Data Uncertainties on Machine Cost," Proc. Specialists' Meeting on Shielding Experiments and Analysis, Garching, Germany, 12-14 February 1990.

3. Chapters 30-40 in P. F. Rose and C. L. Dunford, "Data Formats and Procedures for the Evaluated Nuclear Data File, ENDF-6," Brookhaven National Laboratory-National Nuclear Data Center report BNL-NCS 44945, ENDF-102, Rev. 10/91 (1991).

4. H. Vonach, "Direct Covariance Data for Coupled Energy-Angle Distributions," Proc. Specialists' Meeting on Evaluation and Processing of Covariance Data (COV 92), 7-9 October 1992, Oak Ridge, Tennessee, sponsored by the Nuclear Energy Agency Nuclear Science Committee of the OECD. See related paper by D. W. Muir, "Description of Covariances of Emission Spectra using ENDF File 30," ibid.

5. R. W. Peelle and D. W. Muir, "Extended Covariance Data Formats for the ENDF/B-VI Data Evaluation," Proc. NEACRP Specialists" Meeting on the Applications of Critical Experiments and Operating Data to Core Design via Formal Methods of Data Adjustment, Jackson Hole, Wyoming, 23-24 September 1988, report NEACRP-L-307, pp. 305-315 (1988).

6. S. A. Badikov, N. S. Rabotnov, and K. I. Zolotarev, "Evaluation of Neutron Dosimetry Reaction Cross Sections and Covariance Analysis with Rational Functions," Proc. Specialists' Meeting on Evaluation and Processing of Covariance Data (COV 92), 7-9 October 1992, Oak Ridge, Tennessee, sponsored by the Nuclear Energy Agency Nuclear Science Committee of the OECD.

7. D. W. Muir and R. E. MacFarlane, "The NJOY Nuclear Data Processing System Volume IV: The ERRORR and COVR Modules", Los Alamos National Laboratory report LA-9303-M, Vol. IV (1985).

8. D. C. Dodder, G. M. Hale, and K. Witte, "EDA, An Energy Dependent Analysis Code for Nuclear Reactions" (code available from authors only for use at Los Alamos).

9. N. M. Larson, "Updated Users' Guide for SAMMY: Multilevel R-Matrix Fits to Neutron Data Using Bayes' Equations", ORNL/TM-9179/R2, Oak Ridge National Laboratory, Oak Ridge, Tennessee (1989).

10. F. H. Fröhner, "Principles and Methods of Data Evaluation", KfK-4099, Kernforschungszentrum Karlsruhe GmbH, Karlsruhe, Germany (1986).

11. E. M. Oblow, et al., "Sensitivity Analysis using Computer Calculus: A Nuclear Waste Isolation Application", *Nucl. Sci. Eng.* **94**, 46 (1986).

12. J. E. Horwedel, "GRESS Version 2.0 User's Manual," ORNL/TM-11591, Oak Ridge National Laboratory report (1991).

13. P. G. Young, E. D. Arthur, and M. B. Chadwick, "Comprehensive Nuclear Model Calculations: Introduction to the Theory and Use of the GNASH Code", Los Alamos National Laboratory report LA-12343-MS (July 1992).

14. C. Bischof and A. Griewank, "ADIFOR: A FORTRAN System for Portable Automatic Differentiation," Proc. 4th Symposium on Multidisciplinary Analysis and Optimization, AIAA Paper 92-4744, pp. 433-441 (1992).

15. G. Reffo, O. Bersillon, D. Muir, and A. Pashchenko, "Reference Nuclear Parameter Library for Nuclear Data Computation," Summary Report of a Consultant's Meeting held in Vienna, Austria, 13-15 November 1991, International Atomic Energy Agency report INDC(NDS)-266/G (November 1992).

Experience with Collapsing and Expanding Multi-group Cross-section Covariance Matrices

I. A. Kodeli* E. Sartori**
* CEA-CE-Saclay, France, ** OECD/NEA Data Bank

Abstract

Utility programs to expand and collapse multi-group cross-section covariance data to a user defined energy structure and for verifying the physical consistency of the data and interpolation procedure have been developed. A relative covariance matrix library, designed for fission and fusion reactor shielding has been assembled on which these utilities were applied.

This procedure is simple, practical and flexible in particular as it allows several input and output formats. The main purpose for developing this method is to facilitate and encourage the use of uncertainty analysis as well as to familiarize the newcomer with it.

Comparison with more rigorous procedures has shown good agreement for interpolations and expansions in which the number of groups changed up to a factor of 3.

The program package including the library was successfully applied to: PWR surveillance capsule dosimetry, fusion blanket shielding, adjustment of iron cross-section data using a benchmark experiment.

Description of the library and the interpolation procedure

The ZZ-VITAMIN-J/COVA library [1], [2] was assembled from the covariance data based on JEF-1, ENDF/B-IV, ENDF/B-V and other evaluations. It contains the relative covariance matrices for the important reactions for 11 material and 19 dosimetry cross-sections. The library is provided in the highly compressed BOXER format [3].

To facilitate the use of the covariance data the program ANGELO was developed at the NEA-DB to expand or collaps the matrices from the original energy group structure to the one required by the user. The main purpose was to write a procedure, which is user-friendly, flexible concerning the input and output formats, and conservative.

In this interpolation scheme the standard deviations are interpolated step-wise in lethargy to the new energy groups. The correlation matrix is interpolated separately in the way that the correlation over the group-lethargy domain is kept constant. Conservative assumptions are made near borders of steps and plateaus in order to insure that the uncertainties are not underestimated. In particular the standard deviation of the threshold is used for the whole energy group which includes the threshold, also if it extends below the threshold value.

No cross-section or flux weighting is carried out. This approach can be justified by the fact that most covariance information is coarse. In a rigorous approach one would require detailed information about fine group fluxes, which should include also self-shielding effects.

Verification of the library was done in several ways:

- Checking whether the covariance matrices are positive definite was performed using a code included in the package [2]. Negative eigenvalues indicate some inconsistency in the covariance data information. Nevertheless, when of small relative magnitude, the negative eigenvalues were found to pose no major problems in the uncertainty or adjustment analyses.
- The sensitivity and adjustment analyses were performed in different energy group structures in order to verify the collapsing procedure implemented in the ANGELO code. Excellent agreement was found between the calculations done in such different group structures as 27 and 14 (see Table 1).
- Performance of the ANGELO collapsing procedure was compared to the more rigorous one implemented in the ERRORR module of NJOY [3]. As demonstrated in Figure 1 good matching was found between the two. Some differences appear near thresholds due to different approaches adopted by the two codes. Nevertheless these differences are expected to have little practical impact in the uncertainty analysis since the cross-sections tend to decrease rapidly in magnitude near threshold.

Examples of applications

The ZZ-VITAMIN-J/COVA package [1] was successfully applied to different domains of neutronics calculations, like reactor shielding, installation cost reduction, improved design. Pressurized water reactor shielding as well as fusion neutronics problems have been considered.

This package was incorporated into the computer code system developed at CEA for the uncertainty and adjustment analysis. The data flow scheme of the system is presented in Figure 2. This system was applied to the 900 MWe EDF power plants with the objective of validating the methodology used in the pressure vessel surveillance programme [4]. All major sources of uncertainties were considered. The adjustment analysis was carried out to improve important nuclear data, such as the iron cross-sections, fission spectrum, response functions, source strength. The adjustment factors determined in this way were found to be realistic and consistent with some recent evaluations.

Table 1: Sensitivities and uncertainties in the fast neutron flux (E>1MeV) in the PWR surveillance capsules, obtained from the calculations in 27 and 14 energy groups.

Cross-section term	27 groups (E>1MeV)		14 groups (E>1MeV)	
	Sensitivity (% per %)	Uncertainty (%)	Sensitivity (% per %)	Uncertainty (%)
Fe total /loss	5.92	15.94	5.89	15.92
Fe elastic /loss+gain	0.64	2.68	0.63	2.64
Fe inelastic /loss	1.72	8.37	1.70	8.33
Fe (n,γ) /loss	0.003	0.04	0.003	0.04

To make an independent test of these results, in particular of the iron adjustments, the ASPIS benchmark experiment was examined using the same procedure. This experiment presents the advantage that the uncertainties other than those due to the cross-section data are considerably reduced. The results of these analyses are presented in detail in a separate paper presented at this conference [5].

A chi-square (χ^2) test was applied to check the consistency between the set of the calculated and the measured values on one side and the corresponding covariance matrices on the other. Values of the chi-square per degree of freedom were found to be not far from 1 in the cases of PWR shielding and the ASPIS experiment, indicating no substantial over- or under-estimation of the input covariance data.

A recent paper [6] concludes on the basis of other applications that "error-correlation files significantly overestimate the uncertainties in transport calculations". The conclusion that can be drawn from the present analysis is therefore much more encouraging. It is also based on more recent covariance data evaluations, while other uncertainty estimations are still based on the Drischler - Weisbin compilation [7].

The adjustments for the iron ENDF/B-IV cross-sections, obtained from the surveillance capsules and the ASPIS analyses are presented in Figures 3 and 4, respectively. Concerning the non-elastic cross-sections, very good agreement can be seen between the two. The difference in the adjustments for the elastic cross-sections are due to the different energy ranges covered by the detectors used for the surveillance capsule dosimetry and those used in the ASPIS experiment. This information can be considered as complementary rather than contradictory. These correction factors agree well with the modifications introduced into the more recent evaluated data, like the ENDF/B-V, /B-V-revision, /B-VI.

One should recall that the adjustment is a very good test of the quality of the covariance data, since it only yields meaningful results provided the input covariance data are reasonable.

Similar analyses have been performed in the framework of the NET (Next European Torus) fusion reactor project. The design of the shielding blanket and its cooling impose some severe constraints on the neutron flux and the related parameters. The most limiting design parameters for the superconducting magnets of NET are the fast neutron flux, the neutron heating of the magnet and the atomic displacements in the copper stabilizer. The accuracy of the calculated neutron parameters is therefore crucial, since it determines the safety margins attributed to them, and in this way the design and the cost of the project.

Extensive cross-section sensitivity and uncertainty studies of the NET fusion reactor have been performed by several laboratories in Europe [8], [9], [10]. Different transport and sensitivity code systems (ANISN-SENSIT, ANISN-SUSD, BISTRO), as well as different approximations in the geometry model descriptions have been used for these analyses. Although the results can be considered as basically consistent, they still give evidence of rather important variations depending on the approximations and the methods used. For the neutron heating of the magnets the uncertainties ranging from 10 to 16.5 % are given in the reports [9] and [10].

Some uncertainty estimations were also evaluated on the basis of the Fe-56 covariance data which has recently become available in the ENDF/B-VI files. Comparing the results based on the JEF-1/ENDF/B-V (ZZ-VITANIN-J/COVA library) and the ENDF/B-VI data, the differences in error estimates of the order of ±20 to ±30 % have been observed; the uncertainties in the nuclear heating in the fusion magnet and in the fast neutron flux are, for example, roughly 20 % larger if based on the ENDF/B-VI data [9], [10]. On the other hand, the ENDF/B-VI iron data yielded about 30 % smaller uncertainties for some reaction rates calculated in the PWR surveillance capsules.

Some remarks have been made concerning the small ±1 % JEF-1/ENDF/B-V standard deviations, attributed to the iron elastic cross-sections in the resonance energy range between 1 keV and 600 keV [10], which are thought to be underestimated.

The magnitude of the variations due to the use of different independent uncertainty evaluations seem therefore to be much smaller than the spread of the data originating from using different calculational methods and models. These results can be interpreted as an indication that the cross-section uncertainties can be estimated with better, or at least comparable, precision than the uncertainties due to other sources like modeling, or calculational methods.

Conclusions

The ZZ-VITAMIN-J/COVA covariance data library and the interpolation procedure seem to be adequate for the uncertainty and data adjustment analysis for radiation shielding applications. It has been successfully applied to several uncertainty analyses of fission and fusion neutronics problems.

The ANGELO interpolation procedure proved to give reliable results for interpolations and expansions in which the number of groups changed up to a factor of 3. This can be explained by the fact that most covariance information is coarse and the rank of the matrix is normally much smaller than its order.

For covariance matrices containing detailed short and long range correlations (i.e. resonance regions) flux weighting should give in general a better performance. This can be achieved by using more rigorous methods, like the NJOY-ERRORR module, which provide group cross-section data and covariances in a consistent way. This code requires on the other hand more expertise from the user and has limitations in the input/output format.

The available covariance data for Fe seem already of quite satisfactory quality comparable to the present state-of-the-art of the calculational tools for radiation transport.

References

[1] I. A. Kodeli, E. Sartori, *Covariance Data Library ZZ-VITAMIN-J/COVA*, NEA 1264/03, 1990.

[2] I. Kodeli, E. Sartori, *Neutron Cross-section Covariance Data in Multigroup Form and Procedure for Interpolation to Users' Group Structures for Uncertainty Analysis Applications*, PHYSOR'90 Conference, Marseille (1990).

[3] D.W. Muir and R.E. MacFarlane, *The NJOY Nuclear Data Processing System, Vol. IV; The ERRORR and COVR Modules*, LA-9303-M November 1985.

[4] I. A. Kodeli, J. C. Nimal, J. C. Lefebvre, *Uncertainty and Data Adjustment Analysis for PWR Shielding*, ANS Topical, M&C, Pittsburg, 1991.

[5] S. H. Zheng, I.Kodeli, C. Raepsaet, C. M. Diop, *Qualification and Improvement of Iron ENDF/B VI and JEF-2 Evaluations by Interpretation of ASPIS Benchmark*, presented at this conference

[6] J. Butler, *Beyond the Horizon in Radiation Shielding: A Stochastic World?*, Proc. ANS Topical: "New Horizon in Radiation Protection and Shielding", (1992).

[7] J. D. Drischler, C R. Weisbin, *Compilation of Multi-group Cross-section Covariance Matrices*, ORNL-5318 (October 1977).

[8] H. Hogenbirk et al., *Sensitivity and Uncertainty Analysis of NET/ITER Shielding Blankets*, Fusion Technology (1990).

[9] A. Hogenbirk, *Extensive sensitivity / uncertainty analysis of a NET fusion reactor blanket*, Report EFF-Doc-128, Nov.1991

[10] T. Parish, A. Santamarina, *Sensitivity and uncertainty analysis of the NET magnet neutronic design parameters to uncertainties in cross section data*, EFF-2 Meeting, Dec. 1991

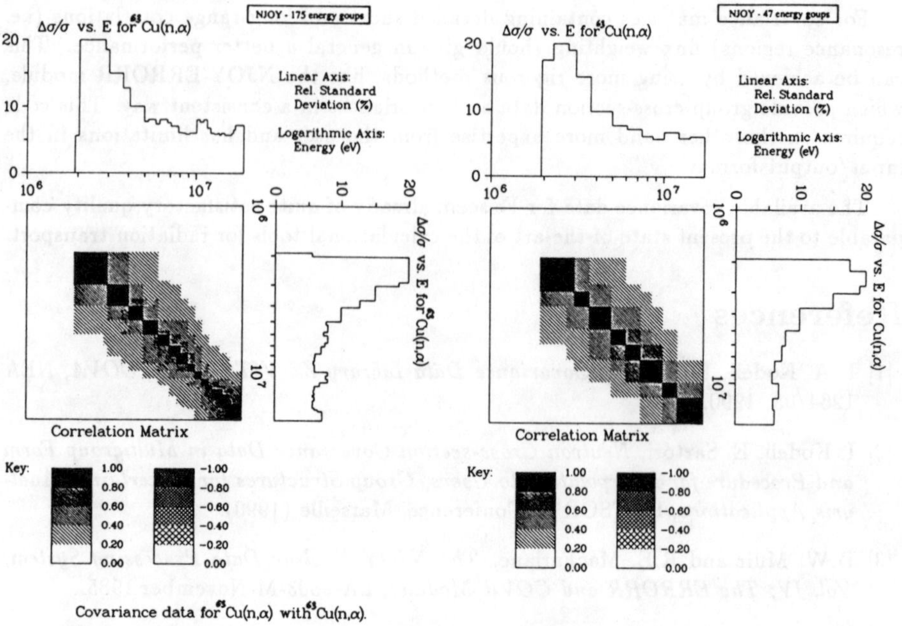

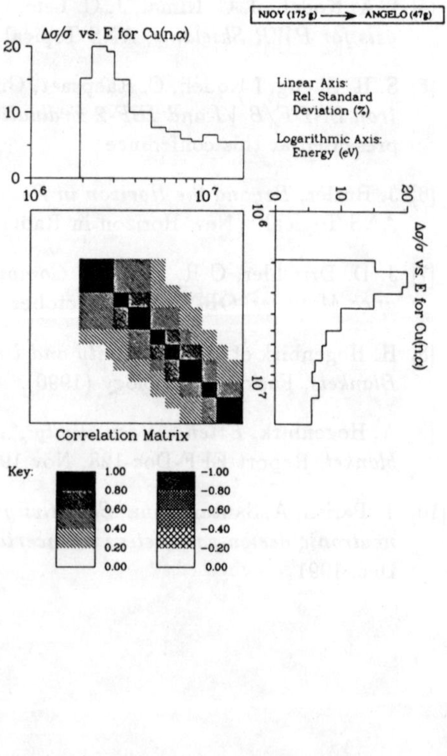

Figure 1: Comparison of the covariance matrices obtained using the codes NJOY and ANGELO. The covariance matrices are practically identical, although contrary to NJOY no flux weighting is used in ANGELO. The differencies appear only near thresholds, and are due to the different approaches in the two codes.

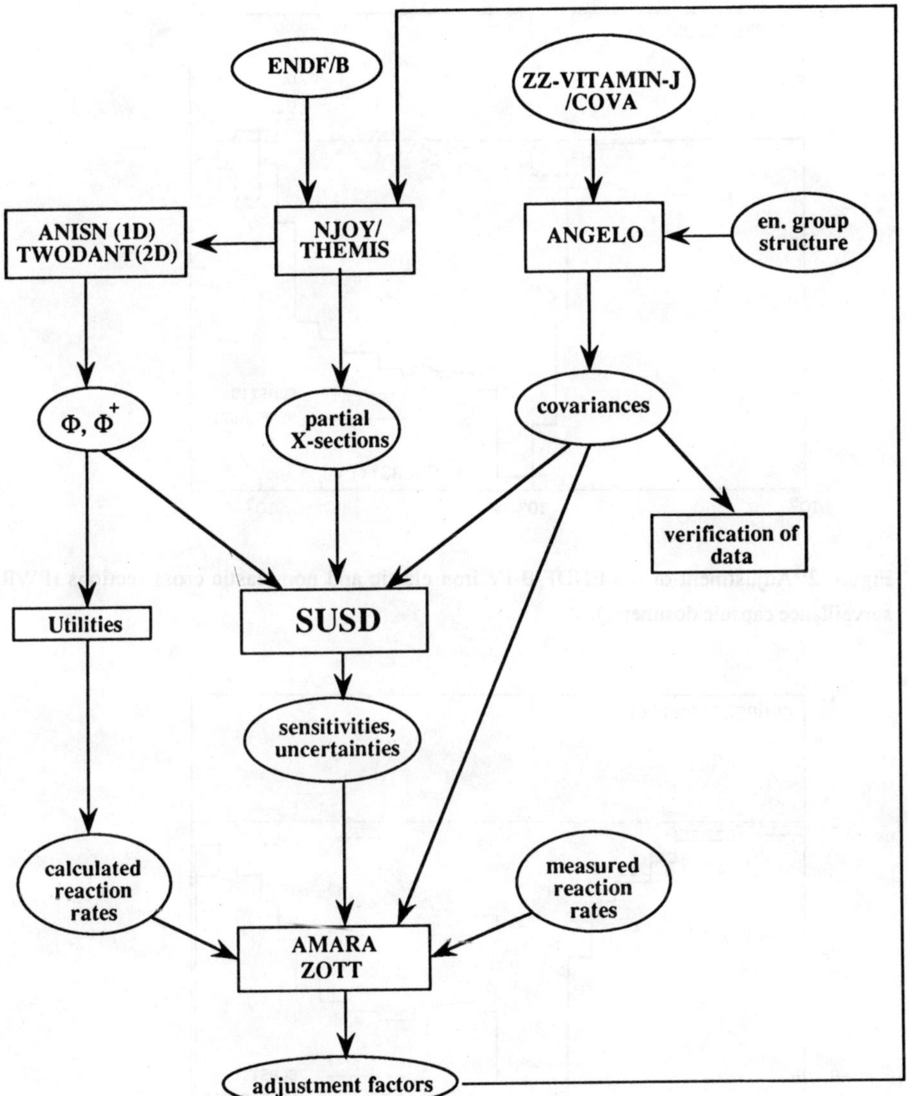

Figure 2: Code system used at CEA Saclay for the sensitivity, uncertainty and adjustment analysis.

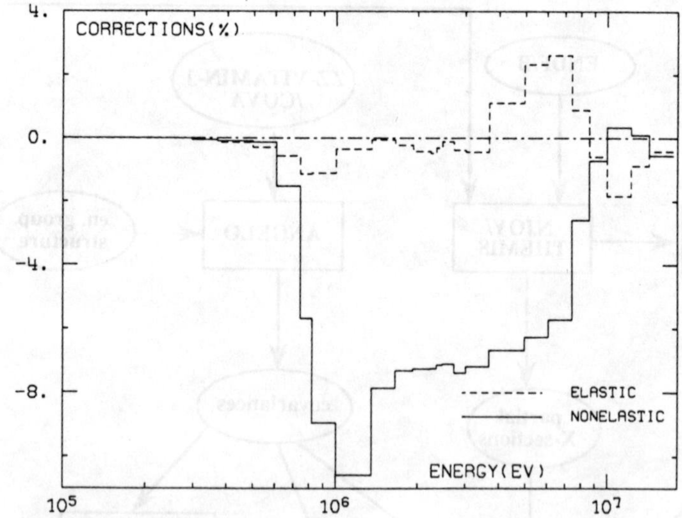

Figure 2: Adjustment of the ENDF/B-IV iron elastic and non-elastic cross sections (PWR surveillance capsule dosimetry).

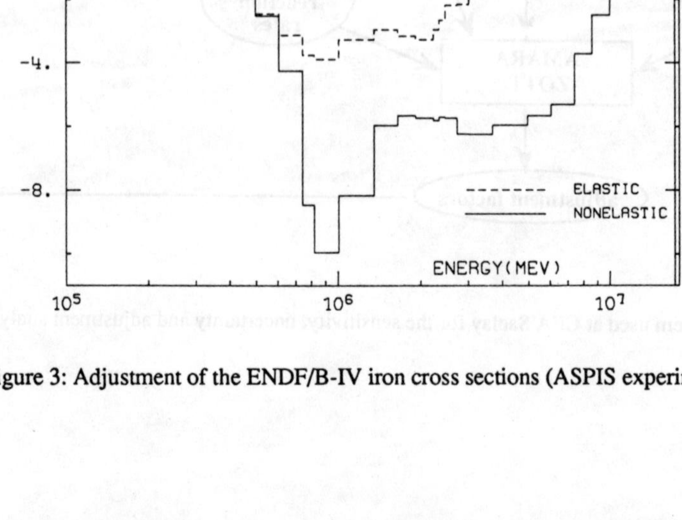

Figure 3: Adjustment of the ENDF/B-IV iron cross sections (ASPIS experiment).

COMPUTER AIDED EVALUATION

INTEGRATED NUCLEAR DATA EVALUATION SYSTEM ON THE BASIS OF KNOWLEDGE ENGINEERING TECHNOLOGY

T. FUKAHORI, T. NAKAGAWA, T. NARITA, Y. KIKUCHI
Nuclear Data Center, Japan Atomic Energy Research Institute, Tokai-mura, Naka-gun, Ibaraki, 319-11 Japan

and

T. ASAMI
Data Engineering Corporation, 1-17-11 Shakujiidai, Nerima-ku, Tokyo, 117 Japan

ABSTRACT

The integrated nuclear data evaluation system (INDES) is being made in order to support the evaluation work. INDES manages databases, for instance, the evaluation data file (EVLDF) storing the basic nuclear physics parameters such as optical model potential parameters, level density parameters, mass excess data, level scheme and gamma transition from excited levels, and various information, for example, names of theoretical calculation codes used in JENDL-3 evaluation work. By using EVLDF and the other databases, INDES automatically creates input data and JCL for several theoretical calculation codes for the nuclear data evaluation. A guidance system in INDES, which is called *'Evaluation Tutor* (ET)', can support users in selecting a set of suitable theoretical calculation codes by applying knowledge engineering technology and the experiences of the evaluation work for JENDL-3. The inference engine of ET is written in FORTRAN77 instead of special computer languages for the artificial intelligence such as LISP and PROLOG so that evaluators can understand what happens in ET easily. The theoretical calculation codes considered in INDES are DWUCKY, ECIS, CASECIS, JUPITOR, EGNASH2, TNG, ALICE-F, PEGASUS, CASTHY2, ELIESE-3, HIKARI, and RESCAL now, and it is easy to add new codes. Furthermore, users can get useful information from data files, which are linked to INDES, such as the ENSDF, evaluated nuclear data libraries, and data files storing experimental data.

1. Introduction

Evaluated nuclear data files are necessary to many applications, for instance, design of nuclear fission and fusion reactors, shielding calculations, etc. Nuclear data evaluation is required to obtain the most reliable data sets for the evaluated nuclear data files. The evaluation must be done by using experimental data and by many complicated theoretical calculations with various basic data such as optical potential parameters, level density parameters, and level scheme. Japanese Evaluated Nuclear Data Library, version 3 (JENDL-3)[1] released in 1989 was accomplished by great efforts of many evaluators with a lot of intricate work. Integrated Nuclear Data Evaluation System (INDES) is being developed to keep their accumulated experiences of JENDL-3, to store basic data of nuclear physics used for the JENDL-3 evaluation, and to support new evaluations,.

Roughly classified, INDES functions are of three categories, which are to select the most suitable set of theoretical calculation codes by applying knowledge engineering technology, to retrieve basic data described above, and to set up input data of theoretical calculation codes automatically. In this paper, functions and databases of INDES are introduced, and especially the guidance system for the suitable code selection installed in INDES is explained.

2. Databases

Parameters for theoretical calculations of nuclear data are stored in Evaluation Data Files (EVLDF) which consist of several files and are controlled by an index file. The data provided in EVLDF are optical potential parameters of nuclei, information of excited levels, level density parameters, other basic information of nuclei, etc. The information of theoretical calculation codes used for the JENDL-3 evaluation are also stored in EVLDF. This information on those codes is gathered into a knowledge-base which is used for code selection for evaluation.

Existing evaluated nuclear data are stored in magnetic disks of the main frame computer system in JAERI. They are compiled in the ENDF format. Almost all of the available evaluated nuclear data in the world are stored. Their MAT numbers, data-set names, etc. are stored in an index file of the evaluated data.

Experimental data on the neutron induced reactions are stored in the NESTOR-2 system by receiving them from the NEA Data Bank and by converting their format from EXFOR. The data on the charged particle induced reactions stored in the CHESTOR system will be installed in INDES with the same manner as those of neutron.

On the nuclear structure data such as level scheme, one can access the Evaluated Nuclear Structure Data File (ENSDF).

3. Function of INDES

INDES is an interactive program available on TSS terminals. Figure 1 shows structure of INDES. It consists of several FORTRAN programs called '*segments*', which are executed by calling them with TSS command procedures from the root segment.

A group of the segments supplies evaluators the useful information for nuclear data evaluation such as existing evaluated data, experimental data, nuclear structure data (from ENSDF), reaction Q-values and threshold energies of the reactions. One segment was prepared for retrieval of the parameters stored in EVLDF. Some segments will be made for evaluation of the level density parameters, the optical model parameters, etc.

Since theoretical calculations take a long CPU time, we have to submit batch jobs for the calculations. The input data and JCL for the theoretical calculation codes are semi-automatically produced in the JCL set-up segments. INDES requires only limited number of input data such as an atomic number and a mass number, name of a parameter set, energies where the calculation will be made, etc. The JCL and input data are set up on a JCL file. The user can submit it as a batch job from INDES. Control of the batch job can be performed by INDES also.

Selection of the suitable theoretical calculation codes can be made with the aid of ET to be described in Chapter 4. The names of codes recommended by ET are shown on the terminal screen. By selecting one of the codes on the screen, one can go into the segment for its JCL and input data setting-up.

4. Automatic Selection of Theoretical Calculation Codes

In order to support users in selecting a set of suitable theoretical calculation codes, a prototype of a nuclear data evaluation guidance system, named '*Evaluation Tutor* (ET)'[12], was developed by applying knowledge engineering technology, which is based on the similar method of 'example-base reasoning' called in the society of the artificial intelligence. The codes considered in ET are DWUCKY[3], ECIS[4], JUPITOR[5], CASECIS[6], EGNASH2[3], TNG[7], PEGASUS[8], ALICE-F[9], CASTHY2[10], ELIESE-3[11], RESCAL[12] and HIKARI[13]. ET consists of an inference engine, frames, a rule-base, two example-bases and calculating modules for certainty factors. A schematic diagram of ET is shown in Fig. 2. The inference engine and calculating modules of ET are written in FORTRAN77, instead of special computer languages for the artificial intelligence such as LISP and PROLOG, since FORTRAN77 is the most familiar computer language to the evaluators of the nuclear data.

The information on the theoretical calculation codes is stored in '*frames*'. A set of the frame consists of '*slots*' of frame name, code type, theories or models used in the code, incident and outgoing particles and energy ranges treated in the code, quantities calculated with the code, required input data, related codes, a code converting the output into the ENDF format, and comments. Values in the slots, except those of code type, quantities and comments, are given as '*a priori certainty factors*'.

The '*rules*' stored in the rule-base are used in the inference engine to select the theoretical calculation codes. The rules are classified into four types according to nuclear reaction processes, which are the direct, preequilibrium, compound and other processes. The each rule consists of an INPUT part requesting input data to the user and an IF-THEN part. If the INPUT part exists, prompt messages specified in the INPUT part are shown on a screen of a terminal to get an answer from the user. The THEN part is executed when the IF part is satisfied. Tasks of the THEN part are to give the information to the user, to set the values, to set the next category of the nuclear reaction process, to go to the next rule, and to search codes satisfying conditions.

The two example-bases are used to obtain '*basic certainty factors*' of the theoretical calculation codes from their frequencies of use. One example-base has been created from the experiences of the JENDL-3 evaluation work. Another is a supplementary example-base, which stores results of code selection performed by ET. A unit in the example-bases consists of target isotope identification, an incident particle, and theoretical calculation codes used in the evaluation.

A flow of ET to select a set of the recommended codes is shown in Fig. 3, and the procedure of code selection is described below. Firstly, the user inputs the basic data such as a target nucleus, a projectile and an incident energy range. Then, basic certainty factors are calculated for all the registered theoretical calculation codes by counting the number of uses in the example-bases. The rules in the rule-base are executed to make a preliminary decision of code selection. All the codes in the frames are classified into four reaction processes. If a process is judged not to be needed, the codes corresponding to the process are omitted from the selection. '*A functional certainty factor*' of each code is calculated from a priori certainty factors stored in the frame. Three candidates having the largest certainty factors are selected for each process. Then, '*frequency certainty factors*' of the three candidates are obtained from the basic certainty factors. The candidate which has the largest frequency certainty factor is selected as the recommended theoretical calculation code for each reaction process. For the codes requiring other auxiliary codes, the auxiliary codes are added to the set of recommended codes. The result of the code selection is stored in the supplementary example–base, and is used for the basic certainty factor calculation to get a learning function.

5. Conclusion

Integrated Nuclear Data Evaluation System, INDES, was introduced, which is being developed to keep experiences and basic data of the nuclear data evaluation for JENDL–3, and to support new evaluations. The prototype of the nuclear data evaluation guidance system, ET, was developed to support users in selecting the theoretical calculation codes, and was installed in INDES. INDES helps not only beginners of the nuclear data evaluation being unfamiliar with using the theoretical calculation codes, but also experienced evaluators to make input data for preliminary calculations. The segments retrieving databases are planned to be enlarged, and several new segments will be produced to make INDES more powerful. The rule–base and the theoretical calculation codes contained in ET will be expanded to treat more detailed procedures.

References

1. K. Shibata, T. Nakagawa, T. Asami, T. Fukahori, T. Narita, S. Chiba, M. Mizumoto, A. Hasegawa, Y. Kikuchi, Y. Nakajima and S. Igarasi, *JAERI*–**1319** (1990).
2. T. Fukahori and T. Nakagawa, *JAERI–M* **92–027** (1992) 214.
3. N. Yamamuro, *JAERI–M* **90–006** (1990).
4. J. Raynal, *IAEA* **SMR–9/8** (1970).
5. T. Tamura, *Rev. Mod. Phys.* **37** (1965) 679.
6. Computer Program CASECIS means CASTHY–ECIS combined code.
7. C. Y. Fu, *ORNL/TM*–**7042** (1980).
8. S. Iijima, T. Sugi, T. Nakagawa and T. Nishigori, *JAERI–M* **87–025** (1987) 337.
9. T. Fukahori, Computer Program ALICE–F for medium energy preequiliblium process calculation, unpublished (1990).
10. S. Igarasi and T. Fukahori, *JAERI*–**1321** (1991).
11. S. Igarasi, *JAERI*–**1224** (1972).
12. K. Shibata and S. Chiba, Computer Program RESCAL for R–matrix theory calculations, unpublished (1987).
13. H. Kitazawa, Computer Program HIKARI for direct–semidirect capture calculations, unpublished (1980).

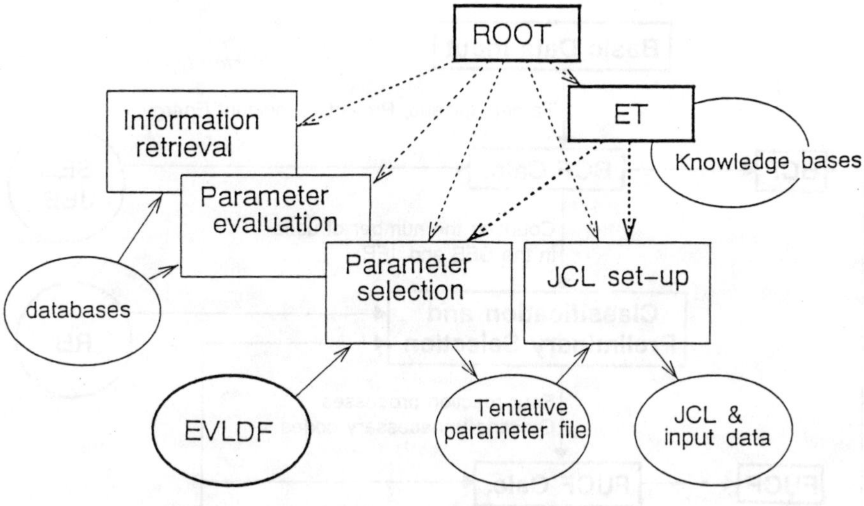

Fig. 1 Structure of INDES

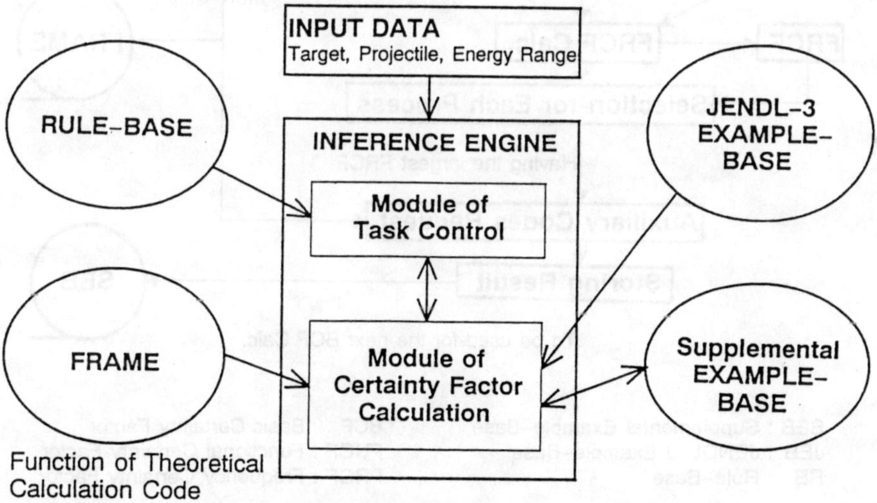

Fig. 2 A schematic diagram of ET

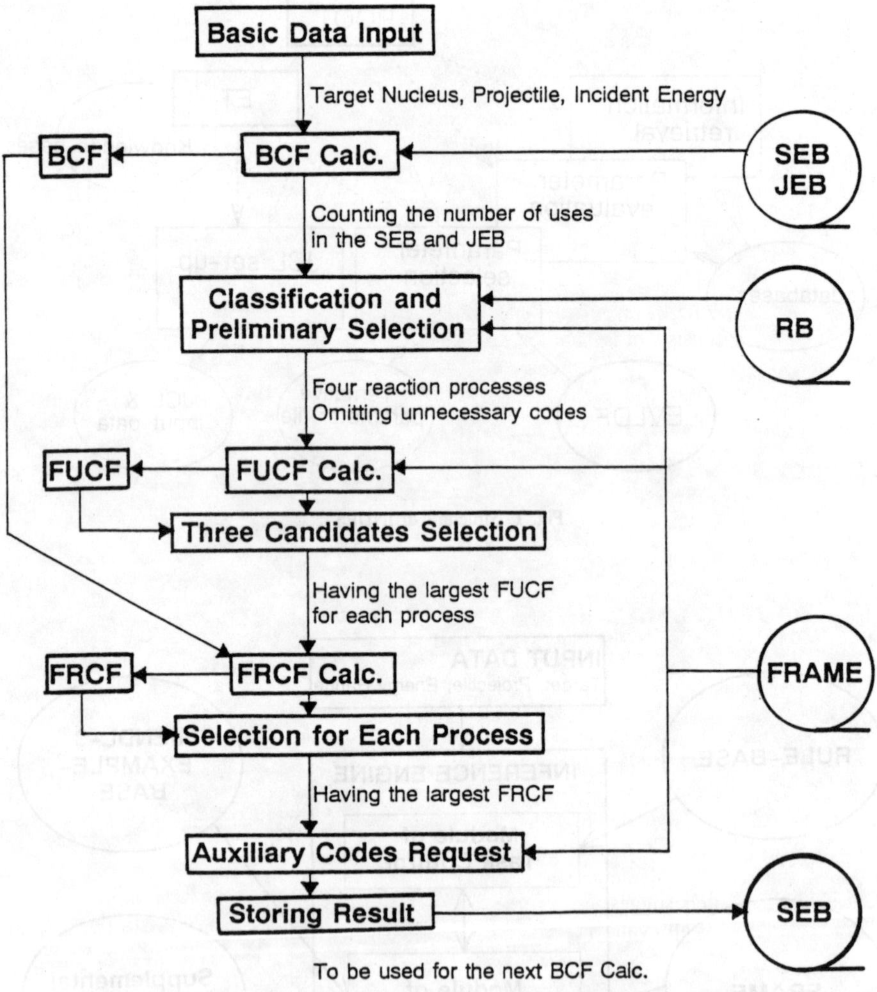

SEB : Supplemental Example–Base
JEB : JENDL–3 Example–Base
RB : Rule–Base

BCF : Basic Certainty Factor
FUCF : Functional Certainty Factor
FRCF : Frequency Certainty Factor

Fig. 3 A Flow of Selection in ET

THE IDA MODULAR SYSTEM OF CODES FOR NUCLEAR DATA EVALUATION

G. REFFO, F. FABBRI, M. HERMAN[1]
ENEA, INN-SVIL Divisione Calcolo
Viale G. B. Ercolani 8, 40138 Bologna, Italy

ABSTRACT

The IDA system of codes was initially designed as a tool for nuclear data evaluation. It consists of linked modules, each corresponding to a different step in the evaluation process. IDA was subsequently enriched by implementation of theoretical developments and evolved into a powerful tool for theoretical investigations as well as for nuclear model testing and intercomparison. IDA contains data tables which are shared by all modules. These tables contain all the information needed for input preparation, including nuclear structure data, and many calculations can be automated by use of default parameters option. According to their use, IDA modules can be grouped for input preparation, for input testing, for extensive calculations and for translation of output into the ENDFB-VI format.

1. Introduction

The IDA modular system of codes became operational in 1979, and many efforts have been made to keep IDA updated according to nuclear model developments. IDA is devoted to calculations of cascading emissions with a maximum of 4 particles and 7 gamma rays emissions in any sequence. IDA works under the restriction of energy, spin and parity conservation, which implies careful consideration of all available nuclear structure information for a number of residual nuclei increasing considerably with projectile energy. Towards the end of the seventies, the increasing lack of personnel involved in nuclear data evaluation and the availability of higher performance computers suggested that we should project and produce a "master code" allowing for the automation of all evaluation procedures which do not need user judgement.

2. Philosophy

When producing theoretical calculations of nuclear cross sections a considerable amount of time is needed for input preparation and testing. Accordingly, first we devoted our attention to input automation by preparation of nuclear parameter data tables. As a result we have enormously

---

[1] Under ENEA contract Prot. n. 1758, 1991.

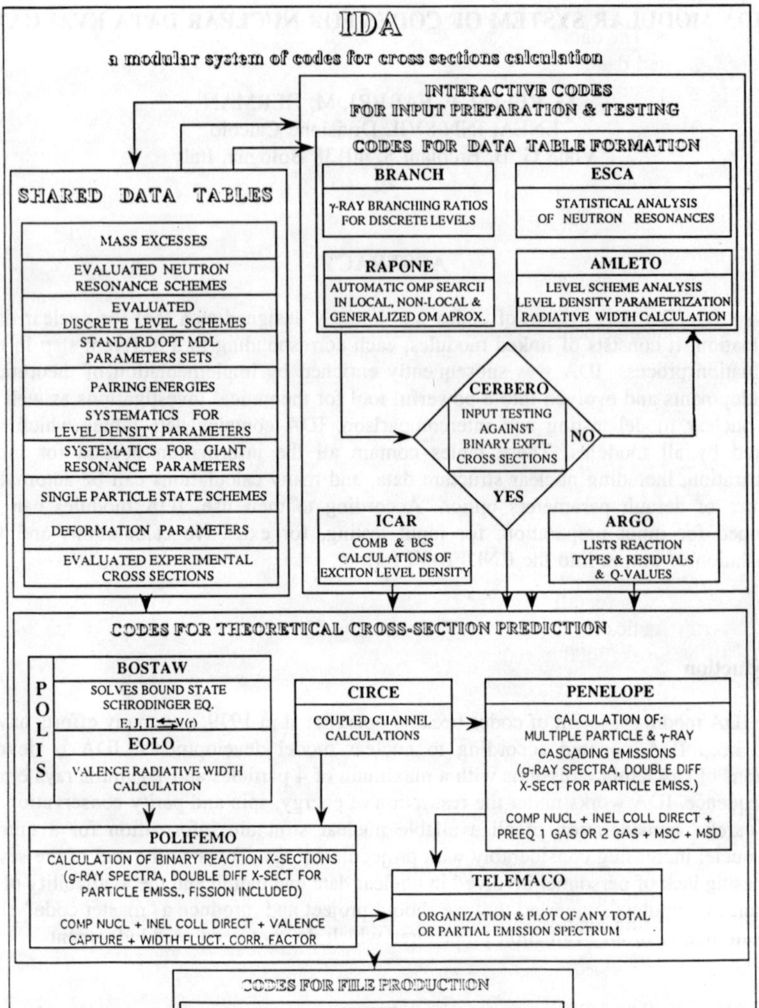
IDA
a modular system of codes for cross sections calculation
INTERACTIVE CODES FOR INPUT PREPARATION & TESTING
CODES FOR DATA TABLE FORMATION
BRANCH
γ-RAY BRANCHING RATIOS FOR DISCRETE LEVELS
ESCA
STATISTICAL ANALYSIS OF NEUTRON RESONANCES
RAPONE
AUTOMATIC OMP SEARCH IN LOCAL, NON-LOCAL & GENERALIZED OM APROX.
AMLETO
LEVEL SCHEME ANALYSIS LEVEL DENSITY PARAMETRIZATION RADIATIVE WIDTH CALCULATION
SHARED DATA TABLES
MASS EXCESSES
EVALUATED NEUTRON RESONANCE SCHEMES
EVALUATED DISCRETE LEVEL SCHEMES
STANDARD OPT MDL PARAMETERS SETS
PAIRING ENERGIES
SYSTEMATICS FOR LEVEL DENSITY PARAMETERS
SYSTEMATICS FOR GIANT RESONANCE PARAMETERS
SINGLE PARTICLE STATE SCHEMES
DEFORMATION PARAMETERS
EVALUATED EXPERIMENTAL CROSS SECTIONS
CERBERO
INPUT TESTING AGAINST BINARY EXPTL CROSS SECTIONS
NO
YES
ICAR
COMB & BCS CALCULATIONS OF EXCITON LEVEL DENSITY
ARGO
LISTS REACTION TYPES & RESIDUALS & Q-VALUES
CODES FOR THEORETICAL CROSS-SECTION PREDICTION
POLIS
BOSTAW
SOLVES BOUND STATE SCHRODINGER EQ.
E, J, Π ⇄ V(r)
EOLO
VALENCE RADIATIVE WIDTH CALCULATION
CIRCE
COUPLED CHANNEL CALCULATIONS
PENELOPE
CALCULATION OF: MULTIPLE PARTICLE & γ-RAY CASCADING EMISSIONS
(g-RAY SPECTRA, DOUBLE DIFF X-SECT FOR PARTICLE EMISS.)
COMP NUCL + INEL COLL DIRECT + PREEQ 1 GAS OR 2 GAS + MSC + MSD
POLIFEMO
CALCULATION OF BINARY REACTION X-SECTIONS
(g-RAY SPECTRA, DOUBLE DIFF X-SECT FOR PARTICLE EMISS., FISSION INCLUDED)
COMP NUCL + INEL COLL DIRECT + VALENCE CAPTURE + WIDTH FLUCT. CORR. FACTOR
TELEMACO
ORGANIZATION & PLOT OF ANY TOTAL OR PARTIAL EMISSION SPECTRUM
CODES FOR FILE PRODUCTION
ULISSE
CROSS-SECTION FILE ORGANIZATION INTO ENDF/B-VI FORMAT

sion

s now about 12 year old. In the meantime progress has been made both in nuclear d in computer science and technology.

attempt to keep up with physics, we have enlarged IDA modules with more and more e many development versions of IDA modules of necessity are being optimized.

calculus and vectorization allow for a considerable reduction in computing time and y to span a much wider energy range without need of dropping details.

that the achievements we are going to reach, fit well the trend of today's nuclear which covers wider and wider energy ranges.

. Fabbri, and R. J. Howertone, this symposium.

mans, H. Gruppelaar, and G. Reffo, Phys. Rev. C22, 73 (1980).

G. Reffo, and C. Costa, Phys. Rev. C 39, 1269 (1989).

, A. Kerman, and S. Koonin, Ann. of Phys. 125, 429 (1980).

J. J. M. Verbaarshot, and H.A. Weidenmuller, Ann. of Phys. 172, 67 (1986).

H. A. Weidenmuller, and S. Yoshida, Ann. of Phys. C26, 166 (1988).

, Reffo and H. A. Weidenmuller Nucl. Phys. A 536, 124(1992).

Udagawa, and H. Lenske, Phys Rev. C26, 397 (1982).

Reffo, H. Lenske and H. H. Wolter, to be published.

J. Berne, and J. M. Miller, Phys. Rev. 165, 1166 (1968).

adioli, E. Gadioli Erba, M. Galmarini, F. Fabbri, and G. Reffo, Phys Rev. 9).

reduced time and typing errors, and optionally, we can produce complete calculations by use of default parameters. The data tables consist of raw data built up from the literature (experimental), as well as evaluated data.

The increasing number of reaction mechanisms to be taken into account and the encreasing complexity of nuclear models, involve more and more nuclear properties which correspond to additional input parameters and option possibilities. An additional simplification has been achieved by writing codes format free. This was facilitated by making the codes interactive, i.e. the codes display the list of the parameters needed by means of symbols or analytical definitions, carefully considered in order to avoid ambiguities. User judgement may be introduced after each run to ascertain wether the results obtained satisfy the expected overall consistency. Finally most of the input preparation codes have been made iterative, i.e., after each run full results are displayed and compared to available experimental data, while the option is offered to repeat the calculation after changement of one or more parameters or of the entire input.

3. The code organization

3.1. *The shared data tables (SDT)*

Ida master code consists of a set of data tables "Shared Data Tables" and four chain of modules, see fig.1. The SDT are shared by all codes. They contain ten files: mass excess table, evaluated neutron resonance schemes, discrete level schemes (1500 isotopes)[1], standard optical model parameter sets for all types of particles, pairing energies, systematics for level density parameters, systematics for the photo-absorbtion giant resonance parameters (E1, E2, M1), single particle level schemes (3000 isotopes), deformation parameters, and evaluated experimental cross sections.

3.2. *Interactive codes for input preparation and testing (ICIPT)*

The codes of IDA are linked together according to sequential logic through the evaluation process. The ICITP contains codes for the determination either of input parameters which are not found in the SDT or which are found in the SDT but need accuracy improved. In addition the ICIPT contains codes for the determination of parameters which are not in the SDT, as well as codes for input testing against experimental cross sections.

3.2.1 *Codes for table formation (CTF)*

AThe group of "Codes for Data Table Formation" are the first to be used when starting an evaluation. These codes are four:

ESCA, iterative and interactive with graphical display (IIG), reads the selected neutron resonance scheme from the SDT and performs the statistical analysis of neutron resonances, according to six different statistical methods, each of which providing with plots and average resonance parameters; runs must be iterated until convergence to the same parameter values is obtained from all methods. Output are the mean observed resonance spacing D and the other average resonance parameters, like the average reduced neutron width, average total radiative width per spin state and s-wawe strength function.

AMLETO, (IIG) reads the selected experimental discrete level scheme from the SDT and from the D-value from ESCA and performs the parametrization of the level density formula choosen. Iterations are needed until a correct fit of known discrete levels is obtained consistent with the mean resonance spacing and with the correct value of total and partial radiative widths. Statistical analysis of spin and parity distributions is also performed. The output is an evaluated discrete level scheme, where the cut energy E, between discrete and continuum regions is determined and all level below E are characterized by known spin parity and branching ratios. The quantities other than experimental being marked with a flag. The level scheme so obtained which still includes all the original experimental information above E is stored back into the SDT.

RAPONE, (IIG) reads from the SDT the selected optical model parameter (OMP) set to be used as a first guess (selection being made by use of 4 digit acronims easily recognizeable). The new OMP set is obtained by a least square minimization procedure fitting the strength function obtained from ESCA, as well as all the other total and differential experimental cross sections available. Numerical and graphical results are displayed. Spherical local and non local approximation can be used as well as generalized optical model. Maximum 200 experimental data distributed in maximum 10 energy points are accepted.

BRANCH, reads the selected discrete level scheme from the SDT and complement it with missing gamma decay branchings up to the cut energy E. This evaluated gamma branchings are flagged, to make sure that they will be easily recognizeable. The new level scheme so obtained is stored into the SDT.

3.2.2. *Codes for input preparation*

The set of CTF is part of the chain of "Interactive Codes for Input Preparation and testing" (ICIPT). The ICITP chain contains 3 more codes:

CERBERO, computes the cross section of any type of reaction in which only one emission can take place. The code, very fast and flexible, is designed for testing the parameters from the chain CTF against experimental cross sections (all the other input parameters needed being taken from the SDT). When the desired cross section accuracy is reached, the parametrization of the selected isotope is ended, if not an additional iteration must be started with one or more modules of the chain CTF. The version CERBERRO computes the uncertainties propagated to cross sections from the uncertainty in the input parameters: a matrix of uncertainties from the individual parameters is printed out as well as the total cross section uncertainty.

ICAR, (IIG), for a selected isotope, reads shell model Hamiltonian eigenvalues from the SDT and by use of combinatorial calculations computes exciton level densities in the frame of BCS theory, spin and parity distributions are also given.

ARGO, given a projectile and a target reads mass exces
produces the list of open reactions together with the respective (
to be parameterized.

3.3. *Codes for theoretical cross section prediction (CTCP)*

Once the complete parametrization of all involved
extensive calculations can be performed by means of th
section prediction". This chain includes 5 codes (see fig

POLIS, consists of two codes BOSTAW and
Schrodinger equation: given the spectroscopic inform
level BOSTAW finds the potential well depth V(r)
radiative width from V(r) obtained by BOSTAW.

CIRCE, performs coupled channel calculatic

POLIFEMO is like CERBERO with incl
allowing targets in any excited state. In coupl
are read from the CIRCE output. POLIFEM
and fission competition is included. The outp
as well as gamma-ray spectra and double

PENELOPE, reads its input from the
is devoted to calculating multiple pa
approximations like POLIFEMO allo
reaction mechanism is included accor
approximations[3]. In the development
according to either Fesbach-Ke
Weidenmuller-Yoshida[5,6,7] theory.
according to the method of Tamur
heavy ion induced reactions are

TELEMACO reads the cro
organizes and plots the data fo
double differential cross sec

3.4. *Codes for file produc*

Finally, for the pur
formats, we have devel
by the chain CTCP a

reduced time and typing errors, and optionally, we can produce complete calculations by use of default parameters. The data tables consist of raw data built up from the literature (experimental), as well as evaluated data.

The increasing number of reaction mechanisms to be taken into account and the encreasing complexity of nuclear models, involve more and more nuclear properties which correspond to additional input parameters and option possibilities. An additional simplification has been achieved by writing codes format free. This was facilitated by making the codes interactive, i.e. the codes display the list of the parameters needed by means of symbols or analytical definitions, carefully considered in order to avoid ambiguities. User judgement may be introduced after each run to ascertain wether the results obtained satisfy the expected overall consistency. Finally most of the input preparation codes have been made iterative, i.e., after each run full results are displayed and compared to available experimental data, while the option is offered to repeat the calculation after changement of one or more parameters or of the entire input.

3. The code organization

3.1. *The shared data tables (SDT)*

Ida master code consists of a set of data tables "Shared Data Tables" and four chain of modules, see fig.1. The SDT are shared by all codes. They contain ten files: mass excess table, evaluated neutron resonance schemes, discrete level schemes (1500 isotopes)[1], standard optical model parameter sets for all types of particles, pairing energies, systematics for level density parameters, systematics for the photo-absorbtion giant resonance parameters (E1, E2, M1), single particle level schemes (3000 isotopes), deformation parameters, and evaluated experimental cross sections.

3.2. *Interactive codes for input preparation and testing (ICIPT)*

The codes of IDA are linked together according to sequential logic through the evaluation process. The ICITP contains codes for the determination either of input parameters which are not found in the SDT or which are found in the SDT but need accuracy improved. In addition the ICIPT contains codes for the determination of parameters which are not in the SDT, as well as codes for input testing against experimental cross sections.

3.2.1 *Codes for table formation (CTF)*

AThe group of "Codes for Data Table Formation" are the first to be used when starting an evaluation. These codes are four:

ESCA, iterative and interactive with graphical display (IIG), reads the selected neutron resonance scheme from the SDT and performs the statistical analysis of neutron resonances, according to six different statistical methods, each of which providing with plots and average resonance parameters; runs must be iterated until convergence to the same parameter values is obtained from all methods. Output are are the mean observed resonance spacing D and the other average resonance parameters, like the average reduced neutron width, average total radiative width per spin state and s-wawe strength function.

AMLETO, (IIG) reads the selected experimental discrete level scheme from the SDT and from the D-value from ESCA and performs the parametrization of the level density formula choosen. Iterations are needed until a correct fit of known discrete levels is obtained consistent with the mean resonance spacing and with the correct value of total and partial radiative widths. Statistical analysis of spin and parity distributions is also performed. The output is an evaluated discrete level scheme, where the cut energy E, between discrete and continuum regions is determined and all level below E are characterized by known spin parity and branching ratios. The quantities other than experimental being marked with a flag. The level scheme so obtained which still includes all the original experimental information above E is stored back into the SDT.

RAPONE, (IIG) reads from the SDT the selected optical model parameter (OMP) set to be used as a first guess (selection being made by use of 4 digit acronims easily recognizeable). The new OMP set is obtained by a least square minimization procedure fitting the strength function obtained from ESCA, as well as all the other total and differential experimental cross sections available. Numerical and graphical results are displayed. Spherical local and non local approximation can be used as well as generalized optical model. Maximum 200 experimental data distributed in maximum 10 energy points are accepted.

BRANCH, reads the selected discrete level scheme from the SDT and complement it with missing gamma decay branchings up to the cut energy E. This evaluated gamma branchings are flagged, to make sure that they will be easily recognizeable. The new level scheme so obtained is stored into the SDT.

3.2.2. *Codes for input preparation*

The set of CTF is part of the chain of "Interactive Codes for Input Preparation and testing" (ICIPT). The ICITP chain contains 3 more codes:

CERBERO, computes the cross section of any type of reaction in which only one emission can take place. The code, very fast and flexible, is designed for testing the parameters from the chain CTF against experimental cross sections (all the other input parameters needed being taken from the SDT). When the desired cross section accuracy is reached, the parametrization of the selected isotope is ended, if not an additional iteration must be started with one or more modules of the chain CTF. The version CERBERRO computes the uncertainties propagated to cross sections from the uncertainty in the input parameters: a matrix of uncertainties from the individual parameters is printed out as well as the total cross section uncertainty.

ICAR, (IIG), for a selected isotope, reads shell model Hamiltonian eigenvalues from the SDT and by use of combinatorial calculations computes exciton level densities in the frame of BCS theory, spin and parity distributions are also given.

ARGO, given a projectile and a target reads mass excesses from the SDT tables and produces the list of open reactions together with the respective Q-values and the residual nuclei to be parameterized.

3.3. *Codes for theoretical cross section prediction (CTCP)*

Once the complete parametrization of all involved residual nuclei has been achieved, extensive calculations can be performed by means of the chain "Codes for Theoretical Cross section prediction". This chain includes 5 codes (see fig. 1):

POLIS, consists of two codes BOSTAW and EOLO. The former solves bound state Schrodinger equation: given the spectroscopic information (energy spin and parity) of a nuclear level BOSTAW finds the potential well depth V(r) and viceversa. The latter computes valence radiative width from V(r) obtained by BOSTAW.

CIRCE, performs coupled channel calculations.

POLIFEMO is like CERBERO with inclusion of valence capture and the possibility of allowing targets in any excited state. In coupled channel approximation the necessary quantities are read from the CIRCE output. POLIFEMO allows for any particle or gamma ray in and out and fission competition is included. The output are total and partial cross sections in any channel, as well as gamma-ray spectra and double differential cross sections for particles.

PENELOPE, reads its input from the output of the ICIPT codes and from the SDT. The code is devoted to calculating multiple particle cascading emissions. It works under the same approximations like POLIFEMO allowing for the same competitions. The preequilibrium reaction mechanism is included according to the exciton model both in one-gas[2] and in two gas approximations[3]. In the development version Multistep Compound calculations can be performed according to either Fesbach-Kerman-Koonin[4] theory or to the Nishioka-Verbaarshot-Weidenmuller-Yoshida[5,6,7] theory. Multistep Direct reaction mechanism has been included according to the method of Tamura-Udagawa-Lenske[8,9]. Particle and gamma ray emissions from heavy ion induced reactions are determined by use of Berne-Harp-Miller master equations[10,11].

TELEMACO reads the cross section files produced by POLIFEMO and by PENELOPE and organizes and plots the data for any total and partial emission spectrum. Angular distribution and double differential cross sections are given for particle only.

3.4. *Codes for file production (CFP)*

Finally, for the purpose of cross section processing and utilization according to international formats, we have developed a rather complex code, ULISSE, which reads cross sections produced by the chain CTCP and translate them into ENDFB-VI format.

4. Conclusion

IDA is now about 12 year old. In the meantime progress has been made both in nuclear physics and in computer science and technology.

In the attempt to keep up with physics, we have enlarged IDA modules with more and more features. The many development versions of IDA modules of necessity are being optimized.

Parallel calculus and vectorization allow for a considerable reduction in computing time and the possibility to span a much wider energy range without need of dropping details.

We hope that the achievements we are going to reach, fit well the trend of today's nuclear data demand, which covers wider and wider energy ranges.

5. References

1. G. Reffo, F. Fabbri, and R. J. Howertone, this symposium.
2. J. M. Akkermans, H. Gruppelaar, and G. Reffo, Phys. Rev. C22, 73 (1980).
3. M. Herman, G. Reffo, and C. Costa, Phys. Rev. C 39, 1269 (1989).
4. H. Feshbach, A. Kerman, and S. Koonin, Ann. of Phys. 125, 429 (1980).
5. H. Nishioka, J. J. M. Verbaarshot, and H.A. Weidenmuller, Ann. of Phys. 172, 67 (1986).
6. H. Nishioka, H. A. Weidenmuller, and S. Yoshida, Ann. of Phys. C26, 166 (1988).
7. M. Herman, G. Reffo and H. A. Weidenmuller Nucl. Phys. A 536, 124(1992).
8. T. Tamura, T. Udagawa, and H. Lenske, Phys Rev. C26, 397 (1982).
9. M. Herman, G. Reffo, H. Lenske and H. H. Wolter, to be published.
10. G. D. Harp, B. J. Berne, and J. M. Miller, Phys. Rev. 165, 1166 (1968).
11. E. Fabrici, E. Gadioli, E. Gadioli Erba, M. Galmarini, F. Fabbri, and G. Reffo, Phys Rev. C40, 2548 (1989).

THE BOLOGNA-LIVERMORE DISCRETE LEVEL SCHEME FILE

G. REFFO AND F. FABBRI
ENEA, INN-SVIL Divisione Calcolo
Viale G. B. ERCOLANI 8, 40138 Bologna, Italy

and

R. J. HOWERTON
Lawrence Livermore National Laboratory
Livermore, CA (USA)

ABSTRACT

Experimental discrete level schemes have been tabulated in Bologna for many years. This collection has recently been enriched by the contribution from Livermore Combined Data Reaction Library (CDRL). The content of the resulting file is limited to commonly needed information (level, energy, spin, parity and gamma-decay schemes) which provide the bulk of data needed for input to compound nucleus calculations. The format is compact and intuitive with special flags noting entries which included other than direct experimental data and contains information for over 1500 nuclei.

1. Purpose and philosophy

Level scheme filing was started over two decades ago, by storage into punched cards. These schemes responded to the needs of the time. During the years a variety of cases have been considered which gave a comprehensive picture of what information is needed. An attempt was made to find a reasonable compromise allowing for maximum information in a form easy to interpret and compact enough to have it clearly displayed in any cross section code output. A very important point was the necessity to be able to distinguish, at a first glance, between experimental and evaluated quantities. As our main intention was to reduce the manual intervention, to start we choose to include only those quantitities which are more often needed in common calculations, dropping the ones that only seldom may be needed. We arrived at the conclusion that the bulk of discrete level information involved in current calculations, whatever the code and the reaction mechanism consists, mostly, of level energy, spin, parity and gamma decay. In a typical 14 Mev nucleon induced reaction several tens of levels are needed, some time hundreds, for about ten residual nuclei involved, becoming a thousand and more for several tens of residual nuclei at 50 MeV incident energies. For each of these levels: energy, spin and parity must be specified in order to be able to account for the details of nuclear structure that affects the hard tails of emission spectra. In addition such information is relevant, as well, for the correct parametrization of the level density in the continuum energy region. Moreover, partial and total gamma ray spectra determination requires the details of gamma decay for each discrete level of each residual nucleus.

2. Information content

The above considerations and practical observation lead the Bologna group to develop a format that in a later time was adopted by other laboratories including Livermore. The latter enriched the library with the majority of the presently available 1500 level schemes. Two representative examples are shown in table 1, the first contains 20 levels and is complete, the second shows 31 out of 191 levels. The illustration of a few lines is sufficient. In the following we describe the line content reading from left to right. Let us consider the first level scheme in table 1. The first line provides the total number of levels (20), the mass A=85 and charge Z=36 followed by compact bibliographic information. Each line which follows contains parity, energy (MeV), spin, A and Z followed by the number of the level in the sequence. The space after the level number is reserved for gamma decay branching ratios. The x placed after the spin value in the 2nd level of the first scheme indicate an isomeric state. All asterisks are flags indicating that the quantity they refer to has been introduced after considerations or information other than experimental. Asterisks in the first column flag the parity, those in the last column before the mass number (see level 10 at 1.611 MeV in the first scheme) flag the spin, the energy flag (meaning doubtful existence of the level) is not shown in the table and would be placed in the middle between the column for the isomeric state flag and for the spin flag.

The branching ratios are expressed with two numbers in fields of 4 digits which start after the Z number. The first two digits give the number of the level reached by the gamma transition under consideration and the next two digits express the branching ratio in percent (00 being 100%). As an example the 15th level of the first scheme, placed at 2.032 MeV decays 17% to the 2nd level, 5% to the 3th, 32% to the 6th, 25% to the 7th and 2% to the 9th. This convention allows for assignement of branching ratios only for levels with numeral <100. In order to flag gamma decay information of origin other than experimental we use alphabetic letters instead of arabic numbers with the simple convention that a=1, b=2, ... , i=9, and z=0, so the 26th level at 6.100 MeV in the second level scheme decays 96% to the 4th level, 2% to the 9th level, 2% to the 10th level.

3. Comment and conclusions

The determination of the cut energy between continuum and discrete region requires a careful analysis of experimental levels that has only been performed for some nuclei. This analysis is based on the use of a recipe for level density, which implies that any result of such an analysis is model dependent. For this reason some people proved somewhat esitant to include too much information which is not strictly amenable to the primary experimental information. Only a few of the level schemes of our file are evaluated ones and complete. Many do not contain branching ratios. It easy to realize the advantage as well as the amount of work needed to provide the user community with an evaluated discrete level scheme file. A considerable step forward in this direction will be taken with the availability of a translation code from ENDSL into our format.

Table 1. An example of filed discrete level schemes for mass and charge numbers A=85, Z=36 and A=19, Z=9 respectively.

```
   20                         85 36  0****lederer 79 and g.w.tepel n.d.s
    1   0.      4.5           85 36  1
   -1    .305    .5     x     85 36  2 100
   -1   1.107  1.5            85 36  3 200
    1   1.141  2.5            85 36  4 100
 *  1   1.167  1.5            85 36  5 200
   -1   1.224  2.5            85 36  6 200
 *  1   1.343  1.5            85 36  7 530 412 305 253
    1   1.417  2.5            85 36  8 176 524
    1   1.427  0.5            85 36  9
 *  1   1.611  4.5          * 85 36 10 azz
 * -1   1.846  3.5          * 85 36 11 azz
    1   1.874  2.5            85 36 12 3zz
   -1   1.939  2.5            85 36 13 346 443 511
 *  1   2.004  2.5            85 36 14 azz
 * -1   2.032  1.5            85 36 15 217 373 508 702
    1   2.055  1.5            85 36 16
 * -1   2.137  1.5            85 36 17 236 305 632 725 902
        2.369                 85 36 18 azz
        2.379  2.5            85 36 19
   -1   2.463  2.5            85 36 20 144 856


  191                         19  9  0****f. ajzenberg-selove:np a329(1983)p.65
    1   0.      0.5           19  9  1
   -1   0.110   0.5           19  9  2 100
    1   0.197   2.5           19  9  3 100
   -1   1.346   2.5           19  9  4 297 303
   -1   1.459   1.5           19  9  5 120 269 311
    1   1.554   1.5           19  9  6 103 205 392
    1   2.780   4.5           19  9  7 300
    1   3.908   1.5           19  9  8 148 217 314 621
   -1   3.999   3.5           19  9  9 318 470 512
   -1   4.033   4.5           19  9 10 400
    1   4.378   3.5           19  9 11 381 719
    1   4.550   2.5           19  9 12 369 405 508 618
   -1   4.556   1.5           19  9 13 136 245 309 404 606
    1   4.648   6.5           19  9 14 700
   -1   4.683   2.5           19  9 15 306 463 531
    1   5.107   2.5           19  9 16 380 520
 *  1   5.337   0.5           19  9 17 137 243 520
   -1   5.418   3.5           19  9 18 471 513 9101006
    1   5.464   3.5           19  9 19 304 432 605 759
    1   5.501   1.5           19  9 20 225 349 416 610
    1   5.535   2.5           19  9 21 107 347 546
   -1   5.621   2.5           19  9 22 339 461
    1   5.938   0.5           19  9 23 107 220 302 563 808
    1   6.070   3.5           19  9 24 353 419 601 7231104
   -1   6.088   1.5           19  9 25 125 261 314
   -1   6.100   4.5           19  9 26 dif 1zbazzb
   -1   6.160   3.5           19  9 27 331 464 501 9021002
    1   6.255   0.5           19  9 28zadhzcdazfaa
    1   6.282   2.5           19  9 29 114 304 436 526 620
    1   6.330   3.5           19  9 30 356 417 6091118
   -1   6.429   0.5           19  9 31zbfazdbazeah
```

THE LLNL INTERACTIVE SYSTEM FOR NUCLEAR DATA EVALUATION

DAVID A. RESLER and ROGER M. WHITE
Nuclear Data Group, Physics Department
Lawrence Livermore National Laboratory
P.O. Box 808, L-298
Livermore, CA 94550, USA

ABSTRACT

LLNL's interactive system for nuclear data evaluation uses a general-purpose two-dimensional graphics code, QPX. With simple commands, we access and plot any information in the Livermore evaluated neutron data library (ENDL), activation library (ACTL), or charged-particle library (ECPL), the Livermore experimental cross section information library (ECSIL), or any evaluated data library using the ENDF format. Features of this system important to evaluation work are discussed. The conversion of evaluations from the ENDF format to the ENDL format for use in our system is discussed.

1. Introduction

In a previous paper,[1] we reviewed specific methods used for evaluation of charged-particle reactions. Specific features of the LLNL interactive evaluation system were described as part of that work. In this paper, we concentrate on generalized tools for nuclear data evaluations using the LLNL interactive system.

2. QPX

QPX is our general-purpose, two-dimensional graphics code and serves as the shell for our interactive system. This code is designed to be a user-friendly visualization and evaluation tool. Using simple commands, data files containing (x,y) or (x,y,Δy) space-delimited data can be accessed and plotted. The limits of the axes can be changed interactively as can be many other properties of the plot. Reasonable default values are picked for all plot properties so that quick plots (QP) can be made instantly. It is designed to run on workstations running the UNIX operating system and employing the X-windowing system (hence the name QPX).

An important capability of this code is that while viewing data, an evaluator can use the computer mouse to define the knots of a cubic-spline curve. In this manner, it is extremely easy to draw a smooth curve. The left mouse button places a knot at the location of the mouse cursor and the middle button removes the knot closest to the location of the cursor. Each time, the resulting cubic-spline curve is redrawn on the screen without removing the previous curves. This allows the user to see exactly how the curve changes each time a knot is added or removed. To refresh the screen with only the last curve, the right button is pushed. In Fig. 1 we show cubic-spline curves that have been drawn

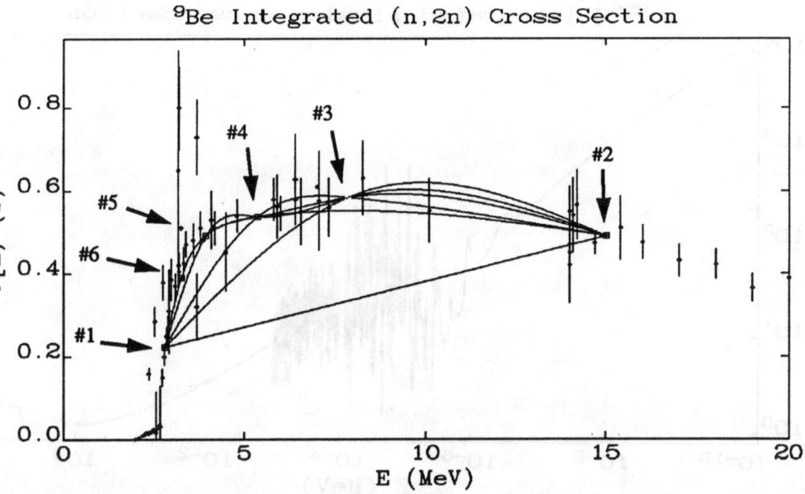

Figure 1: Cubic-spline curves drawn with QPX and the computer mouse. The order of the knot placement is indicated. Note the lowering of the curve near 7 MeV when knot #3 is removed.

through experimental data using QPX and the computer mouse.

2.1. ENDL, ACTL, ECPL

The contents of the Livermore evaluated neutron data library (ENDL), activation library (ACTL), and charged-particle library (ECPL) have been placed into a hierarchial tree structure for easy access on our UNIX workstations. The ENDL library contains evaluations for 109 materials, the ACTL library contains evaluations for 360 materials, and the ECPL library contains evaluations for 14 materials. QPX allows a user to enter simple commands to access and plot any quantities in these libraries. In Fig. 2 we show the integrated fission cross section for ^{235}U that was obtained from the sample QPX session shown in Fig. 3. It is extremely easy for QPX to access this information since *all* of the data in these libraries exist in tabular form using linear-linear interpolation between all points.

2.2. ECSIL

The Livermore experimental cross section information library (ECSIL) is also readily accessible from QPX. This library contains experimental data for neutron-induced reactions from over 6,000 references dating back to 1938. Many of these data were obtained from the National Nuclear Data Center at Brookhaven National Laboratory in the EXFOR format and were translated at LLNL into a common units base and format.

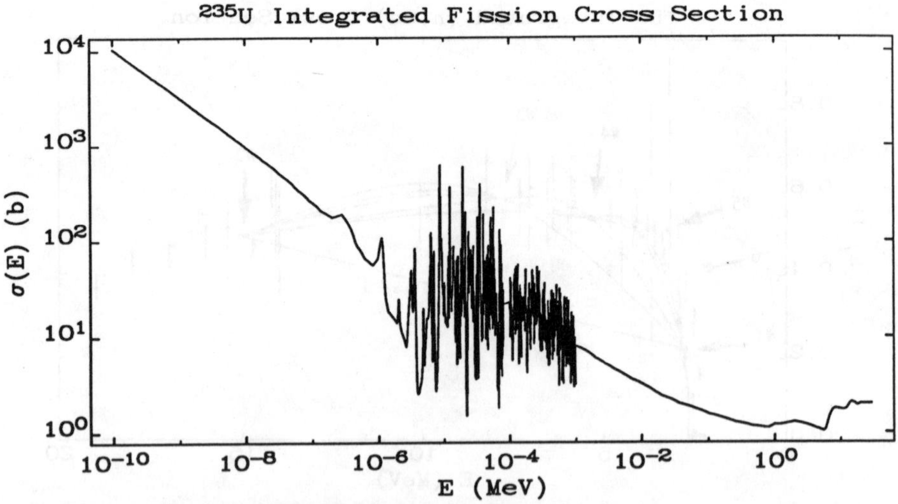

Figure 2: Integrated fission cross section from the LLNL Evaluated Neutron Data Library (ENDL). The QPX commands required to generate this figure are shown in Fig. 3.

3. Conversion of ENDF/B to the ENDL Format

While describing the same basic kinds of data, the formats employed by the Livermore libraries (ENDL) and the U.S. libraries (ENDF/B) are completely different. The ENDL format for a data set consists of two header lines followed by either 2, 3, or 4 columns of linear-linear interpolatable data. The header lines contain the target information and the ENDL 'quantum numbers' (C, I, S, etc.) describing the type of reaction information. An example of the ENDL format is shown in Fig. 4 for the energy distributions of secondary neutrons from ^{113}Cd(n,2n) as translated from ENDF/B-V. The first column gives the incident neutron energy, the second column is the energy of the emitted neutron, and the fourth column is the normalized probability distribution. In the ENDF/B system, the data are expressed in many different formats employing multiple interpolation regions and the possibility of six different interpolation schemes. See Fig. 5 for the original ENDF/B-V information which corresponds to Fig. 4. In this case, the energy distributions are described by a combination of both an arbitrary tabulated function and an evaporation spectrum.

Before using an evaluation in the ENDF/B format, we preprocess the information into pointwise form using D. E. Cullen's suite of codes.[2] The output is then passed through an additional set of codes and translated into the ENDL format. In the ENDL format, the information in ENDF/B formatted evaluations can be accessed and plotted using QPX.

For ENDF/B-V, we have completed a full translation to the ENDL format. For

```
X QP   --  Version:   2.7, 92/08/10 11:12:36

Enter QP command:   endl

     0001 n        14000 Si      40000 Zr       79197 Au197     94239 Pu239
     1001 H1       15031 P31     41093 Nb93     80000 Hg        94240 Pu240
     1002 H2       16032 S32     42000 Mo       82000 Pb        94241 Pu241
     1003 H3       17000 Cl      47107 Ag107    83209 Bi209     94242 Pu242
     2003 He3      18000 Ar      47109 Ag109    90231 Th231     94243 Pu243
     2004 He4      19000 K       48000 Cd       90232 Th232     95241 Am241
     3006 Li6      20000 Ca      49000 In       90233 Th233     95242 Am242
     3007 Li7      22000 Ti      50000 Sn       91233 Pa233     95243 Am243
     4007 Be7      23051 V51     51000 Sb       92233 U233      96242 Cm242
     4009 Be9      24000 Cr      53127 I127     92234 U234      96243 Cm243
     5010 B10      25055 Mn55    54000 Xe       92235 U235      96244 Cm244
     5011 B11      26000 Fe      54134 Xe134    92236 U236      96245 Cm245
     6012 C12      27059 Co59    56138 Ba138    92237 U237      96246 Cm246
     6013 C13      28000 Ni      63000 Eu       92238 U238      96247 Cm247
     7014 N14      28058 Ni58    64000 Gd       92239 U239      96248 Cm248
     7015 N15      29000 Cu      67165 Ho165    92240 U240      97249 Bk249
     8016 O16      30000 Zn      72000 Hf       93235 Np235     98249 Cf249
     9019 F19      31000 Ga      73181 Ta181    93236 Np236     98250 Cf250
    10020 Ne20     33074 As74    74000 W        93237 Np237     98251 Cf251
    11023 Na23     33075 As75    75185 Re185    93238 Np238     98252 Cf252
    12000 Mg       39088 Y88     75187 Re187    94237 Pu237     99120 fp120
    13027 Al27     39089 Y89     78000 Pt       94238 Pu238

Enter isotope descriptor:   u235

  C short &   long reaction type descriptors

   1 tot       total
  10 elas      elastic scattering
  11 n         n
  12 2n        2n
  13 3n        3n
  14 4n        4n
  15 f         fission
  46 g         gamma
  55 xg        xgamma

Enter reaction type descriptor:   f

  I short &  long reaction property descriptors

  0 int      integrated cross-sections and related parameters
  4 ead      energy-angle distributions, normalized legendre coefficients
  7 nubar    average neutrons per fission (prompt or delayed)
  9 mult     multiplicities
 10 aveenyo  average energy deposits to Yo
 11 aveenrn  average energy deposits to residual nucleus

Enter reaction property descriptor:   0

  The following dataset has been selected:

  library(endl)      target(u235)       projectile(n)

   C= 15   fission
   I=  0   integrated cross-sections and related parameters
  Yo=  0   not applicable or none
   S=  0   no x-field data

Enter QP command:  p
```

Figure 3: Sample QPX session showing the four commands required to access and the single command to plot the ^{235}U integrated fission cross section shown in Fig. 2.

```
 48113  1  1  0.000000+0 920930 12  0.000000+0  1.00000+50  2.586000-8
12  4  0  -6.53980+0  0.000000+0  0.000000+0  0.000000+0  0.000000+0
 6.59823000E+00  1.000000E-10  0.000000E+00  1.920040E-10
 6.59823000E+00  1.000000E-02  0.000000E+00  2.919859E+01
 6.59823000E+00  2.000000E-02  0.000000E+00  2.470091E+01
 6.59823000E+00  3.000000E-02  0.000000E+00  1.775450E+01
 6.59823000E+00  4.000000E-02  0.000000E+00  1.365683E+01
 6.59823000E+00  5.000000E-02  0.000000E+00  1.126418E+01
 6.59823000E+00  5.843006E-02  0.000000E+00  1.022340E+01
 7.50000000E+00  1.000000E-10  0.000000E+00  9.235492E-09
 7.50000000E+00  1.000000E-02  0.000000E+00  6.622125E+00
 7.50000000E+00  1.112447E-02  0.000000E+00  7.689136E+00
 7.50000000E+00  1.224894E-02  0.000000E+00  7.180500E+00
...
 7.50000000E+00  4.024014E-01  0.000000E+00  5.573474E-01
 7.50000000E+00  4.400000E-01  0.000000E+00  5.052957E-01
 7.50000000E+00  5.986021E-01  0.000000E+00  1.046834E+00
 7.50000000E+00  7.948028E-01  0.000000E+00  3.747333E-01
 7.50000000E+00  9.602000E-01  0.000000E+00  1.056197E+00
...
 2.00000000E+01  0.000000E+00  0.000000E+00  0.000000E+00
 2.00000000E+01  1.000000E-02  0.000000E+00  2.977437E-01
 2.00000000E+01  2.000000E-02  0.000000E+00  4.266854E-01
 2.00000000E+01  3.000000E-02  0.000000E+00  4.586002E-01
 2.00000000E+01  4.000000E-02  0.000000E+00  4.381352E-01
...
 2.00000000E+01  8.500000E-01  0.000000E+00  9.175333E-02
 2.00000000E+01  8.600000E-01  0.000000E+00  5.192256E-02
 2.00000000E+01  8.700000E-01  0.000000E+00  9.388713E-02
 2.00000000E+01  8.800000E-01  0.000000E+00  5.334506E-02
 2.00000000E+01  1.346020E+01  0.000000E+00  9.602093E-02
                                                                       1
```

Figure 4: Example of the ENDL format for the energy distributions of secondary neutrons for ^{113}Cd(n,2n). The dots indicate lines which have been omitted. Each section of ENDL starts with two header lines and ends with a line containing only a '1' in column 72.

```
 4.81130+ 4 1.11930+ 2          0          0          2          01318 5 16  506
 6.53980+ 6 0.00000+ 0          0          9          1          41318 5 16  507
          4          2                                            1318 5 16  508
 6.59823+ 6 0.00000+ 0 6.59823+ 6 5.00000- 1 2.00000+ 7 5.00000- 11318 5 16  509
 2.00000+ 7 0.00000+ 0                                            1318 5 16  510
 0.00000+ 0 0.00000+ 0          0          0          1          31318 5 16  511
          3          5                                            1318 5 16  512
 6.59823+ 6 6.47153+ 5 1.50000+ 7 9.75750+ 5 2.00000+ 7 9.75750+ 51318 5 16  513
 0.00000+ 0 0.00000+ 0          0          1          1          41318 5 16  514
          4          2                                            1318 5 16  515
 6.59823+ 6 0.00000+ 0 6.59823+ 6 5.00000- 1 2.00000+ 7 5.00000- 11318 5 16  516
 2.00000+ 7 0.00000+ 0                                            1318 5 16  517
 0.00000+ 0 0.00000+ 0          0          0          1          31318 5 16  518
          3          2                                            1318 5 16  519
 0.00000+ 0 6.59823+ 6          0          0          1          61318 5 16  520
          6          2                                            1318 5 16  521
 1.00000- 4 4.33807-16 1.00000+ 4 1.45486- 5 2.00000+ 4 1.79744- 51318 5 16  522
 4.00000+ 4 2.04983- 5 5.84300+ 4 2.07297- 5 5.84300+ 4 0.00000+ 01318 5 16  523
 0.00000+ 0 7.50000+ 6          0          0          1         171318 5 16  524
         17          2                                            1318 5 16  525
 1.00000- 5 1.00000-16 1.00000+ 4 3.35370- 6 2.00000+ 4 4.14340- 61318 5 16  526
 4.00000+ 4 4.72520- 6 6.00000+ 4 4.78310- 6 8.00000+ 4 4.60920- 61318 5 16  527
 1.00000+ 5 4.31890- 6 1.20000+ 5 3.97090- 6 1.40000+ 5 3.59830- 61318 5 16  528
 1.60000+ 5 3.22120- 6 1.80000+ 5 2.85190- 6 2.00000+ 5 2.49810- 61318 5 16  529
 2.50000+ 5 1.70680- 6 3.00000+ 5 1.06400- 6 3.50000+ 5 5.65100- 71318 5 16  530
 4.00000+ 5 1.89300- 7 4.40000+ 5 1.00000-26                      1318 5 16  531
 0.00000+ 0 2.00000+ 7          0          0          1         171318 5 16  532
         17          2                                            1318 5 16  533
 1.00000- 5 1.00000-16 1.00000+ 4 3.35370- 6 2.00000+ 4 4.14340- 61318 5 16  534
 4.00000+ 4 4.72520- 6 6.00000+ 4 4.78310- 6 8.00000+ 4 4.60920- 61318 5 16  535
 1.00000+ 5 4.31890- 6 1.20000+ 5 3.97090- 6 1.40000+ 5 3.59830- 61318 5 16  536
 1.60000+ 5 3.22120- 6 1.80000+ 5 2.85190- 6 2.00000+ 5 2.49810- 61318 5 16  537
 2.50000+ 5 1.70680- 6 3.00000+ 5 1.06400- 6 3.50000+ 5 5.65100- 71318 5 16  538
 4.00000+ 5 1.89300- 7 4.40000+ 5 1.00000-26                      1318 5 16  539
 0.00000+ 0 0.00000+ 0          0          0          0          01318 5  0  540
```

Figure 5: Example of ENDF/B format for the energy distributions of secondary neutrons from ^{113}Cd(n,2n).

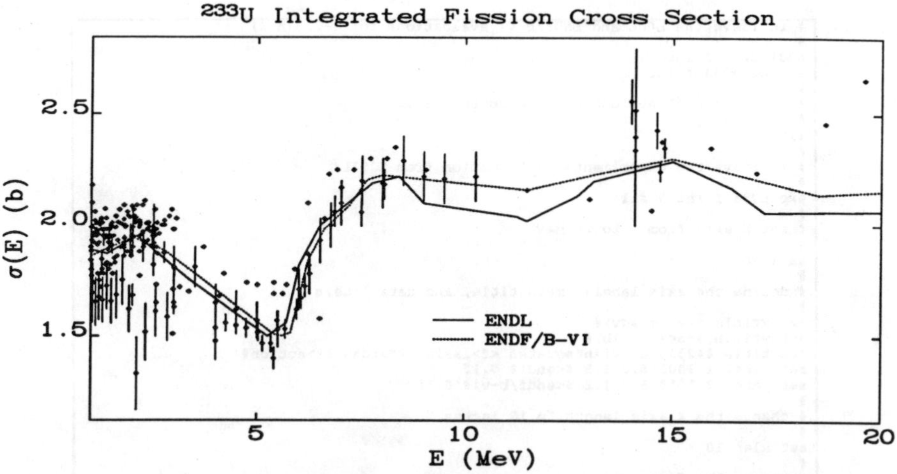

Figure 6: Integrated fission cross section for ^{233}U. Shown are the evaluations from ENDL and ENDF/B-VI along with all of the experimental data available from ECSIL. The QPX commands to generate this plot are shown in Fig. 7.

ENDF/B-VI, we have so far translated the integrated cross sections. In Fig. 6, we show experimental data from ECSIL and the evaluations from ENDL and ENDF/B-VI for the ^{233}U integrated fission cross section. In Fig. 7, we show the QPX command file required to generate the plot shown in Fig. 6. This type of intercomparison of multiple evaluations and the available experimental data will be essential for moving towards a single set of evaluations, i.e., determining the best of the world's evaluations for any isotope.

4. Data Verification—Integral Testing

4.1. Kerma

Effects produced by neutron reactions in matter occur principally by the energy loss via ionization and excitation of the charged particles (including residual nuclei) produced. Kerma is defined as the kinetic energy released in matter (per unit mass) and is the sum of all energy transferred to light charged-particles and residual nuclei in a reaction. The neutron kerma factor is defined as the kerma produced per unit neutron fluence.

Because of the importance of neutron kerma factors for both medical applications and in determining heating of materials in fission or fusion power applications, direct measurements of kerma factors have been made. In addition, kerma factors have been calculated from evaluations of reaction data. This is often difficult since many evaluations do not contain all of the requisite information, e.g., the explicit energy distributions of secondary charged particles. The ENDL library does contains explicit energy distributions

```
# retrieve the ENDL and ENDF/B-VI evaluations for U-233(n,f)
#
endl u233 f int
endfb6 u233 f int
#
# draw the ENDF evaluation as a dotted line
#
line10 2
#
# retrieve the experimental information from ECSIL
#
exp u233 f int 0 all
#
# set X axis from 1 to 20 MeV
#
sa 1 20
#
# define the axis labels, main title, and data labels
#
set xtitle $<e (m>e<v)#
set ytitle $*s<(e)> (b)#
set title $&233$<u> <i>ntegrated <f>ission <c>ross <s>ection#
set label 1 1001 5.0 1.5 $<endl# 0.15
set label 2 1010 5.0 1.2 $<endf/b-vi# 0.15
#
# change the X axis length to 10 inches
#
set xlen 10
#
# change the font
#
set font 3
#
# make the plot
p
```

Figure 7: QPX commands used to make the plot of the ^{233}U integrated fission cross section shown in Fig. 6. The lines starting with '#' are QPX comment lines.

for all secondary particles from all neutron-induced reactions and can therefore be used directly to calculate kerma factors using our utility code KERMA. In Fig. 8 we show the neutron kerma factor for ^{12}C as calculated from ENDL along with measured values.

The calculation of neutron kerma factors is one of the *integral* tests we perform on our evaluated data libraries. When a calculation of neutron kerma does not match the experimental measurements of kerma, it is an indication that there is a potential problem with the basic evaluation. Integral tests such as this are not used as the basis for adjusting the evaluations but are used as a guide to what microscopic data we might need to reevaluate. The success of an integral test is satisfying but may only be fortuitous. Therefore, it is only the differences found in integral tests which are significant.

4.2. *Critical Assemblies and Pulsed-Sphere Experiments*

We also perform calculations using our evaluated libraries to compare with our neutron and gamma-ray pulsed-sphere data. These tests check ENDL in the region of 14 MeV and lower. In addition, we have a suite of 69 critical assemblies with which we validate our evaluated libraries. We have started detailed comparisons of using not only ENDL in these integral tests, but our translations of the ENDF/B-V and ENDF/B-VI libraries as well.

In Fig. 9 is a diagram showing the overall picture of the use of nuclear data in applications codes. The integral tests described above make up only one small part of the application calculations. As part of our evaluations, we feel it is critical to close the loop

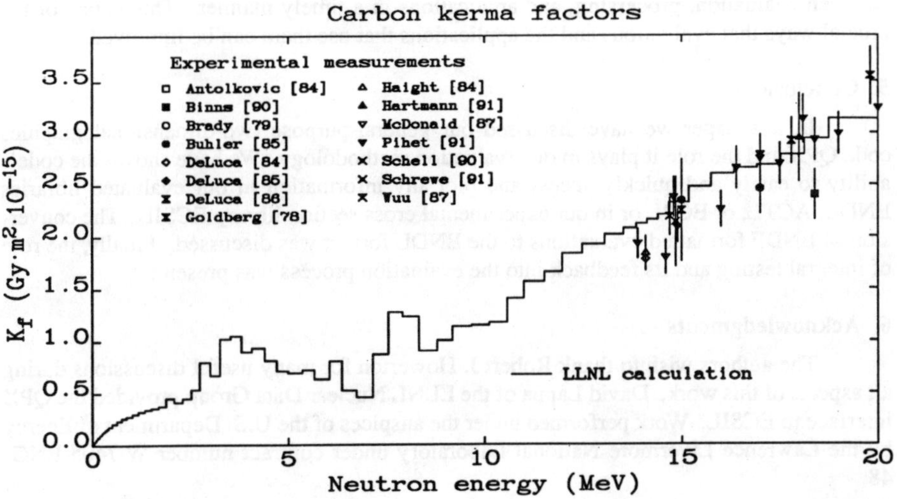

Figure 8: Neutron kerma factors for ^{12}C. The solid line was calculated from ENDL. The references for the experimental data are given elsewhere.[3]

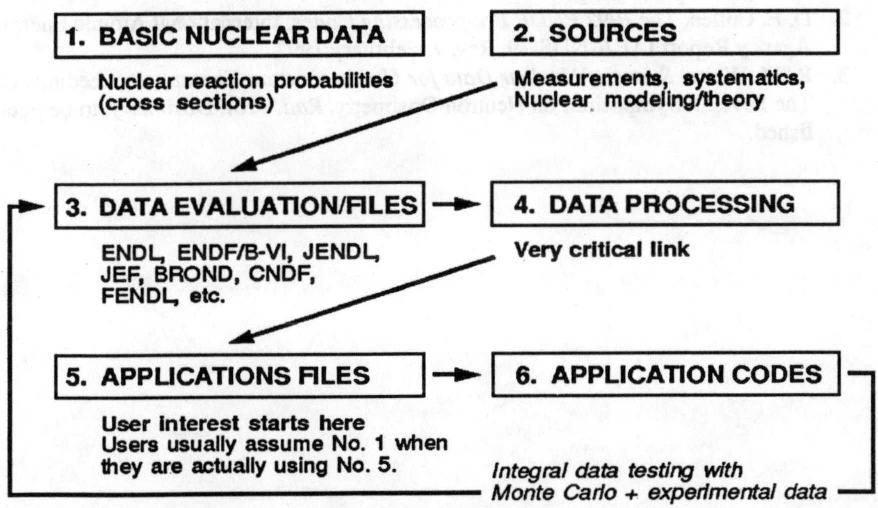

Figure 9: Diagram showing the flow from basic data to applications codes. Our attempt at LLNL has been to close this loop as efficiently as possible so that evaluations and the applications that use them can be improved in a timely manner.

between evaluation, processing, and applications in a timely manner. This is one of the crucial ways that evaluations and the applications that use them can be improved.

5. Conclusions

In this paper we have discussed our general-purpose two-dimensional graphics code QPX and the role it plays in our evaluation methodology. We have shown the code's ability to easily and quickly access and plot any information in our evaluated libraries ENDL, ACTL, or ECPL or in our experimental cross section library ECSIL. The conversion of ENDF formatted evaluations to the ENDL format was discussed. Finally, the role of integral testing and its feedback into the evaluation process was presented.

6. Acknowledgments

The authors wish to thank Robert J. Howerton for many useful discussions during all aspects of this work. David Lappa of the LLNL Nuclear Data Group provided the QPX interface to ECSIL. Work performed under the auspices of the U.S. Department of Energy by the Lawrence Livermore National Laboratory under contract number W-7405-ENG-48.

7. References

1. D. A. Resler and R. M. White, *Charged-Particle Evaluation Methodology at LLNL*, these proceedings.
2. D. E. Cullen, *The 1992 ENDF Pre-processing Codes*, International Atomic Energy Agency Report IAEA-NDS-39, Rev. 7, February 1992.
3. R. M. White, *Status of Nuclear Data for Use in Neutron Therapy*, Proceedings of The Seventh Symposium on Neutron Dosimetry, *Rad. Prot. Dosimetry*, to be published.

List of Participants

Name	Institution	Nation
Audi, Georges	CSNSM, Orsay	France
Bastian, Claude	CBNM, Geel	EURATOM
Bersillon, Oliver	CEN, Bruyeres-le-Chatel	France
Blachot, Jean	CEN, Grenoble	France
Carlson, Allan	Natioinal Institute of Standards and Technology	United States
Caro, Edmund	Knolls Atomic Power Laboratory	United States
Chadwick, Mark	Los Alamos National Laboratory	United States
Cheng, Edward	TSI Research Corp.	United States
Chiba, Satoshi	Japan Atomic Energy Research Institute	Japan
Cho, Nam Zin	KAIST, Taejon	Korea
Cullen, Dermott	Lawrence Livermore National Laboratory	United States
Decher, Ulrich	ABB Combustion Engineering	United States
Delaroche, Jean	CEN, Bruyeres-le-Chatel	France
Difilippo, Felix	Oak Ridge National Laboratory	United States
Dudley, Thomas	Knolls Atomic Power Laboratory	United States
Dunford, Charles	Brookhaven National Laboratory	United States
Filges, Detlef	KFA, Julich	Germany
Fort, Eric	CEN, Cadarache	France
Frederico, Tobias	IEA, Sao Jose dos Campos	Brazil
Froehner, Fritz	KfK, Karlsruhe	Germany
Fu, C.Y.	Oak Ridge National Laboratory	United States
Fukahori, Tokio	Japan Atomic Energy Research Institute	Japan
Ganesan, Srinivasan	Nuclear Data Section	IAEA
Gardner, Donald	Lawrence Livermore National Laboratory	United States
Gardner, Maureen	Lawrence Livermore National Laboratory	United States
Gohs, Uwe	UAV, Dresden	Germany
Griffin, Patrick	Sandia National Laboratory	United States
Guzhovskij, Boris	IEP, Arzamas	Russia
Hale, Gerald	Los Alamos National Laboratory	United States
Hetrick, David	Oak Ridge National Laboratory	United States
Holden, Norman	Brookhaven National Laboratory	United States
Howerton, Robert	Lawrence Livermore National Laboratory	United States
Hwang, Richard	Argonne National Laboratory	United States
Ignatyuk, Anatoly	IPPE, Obninsk	Russia
Ishibashi, Kenji	Kyushu University	Japan
Janeva, Natalia	INR, Sofia	Bulgaria
Kahler, Albert	Bettis Atomic Power Laboratory	United States
Kalka, Harald	Max Planck Institute, Heidelberg	Germany

Kanda, Yukinori	Kyushu University	Japan
Katoh, Toshiro	Nagoya University	Japan
Kawano, Toshihiko	Kyushu University	Japan
Kikuchi, Yasuyuki	Japan Atomic Energy Research Institute	Japan
King, Ming-Ming	Tsing Hua University	Taiwan
Kishida, Norio	Century Research Corp.	Japan
Koning, A.J.	ECN, Petten	Netherlands
Kopecky, Jura	ECN, Petten	Netherlands
Larson, Duane	Oak Ridge National Laboratory	United States
Larson, Nancy	Oak Ridge National Laboratory	United States
Lubitz, Cecil	Knolls Atomic Power Laboratory	United States
Lukyanov, Alexandre	Sofia State University	Bulgaria
MacMahon, Desmond	Imperial College	United Kingdom
Madland, David	Los Alamos National Laboratory	United States
Maekawa, Hiroshi	Japan Atomic Energy Research Institute	Japan
Mannhart, Wolf	PTB, Braunschweig	Germany
Maslov, Vladimir	RPCPI, Minsk	Belarus
Massey, Thomas	Ohio University	United States
McKnight, Richard	Argonne National Laboratory	United States
Michel, Rolf	University of Hannover	Germany
Muir, Douglas	Los Alamos National Laboratory	United States
Nordborg, Claes	NEA Data Bank	NEA
Oblozinsky, Pavel	SAS Bratislava	Slovak Rep.
Pashchenko, Anatoly	Nuclear Data Section	IAEA
Pearlstein, Sol	Brookhaven National Laboratory	United States
Peelle, Robert	Oak Ridge National Laboratory	United States
Pelloni, Sandro	PSI, Wurenlingen	Switzerland
Pevey, Ronald	Savanah River Laboratory	United States
Poenitz, Wolfgang	Argonne National Laboratory, West	United States
Prael, Richard	Los Alamos National Laboratory	United States
Pronyaev, Vladimir	IPPE, Obninsk	Russia
Raepsaet, Caroline	CEN, Saclay	France
Ramstroem, Elizabet	Uppsala University	Sweden
Reffo, Gianni	ENEA, Bologna	Italy
Resler, David	Lawrence Livermore National Laboratory	United States
Rothenstein, Wolfgang	Technion, Haifa	Israel
Roussin, Robert	Radiation Shielding Information Center	United States
Salvatores, Massimo	CEN, Cadarache	France
Sartori, Enrico	NEA Data Bank	NEA

Schmidt, Josef	Vienna	Austria
Schneider, Amiram	Ben Gurion University	Israel
Shamu, Robert	Western Michigan Univ.	United States
Shubin, Yuri	IPPE, Obninsk	Russia
Smith, Donald	Argonne National Laboratory	United States
Sumita, Kenji	Osaka University	Japan
Tagesen, Siegfried	IRK, Vienna	Austria
Tellier, Henry	CEN, Saclay	France
Townsend, Lawrence	NASA Langley Research Center	United States
Varlamov, Vladimir	Moscow State University	Russia
Vonach, Herbert	IRK, Vienna	Austria
Wagschal, Jekutiel	Hebrew University	Israel
Weaver, David R.	University of Birmingham	United Kingdom
White, Roger	Lawrence Livermore National Laboratory	United States
Wilson, William B.	Los Alamos National Laboratory	United States
Winkler, Gerhard	IRK, Vienna	Austria
Wright, R.Q.	Oak Ridge National Laboratory	United States
Young, Philip	Los Alamos National Laboratory	United States
Zhao Zhixiang	CNDC, Beijing	P.R.China
Zhuang Youxiang	CNDC, Beijing	P.R.China
Zucker, Martin	Brookhaven National Laboratory	United States
Zvenigorodskij, Anatoly	IEP, Arzamas	Russia